THE LIBRARY

D0821705

WITHDRAWN

PLANT–ANIMAL INTERACTIONS

The union of a North American *Papilio glaucus* male (right) with a Brazilian *Papilio scamander* female (left). This mating resulted from individuals that were hand paired, and it produced viable progeny. (Drawing by Ali Partridge, based on a photograph by Mark Scriber and Robert Lederhouse. Used with permission.)

PLANT–ANIMAL INTERACTIONS

Evolutionary Ecology in Tropical and Temperate Regions

Edited by

Peter W. Price
Department of Biological Sciences
Northern Arizona University
Flagstaff, Arizona

Thomas M. Lewinsohn
Department of Zoology
Instituto de Biologia
Universidade Estadual de Campinas
Campinas, Brazil

G. Wilson Fernandes
Department of Biological Sciences
Northern Arizona University
Flagstaff, Arizona
and
Department of General Biology
Universidade Federal de Minas Gerais,
Belo Horizonte, Minas Gerais, Brazil

Woodruff W. Benson
Department of Zoology
Instituto de Biologia
Universidade Estadual de Campinas
Campinas, Brazil

A WILEY-INTERSCIENCE PUBLICATION
John Wiley & Sons, Inc.
NEW YORK / CHICHESTER / BRISBANE / TORONTO / SINGAPORE

In recognition of the importance of preserving what has
been written, it is a policy of John Wiley & Sons, Inc. to
have books of enduring value published in the United
States printed on acid-free paper, and we exert our best
efforts to that end.

Copyright © 1991 by John Wiley & Sons, Inc.

All rights reserved. Published simultaneously in Canada.

Reproduction or translation of any part of this work
beyond that permitted by Section 107 or 108 of the
1976 United States Copyright Act without the permission
of the copyright owner is unlawful. Requests for
permission or further information should be addressed to
the Permissions Department, John Wiley & Sons, Inc.

Library of Congress Cataloging in Publication Data:

Plant–animal interactions : evolutionary ecology in tropical and temperate regions /
edited by Peter W. Price ··· [et al.].
p. cm.

Based on papers from an international symposium, held at UNICAMP
(Campinas State University), Campinas, Brazil, 1988.
"A Wiley-Interscience publication."
Includes bibliographical references and index.
ISBN 0-471-50937-X

1. Animal–plant relationships—Congresses. 2. Ecology—
Congresses. 3. Evolution—Congresses. I. Price, Peter W.
QH549.5.P53 1991
574·5'24—dc20   90-39766
                                                                          CIP

Printed in the United States of America

10 9 8 7 6 5 4 3 2 1

CONTRIBUTORS

T. MITCHELL AIDE, Biology Department, University of Utah, Salt Lake City, Utah, U.S.A.

MIGUEL A. ALTIERI, Division of Biological Control, University of California-Berkeley, Albany, California, U.S.A.

WOODRUFF W. BENSON, Department of Zoology, Instituto de Biologia, Universidade Estadual de Campinas, Campinas, São Paulo, Brazil.

BARBARA L. BENTLEY, Department of Ecology and Evolution, State University of New York at Stony Brook, Stony Brook, New York, U.S.A.

KEITH S. BROWN, Jr., Department of Zoology, Instituto de Biologia, Universidade Estadual de Campinas, Campinas, São Paulo, Brazil

DAVID B. CLARK, La Selva Biological Station, Organization for Tropical Studies, San Pedro, Costa Rica.

DEBORAH A. CLARK, La Selva Biological Station, Organization for Tropical Studies, San Pedro, Costa Rica.

PHYLLIS D. COLEY, Department of Biology, University of Utah, Salt Lake City, Utah, U.S.A.

ROBERT K. COLWELL, Department of Ecology and Evolutionary Biology, University of Connecticut, Storrs, Connecticut, U.S.A.

MICHAEL CYTRYNOWICZ, Department of Biology, Universidade Federal de Santa Catarina, Florianopolis, Santa Catarina, Brazil.

DIANE W. DAVIDSON, Department of Biology, University of Utah, Salt Lake City, Utah, U.S.A.

RODOLFO DIRZO, Centro de Ecologia, Universidad National Autonoma de Mexico, Mexico, D.F., Mexico.

DAVID S. DOBKIN, Department of Biology, Rutgers University, Camden, New Jersey, U.S.A.

PAUL FEENY, Section of Ecology and Systematics, Corson Hall, Cornell University, Ithaca, New York, U.S.A.

G. WILSON FERNANDES,   Department of Biological Sciences, Northern Arizona University, Flagstaff, Arizona, U.S.A., and Department of General Biology, Universidade Federal de Minas Gerais, Belo Horizonte, Minas Gerais, Brazil.

THEODORE H. FLEMING,   Department of Biology, University of Miami, Coral Gables, Florida, U.S.A.

ROBIN B. FOSTER,   Department of Botany, Field Museum of Natural History, Chicago, Illinois, U.S.A.

RONALDO B. FRANCINI,   Department of Zoology, Instituto de Biologia, Universidade Estadual de Campinas, Campinas, São Paulo, Brazil.

DOUGLAS J. FUTUYMA,   Department of Ecology and Evolution, State University of New York at Stony Brook, Stony Brook, New York, U.S.A.

MARIA ALICE GARCIA,   Department of Zoology, Instituto de Biologia, Universidade Estadual de Campinas, Campinas, São Paulo, Brazil.

LAWRENCE E. GILBERT,   Department of Zoology, University of Texas, Austin, Texas, U.S.A.

ROBERT H. HAGEN,   Department of Entomology, Michigan State University, East Lansing, Michigan, U.S.A.

BERNARD HALLET,   Quaternary Research Center, University of Washington, Seattle, Washington, U.S.A.

AMY JO HEYNEMAN,   2019 E. Newton St., Seattle, Washington, U.S.A.

NELSON D. JOHNSON,   Department of Ecology and Evolution, State University of New York at Stony Brook, Stony Brook, New York, U.S.A.

MARCOS KOGAN,   Agricultural Entomology, University of Illinois and Illinois Natural History Survey, Champaign, Illinois, U.S.A.

KATHERINE C. LARSON,   Department of Biological Sciences, Northern Arizona University, Flagstaff, Arizona, U.S.A.

JOHN H. LAWTON,   Centre for Population Biology, Imperial College at Silwood Park, Ascot, Berkshire, England.

ROBERT C. LEDERHOUSE,   Department of Entomology, Michigan State University, East Lansing, Michigan, U.S.A.

THOMAS M. LEWINSOHN,   Department of Zoology, Instituto de Biologia, Universidade Estadual de Campinas, Campinas, São Paulo, Brazil.

PEDRO W. LOZADA,   Department of Entomology, Museo de Historia Natural (Javier Prado), Universidad Nacional Mayor de San Marcos, Lima, Peru.

ROBERT J. MARQUIS,   Department of Biology, University of Missouri, St. Louis, Missouri, U.S.A.

JOYCE MASCHINSKI, Department of Biological Sciences, Northern Arizona University, Flagstaff, Arizona, U.S.A.

SAMUEL J. MCNAUGHTON, Biological Research Laboratories, Syracuse University, Syracuse, New York, U.S.A.

ALVARO MIRANDA, Centro de Ecologia, Universidad National Autonoma de Mexico, Mexico, D.F., Mexico.

ANA BEATRIZ BARROS de MORAIS, Department of Zoology, Instituto de Biologia, Universidade Estadual de Campinas, Campinas, São Paulo, Brazil.

PAULO CESAR MOTTA, Department of Zoology, Instituto de Biologia, Universidade Estadual de Campinas, Campinas, São Paulo, Brazil.

SHAHID NAEEM, Department of Biology, University of Michigan, Ann Arbor, Michigan, U.S.A.

JOÃO VASCONCELLOS NETO, Department of Zoology, Instituto de Biologia, Universidade Estadual de Campinas, Campinas, São Paulo, Brazil.

PAULO S. OLIVEIRA, Department of Zoology, Instituto de Biologia, Universidade Estadual de Campinas, Campinas, São Paulo, Brazil.

ARY T. OLIVEIRA-FILHO, Department of Zoology, Instituto de Biologia, Universidade Estadual de Campinas, Campinas, São Paulo, Brazil.

KEN N. PAIGE, Environmental Research Laboratory, Institute of Environmental Studies, University of Illinois, Urbana, Illinois, U.S.A.

PETER W. PRICE, Department of Biological Sciences, Northern Arizona University, Flagstaff, Arizona, U.S.A.

MARIA ROMSTÖCK-VÖLKL, Department of Animal Ecology, University of Bayreuth, Bayreuth, West Germany.

J. MARK SCRIBER, Department of Entomology, Michigan State University, East Lansing, Michigan, U.S.A.

ROY R. SNELLING, Entomology Section, Natural History Museum of Los Angeles County, Los Angeles, California, U.S.A.

JOSÉ ROBERTO TRIGO, Department of Zoology, Instituto de Biologia, Universidade Estadual de Campinas, Campinas, São Paulo, Brazil.

THOMAS G. WHITHAM, Department of Biological Sciences, Northern Arizona University, Flagstaff, Arizona, U.S.A. and Biology Department, Museum of Northern Arizona, Flagstaff, Arizona, U.S.A.

HELMUT ZWÖLFER, Department of Animal Ecology, University of Bayreuth, Bayreuth, West Germany.

PREFACE

This book provides the reader with an unusual blend of authors and topics relating to the ecology of plant–animal interactions in tropical and temperate settings. In North America and Europe scientists are frequently unaware of the important ecological research being undertaken in regions of greater concern throughout the world over conservation issues and questions of global change. Moreover, tropical ecology includes a wide range of non-rainforest environments, many of which are also threatened as well as interesting in their own right. In addition to providing a selection of recent important results on plant–animal relationships, we hope this book will stimulate the emerging body of ecologists in tropical countries with new viewpoints on familiar tropical systems and with fresh ideas from temperature latitudes barely touched on in tropical systems as yet. The continued conservation of tropical habitats and their associated biological communities will ultimately depend on the expertise of indigenous scientists and their ability to command public opinion in critical regions. We believe that the contributions in this book should aid in expanding this expertise.

Indeed, the motivation for this book came in large part from a desire to foster more tangible links between researchers in temperate and tropical regions and to develop ties among established researchers and young scientists alike. The initial step took the form of an international symposium on the Evolutionary Ecology of Tropical Herbivores held at UNICAMP (Campinas State University), Campinas, Brazil, in 1988. Since Campinas is becoming a major center of research on tropical plant–animal relationships, and given the current momentum of evolutionary ecology in Brazil, this path was particularly appropriate and drew a large attendance. Collaborative research and graduate-study opportunities forged during the symposium will certainly have a lasting impact on many participants. It also seems fitting that Brazil should host a meeting of this nature and become more involved in the scientific mainstream with regard to questions essential to its destiny.

This book was not written as a compendium of research on plant–animal interactions. We do however expect that the different approaches displayed here, and the "flavor" of how ecologists deal with their particular research problems, will prove to be valuable and instructive. Collaborative enterprises such as this can only enrich our experiences and our understanding of the fascinating patterns of nature.

The symposium and this book would not have been possible without the generous support of the Brazilian Conselho Nacional de Desenvolvimento

Científico e Technológico (CNPq), the U.S. National Science Foundation Ecology Program and U.S.–Brazil International Program, the Fundação de Amparo à Pesquisa do Estado de São Paulo (FAPESP), the Third World Academy of Sciences, the Fundação MB, UNICAMP, Northern Arizona University, and the Secretarias de Ciência e Tecnologia e do Meio-Ambiente do Estado de São Paulo. To these and to the many colleagues, students, and friends who helped with the symposium and manuscript production, we are most grateful.

April 1990
Flagstaff, Arizona   PETER W. PRICE
Campinas, Brazil   THOMAS M. LEWINSOHN
Flagstaff, Arizona   G. WILSON FERNANDES
Campinas, Brazil   WOODRUFF W. BENSON

CONTENTS

PLANT–ANIMAL INTERACTIONS

1 Introduction: Historical Roots and Current Issues in Tropical Evolutionary Ecology

THOMAS M. LEWINSOHN, G. WILSON FERNANDES,
WOODRUFF W. BENSON, and PETER W. PRICE

Ecology, once it progresses beyond ancient natural history, is largely ignored in accounts of the history of biology (e.g., Théodoridès, 1965; Ronan, 1983; Allen, 1975), but its own historiography is growing steadily (Egerton, 1983; Dajoz, 1984; Kingsland, 1985; McIntosh, 1985; Sheail, 1987; see also Allee et al., 1949, Chapters 2 and 3, and special sections in the Journal of the History of Biology, 1986, Vol. 19 and 1988, Vol. 21). Nonetheless, the development of tropical ecology has never been adequately reviewed and is still difficult to trace.

The following historical notes are intended to call attention to certain issues that help to place the ensuing contributions in a more comprehensive perspective. There is no attempt at completeness; references are largely to neotropical work and especially to Brazil because they are more familiar to the authors, but they have enough in common with other tropical countries to represent them to a certain degree.

This brief outline of the historical background of tropical ecology is followed by an overview of the main concerns and thematic arrangement of the book.

HISTORICAL NOTES ON TROPICAL ECOLOGY

Beginnings—Priests and Travelling Collectors

The discovery of the New World and the increasing explorations of the Old World tropics opened new horizons for 16th century Europeans. Opportunities for new scientific discoveries were dramatically enhanced. Dozens of expeditions, motivated chiefly by economic and religious concerns, followed on the discovery of America (Eiseley, 1958; Goodman, 1972).

Plant-Animal Interactions: Evolutionary Ecology in Tropical and Temperate Regions, Edited by
Peter W. Price, Thomas M. Lewinsohn, G. Wilson Fernandes, and Woodruff W. Benson.
ISBN 0-471-50937-X © 1991 John Wiley & Sons, Inc.

Dedicated primarily to the conversion of native peoples, some of the Jesuit missionaries had a particular interest in ethnography and natural history, which qualifies them as the first naturalists, geographers, and anthropologists among the European conquerors (Goodman, 1972). After political troubles put an end to the Catholic missions, many remote regions became virtually "terrae incognitae" until penetrated once again by 19th century scientists (Holanda, 1963).

During the 16th and 17th centuries, several invasions of the Spanish and Portuguese colonies provided naturalists of other European countries with the opportunity to gather specimens and information. Marcgraf and Piso's efforts during the Dutch takeover of northeastern Brazil produced the first detailed account of South American natural history and indigenous medicine (Piso, 1658; on the cover this edition sports tropical fruit, a sloth, a macaw, a jaguar, and an Indian, in the unexpected company of a rhinoceros, a dodo, and an Arab).

One would do well to remember that although historical accounts usually stand mute about the "savages," Indian guides and companions were the direct source of the accounts of the natural history, behavior, and properties of the strange new plants and animals given by most naturalists and collectors up to the 18th century, and some latter-day travellers as well. Indeed, many indigenous names were assimilated into the conquerors' languages, and they provided indications for the taxonomical relationships of the organisms they designated.

The firm control of educational institutions by the Roman Catholic Church, and its hostility to the natural sciences, stifled scientific growth in Spain and Portugal and even more so in their colonies (Cunha, 1980; Gómez and Savage, 1983). Only in the late 18th century were the natural sciences included in the curricula of Coimbra, Portugal's foremost university (Cunha, 1980).

At this time, the Brazilian Alexandre R. Ferreira was commissioned by the Portuguese Crown to do a natural history survey in the Amazon region; he returned to Portugal in 1793 with seven years' collections of specimens. This collection was coveted by leading European scientists and, in keeping with the times, when Portugal was taken over by Napoleon's army in 1808, Geoffroy de Saint-Hilaire immediately confiscated it—together with 554 still unpublished plates prepared for Friar Velloso's *Flora of Rio de Janeiro* (Neiva, 1929). The Saint-Hilaire brothers described many of Ferreira and Velloso's new species. Ferreira only saw fragments of his work published during his lifetime and his *Philosophical Voyage* was only recently printed (Ferreira, 1971). Velloso, author of 400-odd new species, was already dead when his major opus went to press under the auspices of the first independent Brazilian government. However, after Brazilian Emperor Dom Pedro I was ousted in 1830, the new Regency refused to pay for the plates and the prints were sold as scrap paper (Neiva, 1929). These tragicomic episodes illustrate well the struggles of the first Latin American scientists and the troubled relations with their European peers (Schwartzman, 1979).

The exploration of the neotropical region in the 19th century was inaugurated by the expeditions of Humboldt and Bonpland between 1799 and 1804

(Goodman, 1972). Humboldt's publications, especially the narrative of his travels (Humboldt and Bonpland, 1818) were immensely popular and influenced naturalists and explorers for a long time (see Darwin, 1860).

Continued explorations of the tropics by early naturalists were undoubtedly prompted by the outstandingly diverse flora and fauna, although travel accounts also included descriptions of the local habits, politics, economy, and even of military installations (e.g., Spix and Martius, 1824). Naturalists obtained thousands of specimens of neotropical plant and animal life for the collections and museums of Europe and later of North America. The exploits and adventures of the 19th century naturalists are described by Cutright (1940), Goodman (1972), and Lloyd (1985) and of course in their own accounts (e.g., Spix and Martius, 1824; Wallace, 1853; Darwin, 1860; Bates, 1863; Belt, 1888).

Although many of the expeditions were officially authorized and even partially supported by local governments, they showed little concern for fostering local science and even less for the improvement of local collections. A little-noticed episode serves to illustrate this. Louis Agassiz, zoologist and geologist notorious for his antievolutionary views, came to Brazil in 1865 to improve the collections of Harvard's new Museum of Comparative Zoology. In the journey diary, he disposes of the Museum in Rio de Janeiro with the scathing comment that a morning at the town market would produce a better fish collection. However, only a few pages before Mrs. Agassiz tells us that her husband dispatched the many specimens accumulated during their absence...including a complete and considerable fish collection containing a number of new species, obtained by order of the Brazilian Emperor (Agassiz and Agassiz, 1868).

In the Old World tropics a somewhat similar progression took place. Explorations for spices, minerals, and slaves were followed by missionaries, some of whom had interests in natural history (e.g., Crowter and Taylor 1859; Livingstone and Livingstone, 1865). These were followed by collectors, and later by the first professional scientists of the 19th century (e.g., Wallace, 1869 and 1878). However, exploration in Asia and Africa was concentrated in tropical colonies of European countries, so that this activity eventually took on a more permanent character; later this led to long-term surveys and even to the establishment of some research institutions (Worthington, 1983).

Tropical Nature, Evolution, and the Beginnings of Ecology

The discovery of tropical wildlife has had a lasting influence on the formation of modern biology (Eiseley, 1958; Mayr, 1982). Taxonomy was affected at once, but tropical exploration produced more than the description and classification of novel organisms; it was decisive, though in somewhat dissimilar ways, for both modern ecology and evolutionary theory.

As early as 1838, Lacordaire recorded insect shifts onto introduced plants, stressing the role of native congeners as sources, while Alphonse de Candolle's *Botanical Geography* of 1855 contains a surprisingly modern description of latitudinal gradients in diversity and also of the basic tenets of dynamic island biogeography (quoted in Dajoz, 1984). Both authors make strong use of tropical examples, attesting their importance to the elaboration of these theories.

Tropical nature provided a dramatic contrast to familiar environments and organisms. Darwin and Wallace were deeply stirred by it, and the gestation of the theory of Natural Selection is partly a result. Other important 19th century evolutionists were equally impressed by their tropical experiences. Mimicry theory, Batesian (Bates, 1863) and Müllerian, (Müller 1879), was first perceived and formulated for tropical organisms. Batesian mimicry was the first discovery to lend direct support to the hypothesis of natural selection (Bowler, 1984) and, according to Mayr (1982) this "came as a godsend;" further instances of mimetic relationships among tropical butterflies were soon described by Wallace. Fritz Müller, a German naturalist, had already emigrated to Brazil when he wrote his influential *Für Darwin* (Müller, 1864), and all his ulterior field work was done in tropical and subtropical regions as one of the first collectors on the staff of the Museu Nacional of Rio de Janeiro.

From the 1880s on, many European naturalists worked in their countries' tropical colonies, not as travelling collectors, but as academic field researchers. German botanists, like Schimper, already had a background in evolution and were not interested in mere "botanizing," but concerned themselves particularly with plant adaptations to environmental features. As McIntosh (1985) says, quoting from Cittadino (1981), "almost all significant ecological work of German botanists of the time was based on their tropical field experience and very little on their European experience."

Eugen Warming, the initiator of modern plant ecology, spent three years studying the cerrado plants of Lagoa Santa in central Brazil, which led to his classic study on their adaptations to fire, drought, and other environmental factors (Warming, 1892). This was followed by the first general treatise of plant ecology (Warming, 1895), whose seminal effect was acknowledged by European plant ecologists of the early 20th century (Sheail, 1987). Thus, plant ecology may legitimately be said to have a tropical origin (Goodland, 1975; Coleman, 1986). It is worth noting that Warming, who worked in a savanna-like community, emphasized the abiotic environment while Schimper, who worked in the humid tropics, was especially impressed by biotic interactions.

Warming, Schimper, and other plant ecologists of the time were primarily interested in unraveling the adaptation of plants to their environment. Such adaptations were presumably evolved, but the underlying evolutionary mechanisms were not directly addressed. Although these ecologists aligned themselves with evolutionary theory, natural selection was of less import and many preferred straightforward Lamarckian mechanics to account for the adaptations of organisms to their environment (McIntosh, 1985; Coleman, 1986).

In contrast, animal adaptation to environmental conditions was mostly studied in stricter temperate or arctic conditions (Elton, 1927), while zoologists working at the turn of the century in tropical regions had different concerns. Following Bates', Wallace's, and Müller's work, the documentation of natural selection and its products under natural conditions was a key objective for British naturalists such as Poulton, Carpenter, and Marshall (Fisher, 1935; Kimler, 1986).

Although tropical animals served more for investigations of natural selection, and tropical plants of adaptations to the environment, this is not an absolute distinction. It does not hold, for instance, for the plant–animal mutualisms which were being profusely described and systematized in the tropics. Interactions of plants with animal pollinators, dispersers, and bodyguards formed much of the substance of tropical natural history till the early 20th century.

It seems clear that by the beginning of this century tropical organisms and communities had contributed decisively to the origin and expansion of ecological and evolutionary thought. Paradoxically, however, tropical nature all but disappeared from the ecological horizon in the ensuing decades.

Eclipse and Fresh Start

From about 1920 to 1950, while ecology acquired the features of an institutionalized (but not necessarily mature) field of science, it suffered profound changes.

The cleavage between plant and animal ecology deepened. Plant ecology was more directed toward the description of vegetation. In Europe the Zürich–Montpellier and Uppsala schools disputed methods of surveying vegetation units; in the United States, ecologists were divided among Clements', Cowles', and Gleason's concepts of community (McIntosh, 1985). None of them, however, paid particular attention to tropical communities. Textbooks such as Weaver and Clements' (1938) influential *Plant Ecology* hardly mentioned an example from the tropics and neither did those of the next decade (Egler, 1951, quoted in McIntosh, 1985).

Tropical animal ecology fared slightly better. In his powerful little book, Elton (1927) included a number of tropical species among the ecological equivalents which illustrate the niche concept, and also noted some particularities, such as the outstanding richness of the tropical night fauna. Allee et al. (1949), the major compendium of animal ecology of its time, contains a fair number of tropical examples, due to the personal tropical experience of its authors; for example, Allee had surveyed rain-forest faunas on Barro Colorado Island (Allee, 1926) and Emerson's interest in termite social organization took him to several neotropical areas.

In general, ecology distanced itself from the tropics, perhaps because of the gradual dissolution of the predominantly tropical colonial empires. Whatever the causes, no doubt P. W. Richards was overstating the "great development of interest in tropical vegetation during the last 15 years" (Richards, 1952); but his own vivid description of tropical rain forests was one of the factors which helped to rekindle this interest.

Tropical research picked up again in the early 1960s not only due to spontaneously growing interest among biologists, but because opportunities for field research improved. The establishment in 1966 of the Smithsonian Tropical

Research Institute on Barro Colorado Island, Panama Canal, a reserve with an already longstanding research tradition, funnelled a steady stream of pre- and postdoctoral American researchers to the area, free from any commitment to a central research program (Leigh, 1982). The Organization of Tropical Studies, a consortium of mainly North American universities founded in 1963 to promote field courses for their students in Costa Rica, became even more important; in its first 15 years alone, over 1200 students took an OTS field course (Janzen, 1977a) and many of them developed a continuing interest in tropical ecology. Elsewhere research also centered on a restricted set of field sites and institutes, though with fewer ecologists than BCI and the OTS sites received.

Ecosystem-directed research had its own tropical upwelling (see Golley, 1983) which even attained a "big science" level in programs such as the El Verde study in Puerto Rico (Odum and Pigeon, 1970) and the Darién project in Panama (Golley et al., 1975). However, even though they drew a large share of resources available for tropical ecological research, long-term ecosystem programs still lag far behind their temperate and boreal counterparts; for instance, tropical localities amount to less than 10% of the International Biological Progam Woodlands data set (Burgess, 1981).

Janzen (1977a) found the space given to tropical ecology in major American and British journals and textbooks strikingly deficient. Since then there has been no clear improvement. Introductory books on specific themes are scarce (e.g., Janzen, 1975; Owen, 1976; Longman and Jenik, 1987) and only one general textbook has a tropical focus (Desmukh, 1986). A sampling of texts on tropical ecology suggests a prevailing interest in relatively large-scale biogeographical faunistic and floristic analyses emphasizing historical and phylogenetic aspects (e.g., Fittkau et al., 1968; Meggers et al., 1973; Prance, 1982). Ecosystem-level studies are also well treated (e.g., Odum and Pigeon, 1970; Golley and Medina, 1975). Even though several of the references given above include some population and community ecology studies, these receive adequate treatment in few compilations (e.g., Montgomery, 1978; Sutton et al., 1983; Clark et al., 1987).

Concern over the disappearance of the tropical rain forests, especially in the Amazon, has produced a large literature of its own (e.g., Goodland and Irwin, 1975; Myers, 1979) and led to collective evaluations of research priorities and recommendations (Farnworth and Golley, 1975; NRC, 1980). Some support has been accordingly diverted to projects such as the Minimal Critical Size of Ecosystems Project in central Amazonia (see Lovejoy et al., 1983), but this is still far from sufficient.

Endogenous Research

Until the 1950s, the science of ecology was almost unheard of in tropical countries. Picado's (1913) work on communities in bromeliad tanks is exceptional for its quality and for being noted outside Costa Rica, but typical for having had little impact in his home country (Gómez and Savage, 1983). In Latin

America native or immigrant scientists, few though they were, gathered a wealth of natural history information; forest and agricultural entomologists, such as Gregório Bondar in northeast Brazil, traced the bionomics of many phytophagous species (cf. Silva et al., 1968). However, this information had restricted circulation and was not organized into an ecological framework, and therefore temperate ecologists were mainly unaware of it.

The extraordinary development of parasitology in Brazil and other tropical countries (Stepan, 1976) entailed extensive research on animal vectors and intermediate or alternate hosts of human and animal diseases. This could have been a powerful input for tropical ecology; but medically-based parasitology developed independently from ecology and, until recently, was disregarded in return (McIntosh, 1985).

From the 1950s on, ecology started to appear in universities and research institutes. In Brazil, it was initially committed to physiological aspects of adaptation: a key subject of plant ecology was the ecophysiological adaptation of cerrado plants to soil, water, climate, and fire, while other work centered on nutrient cycling or phytogeographical and phytosociological surveys. Field research on terrestrial animals mostly addressed faunistic and zoogeographic studies, with bionomical information sometimes added. Meanwhile, only a few budding geneticists influenced by Theodosius Dobzhansky's visits and courses (Schwartzman, 1979) took up some field work which touched on the interface of ecology and evolution.

It is fairly clear that in many Latin American countries modern ecology tended strongly toward ecosystem-centered approaches, in terrestrial and even more in freshwater ecology. This trend dovetailed naturally with preexisting ecophysiological experience, and also fit in with the traditional interest in typification of communities at the regional level.

These trends were reinforced by much of the recent international collaborative research with American, French, British, German, and other research groups, developed in the form of short or extended expeditions, joint projects, or permanent programs. At their best, in these collaborative enterprises local researchers have had equal standing, and reference collections as well as other permanent results have been reasonably shared with local institutions. At worst, "collaboration" has meant using local facilities for logistic support and local participants as unskilled labor: a contemporary version of the 19th century expeditions which contributed nothing to foster science in the countries in which they worked.

Most tropical countries revised university curricula and created or enlarged graduate courses over the last 25 years. Ecology too has acquired a larger, albeit still precarious, standing in university-based research. In exceptional cases a permanent program of field research has been undertaken, such as the Mexican Universidad Nacional Autonoma de Mexico group has carried on successfully at the Las Tuxtlas site for more than two decades (see Gómez-Pompa et al., 1976; Gómez-Pompa and Del Amo, 1985). Within this recent expansion, ecology in tropical countries has become increasingly diversified. Population and community studies having evolutionary–ecological frameworks are coming into their own.

THE PERVADING THEME: SPECIES DIVERSITY

"Now, the greater diversity of living beings found in the tropical compared to the temperate and cold zones is the outstanding difference which strikes the observer" (Dobzhansky, 1950). It was the variety of plants and animals which, without exception, first impressed naturalists in the tropics. With all the problems entailed by enthroning a single subject, the best candidate for a unifying theme in tropical ecology is species diversity and its corollaries.

One can recognize two sides to the question of tropical diversity: first, how and why did so many species come to be? Second, how can so many species coexist continuously? For convenience these may be thought of, respectively, as evolutionary and ecological questions of tropical diversity. Both are closely linked insofar as answers to the evolutionary question invoke past ecological settings, inferred from present-day ones; and, conversely, insofar as evolutionary processes are presumed to underlie and explain present coexistence.

In his pioneering paper on speciation in the tropics, Dobzhansky (1950) proposed that under mild abiotic circumstances the major adaptive challenges would stem from interspecific interactions. Temperate regions were seen as harsher and requiring primarily adaptation to physical stresses, while in the humid tropics selective pressures derive mostly from biotic interactions; the same concept has appeared under different guises before and since (e.g., Wallace, 1878; MacArthur, 1969; Robinson, 1978). Fine-tuned mutualisms, greater trophic and habitat specialization (and for phytophagous insects, host–plant shifts as well) would be involved in tropical species formation.

Shifting conditions instead of stable ones (especially climatic), may equally be taken to drive tropical speciation. The most familiar argument for this is the refuge hypothesis, which proposes allopatric speciation in forest tracts isolated during Pleistocene cold and dry periods (Moreau, 1966; Haffer, 1969; Vanzolini and Williams, 1970; Prance 1982). Of course, nonforest islands within continuous forests can play an equivalent role (Brown and Benson, 1977; Huber, 1982). Moreover, Baker (1970), following an idea of G. L. Stebbins', speculated that because of their milder conditions, tropical forests could accumulate species orginated in savannas, dry forests, or mountains, where greater selective stresses and rapidly shifting distributions would facilitate speciation.

The question of maintenance of contemporary species-rich communities was reviewed by Pianka (1966), who distinguished 10 explanations for latitudinal gradients in diversity. Some of these are purely evolutionary, such as the idea that tropical regions are older and had more opportunity for speciation. Others are derived from structural features: climatic stability, higher productivity, greater spatial heterogeneity. Biotic interactions too are prominent in Pianka's roster,

but their roles are not clear-cut: thus, competition could drive higher specializ-ation and coexistence based on resource partitioning; but, on the contrary, it could result in exclusion. In the latter case, predators or catastrophes are invoked as factors which can depress competitive exclusion and diversity reduction (see Huston, 1979).

Many other theoretical approaches can be brought into this general framework. Thus, according to Elton's view, complex communities are more stable (Elton, 1958), implying that tropical communities should be safer from disruption than temperate ones. However, model analyses such as those by May (1973) lead to a precisely opposite expectation. The evolution of specialization and of food-web structure are also directly implicated in this discussion.

One further concept which should be mentioned is equilibrium. Many theoretical population and community models presuppose, explicitly or not, an equilibrium condition (see MacArthur, 1972). Often this is intended only as a simplifying assumption to make models more tractable. However, to many minds tropical systems, and especially rain forests, epitomize equilibrium communities, to which theoretical equilibrium models should therefore provide a reasonable approximation, particularly to the importance, magnitude, and structural consequences of biotic interactions.

However disparate the viewpoints and hypotheses on tropical species diversity, there is little disagreement on the pervasive role of biotic interactions (MacArthur, 1969) and among these plant–animal interactions are clearly of high interest (Strong et al., 1984a; Howe and Westley, 1988). Different reasons lead Farnworth and Golley (1975) to submit that "plant–herbivore interactions are of great importance because herbivores, as consumers of primary productivity, are able to exert powerful influences on the structure and functioning of ecosystems...many of the most important economic problems in disturbed tropical ecosystems relate to herbivore infestations of crops."

THEMATIC ARRANGEMENT OF THE BOOK

The six sections of this book are arranged around largely interconnected themes. The questions to which each section is addressed are outlined below in general terms. No reference is made to particular chapters, since these are presented in the introductions which precede each section.

Tropical—Temperate Comparisons

The contrast between tropical and temperate species, species interactions, and communities, is a recurrent subject throughout this book, but in the first section this contrast is directly addressed at a more general level.

The prevailing view of tropical communities as generally more diverse than temperate ones (Fischer, 1960) is presented by most current textbooks of ecology and evolution (e.g., Pianka, 1984; Krebs, 1985). Although some excep-tional taxa are noted, these do not affect the general pattern because they are

counterbalanced by exclusively tropical taxa, such as palms and fruit bats (Brown and Gibson, 1983; Begon et al., 1986). However, there is now enough evidence that more groups of organisms do not fit the general rule (e.g., parasitoids, Owen and Chanter, 1970; aphids, Dixon et al., 1987; galling insects, Fernandes and Price, 1988; but see Stevens, 1989) to justify its reevaluation. In other words, does the rule still hold as a general statement on species diversity; and are there unifying features to the exceptions—or is each one a case in itself?

Theoretical explorations of the relationship of body size to population abundance (Van Valen, 1973) and to diversity (May, 1978) have signaled the usefulness of investigating variation in body size in different groups of organisms and communities (see Peters, 1983; Dial and Marzluff, 1988). Even very rough comparisons of temperate and tropical communities (e.g., Schoener and Janzen, 1968) have shown that differences do indeed exist. The theoretical integration of gradients in diversity, body size, and population abundances is one of the more challenging contemporary problems. The empirical verification of proposed hypotheses is a difficult but certainly rewarding task.

Other approaches to tropical–temperate contrasts are directly relevant to plant–animal interactions. Qualitative differences in plant defense modes have been suggested since Kerner (1878), and differential responses by tropical herbivores to plant defenses were recorded by Spruce (1908). Differences in alkaloid production between temperate and tropical plants, found by Levin (1976), raise the possibility of other gradients in defensive features.

Together, these comparative studies signal the feasibility of a summarization of large-scale variation not of species richness alone, but of suites of animal and plant features, which may in time allow a more thorough understanding of the processes and mechanisms which foster greater diversity of many taxa in the tropics, and of others in temperate regions.

Mutualistic Relationships Between Plants and Animals

After attracting much interest in the 19th and early 20th century (see Allee et al., 1949; Mitman, 1988; Keller, 1988), mutualistic interactions were dislodged from mainstream theoretical ecology. This has now changed entirely: mutualism has never drawn so much attention as in the last few years (e.g., Addicott, 1984; Boucher, 1985; Beattie, 1985) and its importance in macroevolution is being increasingly recognized (Price, 1988; Margulis and Fester, 1990).

Plant–animal mutualisms encompass an enormous spectrum of relationships (Abrahamson, 1989) whose benefits to the plant may include protection from herbivores and competitors (Janzen, 1966; Bentley, 1977), supply of essential nutrients (Rickson, 1974), pollination (Faegri and Pijl, 1979; Real, 1983), and seed dispersal (Pijl, 1972; Fleming et al., 1987).

Mutualistic relationships are especially diversified in the tropical region, and some of the more specialized ones are unique to it. Ant–myrmecophyte associations are outstanding and intensively studied examples (Spruce, 1908; Wheeler, 1910; Janzen, 1966; Rickson, 1974; Beattie, 1985; Jolivet, 1986). Others,

such as seed dispersal by birds (e.g., Herrera, 1982), bats (Fleming and Heithaus, 1981), or fishes (Goulding, 1980) are also conspicuous features of tropical systems, but curiously their importance to plant populations and communities are still little understood.

The studies in this section address a range from comparatively loose associations to almost obligate and species-specific plant–animal partnerships. Rather than focusing on single cases, they all undertake broad comparisons along habitat gradients or over morphological, behavioral, and phenological character sequences.

Although it has long been accepted that mutualisms are most common in the tropics, other patterns are only recently being perceived. Why are mutualistic species especially diversified or abundant in certain kinds of communities? Under what circumstances do more specialized associations arise, and what environmental conditions enhance them? What community patterns result from combinations of sets of potential plant and animal partners?

These and other questions emerge from the newly-recognized patterns. Although provisional, the emerging answers promise a quantum advance in our grasp of the evolution of interspecific interactions and their community consequences. It is worth noting that the evolution of specialization, which has been especially investigated within antagonistic relationships, is nowadays equally discussed in relation to mutualisms.

Antagonistic Relationships Between Plants and Animals

A seemingly straightforward subject, this is one of the central areas of concern of contemporary animal–plant studies (Crawley, 1983; Denno and McClure, 1983; Strong et al., 1984a; Howe and Westley, 1988). The contributions included in this section show very clearly that there are tremendous gaps in our understanding of factors which determine herbivore occurrence and activity on certain plants, let alone their effect on attacked plants and plant populations (Crawley, 1989).

Animal occurrence and feeding has been shown to respond to plant geographical range and local abundance; to their size and structural complexity; to their nutritional quality, secondary chemistry, and phenology, among other factors (for reviews see Montgomery, 1978; Crawley, 1983; McDowell, 1984; Strong et al., 1984a; Slansky and Rodriguez 1987). Any or all of these are important in determining the consequences of animal feeding patterns and intensity (Crawley, 1989).

More herbivore species in concert are supposed to cause greater aggregate damage to a plant. This, and other similar simplifications, is an unwarranted assumption (Rhoades and Cates, 1976). Therefore, herbivore impact on plant growth, survival, and reproduction has to be ascertained directly. Nonetheless, often it will prove very difficult to find in nature the required herbivore species combinations, and also to control them in field experiments. Obviously, difficulties are even greater in assessing the consequences of herbivory on population features or on multispecies plant assemblages.

Studies that contribute data sets on this subject are few and additions are valuable, doubly so when they refer to tropical systems.

Janzen (1988) has emphasized that aseasonal tropical systems may have no fixed yearly population peak of herbivores and herbivore activity. Moreover, for a thorough assessment of herbivore effects, studies over several years are essential, as the most crucial effects are not necessarily immediate and between-year variation may show surprising patterns. Such studies are among the more urgent ones to be undertaken because of the rapid disappearance of areas sufficiently large to study the required combinations of factors and estimate their effects under natural conditions.

Plant–Butterfly Interactions

Butterflies and large moths are aesthetically appealing organisms which have always attracted collectors the world over; furthermore, their immatures are usually ectophagous on plants and so are comparatively easy to collect and rear. This unique combination of features has made the Macrolepidoptera conceivably the only major insect group which is reasonably well known in all geographical regions (Vane-Wright and Ackery, 1984). Among tropical phytophagous insects, they stand out even more from other taxa, whose study is still encumbered by considerable alpha-taxonimical and life-historical gaps. This solid background has allowed tropical butterfly ecology to advance into subjects scarcely touched by studies on other phytophagous groups.

The contributions to this section emphasize especially the chemical and behavioral components of butterfly and plant evolution. Chemical ecology which flourished with plant–lepidopteran studies (e.g., Brower and Brower, 1964; Feeny, 1970, 1976) has provided powerful new tools to examine proximal insect responses to plants. Recognition and choice of host plants by females, and their acceptance and tolerance by larvae, are sufficiently understood in some groups to make sense of their host–plant use patterns.

The evolution of present plant–butterfly patterns of association is a more complex matter. Plant chemistry has offered a number of valuable clues to explain some puzzling combinations and transferences among host taxa, but other factors are necessarily involved. In combination, the contributions to this section offer a spectrum of such factors: large-scale changes in plant distribution and diversification over different time scales; habitat distributions and phenologies of local potential host-plant sets; interspecific effects of mating behavior; the parasitoid matrix; and so on.

Coevolution is still very much an open question. Though there are indications of its operation, the evidence for selection of plant features by a particular phytophagous species or species combination is mostly circumstantial, with some rare exceptions such as egg-mimicking plant structures (Gilbert, 1975; Shapiro, 1981), and the matching of plant and butterfly phylogenies is not so close as originally expected (Ehrlich and Raven, 1964; see Miller, 1987). Further investigation of coevolution should be especially rewarding in tropical communities, where butterflies attain maximum diversification and all kinds of host use and association are easily found.

Specificity in Plant Utilization

Specialization in tropical communities has been regarded as a cause and/or a consequence of supposedly tighter species packing; accordingly, it has been most investigated with respect to interspecific competition.

Other things being equal, generalists should hold an advantage over specialists simply by tolerating a wider range of conditions; thus, specialization is not self-explanatory (Futuyma and Moreno, 1988). The common assumption of greater specialist efficiency would seem to account for a general drive toward specialization; but studies which tested this directly often failed to find the expected greater survival or growth efficiency of specialists over generalists, or even of specialists on their preferred compared to unused plants (e.g., Smiley, 1978; Smiley and Wisdom, 1985). The issue is as polemical now as ever before (see Bernays and Graham, 1988 and ensuing responses in Ecology 69, pp. 893–915).

Contributions to this section emphasize less familiar features associated with specialization, for example, host restriction as a mate-finding evolutionary strategy. Less straightforward functions of specialization may turn out to be at least as important as the presumed gains in immediate survival and growth efficiency, on which much of the previous theory rested.

Phytophagous insects are prime subjects for simultaneous studies of the genetical, physiological, and behavioral components of trophic specialization. Furthermore, they lend themselves excellently to investigations of the geographical differentiation of these components. The chapters in this section, together with several from the two preceding, typify a new run of studies which should be especially important for the understanding of the evolution of tropical communities, and should help us to make sense of the bewildering array of life history patterns found among their component species.

Community Patterns in Natural and Agricultural Systems

Do communities have discernible patterns, and if so, do these patterns reflect "real" structure, or are they conceivably derived from data-ordering and analytical procedures? Furthermore, is there strong evidence that particular perceived patterns result from certain ecological or evolutionary conditions, or are any such causal implications arguable at best?

Dissent on these questions escalated to a heated dispute during the last decade (see Salt, 1984; Strong et al., 1984b). Although there is less furor now, no full agreement has been attained. More recently, several voices have argued for a pluralistic stance in ecology, conceding the intrinsic complexity and the unreality of simultaneously simple and universal determinants of community structure (Diamond and Case, 1986; McIntosh, 1987). Within such an approach, one may ask what factors should be especially important in governing diversity and abundance of which kinds of organisms in what environmental settings? This leads to a more restricted set of predictions which are more amenable to direct testing (Schoener, 1986).

Tropical community ecology suffers first of all from a scarceness of basic data on different organisms in various communities and in various geographical regions and scales. Tropical plant–animal interfaces are so little surveyed that any present assessment of pattern in phytophagous assemblages is tentative at best. Still, current research is promising and, given the necessary support, substantial advances are to be expected.

The rapid expansion of intensive large-scale farming in the tropics places agroecology among the most pressing research areas. The closing chapters of this book address several aspects of agroecological systems, from the local scale whose short-term dynamics can be investigated through direct experimentation, to larger-scale questions where historical processes are more conspicuous. Hopefully, such studies will attract the interest of more workers in applied subjects, especially agronomy, forestry, and land management.

TROPICAL ECOLOGY IN PROSPECT

Recent tropical research has had a much needed myth-dispelling character. Widely accepted misconceptions, maintained by lack of familiarity with tropical communities, are being disproved by field work. These range over every aspect of ecology, from the smaller variability of population abundances (not so for insects, Bigger, 1976; Wolda, 1978) to the greater production of timber by lowland tropical forests (negated by Jordan, 1983).

The contributions to this book will add some substance to a less stylized conception of the tropics. It is a more complex view, with life-historical, local, and historical features added so that perceived patterns make better sense; in short, a view with less simple rules and more exceptions to them than ecologists may once have hoped for. However, this is not to say that no general trends exist nor that predictions are impossible. MacArthur (1972) stated that preoccupation with natural history could lead biologists to renounce searching out general patterns (or from doing science, as he put it). The reference to natural history, which is common to most contributions in this book, should be taken as part of a trend toward more restricted theory and more contingent predictions.

Faced with the alarmingly accelerating disappearance of tropical natural systems, some science administrators and scientists themselves are coming to consider so-called "pure" research as an unaffordable luxury in an emergency situation; botanizing while the Amazon burns, as it were. This should not be accepted uncritically. It is certainly true that a lot of current research is irrelevant to the most pressing practical problems in tropical countries, and that a concerted effort of strategical goal-oriented research is urgently needed. However, so little is known about the ecology of natural tropical systems that any morsel of fact is soon circulated, simplified, generalized, and converted into fanciful policy proposals. This precarious situation can only be changed by a substantially improved understanding of tropical systems, and to that end fundamental research simply cannot be dispensed with. We are a long way still from the minimally necessary ecological knowledge of tropical systems.

The future of tropical ecology depends on how much tropical nature remains after the next few decades. This is too obvious to need more elaboration than Janzen's (1977b) quip: "You cannot collect an interaction and keep a specimen on display in a museum." We can only hope that, given the opportunity, the prospects of tropical natural systems will also be improved by ecological research.

REFERENCES

Abrahamson, W. G. (Ed.). 1989. *Plant–animal interactions.* McGraw-Hill, New York.

Addicott, J. F. 1984. *Mutualistic interactions in population and community processes.* Pp. 437–455 in P. W. Price, C. N. Slobodchikoff, and W. S. Gaud (Eds.). A new ecology; novel approaches to interactive systems. Wiley, New York.

Agassiz, L. and E. C. Agassiz. 1868. *A journey in Brazil.* Ticknor and Fields, Boston (Portuguese translation, 1975, Itatiaia, Belo Horizonte).

Allee, W. C. 1926. Distribution of animals in a tropical rain-forest with relation to environmental factors. *Ecology* 7: 445–468.

Allee, W. C., A. E. Emerson, O. Park, T. Park, and K. P. Schmidt. 1949. *Principles of animal ecology.* Saunders, Philadelphia.

Allen, G. 1975. Life science in the twentieth century. Cambridge University Press, Cambridge (1981 reprint).

Baker, H. G. 1970. Evolution in the tropics. *Biotropica* 2: 101–111.

Bates, H. W. 1863. *The naturalist on the river Amazons.* 2nd ed., 1864. Murray, London.

Beattie, A. J. 1985. *The evolutionary ecology of ant–plant mutalisms.* Cambridge University Press, Cambridge.

Begon, M., J. L. Harper, and C. R. Townsend. 1986. *Ecology.* Blackwell, Oxford.

Belt, T. 1888. *The naturalist in Nicaragua.* 2nd ed. Bumpus, London.

Bentley, B. 1977. Extrafloral nectaries and protection by pugnacious bodyguards. *Annual Review of Ecology and Systematics* 8: 407–427.

Bernays, E. and M. Graham. 1988. On the evolution of host specificity in phytophagous arthropods. *Ecology* 69: 886–892.

Bigger, M. 1976. Oscillations of tropical insect populations. *Nature* 259: 207–209.

Boucher, D. H. (Ed.). 1985. *The biology of mutualism: Ecology and evolution.* Oxford University Press, Oxford.

Bowler, P. J. 1984. *Evolution; the history of an idea.* University of California Press, Berkeley, CA.

Brower, L. P. and J. V. Z. Brower. 1964. Birds, butterflies and plant poisons: A study in ecological chemistry. *Zoologica* 49: 137–159.

Brown, J. H. and A. C. Gibson. 1983. *Biogeography.* Mosby, Saint Louis, MI

Brown, K. S., Jr. and W. W. Benson. 1977. Evolution in modern Amazonian non-forest islands: *Heliconius hermathena*. *Biotropica* **9**: 95–117.

Burgess, R. L. 1981. Physiognomy and phytosociology of the international woodlands research sites. Pp. 1–35 in D. E. Reichle (Ed.). *Dynamic properties of forest ecosystems.* Cambridge University Press, Cambridge.

Cittadino, E. 1981. *Plant adaptation and natural selection after Darwin: ecological plant physiology in the German Empire, 1880–1900.* Ph.D. dissertation, University of Wisconsin, Madison, WI.

Clark, D. A., R. Dirzo, and N. Fetcher (Eds.). 1987. Ecología y ecofisiología de plantas en los bosques mesoamericanos. *Revista de Biología Tropical 35*, suplemento 1.

Coleman, W. 1986. Evolution into ecology? The strategy of Warming's ecological plant geography. *Journal of the History of Biology* **19**: 181–196.

Crawley, M. J. 1983. *Herbivory; the dynamics of animal–plant interactions.* Blackwell, Oxford.

Crawley, M. J. 1989. Insect herbivores and plant population dynamics. *Annual Review of Entomology* **34**: 531–564.

Crowter, S. and J. C. Taylor. 1859. *The gospel on the banks of the Niger.* Dawson of Pall Mall, London.

Cunha, L. A. 1980. A universidade temporã. Editora Civilização Brasileira, Rio de Janeiro.

Cutright, P. R. 1940. *The great naturalists explore South America.* MacMillan, New York.

Dajoz, R. 1984. Eléments pour une histoire de l'écologie. La naissance de l'écologie moderne au XIXe. siècle. *Histoire et Nature* (**24–25**): 5–111.

Darwin, C. 1860. *The voyage of the Beagle.* Natural History Library, New York.

Denno, R. F. and M. S. McClure (Eds.). 1983. *Variable plants and herbivores in natural and managed systems.* Academic, New York.

Desmukh, I. 1986. *Ecology and tropical biology.* Blackwell, Palo Alto.

Dial, K. P. and J. M. Marzluff. 1988. Are the smallest organisms the most diverse? *Ecology* **69**: 1620–1624.

Diamond, J. M. and T. J. Case (Eds.). 1986. *Community ecology.* Harper and Row, New York.

Dixon, A. F. G., P. Kindlman, J. Leps, and J. Holman. 1987. Why are there so few species of aphids, especially in the tropics. *American Naturalist* **129**: 580–592.

Dobzhansky, T. 1950. Evolution in the tropics. *American Scientist* **38**: 209–221.

Egerton, F. E. 1983. The history of ecology: Achievements and opportunities, part one. *Journal of the History of Biology* **16**: 259–310.

Egler, F. E. 1951. A commentary on American plant ecology based on the textbooks of 1947–1949. *Ecology* **32**: 673–695.

Ehrlich, P. R. and P. H. Raven. 1964. Butterflies and plants: A study in coevolution. *Evolution* **18**: 586–608.

Eiseley, L. 1958. *Darwin's century.* Doubleday, New York.

Elton, C. 1927. *Animal ecology.* Sidgwick and Jackson, London (1971 reprint, Science Paperbacks, London).

Elton, C. 1958. *The ecology of invasions by animals and plants.* Methuen, London.

Faegri, K. and L. van der Pijl. 1979. *The principles of pollination ecology.* Pergamon, Oxford.

Farnworth, E. G. and F. B. Golley (Eds.). 1975. *Fragile ecosystems; evaluation of research and applications in the Neotropics.* Springer, New York.

Feeny, P. 1970. Seasonal changes in oak leaf tannins and nutrients as a cause of spring feeding by winter moth caterpillars. *Ecology* **51**: 565–581.

Feeny, P. 1976. Plant apparency and chemical defense. *Recent Advances in Phytochemistry* **10**: 1–40.

Fernandes, G. W. and P. W. Price. 1988. Biogeographical gradients in galling species richness: Test of hypotheses. *Oecologia* **76**: 161–167.

Ferreira, A. R. 1971–1972. *Viagem filosófica pelas capitanias do Grão Pará, Rio Negro, Mato Grosso e Cuiabá.* Conselho Federal de Cultura, Rio de Janeiro.

Fischer, A. G. 1960. Latitudinal variations in organic diversity. *Evolution* **14**: 64–81.

Fisher, R. A. 1935. *The genetical theory of natural selection.* 2nd ed. 1958 reprint, Dover, New York.

Fittkau, E. J., J. Illies, H. Klinge, G. H. Schwabe, and H. Sioli (Eds.). 1968. *Biogeography and ecology in South America.* 2 vols. Dr. W. Junk, The Hague.

Fleming, T. H. and E. R. Heithaus. 1981. Frugivorous bats, seed shadows, and the structure of tropical forests. *Biotropica* (suppl.) **13**: 45–53.

Fleming, T. H., R. Breitwisch, and H. Whitesides. 1987. Patterns of tropical vertebrate frugivore diversity. *Annual Review of Ecology and Systematics* **18**: 91–109.

Futuyma, D. J. and G. Moreno. 1988. The evolution of ecological specialization. *Annual Review of Ecology and Systematics* **19**: 207–233.

Gilbert, L. E. 1975. Ecological consequences of a coevolved mutualism between butterflies and plants. Pp. 210–240 in L. E. Gilbert and P. R. Raven (Eds.) *Coevolution of animals and plants.* University of Texas Press, Austin.

Golley, F. B. 1983. Introduction. Pp. 1–8 in F. B. Golley (Ed.). *Tropical rain forest ecosystems.* Elsevier, Amsterdam.

Golley, F. B. and E. Medina (Eds.). 1975. *Tropical ecological systems.* Springer, Berlin.

Golley, F. B., J. T. McGinnis, R. G. Clements, G. I. Child, and M. J. Duever. 1975. *Mineral cycling in a tropical moist forest ecosystem.* University of Georgia Press, Athens, Georgia.

Gómez, L. D. and J. M. Savage. 1983. Searchers on that rich coast: Costa Rican field biology, 1400–1980. Pp. 1–11 in D. H. Janzen (Ed.). *Costa Rican natural history.* University of Chicago Press, Chicago.

Gómez-Pompa, A. and S. Del Amo (Eds.). 1985. *Investigaciones sobre la regeneración de selvas altas on Veracruz, México. II.* Editorial Alhambra Mexicana, México.

Gómez-Pompa, A., C. Vázquez-Yanes, S. Del Amo and A. Butanda (Eds.). 1976. *Investigaciones sobre la regeneración de selvas altas en Veracruz, México.* Compañia Editorial Continental, México.

Goodland, R. J. 1975. The tropical origins of ecology: Eugen Warming's jubilee. *Oikos* **26**: 240–245.

Goodland, R. J. and H. Irwin. 1975. *Amazon jungle: Green hell to red desert?* Elsevier, Amsterdam.

Goodman, E. J. 1972. *The explorers of South America.* Macmillan, New York.

Goulding, M. 1980. *The fishes and the forest.* University of California Press, Berkeley.

Haffer, J. 1969. Speciation in Amazonian forest birds. *Science* **165**: 131–137.

Herrera, C. M. 1982. Seasonal variation in the quality of fruits and diffuse coevolution between plants and avian dispersers. *Ecology* **63**: 773–785.

Holanda, S. B. 1963. *História geral da civilização brazileira. Difusão Europeia do Livro,* São Paulo.

Howe, H. F. and L. C. Westley. 1988. *Ecological relationships of plants and animals.* Oxford University Press, New York.

Huber, O. 1982. Significance of savanna vegetation in the Amazon territory of Venezuela. Pp. 221–244 in G. T. Prance (Ed.). *Biological diversification in the tropics.* Columbia University Press, New York.

Humboldt, A. and A. Bonpland. 1818. Personal narrative of travels to the equinoctial regions of the new continent during the years 1799–1804. 7 Vols. Translated from the French by H. M. Williams. Longman, London.

Huston, M. 1979. A general theory of species diversity. *American Naturalist* **113**: 81–101.

Janzen, D. H. 1966. Coevolution of mutualism between ants and acacias in Central America. *Evolution* **20**: 249–275.

Janzen, D. H. 1975. *Ecology of plants in the tropics.* Edward Arnold, London.

Janzen, D. H. 1977a. The impact of tropical studies on ecology. *Academy of Natural Sciences Special Publication* **12**: 159–187.

Janzen, D. H. 1977b. Promising directions of study in tropical animal–plant interactions. *Annals of the Missouri Botanical Garden* **64**: 706–736.

Janzen, D. H. 1988. Ecological characterization of a Costa Rican dry forest caterpillar fauna. *Biotropica* **20**: 120–135.

Jolivet, P. 1986. *Les fourmis et les plantes: Un exemple de coévolution.* Boubée, Paris.

Jordan, C. F. 1983. Productivity of tropical rain forest ecosystems and the implications for their use as future wood and energy sources. Pp. 117–136 in F. B. Golley (Ed.). *Tropical rain forest ecosystems.* Elsevier, Amsterdam.

Keller, E. F. 1988. Demarcating public from private values in evolutionary discourse. *Journal of the History of Biology* **21**: 195–211.

Kerner, A. 1878. *Flowers and their unbidden guests.* Kegan Paul, London.

Kimler, W. C. 1986. Advantage, adaptiveness, and evolutionary ecology. *Journal of the History of Biology* **19**: 215–233.

Kingsland, S. E. 1985. *Modeling Nature; episodes in the history of population ecology.* University of Chicago Press, Chicago.

Krebs, C. J. 1985. *Ecology; the experimental analysis of distribution of abundance.* 3rd ed. Harper and Row, New York.

Leigh, E. G., Jr. 1982. Introduction. Pp. 11–17 in E. G. Leigh, A. S. Rand, and D. M. Windsor (Eds.). *The ecology of a tropical forest; seasonal rhythms and long-term changes.* Smithsonian Institution Press, Washington, D. C.

Levin, D. 1976. *Alkaloid-bearing plants: An Ecogeographic perspective. American Naturalist* **110**: 261–284.

Livingstone, D. and C. Livingstone. 1865. *Narrative of an expedition to the Zambesi and its tributaries.* Murray, London.

Lloyd, C. 1985. *The travelling naturalists.* Croom Helm, London.

Longman, K. A. and J. Jeník. 1987. *Tropical forest and its environment.* 2nd ed. Longman, London.

Lovejoy, T. E., R. O. Bierregaard, J. M. Rankin, and H. O. R. Schubart. 1983. Ecological dynamics of tropical forest fragments. Pp. 377–384 in Sutton, S. L., T. C. Whitmore, and A. C. Chadwick (Eds.). *Tropical rain forest: Ecology and management*. Blackwell, Oxford.

MacArthur, R. H. 1969. Patterns of communities in the tropics. *Biological Journal of the Linnean Society* 1: 19–30.

MacArthur, R. H. 1972. *Geographical ecology*. Harper and Row, New York.

Margulis, L. and R. Fester (Eds.). 1990. *Evolution and speciation: Symbiosis as a source of evolutionary innovation*. MIT Press, Boston (in press).

May, R. M. 1973. *Stability and complexity in model ecosystems*. Princeton University Press, Princeton.

May, R. M. 1978. The dynamics and diversity of insect faunas. Pp. 188–204 in L. A. Mound and N. Waloff (Eds.). *Diversity of insect faunas*. Blackwell, Oxford.

Mayr, E. 1982. *The growth of biological thought*. Belknap. Cambridge, MA.

McDowell, L. R. 1984. *Nutrition of grazing ruminants in warm climates*. Academic, New York.

McIntosh, R. P. 1985. *The background of ecology*. Cambridge University Press, Cambridge.

McIntosh, R. P. 1987. Pluralism in ecology. *Annual Review of Ecology and Systematics* 18: 321–341.

Meggers, B. J., E. S. Ayensu, and W. D. Duckworth (Eds.). 1973. *Tropical forest ecosystems in Africa and South America: A comparative review*. Smithsonian Institution Press, Washington, D. C.

Miller, J. S. 1987. Host–plant relationships in the Papilionidae (Lepidoptera): Parallel cladogenesis or colonization? *Cladistics* 3: 105–120.

Mitman, G. 1988. From the population to society: The cooperative metaphors of W. C. Allee and A. E. Emerson. *Journal of the History of Biology* 21: 173–194.

Montgomery, G. G. (Ed.). 1978. *The ecology of arboreal folivores*. Smithsonian Institution Press, Washington, D. C.

Moreau, R. E. 1966. The bird faunas of Africa and its islands. Academic, London.

Müller, F. 1864. *Für Darwin*. Engelmann, Leipzig.

Müller, F. 1879. *Ithuna* and *Thyridia*: a remarkable case of mimicry in butterflies. *Transactions of the Royal Entomological Society of London* 1879: XX–XXIX.

Myers, N. 1979. The sinking ark. Pergamon, Oxford.

Neiva, A. 1929. *Esboço histórico sobre a Botânica e Zoologia no Brazil*. Sociedade Impressora Paulista, São Paulo.

NRC (National Research Council). 1980. *Research priorities in tropical biology*. National Academy of Sciences, Washington.

Odum, H. T. and R. Pigeon (Eds.). 1970. *A tropical rain forest*. U. S. Atomic Energy Commission, Springfield.

Owen, D. F. 1976. *Animal ecology in tropical Africa*. 2nd ed. Longman, London.

Owen, D. F. and D. O. Chanter. 1970. Species diversity and seasonal abundance in tropical Ichneumonidae. *Oikos* 21: 142–144.

Peters, R. H. 1983. *The ecological implications of body size*. Cambridge University Press, Cambridge.

Pianka, E. R. 1966. Latitudinal gradients in species diversity: A review of concepts. *American Naturalist* **100**: 33–46.

Pianka, E. R. 1984. *Evolutionary ecology*. 4th ed. Harper and Row, New York.

Picado, C. 1913. Les Bromeliacées épiphytes considerées comme milieu biologique. *Bulletin de la Société d'Entomologie de France et Belgique* **7**: 215–360.

Pijl, L. van der. 1972. *Principles of dispersal in higher plants*. Springer, Berlin.

Piso, G. 1658. *De Indiae utriusque re naturali et medica*. Elsevier, Amsterdam (Portuguese translation, Instituto Nacional do Livro, Rio de Janeiro, 1957).

Prance, G. T. (Ed.). 1982. *Biological diversification in the tropics*. Columbia University Press, New York.

Price, P. W. 1988. An overview of organismal interactions in ecosystems in evolutionary and ecological time. *Agriculture, Ecosystems and Environment* **24**: 369–377.

Real, L. (Ed.) 1983. *Pollination biology*. Academic, New York.

Rhoades, D. F. and R. G. Cates. 1976. Toward a general theory of plant antiherbivore chemistry. *Recent Advances in Phytochemistry* **10**: 168–213.

Richards, P. W. 1952. *The tropical rain forest; an ecological study*. Cambridge University Press, Cambridge.

Rickson, F. R. 1974. Absorption of animal tissue breakdown products into a plant stem—the feeding of a plant by ants. *American Journal of Botany* **66**: 87–90.

Robinson, M. H. 1978. Is tropical biology real? *Tropical Ecology* **19**: 30–50.

Ronan, C. A. 1983. *The Cambridge illustrated history of the World's science*. Cambridge University Press, Cambridge.

Salt, G. G. 1984. Ecology and evolutionary biology: A roundtable on research. *Reprint of American Naturalist* **122**(5), 1983. Chicago University Press, Chicago.

Schoener, T. W. 1986. Patterns in terrestrial vertebrate versus arthropod communities: Do systematic differences in regularity exist? Pp. 556–586 in J. Diamond and T. J. Case (Eds.). *Community ecology*. Harper and Row, New York.

Schoener, T. W. and D. H. Janzen 1968. Notes on environmental determinants of tropical versus temperate insect size patterns. *American Naturalist* **102**: 207–224.

Schwartzman, S. 1979. *Formação da comunidade científica no Brasil*. Editora Nacional, São Paulo.

Shapiro, A. M. 1981. The pierid red-egg syndrome. *American Naturalist* **117**: 276–294.

Sheail, J. 1987. *Seventy-five years in ecology: The British Ecological Society*. Blackwell, Oxford.

Silva, A. G. A., C. R. Gonçalves, D. M. Galvão, A. J. L. Gonçalves, J. Gomes, M. N. Silva, and L. Simoni. 1968. *Quarto catálogo dos insetos que vivem nas plantas do Brasil*. 4 Vols. Ministério da Agricultura, Rio de Janeiro.

Slansky, F., Jr. and J. G. Rodriguez (Eds.). 1987. *Nutritional ecology of insects, mites, spiders, and related invertebrates*. Wiley, New York.

Smiley, J. T. 1978. Plant chemistry and evolution of host specificity: New evidence from *Heliconius* and *Passiflora*. *Science* **201**: 745–747.

Smiley, J. T. and C. S. Wisdom. 1985. Determinants of growth rate on chemically heterogeneous host plants by specialist insects. *Biochemical Systematics and Ecology* **13**: 305–312.

Spix, J. B. and C. F. P. von Martius. 1824. *Reise in Brasilien.* Friedrich Fleischer, München (Portuguese translation, Itatiaia, Belo Horizonte, 1981).

Spruce, R. 1908. *Notes of a botanist on the Amazon and Andes.* 2 vols. Macmillan, London.

Stepan, N. 1976. *Beginnings of Brazilian science.* Science History Publications, New York (Portuguese translation, Artenova, Rio de Janeiro, 1981).

Stevens, G. C. 1989. The latitudinal gradient in geographical range: How so many species coexist in the tropics. *American Naturalist* 133: 240–256.

Strong, D. R., J. H. Lawton, and R. Southwood. 1984a. *Insects on plants: Community patterns and mechanisms.* Blackwell, Oxford.

Strong, D. R., D. Simberloff, L. G. Abele, and A. B. Thistle (Eds.). 1984b. *Ecological communities: Conceptual issues and the evidence.* Princeton University Press, Princeton.

Sutton, S. L., T. C. Whitmore, and A. C. Chadwick (Eds.). 1983. *Tropical rain forest: Ecology and management.* Blackwell, Oxford.

Théodoridès, J. 1965. *Histoire de la Biologie.* Presses Universitaires de France, Paris.

Van Valen, L. 1973. Body size and numbers of plants and animals. *Evolution* 27: 27–35.

Vane-Wright, R. I. and P. R. Ackery (Eds.). 1984. *The biology of butterflies.* Academic Press, London.

Vanzolini, P. E. and E. E. Williams. 1970. South American anoles: The geographic differentiation and evolution of the *Anolis chrysolepis* species group (Sauria, Iguanidae). *Arquivos de Zoologia* 19: 1–298.

Wallace, A. R. 1853. *A narrative of the travels on the Rio Negro.* Haskell, New York.

Wallace, A. R. 1869. *The Malay Archipelago.* Macmillan, London.

Wallace, A. R. 1878. *Tropical nature and others essays.* Macmillan, London.

Warming, E. 1892. Lagoa Santa. Et bidrag til den biologiske plantegeografi. *Kongel. Danske Vidensk. Selsk. Skr.* 6: 153–488 (Portuguese translation by A. Loefgren, 1908; 1973 reprint, Itatiaia, Belo Horizonte).

Warming, E. 1895. *Plantesamfund; Grundtraek af den oekologiska plantegeografi.* Philipsen, Copenhagen (English translation: Oecology of plants—An introduction to the study of plant communities. Clarendon, Oxford, 1909).

Weaver, J. E. and F. E. Clements 1938. *Plant ecology.* McGraw-Hill, New York.

Wheeler, W. M. 1910. *Ants, their structure, development and behavior.* Columbia, New York.

Wolda, H. 1978. Fluctuations in abundance of tropical insects. *American Naturalist* 112: 1017–1045.

Worthington, E. B. 1983. *The ecological century; a personal appraisal.* Clarendon Press, Oxford.

SECTION I
Tropical and Temperate Comparisons

Collections and natural history accounts of the great voyagers of the 18th century, especially in tropical regions, had a momentous effect on the development of modern taxonomical schemes (Eiseley, 1958). The personal confrontation of 19th-century naturalists with tropical nature was equally decisive for the gestation of ecology. Avid and perceptive collectors, they referred time and again to the staggering variety of tropical organisms; thus, in an often-quoted passage, Bates (1863) contrasted the 700-odd species of butterflies to be found "within an hour's walk" of the town of Belém, Pará with the 66 British or 321 European species then known. Although overwhelmed by this grand diversity, Bates as well as his predecessors noticed that some groups, such as the ground beetles, were much scarcer and less diverse in tropical than in temperate countries: ecological, evolutionary, and biogeographical explanations for these differences were alluded to. Indeed, the impact of tropical diversity on both Darwin and Wallace was essential to their development of the theory of evolution by natural selection.

The first general treatments of ecology were produced in the late 19th century by Warming (1895) and Schimper (1898), both of whom had worked long and fruitfully in the tropics. However, as McIntosh (1985) points out, these early tropical roots of ecology were almost forgotten during the first half of the 20th century, when evolution was considered of little concern and most investigators concentrated on boreal and temperate communities. Nonetheless, a number of generalizations on tropical versus temperate ecological systems became established in the literature, sometimes with little indication of their origin or factual basis. Indeed, the more recent flourishing of tropical studies has shown that conventional wisdom on the features and mechanics of tropical communities needs to be reexamined from the very beginning.

In our first section tropical and temperate communities are contrasted in various ways meant to pinpoint essential features in which they differ. In Chapter 2, two linked aspects are broadly reviewed: where there are major differences in plant antiherbivore features, on one hand, and in the intensity of herbivory, on the other. Chapter 3 is devoted to questioning often-repeated statements on the greater diversity in the tropics, their correlates, and implications. In Chapter 4, diversity, abundance, and size patterns are meshed together in an effort toward a difficult but rewarding synthesis. Chapter 5 follows a different blueprint, with a more restricted but robust comparison of diversity gradients in tropical and temperate settings.

An issue that pervades this section is whether differences between tropical and temperate systems are mostly a matter of degree, or whether qualitative and structural differences are such as to imply different governing processes: to what extent, indeed, are tropical and temperate ecologies distinct?

REFERENCES

Bates, H. W. 1863. *The naturalist on the river Amazons.* 2nd ed. 1864. Murray, London.

Eiseley, L. 1958. *Darwin's Century.* Anchor Books, New York.

McIntosh, R. P. 1985. *The background of ecology; concept and theory.* Cambridge University Press, Cambridge.

Schimper, A. F. W. 1898. *Pflanzen-Geographie auf physiologischer Grundlage.* Fischer, Jena (English translation: *Plant ecology upon a physiological basis.* Clarendon, Oxford, 1903).

Warming, E. 1895. *Plantesamfund; Grundtraek af den oekologiska plantegeografi.* Philipsen, Copenhagen (English translation: *Oecology of plants: An introduction to the study of plant communities.* Clarendon, Oxford, 1909).

2 Comparison of Herbivory and Plant Defenses in Temperate and Tropical Broad-Leaved Forests

PHYLLIS D. COLEY and T. MITCHELL AIDE

One important component of plant–herbivore interactions over both ecological and evolutionary time scales is plant defense. Defenses influence the population dynamics and distribution of plants and herbivores and, in turn, this complicated interplay influences the evolution of defenses. Variation among species or communities in defense levels therefore indicates that they differ fundamentally in the role of plant–herbivore interactions. The focus of this chapter is to determine if there are differences in antiherbivore defenses between temperate and tropical communities, and if so, why? First we will provide data comparing levels of herbivory and plant defense; then we will examine selective pressures which may be responsible for these patterns; and finally we will discuss some of the implications for plant–herbivore interactions. Although quantitative data with which to make these comparisons are still difficult to find, our preliminary results suggest that this is a fruitful area for comparative research. The evidence we present is not meant to provide definitive answers, but rather to identify intriguing patterns and encourage future research.

Forest Communities

It is clearly naive to make sweeping generalities about differences between temperate and tropical communities without acknowledging the enormous variation within a single latitude. To partially control for this variation, we focused our latitudinal comparison on broad-leaved forests, exclusive of Australia. Whenever possible we subdivided the tropical data into three smaller categories: seasonal dry forest species, wet forest shade-tolerant species, and wet forest light-gap species. Although the latter two both occur together in wet forests, they are separated because of very different microhabitat preferences.

Plant-Animal Interactions: Evolutionary Ecology in Tropical and Temperate Regions, Edited by Peter W. Price, Thomas M. Lewinsohn, G. Wilson Fernandes, and Woodruff W. Benson. ISBN   0-471-50937-X   © 1991 John Wiley & Sons, Inc.

Shade-tolerant species can exist in the shaded understory for many years, while gap species are found only in the high light environments of gaps made by fallen trees (Brokaw 1985; Denslow 1987; Hartshorn 1980). Although vertebrate herbivores are present in all four forest categories, the major herbivores are probably insects. Our conclusions are therefore based on a subset of the possible plant and herbivore communities in the world.

HERBIVORY

One estimate of the ecological importance of plant–herbivore interactions within a community is the level of herbivory. We obtained estimates of annual rates of herbivory (percent leaf area damaged per year) for 23 forest communities. Annual rates of herbivory in tropical forests (10.9%) were significantly greater than in temperate forests (7.5%) (Fig. 2-1, $P < 0.05$).

Unfortunately, different methods were employed in the different studies, so there is a large degree of uncertainty associated with these estimates. Furthermore, most studies made one-time measurements of holes in leaves. This gives an underestimate of herbivory since it does not include leaves that were completely eaten (Coley 1982; Lowman 1984). We therefore encourage additional quantitative measures of rates of herbivory in different communities.

Significantly higher herbivory in the tropics could result from greater herbivore pressure, or from lower levels of defenses. Since in the following sections we will show evidence for higher defenses in the tropics, we suggest that these results reflect greater overall herbivore pressure.

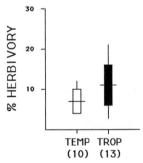

Figure 2-1. Annual rates of herbivory in temperate and tropical broad-leaved forests. Plots indicate mean ± sd and range. Sample sizes are in parentheses. $P < 0.05$, T-test. Data are from Adis et al. 1979, Bray 1964, Coley 1982, de la Cruz and Dirzo 1987, Dirzo 1984, Edwards 1977, Funke 1973, Jordan and Uhl 1978, Leigh and Smythe 1978, Nielson 1978, Odum and Ruiz-Reyes 1970, Proctor et al. 1983, Wint 1983, Woodwell and Whittaker 1968, Reichle and Crossley 1967, Reichle et al. 1973, Stanton 1975.

DEFENSES OF MATURE LEAVES

The diversity of plant characteristics which have been shown to deter herbivores is enormous (Rosenthal and Janzen 1979; Whittaker and Feeny 1971). We were able to obtain comparative data on several chemical, physical, and nutritional aspects of mature leaves which can function as defenses. These include alkaloids, simple phenols, condensed tannins, acid detergent fiber, toughness, nitrogen, and water.

Alkaloids

In a review of the literature, Levin (1976) found that the frequency of species with alkaloids is significantly greater in tropical than temperate regions, whether

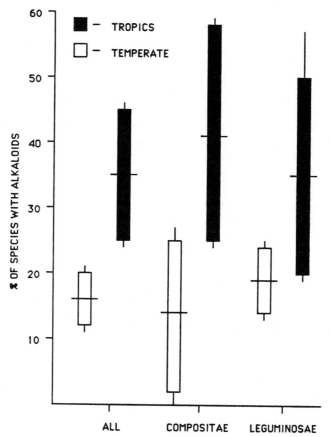

Figure 2-2. A latitudinal comparison of the percentage of species with alkaloids in the flora of 13 countries (7 tropic and 6 temperate), or in the Compositae or Leguminosae separately. Data are from Levin 1976 (Tables 2 and 4, data from Australia were excluded). Plots indicate mean ± sd and range.

considering the entire data set or only the families Leguminosae and Compositae (Fig. 2-2). On average, 16% of the temperate species contain alkaloids, as compared with over 35% of the tropical species. A more detailed analysis of over 500 genera indicated that alkaloids of tropical genera are more toxic and occur in higher concentrations (Levin and York 1978).

These data (Levin 1976; Levin and York 1978) show that tropical alkaloids are both significantly more common and more toxic than temperate ones. However, these may be overestimates, since tropical species were not sampled at random, but rather with a bias toward families from which alkaloids were known. Nonetheless, the latitudinal differences in alkaloids are so enormous that we suggest there is a distinctly greater probability of defense by alkaloids in the tropics.

Phenolic Compounds

Phenolic compounds are essentially ubiquitous in the plant kingdom and can be present in very large concentrations in leaf tissue (Swain 1979). Although simple phenols have been used as an index of defense, they serve many functions in the plant and are difficult to quantify accurately (Gershenzon 1984; Zucker 1983). Data on simple phenols should therefore be interpreted cautiously. Concentrations of phenols averaged less than 8% dw (Fig. 2-3a; Table 2-1; total $n = 286$ species) with no significant differences among forests types.

Condensed tannins, or polyphenols, have been suggested as one of the most effective types of defensive compounds (Feeny 1976, Rhoades and Cates 1976). Despite conflicting data on effectiveness and mode of action of tannins (Bernays 1981; Martin and Martin 1982; Martin et al. 1987; Zucker 1983), it is likely that they function as defenses against herbivores or pathogens (Coley 1986). Data from 243 species using the BUOH/HCl assay for condensed tannins shows significant differences among forest types (Fig. 2-3b; Table 2-1). Leaves of temperate deciduous species have the lowest concentrations (1.9% dw), and shade-tolerant tropical species have highest (6.0% dw).

In all of these studies quebracho tannin was used as a standard. Since different tannins have different reactivities in this assay (Wisdom et al. 1987), quebracho tannin may not give a reliable estimate of the amount of tannin in all species. There is no particular reason to assume that temperate tannins are less reactive, so it is unlikely that the low levels in deciduous temperate species are an artifact. Nonetheless, we suggest that future surveys use tannins extracted from each species as its own standard.

Toughness and Fiber Content

Leaf toughness or high fiber content provide structural support for the leaf, but have also been shown to be extremely effective as antiherbivore defenses (Coley 1983, 1987; Feeny 1970; Raupp 1985; Tanton 1962). Toughness of temperate species is similar to that of species from the seasonal dry forests of Costa Rica,

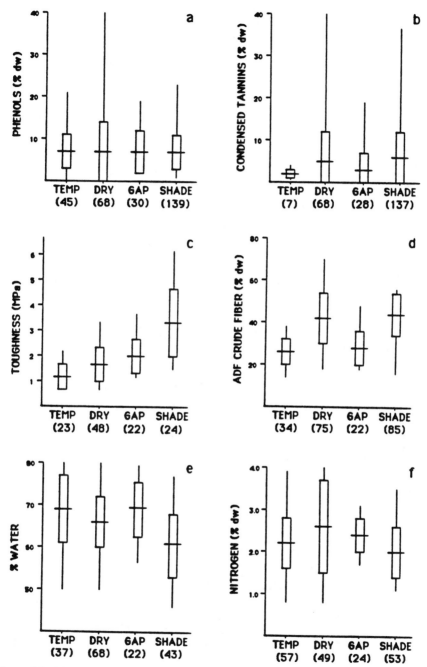

Figure 2-3. Mature leaf characteristics for broad-leaved tree species from temperate and tropical forests. See Table 2-1 for details. Plots indicate mean ± sd and range.

TABLE 2-1. Mature leaf characteristics for broad-leaved tree species from temperate and tropical forests. See text for a description of forest categories. Means followed by different letters are significantly different at $P < 0.05$, Tukey's studentized range test

Variable		Temperate	Tropical		
			Dry	Gap	Shade
Simple phenols[a]	Mean	6.51a	6.59a	7.40a	7.06a
	N	45	68	30	139
	sd	4.4	6.6	4.5	4.3
	Range	0.2–21.0	0.9–42.4	1.9–19.4	0.6–22.8
Condensed tannins[b]	Mean	1.94ab	5.27ab	2.93a	6.03b
	N	7 •	68	28	137
	sd	1.2	7.5	4.4	6.2
	Range	0.5–3.5	0–43.8	0–19.1	0–37.0
Toughness[c]	Mean	0.12a	0.17ab	0.20b	0.33c
	N	23	48	22	24
	sd	0.04	0.07	0.07	0.13
	Range	0.07–0.21	0.04–0.33	0.13–0.36	0.14–0.61
Crude fiber (ADF)[d]	Mean	25.2a	42.4b	29.2a	44.6b
	N	34	75	22	85
	sd	5.5	12.8	7.6	10.0
	Range	14.0–37.4	18.0–70.3	19.7–47.7	15.6–55.5
Water[e]	Mean	69.0a	66.3a	69.9a	60.5b
	N	37	68	22	43
	sd	7.5	6.3	6.4	6.7
	Range	50.0–80.2	50.0–80.0	57.0–82.0	46.0–77.0
Nitrogen[f]	Mean	2.25a	2.58	2.42a	1.99b
	N	57	49	24	54
	sd	0.6	1.1	0.4	0.6
	Range	0.8–3.9	0.8–6.4	1.7–3.1	1.1–3.5
Leaf lifetimes[g]	Mean	6.0a	9.6a	6.9a	32.2b
	N	7	8	24	43
	sd	2.8	3.0	3.9	16.1
	Range	3.5–12.0	6.0–12.0	2.0–21.0	7.5–66.0

[a]Simple phenols (mg/g dw) are measured by the Folin Denis assay, ANOVA $P > 0.4$. Data are from Baldwin and Schultz 1988, Coley 1983, Crankshaw and Langenheim 1981, Faeth 1985, Gartlan et al. 1978, Hanley et al. 1987, Haukioja et al. 1978, Janzen and Waterman 1984, Langenheim et al. 1986, McKey et al. 1978, Meyer and Montgomery 1987, Mole et al. 1988, Montgomery (unpublished data), Oates et al. 1980, Potter and Kimmerer 1986, Ricklefs and Matthew 1982, Schultz et al. 1982, Waterman 1983, Waterman et al. 1984.

[b]Condensed tannins (mg/g dw) are measured by the BUOH/HCl assay with quebracho tannin as a standard. ANOVA was performed on an arcsin squareroot transformation, $P < 0.01$. Data are from Aide and Zimmerman 1990, Baldwin et al. 1987, Baldwin and Schultz 1988, Coley 1983, Crankshaw and Langenheim 1981, Faeth 1985, Gartlan et al. 1978, Haukioja et al. 1978, Janzen and Waterman

though gap and shade-tolerant species of wet tropical forests are significantly tougher (Fig. 2-3c; Table 2-1). Crude fiber content, measured by the acid-detergent method (Van Soest 1975), is also higher in tropical species (Fig. 2-3d; Table 2-1).

Nitrogen and Water Content

Two important components of leaf nutritional quality for herbivores are water and nitrogen contents (McNeill and Southwood 1979; Scriber and Slansky 1981). Water contents for shade-tolerant wet forest species average 60.5% and are significantly lower than for the other three forest types (66.3–69.9%; Fig. 2-3e; Table 2-1). Similar patterns are seen for nitrogen content, with shade-tolerant species having the lowest nitrogen concentrations (Fig. 2-3f; Table 2-1). These data suggest that shade-tolerant tropical species are significantly less nutritious for herbivores than species of other forest types.

HERBIVORY ON YOUNG LEAVES

The data presented in the preceding sections indicate greater herbivory and defense of mature leaves in tropical as compared with temperate communities. Do young leaves show similar latitudinal trends? Quantitative data on young leaves are exceedingly difficult to obtain; however, in the following sections we present evidence which suggests that young leaves also suffer higher levels of herbivory and have greater defenses in the tropics.

In most species, young leaves are more vulnerable to herbivores than are mature leaves. They have higher nitrogen and water contents and lower

---

1984, McKey et al. 1978, Meyer and Montgomery 1987, Mole et al. 1988, Montgomery (unpublished data), Oates et al. 1980, Palo 1984, Waterman 1983, Waterman et al. 1984.
[c]Toughness (MegaPascals) is expressed as the force necessary to punch a 5-mm diameter rod through a leaf. ANOVA $P < 0.0001$. Data are from Coley 1983, Howard (unpublished data), Lechowicz (unpublished data).
[d]Crude fiber (mg/g dw) is measured as acid detergent fiber. ANOVA $P < 0.001$. Data are from Coley 1983, Janzen and Waterman 1984, Oates et al. 1980, Ricklefs and Matthew 1982, Waterman 1983.
[e]Water is a percentage. ANOVA $P < 0.0001$. Data are from Ayres and MacLean 1987, Coley 1983, Crankshaw and Langenheim 1981, Denno and McClure 1983, Janzen and Waterman 1984, Kraft and Denno 1982, Langenheim et al. 1986, Lechowicz (unpublished data), Meyer and Montgomery 1987, Montgomery (unpublished data), Oates et al. 1980, Potter and Kimmerer 1986, Schroeder 1986, Schultz et al. 1982.
[f]Nitrogen is in mg/g dw. ANOVA $P < 0.002$. Data are from Aide and Zimmerman 1990, Ayres and MacLean 1987, Coley 1983, Denno and McClure 1983, Faeth 1985, Faeth et al. 1981, Hanley et al. 1987, Haukioja et al. 1978, Howard (unpublished data), Langenheim et al. 1986, Lechowicz (unpublished data), Meyer and Montgomery 1987, Milton 1979, Mole et al. 1988, Muller et al. 1987, Montgomery (unpublished data), Oates et al. 1980, Potter and Kimmerer 1986, Schroeder 1986, Schultz et al. 1982, Waterman et al. 1984.
[g]Leaf lifetimes are in months. ANOVA $P < 0.0001$. Data are from Aide and Zimmerman 1990, Coley 1988, Hladik and Blanc 1987, Kozlowski 1971, Marin and Medina 1981, Marquis 1984, Potter and Kimmerer 1986, Southwood et al. 1986, Uhl 1982, Waltz 1984.

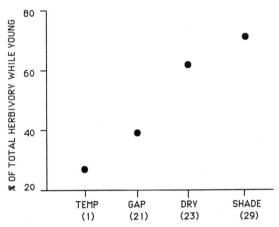

Figure 2-4. The percentage of total lifetime herbivory that occurs while the young leaves are expanding. Temperate data are a pooled estimate for an undetermined number of species in a *Liriodendron* forest (Reichle et al. 1973). Dry forest data are for 23 species (Stanton 1975, Ribeiro et al. unpublished data). Wet forest data are for 21 gap species (Coley 1983) and 29 shade-tolerant species (Coley 1983; Cooke et al. 1984; Wint 1983).

toughness, all characteristics which have been correlated with higher herbivory in laboratory and field studies (Raupp and Denno 1983; Scriber and Slansky 1981). The percentage of the total lifetime damage to a leaf that occurs while it is young (expansion phase) is less for temperate deciduous species compared with all three classes of tropical species (Fig. 2-4). This difference is even more amazing since the expansion phase is similar across forests, but total leaf lifespan is increasing along the *x* axis (Fig. 2-4). Therefore, in shade-tolerant tropical species, 70% of the herbivory during the 2.5-year lifespan occurs during the first month, while the leaf is young.

These data indicate that not only are young tropical leaves suffering higher absolute rates of damage than temperate leaves, they also have higher relative amounts. Furthermore, herbivory on young leaves can have a high impact on the plant since most resources have already been imported to the leaf but it has not yet begun to export photosynthate. This suggests that it is particularly important to both ecological and evolutionary questions to understand herbivory and defensive patterns of young tropical leaves.

DEFENSES OF YOUNG LEAVES

Physical and Chemical Defenses

Data on phenolic compounds and toughness suggest that young leaves are better defended in the tropics than in the temperate zone. Figure 2-5 expresses defense investment as the ratio of defenses in young and mature leaves for species in which both were measured. A ratio of one indicates that concentrations were similar in

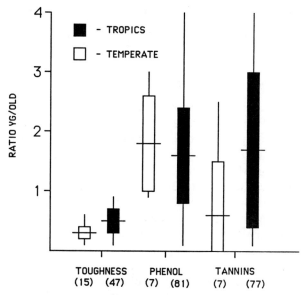

Figure 2-5. The ratio of defenses in young and mature leaves. T-tests of temperate/tropic comparisons are significant for toughness ($P < 0.001$) and tannins ($P < 0.04$), but not for phenols ($P > 0.6$). Data are from Ayres and MacLean 1987, Becker 1981, Becker and Martin 1982, Coley 1983, Crankshaw and Langenheim 1981, Dement and Mooney 1974, Faeth 1985, Feeny 1970, Gartlan et al. 1978, Hall and Langenheim 1986, Kraft and Denno 1982, Langenheim et al. 1986, Meyer and Montgomery 1987, Mole et al. 1988, Montgomery (unpublished data), Mooney et al. 1983, Oates et al. 1980, Potter and Kimmerer 1986, Raupp 1985, Rhoades and Cates 1976, Schroeder, 1986, Schultz et al. 1982. Plots indicate mean $\pm$ sd and range.

both aged leaves. For simple phenols measured by the Folin Denis assay, young leaves had almost twice the concentrations of mature leaves, however there is no significant latitudinal difference. Young tropical leaves are significantly tougher than temperate leaves. Condensed tannins show the largest differences between temperate and tropical species. Young tropical leaves have almost twice the concentrations of mature leaves, while young temperate leaves have only half as much. Because most early work on tannins was done in the temperate zone, it was thought to be physiologically impossible to have significant levels of condensed tannins in expanding leaves (McKey 1979; Orians and Janzen 1974). These data show that tropical young leaves do have condensed tannins and at levels greater than mature leaves.

Cost–benefit models of defense suggest that young leaves should be defended by small-molecular-weight mobile compounds (Feeny 1976; McKey 1974, 1978; Rhoades and Cates 1976). These have the advantage of only incurring turnover costs during the short period while the leaf is young (Coley et al. 1985). Furthermore, they can be reclaimed when the leaf is better defended by physical defenses (McKey 1979). It is therefore likely that much of the chemical defense of young leaves is in these mobile compounds about which we have no data. We suggest that the discrepancy in defense between young temperate and tropical leaves will be even greater when mobile compounds are considered.

Extra-floral Nectaries and Food Bodies

Many plants produce extra-floral nectaries (EFN) and food bodies which attract ants to the plant. There is strong support for the claim that the ants in turn provide the plant with partial protection from herbivores (Beattie 1985; Bentley 1977). Ant rewards are usually only active for a short period while the leaves are expanding or young (McKey 1989).

In a taxonomic review, Elias (1983) identified 66 families with EFN. Following the descriptions of Heywood (1978), we categorized families into three distributional classes: temperate, tropical, and cosmopolitan. For the analysis we only used the 48 families in the three subclasses (Dillenidae, Rosidae, and Asteridae) in which EFN are relatively common (Elias 1983).

There is a significant latitudinal effect on the distribution of families with EFN (Table 2-2, $P < 0.01$). Thirty-nine percent of the families with EFN are tropical, 27% are cosmopolitan, and 12% are temperate. Although the Dillenidae and Rosidae have a greater proportion of tropical families ($P < 0.05$), the latitudinal distribution of EFN was independent of subclass ($P > 0.1$), suggesting no phylogenetic bias.

Two potential sampling problems with this analysis are the effect of family size and the intensity of study. First, if tropical families have more species, there would be an increased probability of discovering a species with EFN. However, family size is greatest in the cosmopolitan group, so it is apparently not an overriding effect. Second, tropical floras are less well studied, decreasing the likelihood of discovering species with EFN. We know that nine tropical families were erroneously classified by Elias (1983) as not having EFN. Further study of tropical species will undoubtedly strengthen the observed trend of greater occurrence of EFN in tropical species.

TABLE 2-2. **The latitudinal distribution of extrafloral nectaries (EFN). The proportion of families known to have EFN is compared between three latitudinal classes (tropical, cosmopolitan, and temperate) within three major subclasses. Values are the number of families with EFN, total number of families, and the percent with EFN in parentheses. Log linear models (Systat 1987) were constructed to determine the interactions between subclass (taxonomic constraints), family distribution (latitudinal comparison), and the presence of EFN**

	Tropical	Cosmopolitan	Temperate	Total
Dillenidae	14/32 (43.8)	1/8 (12.5)	2/15 (13.3)	17/55 (30.9)
Rosidae	13/40 (32.5)	4/16 (25.0)	3/23 (13.0)	20/79 (25.3)
Asteridae	5/10 (50.0)	5/13 (38.5)	1/11 (9.1)	11/34 (32.3)
Total	32/82 (39.0)	10/37 (27.0)	6/49 (12.2)	48/168 (28.6)

TABLE 2-3. Percent of species and stems with ant rewards in different temperate and tropical communities. Ant rewards include food bodies or extrafloral nectaries

Community	Occurence of Ant Rewards		Reference
	Species (%)	Stems (%)	
Tropics			
Barro Colorado Island, Panama			
Total	37	16	Schupp and
Gap	39	22	Feener
Shade	15	5	(unpublished)
Lianas	45		
Guanacaste, Costa Rica			
Edge		80	Bentley 1976
Interior		40	
Jamaica			
Low elevation		28	Keeler 1979
High elevation		0	
Cerrado, Brazil			
Itirapina	15.4	17.5	Oliveira and
Sao Simao	20.0	18.6	Leitao–Filho
Mogi-Guacu	20.2	18.6	1987
Mogi-Mirim	16.5	15.1	
Luis Antonio	21.9	7.6	
Temperate			
Iowa, USA			
Deciduous forest	25.9		Schupp (unpublished)
Nebraska, USA			
Total	3		Keeler 1980
Riparian		1.3	
Deciduous forest		1.8	
Tall grass Prairie		0	
Sandhill Prairie		8.3	
California, USA			
Total	0	0	Keeler 1981
Colorado, USA			
Alpine	< 1		D. Inouye (personal communication)

Community studies on the distribution of ant rewards provide further evidence for their relatively greater occurrence in tropical regions (Table 2-3). Analyses of floristic composition as well as relative abundance suggest that ant defense is significantly more common in the tropics. Approximately 30% of the species and individual plants in lowland tropical forests contain ant rewards. This percentage is somewhat less in the cerrado (tropical savannah) and drops considerably in temperate and alpine communities. Reduced ant defense in these areas may be partially explained by a reduced abundance and diversity of ants (Kusnezov 1952). Furthermore, the more obligate, species-specific myrmecophytic associations such as *Acacia–Psuedomyrmex* (Janzen 1967) and *Cecropia–Azteca* (Schupp 1986) are apparently restricted to tropical regions.

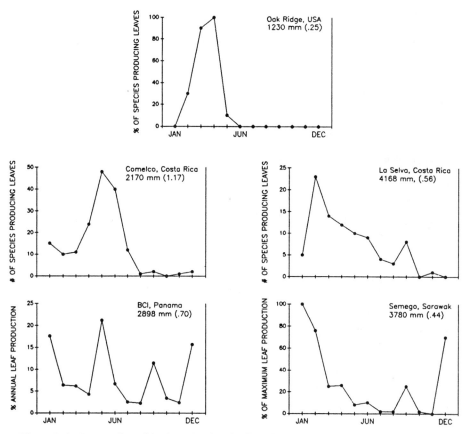

Figure 2-6. Seasonality of leaf production in five sites with differing degrees of abiotic seasonality. Annual rainfall in millimeters and the coefficient of variation in rainfall (in parentheses) is presented for each site. Data are from Frankie et al. 1974 (Comelco and La Selva), Fogden 1972 (Semego), Aide 1988 (BCI), and Taylor 1974 (Oak Ridge).

Phenological Escape

Young leaves are exceptionally vulnerable to herbivores, but only for a relatively short period before they toughen and become better defended. It has been suggested that phenological patterns of leaf production can help protect young leaves by temporally altering this period of vulnerability (Aide 1988; Crawley 1983; McKey 1974). Damage may be reduced if leaves are produced in synchronous flushes which satiate herbivores, or in seasons when herbivore populations are low. Is the timing of leaf production merely a reflection of abiotic constraints, or is there evidence that phenological patterns may be responding to selection by herbivores?

Figure 2-6 gives patterns of leaf production for five forests with differing degrees of abiotic seasonality: (a) Oak Ridge is a temperate broad-leaved forest with a severe winter; (b) Comelco is a dry forest site in Costa Rica; (c) Barro Colorado Island in Panama is a moist forest with a 4-month dry season; (d) La Selva in Costa Rica has relatively uniform rainfall throughout the year; and (e) Semego in Sarawak on the equator has uniform rainfall and temperature regimes year-round. In all sites, leaf production is highly synchronous despite the fact that there may be few abiotic constraints in some of the tropical sites.

A more detailed study of 32 species of saplings in the understory of Barro Colorado Island elucidates some of the defensive advantages of seasonal leaf production (Aide 1988). Thirty-four percent of the annual leaf production occurs in the dry season when water stress is severe (Rundel and Becker 1987), but herbivore abundance is low (Wolda 1978). Herbivore pressure may be driving the peak of leaf production toward an abiotically unfavorable time of year. Furthermore, individuals that produce leaves during the population peaks suffer significantly less damage than individuals that produce leaves during the troughs (Aide 1988). Thus there is an advantage to synchrony, regardless of season, due to satiation of herbivores.

SELECTIVE PRESSURES FOR LATITUDINAL DIFFERENCES IN DEFENSE

For all the chemical, physical, nutritional, and phenological defenses which we have reported, there are significant or suggestive trends for greater defense in tropical communities. In the following sections we examine several selective factors which have been suggested as being important for the evolution of defenses, and discuss how they may differ latitudinally.

Inherent Growth Rate

Theoretical and empirical data have argued that the inherent growth rate of a species is an important evolutionary determinant of defense investment (Coley et al. 1985; Grime 1977; Gulmon and Mooney 1986; Janzen 1974; McKey 1979). Slow-growing species tend to have higher levels of defense and to suffer lower

TABLE 2-4. **Studies comparing herbivory on fast- and slow-growing plant species in temperate and tropical communities**[a]

Herbivory	Temperate	Tropical	Both
Fast > slow	14	6	20
Slow > fast	0	1	1

[a]Data are from Batzli and Jung 1980, Bryant et al. 1983, Cates and Orians 1975, Chapin 1980, Coley 1983, de la Cruz and Dirzo 1987, Dirzo 1984, Grime et al. 1968, Kuropat 1984, MacLean and Jensen 1984, McKey and Gartlan 1981, Pimental 1976, Rathcke 1985, Reader and Southwood 1981, Robus 1981, Struhsaker 1978, Trudell and White 1981, Waltz 1984, Weeken 1969, Whitten et al. 1987, Williams et al. 1984.

rates of herbivory. In 20 of 21 community studies, fast-growing species had higher herbivory than slow-growing species (Table 2-4). Although there are large differences in growth rates for species within and among communities, we were unable to find sufficient evidence to show latitudinal differences. So at present, there is no reason to expect tropic–temperate differences in defenses to be due to differences in inherent growth rates.

Leaf Lifetimes

Many theories have considered the risk of herbivory and the value of a plant tissue to be two important factors influencing the evolution of defense (Feeny 1976; Janzen 1974; McKey 1979; Rhoades and Cates 1976; Southwood et al. 1986). We suggest that leaf lifetime will influence both the risk and value of leaves. First, long-lived leaves are at risk to herbivores for longer times than short-lived leaves. Second, long-lived leaves may be more valuable to the plant because damage to a long-lived leaf represents a greater reduction in future productivity than would damage to a short-lived leaf. Third, species with long-lived leaves are common in resource-poor environments, where replacement of damaged leaves is costly or impossible. Therefore, because of both risk and value considerations, it is hypothesized that long-lived leaves should be better defended against herbivores.

Leaf lifetime is also thought to influence the type of defense which is most cost-effective (Coley et al. 1985; Coley 1988; McKey 1979, 1984). Quantitative or immobile defenses, such as tannins and lignins, can be present in large concentrations, which represent considerable initial construction costs. However, they are extremely stable metabolically, so the maintenance costs associated with turnover are low. Small-molecular-weight mobile defenses, such as alkaloids, cardiac glycosides, and monoterpenes, are present in low concentrations, but may have continued costs associated with rapid rates of turnover. Because of these differences in cost schedules, small-molecular-weight compounds may be

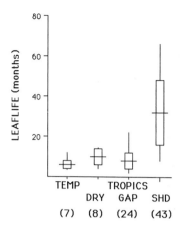

Figure 2-7. A comparison of leaf lifetimes for the four forest categories. See Table 2-1 for details. Plots indicate mean ± sd and range.

most cost-effective in short-lived leaves, and quantitative defenses in long-lived leaves.

Does leaf lifetime show a latitudinal trend, and could this explain some of the defense differences between temperate and tropical communities? Data on 87 species show that leaf lifetimes differ significantly among forest types (Figure 2-7). Broad-leaved temperate forest leaves have the shortest leaf lifetimes, though they are not significantly different from the seasonal dry tropical forests with long periods of drought and a high abundance of deciduous species. Nor are they significantly different from light-gap species which grow in high-resource microhabitats and can maintain rapid growth and high leaf turnover rates. Leaf lifespans of these three groups are significantly shorter than the 2.5 years typical for saplings of shade-tolerant tropical species. Lifespans for canopy individuals may be slightly shorter.

Differences among forest types in average leaf lifetimes suggest that there should be differences in defenses, with greater quantitative defense and lower qualitative defense in longer-lived leaves. Two important quantitative defenses, condensed tannin and crude fiber, support this prediction, showing a significant positive relationship with leaf lifetime (Fig. 2-8). If the data are transformed to approach a linear relationship, as much as 90% of the variation among forests in these two defenses can be explained by differences in leaf lifetime. Latitudinal differences in quantitative defenses may therefore be primarily due to latitudinal differences in leaf lifetimes. Nonetheless, it is interesting that although temperate, gap, and dry-forest species do not differ significantly in leaf lifetimes, the tropical species are much better defended.

Qualitative defenses such as alkaloids and EFN are predicted to show a negative correlation with leaf lifetime (Coley et al. 1985; McKey 1974, 1984). The higher frequency of both in tropical communities does not support this prediction. These data suggest that although leaf lifetime may explain most of the latitudinal differences in quantitative defense, there appear to be other factors which have selected for increased qualitative defense in tropical species.

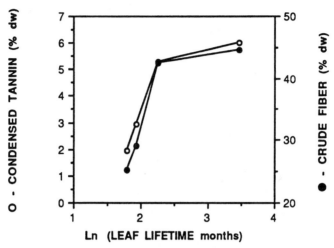

Figure 2-8. Condensed tannin and crude fiber (ADF) contents as a function of the natural log of leaf lifetime for the 4 forest categories. Data are from Table 2-1.

HERBIVORE PRESSURE

Latitudinal differences in herbivore pressure could lead to the evolution of greater defense in the tropics. As was shown earlier (Figure 2-1), actual rates of herbivory are significantly higher in the tropics. Higher herbivory coupled with higher defense suggests greater herbivore pressure. Clearly, greater herbivore pressure will increase the risk of damage and the potential benefit of defenses, favoring selection for enhanced defense of any type (Coley et al. 1985; Feeny 1976; Levin 1976; Rhoades and Cates 1976). Greater herbivore pressure may therefore help explain the generally greater defense levels in the tropics, and particularly the higher frequency of alkaloids and EFN. Why herbivore pressure should show a latitudinal trend is not entirely clear. Plausible explanations include less severe abiotic control of herbivore populations and greater predictability and productivity of plants.

A second related factor which may have influenced the evolution of higher defenses in young tropical leaves is their predictability or availability to herbivores. In highly seasonal environments, such as temperate forests with periods of cold and dry forests with periods of drought, insect herbivore populations may be severely lowered. Leaf production at the onset of favorable growth periods thus coincides with a period of greatly reduced herbivore pressure. In the more benign climate of the humid tropics, there may be less abiotic regulation of herbivore abundance and at no time of year do populations drop so low. So young tropical leaves not only face the higher herbivore pressure experienced by tropical plants in general, but also a higher seasonal availability to herbivores due to a less severe climate.

IMPACT OF PLANT DEFENSES ON HERBIVORES

The evidence presented in previous sections argues that mature tropical leaves are both better defended physically and chemically, and are less nutritious than temperate leaves. Does this have any effect on their interactions with herbivores? Although the question is beyond the scope of the chapter, we would like to suggest that it does, and raise a few possibilities.

Higher defense and lower nutritional quality may lead to longer developmental times for herbivores in the tropics. This could in turn increase their vulnerability to predators and parasitoids. Predator avoidance techniques such as chemical defense, construction of physical barriers, and crypsis may therefore play a larger role in the tropics. This could translate into a stronger effect of third trophic level interactions on population regulation of tropical herbivores (Price 1980; Price et al. 1980, 1986).

Greater defense of tropical leaves may also have implications for the degree of host-plant specialization. Theoretical considerations suggest that greater defense should lead to a higher degree of diet specialization. Although sufficient data to test this are not available, preliminary indications do not support the suggestion (Marquis and Braker 1991).

Another latitudinal distinction between plant–herbivore interactions is that in tropical communities more of the herbivory happens to young leaves. Because young leaves are available for a shorter period than mature leaves, this may put ecological and evolutionary constraints on host-finding by herbivores. Herbivores must find the right plant at the right stage of development (Aide and Londoño 1989). Their life cycles may therefore show sophisticated fine tuning to the periodicity of leaf production in their host plants.

CONCLUSIONS

The preliminary evidence we have presented suggests that there are significant differences in herbivory and defenses among temperate and tropical broad-leaved forests. Tropical species have apparently had an evolutionary history of higher herbivore pressure, and have responded with a battery of physical, chemical, mutualistic, and phenological defenses. The apparently greater importance of plant–herbivore interactions in these tropical communities suggests that it is an important area for future studies examining both ecological as well as evolutionary questions.

ACKNOWLEDGMENTS

We thank W. Montgomery, M. Lechowicz, E. Schupp, D. Feener, K. Keeler, and J. Howard for generously allowing us to use their unpublished data. We are also grateful to R. Marquis and T. Kursar for constructive ideas, and to M. Geber, E.

Schupp, E. G. Leigh, Jr., and J. Bryant for comments on the manuscript. This research was supported by NSF grant BSR 8407712 to PDC.

REFERENCES

Adis, J., K. Furch, and U. Irmler. 1979. Litter production of a Central-Amazonian blackwater inundation forest. *Tropical Ecology* **20**: 236–245.

Aide, T. M. 1988. Herbivory as a selective agent on the timing of leaf production in a tropical understory community. *Nature* **336**: 574–575.

Aide, T. M. and C. Londoño. 1989. The effects of rapid leaf expansion on the growth and survivorship of a lepidopteran herbivore. *Oikos* **55**: 66–70.

Aide, T. M. and J. K. Zimmerman. 1990. Patterns of insect herbivory, growth, and survivorship in juveniles of a neotropical liana, *Connarus turczaninowii* (Connaraceae). *Ecology* **71**: 1412–1421.

Ayres, M. P. and S. F. MacLean, Jr. 1987. Development of birch leaves and the growth energetics of *Epirrita autumnata* (Geometridae). *Ecology* **68**: 558–568.

Baldwin, I. T., J. C. Schultz, and D. Ward. 1987. Patterns and sources of leaf tannin variation in yellow birch (*Betula alleghaniensis*) and sugar maple (*Acer saccharum*). *Journal of Chemical Ecology* **13**: 1069–1078.

Baldwin, I. T. and J. C. Schultz. 1988. Phylogeny and the patterns of leaf phenolics in gap- and forest-adapted *Piper* and *Miconia* understory shrubs. *Oecologia* **75**: 105–109.

Batzli, G. O. and J. G. Jung 1980. Nutritional ecology of microtine rodents: resource utilization near Atasook, Alaska. *Arctic and Alpine Research* **12**: 483–499.

Beattie, A. J. 1985. *The evolutionary ecology of ant–plant mutualisms.* Cambridge University Press, Cambridge.

Becker, P. 1981. Potential physical and chemical defenses of *Shorea* seedling leaves against insects. *Malyasian Forester* **2&3**: 346–356.

Becker, P. and J. S. Martin. 1982. Protein precipitating capacity of tannins in *Shorea* Dipterocarpaceae seedling leaves. *Journal of Chemical Ecology* **8**: 1353–1368.

Bentley, B. L. 1976. Plants bearing extrafloral nectaries and the associated ant community: Interhabitat differences in the reduction of herbivore damage. *Ecology* **57**: 815–820.

Bentley, B. L. 1977. Extrafloral nectaries and protection by pugnacious bodyguards. *Annual Review of Ecology and Systematics* **8**: 407–427.

Bernays, E. A. 1981. Plant tannins and insect herbivores: An appraisal. *Ecological Entomology* **6**: 353–360.

Bray, J. R. 1964. Primary consumption in three forest canopies. *Ecology* **45**: 165–167.

Brokaw N. V. L. 1985. Gap-phase regeneration in a tropical forest. *Ecology* **66**: 682–687.

Bryant, J. P., F. S. Chapin III and D. R. Klein. 1983. Carbon/nutrient balance of boreal plants in relation to vertebrate herbivory. *Oikos* **40**: 357–368.

Cates, R. G. 1975. The interface between slugs and wild ginger: Some evolutionary aspects. *Ecology* **56**: 391–400.

Cates, R. G. and G. H. Orians. 1975. Successional status and the palatability of plants to generalized herbivores. *Ecology* **56**: 410–418.

Chapin, F. S. III. 1980a. The mineral nutrition of wild plants. *Annual Review of Ecology and Systematics* **11**: 261–285.

Chapin, F. S. III. 1980b. Nutrient allocation and responses to defoliation in tundra plants. *Arctic and Alpine Research* **12**: 553–563.

Coley, P. D. 1982. Rates of herbivory on different tropical trees. In E. G. Leigh, A. S. Rand, and D. M. Windsor (Eds.). *The ecology of a tropical forest: Seasonal rhythms and long-term changes.* Smithsonian Institution Press, Washington D. C. Pp. 123–132.

Coley, P. D. 1983. Herbivory and defensive characteristics of tree species in a lowland tropical forest. *Ecological Monographs* **53**: 209–233.

Coley, P. D. 1986. Costs and benefits of defense by tannins in a neotropical tree. *Oecologia* **70**: 238–241.

Coley, P. D. 1987. Patrones en las defensas de las plantas: Por que los herbivoros prefieren ciertas especies? *Revista de Biologia Tropical* **35** (Supl. 1): 151–164.

Coley, P. D. 1988. Effects of plant growth rate and leaf lifetime on the amount and type of anti-herbivore defense. *Oecologia* **74**: 531–536.

Coley, P. D., J. P. Bryant, and F. S. Chapin III. 1985. Resource availability and plant anti-herbivore defense. *Science* **230**: 895–899.

Cooke, F. P., J. P. Brown, and S. Mole. 1984. Herbivory, Foliar enzyme inhibitors, nitrogen, and leaf structure of young and mature leaves in a tropical forest. *Biotropica* **16**: 257–263.

Crankshaw, D. R. and J. H. Langenheim. 1981. Variation in terpences and phenolics through leaf development in *Hymenea* and its possible significance to herbivory. *Biochemical Systematics and Ecology* **9**: 115–124.

Crawley, M. J. 1983. *Herbivory: The dynamics of animal-plant interactions.* University of California Press, Berkeley, CA.

de la Cruz, M. and R. Dirzo. 1987. A survey of the standing levels of herbivory in seedling from a Mexican rain forest. *Biotropica* **19**: 98–106.

Denno, R. F. and M. S. McClure. 1983. *Variable plants and herbivores in natural and managed systems.* Academic Press, New York.

Dement, W. A. and H. A. Mooney. 1974. Seasonal variation in the production of tannins and cyanogenic glycosides in the chaparral shrub, *Heteromeles arbutifolia*. *Oecologia* **41**: 65–76.

Denslow, J. S. 1987. Tropical rainforest gaps and tree species diversity. *Annual Review of Ecology and Systematics* **18**: 431–451.

Dirzo, R. 1984. Insect–plant interactions: Some ecophysiological consequences of herbivory. In E. Medina, H. A. Mooney, and C. Vazquez-Yanes (Eds.). *Physiological ecology of plants in the wet tropics.* Junk, The Hague, pp. 209–224.

Edwards, P. J. 1977. Studies of mineral cycling in a montane rain forest in New Guinea II. The production and disappearance of litter. *Journal of Ecology* **65**: 971–992.

Elias, T. S. 1983. Extrafloral nectaries: Their structure and distribution. In B. L. Bentley and T. S. Elias (Eds.). *The biology of nectaries.* Columbia University Press, New York, pp. 174–203.

Faeth, S. H. 1985. Quantitative defense theory and patterns of feeding by oak insects. *Oecologia* **68**: 34–40.

Faeth, S. J., S. Mopper, and D. Simberloff. 1981. Abundances and diversity of leaf-mining insects on three oak host species: Effects of host–plant phenology and nitrogen content of leaves. *Oikos* **37**: 238–251.

Feeny, P. 1970. Seasonal changes in oak leaf tannins and nutrients as a cause of spring feeding by winter moth caterpillar. *Ecology* **51**: 565–581.

Feeny, P. 1976. Plant apparency and chemical defense. In J. Wallace and R. L. Mansell (Eds.). *Biochemical interaction between plants and insects.* Recent Advances in Phytochemistry. Volume 10. Plenum Press, New York, pp. 1–40.

Fogden, M. O. L. 1972. The seasonality and population dynamics of equatorial forest birds in Sarawak. *Ibis* **114**: 307–343.

Frankie, G. W., H. G. Baker, and P. A. Opler 1974. Comparative phenological studies of trees in tropical lowland wet and dry forest sites of Costa Rica. *Journal of Ecology* **62**: 881–919.

Funke, W. 1973. Rolle der tiere. In H. Ellenberg (Ed.). *Wald-okosystemen des solling okosystemforschung.* Springer, Berlin, pp. 143–164.

Gartlan, J. S., D. B. McKey, and P. G. Waterman, 1978. Soils, forest structure and feeding behavior of primates in a Cameroon costal rainforest. In D. J. Chivers and J. Herbert (Eds.). *Recent advances in primatology.* Volume 1. London Academic Press, London, England, pp. 259–267.

Gershenzon, J. 1984. Changes in the levels of plant secondary metabolites under water and nutrient stress. In B. Timmermann, C. Steelink, and F. Loewus (Eds.). *Phytochemical adaptations to stress.* Plenum Press, New York, pp. 273–320.

Grime, J. P. 1977. Evidence for the existence of three primary stategies in plants and its relevance to ecological and evolutionary theory. *American Naturalist* **111**: 1169–1194.

Grime, J. P., S. F. MacPherson-Stewart, and R. S. Dearman. 1968. An investigation of leaf palatability using the snail *Cepaea nemoralis. Journal of Ecology* **56**: 405–420.

Gulmon, S. L. and H. A. Mooney 1986. Costs of defense on plant productivity. In T. J. Givnish (Ed.). *On the economy of plant form and function.* Cambridge University Press, Cambridge, pp. 681–698.

Hall, G. D. and J. H. Langenheim. 1986. Temporal changes in the leaf monoterpenes of *Sequoia sempervirens. Biochemical Systematics and Ecology* **14**: 61–70.

Hanley, T. A., R. G. Cates, B. Van Horne, and J. D. McKendrick. 1987. Forest stand-age-related differences in apparent nutritional quality of forage for deer in southeastern Alaska. In F. D. Provenza, J. T. Flinders, and E. D. McArthur (Eds.). *Proceedings-symposium on plant–herbivore interactions.* Forest Service, USDA, Ogden, pp. 9–17.

Hartshorn, G. S. 1980. Neotropical forest dynamics. *Biotropica* **12** (supplement): 23–30.

Haukioja, E., P. Niemela, L. Iso-Iivari, H. Ohala, and E. Aro. 1978. Birch leaves as a resource for herbivores. I. Variation in the suitability of leaves. *Report of Kevo Subarctic Research Station* **14**: 5–12.

Heywood, V. H. 1978. *Flowering plants of the world.* Mayflower Books, New York.

Hladik, A. and P. Blanc. 1987. Croissance des plantes en Sous-Bois de foret dense humide (Makokou, Gabon). *La Terre et La Vie* **42**: 209–234.

Janzen, D. H. 1967. Interaction of the bull's-horn acacia (*Acacia cornigera* L.) with an ant inhabitant (*Pseudomyrmex ferruginea* F. Smith) in Eastern Mexico. *University of Kansas Science Bulletin* **47**: 315–558.

Janzen, D. H. 1974. Tropical blackwater rivers, animals and mast fruiting by the Dipterocarpaceae. *Biotropica* **6**: 69–103.

Janzen, D. H. and P. G. Waterman. 1984. A seasonal census of phenolics, fiber and alkaloids in foliage of forest trees in Costa Rica: Some factors influencing their distribution and relation to host selection by Sphingidae and Saturniidae. *Biological Journal of the Linnean Society* **21**: 439–454.

Jordan, C. F. and C. Uhl. 1978. Biomass of a 'tierra firme' forest of the amazon Basin. *Oecologia Plantarum* **13**: 387–400.

Keeler, K. H. 1979. Distribution of plants with extrafloral nectaries and ants at two elevations in Jamaica. *Biotropica* **11**: 152–154.

Keeler, K. H. 1980. Distribution of plants with extrafloral nectaries in temperate communities. *American Midland Naturalist* **104**: 274–280.

Keeler, K. H. 1981. Cover of plants with extrafloral nectaries at four northern California sites. *Madrono* **28**: 26–29.

Kozlowski, T. T. 1971. *Growth and development of trees*. Vol I. Academic Press, New York.

Kraft, S. K., and R. F. Denno. 1982. Feeding responses of adapted and non-adapted insects to the defensive properties of *Baccharis halimifolia* L. (Compositae). *Oecologia* **52**: 156–163.

Kuropat, P. J. 1984. *Foraging behavior of caribou on a calving ground in northwestern Alaska*. Thesis, University of Alaska, Fairbanks, 94 pp.

Kusnezov, N. 1952. Numbers of species of ants in faunae of different latitudes. *Evolution* **11**: 298–299.

Langenheim, J. H., C. A. Macedo, M. K. Ross and W. H. Stubblebine. 1986. Leaf development in the tropical leguminous free *Copaifera* in relation to microlepidopteran herbivory. *Biochemical Systematics and Ecology* **14**: 51–59.

Leigh, E. G. Jr. and N. Smythe. 1978. Leaf production, leaf consumption and the regulation of folivory on Barro Colorado Island. In G. G. Montgomery (Ed.). *The ecology of arboreal folivores*. Smithsonian Institution Press, Washington, D. C., pp. 35–50.

Levin, D. A. 1976. Alkaloid-bearing plants: An ecogeographic perspective. *American Naturalist* **110**: 261–284.

Levin, D. A. and B. M. York, Jr. 1978. The toxicity of plant alkaloids: An ecogeographic perspective. *Biochemical Systematics and Ecology* **6**: 61–76.

Lowman, M. D. 1984. An assessment of techniques for measuring herbivory: Is rainforest defoliation more intense than we thought? *Biotropica* **16**: 264–268.

MacLean, S. F. and T. J. Jensen. 1984. Food plant selection by insect herbivores: The role of plant growth form. *Oikos* **44**: 211–221.

Marin, D. and E. Medina. 1981. Leaf duration, nutrient content and sclerophylly of very dry tropical forest trees. *Acta Cientifica Vemezp'ama* **32**: 508–514.

Marquis, R. J. 1984. Leaf herbivores decrease fitness of a tropical plant. *Science* **226**: 537–539.

Marquis, R. J. and H. E. Braker. 1991. Plant-herbivore interactions at La Selva: Diversity, specificity and impact. In L. M. McDade, K. S. Bawa, G. S. Hartshorn, and H. E. Hespenheide (Eds.). *La Selva: Ecology and Natural History of a Neotropical Rain Forest*. University of Chicago, Chicago.

Martin, J. S., M. M. Martin, E. A. Bernays. 1987. Failure of tannic acid to inhibit digestion or reduce digestibility of plant protein in gut fluids of insect herbivores: Implications for theories of plant defense. *Journal of Chemical Ecology* **13**: 605–622.

Martin, J. S. and M. M. Martin. 1982. Tannin assays in ecological studies: Lack of correlation between phenolics, proanthocyanidins and protein-precipitating constituents in mature foliage of six oak species. *Oecologia* **54**: 205–211.

McKey, D. 1974. Adaptive patterns in alkaloid physiology. *American Naturalist* **108**: 305–320.

McKey, D. B. 1979. The distribution of secondary compounds within plants. In G. A. Rosenthal and D. H. Janzen (Eds.). *Herbivores: Their interactions with secondary plant metabolites.* Academic Press, New York, pp. 55–133.

McKey, D. B. 1984. Interaction of the ant–plant *Leonardoxa africana* (Caesalpiniaceae) with its obligate inhabitants in a rainforest in Cameroon. *Biotropica* **16**: 81–99.

McKey, D. 1989. Interactions between ants and leguminous plants. In J. Zarucchi and C. Stirton (Eds.). *Advances in legume biology.* Monographs in Systematic Botany of the Missouri Botanical Garden, **29**: 673–718.

McKey, D. B., P. G. Waterman, C. N. Mbi, J. S. Gartlan, and T. T. Strusaker. 1978. Phenolic content of vegetation in two African rain-forests: Ecological implications. *Science* **202**: 61–64.

McKey, D. B. and J. S. Gartlan. 1981. Food selection by black colobus monkeys (*Colobus satanus*) in relation to plant chemistry. *Biological Journal of the Linnean Society* **16**: 115–146.

McNeill, S. and T. R. E. Southwood. 1979. The role of nitrogen in the development of insect/plant relationships. In J. B. Harborne (Ed.). Biochemical aspects of plant and animal coevolution. *Phytochemical and Chemical Society European Symposium Series* **15**: 77–98.

Meyer, G. A. and M. E. Montgomery. 1987. Relationships between leaf age and the food quality of cottonwood foliage for the gypsy moth, *Lymantria dispar. Oecologia* **72**: 527–532.

Milton, K. 1979. Factors influencing leaf choice by howler monkeys: A test of some hypotheses of food selection by generalist herbivores. *American Naturalist* **114**: 362–378.

Mole, S., J. A. M. Ross, and P. G. Waterman. 1988. Light-induced variation in phenolic levels in foliage of rain-forest plants. I. Chemical changes. *Journal of Chemical Ecology* **14**: 1–21.

Mooney, H. A., S. L. Gulmon and N. D. Johnson. 1983. Physiological constraints on plant chemical defenses. In. P. A. Hedin (Ed.). *Plant resistance to insects.* American Chemical Society, Washington, D.C.

Muller, R. N., P. J. Kalisz, and T. W. Kimmerer. 1987. Intraspecific variation in production of astringent phenolics over a vegetation-resource availability gradient. *Oecologia* **72**: 211–215.

Nielson, B. O. 1978. Above ground food resources and herbivory in a beech forest ecosystem. *Oikos* **31**: 273–279.

Oates, J. F., P. G. Waterman, and G. M. Choo. 1980. Food selection by the south Indian leaf-monkey, Presbytis Johnii, in relation to leaf chemistry. *Oecologia* **45**: 45–56.

Odum, H. T. and J. Ruiz-Reyes. 1970. Holes in leaves and the grazing control mechanism. In H. T. Odum (Ed.). *A tropical rain forest.* Div of Tech Info, U.S. Atomic Energy Commission.

Oliveira, P. S. and H. F. Leitao-Filho. 1987. Extrafloral nectaries: Their taxonomic distribution and abundance in the woody flora of Cerrado vegetation in southeast Brazil. *Biotropica* **19**: 140–148.

Orians, G. H. and D. H. Janzen. 1974. Why are embryos so tasty? *American Naturalist* **108**: 581–592.

Palo, R. T. 1984. Distribution of birch (*Betula* spp.), Willow (*Salix* spp.) and poplar (*Populus* spp.) secondary metabolites and their potential role as chemical defense against herbivores. *Journal of Chemical Ecology* **10**: 499–520.

Pimentel D. 1976. World food crisis: Energy and pests. *Bulletin of the Entomogical Society of America* **22**: 20–26.

Potter, D. A. and T. W. Kimmerer. 1986. Seasonal allocation of defense investment in *Ilex opaca* and constraints on a specialist leafminer. *Oecologia* **69**: 271–224.

Price, P. W. 1980. Evolutionary biology of parsites. Princeton University Press, Princeton.

Price, P. W., C. E. Bouton, P. Gross, B. A. McPheron, J. N. Thompson, and A. E. Weis. 1980. Interactions among three tropic levels: Influence of plants on interactions between insect herbivores and natural enemies. *Annual Review of Ecology and Systematics* **11**: 45–65.

Price, P. W., M. Westoby, B. Rice, P. R. Atsatt, R. S. Fritz, J. N. Thompson, and K. Mobley. 1986. Parasite mediation in ecological interactions. *Annual Review of Ecology and Systematics* **17**: 487–505.

Proctor, J., J. M. Anderson, S. C. L. Fogden, and H. W. Vallack. 1983. Ecological studies in four contrasting lowland rain forests in Gunung Mulu National Park, Sarawak II. Litterfall, litter standing crop and preliminary observation on herbivory. *Journal of Ecology* **71**: 261–284.

Rathcke, B. 1985. Slugs as generalist herbivores: Test of three hypotheses on plant choices. *Ecology* **66**: 828–836.

Raupp, M. J. 1985. Effects of leaf toughness on mandibular wear of the leaf beetle *Plogiaodera versicolora*. *Ecological Entomology* **10**: 73–79.

Raupp, M. J. and R. F. Denno. 1983. Leaf age as a predictor of herbivore distribution and abundance. In R. F. Denno and M. S. McClure (Eds.). *Variable plants and herbivores in natural and managed systems*. Academic Press, New York, pp. 91–124.

Reader, P. M. and T. R. E. Southwood. 1981. The relationship between palatability to invertebrates and the successional status of a plant. *Oecologia* **51**: 271–275.

Reichle, D. E., R. A. Goldstein, R. I. Van Heok, Jr., and G. J. Dodson. 1973. Analysis of insect consumption in a forest canopy. *Ecology* **54**: 1076–1084.

Reichle, D. E. and D. A. Crossley. 1967. Investigation of heterotropic productivity in forest insect communities. In K. Petrusewicz (Ed.). *Secondary productivity of terrestrial ecosystem*. Panstwowe Wydawnietwo Naukowe, Warsaw, pp. 563–587.

Rhoades, D. F. and R. G. Cates, 1976. Toward a general theory of plant antiherbivore chemistry. In J. Wallace and R. L. Mansell (Eds.). Biochemical interactions between plants and insects. *Rec. Adv. Phytochem.* Vol. 10, Plenum Press, New York, pp. 168–213.

Ricklefs, R. E. and K. K. Mathew. 1982. Chemical characteristics of the foliage of some deciduous trees in southeastern Ontario. *Canadian Journal of Botany* **60**: 2037–2045.

Robus, M. 1981. Foraging behavior of muskoxen in artic Alaska. Thesis, University of Alaska, Fairbanks, 78 pp.

Rosenthal, G. A. and D. H. Janzen. 1979. *Herbivores: Their interaction with secondary plant metabolites*. Academic Press, New York.

Rundel, P. W. and P. F. Becker. 1987. Cambios estacionales en las relaciones hidricas y en la fenologia vegetativa de plantas del estrato bajo del bosque tropical de la Isla de Barro Colorado, Panama. *Revista de Biologia Tropical* **35** (Supl. 1): 71–84.

Schroeder, L. A. 1986. Changes in tree leaf quality and growth performance of Lepidopteran larvae. *Ecology* **67**: 1628–1636.

Schultz, J. C., P. J. Nothnagle, and I. J. Baldwin. 1982. Seasonal and individual variation in leaf quality of two northern hardwoods tree species. *American Journal of Botany* **69**: 753–759.

Schupp, E. W. 1986. *Azteca* protection of *Cecropia*: Ant occupation benefits juvenile trees. *Oecologia* **70**: 379–385.

Scriber, J. M. and F. Slansky, Jr. 1981. The nutritional ecology of immature insects. *Annual Review of Entomology* **26**: 183–211.

Southwood, T. R. E., V. K. Brown, and P. M. Reader. 1986. Leaf palatablility, life expectancy and herbivore damage. *Oecologia* **70**: 544–548.

Stanton, N. 1975. Herbivore pressure on two types of tropical forests. *Biotropica* **7**: 8–12.

Struhsaker, T. T. 1978. Interrelations of red colobus and rain-forest trees in the Kibale Forest, Uganda. In G. G. Montgomery (Ed.). *The ecology of arboreal folivores.* Smithsonian Institution Press, Washington, D.C., pp. 397–422.

Swain, T. 1979. Tannins and lignins. In G. A. Rosenthal, and D. H. Janzen (Eds.). *Herbivores: Their interaction with secondary plant metabolites.* Academic Press, New York, pp. 657–682.

Systat. 1987. *Systat: The system for statistics.* Systat, Inc., Evanston, IL.

Tanton, M. T. 1962. The effect of leaf "toughness" on the feeding of the larvae of the mustard beetle, *Phaedon cochleariae* Fabricus. *Entomologia Experimentalis et Applicata* **5**: 74–78.

Taylor, F. G. Jr. 1974. Phenodynamics of production in a mesic deciduous forest. In H. Lieth (Ed.). *Phenology and seasonality modeling.* Springer-Verlag, New York, pp. 237–254.

Trudell, J. and R. G. White. 1981. The effect of forage structure and availability of food intake, biting rate, bite size and daily eating time of reindeer. *Journal of Applied Ecology* **18**: 63–81.

Uhl, C. 1982. Recovery following disturbances of different intensities in the Amazon rain forest of Venezuela. *Interciencia* **7**: 19–24.

Van Soest, P. J. 1975. Physico-chemical aspects of fiber digestion. In I. W. McDonald, and A. C. I. Warner (Eds.). *Proceedings of the IV International Symposium on Ruminant Physiology.* University of New England Publishing Unit, Armidale, New South Wales, Australia, pp. 351–365.

Waltz, S. A. 1984. *Comparative study of predictability, value, and defenses of leaves of tropical wet forest trees.* Dissertation. University of Washington, Seattle, Washington.

Waterman, P. G. 1983. Distribution of secondary metabolites in rainforest plants: Toward an understanding of cause and effect. In S. L. Sutton, T. C. Whitmore, and C. Chadwick (Eds.). *Tropical rain forest: Ecology and management.* Blackwell Scientific, Oxford, pp. 167–179.

Waterman, P. G., J. A. M. Ross, and D. B. McKey. 1984. Factors affecting levels of some phenolic compounds, digestibility, and nitrogen content of the mature leaves of *Barteria fistulosa* (Passifloraceae). *Journal of Chemical Ecology* **10**: 387–401.

Weeken, R. B. 1969. Foods of rock and willow ptarmigan in central Alaska with comments on interspecific competition. *Auk* **86**: 271–281.

Whittaker, R. H., and P. P. Feeny. 1971. Allelochemics: Chemical interactions between species. *Science* **171**: 757–770.

Whitten, A. J., J. Mustafa, and G. S. Henderson. 1987. *The ecology of Sulawesi.* Gadjah Mada University Press, Bulaksumur, Indonesia.

Williams, J. B., D. Best, and C. Warford. 1984. Foraging ecology of ptarmigan at Meade River, Alaska. *Wilson Bulletin* **92**: 341–351.

Wint, G. R. W. 1983. Leaf damage in tropical rain forest canopies. In S. L. Sutton, T. C. Whitmore, and A. C. Chadwick (Eds.). *Tropical rainforest: Ecology and management.* Blackwell Publishers, Oxford, pp. 229–239.

Wisdom, C. S., A. Gonzalez-Coloma, and P. W. Rundel. 1987. Ecological tannin assays. Evaluation of proanthocyanidins, protein binding assays and protein precipitating potential. *Oecologia* **72**: 395–401.

Wolda, H. 1978. Seasonal fluctuations in rainfall, food and abundance of tropical insects. *Journal of Animal Ecology* **47**: 369–381.

Woodwell, G. and R. H. Whittaker. 1968. Primary production in terrestrial ecosystems. *American Zoologist* **8**: 19–30.

Zucker, W. V. 1983. Tannins: Does structure determine function? An ecological perspective. *American Naturalist* **121**: 335–365.

3 Patterns in Communities Along Latitudinal Gradients

PETER W. PRICE

Ecology has tended to develop with a typological approach to patterns in nature with emphasis on finding one pattern sufficient to account for the majority of organisms. Such patterns have relied heavily on a limited number of taxa, particularly plants, birds, and mammals. Therefore, we can frequently read in the general ecological literature about pervasive patterns such as the following. "In virtually all groups of organisms, the number of species increases markedly towards the equator" (Ricklefs 1979, p. 687). "Competition is keener in the tropics, and niches are smaller" (Krebs 1985, p. 532). Most explanations for high diversity in the tropics are "based on the assumption that diversity is near equilibrium in most regions" (Futuyma 1986, p. 391). "In many diverse communities, such as tropical rain forests, populations are thought to be often near their maximal sizes (equilibrium populations ...), with the result that intraspecific and interspecific competition are frequently keen... and the resulting small niches make high diversity possible" (Pianka 1974, p. 241). "In predictable, stable environments populations will frequently reach the carrying capacity of the environment and selection will operate by improving adaptations for living in these crowded, competitive conditions, where enemies such as predators and parasites will be very effective—K selection" (Price 1984, p. 252).

These quotations are taken out of a context which is usually more equivocal, stating problems and exceptions, and lacking good empirical evidence for some patterns. We all attempt to generalize and this is justified, and I cite my own book to ensure that I am not critical of the ways textbooks on ecology are written. However, in the long run I think it is crucial to reduce the emphasis on typology in ecology, and accept the inevitable conclusion that there are many patterns to be understood. The sooner we move from broad generalizations based on scanty data to specific patterns, the sooner will a firm basis for comparative mechanistic understanding in ecology be established. Then generalizations can be derived from empirical data as far as those data allow. As a more mechanistic ecology

Plant-Animal Interactions: Evolutionary Ecology in Tropical and Temperate Regions, Edited by Peter W. Price, Thomas M. Lewinsohn, G. Wilson Fernandes, and Woodruff W. Benson. ISBN 0-471-50937-X © 1991 John Wiley & Sons, Inc.

develops there is a focus on specific empirical studies, from which we build to generality. This approach contrasts with much of ecological thinking relating to broad latitudinal patterns which emphasized erection of hypotheses, without adequate empirical tests (e.g., Dobzhansky 1950; MacArthur 1962; see also references in Pianka's 1966 review).

My object in this chapter, therefore, is to take another look at some of the broad generalizations about patterns along latitudinal gradients, emphasizing herbivorous insects. I would like to think that this will stimulate more empirical tests of general theory, particularly by scientists native to the tropics. After much field study we may then build on a sound data-based foundation to understanding mechanistically the range of patterns that truly capture the essence of nature. I believe this is the major challenge for ecologists.

The chapter is arranged as a set of questions about tropical and temperate comparisons, and my attempt at answers: (1) Are there more species in the tropics? (2) Are species more specialized in the tropics? (3) Is there more competition in the tropics? (4) Are there vacant niches in the tropics? The first question looks at species richness within taxa along latitudinal gradients. The other questions relate to mechanisms of interaction, with competition as a focus which unites the three points. Of course many other aspects of mechanisms driving tropical diversity could be discussed, but the ones covered are central issues in ecology, and are most in need of reexamination in my opinion. My conclusions will be unconventional and no doubt controversial, and therefore hopefully will encourage more careful empirical studies.

ARE THERE MORE SPECIES IN THE TROPICS?

"Perhaps the most widely recognized pattern in species diversity is the increase that occurs from the poles to the tropics" (Begon et al. 1986, p. 800). I cannot dispute this as a broad generalization, but when we look at specific taxa the exceptions are more numerous than anybody has been willing to admit. For herbivorous insects it is easy to find examples of higher diversity in temperate regions (Table 3-1). This is not a trivial sample either, representing more species than all birds and mammals from which so much ecological theory is derived. Indeed, this small sample is probably larger than all terrestrial vertebrates.

There is an encouraging development of theory to account for the empirical pattern seen in the aphids (Fig. 3-1) based on Eastop's (1972, 1978) studies. Dixon et al. (1987) argue that these insects cannot live long without food, are highly specific to host-plant species, but they locate hosts with low efficiency. Therefore, they cannot exploit rare plants, which per unit area are more common in the tropics. Janzen (1975, 1981) made equivalent arguments for parasitic wasps on insect herbivores in the tropics, which are also better represented in temperate regions (Table 3-1) (see also Gauld 1986a, b).

The global pattern for galling species richness is becoming clear (Fig. 3-2), in which samples in the lowland tropics are usually no more diverse than in the arctic, and really high diversity is associated with sclerophyll vegetation. This

TABLE 3-1. Examples of Insect Groups Better Represented in Temperate Regions than in the Lowland Tropics[a]

Taxon	Numer of Species in World	Sources
Herbivores		
Agromyzid leaf-mining flies	c. 1800	Spencer 1972
Aphids or plant lice	4064	Eastop 1972, 1978; Dixon et al. 1987 (Fig. 1)
Psyllids or jumping plant lice	1728	Eastop 1972, 1978 (Fig. 1)
Tenthredinid sawflies	c. 5000	Benson 1950; Smith 1979
Galling insects	7930	This chapter (Fig. 3-2); Felt 1940
Total	20,522	
Parasitoids		
Ichneumonid wasps	14,816	Townes 1969; Janzen 1981; Gauld 1986a

[a]Estimates of valid described species are used, not estimates of total species present. Townes (1969) estimates a probable total of 60,464 species of ichneumonids.

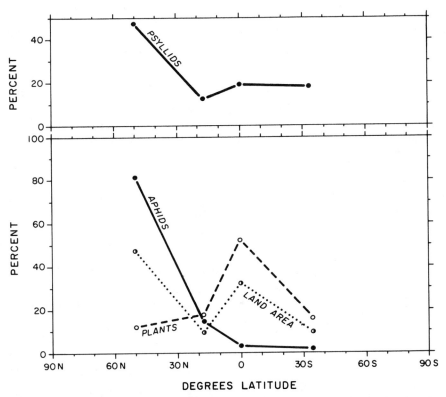

Figure 3-1. The distribution of aphids and psyllids in relation to latitude, showing the preponderance of species in the north temperate. The distribution of plant species and land area do not account for the pattern in the herbivore species (all data from Eastop 1978).

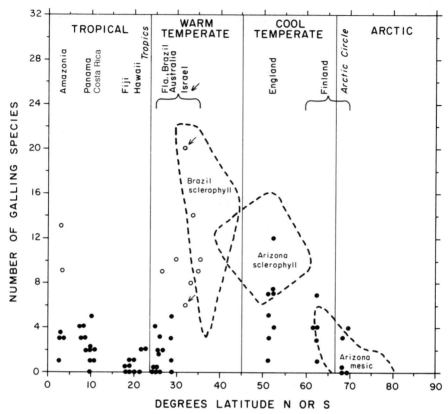

Figure 3-2. The distribution of galling insect species on the latitudinal gradient, based on samples of 45 trees and 100 shrubs per sample, or 1-hr censuses. Locations are given above; open circles are from sites with sclerophyll vegetation and closed circles are from mesic or tropical vegetation. The dashed lines for Brazil and Arizona enclose the sample distributions obtained by Wilson Fernandes, and are corrected for altitude. Data from Panama are from Jeffrey Brawn, and for Israel, indicated by arrows, from Dan Gerling. Note the two high records of galling species richness from Amazonian sclerophyllous campina vegetation.

example (Table 3-1) is not based on a global catalog of species, for this does not exist, but on samples on a very local scale usually using 45 trees and 100 shrubs as the basis or a 1-hr census. The advantage is that every site is equally sampled so there is no bias in research intensity. Sclerophyll vegetation does occur in the tropics, for example, in the cerrado of Brazil, sampled by Wilson Fernandes (1987). But when latitude is adjusted for altitude (4° increased latitude for every 305 m elevation), the Brazil sclerophyll samples overlap those of sclerophyll vegetation from Arizona, Australia, and Israel.

These patterns, documented clearly in Table 3-1, illustrate that the tropics are not the universal centers of speciation and radiation we sometimes believe. They are even repellent to species invasion, as with aphids. Many radiations of herbivorous taxa have occurred in cool temperate latitudes, on cool-adapted plant taxa such as conifers [many aphids (Eastop 1978), all diprionid sawflies (Smith 1979)], and the often associated willows and poplars (such as tenthredinid sawflies). The radiations of insect gallers in the drier vegetations of the Mediterranean, chapparal, desert, and cerrado, are often associated with poor soils as in Australia. They indicate some special relationships between gall formers and plants not found commonly in the wet tropics, which Wilson Fernandes is investigating. One exception which proves the rule is the campina sclerophyllous shrub vegetation on the poor white sands along the Rio Negro, Amazonia, where gallers are diverse (Fig. 3-2).

If speciation and radiation of herbivore taxa can be so extensive outside the tropics, it reinforces the question of why so many other taxa have had the common response of radiation in the tropics. Is it simply that there are more plant species present? Or is it also the additive effects of all the climatic zones represented in tropical latitudes from rain forest to alpine? These factors could be resolved easily with a good field study. I think the answers will come more readily by considering radiations of many groups in relation to their specific biologies as they relate to tropical and temperate climates.

ARE SPECIES MORE SPECIALIZED IN THE TROPICS?

"Entry into a smaller fauna is often accompanied by ecological release" (MacArthur and Wilson 1967, p. 105). I do not know of a single insect herbivore taxon which is well documented as more specialized in the tropics than in temperate regions. This contrasts with the examples used by MacArthur and Wilson (1967) including ants, birds, and mammals. Admittedly the "Niche Compression Hypothesis" states that "as more species invade and are packed in, the occupied habitat shrinks... but not the range of acceptable food items within the occupied habitat" (MacArthur and Wilson 1967, p. 108). But for insect herbivores food defines habitat, so feeding specialization equates with habitat specialization, so if there is truly more competition and niche compression in the tropics, this should be seen in narrower specialization in the number of host plants utilized by herbivores. This cause and effect relationship has not been demonstrated for any insect herbivore taxon.

The fact is that most parasites are very specialized, including insect herbivores (Price 1980), and they are perhaps equally specialized throughout the latitudinal range over which the taxon exists. Bruchid beetles show no trend in specificity on a latitudinal gradient (Table 3-2). A comparison of tropical and temperate butterflies also illustrates this point (Table 3-3). The North American and British faunas are well studied and represent accurate pictures of the number of host plants utilized. In the tropics only the *Heliconius* species have been studied enough for a good comparison, and specificity is similar to that in temperate regions. This is in spite of the many more butterfly species in the tropics. In the

TABLE 3-2. The Number of Host-Plant Species Utilized by Bruchid Beetles in the Genus *Acanthoscelides* in Tropical and Temperate North America[a]

	Tropical	Temperate
Number of Hosts	Southern Mexico and Central America	Arizona, California, Oregon
1	35.9	35.1
2	23.1	22.7
3	10.3	10.3
4	10.3	6.2
5	10.3	5.2
6		3.1
7		5.2
8	2.6	2.1
9	2.6	
10		1.0
Number of species	97	39

[a]The percentage of species that fall into each class of number of host-plant species utilized is provided [Johnson (1970, 1983) provides the host records].

TABLE 3-3. Specificity of Some Tropical and Temperate Butterflies, Showing the Percentage of Species with Known Food Plants that Fall into Each Class of Number of Host-Plant Species Utilized[a]

	Tropical			Temperate	
Number of Hosts	*Heliconius* species Mid and South America	Malay Peninsula	East Africa	North America	Britain
1	24	26	31	22	21
2	14	22	31	21	24
3	16	15	11	12	13
4	4	5	5	11	4
5	6	2	3	7	16
6	2		2	3	9
7	2	1	1	7	3
8	4			1	4
9	2		< 1	1	
10–19	12		1	12	4
20–29	8			1	
30–39	8			2	
Unknown but more than one		30	15		
Number of species	51	144	292	214	67

[a]From Price 1980, for sources see this reference.

other two tropical examples many host-plant records were given as genera, or not all host species were listed, so specificity is overestimated. Scriber (1973) analyzed specificity of papilionid butterflies by using the number of host-plant families utilized, and found more specificity of this kind in the tropics where more butterfly species exist. However, this tells us little about community structure, as the scale is too coarse to understand the mechanisms involved. It is also difficult to use DeVries (1987) for a comparative estimate of butterfly specificity in Costa Rica, because so few host ranges are reported at the plant species level. For example, in the Pieridae about 33% of species have unknown hosts, and only 38% have hosts listed by species only, without higher taxa included in the list. If specificity of butterflies (and grasshoppers) is higher in the tropics as Marquis and Braker (1989) report, we still cannot argue that competition is the cause, for most probably phytochemical diversity would play a stronger role.

Many other insect herbivore taxa are as specific as the butterflies or much more specific, even in England around 53°N. Mirid bugs have 22% of species specific to one host, gelechiid moths have 53% on one host, and agromyzid leaf mining flies have 57% on one host (Price 1980). Parasitoids seem to be equally specialized, with 53% of ichneumonids on one host, and 60% of braconids on one host in temperate faunas (Price 1980), but with reduced specificity in the tropics because specialized larval parasitoids decline in species number with reduced latitude and the more generalized pupal parasitoids increase (Gauld 1986b). Hawkins (1990) also supported these patterns when the smaller scale of number of parasitoid species per host species was utilized. Some additional taxa seem to be less specialized in the tropics than in temperate regions, for example, the aphids (Eastop 1972), bark and ambrosia beetles (Beaver 1979a), and membracids (Wood 1984).

Colwell's (1986; Chapter 20 of this book) detailed studies show that humming-bird flower mites are more specialized in wet tropical forests than in tropical highlands and temperate latitudes. Here specificity seems to be driven by sexual selection for increased mate finding and reduced congeneric courtship and mating, rather than increased interspecific competition for resources.

Specialization is a property of certain ways of life, such as those of parasites, and not a property that usually varies along latitudinal gradients. This is well illustrated by Rohde's (1978a) data on monogenean parasites on marine fishes. At all latitudes from 0° to 70°N species are equally specialized with about 83% of parasite species recorded from only one fish species (Fig. 3-3). Specialization does not change even though species richness of parasites is high in tropical waters and very low in temperate and arctic waters. The niche compression hypothesis does not hold for these kinds of organisms (see also Lawton 1984). Even with an increase in number of parasites per host of 6.5 times, there is no evidence of change in specificity. This is a greater change than in many examples from birds and mammals which do illustrate compression (Puerto Rico vs. Panama birds, 3.2x the number of species on the mainland compared to the island, MacArthur et al. 1966; Caribbean islands vs. Panama birds, 1.3x, 2.4x, 4.1x, 6.8x, Cox and Ricklefs 1977; Jamaica vs. Honduras birds, 6x, Lack 1976; hares on Newfound-

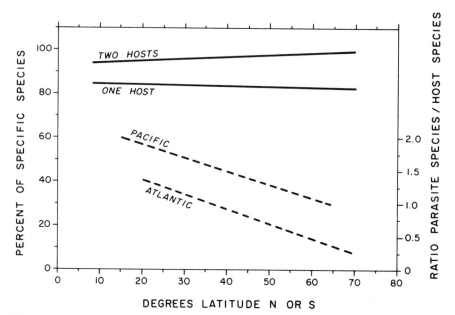

Figure 3-3. Specificity of monogenean parasites on a latitudinal gradient (solid lines), and the number of parasite species per host species (dashed lines). Two indices for specificity were used: the percent of species utilizing only one host species, and the percent of species known from only two host species. The ratio of parasite to host species is given for Pacific and Atlantic oceans and varies from a low of 0.31 in the Atlantic to a high of 2.0 in the Pacific, an increase in parasite diversity of 6.5 times (all data are from Rohde 1978a, 1986, adjusted approximately for latitude instead of sea temperature which Rohde used as the abscissa). From top line down r^2 values are respectively: two hosts, 0.01 NS; one host, 0.17 NS; Pacific, 0.89 $p < .01$; Atlantic, 0.98 $p < .01$.

land, 2x, Cameron 1958). Wilson and Taylor (1967) estimated that evolved niche compression and assortative species accumulation would result in increases of 1.5–2 times in the number of ant species on Polynesian islands (their Fig. 5). A tropical fish species may support up to nine monogenean species, and a host individual may have seven species present (Rohde 1978b). In colder waters over 60% of fish species have no monogenean parasites and about 20–30% of the remainder have only one species. Many parasites therefore live in a vacuum of interaction from other species on the same trophic level, and yet show no change in host specificity or specificity of location within the habitat provided by a single fish (Rohde 1978a, b, 1979, 1980). Competition has had no evident impact on host or site specificity in these organisms.

Insect herbivores and animal parasites alike are generally constrained to be specific because of the host–parasite interaction and not because of competition which forces niche compression (Price 1980). Their ways of life are very different from the birds and mammals from which we have attempted to generalize about nature. But with a less typological approach we can recognize substantially different patterns in nature because substantially different ecological forces impinge on organisms with different ways of life. For insect herbivores specialization is very common universally and not a special attribute of tropical species.

IS THERE MORE COMPETITION IN THE TROPICS?

"Competition was found in 90% of the studies and 76% of their species, indicating its pervasive importance in ecological systems" (Schoener 1983, p. 276). But as to its relative importance in tropical and temperate regions, the data from experimental studies on interspecific competition are so skewed to the temperate that comparisons are unjustified. Taking Schoener's (1983) list of studies and plotting them in relation to latitude for studies on plants, aquatic herbivores, and terrestrial herbivores plus omnivores, only 8 studies out of 100 are in the tropics (1 on plants, 5 on aquatic herbivores, 2 on terrestrial herbivores) (Fig. 3-4). Indeed, if

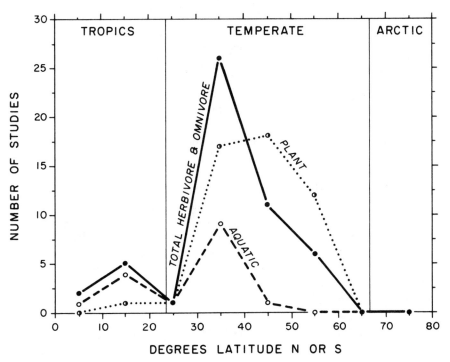

Figure 3-4. Distribution of 100 studies on competition in relation to latitude. Studies were classed in 10° latitude classes and totals are plotted at the midpoint of each class. Totals are given for terrestrial plants (49 studies), aquatic herbivores (16 studies), and total herbivores and omnivores (51 studies), with the difference between aquatic and total being terrestrial herbivores (23 studies) plus terrestrial omnivores (12 studies). All studies in these categories are totals listed by Schoener (1983).

any of the theory on competition being stronger in the tropics were correct, how is it that so much competition can be observed in temperate regions? So some of the basic tenets of r- and K-selection, and the Niche Compression Hypothesis, do not seem to fit broad-scale latitudinal patterns.

Connell (1983) also found much evidence of competition in good experimental studies: "Competition was found in most of the studies, in somewhat more than half of the species, and in about two-fifths of the experiments" (p. 682). But Schoener found less evidence for competition in phytophagous herbivores than in plants, granivores, nectarivores, carnivores, and scavengers. Connell found similar frequencies of competition among plants, herbivores, and carnivores.

So anybody claiming that there is more competition in the tropics than in temperate regions had better base their assertion on their own empirical field studies using well-controlled experiments on a comparative basis both in tropical and temperate locations. There is a serious need for such studies for they would be the first good tests of much basic ecological theory.

ARE THERE VACANT NICHES IN THE TROPICS?

"We have already seen how very hard it is for a second species to colonize an island containing a reasonably close competitor" (MacArthur 1972, p. 247). The concepts that there is more biotic interaction in the tropics, that populations are close to carrying capacity most of the time, and that communities are saturated with species with narrow niches so that invasion by new species is prevented, run through much of our thinking in ecology on latitudinal gradients. And yet there is substantial evidence that many tropical communities exist in a nonequilibrium state: coral reef fish (Sale 1977), corals (Connell 1979), orchid and bromeliad epiphytes (Benzing 1978a, b), tropical trees (Connell 1979), and some tropical bark beetles (Beaver 1979b), to name a few. In addition, I have argued above that tropical herbivores and other species of invertebrates are not more specialized than their equivalents in temperate regions. The obvious deduction is that niche compression has not occurred in the tropics, and there may well be niche space available for colonization by new species. This point is central to much of ecology for it examines the degree to which communities are saturated with species. The quote from MacArthur (1972) is an accurate statement no doubt, but the question we need to address now is "How often is a close competitor present in a community invaded by another species?"

The evidence for empty niches in the tropics is clear enough in some cases. The Pleistocene extinctions of large mammals left only the tapir in Costa Rica (Janzen 1979, 1983; Janzen and Martin 1982). None has been replaced naturally, but horses and cattle survive well and are rather specialized in some of the seeds they disperse, suggesting a flora with relict adaptations to the large herbivores of the Pleistocene: horses are greatly attracted to *Crescentia alata* fruits, cattle to *Pithecellobium saman* fruits, and perhaps mastodonts were greedy for *Simaba cedron* fruits (Janzen 1979). Indeed, the reestablishment of Guanacaste dry-tropical forest over cleared land depends heavily on introduced large mammals. There can be no doubt that many large herbivores could live in Central America, as they did 10,000 years ago, and in their absence empty niches remain. We do not see a saturated or equilibrium community here.

Another case in provided by Lawton's (1984) studies of insect herbivores on bracken fern. All evidence suggests many vacant niches in both tropical and temperate assemblages, no niche compression when many species are present, and virtually no structuring of the community through competition. At peak richness temperate bracken communities included 15 species (Skipwith Common, 54°N) and in Papua New Guinea 14 species (Hombrom Bluff, 5°S). There were totally unoccupied niches in each location, relative to occupation at the other site: no rachis miners at Skipwith Common, 5 at Hombrom Bluff; 2 pinna gallers at Skipwith, none at Hombrom; no rachis chewers at Skipwith, 2 at Hombrom, and so on. Lawton (1984, p. 93) emphasized the generality of his findings on bracken fern communities relative to other phytophagous insect communities, including "the large number of apparently empty niches."

Yet another source of evidence for vacant niches comes from the many other introductions of exotic species and their effect, or lack of effect, on the resident biota (see also Mayr 1963 for the same argument). Simberloff (1981) reviewed much of the literature on this subject and concluded that a very small proportion of studies demonstrated extinction of a resident species by an introduced species. Extinction of a resident species would mean there was no vacant niche present in the recipient community. But if no extinction occurs, then a vacant niche for the introduced species must be present. No extinction based on competition occurred in over 99% of the 854 cases reviewed by Simberloff, and in 79% of cases no effect of the introduced species on the resident community was reported. There is little evidence of latitudinal trends in resistance to colonizers from competition, based on the more regional studies (Table 3-4). Extinction did occur owing to introductions, but these were mostly the result of introduced predators affecting prey species or because of habitat change.

TABLE 3-4. Regional Studies on Species Introductions and the Effect on Resident Species[a]

			Extinction Caused By			
Latitude	Location	Number of Introductions	Predation	Habitat Change	Competition	Source
50–60°N	England	72	1(1.4%)	0	1(1.4%)	Lever 1979
20–30°N	Florida	44	0	0	2(4.5%)	Courtenay and Robbins 1973; Courtenay et al. 1974
40–60°S	Southern Islands	14	3 (21.4%)	4 (28.6%)	0	Holdgate and Wace 1961

[a]Based on Simberloff 1981

The literature on biological control of insects is particularly valuable and interesting because much effort has been expended in both tropical and temperate latitudes (DeBack 1974b). Simberloff (1981) concluded from the more than 100 introductions that no extinctions occurred from competition, or any other cause. Admittedly, many cases of biological control concern pest species introduced with introduced plants, so depauperate communites could be involved. However, where collecting has occurred in the pest's original location, as for the cottony cushion scale, the community is hardly richer, for all the available biological control agents are usually imported, and communities of parasitoids may even be enriched on introduced host herbivores because of imported enemies from several regions of the world, as with *Aphytis* species on the California red scale (DeBach 1974a). Therefore, the results of biological control attempts on insect herbivores should be a valid approach to testing for empty niches, especially considering all the other disturbed communities which we study and should be interested in.

If there really is more biotic interaction in the tropics and more physical effects in temperate regions, as conventional wisdom assumes, then we should expect more successes with biological control in tropical regions. "Success has not occurred more frequently in tropical and subtropical (mild climate) areas... The proportion of successes to introductions or of introductions to establishment has probably been as high or higher in British Columbia... and New Zealand... as in any of the tropical areas" (DeBach 1974a, 694).

Two deductions are possible from this large body of research work: (1) Introductions of new species are just as easy in tropical latitudes as in temperate latitudes (Fig. 3-5), indicating that vacant niches existed for parasitoids and predators of insect herbivores in all latitudes. No evident significant latitudinal gradient exists. (2) Biotic interactions between predators or parasitoids with their prey or hosts are as readily established in tropical and temperate latitudes, indicating no latitudinal gradient in the importance of biotic interactions between trophic levels.

Four lines of empirical evidence all support the same conclusion that empty niches exist in the tropics on a large scale: many more large mammals could exist there than are now present; tropical insect herbivore communities have as many vacant niches as in the temperate; many species can be introduced without affecting the resident flora and fauna; many natural enemies can be introduced to regulate insect herbivore pests. I know of no empirical evidence based on manipulative studies contrary to the argument that tropical communities are usually undersaturated and not at equilibrium.

Some would argue that a niche is a property of a species or population (e.g., Colwell and Fuentes 1975), so empty niches cannot exist (e.g., Herbold and Moyle 1986; Colwell 1989). However, the approach taken in this chapter in which the niche is a property to the habitat, and similar habitats should support similar kinds of niches, is well established in the literature. In Hutchinson's definition (1957, p. 416) "the state of the environment which would permit the species... to exist indefinitely" is an essential part of the concept. The idea of empty niches has

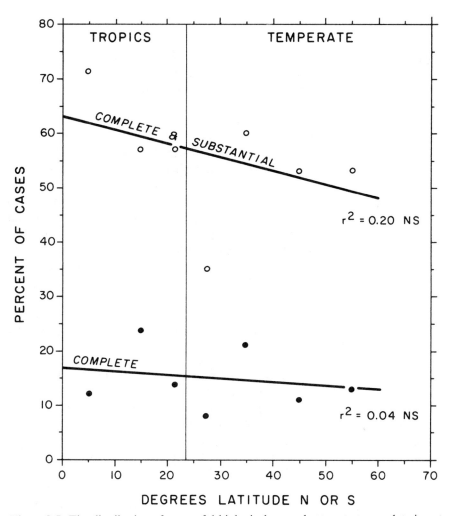

Figure 3-5. The distribution of successful biological control attempts to regulate insect herbivore pests along latitudinal gradients (based on data in DeBach 1974b). DeBach classified 158 cases in the categories complete, substantial or partial control. The percent of cases for complete (solid circles) and complete plus substantial control (open circles) are given, and in neither case is there a significant latitudinal trend. Points are plotted at the midpoint of each latitudinal range and cases between 20 and 30° are divided into those in the tropics and those in temperate regions.

been used by many authors, including Grinnell (1914), Hutchinson (1957), Mayr (1963), Colbert (1980), and Strong et al. (1984), and the comparative approach to recognizing niche vacancy in host habitats is clearly defined (e.g., Lawton and Price 1979; Rohde 1979; Price 1980; Lawton 1984). Therefore, further comparative studies on frequency of vacant niches in tropical and temperate systems is fully justified and should prove illuminating.

CONCLUSIONS

I have addressed four simple questions fundamental to much theory on variation in community structure along latitudinal gradients, emphasizing insect herbivores and other small organisms when possible. To the question "Are there more species in the tropics?" the answer is yes, but with many taxa of insect herbivores not showing this trend, suggesting the need for a much more careful look at tropical diversity and what kinds of organisms really contribute to it. The question of specialization in the tropics was answered by showing equally narrow niche occupation among organisms, like insect herbivores and animal parasites, along latitudinal gradients. "Is there more competition in the tropics?" We simply do not have enough good experimental studies to say, but given the strong and extensive evidence for competition in temperate regions it is doubtful that competition will be found to be stronger in the tropics. Finally, empirical evidence suggests that many vacant niches persist in tropical areas, just as they do in temperate regions, and no latitudinal trends are evident. Both the evidence for universal specialization in general, and the presence of empty niches, combine to support the argument that competition will not be found to be more concentrated in tropical latitudes. This together with the empirical evidence from biological control of insect herbivores suggests that biotic interactions in general are probably no more intense in the tropics than in temperate zones. They may be more diverse, but not cause any stronger selection pressure.

The theory on latitudinal gradients of species diversity was founded largely on more generalized species like birds, mammals, and ants, using observation in the absence of controlled experiments or any manipulative approaches, and only the simplest of empirical data, such as provided in Fischer's (1960) analysis. In spite of this theory being with us for almost 40 years, since Dobzhansky's (1950) early arguments, very little progress has been made toward a mechanistic understanding of latitudinal gradients in species diversity, certainly not for insect herbivores.

I have argued that in the three areas of specialization, competition, and niche vacancy, the empirical evidence indicates no real differences between tropical and temperate zones. Clearly different are the physical factors which set environments: temperature, humidity, light intensity, environmental stability, area, and altitudinal gradients which define the number of life zones and habitats in a region. These differences have clear consequences for the biota, as reviewed in Figure 3-6, resulting in high species number of plants and animals. The scheme was developed in 1975 (Price 1975) before I had started wondering about specialization in tropical herbivores and parasites (Price 1980). It now seems that specialization by animals should be deleted as a relevant part of the process, but other aspects of the scheme may still be valid.

The challenge for the future, as I see it, is to plan for directly comparable studies, preferably by one investigator, to look in detail at tropical and temperate sites. The number of species present and their specificity or niche breadth should be measured. Competition experiments at both latitudes should be conducted.

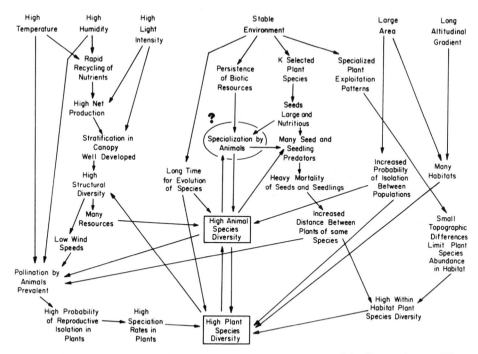

Figure 3-6. Starting with six physical factors listed at the top of the figure, the possible consequences for high plant and animal diversity in the tropics are displayed. This composite scheme is derived from many sources (from Price 1975, 1984). A ring around "specialization by animals" has been added with a question mark to indicate that present evidence does not support this aspect of the diagram.

Both positive and negative results should be reported. For insect herbivores the number of species per plant species should be evaluated, and population dynamics studies undertaken. In 10 years, with this kind of study, I think we could advance the understanding of community structure along latitudinal gradients more than in the past 40 years.

ACKNOWLEDGMENTS

I am grateful to John N. Thompson and Timothy P. Craig for helpful comments on this chapter. Financial support was provided by a grant from the National Science Foundation BSR-8705302. I deeply appreciate the samples on galling insects taken by Jeffrey Brawn in Panama, and Dan Gerling in Israel, and Barbara Bentley's suggestion that I should sample in campina vegetation along the Rio Negro.

REFERENCES

Beaver, R. A. 1979a. Host specificity in temperate and tropical animals. *Nature* **281**: 139–141.

Beaver, R. A. 1979b. Non-equilibrium "island" communities: A guild of tropical bark beetles. *Journal of Animal Ecology* **48**: 987–1002.

Begon, M., J. L. Harper, and C. R. Townsend. 1986. *Ecology: Individuals, populations, and communities.* Sinauer, Sunderland, Massachusetts.

Benson, R. B. 1950. An introduction to the natural history of British sawflies. *Transactions of the Society for British Entomology* **10**: 45–142.

Benzing, D. H. 1978a. Germination and early establishment of *Tillandsia circinnata* Schlecht. (Bromeliaceae) on some of its hosts and other supports in southern Florida. *Selbyana* **2**: 95–106.

Benzing, D. H. 1978b. The life history profile of *Tillandsia circinnata* (Bromeliaceae) and the rarity of extreme epiphytism among the angiosperms. *Selbyana* **2**: 325–337.

Cameron, W. A. 1958. *Mammals of the islands in the Gulf of St. Lawrence.* National Museum of Canada Bulletin 154.

Colbert, E. H. 1980. *Evolution of the vertebrates.* Wiley, New York.

Colwell, R. K. 1986. Community biology and sexual selection: Lessons from Hummingbird flower mites. Pp. 406–424. In J. Diamond and T. J. Case (Eds.). *Community ecology.* Harper and Row, New York.

Colwell, R. K. 1989. Ecology and biotechnology: Expectations and outliers. In J. Fiksel and V. T. Covello (Eds.). *Risk analysis approaches for environmental releases of genetically engineered organisms.* Springer-Verlag, Berlin.

Colwell, R. K. and E. R. Fuentes. 1975. Experimental studies of the niche. *Annual Review of Ecology and Systematics* **6**: 281–310.

Connell, J. H. 1983. On the prevalence and relative importance of interspecific competition: Evidence from field experiments. *American Naturalist* **122**: 661–696.

Connell, J. H. 1979. Tropical rain forests and coral reefs as open nonequilibrium systems. *Symposium of the British Ecological Society* **20**: 141–163.

Courtenay, W. R. and C. R. Robins. 1973. Exotic aquatic organisms in Florida with emphasis on fishes: A review and recommendations. *Transactions of the American Fisheries Society* **102**: 1–12.

Courtenay, W. R., H. F. Sahlman, W. W. Miley and D. J. Herrema. 1974. Exotic fishes in fresh and brackish waters of Florida. *Biological Conservation* **6**: 291–302.

Cox, G. W. and R. E. Ricklefs. 1977. Species diversity, ecological release, and community structuring in Caribbean land bird faunas. *Oikos* **29**: 60–66.

DeBach, P. (Ed.). 1974a. *Biological control of insect pests and weeds.* Reinhold, New York.

DeBach, P. 1974b. Successes, trends, and future possibilities. Pp. 673–713. In P. DeBach (Ed.). *Biological control of insect pests and weeds.* Reinhold, New York.

DeVries, P. J. 1987. *The butterflies of Costa Rica and their natural history.* Princeton University Press, Princeton, New Jersey.

Dixon, A. F. G., P. Kindlmann, J. Leps, and J. Holman. 1987. Why there are so few species of aphids, especially in the tropics. *American Naturalist* **129**: 580–592.

Dobzhansky, T. 1950. Evolution in the tropics. *American Scientist* **38**: 209–221.

Eastop, V. F. 1972. Deductions from the present day host plants of aphids and related insects. *Symposium of the Royal Entomological Society of London* **6**: 157–178.

Eastop, V. F. 1978. Diversity of the Sternorrhyncha within major climatic zones. *Symposium of the Royal Entomological Society of London* **9**: 71–88.

Felt, E. P. 1940. *Plant galls and gall makers.* Comstock, Ithaca.

Fernandes, G. W. 1987. *Tropical and temperature altitudinal gradients in galling species richness.* Master of Science Thesis. Northern Arizona University.

Fischer, A. G. 1960. Latitudinal variations in organic diversity. *Evolution* **14**: 64–81.

Futuyma, D. J. 1986. *Evolutionary biology.* 2nd ed. Sinauer, Sunderland, Massachusetts.

Gauld, I. D. 1986a. Latitudinal gradients in ichneumonid species-richness in Australia. *Ecol. Entomol.* **11**: 155–161.

Gauld, I. D. 1986b. Taxonomy, its limitations and its role in understanding parasitoid biology. Pp. 1–21. In J. Waage and D. Greathead (Eds.). *Insect parasitoids.* Academic Press, London.

Grinnell, J. 1914. An account of the mammals and birds of the Lower Colorado Valley. *University of California Publications in Zoology* **12**: 51–294.

Hawkins, B. A. 1990. Global patterns of parasitoid assemblage size. *J. Anim. Ecol.* **59**: 57–72.

Herbold, B. and P. B. Moyle. 1986. Introduced species and vacant niches. *American Naturalist* **128**: 751–760.

Holdgate, M. W. and N. M. Wace. 1961. The influence of man on the floras and faunas of southern islands. *Polar Record* **10**: 473–493.

Hutchinson, G. E. 1957. Concluding remarks. *Cold Spring Harbor Symposium in Quantitative Biology* **22**: 415–427.

Janzen, D. H. 1975. Interactions of seeds and their insect predators/parasitoids in a tropical deciduous forest. Pp. 154–186. In P. W. Price (Ed.). *Evolutionary strategies of parasitic insects and mites.* Plenum, New York.

Janzen, D. H. 1979. New horizons in the biology of plant defenses. Pp. 331–350. In G. A. Rosenthal and D. H. Janzen (Eds.). *Herbivores: Their interaction with secondary plant metabolites.* Academic, New York.

Janzen, D. H. 1981. The peak of North American ichneumonid species richness lies between 38° and 41°N. *Ecology* **62**: 532–537.

Janzen, D. H. 1983. Dispersal of seeds by vertebrate guts. Pp. 232–262. In D. J. Futuyma and M. Slatkin (Eds.). *Coevolution.* Sinauer, Sunderland, Massachusetts.

Janzen, D. H. and P. S. Martin. 1982. Neotropical anachronisms: The fruits the gomphotheres ate. *Science* **215**: 19–27.

Johnson, C. D. 1970. Biosystematics of the Arizona, California, and Oregon species of the seed beetle genus *Acanthoscelides*, Schilsky (Coleoptera: Bruchidae). *Univ. California Pubs. Entomol.* **59**: 1–116.

Johnson, C. D. 1983. Ecosystematics of *Acanthoscelides* (Coleoptera: Bruchidae) of Southern Mexico and Central America. *Entomol. Soc. Amer. Misc. Pub.* **56**: 1–370.

Krebs, C. J. 1985. *Ecology: The experimental analysis of distribution and abundance.* Harper and Row, New York.

Lack, D. 1976. *Island biology illustrated by the land birds of Jamaica.* University of California Press, Berkeley.

Lawton, J. H. 1984. Non-competitive populations, non-convergent communites, and vacant niches: The herbivores of bracken. Pp. 67–100. In D. R. Strong, D. Simberloff, L. G. Abele, and A. B. Thistle (Eds.). *Ecological communities: Conceptual issues and the evidence.* Princeton University Press, Princeton, N.J.

Lawton, J. H. and P. W. Price. 1979. Species richness of parasites on hosts: Agromyzid flies of the British Umbelliferae. *J. Anim. Ecol.* **48**: 619–637.

Lever, C. 1979. *The naturalized animals of the British Isles.* Granada, St. Albans, Hertfordshire.

MacArthur, R. H. 1962. Some generalized theorems of natural selection. *Proceedings of the National Academy of Sciences* **48**: 1893–1897.

MacArthur, R. H. 1972. *Geographical ecology: Patterns in the distribution of species.* Harper and Row, New York.

MacArthur, R. H., H. Recher, and M. Cody. 1966. On the relation between habitat selection and species diversity. *American Naturalist* **100**: 319–327.

MacArthur, R. H. and E. O. Wilson. 1967. *The theory of island biogeography.* Princeton University Press, Princeton, New Jersey.

Marquis, R. J. and H. E. Braker. 1989. Plant/herbivore interactions at La Selva: Diversity, specialization and impact on plant populations. In K. Bawa, G. S. Hartshorn, and H. Hespenheide. *La Selva: Ecology of a lowland rain forest.* Sinauer, Sunderland, Mass.

Mayr, E. 1963. *Animal species and evolution.* Belknap Press of Harvard University Press, Cambridge, Mass.

Pianka, E. R. 1966. Latitudinal gradients in species diversity: A review of concepts. *American Naturalist* **100**: 33–46.

Pianka, E. R. 1974. *Evolutionary ecology.* Harper and Row, New York.

Price, P. W. 1975. *Insect ecology.* Wiley, New York.

Price, P. W. 1980. *Evolutionary biology of parasites.* Princeton University Press, Princeton, New Jersey.

Price, P. W. 1984. *Insect ecology.* 2nd ed. Wiley, New York.

Ricklefs, R. E. 1979. *Ecology.* 2nd ed. Chiron, New York.

Rohde, K. 1978a. Latitudinal differences in host-specificity of marine Monogenea and Digenea. *Marine Biology* **47**: 125–134.

Rohde, K. 1978b. Latitudinal gradients in species diversity and their causes. II. Marine parasitological evidence for a time hypothesis. *Biologisches Zentralblatt* **97**: 405–418.

Rohde, K. 1979. A critical evaluation of intrinsic and extrinsic factors responsible for niche restriction in parasites. *American Naturalist* **114**: 648–671.

Rohde, K. 1980. Comparative studies on microhabitat utilization by ectoparasites of some marine fishes from the North Sea and Papua New Guinea. *Zoologischer Anzeiger* **204**: 27–63.

Rohde, K. 1986. Differences in species diversity of Monogenea between the Pacific and Atlantic Oceans. *Hydrobiologia* **137**: 21–28.

Sale, P. F. 1977. Maintenance on high diversity in coral reef fish communities. *American Naturalist* **111**: 337–359.

Schoener, T. W. 1983. Field experiments on interspecific competition. *American Naturalist* **122**: 240–285.

Scriber, J. M. 1973. Latitudinal gradients in larval feeding specialization of the World Papilionidae (Lepidoptera). *Psyche* **80**: 355–373.

Simberloff, D. 1981. Community effects of introduced species. Pp. 53–81. In M. H. Nitecki (Ed.). *Biotic crises in ecological and evolutionary time.* Academic, New York.

Smith, D. R. 1979. Suborder Symphyta. Pp. 3–137. In K. V. Krombein, P. D. Hurd, D. R. Smith, and B. D. Burks (Eds.). *Catalog of Hymenoptera in America north of Mexico.* Smithsonian Institution Press, Washington, D.C.

Spencer, K. A. 1972. *Handbook for the identification of British insects: Diptera, Agromyzidae.* Royal Entomological Society of London, London.

Strong, D. R., J. H. Lawton, and R. Southwood. 1984. *Insects on plants: Community patterns and mechanisms.* Harvard University Press, Cambridge.

Townes, H. 1969. The genera of Ichneumonidae. *Memoirs of the American Entomological Institute* **11**: 1–300.

Wilson, E. O. and R. W. Taylor. 1967. An estimate of the potential evolutionary increase in species density in the Polynesian ant fauna. *Evolution* **21**: 1–10.

Wood, T. K. 1984. Life history patterns of tropical membracids (Homoptera: Membracidae). *Sociobiology* **8**: 299–344.

4 Species Richness, Population Abundances, and Body Sizes in Insect Communities: Tropical Versus Temperate Comparisons

JOHN H. LAWTON

The structure of animal communities can be described in a number of ways; the simplest is the number of coexisting species. More subtle descriptions include estimates of population abundance and fluctuations shown by component species, the number of species in each abundance category, and the body size distribution of members of the community. This chapter explores these and related patterns in insect assemblages with particular emphasis on phytophagous species. It has two main aims. The first is to show how a number of population and community characteristics are interrelated, despite having separate lives in most of the current literature. The second is to speculate on similarities and differences between tropical and temperate communities. So broad and general a theme echoes the observation by Young (1982) "..that the tropics generally contain a greater number of closely related species for many genera, and also more genera in different families of insects. Somehow, therefore, there must be relationships between the population structure and dynamics of individual species and the ability for many more species to occupy the same environment or habitat.... As the great majority of insects are herbivores or plant predators, it is the interrelationships of plants and insects that probably account for the spatial and temporal structuring of insect populations in the tropics."

The chapter does not provide definitive answers to specific questions within these broad themes; rather, it tries to provide a framework that may be useful for comparing tropical and temperate communities in the future. In so doing, it raises many more problems than it solves.

Plant-Animal Interactions: Evolutionary Ecology in Tropical and Temperate Regions, Edited by Peter W. Price, Thomas M. Lewinsohn, G. Wilson Fernandes, and Woodruff W. Benson. ISBN 0-471-50937-X © 1991 John Wiley & Sons, Inc.

PATTERNS IN THE DYNAMICS AND DISTRIBUTIONS
OF ANIMAL POPULATIONS

Theoretical Predictions

The local population dynamics and regional distributions of animal species from a variety of taxa, including phytophagous insects, are now known to be interrelated in a number of interesting ways (Gaston and Lawton 1988a,b). The great majority of studies, however, have focused on particular pairs of attributes. But as Figure 4-1 makes plain, they are each part of a complex web of dependent and independent variables, making it difficult to sort out "cause and effect." Nevertheless, arguments can be advanced to explain nine pairs of relationships. The outer edge of the triangle in Figure 4-1 defines the first three of these.

(1) *Local Population Abundance and Regional Distribution.* A positive relationship between average population size measured at occupied sites, and geographic range measured by the number of occupied sites, has been demonstrated for several taxa, including plants, insects, and birds (Hanski 1982; Fuller 1982; Bock and Ricklefs 1983; Brown 1984; O'Connor and Shrubb 1986; Bock 1987; Brown and Maurer 1987). A positive relationship is expected if widespread species are more flexible in their use of resources (McNaughton and Wolf 1970; Brown 1984; Brown and Maurer 1987).

(2) *Local Population Variability and Regional Distribution.* Species with restricted regional distributions and high local population variability should rarely be encountered, because they have a high probability of extinction (Glazier 1986). Thus we expect a positive correlation between the variability shown by local populations and size of geographic range.

(3) *Local Population Abundance and Variability.* Means and variances are usually positively correlated in population data (Taylor 1961). Rare, highly variable populations have high probabilities of extinction.

Inside the triangle defined by these first three relationships are three more, linking local population dynamics and regional distributions to species body sizes (Figure 4-1) as follows.

(4) *Local Population Abundance and Body Size.* A negative relationship between these variables is well documented (Peters 1983; Peters and Wassenberg 1983; Brown and Maurer 1986; Damuth 1987; Robinson and Redford 1986). It may be generated by an increase in *per capita* resource use with increasing body size. However, as we shall discover later, this relationship may not be as robust as the references given above and plausible theoretical explanation imply.

(5) *Local Population Variability and Body Size.* Smaller species have higher intrinsic rates of increase, r, than larger species (Smith 1954; Southwood 1981; Gaston 1988), which will tend to generate greater deterministic population oscillations (May 1981). Variability may be reinforced if smaller species are more susceptible to environmental fluctuations (Bonner 1965; Southwood 1977; Lindstedt and Boyce 1985). A negative correlation between population varia-

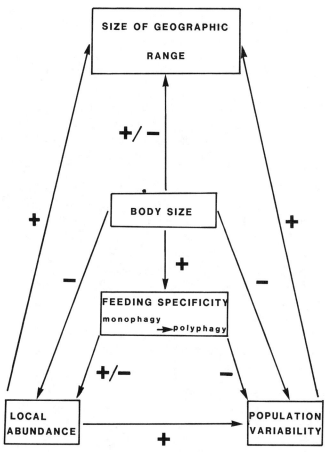

Figure 4-1. Interrelationships between population dynamics, size of geographic range, body size, and feeding specificity (from few to many host plants). The arrows run from presumed independent to dependent variables with the expected sign of the correlation.

bility and body size is therefore predicted, but until recently (see below) has not been tested, except in comparisons of fluctuations shown by (large-bodied) vertebrate and (small-bodied) invertebrate populations (Schoener 1985).

(6) *Regional Distribution and Body Size.* To be consistent with the other relationships depicted in Figure 4-1, a negative correlation between body size and size of geographic range is expected, and may be generated if small-bodied species with high *r* are better colonists than large-bodied species. This may be reasonable given that extinction rates are higher for small populations (MacArthur and Wilson 1967; Leigh 1981), and that high *r* values carry newly founded populations out of the "danger zone" more readily than low *r*'s (Lawton and Brown 1986). However, counter arguments suggest that species of large body size and hence low population density will have high probabilities of extinction if

they have small geographic ranges (because total population sizes will be small). These arguments predict a positive relationship between number of sites occupied and body size (Brown and Maurer 1987).

Finally, we can trace relationships between feeding specificity, body size, and population dynamics.

(7) *Body Size and Feeding Specificity.* Both for mammalian (Jarman 1974) and insect herbivores (Wasserman and Mitter 1978; Niemela et al. 1981) large-bodied species tend to be more polyphagous than small-bodied species. At least for insects, a satisfactory theoretical explanation for the pattern is hard to find.

(8) *Local Population Variability and Feeding Specificity.* MacArthur (1955) predicted that polyphagous species should have more stable populations than more specialized species, because generalists are less susceptible to fluctuations in the abundance of individual resources. Although Watt (1964) presents counter arguments, recent analyses by Redfearn and Pimm (1988) for a range of phytophagous insect populations generally support MacArthur's prediction.

(9) *Feeding Specificity and Local Abundance.* To conform with the other relationships in Figure 4-1, trophic specialists should be locally more abundant than polyphagous species. Such a correlation would be consistent with the "jack of all trades, master of none" principle. However, this prediction is clearly at variance with Brown's (1984) argument which underpins pattern (1), namely that generalists will be locally more abundant than specialists. The expected correlation is therefore unclear.

Figure 4-1 is complex, and nine separate relationships are difficult to keep in mind all at once. Accordingly, the theoretical arguments just outlined have been kept as simple as possible. For many of them, there are uncertainties and subtleties that I have deliberately avoided, without I hope distorting the picture too badly. The problems are discussed in more detail by Brown and Maurer (1987), Redfearn and Pimm (1988), Gaston (1988a), and Gaston and Lawton (1988a,b).

Tests of Theoretical Predictions: Data for British Phytophagous Insects

Table 4-1 gathers together tests of these predictions, based on data for British phytophagous insects in several recent studies. The first two (Gaston and Lawton 1988a,b) examine the feeding stages of a taxonomically heterogeneous group of 21 species on a single host plant, bracken (*Pteridium aquilinum*), growing in open and woodland sites [there are small, but consistent differences both in the species composition of the insects and their abundances in the two habitats (MacGarvin et al. 1986)]. The remaining studies deal with more closely related taxa, and differ markedly from the bracken data and from one another in the nature and geographic scale of the samples. Gaston and Lawton (1988a) and Gaston (1988a) analyze information on 263 species of adult moths and 97 species of alate aphids (sampled by light traps and suction traps, respectively) from a wide variety of

TABLE 4-1. Summary of Theoretically Predicted and Empirically Observed Correlations Between the Nine Pairs of Relationships Depicted in Figure 4-1.

	Predicted Correlation	Observed Correlation			
		Bracken Insects Open	Bracken Insects Woodland	Adult Moths	Alate Aphids
Size of geographic range vs. local population abundance	+	+	+ +	+ + +	+ + +
Size of geographic range vs. population variability	+	0	+ +	+ + +	+ +
Population variability vs. local population abundance	+	+ + +	+ + +	+ + + (0)	+
Local population abundance vs. body size	−	− − −	− − −	0	Not tested
Population variability vs. body size	−	− − −	− − −	− − −	Not tested
Size of geographic range vs. body size	+/−	− − −	− − −	+	Not tested
Feeding specificity (mono + polyphagy) vs. body size	+	0	+	+	Not tested
Population variability vs. feeding specificity (mono + polyphagy)	−	−	−	+ (−)[b]	Not tested (−)[c]
Local population abundance vs feeding specificity (mono + polyphagy)	+/−	−	− (0)	0 (+)[d]	Not tested

[a]The majority of the data are from Gaston (1988a) and Gaston and Lawton (1988a, b). Results in parentheses are for similar tests carried out by Redfearn and Pimm (1988); "not tested" refers to Gaston and Lawton (1988a). In the list of pairwise relationships, the presumed dependent variable is given first. Predicted and observed correlations with feeding specificity are on a scale from few to many host plants (i.e., monophagy to polyphagy). +, −, and 0 refer to positive, negative, and no correlations, respectively, with the statistical significance of observed relationships being indicated by one ($0.01 < P < 0.05$), two ($0.001 < P < 0.01$) and three ($P < 0.001$) symbols, respectively.

[b]This result, though statistically significant, it biologically fragile, because only three species in Redfearn and Pimm's sample are host specialists.

[c]Six different sites are analyzed separately by Redfearn and Pimm, and there are significant negative relationships in non-host alternating aphids at three of them.

[d]Not statistically significant ($P = 0.08$).

plant species and habitats throughout Britain in the Rothamsted Insect Survey (Taylor and Woiwod 1980). Redfearn and Pimm (1988) also look at moths and aphids in the Rothamsted Survey, but use many fewer species, fewer sites, and different statistical techniques to those employed by Gaston and myself. The different sample sizes in particular seem to have led to different conclusions, and underline the caution that must be exercised when interpreting the results in Table 4-1.

There is evidence in Table 4-1 to support all the theoretical predictions. Despite the heterogeneous nature of the data, only one significant correlation has the wrong sign [monophagous moths have less variable populations than polyphagous species in Gaston's (1988a) analysis; theory predicts the reverse and is supported, albeit feebly, by Redfearn and Pimm (1988)]. Moreover, where theoretical predictions are uncertain (plausible arguments exist for both positive and negative correlations), both are observed, and in the case of the relationship between local population abundance and feeding specificity seem certain to be due to different methodologies. The bracken data refer to specialist and generalist insects sampled on one species of plant; here generalists are rarer than specialists. The data for British moths refer to insects drawn from many different species of host plants, and one would not be surprised if at least sometimes, the summed abundances of polyphages was higher than that of monophages.

It would be unwise to pay too much attention to the statistical significance of the correlations in Table 4-1, partly because the data have been gathered in very different ways in different studies, creating very different errors and biases, and partly because high statistical significance tends to be associated with independent variables that have wide absolute ranges. For this reason, because adult aphids cover a very narrow range of body sizes, correlations between population characteristics and aphid body sizes have not been calculated (Gaston and Lawton 1988a).

At the present time it is impossible to say which of the statistically significant pairs of relationships in Table 4-1 are not biologically related, but are instead the products of shared correlations with other variables. It has proved to be very difficult to sort out cause and effect in these data (Gaston and Lawton 1988a,b). But I see no reason in principle why most of the patterns should not reinforce one another in the manner envisaged in Figure 4-1.

The Consequences for Populations of Tropical Phytophagous Insects

There are no data comparable to those summarized in Table 4-1 for any group of tropical insects. Nor is it sensible to discuss "tropical insects" as if they were all the same. Instead, I propose to narrow the argument and consider insect herbivores in tropical rain forest. To work out the possible consequences of the patterns in Figure 4-1 for such organisms, two generalizations seem particularly relevant.

Compared with temperate forests, herbivores exploiting the canopies of rain forest trees live in a bewilderingly varied mosaic of plants, often with long distances separating individuals of the same species of tree (Hubbell and Foster

1983; Grubb 1987; Gentry 1988). In secondary vegetation and forest understories, foliage of one type occurs in much smaller patches, heavily mixed with many other species (Janzen 1973a). Obviously, vegetation also tends to be distributed in a more fine-grained way in the understory caompared with the canopy of temperate forests, but with much lower overall diversity than in tropical habitats. Consequently, as Elton (1973) noted, compared with phytophagous insects from temperate regions, populations of tropical herbivores tend to be small and scattered among highly diverse vegetation.

A second broad generalization is that contrary to early expectations, populations of insects in tropical forests are not more stable than populations of similar taxa in temperate regions (Bigger 1976; Wolda 1978, 1983); populations fluctuate more or less to the same degree in both regions.

Taken together with the relationships depicted in Figure 4-1, these generalizations lead to a number of tentative and very speculative predictions. For example, populations of host–specialist insects tend to be rarer (per standard unit of plant) in mixed than in pure stands of vegetation (Strong et al. 1984), not least because of the difficulties of finding and staying on suitable hosts in highly diverse vegetation. These problems must be formidable in a rain forest, particularly for small insects with limited powers of dispersal. Hence, if Bigger and Wolda's observations are not somehow misleading us, and other things being equal, I would expect populations of rare, small-bodied, host-specialized phytophagous insects to be incredibly vulnerable to the risk of at least local stochastic extinction in rain forest habitats, compared with their temperate region counterparts.

Over evolutionary time, differential extinctions should create several differences between assemblages of tropical and temperate region herbivores. They include proportionately more large-bodied species in the tropics, because being large should increase the probability of living long enough to discover highly scattered food plants, as well as increasing powers of controlled dispersal. Large-bodied species may also be favored because they tend to be more polyphagous than small-bodied species (an obvious advantage in highly heterogeneous vegetation), and to have more stable populations. The difficulties of findings hosts in highly diverse vegetation also suggest that comparing related taxa of similar body size, tropical herbivores will be more polyphagous than their temperate cousins.

It is not difficult to continue with these speculations. For example, I have said nothing about range sizes. Rapoport (1982) suggests that the proportion of bird and mammal species with restricted geographic ranges is higher in the tropics than in temperate regions. If this turns out to be true for insects, it would be consistent with a higher proportion of locally rare species, and could also be related to body size, if range sizes and body sizes are interdependent.

Partial Tests of These Predictions

The notion that tropical faunas may be more polyphagous than relatives from temperate habitats flies in the face of received wisdom (e.g., Janzen 1973a). Yet as

Price (Chapter 3) shows, for almost all taxa where comparisons have been made, tropical species are certainly not more host-specific than similar temperate species, and are often less specific, exactly as predicted for species exploiting rare and highly fragmented resources. Aphids (Dixon et al. 1987), bark and ambrosia beetles (Beaver 1979), and ichneumonid parasitoids (Gauld 1986) are all more polyphagous in the tropics. Butterflies (Price 1980) and geometrid moths in the subfamily Ennominae (Wasserman and Mitter 1978), in contrast, do not appear to become more generalized in their use of host plants at lower latitudes, but nor do they become more specialized. Without further information on the spatial distribution and abundance of hosts, and insect dispersal abilities, such data will be difficult to interpret (for further discussions, see Young 1982).

Aphids are also intriguing in one other way. Most of the 4000 known species occur only in temperate regions, with the ratio of aphid species to plant species declining steadily as plant species richness rises. In contrast, the ratio of butterfly species to plant species is unaffected by floral diversity (Dixon et al. 1987). Dixon and his colleagues attribute the dearth of aphids in the tropics to their small size, high host specificity, and the short period of time they are able to survive without food, exacerbated by the low efficiency with which they find suitable hosts. Butterflies are much larger, live longer, and have more control over host finding. In other words, selection works against being a small, short-lived host specialist in the tropics. Larger taxa face fewer problems.

Although these data are not incompatable with the tentative speculations oulined earlier, more useful would be information on population abundance, body size, and number of species in entire assemblages of insects. It is to such data that we must now turn.

POPULATION ABUNDANCE, BODY SIZE, AND NUMBER OF SPECIES IN TROPICAL AND TEMPERATE INSECT COMMUNITIES

It is not possible to compare all the components of Figure 4-1 for any pair of insect communities, one temperate and the other tropical. It is possible, however, to compare linked relationships between body size, abundance, and number of species. The data then force us to think about the consequences for population variability and feeding specificity. We shall have nothing further to say about size of geographic range, not because it is unimportant, but because useful data are almost entirely nonexistent.

Theoretical Relationships

Three interrelated patterns provide a powerful way of describing the structure of ecological communities. They are the frequency distribution of species abundances (Preston 1948; May 1975; Sugihara 1980; Gray 1987); the number of species of different body sizes (Hemmingsen 1934; May 1978, 1986; Griffiths 1986); and one of the relationships already explored in Figure 4-1, namely the

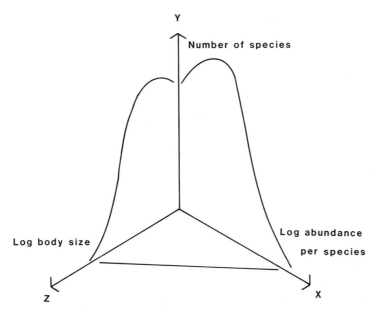

Figure 4-2. The diagram illustrates how three pairs of relationships define a surface in three dimensions. Each pair of relationships has its own body of literature and associated theory. They are the frequency distribution of species abundances (YX plane), the number of species of different body sizes (the YZ plane), and the abundance of individual species as a function of body size (the ZX plane) (see text for details).

abundance of individual species as a function of body size. Their interdependence is illustrated in Figure 4-2; clearly, fixing any two of the basic pairs of relationships automatically defines the limits of the third (Harvey and Lawton 1986). A growing number of papers consider relationships between two of the three faces of Figure 4-2 (e.g., Janzen 1973b; Griffiths 1986; Strayer 1986; Harvey and Godfray 1987), but because data are almost always presented in two dimensions (e.g., frequency distributions of species abundances are presented by summing across individuals of all body sizes), development of a unifying theory is hampered by ignorance about the shape of the full three-dimensional surface. May (1978) is one of the few to have realized that these community patterns are interrelated. However, we now have data on the full surface for an assemblage of tropical insects and a roughly comparable one for temperate insects.

Data for Assemblages of Tropical and Temperate Insects

The tropical data were obtained by fogging the canopies of 10 trees of 5 species in a lowland rain forest near Bukit Sulang, Brunei (Morse et al. 1988). So far, only data on adult Coleoptera have been sorted to species, counted and measured; nevertheless, this subsample contains 859 species of beetles and nearly 4000 individuals. As Morse et al. point out, confining the study to adult beetles has both advantages and disadvantages. One clear advantage is that unlike nymphs and larvae, adults vary very little in size. A major problem is that beetles may not be representative of the entire insect assemblage [e.g., the mean body lengths of beetles in three of the four areas sampled by Schoener and Janzen (1968) are less than the mean body lengths of all insect taxa in the same samples]. Keep this important caveat in mind throughout the ensuing discussions. Nevertheless, adult beetles constitute a significant portion of both individuals and species of insects in the canopies of tropical forest (Erwin 1983).

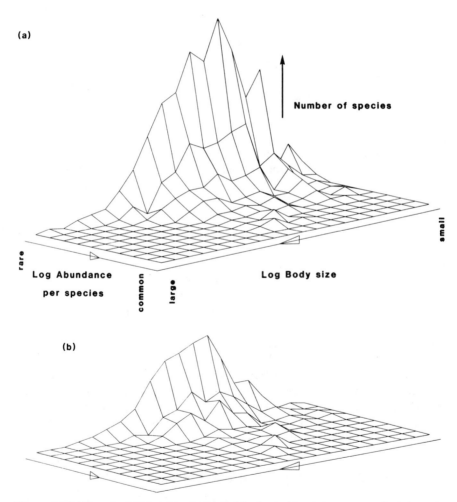

Figure 4-3. (*a*) Data for 859 species of adult beetles fogged from the canopy of rainforest in Brunei, plotted on the axes in Figure 4-2, to reveal the full, three dimensional surface. (*b*) The subsample of 384 species of beetles classified as herbivores is illustrated (from Morse et al. 1988).

Figure 4-3 shows the full three-dimensional surface for all beetles, together with a subsample of the data for 384 species classified as herbivores (see Morse et al. 1988 for details). Herbivores have proportionately fewer very small, very rare species in the "back corner" of the graph than are found in the entire assemblage, but otherwise the basic shapes of the two surfaces are similar.

Projecting the data onto each of the faces of Figure 4-3 to yield more familiar pairs of relationships generates no particular surprises in two cases. The full data are in Morse et al. (1988). Here we need simply note that the species frequency distribution (the YX plane in Fig. 4-2) shows no sign of having reached a mode, despite the large number of species in the collection; 496 species (58% of the sample) are represented by single individuals. Janzen and Schoener (1968) found a similar pattern in sweep net samples of tropical insects (see also the rich compendium of data tabulated in Janzen 1973b). The number of species in different body-size classes (the YZ plane) is also fairly unremarkable, with a mode at body lengths of about 2 mm. However, the proportion of small species is higher than in comparable samples taken by Erwin (1983), and lies close to theoretically expected maxima (for further discussions see Lawton 1986; May 1986; and Morse et al. 1988).

The most interesting feature of Figure 4-3a is the ZX plane. Projecting the three-dimensional surface onto the "floor" of this graph, to yield the supposedly familiar relationship between the abundances of individual species and their body sizes (Fig. 4-1 and row 4, Table 4-1) reveals a very unexpected picture (Fig. 4-4). Average population abundances do not decline significantly with body size, for all beetles or just herbivores. The expected relationship is destroyed by large numbers of rare species with body lengths less than 2 mm.

Two points are worth making at this juncture. First, as Figures 4-2 and 4-3 make clear, if there is a simple inverse correlation between population abundance and body size, there must be a complete lack of data points close to the Y axis on the ZX plane of Figure 4-2; that is, there will be a "hole" in the back corner of the full surface, furthest from the reader. It then follows that rare species in the species-frequency distribution (the YX plane) will all be large, and that small species on the YZ plane will all be common. This is not the case with the Brunei beetles, and the surface of Figure 4-3 is intact. Second, these are not the only data that find no, or only a very feeble, correlation between species abundance and body size (Juanes 1986; Brown and Maurer 1987). It remains to be seen whether differences in the strength of the relationship between body size and population abundances are due to zoogeographical region, taxonomy, the range of species body sizes in the sample, or simply to sampling methods. Fogging, for example, must inevitably collect "tourists," rare individuals that happen to be passing through the canopy at the time. These, and related uncertainties and problems highlight how little we actually know about an apparently very simple relationship (see Morse et al. 1988 for further discussions).

Data exactly comparable to Figure 4-3 are not available for temperate insects. However, very tentative comparisons can be made for arthropods fogged in July from the canopies of three birch (*Betula pendula*) trees at Skipwith Common, in

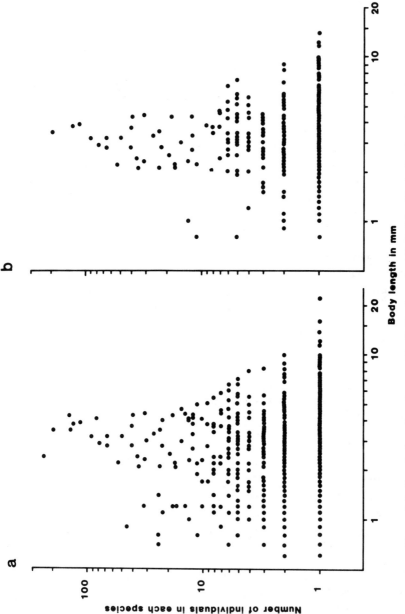

Figure 4-4. Population abundances of individual species of adult beetles in the Brunei samples, as a function of body size (i.e., the ZX plane of Fig. 4-2), for (*a*) all beetles and (*b*) herbivores. Unlike the temperate region data in row 4 of Table 4-1, there is no statistically significant inverse correlation for these tropical insects.

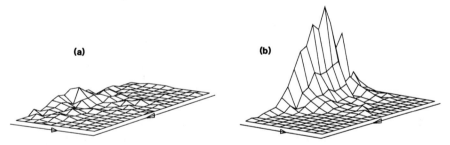

Figure 4-5. (*a*) Data for all arthropods fogged from the canopy of birch trees in northern England, plotted on the same scale and axes as the Brunei beetle data in Figure 4-3*a*, here reproduced as (*b*) to facilitate direct comparison.

the north of England, by David Morse (personal communication). Although there are nearly 5000 individuals in the combined sample, it contains too few beetles to justify separate analysis. Nor have the species been sorted into guilds. Figure 4-5*a* therefore represents the entire assemblage of arthropods, with the data for Brunei beetles reproduced in Figure 4-5*b* on the same scale for ease of comparison.

Two things are immediately obvious from this figure. First, there are proportionately many more rare species in the tropical system; and second, large-bodied species are proportionately commoner in the temperate assemblage. Both patterns are at variance with the predictions made earlier, based on population dynamics.

DISCUSSION AND DIRECTIONS FOR FUTURE RESEARCH

Interesting as it is, comparing the two graphs in Figure 4-5 is extremely hazardous. I have done it more in the hope that it will provoke better tropical–temperate comparisons in the future than in the belief that this particular effort is reliable. Not only are the two figures based on very different taxa, but as Janzen and Schoener (1968), Schoener and Janzen (1968), Janzen (1973a), Erwin (1983), and Rees (1983) have shown, the size structure of samples of insects from tropical and temperate habitats is markedly influenced by such things as season, moisture, and habitat. For example, the modal weight of Homoptera and Heteroptera caught at light and suction traps in a rainforest in Sulawesi increased as rainfall during the trapping period increased (Rees 1983). Contrary to the results in Figure 4-5, Schoener and Janzen's (1968) sweep net samples of insects from forest understories generally had higher mean body lengths is tropical than in temperate collections. However, the proportion of small insects decreased markedly in drier, hotter tropical habitats (Janzen and Schoener 1968). I would therefore expect pictures like Figure 4-5 to look very different in different types of tropical habitats, and at different seasons within one habitat. It remains to be seen whether these differences are nevertheless smaller than those that exist between equivalent tropical and temperate communities.

Figure 4-5 also raises a more general issue. The foundations for making large-scale comparisons of this type were laid by Janzen and Schoener over 20 years ago, in a sweep-sample program of herculean proportions, and great vision. Unfortunately, this type of work is no longer fashionable; it has been replaced by reductionism, experiments, and detailed studies on one or two species. I have nothing against such work. Indeed it is essential. But it is monumentally misleading to suggest that you can reassemble a rain forest by studying small groups of species. Models and experiments are useful ways to understand tiny bits of the forest that make up tractable "small number systems" (Harris 1985). But reductionist attempts to build up detailed understanding of the structure and dynamics of even a few tens of species, by combining individual studies on one or a few, will rapidly run into horrendous difficulties; predicting the dynamics of "middle number systems" is a hopelessly intractable task. One way out of this dilemma is therefore to seek large-scale, statistical patterns that emerge in "large number systems" involving many species (Harris 1985). Ecologists ought to know much more about the size structure and relative species abundances of tropical and temperate animal communities, to name but two large-scale attributes (see, for example, May 1978, 1986; Peters 1983). Without patterns in nature to guide us, we risk being overwhelmed by detail (Southwood 1988).

Figure 4-1 and 4-2 are both small attempts to organize information and to seek patterns in nature. They differ to the extent that Figure 4-1 is underpinned by population dynamics theory, albeit much of it imperfect and in need of refinement. Each face of Figure 4-2 has a separate body of theory associated with it, but nothing to predict the shape of the full three-dimensional surface. Blindly fitting statistical distributions to the data in Figure 4-5 would not be very interesting; trying to explain why there may be more large species and fewer rare species in the temperate assemblage is.

Undoubtedly one of the most interesting features of the Brunei data is the large number of rare species in the sample; many of these are small-bodied, and many of them are herbivores. Elton (1973), amongst others, also comments on the large number of small, very rare species of insects in tropical forests. Yet earlier speculations, and Dixon et al.'s (1987) data on aphids both point to the fact that it is more difficult to be a small, rare herbivore in the tropics than in temperate regions. So why are there apparently so many exceptions in the Brunei beetles?

Sampling artifacts aside, there are many possibilities. One is that species represented by only one or two individuals in the fogging samples nevertheless sustain quite large local populations in the canopy. In which case, if we continued to sample, the species-frequency distribution predicts that we should find still more, even rarer species! This explanation also begs several questions. What is the minimum viable population size for a small insect (Soulé 1987)? What effect would even further sampling have on the graph of species abundance against body size? (the most likely answer is to destroy even further any semblance of an inverse correlation). And how many species of beetles are there in the canopy? (the answer is certainly a vast number—see Erwin 1983).

Another possibility is that small canopy beetles are not like aphids; rather they live longer, and can therefore more easily locate rare, highly scattered food plants. Dixon et al. (1987) suggest, for example, that psyllids are much more successful in the tropics than aphids because they survive longer away from the host plant. A third explanation, already touched upon is that most of these small, herbivorous beetles are not oligophagous or monophagous (as Fig. 4-1 predicts they should be), but polyphagous. In other words, patterns of host-plant specificity as a function of body size, suggested for temperate species, do not apply to tropical beetles. Dr. C. J. C. Rees has also suggested to me that many of the species classified as herbivores may not, in fact, be exploiting the foliage of the sample tree, but might feed on epiphytes, or algae.

If patterns of host-plant specificity differ, how else might the relationships depicted in Figure 4-1 change with latitude? Rare, fragmented populations have a greater chance of persisting if they are stable. It is unclear from Figure 4-1 and the associated theory whether populations of small but rare tropical species will have very stable numbers because they are rare, or be very unstable because they are small. It would also be interesting to know whether amplitudes of population fluctuations found in tropical and temperate insects not only cover the same range when related taxa are compared (Wolda 1978, 1983), but whether they are the same for species of similar body sizes.

Obviously, there is no shortage of interesting questions, even if they arise from very limited data of uncertain generality. The problem now is to get more and better data, using comparable sampling techniques in tropical and temperate systems.

ACKNOWLEDGMENTS

I am very grateful to the symposium organizers for the invitation to take part, and to their sponsors, together with the University of York and the Royal Society of London for the financial support that made it possible. The ideas in this chapter have benefited greatly from discussions with Kevin Gaston, Paul Harvey, David Morse, Stuart Pimm, and Nigel Stork. Brad Hawkins, Kevin Gaston, and Steve Juliano made valuable comments on the manuscript.

REFERENCES

Beaver, R. A. 1979. Host specificity of temperate and tropical animals. *Nature* **281**: 139–141.

Bigger, M. 1976. Oscillations of tropical insect populations. *Nature* **259**: 207–209.

Bock, C. E. 1987. Distribution–abundance relationships of some Arizona landbirds: A matter of scale? *Ecology* **68**: 124–129.

Bock, C. E. and R. E. Ricklefs. 1983. Range size and local abundance of some North American songbirds: A positive correlation. *American Naturalist* **122**: 295–299.

Bonner, J. T. 1965. *Size and Cycle: An Essay on the Structure of Biology*. Princeton University Press, Princeton, New Jersey.

Brown, J. H. 1984. On the relationship between abundance and distribution of species. *American Naturalist* **124**: 255–279.

Brown, J. H. and B. A. Maurer. 1986. Body size, ecological dominance and Cope's rule. *Nature* **324**: 248–250.

Brown, J. H. and B. A. Maurer. 1987. Evolution of species assemblages: Effects of energetic constraints and species dynamics on the diversification of the North American avifauna. *American Naturalist* **130**: 1–17.

Damuth, J. 1987. Interspecific allometry of population density in mammals and other animals: The independence of body mass and population energy-use. *Biological Journal of the Linnean Society* **31**: 193–246.

Dixon, A. F. G., P. Kindlmann, J. Leps, and J. Holman. 1987. Why are there so few species of aphids, especially in the tropics. *American Naturalist* **129**: 580–592.

Elton, C. S. 1973. The structure of invertebrate populations inside neotropical rain forest. *Journal of Animal Ecology* **42**: 55–104.

Erwin, T. L. 1983. Beetles and other insects of tropical forest canopies at Manaus, Brazil, sampled by insecticidal fogging. *Tropical Rain Forest: Ecology and Management*. S. L. Sutton, T. C. Whitmore, and A. C. Chadwick (Eds.)., pp. 59–75. Special Publications series of the British Ecological Society 2. Blackwell Scientific, Oxford.

Fuller, R. J. 1982. *Bird Habitats in Britain*. Poyser, Calton, Staffs.

Gaston, K. J. 1988a. Patterns in the local and regional dynamics of moth populations. *Oikos*, **53**: 49–57.

Gaston, K. G. 1988b. The intrinsic rates of increase of insects of different sizes. *Ecological Entomology* **13**: 399–409.

Gaston, K. G. and J. H. Lawton. 1988a. Patterns in the distribution and abundance of insect populations. *Nature* **331**: 709–712.

Gaston, K. J. and J. H. Lawton. 1988b. Patterns in body size, population dynamics and regional distribution of bracken herbivores. *American Naturalist*, **132**: 662–680.

Gauld, I. D. 1986. Latitudinal gradients in ichneumonid species-richness in Australia. *Ecological Entomology* **11**: 155–161.

Gentry, A. H. 1988. Tree species richness of upper Amazonian forests. *Proceedings of the National Academy of Sciences. USA* **85**: 156–159.

Glazier, D. S. 1986. Temporal variability of abundance and the distribution of species. *Oikos* **47**: 309–314.

Gray, J. S. 1987. Species-abundance patterns. *Organisation of Communities Past and Present*. J. H. R. Gee and P. S. Giller (Eds.)., pp. 53–67. Blackwell Scientific, Oxford.

Griffiths, D. 1986. Size-abundance relations in communities. *American Naturalist* **127**: 140–166.

Grubb, P. J. 1987. Global trends in species-richness in terrestrial vegetation: A view from the northern hemisphere. *Organisation of Communities Past and Present* J. H. R. Gee and P. S. Giller (Eds.), pp. 98–118. Blackwell Scientific, Oxford.

Hanski, I. 1982. Dynamics of regional distribution: The core and satellite species hypothesis. *Oikos* **38**: 210–221.

Harris, G. P. 1985. The answer lies in the nesting behaviour. *Freshwater Biology* **15**: 375–380.

Harvey, P. H. and H. C. J. Godfray 1987. How species divide resources. *American Naturalist* **129**: 318–320.

Harvey, P. H. and J. H. Lawton. 1986. Patterns in three dimensions. *Nature* **324**: 212.

Hemmingsen, A. M. 1934. A statistical analysis of the differences in body size of related species. *Videnskabelige Meddelelser Dansk Naturhistorisk Forening Kobenhaven* **98**: 125–160.

Hubbell, S. P. and R. B. Foster. 1983. Diversity of canopy trees in a neotropical forest and implications for conservation. *Special Publications Series of the British Ecological Society* **2**: 25–41. Blackwell Scientific, Oxford.

Janzen, D. H. 1973a. Sweep samples of tropical foliage insects: Effects of seasons, vegetation types, elevation, time of day, and insularity. *Ecology* **54**: 688–707.

Janzen, D. H. 1973b. Sweep samples of tropical foliage insects: Description of study sites, with data on species abundances and size distributions. *Ecology* **54**: 659–685.

Janzen, D. H. and T. W. Schoener. 1968. Differences in insect abundance and diversity between wetter and drier sites during a tropical dry season. *Ecology* **49**: 96–110.

Jarman, P. J. 1974. The social organisation of antelopes in relation to their ecology. *Behaviour* **48**: 215–267.

Juanes, F. 1986. Population density and size in birds. *American Naturalist* **128**: 921–929.

Lawton, J. H. 1986. Surface availability and insect community structure: The effects of architecture and fractal dimension of plants. *Insects and the Plant Surface*. B. Juniper and T. R. E. Southwood (Eds.), pp. 317–331. Edward Arnold, London.

Lawton, J. H. and K. C. Brown. 1986. The population and community ecology of invading insects. *Philosophical Transactions of the Royal Society of London B* **314**: 607–617.

Leigh, E. G. Jr. 1981. On the average lifetime of a population in a varying environment. *Journal of Theoretical Biology* **90**: 213–239.

Lindstedt, S. L. and M. S. Boyce. 1985. Seasonality, fasting endurance, and body size in mammals. *American Naturalist* **125**: 873–878.

MacArthur, R. H. 1955. Fluctuations of animal populations, and a measure of community stability. *Ecology* **36**: 533–536.

MacArthur, R. H. and E. O. Wilson. 1967. *The Theory of Island Biogeography*. Princeton University Press, Princeton, New Jersey.

MacGarvin, M., J. H. Lawton, and P. A. Heads. 1986. The herbivorous insect communities of open and woodland bracken: Observations, experiments and habitat manipulations. *Oikos* **47**: 135–148.

May, R. M. 1975. Patterns of species abundance and diversity. *Ecological and Evolution of Communities*. M. L. Cody and J. M. Diamond (Eds.), pp. 81–120. Harvard University Press, Cambridge, MA.

May, R. M. 1978. The dynamics and diversity of insect faunas. *Diversity of Insect Faunas* L. A. Mound and N. Waloff (Eds.), pp. 188–204. Symposium of the Royal Entomological Society of London 9. Blackwell Scientific, Oxford.

May, R. M. 1981. Models for single populations. I. *Theoretical Ecology: Principles and Applications*, R. M. May (Ed.), pp. 5–29. Blackwell Scientific, Oxford.

May, R. M. 1986. The search for patterns in the balance of nature: Advances and retreats. *Ecology* **67**: 1115–1126.

McNaughton, S. J. and L. L. Wolf. 1970. Dominance and the niche in ecological systems. *Science* **167**: 131–139.

Morse, D. R., N. E. Stork, and J. H. Lawton. 1988. Species number, species abundance and body length relationships of arboreal beetles in Bornean lowland rain forest trees. *Ecological Entomology* **13**: 25–37.

Niemela, P., S. Hanhimaki, and R. Mannila. 1981. The relationship of adult size in noctuid moths (Lepidoptera, Noctuidae) to breadth of diet and growth form of host plants. *Annals Entomologica Fennici* **47**: 17–20.

O'Connor, R. J. and M. Shrubb. 1986. Farming and Birds. Cambridge University Press, Cambridge.

Peters, R. H. 1983. *The Ecological Implications of Body Size.* Cambridge University Press, Cambridge.

Peters, R. H. and K. Wassenberg. 1983. The effect of body size on animal abundance. *Oecologia* **60**: 89–96.

Preston, F. W. 1948. The commonness and rarity of species. *Ecology* **29**: 254–283.

Price, P. W. 1980. *Evolutionary Biology of Parasites.* Princeton University Press. Princeton, New Jersey.

Rapoport, E. H. 1982. *Areography.* Pergamon Press, Oxford.

Redfearn, A. and S. L. Pimm. 1988. Population variability and polyphagy in herbivorous insect communities. *Ecological Monographs* **58**: 39–55.

Rees, C. J. C. 1983. Microclimate and the flying Hemiptera fauna of a primary lowland rain forest in Sulawesi. *Tropical Rain Forest: Ecology and Management.* S. L. Sutton, T. C. Whitmore, and A. C. Chadwick (Eds.), pp. 121–136. Special Publications Series of the British Ecological Society 2. Blackwell Scientific, Oxford.

Robinson, J. G. and K. H. Redford. 1986. Body size, diet, and population density of neotropical forest mammals. *American Naturalist* **128**: 665–680.

Schoener, T. W. 1985. Patterns in terrestrial vertebrate versus arthropod communities: Do systematic differences in regularity exist? In *Community Ecology*, J. Diamond and T. S. Case (Eds.), pp. 556–586. Harper and Row, New York.

Schoener, T. W. and D. H. Janzen. 1968. Notes on environmental determinants of tropical versus temperature insect size patterns. *American Naturalist* **102**: 207–224.

Smith, F. E. 1954. Quantitative aspects of population growth. In *Dynamics of Growth Processes*, E. Boell (Ed.), pp. 274–294. Princeton University Press, Princeton, New Jersey.

Soulé, M. E. 1987 (Ed.). *Viable Populations for Conservation.* Cambridge University Press, Cambridge.

Southwood, T. R. E. 1977. Habitat, the templet for ecological strategies? *Journal of Animal Ecology* **46**: 337–365.

Southwood, T. R. E. 1981. Bionomic strategies and population parameters. In *Theoretical Ecology: Principles and Applications*, R. M. May (Ed.), pp. 30–52. Blackwell Scientific, Oxford.

Southwood, T. R. E. 1988. Tactics, strategies and templets. *Oikos* **52**: 3–18.

Strayer, D. 1986. The size and structure of a lacustrine zoobenthic community. *Oecologia* **69**: 513–516.

Strong, D. R., J. H. Lawton, and T. R. E. Southwood. 1984. *Insects on Plants: Community Patterns and Mechanisms.* Blackwell Scientific, Oxford.

Sugihara, G. 1980. Minimal community structure: An explanation of species abundance patterns. *American Naturalist* **116**: 770–787.

Taylor, L. R. 1961. Aggregation, variance and the mean. *Nature* **189**: 732–735.

Taylor, L. R. and I. P. Woiwod. 1980. Temporal stability as a density-dependent species characteristic. *Journal of Animal Ecology* **49**: 209–224.

Wasserman, S. S. and C. Mitter. 1978. The relationship of body size to breadth of diet in some Lepidoptera. *Ecological Entomology* **3**: 155–160.

Watt, K. E. F. 1964. Comments on fluctuations of animal populations and measures of community stability. *Canadian Entomologist* **96**: 1434–1442.

Wolda, H. 1978. Fluctuations in abundance of tropical insects. *American Naturalist* **112**: 1017–1045.

Wolda, H. 1983. "Long-term" stability of tropical insect populations. *Researches on Population Ecology Supplement* **3**: 112–126.

Young, A. M. 1982. *The Population Biology of Tropical Insects*. Plenum, New York.

NOTE ADDED IN PROOF

Since this paper was written, New York (B. H. McArdle, K. G. Gaston and J. H. Lawton (1990) *Journal of Animal Ecology* **59**: 439–454) shows that most of the estimates of population variation on which the tests in Table 4-1 are based are in error because of the way in which data were transformed prior to analysis. The theoretical framework discussing population variation is correct; empirical tests of this part of the theoretical framework, carried out by Gaston and Lawton can no longer be considered secure.

5 Comparison of Tropical and Temperate Galling Species Richness: The Roles of Environmental Harshness and Plant Nutrient Status

G. WILSON FERNANDES and PETER W. PRICE

Species richness has inspired thinking in biology since the beginning of the Victorian age when travellers first came in contact with the extravagance of life in the tropics. Since then, species richness and diversity phenomena have themselves become subject to scrutiny and debate.

The increase of species diversity with decreasing latitude is well established. A few examples will illustrate this global pattern in species richness. The number of ant species increases steadily from the arctic part of Alaska (65–70°N; 7 species) to Trinidad (10–11°N; 134 species), and from Tierra del Fuego (55–57°S; 2 species) to São Paulo (20–25°S; 222 species) (see Kusnezov 1957; Fischer 1960; Darlington 1965). Beetles (Clarke 1954), and dragonflies (Williams 1964) also show the same trend of increasing species richness with decreasing latitude. Several other animal taxa (see Simpson 1964; MacArthur 1972; Horn and Allen 1978; Brown and Gibson 1983), and plant taxa (e.g., Richards 1957) show the same trends in the distribution of species richness.

However some taxa are exceptions. For example, ichneumonids are richest in mid-latitudes in North America at about 40°N (Janzen 1981). Bees (Michener 1979), aphids (Eastop 1972; Dixon et al. 1987), and psyllids (Eastop 1978) are also richer in nontropical areas (see also Price, Chapter 3).

The more widespread geographic patterns of animal distribution and diversity are somewhat mimicked by altitudinal gradients (e.g., Holdridge 1967; Holdridge et al. 1971; Claridge and Singharo 1978; Lawton et al. 1987; Wolda 1987).

The southwestern United States offers excellent conditions for studies of

Plant-Animal Interactions: Evolutionary Ecology in Tropical and Temperate Regions, Edited by Peter W. Price, Thomas M. Lewinsohn, G. Wilson Fernandes, and Woodruff W. Benson.
ISBN   0-471-50937-X © 1991 John Wiley & Sons, Inc.

factors influencing species richness. For example in Arizona, habitat types range from harsh desert to alpine-tundra vegetation in a few hundred kilometers, passing through several vegetational belts (Merriam 1890). Casual observations suggested a strong increase in galling insect species diversity with decreasing altitude. Even with fewer vegetational belts than in Arizona, the same pattern was evident on the altitudinal gradient of the Serra do Cipó in southeastern Brazil. Therefore, we undertook a study to verify if the pattern of increasing species richness with decreasing elevation existed for gall-forming insects, and whether the trend was more related to ecological factors other than altitude. Insect galls are good subjects for ecological studies owing to their abundance, diversity, and sessile habit, which makes them easier to census than unconcealed herbivores. We studied galling species richness along geographic gradients that historically have yielded the most dramatic contrasts in species diversity. These gradients include latitude, altitude, moisture, and temperature.

The following hypotheses were tested: (1) the altitudinal–latitudinal hypothesis, which predicts that as altitude or latitude decline species diversity increases; (2) the harsh environment hypothesis, which predicts that galling species richness will be higher in dry, hygrothermally stressed habitats; (3) the plant species richness hypotheses, which predicts a positive correlation between plant species number and galling species number; (4) the species-area hypothesis, which predicts a positive correlation between host species area and galling species number; (5) the plant architecture (structural complexity) hypothesis, which predicts a positive correlation between plant structural complexity and galling species richness. The present paper develops from Fernandes (1987) and Fernandes and Price (1988), adding new data and new analyses on the richest galling taxa, the cecidomyiids. We also begin to investigate the mechanisms creating the patterns we observed.

The harsh environment hypothesis predicts that galling species diversity is better related to dry habitats than to any of the other hypotheses. We have argued elsewhere (Price et al. 1986, 1987) that a major adaptive advantage to the galling habit is protection for the insect from hygrothermal stress. If this were correct, galling species diversity should be higher in dry habitats than in mesic habitats independent of altitude, plant species richness, plant structural diversity, and host species area.

Our temperate sites were taken on an altitudinal gradient ranging from the lower Sonoran Desert at 0 m (sea level) in Bahía Kino (27°N), Mexico to the top of the San Francisco Peaks (36°N) at 3843 m in the United States. We sampled 27 sites distributed in pairs, on north and south slopes, at 13 elevational locations. An additional site was at the mountain peak (3843 m). Samples were taken at each 305 m (1000 ft) from sea level. This altitudinal gradient comprised six different plant communities, from the bottom to the top: desert, chaparral, pinyon–juniper woodland, ponderosa pine forest, spruce–fir forest, and alpine–tundra vegetation. Details concerning each vegetation type can be found in Merriam (1890) and Lowe (1964).

Our samples in the tropics were on Serra do Cipó, Minas Gerais state, Brazil.

The altitudinal gradient sampled ranged from 650 m (12°S) to 1350 m (19°S) and was dominated by cerrado vegetation. Details concerning the Serra do Cipó vegetation can be found in Warming (1908), Joly (1970), Goodland and Ferri (1979), and Giulietti et al. (1987). The sample sites were at 650, 900, 1050, 1100, 1200, and 1350 m above sea level.

We decided a priori to study pairs of habitats selected for hygrothermal distinctness and made comparisons of galling species richness between habitats. We also studied the change in free-feeding insect herbivore richness along the elevational gradient to further test the prediction above that galling habit is associated with protection from hygrothermal stress. The two habitat types were defined as predominantly mesic or xeric, with riparian habitats as mesic, and sites distant from surface water as xeric sites. We sampled 16 sites at six elevations in Arizona at every 305 m from 915 to 2440 m above sea level, and the six altitudinal sites in Serra do Cipó, Brazil.

The host-plant species area hypothesis was tested using distributional maps of the host plants in the United States (Benson and Darrow 1944; Little 1971, 1976; Anderson 1986). In the statistical analysis we used the continuous area which covered the altitudinal gradient surveyed.

For direct comparison between tropical and temperate sites, altitude was adjusted to the equivalent latitude by the standard conversion: 305 m altitude increase is equivalent to 4° increase in latitude (Merriam 1899; also see Lowe 1964).

For more details on the sampling strategy and sample areas see Fernandes (1987) and Fernandes & Price (1988).

TRENDS IN TEMPERATE AND TROPICAL GALLING INSECT DISTRIBUTION

Test of Hypotheses

Galling species richness on herbs, shrubs, woody plants, and all plants increased with decreasing altitude in dry sites in the temperate region. The decrease in altitude accounted for 82% of the variation in galling species richness on shrubs (Fig. 5-1a). The same pattern was found for galling species richness on herbs, accounting for 37% of the variation (Fig. 5-1b). Likewise, decreasing altitude accounted for 75% of variation in galling species richness on woody palnts (= shrubs + tress; Fig. 5-1c) and 78% on all plants (= herbs + shrubs + trees; Fig. 5-1d). However, galling species richness on trees and altitude were not significantly correlated.

The pattern of increasing galling species richness with decreasing altitude held for trees, woody plants, and all plant groups in the tropical sites (Fig. 5-2). Stepwise multiple regression analysis of galling species richness on altitude and tree species richness showed that altitude was the best predictor of galling species richness. Again, galling species richness on woody and all plants was primarily explained by altitude. However, galling species richness on herbs was mostly

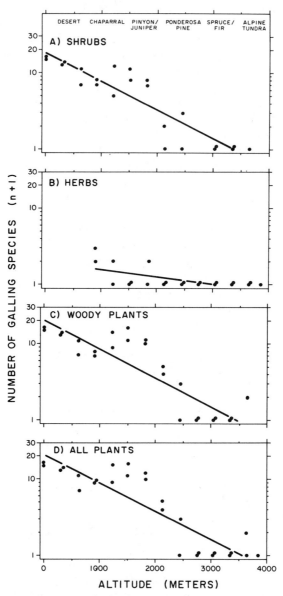

Figure 5-1. Regression of galling species richness on shrubs, herbs, woody plants (shrubs + trees), and all plants (herbs + shrubs + trees), as related to altitude in sites away from riparian habitats in the temperate region (from sea level to 3843 m). Vegetation types along the altitudinal gradient are given at the top of the figure. Mesic sites were not included. (A) shrubs: $r^2 = 0.82$; $p < 0.0005$; $\log_{10}(y + 1) = 1.26 - 0.000377x$; (B) herbs: $r^2 = 0.37$; $p < 0.005$; $\log_{10}(y + 1) = 0.284 - 0.000092x$; (C) woody plants: $r^2 = 0.75$; $p < 0.0005$; $\log_{10}(y + 1) = 1.31 - 0.000367x$; all plants: $r^2 = 0.78$; $p < 0.0005$; $\log_{10}(y + 1) = 1.32 - 0.000366x$. Note that the y axis is on a logarithmic scale.

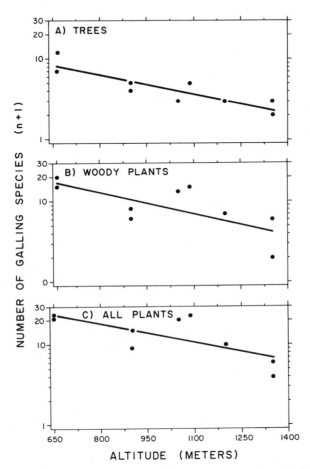

Figure 5-2. Regression of galling species number on trees, woody, and all plants on the altitudinal gradient from 650 to 1350 m above sea level in Serra do Cipó (Brazil). The regressions were: (A) trees: $r^2 = 0.78$; $p < 0.001$; $\log_{10} (y + 1) = 1.42 - 0.000779x$; (B) woody plants: $r^2 = 0.50$; $p < 0.025$; $\log_{10} (y + 1) = 1.76 - 0.000812x$; (C) and all plants: $r^2 = 0.54$; $p < 0.01$; $\log_{10} (y + 1) = 1.88 - 0.000772x$. Note that the y axis is on a logarithmic scale.

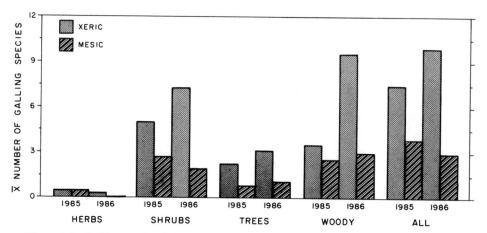

Figure 5-3. Galling species richness on plants of different structural complexity between xeric (stippled bars) and mesic (dashed bars) sites along the altitudinal gradient of 915 and 2440 m in Arizona both in 1985 and 1986. Wilcoxon text (Zar 1984) for: herbs (1985: NS; 1986: NS); shrubs (1985: $p < 0.05$; 1986: $p < 0.05$); trees (1985: $p < 0.01$; 1986: $p < 0.01$); woody plants (1985: $p < 0.01$; 1986: $p < 0.04$); all plants (1985: $p < 0.01$; 1986: $p < 0.04$).

explained by variation in herb species number. The increase in herb species number alone accounted for 56% of the variation in galling species richness on herbaceous plants. Neither of the entered variables, elevation and shrub species richness, explained the variation of galling species richness on shrubs.

Temperate galling species distribution was influenced by hygrothermal stress. Galling species richness on shrubs, trees, woody plants, and all plants was greater in xeric sites than in mesic sites, in both 1985 and 1986 field seasons (Fig. 5-3). In addition, in xeric sites altitude accounted for significant levels of galling species richness on all plant groups except trees, but in mesic sites no significant relationships were seen (Fig. 5-4). There were clear distributional trends in galling insect richness in xeric sites, with galling species number increasing as hygrothermal stress increased. However, no statistically significant trends were observed for galling insects in mesic sites. Therefore, the effect of altitude was best explained by the increasing hygrothermal stress in dry sites at lower elevations. Analyses of covarince (Snedecor and Cochran 1980) were performed for the data of 1985 and 1986 collected between 915 and 2440 m for galling species richness on elevation and the xeric–mesic factor (indicator variable) for all plant growth forms. Only the xeric–mesic factor accounted for significant variation in galling species richness. We therefore rejected the altitudinal gradient hypothesis based on the strength of the association between increasing hygrothermal stress and decreasing elevation.

Tropical galling species distribution was, likewise, influenced by hygrothermal stress. Galling species richness on all plant architectural types was significantly higher in xeric sites compared to mesic sites (Fig. 5-5). In addition, on the altitudinal gradient there were clear trends in galling species richness in dry sites

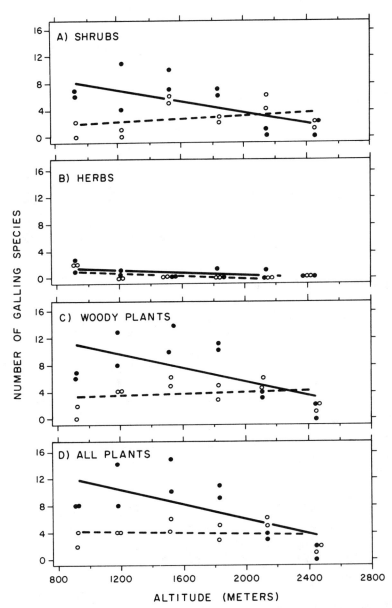

Figure 5-4. Regression of galling species richness on altitude in xeric (closed circles and solid lines) and mesic (open circles and dashed lines) sites during the field season of 1985 for the altitudinal gradient between 915 and 2440 m in Arizona. (A) shrubs (xeric: $r^2 = 0.48$, $p < 0.025$, $y = 13.0 - 0.00473x$); mesic: $r^2 = 0.10$, $p > 0.25$, $y = 0.62 + 0.00122x$); (B) herbs (xeric: $r^2 = 0.42$, $p < 0.025$, $y = 1.75 - 0.000796x$; mesic: $r^2 = 0.30$, $p > 0.25$, $y = 1.75 - 0.000796x$); (C) woody plants (xeric: $r^2 = 0.34$, $p < 0.05$, $y = 15.3 - 0.00482x$; mesic: $r^2 = 0.01$, $p > 0.25$, $y = 2.87 + 0.00037x$); (D) all plants (xeric: $r^2 = 0.43$, $p < 0.025$, $y = 17.1 - 0.0562x$; mesic: $r^2 = 0.02$, $p > 0.25$, $y = 4.62 - 0.00042x$). There were no clear trends in galling species richness distribution on trees in both xeric and mesic sites in 1985 and 1986 ($p > 0.25$ in both sites and both years).

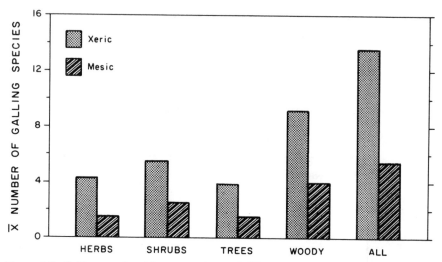

Figure 5-5. Galling species richness on plants of different structural complexity between xeric (stippled bars) and mesic (dashed bars) sites along the altitudinal gradient of 650 and 1350 m in Serra do Cipó, Brazil. Wilcoxon test performed for: herbs ($p < 0.007$); shrubs ($p < 0.025$); trees ($p < 0.03$), woody plants ($p < 0.019$), and all plants ($p < 0.006$).

for trees, woody, and all plants (Fig. 5-6), with galling species number increasing as hygrothermal stress increased. On the other hand, no clear distributional patterns were observed in mesic sites. These tropical data support the conclusion above that galling insects are more species-rich in hygrothermally stressed habitats.

Temperate free-feeding insects were more numerous in mesic areas than in xeric areas (Fig. 5-7A). The opposite pattern was observed for gall-making insects in the same time period and on the same altitudinal gradient. The galling fauna was more numerous in xeric sites (Fig. 5-7B).

Tropical free-feeding insect herbivore species richness was significantly higher in mesic sites than in xeric sites on the altitudinal gradient of Serra do Cipó (Fig. 5-8A). On the other hand, gallers richness was significantly higher in dry sites (Fig. 5-8B). These differences between free feeders and gallers support the conclusion that gallers are strongly associated with habitats of high hygrothermal stress and have distributions largely distinct from the free feeders. Furthermore, the data indicate that this phenomenon is broadly distributed. We therefore accepted the harsh environment hypothesis.

Testing the plant species richness hypothesis, the increase of galling species richness on shrubs and woody plants was highly correlated with increasing shrub species richness (Fig. 5-9). However, variation in herbs, trees, and all plant species number did not account for any of the variation in galling species richness.

Plant species richness does not offer an adequate explanation for the trends in

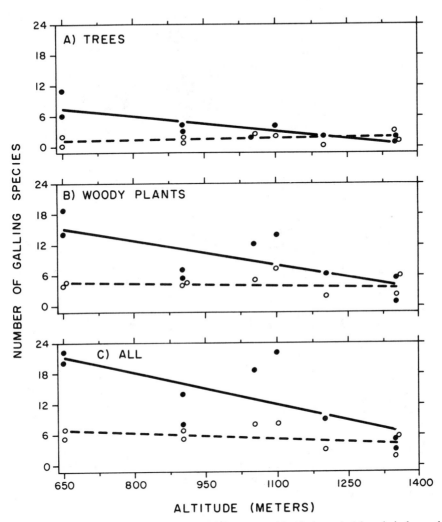

Figure 5-6. Regression of galling species richness on altitude in xeric (closed circles and solid lines) and mesic (open circles and dashed lines) sites along the altitudinal gradient of 650 and 1350 m in Serra do Cipó, Brazil. (A) trees (xeric: $r^2 = 0.67$, $p < 0.01$, $y = 13.5 - 0.00943x$; mesic: $r^2 = 0.05$, $p > 0.25$, $y = 0.57 + 0.00086x$); (B) woody plants (xeric: $r^2 = 0.52$, $p < 0.025$, $y = 25.3 - 0.0158x$; mesic: $r^2 = 0.03$, $p > 0.25$, $y = 4.91 - 0.00089x$); (C) all plants (xeric: $r^2 = 0.50$, $p < 0.05$, $y = 33.9 - 0.0200x$; mesic: $r^2 = 0.19$, $p > 0.10$, $y = 9.16 - 0.00354x$).

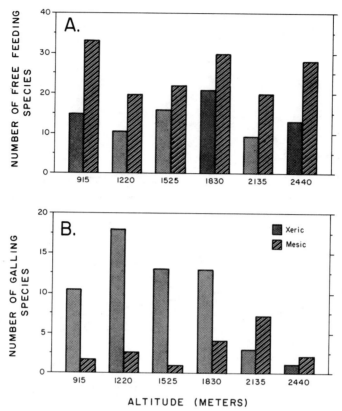

Figure 5-7. Free-feeding and galling insect herbivore distributions in xeric (stippled bars) and mesic (dashed bars) habitats on an altitudinal gradient ranging from 915 to 2440 m in Arizona. (A) Distributional trends in free-feeding insect species richness. Free feeders were significantly more species rich in mesic than in xeric sites (Wilcoxon test = 36.0, $p < 0.0005$). Except at 915 and 1220 m, which were sampled twice, all other sites were sampled only once. (B) Distributional trends in galling insect species richness. Galling species were significantly more species rich in xeric sites compared to mesic sites (Wilcoxon test = 3.0, $p < 0.05$). Except at 915 m site, which was sampled twice, all other sites were sampled only once.

galling species richness. The pattern of galling species richness was more associated with plant taxa than with plant richness when galling species richness was observed within host species or genera. For example, within the genera *Juniperus*, *Pinus*, and *Quercus*, galling species richness increased with declining altitude without an increase in number of host-plant species present. In addition, elevation was the most important factor explaining galling species richness on several plant species or genera (see Fernandes and Price 1988). Furthermore, most galler richness was supported by one species of plant at many sites. For example, at 305 m *Larrea tridentata* supported 92% of all galls, although

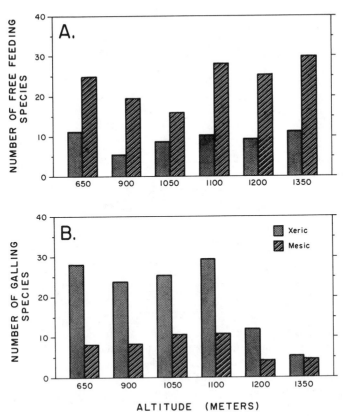

Figure 5-8. Free-feeding and galling insect herbivore distributions in xeric (stippled bars) and mesic (dashed bars) habitats on an altitudinal gradient ranging from 650 to 1350 m in Serra do Cipó (Brazil). (A) Distributional trends in free-feeding insect species richness. Free feeders were significantly more species rich in mesic than in xeric sites (Wilcoxon test = 171.0, $p < 0.0000$). All sites were sampled three times. (B) Distributional trends in galling insect species richness. Galling species were significantly more species rich in xeric sites compared to mesic sites (Wilcoxon test = 1.0, $p < 0.006$).

representing only 23% of all plant species (Fernandes and Price 1988). We concluded that plant species number is purely a factor correlated with galling species richness, but with no explanatory power, and therefore we rejected this hypothesis.

Species area occupied by nine host plants of galling insects was a poor predictor of gall-forming insect richness. We used the number of gall-forming insects on shrubs, the plant architectural type most heavily attacked by gallers in Arizona, to test the hypothesis that galling species richness would increase with increasing host-plant area. Only 3% of the variation in galling species richness was accounted for by host-plant area, and the relationship was not significant at the 0.05 level (Fig. 5-10). Thus, this hypothesis was not supported.

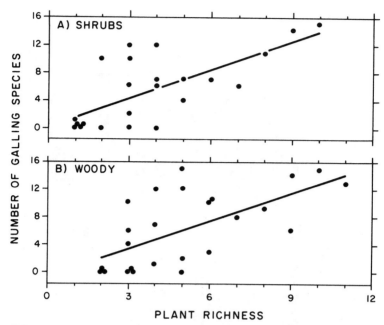

Figure 5-9. Regression of galling species number on shrub species number and on woody species number for the temperate region gradient (from sea level to 3843 m). (A) Change in shrub species number accounted for 49% of the variation in galling species richness ($r^2 = 0.49$, $p < 0.001$, $y = 0.29 + 1.37x$). (B) Change in woody plant species number accounted for 42% of the variation in galling species richness ($r^2 = 0.42$, $p < 0.001$, $y = 0.05 + 1.35x$).

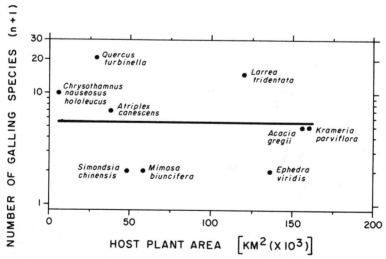

Figure 5-10. Regression of galling species number of host plant area (only shrub species). Three percent of the variation in galling species richness was accounted for by host plant area; however, the relationship was not statistically significant ($r^2 = 0.03$, $p > 0.25$, $\log_{10}(y + 1) = 0.713 - 0.000001x$). Note that the y axis is on a logarithmic scale.

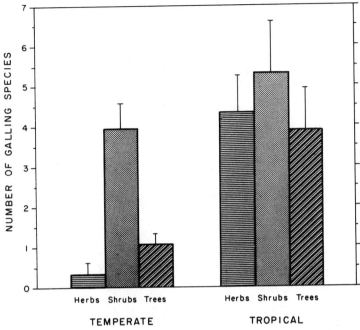

Figure 5-11. Average number of galling species on plants of different structural type in temperate and tropical regions; herbs, shrubs, and trees. Temperate shrubs supported significantly more galling species than herbs and trees ($\bar{X} = 3.94$, $P < 0.05$). Temperate herbs ($\bar{X} = 0.31$) and trees ($\bar{X} = 1.06$) were not statistically different in number of galling species. Tropical plant growth forms supported statistically equal numbers of galling species ($p > 0.50$): herbs ($\bar{X} = 4.33$), shrubs ($\bar{X} = 5.33$), and trees ($\bar{X} = 3.89$).

Increasing plant structural complexity did not correlate with increased galling species richness. Shrubs supported significantly more galling species than trees and herbs (Fig. 5-11). In the tropical sites the number of galling species on each plant group was not significantly different between architectural types (Fig. 5-11). Therefore, we rejected the hypothesis that number of gallers would increase as plant structural complexity increased.

TROPICAL VS. TEMPERATE GALLING SPECIES RICHNESS

A simple linear regression of the number of galling species on latitude adjusted for altitude (see Fernandes 1987) clearly shows a trend of increasing galling species frequency with decreasing "latitude" for both tropical and temperate data (Fig. 5-12). Overall, samples were taken from the equivalents of "latitudes" from 28°30′ through 85°30′. The decrease in the "latitude equivalents" explained 53% of the variation in tropical galling species richness and 78% of the variation in temperate galling species number (Fig. 5-12).

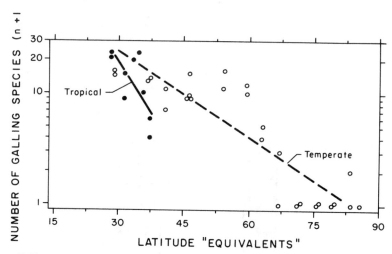

Figure 5-12. Tropical (closed circles and solid line) and temperate (open circles and dashed line) linear regressions of galling species number on latitude corrected for altitude ("latitude equivalents"). The equations are: Tropical [$\log_{10} (y + 1) = 3.09 - 0.0607x$, $r^2 = 0.53$, $P < 0.025$], and Temperate [$\log_{10} (y + 1) = 2.12 - 0.0253x$, $r^2 = 0.78$, $p < 0.0000$]. Note that the y axis is on a logarithmic scale.

Although there were no statistically significant differences in galling species richness between tropical and temperate sites, a trend existed in each plant type for more galling species in tropical than in temperate regions (see Fig. 5-11). We took a conservative approach to compare our tropical and temperate data, using data from equivalent altitudes, between 610 and 1525 m.

TRENDS IN THE DISTRIBUTION OF TEMPERATE CECIDOMYIID GALLERS

All galls on herbaceous plants in dry sites were caused by cecidomyiids. With the exception of *Quercus gambelli* (at 2135 m altitude), and *Picea engelmani* (at 3660 m altitude) which were galled by cynipid and adelgid species, respectively, all other tree species were galled by cecidomyiids. Thus the distributional trends in temperate insect gall richness on these two plant architectural types were primarily dictated by the cecidomyiids. However, shrubs are galled by a more diverse array of insect gallers, such as cynipids, psyllids, tephritids, and cecidomyiids. A simple linear regression showed that 86% of the variation in the cecidomyiid gall richness was explained by changes in altitude (Fig. 5-13).

The cecidomyiids are possibly the only taxon which could furnish sound comparative data worldwide. They are well represented in all biogeographical regions (Gagné 1984; Skuhravá et al. 1984 and references therein). In our study, the cecidomyiids were responsible for causing all galls on herbaceous plants, and

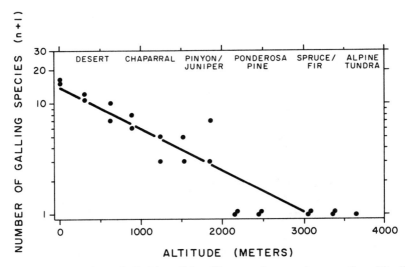

Figure 5-13. Regression of Cecidomyiid galling species number on the altitudinal gradient ranging from sea level in Bahía Kino (Mexico) to 3843 m at the top of the San Francisco Peaks in northern Arizona. The decrease in altitude explained 86% of the variation in Cecidomyiidae galling species richness [$r^2 = 0.86$, $p < 0.0000$, $\log_{10}(y + 1) = 1.14 - 0.000375x$]. Note that the y axis is on a logarithmic scale.

the majority of tree galls. On shrubs, when the cynipids were removed from the analyses, the trends in species richness were even clearer. Thus, the distribution of gall-forming species richness on shrubs was primarily explained by the distribution of cecidomyiid galls. While cynipids, tenthredinid sawflies, and aphids are concentrated on northern temperate hosts and psyllids on Australian hosts, the cecidomyiid gallers are well represented in all major biogeographical regions.

MECHANISMS PRODUCING THE PATTERNS

The question raised by our studies (Fernandes 1987; Fernandes and Price 1988) were numerous. For example, why should gallers be better adapted to live in hygrothermally stressed habitats? Why should they not be as abundant in mesic habitats? Are they less effective in colonizing plants in mesic sites? Why did gallers that successfully colonized mesic sites not speciate and radiate as extensively as those that colonized xeric sites?

Research on the mortality factors of gallers on the same host plant in xeric sites and mesic sites strengthens the view that gallers survive better in xeric compared to mesic sites. For example, a psyllid leaf galler [probably *Pachypsylla venusta* (Felt 1940)] on *Celtis reticulata* in northern Arizona suffers lower rates of parasitism and fungal attack in xeric sites than in mesic sites (Fig. 5-14). Parasitism and diseases caused higher mortality of two cecidomyiid leaf gallers in southeastern Bazil in mesic sites than in xeric sites (Fig. 5-15). Thus, those

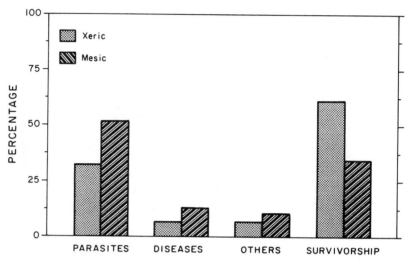

Figure 5-14. Differential mortality and survivorship of a psyllid leaf galler (*Pachypsylla venusta*) on *Celtis reticulata* (Ulmaceae) in northern Arizona. Parasitism and fungal attack were significantly lower in xeric (stippled bars) habitats than in mesic (dashed bars) sites ($P < 0.000$, and $p < 0.004$, respectively). Consequently, survivorship was higher in xeric sites than in mesic sites (Hotellings $t^2 = 1.277$, $p < 0.000$).

gallers that colonized plants in xeric sites were more successful than those that colonized plants in mesic sites, and selection may have worked on female oviposition behavior for preference of xeric sites.

These results indicate what may be part of a general pattern. Insect gallers that colonize plants in xeric habitats survive better than those that colonize plants in mesic habitats. Parasites and fungal attack keep gallers from becoming more abundant in mesic sites. The differential survivorship in xeric and mesic habitats may be the general proximate factor influencing the distributional patterns of galling species richness in both temperate and tropical regions. Carroll (1988) reported that invasion of the gall by the endophytic fungus, *Rhabdocline parkeri*, is the most important mortality factor of cecidomyiid needle gallers (*Contarinia* spp.) on Douglas-fir during wet years (see also Carroll and Carroll 1978). *Rhabdocline parkeri* is found in virtually 100% of sites where annual rainfall exceeds 100 cm. Furthermore, Taper et al. (1986) found that fungal infestation and parasitism were the most important mortality factors for the cynipid, *Dryocosmus dubiosus*, a galler on *Quercus agrifolia* and *Q. wislizenii*. Fungal growth is very common on insect galls growing on plants in humid habitats in Brazil, and when gall walls are damaged the larval chamber quickly fills with fungal hyphae. In addition, fungal galls (witch's brooms) are common in wet years in the cerrado vegetation of Brazil but not as common during dry years (personal observation). This is not surprising because fungal reproduction and growth is dependent on humidity. The ultimate factor influencing distribution of gallers

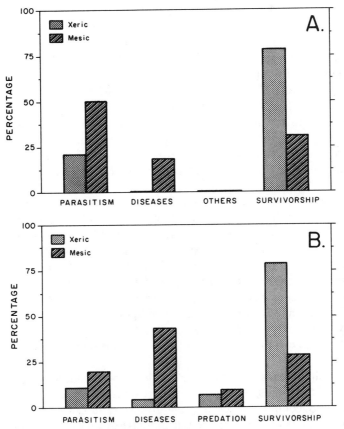

Figure 5-15. Differential mortality and survivorship of two gallers on tropical host plants in southeastern Brazil. (A) Cecidomyiid leaf galler on *Copaifera langsdorffii* (Leguminosae). Parasitism and fungal attack were significantly lower in xeric (stippled bars) habitats than in mesic (dashed bars) habitats ($P < 0.001$ in both cases). Survivorship was higher in xeric habitats than in mesic habitats (Hotellings $t^2 = 2.229$, $p < 0.000$). (B) Cecidomyiid leaf galler on *Bauhinia* sp. (Leguminosae). Parasitism and diseases were significantly lower in xeric (stippled bars) habitats than in mesic (dashed bars) habitats ($P < 0.023$, $p < 0.000$, respectively). Predation was not statistically different between habitats. Consequently, survivorship was higher in xeric sites than in mesic sites (Hotellings $t^2 = 3.174$, $p < 0.000$).

and richness may well be an evolved response of females, which select xeric sites in preference to mesic sites.

Plant allelochemical concentration and diversity may also drive richness of the galling insect fauna. Taper and Case (1987) showed positive correlations among leaf tannin levels and leaf–cynipid–galler richness and abundance. They argued that tannins may aid in the defense of the cynipid larvae from either hyperparasites, herbivores, pathogens, or some combination of these factors (see also Cornell 1983). Fungal growth on the cynipid galls was negatively correlated with leaf tannin content. Müller et al. (1987) reported higher astringent phenol content in plants inhabiting stressful sites.

Fernandes and Price (1988) and Price (Chapter 3) stated that galling seems to be strongly associated with sclerophylly: galls are very common in the chaparral of Arizona, in eastern Australia, dominated by sclerophyllous vegetation, and even in mesic Australian latitudes in coastal areas where sclerophylly remains common. We observed that galling species are common in the campina vegetation of sclerophyllous shrubs on the depauperate sands along the Rio Negro and in northern Amazonia (Brazil), and are three or four times richer than in adjacent rain forest.

The relationship between galling species richness and sclerophylly is worthy of more study. Sclerophylly is common in plants that can tolerate low levels of nutrients, particularly low phosphorous (Beadle 1954; Loveless 1961, 1962; Mooney and Dunn 1970; Orians and Solbrig 1977; Grubb 1977). Phosphate deficiency is not only characteristic of physically dry soils, but it "also characterizes acid soils in areas of high rainfall but especially when these are rich in organic matter since phosphorous in organic combinations is even less readily available to the plant than from inorganic compounds such as iron–phosphate" (Salisbury 1959, p. 139; see also Kruger et al. 1983). The species richness of gallers in sclerophyllous campina vegetation fits this pattern in terms of acid soil and high rainfall.

Sclerophyllous plants generally have high levels of secondary compounds also. Water and phosphorous-deficient plants have a higher content than unstressed plants of many compounds: cyanogenic glycosides (e.g., Cooper-Driver et al. 1977; Clark et al. 1979), glucosinolates (e.g., Gershenzon 1984), alkaloids (Hsiao 1973; Hsiao et al. 1976; Stewart and Lahrer 1980; Bradford and Hsiao 1982), phenolic compounds (McClure 1975, 1979; McKey et al. 1978; Gershenzon 1984; Müller et al. 1987), and terpenoids (Wilson and Ng 1975; Gershenzon 1984).

Although studies on the dynamics and physiology of plant secondary compounds in wild plants are still fragmentary, there are indications that secondary compounds increase under water and/or nutrient stress both between habitats and within a habitat. Levels of secondary compounds are in general higher in individuals growing in dry and/or nutrient-poor sites than in wet and/or nutrient-rich sites (e.g., Majak et al. 1980a; Gershenzon 1984). Increases in secondary compound levels during dry periods have also been observed in some plant species (e.g., Blohm 1962; Hegnauer 1966; Cooper-Driver et al. 1977; Majak et al. 1980b; Kennedy and Bush 1983; Mattson and Haack 1987), or with increasing plant age and/or phenology (e.g., Feeny 1970).

The correlated traits of sclerophylly, low nutrient status plants, high chemical defense, and galling suggests that the distribution of gallers may be causally related to these traits. We explore this idea in the next section.

THE EVOLUTION OF GALLERS ON PLANTS IN
LOW NUTRIENT SOILS

Plants growing on nutrient-poor soils have low levels of nutrients (e.g., Janzen 1974; Grime 1977; McKey et al. 1978) (Fig. 5-16). Because nutrients are limiting, these plants are adapted to conserve nutrients by developing long-lived leaves (Chapin 1980; Kruger et al. 1983; Coley 1983, 1988; Coley et al. 1985). Long-lived leaves have a reduced probability of abscission and are tougher (Monk 1966;

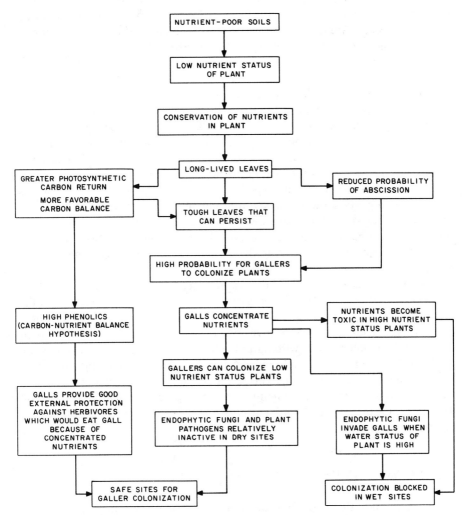

Figure 5-16. Factors potentially involved in the evolution of gallers on low-nutrient status plants.

Grubb 1986). Thus, sclerophyllous plants are likely to be predictable in time and space, increasing the probability that gallers will colonize such plants and survive on their long-lived leaves.

Nutrients are concentrated in gall tissue (e.g., Fourcroy and Braun 1967; Rohfritsch and Shorthouse 1982; Abrahamson and McCrea 1985; McCrea et al. 1985; Weis et al. 1988) compensating for low nutrient status of leaves. Nutrients become toxic to gallers in high nutrient status plants, and colonization is blocked in these plants. In addition, endophytic fungi invade galls when water status of the plant is high (Carroll 1988). Thus, galler colonization is prevented once again. However, endophytic fungi and plant pathogens are relatively inactive in dry habitats (Preece and Dickinson 1971; Levin 1976). Although plants inhabiting stressful habitats are more likely to contain high allelochemical content, especially phenolics (e.g., Janzen 1974; Gershenzon 1984; Müller et al. 1987), gallers are apparently able to circumvent such plant constitutive defenses. The galled tissue in which the gall larvae feed (the nutritive tissue) is phenol-free (e.g., Larew 1982). Hence, gallers were able to circumvent plant defenses and manipulate their host plants to obtain a rich, nontoxic food source, and protection against environmental harshness. Because their sclerophyllous host plants contained high phenolic levels, gallers occupied a niche relatively free of diseases and enemies (especially herbivores) which would eat the gall because of its concentrated nutrients (but see Fernandes et al. 1987).

Plants supporting high galling species richness in northern latitudes, such as willows and poplars, grow in wet, low-nutrient-status soils such as sandy alluvial river banks. Bog vegetation also grows in low nutrient status soils (Chapin 1980). Therefore, northern plants may present similar conditions for gallers as sclerophyllous plants, only lacking the characteristic tough leaves in some but not all cases.

In summary, the evolution of gall-forming insects on their sclerophyllous host plants involved at least three major classes of characters: (a) physiological traits that affected offspring growth and reproductive success on a particular host plant; (b) predictability of a safe environment, evergreen leaves, in which a food plant mediated interactions with natural enemies, such that pathogens and gall herbivores were negatively affected; (c) female behaviors that influenced the choice of a plant for its offspring. The pattern of galling species richness seems to be clear, and is summarized in Figure 5-16. However, more research is needed on the interacting factors which force patterns, and the exceptions to the scenario we propose.

ACKNOWLEDGMENTS

We are pleased to thank J. Gershenzon, K. Paige, B. A. Hawkins, D. Simberloff, and P. D. Coley for their criticisms on this manuscript. Logistical support was provided by the Departamento de Biologia Geral (Universidade Federal de Minas Gerais—Brazil), the United States Forestry Department, and M. Wong and P. Becker of the World Wildlife Fund in Manaus, Brazil. The authors thank Angela Fernandes for her assistance during field collection and laboratory work. Financial support was provided by the U.S. National Science Foundation (Grants N° BSR-8314594, and BSR-8705302), and by the Conselho Nacional de Pesquisas (CNPq–n° 200.747/84-3-Z0). We gratefully acknowledge a Sigma Xi Grants-in-Aid to GWF.

REFERENCES

Abrahamson, W. G. and K. D. McCrea. 1985. Nutrient and biomass allocation in *Solidago altissima* stem gall insect–parasitoid guild food chain. *Oecologia* **68**: 174–180.

Anderson, L. C. 1986. An overview of the genus *Chrysothamnus* (Asteraceae). Pp. 29–45 in E. D. McArthur and B. L. Welch (Eds.). *Proceedings—symposium on the biology of Artemisia and Chrysothamnus*. United States Department of Agriculture, General Technical Report INT-200. Ogden, Utah, USA.

Beadle, N. C. W. 1954. Soil phosphate and the delimitation of plant communities in eastern Australia. *Ecology* **35**: 370–375.

Benson, L. and R. A. Darrow. 1944. *A manual of southwestern desert trees and shrubs*. University of Arizona Press, Tucson, AZ.

Blohm, H. 1962. *Poisonous plants of Venezuela*. Harvard University Press, Cambridge, MA.

Bradford, K. J. and T. C. Hsiao. 1982. Physiological responses to moderate water stress. Pp. 264–324 in O. L. Lange, P. S. Nobel, C. B. Osmond, and H. Ziegler (Eds.). *Physiological plant ecology. II. Water relations and carbon assimilation*. Springer-Verlag, Berlin, Germany.

Brown, J. H. and A. C. Gibson. 1983. *Biogeography*. Mosby, St. Louis, MO.

Carroll, G. C. 1988. Fungal endophytes in stems and leaves: From latent pathogen to mutualistic symbiont. *Ecology* **69**: 2–9.

Carroll, G. C. and F. E. Carroll. 1978. Studies on the incidence of coniferous needle endophytes in the Pacific Northwest. *Canadian Journal of Botany* **56**: 3034–3343.

Chapin, F. S. 1980. The mineral nutrition of wild plants. *Annual Review of Ecology and Systematics* **11**: 233–260.

Claridge, M. F. and J. S. Singharo. 1978. Diversity and altitudinal distribution of grasshoppers (Acridoidea) on a Mediterranean mountain. *Journal of Biogeography* **5**: 239–250.

Clark, R. B., H. J. Gorz, and F. A. Haskins. 1979. Effects of mineral elements on hydrocyanic acid potential in sorghum seedlings. *Crop Science* **19**: 757–761.

Clarke, G. L. 1954. *Elements of ecology*. Wiley, New York.

Coley, P. D. 1983. Herbivory and defensive characteristics of tree species in a lowland tropical forest. *Ecological Monographs* **53**: 209–233.

Coley, P. D. 1988. Effects of plant growth rate and leaftime on the amount and type of anti-herbivore defense. *Oecologia* **74**: 531–536.

Coley, P. D., J. P. Bryant, and F. S. Chapin III. 1985. Resource availability and plant anti-herbivore defense. *Science* **230**: 895–899.

Cooper-Driver, G., S. Finch, T. Swain, and E. Bernays. 1977. Seasonal variation in secondary plant compounds in relation to the palatability of *Pteridium aquilinum*. *Biochemical Systematics and Ecology* **5**: 177–183.

Cornell, H. V. 1983. The secondary chemistry and complex morphology of galls formed by the Cynipidae (Hymenoptera): Why and how? *American Midland Naturalist* **110**: 225–234.

Darlington, P. J. 1965. *Biogeography of the southern end of the world*. Harvard, Cambridge, MA.

Dixon, A. F. G., P. Kindlman, J. Leps, and J. Holman. 1987. Why there are so few species of aphids, especially in the tropics. *American Naturalist* **129**: 580–592.

Eastop, V. F. 1972. Deductions from the present day host plants of aphids and related insects. *Symposium of the Royal Entomological Society of London* **6**: 157–178.

Eastop, V. F. 1978. Diversity of the Sternorrhyncha within major climatic zones. *Symposium of the Royal Entomological Society of London* **9**: 71–88.

Feeny, P. 1970. Seasonal changes in oak leaf tannins and nutrients as a cause of spring feeding by winter moth caterpillars. *Ecology* **51**: 565–581.

Felt, E. P. 1940. *Plant galls and gall makers*. Comstock, New York.

Fernandes, G. W. 1987. *Tropical and temperate altitudinal gradients in galling species richness*. MSc Thesis. Northern Arizona Universiy. Flagstaff, AZ.

Fernandes, G. W., R. P. Martins, and E. Tameirão Neto. 1987. Food web relationships involving *Anadiplosis* sp. (Diptera: Cecidomyiidae) on *Machaerium aculeatum* (Leguminosae). *Revista Brasileira de Botânica* **10**: 117–123.

Fernandes, G. W. and P. W. Price. 1988. Biogeographical gradients in galling species richness: tests of hypotheses. *Oecologia* **76**: 161–167.

Fischer, A. G. 1960. Latitudinal variations in organic diversity. *Evolution* **14**: 64–81.

Fourcroy, M. and C. Braun. 1967. Observations sur la galle de l'*Aulax glechomae* L. sur *Glechoma heredacea* L. II. Histologie et role physiologique de la coque sclerifée. *Marcellia* **34**: 3–30.

Gagné, R. J. 1984. The geography of gall insects. Pp. 305–322 in T. N. Ananthakrishnan (Ed.). *Biology of gall insects*. Oxford & IBH, New Delhi, India.

Giulietti, A. M., N. L. Menezes, J. R. Pirani, M. Meguro, and M. G. L. Wanderley. 1987. Flora da Serra do Cipó: caracterizacão e lista de espécies. *Boletin de Botânica* **9**: 1–151.

Goodland, R. and M. G. Ferri. 1979. *Ecologia do cerrado*. Edusp & Itatiaia, Belo Horizonte, Brazil.

Gershenzon, J. 1984. Changes in the levels of plant secondary metabolites under water and nutrient stress. Pp. 273–320 in B. N. Timmermann, C. Steelink, and F. A. Loewus (Eds.). *Phytochemical adaptations to stress*. Plenum, New York.

Grime, J. P. 1977. Evidence for the existence of three primary strategies in plants and its relevance to ecological and evolutionary theory. *American Naturalist* **111**: 1169–1194.

Grubb, P. J. 1977. Control of forest growth and distribution on wet tropical mountains: With special reference to mineral nutrition. *Annual Review of Ecology and Systematics* **8**: 83–107.

Grubb, P. J. 1986. Sclerophylls, pachyphylls and pycnophylls: The nature and significance of hard leaf surfaces. Pp. 137–150 in B. E. Juniper and T. R. E. Southwood (Eds.). *Insects and the plant surface*. Edward Arnold, London, England.

Hegnauer, R. 1966. *Chemotaxonomie der Pflanzen*. Birkhauser Verlag, Basel, Germany.

Holdridge, L. R. 1967. *Life zone ecology* (2nd ed.). Tropical Research Center, San José, Costa Rica.

Holdridge, L. R., W. C. Grenke, W. H. Hatheway, T. Liang, and J. A. Tosi, Jr. 1971. *Forest environments in tropical life zones: A pilot study.* Pergamon, Oxford, England.

Horn, M. L. and L. G. Allen. 1978. A distributional analysis of California coastal marine fishes. *Journal of Biogeography* **5**: 23–42.

Hsiao, T. C. 1973. Plant responses to water stress. *Annual Review of Plant Physiology* **24**: 519–570.

Hsiao, T. C., E. Acevedo, E. Fereres, and D. W. Henderson. 1976. Stress metabolism, water stress, growth, and osmotic adjustment. *Phylosophical Transactions of the Royal Society of London* (B) **273**: 479–500.

Janzen, D. H. 1974. Tropical blackwater rivers, animals and mast fruiting by the Dipterocarpaceae. *Biotropica* **6**: 69–103.

Janzen, D. H. 1981. The peak in North American ichneumonid species richness lies between 38° and 42°N. *Ecology* **62**: 532–537.

Joly, A. B. 1970. *Conheca a vegetacão brasileira.* Edusp, São Paulo, Brazil.

Kennedy, C. W. and L. P. Bush. 1983. Effect of environmental and management factors on the accumulation of *N*-formyl loline alkaloids in tall fescue. *Crop Science* **23**: 547–552.

Kruger, F. J., D. T. Mitchell, and J. Jarvis. 1983. *Mediterranean-type ecosystems.* Springer-Verlag, Berlin, Germany.

Kusnezov, N. 1957. Number of species of ants in faunae of different latitudes. *Evolution* **11**: 298–299.

Larew, H. 1982. *A comparative anatomical study of galls caused by the major cecidogenetic group, with special emphasis on the nutritive tissue.* PhD dissertation, Oregon State University, Corvallis, OR.

Lawton, J. H., M. McGarvin, and P. A. Heads. 1987. Effects of altitude on the abundance and species richness of insect herbivores on bracken. *Journal of Animal Ecology* **56**: 147–160.

Levin, D. A. 1976. The chemical defenses of plants to pathogens and herbivores. *Annual Review of Ecology and Systematics* **7**: 121–159.

Little, E. L. 1971. *Atlas of United States trees. Coniferous and important hardwoods.* Vol. 1. United States Department of Agriculture, Miscellaneous Publications n° 1146.

Little, E. L. 1976. *Atlas of United States trees. Minor western hardwoods.* Vol. 3. United States Department of Agriculture, Miscellaneous Publications n° 1314.

Loveless, A. R. 1961. A nutritional interpretation of sclerophylly based on differences in the chemical composition of sclerophyllous and mesophytic leaves. *Annals of Botany* (*N.S.*) **25**: 168–184.

Loveless, A. R. 1962. Further evidence to support a nutritional interpretation of sclerophylly. *Annals of Botany* (*N.S.*) **26**: 551–561.

Lowe, C. H. 1964. *Arizona's natural environment.* University of Arizona Press, Tucson, AZ.

Majak, W., D. A. Quinton, and K. Broersma. 1980a. Cyanogenic glycoside levels in Saskatoon serviceberry. *Journal of Range Management* **33**: 197–199.

Majak, W., R. E. McDiarmid, J. W. Hall, and A. L. Van Ryswyk. 1980b. Seasonal variation in the cyanide potential of arrowgrass (*Triglochin maritima*). *Canadian Journal of Plant Science* **60**: 1235–1241.

Mattson, W. J. and R. A. Haack. 1987. The role of drought stress in provoking outbreaks of phytophagous insects. Pp. 365–407 in P. Barbosa and J. C. Schultz (Eds.). *Insect outbreaks*. Academic, New York.

MacArthur, R. H. 1972. *Geographical ecology: Patterns in the distribution of species*. Harper and Row, New York.

McClure, J. W. 1975. Physiology and functions of flavonoids. Pp. 970–1055 in J. B. Harborne, T. J. Mabry, and H. Mabry (Eds.). *The Flavonoids*, Chapman and Hall, London, England.

McClure, J. W. 1979. The physiology of phenolic compounds in plants. *Recent Advances in Phytochemistry* **12**: 525–556.

McCrea, K. D., W. G. Abrahamson, and A. E. Weis. 1985. Goldenrod ball gall effects on *Solidago altissima*: ^{14}C translocation and growth. *Ecology* **66**: 1902–1907.

McKey, D., P. T. Waterman, C. N. Mbi, J. S. Gartlan, and T. T. Struhsaker. 1978. Phenolic content of vegetation in two African rain forests: ecological implications. *Science* **202**: 61–64.

Merriam, C. H. 1890. Results of a biological survey of the San Francisco Mountains region and desert of the Little Colorado in Arizona. *United States Department of Agriculture, North America Fauna* **3**: 1–136.

Merriam, C. H. 1899, Zone temperatures, *Science* **9**: 116.

Michener, C. O. 1979. Biogeography of the bees. *Annals of the Missouri Botanical Garden* **66**: 277–347.

Monk, C. D. 1966. An ecological significance of evergreenness. *Ecology* **47**: 504–505.

Mooney, H. A. and E. L. Dunn. 1970. Convergent evolution of Meditteranean-climate evergreen sclerophyll shrubs. *Evolution* **24**: 292–303.

Müller, R. N., P. J. Kalisz, and T. W. Kimmerer. 1987. Intraspecific variation in production of astringent phenolics over a vegetation–resource availability gradient. *Oecologia* **72**: 211–215.

Orians, G. and D. Solbrig. 1977. A cost-income model of leaves and roots with special reference to arid and semi-arid areas. *American Naturalist* **111**: 677–690.

Preece, T. F. and C. H. Dickinson (Eds.). 1971. *Ecology of leaf surface micro-organisms*. Academic, New York.

Price, P. W., G. W. Fernandes, and G. L. Waring. 1987. Adaptive nature of insect galls. *Environmental Entomology* **16**: 15–24.

Price, P. W., G. L. Waring, and G. W. Fernandes. 1986. Hypothesis on the adaptive nature of galls. *Proceedings of the Entomological Society of Washington* **88**: 361–363.

Richards, P. W. 1957. *The tropical rain forest*. Cambridge Press, Cambridge, England.

Rohfritsch, O. and J. D. Shorthouse. 1982. Insect galls. Pp. 131–152 in G. Kahl and J. S. Schell (Eds.). *Molecular biology of plant tumors*. Academic, New York.

Salisbury, E. J. 1959. Casual plant ecology. Pp. 124–144 in W. B. Turrill (Ed.). *Vistas in botany* (Vol. I). Pergamon, New York.

Simpson, G. G. 1964. Species density of North American recent mammals. *Systematic Zoology* **13**: 57–73.

Skuhravá, M., V. Skuhravý, and J. W. Brewer. 1984. Biology of gall midges. Pp. 169–222 In T. N. Ananthakrishnan (Ed.). *Biology of gall insects*. Oxford and IBH, New Delhi, India.

Snedecor, G. W. and W. G. Cochran. 1980. *Statistical methods* (7th ed.). Iowa State University Press, Ames, IO.

Stewart, G. R. and F. Lahrer. 1980. Accumulation of amino acids and related compounds in relation to environmental stress. Pp. 609–635 in B. J. Mifflin (Ed.). *The biochemistry of plants. Amino acids and derivatives* (Vol. 5). Academic, New York.

Taper, M. L. and T. J. Case. 1987. Interactions between oak tannins and parasite community structure: Unexpected benefits of tannins to cynipid gall-wasps. *Oecologia* **71**: 254–261.

Taper, M. L., E. M. Zimmerman, and T. J. Case. 1986. Sources of mortality for a cynipid gall-wasp [*Dryocosmus dubiosus* (Hymenoptera: Cynipidae)]: The importance of the tannin/fungus interaction. *Oecologia* **68**: 437–445.

Warming, E. 1908. *Lagoa Santa*. Portuguese translation by A. Löfgren. Imprensa Oficial do Estado de Minas Gerais, (Reprinted 1973). Edusp and Itatiaia, Belo Horizonte, Brazil.

Weis, A. E., R. Walton, and C. L. Crego. 1988. Reactive plant tissue sites and the population biology of gall makers. *Annual Review of Ecology and Systematics* **33**: 467–486.

Williams, C. B. 1964. *Patterns in the balance of nature and related problems in quantitative ecology*. Academic, London, England.

Wilson, J. R. and T. T. Ng. 1975. Influence of water stress on parameters associated with herbage quality of *Panicum maximum* var. *trichoglume*. *Australian Journal of Agriculture Research* **26**: 127–136.

Wolda, J. R. 1987. Altitude, habitat and tropical insect diversity. *Biological Journal of the Linnean Society* **30**: 313–323.

Zar, J. H. 1984. *Biostatistical analyses*. (2nd ed.). Prentice-Hall, Englewood Cliffs, NJ.

SECTION 2
Mutualistic Relationships Between Plants and Animals

Some of the outstanding mutualistic interactions in the tropics came early to the attention of the travelling naturalists. Thus, Martius commented on "those weird melastomes, in whose turgid petioles certain minute ants make their nests" (Spix and Martius 1831). Ant–plant associations were commented upon by several other travellers. While some descriptions emphasized only their more curious features, Belt (1888) and Schimper (1888), however, were quite aware of the intricate character of the reciprocal adaptations of the mutualists, and of their relevance for the theory of evolution and natural selection. Spruce (1908) devoted a chapter to "Ants as modifiers of plant-structure" after his travels in Brazil during the 1850s.

Although interactions of plants and animal dispersers are often less striking and easy to observe than those of pollinators, the ample survey presented in Chapter 6 shows their far-reaching consequences for the ecology and evolution of many tropical plant and vertebrate taxa. In the following two chapters, ant–plant relationships, possibly the most important among mutualisms involving terrestrial invertebrates, are examined in two complementary ways. In Chapter 7 the evolutionary intricacies of a particular set of ant–plants and their associates are presented in detail. Plants with extra floral nectaries represent a large share of the woody vegetation in the open tropical communities of western Brazil, as is shown in Chapter 8.

Beyond the bizarre features of the more extreme cases, mutualistic ties of plants and animals are thus pervasive and influential in tropical communities, and their ecological and evolutionary consequences are far from being fully recognized.

REFERENCES

Belt, T. 1888. *The naturalist in Nicaragua.* 2nd ed. Bumpus, London.

Schimper, A. F. W. 1888. Die Wechselbeziehungen zwischen Pflanzen und Ameisen im tropischen Amerika. *Botanische Mitteilungen aus den Tropen, Jena,* **5**: 1–95.

Spix, J. B. von and C. F. P. von Martius. 1831. *Reise in Brasilien; Dritter Theil.* Friedrich Fleischer, Muenchen (Portuguese translation, Itatiaia, Belo Horizonte, 1981).

Spruce, R. 1908. *Notes of a botanist on the Amazon and Andes.* Vol. 2. Macmillan, London.

6 Fruiting Plant–Frugivore Mutualism: The Evolutionary Theater and the Ecological Play

THEODORE H. FLEMING

The title of this paper is a modification of the title of G. E. Hutchinson's 1965 book *The Ecological Theater and the Evolutionary Play*, in which he discusses four rather disparate topics that provide insight into his philosophy concerning the evolution of life. I have modified Hutchinson's metaphor to indicate that in this paper I will be dealing with the historical (evolutionary) aspects of an important ecological interaction—the mutualism between fruit-eating verte-brates and their food plants. Until recently (Howe and Westley 1986, 1988) this mutualism has been virtually ignored in ecological textbooks, despite the fact that in the tropics at least it has had a profound influence on the evolution of plants and animals. A high proportion of tropical shrubs and trees, for example, produce fleshy fruits and rely on a wide variety of vertebrates to disperse their seeds. These fruits support a substantial portion of the biomass of forest-dwelling tropical vertebrates (e.g., Terborgh 1986).

In this review I will address three major questions: (1) What are the diversity patterns of fruits and frugivores in time and space; (2) What evolutionary processes lie behind these patterns; and (3) To what extent have the morpholog-ical and ecological traits of plants and animals influenced each other's evolution? I do not intend this paper to be an exhaustive review of the literature. Instead, it will concentrate on the broad historical picture and will place less emphasis on details of one end product of this mutualism, namely, seed dispersal. Various aspects of seed dispersal have been well-reviewed recently (e.g., Howe and Smallwood 1982; Janzen 1983a,b, 1985; Howe and Westley 1986, 1988; papers in Estrada and Fleming 1986 and Murray 1986; Fleming 1988). Throughout this paper I use the term "fruit" to denote any fleshy diaspore consumed by vertebrates.

Perhaps the main reason why frugivory has escaped the attention of textbook

Plant-Animal Interactions: Evolutionary Ecology in Tropical and Temperate Regions, Edited by Peter W. Price, Thomas M. Lewinsohn, G. Wilson Fernandes, and Woodruff W. Benson. ISBN   0-471-50937-X   © 1991 John Wiley & Sons, Inc.

writers is that, prior to 1971, the study of fruits and their vertebrate consumers fell into the realm of natural history and was largely conducted in a theoretical vacuum. Snow (1971) began constructing a theoretical framework by suggesting that a dichotomy exists among avian frugivores in which many species eat fruit in an opportunistic fashion while a few species are specialized frugivores. Opportunists were thought to eat mainly "low-quality" fruits containing many small seeds embedded in a pulp that was rich in carbohydrates and water but poor in lipids and protein. Specialists, in contrast, concentrated on large-seeded, "high-quality" fruits whose pulp was rich in lipids.

McKey (1975) expanded upon this theme by postulating that opportunistic frugivores provided low-quality seed dispersal in terms of treatment of seeds in the gut, reliability of visitation to fruiting plants, and deposition of seeds in appropriate germination sites, whereas specialists provided high-quality seed dispersal and hence were likely to be more closely coevolved with their food plants than were opportunists. Howe and Estabrook (1977) contributed to this coevolutionary framework by hypothesizing that trees whose fruits are "aimed" at opportunistic frugivores are under directional selection favoring the production of ever-larger fruit crops, whereas trees whose fruits are "aimed" at specialized frugivores are under stabilizing selection for intermediate crop sizes to avoid satiating their smaller pool of potential dispersers. The Howe–Estabrook scheme also envisioned differences in the fruiting phenology of these two groups of trees with "opportunistic" species producing large fruit crops in short ("peaked") seasons and "specialized" species producing smaller crops over extended fruiting seasons.

Beginning with Wheelwright and Orians (1982), researchers working with fruits and frugivores questioned the degree to which this mutualism, whose principal benefits are seed mobility for plants and energy and nutrients for animals, is the product of tight coevolution between plants and their dispersers. Current opinion is that the traits of fruits and their frugivores are the product of diffuse coevolution in which groups of plants interact with groups of animals. The existence of fruit "syndromes" (van der Pijl 1982, Janson 1983) suggests that this coevolution has occurred between taxonomically restricted groups of plants and animals. The end products of this process are the subject of this chapter.

EVOLUTIONARY PATTERNS IN PLANTS AND FRUGIVOROUS VERTEBRATES

General Overview

Effective seed dispersal has been an adaptive concern of plants ever since the seed habit evolved in land plants of the Devonian and Mississippian periods. As reviewed by Tiffney (1986a), fossil evidence indicates that biotic dispersal began in the Pennsylvanian with fish and amphibians possibly being early vertebrate seed dispersers. In the late Paleozoic and Mesozoic eras, various plant lineages, including medullosan pteridosperms, cordaitaleans, seed ferns, cycads, and ginkgos, evolved flesh-covered diaspores. Herbivorous dinosaurs undoubtedly ate and dispersed seeds in the Cretaceous, but, according to Tiffney (1986a), prior to the rise of the angiosperms the fruiting plant–herbivore interaction remained generalized. Specialized mutualisms did not exist until the demise of the dinosaurs and the evolution of fruit-eating birds and mammals.

The earliest angiosperms possessed small, abiotically dispersed seeds (Tiffney 1984, 1986a). Prior to the Tertiary, the most common fruit types in fossil angiosperm floras were dehiscent follicles or capsules bearing small (mean volume = 1.7 mm^3), thin-walled seeds which lacked evidence of distinctive dispersal adaptations. Large seeds (with a volume of $\geqslant 1000$ mm^3) first appeared in the Eocene. Tiffney (1984) postulated that the change in range of seed sizes in angiosperms resulted from two factors: (1) a shift from weedy, second growth and forest understory habits to larger-statured shrub and tree habits and (2) a greater reliance on vertebrates for seed dispersal.

Various authors (e.g., Stebbins 1974, 1981; Regal 1977; Doyle 1977; Crepet 1984; Tiffney 1984, 1986a,b) have suggested that biotic dispersal was one of the key adaptations that led to the "explosive" radiation of angiosperms beginning in the late Cretaceous. Herrera (1989), however, disputed this view by pointing out that endozoochory is proportionately more common in gymnosperm families (64.3% are endozoochorous) than in angiosperm families (27.1% endozoochorous) and that in angiosperms, endozoochorous families contain no higher species diversity than families with other dispersal modes. He concluded that "The evolution of fleshy diaspores (and endozoochorous dispersal) apparently has not constituted a sufficiently important evolutionary breakthrough to promote diversification in and of itself."

Evolutionary Patterns in Contemporary Angiosperms

Our biosphere presently contains approximately 235×10^3 species of angiosperms and 700 species of gymnosperms. In the case of gymnosperms, this total represents the culmination of about 290 million years of erolution, whereas it represents about 125–130 million years of angiosperm evolution. Current diversity in angiosperms is probably higher than at any time in the past, but gymnosperms clearly are a declining group whose diversity peaked in the middle of the Mesozoic era (Doyle 1977; Niklas et al. 1985). Below I discuss several patterns of diversity and adaptation in contemporary angiosperms that collectively represent the botanical template upon which the evolution of modern fruit-eating vertebrates has taken place.

Although gymnosperms have historically been an important source of fleshy "fruit" for vertebrates (van der Pijl 1982; Herrera 1989), I emphasize angiosperms in this review because they are by far the most important sources of fruit in contemporary habitats and have likely been so throughout the Tertiary. I obtained data for the following analyses from two principal sources. Data on taxonomic arrangements and species richness, geographic distributions, growth habits, and fruit types by family come from Heywood (1978). Data on the fossil ages of families, with the arrangement of families in subclasses modified to follow Heywood (1978), come from Muller (1981).

Diversity of Fruit Types. I classified each of 281 angiosperm families (15 families of aquatic monocots in the Alismatidae were excluded from this and other analyses) in one of six classes of fruit (Table 6-1A). About 28% of these families produce fleshy fruits, 20% contain species producing either fleshy or dry fruits ("mixed" families), and 53% produce only dry fruits. Drupes and berries are the most common types of fleshy fruits, and dry capsules are the most common type of nonfleshy fruits.

The distribution of fruit types differs among the six dicot subclasses ($X^2 = 28.10$, $df = 10$, $p = 0.0017$) (Table 6-1B). Except in the Magnoliidae and Rosidae (which contains the highest proportion of mixed families), dry fruits dominate each subclass. The Rosidae has the most even representation of fruit types; the Hamamelidae has the most "polarized" distribution; and the Asteridae has the lowest representation of fleshy fruits in its families.

The distribution of fruit types differs significantly among three growth habit classes ($X^2 = 42.12$, $df = 4$, $p \ll 0.0001$) (Table 6-2A). Dry fruits predominate (72.3% of the families) in herbaceous and "mixed" families. Woody families produce a more even mix of fruit types with fleshy fruits predominating (39%).

TABLE 6-1. Distribution of Angiosperm Families among Fruit Types

Fruit Type	A. Terrestrial Dicots and Monocots Number of Families (proportion)
Berry or drupe	67 (0.238)
Other fleshy	11 (0.039)
Mixed[a]	55 (0.196)
Dry capsule	80 (0.285)
Dry non-capsule	68 (0.242)

B. Dicots by Subclass
Proportion of Families by Fruit Type

Subclass (No. of families)	Fleshy	Mixed[a]	Dry
Magnoliidae (28)	0.571	0.107	0.321
Hamamelidae (11)	0.273	0	0.727
Caryophyllidae (13)	0.231	0.154	0.616
Dilleniidae (60)	0.233	0.200	0.567
Rosidae (87)	0.298	0.275	0.425
Asteridae (43)	0.116	0.163	0.721

[a]Includes families with fleshy and dry fruits.

TABLE 6-2. Distributions of Angiosperm Families by Plant Habit and Species Richness

Habit (No. of families)	A. Plant Habit Proportion of families by fruit type			
	Fleshy	Mixed[a]	Capsule	Other Dry
Herbaceous (75)	0.133	0.093	0.427	0.347
Mixed[b] (52)	0.096	0.231	0.346	0.327
Woody (146)	0.391	0.240	0.212	0.157

Number of Species per Family	B. Species Richness Proportion of Families by Fruit Type		
	Fleshy ($n = 78$ families)	Mixed[a] ($n = 55$)	Dry ($n = 148$)
1–10	0.231	0.036	0.149
11–100	0.244	0.127	0.358
101–1000	0.397	0.509	0.351
> 1000	0.128	0.327	0.142

[a]Includes families with fleshy and dry fruits.
[b]Includes families with herbaceous and woody species.

Species Richness and Fruit Types. I examined the question—Does the number of species per family differ among fruit types?—by tallying each of the 281 families in one of four diversity classes by fruit type (Table 6-2B). Distributions differ significantly among fruit types ($X^2 = 28.0$, $df = 6$, $p \ll 0.0001$). Mixed families tend to have more species per family than do the other two fruit types. The two largest angiosperm families (Leguminosae and Orchidaceae) produce non-capsular dry fruits. Families producing fleshy fruits do not necessarily contain more species, on average, than those producing dry fruits. Herrera (1989) has used this to infer that biotic dispersal does not necessarily result in higher rates of speciation. In contrast, Tiffney (1986b) noted that in the Hamamelidae [which is constituted differently than the arrangement in Heywood (1978)], families producing biotically dispersed (but not necessarily fleshy) fruits (e.g., Fagaceae, Moraceae) contain more species than families that are abiotically dispersed.

Geographic Distributions of Families. I classified each of the 281 families into one of seven broad geographic classes by fruit type (Table 6-3). The distribution of fruit types differs significantly among geographic regions ($X^2 = 50.04$, $df = 12$, $p \ll 0.0001$). Most (73.8%) cosmopolitan and temperate families produce dry fruits, whereas fleshy fruits (or a "mixed" situation) predominate in pantropical families of which only 27.3% produce dry fruits. In families restricted to either the New or Old World tropics, dry fruit families tend to predominate over fleshy-fruited or mixed families (58.0% vs. 42.0% respectively).

TABLE 6-3. Geographic Distributions of Angiosperm Families by Fruit Type

Distribution	Proportion of Families by Fruit Type			
(No. of families)	Fleshy	Mixed[a]	Capsule	Other Dry
Cosmopolitan (46)	0.131	0.261	0.326	0.283
N and S temperate (22)	0.091	0.136	0.545	0.227
N temperate (44)	0.181	0.159	0.295	0.364
S temperate (11)	0.091	0	0.364	0.545
Pantropical (88)	0.409	0.318	0.182	0.091
Neotropics (22)	0.318	0.091	0.364	0.227
Paleotropics (40)	0.355	0.075	0.275	0.295

[a]Includes families with fleshy and dry fruits.

Geological Distributions of Dicot Subclasses and Families. As expected, the first fossil record (of pollen, not fruits or seeds) tends to be earlier in families of Magnoliidae and Hamamelidae than in the other four dicot subclasses (Table 6-4A). The temporal distributions of Magnoliidae and Asteridae (in two time blocks—Eocene or before, Oligocene and later) differ significantly ($X^2 = 5.58$, $df = 1$, $p = 0.018$). When families are classified by (contemporary) fruit types (Table 6-4B), their distributions in the two time blocks above do not differ

TABLE 6-4. Geological Distributions of First Pollen Appearances of Dicot Angiosperms

No. of Families	Proportion of Families				
	Cretac.	Paleoc.	Eoc.	Oligoc.	Mioc. on
A. By Subclass					
Subclass					
Magnoliidae (11)	0.455	0.091	0.182	0.091	0.182
Hamamelidae (8)	0.625	0.250	0.125	0	0
Caryophyllidae (6)	0.167	0.167	0.167	0.167	0.333
Dilleniidae (24)	0.209	0.083	0.250	0.250	0.208
Rosidae (43)	0.210	0.116	0.256	0.209	0.210
Asteridae (18)	0	0.056	0.222	0.278	0.445
B. By Fruit Types[a]					
Fruit types					
Fleshy (31)	0.258	0.129	0.323	0.129	0.162
Mixed[b] (34)	0.235	0.059	0.235	0.324	0.147
Dry (44)	0.182	0.136	0.159	0.159	0.364

[a]Based on contemporary members of the families.
[b]Includes families with fleshy and dry fruits.

significantly ($X^2 = 4.17$, $df = 2$, $p = 0.12$). However, the combined geological records of fleshy-fruited and mixed families differ significantly from those of dry-fruited families ($X^2 = 4.02$, $df = 1$, $p = 0.045$) with families in the former group appearing earlier in the geologic record, on average, than those in the latter group.

The "Popular" Fruit Families. Not all plant families producing fleshy fruits are equally "popular" with frugivorous vertebrates. For example, Snow (1971, 1981), Wheelwright et al. (1984), and Moermond and Denslow (1985) have noted that only a handful of plant families provide the bulk of the fruit eaten by frugivorous tropical birds. Likewise, studies of the diets of tropical primates and bats indicate that these groups concentrate on a small subset, often different from that of birds and each other, of fruiting families (for primates: Chivers 1980; Terborgh 1983; Gautier-Hion et al. 1985b; for bats: Gardner 1977; Marshall 1983, 1985; Fleming 1988). These families are listed in Table 6-5A.

It is of interest to know whether these families represent a random draw from the pool of all plant families or whether they represent a biased subset regarding their species richness, geographic distributions, geological age, and phylogenetic (subclass) affinities. Results summarized in Table 6-5B indicate that, compared with the "average" angiosperm family, the popular families are (1) larger, (2) more likely to be pantropical, and (3) geologically older. Although they represent a random draw from dicot subclasses, they tend to be more concentrated in the Magnoliidae (the oldest extant subclass) and in the Rosidae (the largest subclass) than expected by chance. Restricting the comparison just to fleshy-fruited and mixed families vs. popular families yields the same results, except that differences in the distributions of the two groups among geological epochs are not significant ($X^2 = 3.44$, $df = 1$, $p = 0.064$).

Many of these families are dominant members of local tropical floras in terms of numbers of species and sometimes numbers of individuals (Gentry 1982a,b). The understories of lowland and mid-elevation neotropical forests, for example, are dominated by shrubs of the families Melastomataceae, Rubiaceae, Piperaceae, and Solanaceae (Gentry and Emmons 1987). It is tempting to speculate that the ecological success of these groups is a direct result of their attractiveness to vertebrate frugivores.

The deviations from chance expectations discussed above raise interesting evolutionary questions that currently lack satisfactory answers. For example, are these families rich in numbers of species (and sometimes genera) and widespread geographically because they are popular or vice versa? Is their popularity due to their geological age and hence the availability of their fruits when modern vertebrate frugivores were beginning to evolve? Do they have special characteristics of their fruiting biology that make them more attractive to frugivores than fruits of other families? And finally, do these families appear to be more "coevolved" with their dispersers than other families?

Summary. Major results of the preceding analyses can be summarized as follows.

TABLE 6-5. The Popular Angiosperm Families whose Fruits are Eaten by Birds, Monkeys, and Bats

A. The Families, by Subclass, and Their Major Dispersers

	Birds	Monkeys	Bats
Magnoliidae			
Annonaceae	x	x	
Lauraceae	x		
Menispermaceae		x	
Myristicaceae	x	x	
Piperaceae			x
Dilleniidae			
Flacourtiaceae		x	
Moraceae	x	x	x
Sapotaceae		x	x
Rosidae			
Anacardiaceae	x	x	
Araliaceae	x		
Burseraceae	x	x	
Euphorbiaceae	x		
Leguminosae		x	
Melastomataceae	x		
Meliaceae	x		
Myrtaceae	x	x	x
Sapindaceae	x	x	
Asteridae			
Rubiaceae	x		
Solanaceae	x		x
Arecidae			
Palmae	x	x	x

B. Major Characteristics of These Families

1. Species richness	No. species per family	Number of families All	Popular	X^2, df, P
	1–10	42	0 ⎫	
	11–100	79	0 ⎬	
	101–1,000	111	7 ⎭	29.21, 1, ≪0.0001
	1,001–10,000	46	12 ⎫	
	> 10,000	3	1 ⎭	
2. Geographic distributions	Worldwide	46	3	
	Pantropical	88	17	25.76, 2, ≪0.001
	Other	146	0	
3. Geological distributions	First record:			
	Cretaceous	31	6 ⎫	
	Paleocene	17	4 ⎬	
	Eocene	25	3 ⎭	4.52, 1, 0.034
	Oligocene	23	2 ⎫	
	Miocene and later	29	0 ⎭	

Fleshy fruits are produced by about 28% of the present angiosperm families (or nearly 50% if families of mixed fruit types are included). Fleshy fruits are more likely to be produced by woody plants than by herbs, and they are more common in tropical families than in temperate families. Families producing fleshy fruits are no richer in species than are other families, but they are more likely to appear earlier in the fossil record (especially those in the Magnoliidae) than other families. Families that are popular with tropical frugivores have been especially successful evolutionarily and ecologically.

Ecological Patterns in Contemporary Fleshy Fruits

Fleshy fruits differ widely in their spatial and temporal availability, in their "design" features, and in the nutritional characteristics of their pulp. All of these features are or have been potential candidates for the action of natural selection as mediated by the foraging behavior of frugivorous vertebrates.

Patterns in Space. Herrera (1984a) pointed out that the spatial and successional availability of fleshy fruits changes with latitude. In temperate forests, fleshy fruits occur mainly in early successional plants and in disturbed areas (e.g., light gaps). In Mediterranean scrublands, fleshy-fruited plants dominate the ends of successional sequences. They are common throughout successional seres in tropical forests with the diversity and density of fleshy-fruited plants being positively correlated with annual precipitation and soil fertility (Gentry 1982b; Gentry and Emmons 1987). Since seed sizes tend to increase along successional sequences and canopy trees tend to produce larger seeds (and larger fruits) than understory plants (Salisbury 1942; Baker 1972; Foster and Janson 1985; Foster 1986), these latitudinal trends suggest that frugivorous vertebrates will be faced with a wider array of seed and fruit sizes in tropical habitats than in temperate habitats. To the extent that consumer body size is correlated with food particle size (it is in frugivorous birds and bats; Wheelwright 1985; Fleming 1988), we should expect to find that, on average, tropical frugivores are larger than temperate frugivores.

Patterns in Time The temporal availability of fruits also varies latitudinally. In mid-latitude temperate forests, the greatest abundance and diversity of fruits occurs in late summer and early autumn—times when the diversity of migrant, facultatively frugivorous birds is highest (Thompson and Willson 1979; Stiles 1980). Peak fruit abundance and diversity occurs in late autumn and winter in Mediterranean scrublands (Herrera 1982, 1984a). Some fleshy fruits are available year round in dry and wet tropical forests, but strong seasonal peaks in abundance and diversity occur in most forests (e.g., Foster 1982; Baker et al. 1983). Multiannual peaks in fruiting are especially prominent in West Malaysia (Appanah 1985).

Morphological/Nutritional Patterns. As described in detail by van der Pijl (1982), fleshy fruits come in a wide variety of sizes, shapes, and degree of external protection. They also display considerable diversity in such design features as number and sizes of seeds, pulp mass:seed mass ratio, and method of presentation as well as in the nutritional composition of their pulp. A comprehensive review of this diversity is beyond the scope of this paper (see discussions in Herrera 1981, 1987 and Moermond and Denslow 1985). The point I wish to make here is that fruits differ considerably in their "profitability" to frugivores (Martin 1985) and that these differences should influence the food choice decisions of frugivores (Moermond and Denslow 1985). In theory, these decisions could have a selective effect on plant traits (Howe and Westley 1986; Tiffney 1986a).

Fruit size is a basic morphological parameter that influences frugivore food choice. In general, mean and maximum (but not minimum) fruit size tends to be positively correlated with body size, at least in volant frugivores (see references above). Although no comprehensive review has yet been published, it is likely that the size-frequency distributions of fruits vary latitudinally with median and modal fruit sizes increasing with decreasing latitude. Data for bird-fruits presented in Herrera (1987) and Wheelwright (1985) are consistent with such a pattern (Fig. 6-1). Fruit widths in the Costa Rican sample were bimodally distributed with modes at 7–12 mm and 17 mm, whereas they were unimodally

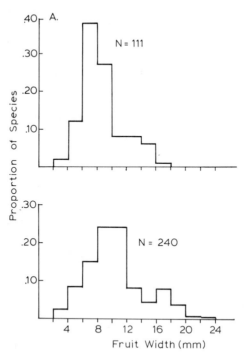

Figure 6-1. Size distributions of fruits at two localities: (A) Mediterrean scrubland; (B) mid-elevation Costa Rican forest. Redrawn with permission from Herrera (1987) and Wheelwright (1985).

distributed with a mode at 7 mm in the Spanish sample. Temperate–tropical differences in fruit sizes undoubtedly reflect the effects of selection for seed size that are independent of fruit–frugivore coevolution.

Patterns in the nutritional composition of fruits are complex. Only in mid-latitude temperate regions and Mediterranean scrublands have nutritional characteristics of fruit floras been systematically examined. Results of four North American studies are contradictory regarding the presence of seasonal patterns in the chemical composition of fruit pulps. Stiles (1980) and Stapanian (1982) found that bird-fruits produced in summer tend to have higher water and carbohydrate contents than those produced in the fall, whereas Johnson et al. (1985) and Piper (1986) failed to find similar trends in the summer and fall fruits they analyzed. The only seasonal trends noted by Johnson et al. and Piper were an increase in potassium and protein from summer to fall and an increase in mean seed mass and its variance, respectively. In contrast, Herrera (1982) and Debussche et al. (1987) reported a decrease in water and carbohydrates and an increase in lipids in autumn–winter (wet season) fruits compared with summer (dry season) fruits in Mediterranean scrublands. Herrera's (1987) analysis of 111 species of Mediterranean fruits further showed that fruit size, seed size, and pulp:seed mass ratios tended to vary independently of each other and that plant growth form influenced interspecific variation in pulp composition. Debussche et al. (1987) found that relative concentrations of water, carbohydrates, and protein in fruit pulp tend to decrease and that lipid content increases in the sequence herbs, shrubs < 2 m tall, and shrubs > 2 m tall. Similar analyses of interspecific variation in the fruit floras of tropical forests are badly needed before we can rigorously address the question, Do distinct nutritional strategies exist in fleshy-fruited plants (e.g., Herrera 1981)?

In summary, fleshy fruits represent a conspicuous and readily accessible but highly variable resource to vertebrate consumers. This variability manifests itself in many ecological dimensions, including historical and contemporary time and

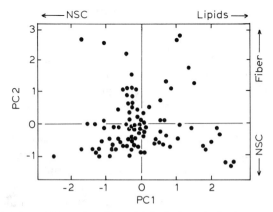

Figure 6-2. Ordination of 91 species of Spanish fruits by principal components analysis on the basis of the water and organic nutrients in their pulp. NSC = nonstructural carbohydrates. Redrawn with permission from Herrera (1987).

space, "chemistry," plant growth form, and phylogeny. A multivariate plot of this variability, such as that conducted by Herrera (1987) (Fig. 6-2) for nutritional characteristics, that encapsulated all dimensions of the variability of extant fruits would be an instructive way of depicting the contemporary frugivory adaptive zone—that is, the set of plant traits that serves as the template upon which frugivore evolution has taken place. It would be interesting to know whether, as is the case in Herrera's plot, fruit traits are tightly clustered around one central "peak" or whether a series of "nodes" consisting of specific plant fruiting "strategies" exist, as multivariate plots of the morphological characteristcs of Gabon fruits (Gautier–Hion et al. 1985b) seem to suggest. Given the mosaic nature of much of evolution, it seems reasonable to predict that the actual adaptive landscape of contemporary fruits will lie somewhere between a single tight cluster of taxa and a series of discrete "minipeaks" onto which specific animal taxa or ecological guilds are mapped. Finally, it would be interesting to know to what extent this landscape has changed through time as a result of fruit–frugivore coevolution.

Evolutionary Patterns in Frugivorous Vertebrates

Contemporary bird and mammal faunas contain a total of 8600 and 4000 species, respectively. In the following analyses I will examine the effect that frugivory has had on the structure of these faunas. Data on taxonomy and species richness, major food types, and geographic distributions of terrestrial families come from two principal sources: Van Tyne and Berger (1975) for birds and Walker (1975) for mammals. I am aware that the familial boundaries of birds, especially in the oscine passerines, are in a state of flux (e.g., A. O. U. 1983), but have accepted Van Tyne and Berger's somewhat outdated classification as being a reasonable reflection of the adaptive diversity of birds. To lump warblers, icterid blackbirds, tanagers, and finches into one family (the Emberizidae) seems inappropriate for my purposes because it combines a wide variety of adaptive types into one taxonomic unit.

Diversity of Food Habits. The distribution of 135 families of terrestrial birds and 107 families of terrestrial mammals among six major diet classes is shown in Table 6-6A. These distributions differ significantly ($X^2 = 33.8$, $df = 3$, $p = \ll 0.0001$). A higher proportion of bird families are mostly frugivorous (16.3% vs. 0.9%), and 35.5% of avian families contain frugivores compared with 19.6% of mammalian families. These families are listed in the Appendix. Insectivory is the most common diet class in birds, whereas insectivory and herbivory are codominant in mammals (Table 6-6A). Vertebrate carnivory is an uncommon diet class in both birds and mammals.

Within birds, frugivory appears to be somewhat more common in passerine families than in nonpasserines, although our impressions of "classic" avian frugivores tend to be of nonpasserines such as oilbirds, toucans, trogons, hornbills, and fruit pigeons. Nineteen of 69 nonpasserine families (27.5%) include

TABLE 6-6. Distributions of Bird and Mammal Families by Diet and Species Richness

A. By Diet Classes
Proportion of Families

Taxon (No. of Families)	Mostly fruit	Fruit + other	Insect	Vegeta-tion	Verte-brates	Other
Birds (135)	0.163	0.193	0.356	0.059	0.059	0.170
Mammals (107)	0.009	0.187	0.336	0.327	0.037	0.103

B. By Species Richness and Diet Class
Proportion of Families

Number of Species per Family	Fruit		Insects or Invertebrates		Other (excluding vertebrates)	
	B^a (48)	M (21)	B (48)	M (36)	B (39)	M (38)
1–40	0.604	0.810	0.708	0.833	0.615	0.763
41–80	0.167	0.095	0.167	0.111	0.103	0.157
81–120	0.063	0	0	0	0.103	0.026
121–160	0.021	0.095	0.021	0	0.051	0
> 160	0.145	0	0.104	0.056	0.128	0.053

^{a}Abbreviations: B = birds, M = mammals; number of families in parentheses.

fruit in their diets, compared with 29 of 67 passerine families (43.3%); differences in these proportions just miss statistical significance ($X^2 = 3.68$, $df = 1$, $p = 0.055$). Of the frugivorous families, four of 19 nonpasserines (21.1%) and 11 of 29 passerines (37.9%) are strongly frugivorous; these proportions do not differ significantly ($X^2 = 1.52$, $df = 1$, $p = 0.22$).

Species Richness and Diet. The distribution of bird and mammal families, classified into three major diet groups, among five classes of species richness are shown in Table 6-6B. Small families containing less than 40 species predominate in both vertebrate classes. Within both classes, distributions among diet groups do not differ significantly ($p \gg 0.05$ in X^2 tests based on two diversity groupings, $\leqslant 40$ species per family and > 40 species per family). Frugivorous families, on average, are no larger (or smaller) than are families in other diet groups (cf. Snow 1971). As expected, given their overall species richness, bird families tend to contain more species than do mammal families; the proportion of bird families containing $\leqslant 40$ species (65.4%) is lower than it is in mammals (80.0%) ($X^2 = 5.79$, $df = 1$, $p = 0.016$).

Geographic Distribution of Frugivorous Families. As expected, frugivorous families are mostly tropical in both birds and mammals (Table 6-7). The degree of tropical concentration in frugivores is twice as strong in mammals (6.3 tropical families: 1 temperate family) as in birds (3.5:1), but this difference is not

TABLE 6-7. Geographic Distributions of Fruit-Eating
Families of Birds and Mammals[a]

Number of Families	Birds	Mammals
Tropics vs. non-tropics	45 vs. 13	20 vs. 3
Neotropics vs. Paleotropics	25 vs. 27	6 vs. 14
New World vs. Old World	26 vs. 29	7 vs. 16
Pantropical	4	0

[a]Data Include Families that are Facultative as well as Obligate
Frugivores.

Note: Families with worldwide or hemisphere-wide distributions are
counted more than once in some comparisons so that certain sums are
not equal (e.g., in birds, number of tropical families do not equal the
number of Neotropical plus Paleotropical families).

statistically significant ($X^2 = 0.77$, $df = 1$, $p = 0.38$). There are twice as many
frugivorous mammal families in the Paleotropics than in the Neotropics but not
in birds; again, this difference is not significant ($X^2 = 1.54$, $df = 1$, $p = 0.21$). There
are four pantropical families of (partially) frugivorous birds (Columbidae,
Psittacidae, Trogonidae, and Capitonidae) but none in mammals.

Geological Distributions. An excellent fossil record enables me to examine the
temporal distributions of the first appearances of many mammal families by diet
class. A poor fossil record (Feduccia 1980) precludes a similar analysis in birds. As
summarized in Table 6-8, first appearances of frugivorous and herbivorous
mammals tend to be later (in the Oligocene or Miocene) than those of insectivores
or carnivores/omnivores. These distributions for three diet classes (fruit,
vegetation, insects) and two time blocks (pre-Oligocene, Oligocene and later)
differ significantly ($X^2 = 8.78$, $df = 2$, $p = 0.012$). Despite the absence of a detailed
fossil record, it is likely that the adaptive radiation of frugivorous birds was
coeval with that of mammals (Feduccia 1980; Houde 1987). If this is true, then the
major radiation of modern frugivorous vertebrates has occurred during the last

TABLE 6-8. Geological Distributions of First Appearances in the Fossil Record of
Mammals by Diet Type

Diet Class (No. of Families)	Proportion of Families				
	Cretac.	Paleoc.	Eoc.	Oligoc.	Mioc. on[b]
Fruit[a] (15)	0	0.066	0.133	0.333	0.467
Vegetation (31)	0	0.032	0.129	0.354	0.485
Insects + invertebrates (21)	0.048	0.048	0.429	0.095	0.381
Vertebrates + omnivores (12)	0	0	0.250	0.417	0.333

[a]Includes facultative and obligate frugivores.
[b]Excludes recent records.

30–40 million years. The evolution of frugivorous passerine birds in the late Tertiary was a relatively late event in this process (Feduccia 1980).

Morphological Patterns. Compared with such adaptive zones as herbivory and carnivory, frugivory does not require drastic morphological and physiological modifications from an insectivorous avian or mammalian ancestor. Indeed, it is likely that morphological adaptations for insectivory preadapted certain groups of birds (e.g., sylviid warblers in Europe) to become effective frugivores (Herrera 1984b). Herrera points out that frugivorous sylviids do not differ from sympatric insectivorous or granivorous passerines in their ratios of gizzard mass, liver mass, and intestinal length to body mass, but they do differ significantly in bill dimensions and mouth widths. Similarly, among tyrannid flycatchers, highly frugivorous species have shorter and broader bills than do non-frugivorous species (Traylor and Fitzpatrick 1982). Moermond and Denslow (1985) review avian morphological adaptations for frugivory and point out that the method of taking fruit (e.g., on the wing or from perches) has influenced bill size and shape, wing and limb dimensions, and the relative mass of limb musculature.

Analogous morphological trends probably also exist in mammals. An example of morphological trends within the microchiropteran family Phyllostomidae (the New World leaf-nosed bats) will serve to illustrate the extent to which frugivores differ in cranial and wing characteristics from insectivorous, carnivorous, nectarivorous, and sanguinivorous members of this ecologically diverse family. Plotted in Fig. 6-3 is an ordination of phyllostomid genera in 2-space as generated by the method of non-metric multidimensional scaling (Mather 1976). I used nine quantitative characters [forearm length, greatest skull length, breadth of braincase, length of maxillary toothrow, breadth across upper molars, alpha angle (an indication of the breadth of the wing relative to its length), wing-tip index, overall aspect ratio, and wing loading] in this analysis. Forearm length and cranial measurements represent generic means calculated from data in Swanepoel and Genoways (1979). Data on wing parameters are generic means taken from Smith and Starrett (1979). The ordination shows that many frugivorous genera in the subfamilies Carolliinae and Stenoderminae are morphologically closer to "basal" insectivorous phyllostomines (e.g., *Macrotus* and *Micronycteris*) (Honeycutt and Sarich 1987) than are carnivorous phyllostomines, nectarivores (subfamily Glossophaginae), and vampires (subfamily Desmodontinae).

In summary, there are similarities and differences in the structure of frugivorous bird and mammal faunas. Similarities include: (1) no differences in patterns of species richness among diet classes; (2) concentration of frugivorous families in the tropics; and (3) more recent appearances of frugivorous families in the fossil record than families in other diet classes. Differences include: (1) a higher proportion of frugivorous families in birds; (2) a higher proportion of temperate frugivorous families in birds, and (3) more families of frugivorous mammals but not birds in the Paleotropics than in the Neotropics. These similarities and differences suggest that the evolutionary theater for the evolution of frugivorous

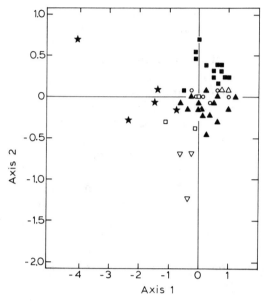

Figure 6-3. Ordination of 48 genera of phyllostomid bats by nonmetric multidimensional scaling based on nine morphological characters (forearm length plus four skull characters and four wing characters; see text). Diets of the subfamilies are as follows: Phyllostominae, insects and fruit (open circles); vertebrates (stars); Glossophaginae, mostly nectar (closed squares); Carolliinae, mostly fruit (open triangles); Stenoderminae, mostly fruit (closed triangles); Brachyphyllinae, nectar and fruit (open squares); Desmodontinae, blood (inverted triangles).

birds and mammals has primarily been the tropics (especially the Paleotropics in mammals), that the tempo of this adaptive radiation (i.e., rate of species formation) has been higher in birds than in mammals, and that birds have been involved to a greater extent than mammals in temperate frugivory.

Summary of the Plant and Animal Patterns

The evolutionary radiations of fleshy-fruited angiosperms and their vertebrate consumers also involve similarities and differences. Principal similarities include: (1) similar proportions of families (ca. 33%) that produce or consume fleshy fruits; (2) concentration of fruits and frugivores in the tropics; and (3) patterns of species richness within families that are similar to those of families of other fruit or diet types. The principal difference is the earlier appearance in the fossil record of many families of fleshy-fruited plants compared with families of avian and mammalian frugivores. This difference raises the question, Who were the consumers of the fleshy fruits of late Cretaceous and early Tertiary angiosperms before their modern consumers evolved? What vertebrate groups were initially involved in the early differentiation of angiosperm families and subclasses, many of whose diagnostic characteristics are based on gynoecial (ovarian and ovular) rather than androecial features (Stebbins 1974, 1981)? As Raven and Axelrod (1974) point out, numerous modern angiosperm genera and families were present 70 million years ago. If these taxa were biotically dispersed, who were their dispersers? Alternatively, perhaps many of these taxa initially were abiotically dispersed and later evolved biotic dispersal methods as modern groups of birds and mammals began to evolve. Such a scenario appears to exist in the Hamamelidae in which the primitive dispersal mode was abiotic and more derived families (e.g., Fagaceae and Moraceae) are biotically dispersed (Tiffney 1986b).

FRUITING PLANT–FRUGIVORE COEVOLUTION

Current opinion (Wheelwright and Orians 1982; Howe 1984; Herrera 1985, 1986) holds that the evolutionary interplay between fruiting plants and frugivores has involved diffuse rather than highly specialized coevolution as a result of numerous evolutionary and ecological constraints. Herrera (1985), for example, points out that a "fine-tuned" relationship between fruits and frugivores is not to be expected because of the dissimilar rates of evolution of animals and plants and the low selection intensities that plants experience from frugivores. Compared with animals, especially birds and mammals (Wyler et al. 1983), the morphological and probably phenological traits of plants evolve slowly (Herrera 1986, 1987; Kochmer and Handel 1986). Frugivores provide low selection intensities because plants often cannot restrict the range of vertebrates eating their fruit, and hence their seeds are handled by animals that vary greatly in their disperser effectiveness (e.g., Howe 1980; Howe and Vande Kerckhove 1981; Pratt and Stiles 1983). In addition, selection pressures provided by other plants, fruit "parasites" such as seed predators or non-disperser invertebrates and vertebrates, and vagaries of the abiotic environment are often more intense than those provided by legitimate dispersers (Jordano 1987a,b). As a result, adaptations for successful dispersal in fleshy-fruited plants should be generalized and "coarse," providing a match with broad taxonomic or ecological groups of dispersers rather than with narrowly restricted sets of species (Janson 1983; Gautier-Hion et al. 1985b; Herrera 1985).

Two recent studies will serve to illustrate the coarse fit between frugivore foraging behavior and plant traits. Contrary to the theoretical predictions of McKey (1975) and Howe and Estabrook (1977), Herrera (1984a) found that the proportion of fruit crops removed in various species of bird-dispersed Mediterranean scrubland plants was independent of fruit nutritional quality, ripening rate and time, and within-plant fruit density. Similarly, Bronstein and Hoffmann (1987) reported no correlations between either the number of bird species or their visitation frequencies and total crop size and average fruit size in Costa Rican *Ficus pertusa*. Caution must be used in interpreting these results, however, because they come from short-term investigations relative to the life spans of the plants being studied. Only if these results hold throughout the life spans of (long-lived) plants would it be fair to conclude that avian frugivores appear not to discriminate among plants on the basis of their phenological or fruit characteristics.

Despite the rather coarse nature of the ecological and evolutionary interplay between fruits and frugivores, some plant and animal traits appear to be the products of the frugivory mutualism. Coevolved plant traits, described in Table 6-9A, include aspects of phenology, infructescence design, and various characteristics of fruit. Perhaps the clearest examples of apparently coevolved

TABLE 6-9. Summary of Plant Traits that are Thought to be Coevolved or Non-coevolved with Frugivorous Seed-Dispersing Vertebrates

A. Coevolved Traits

1. Phenological traits
    a. Timing of peak numbers of fruiting species corresponds with peak abundance and diversity of avian migrants in eastern U. S. temperate forests and Mediterranean scrublands (Thompson and Willson 1979; Stiles 1980; Herrera 1984a; Skeate 1987).
    b. Degree of ripening synchrony and length of ripening period within species reflects the seasonal abundance of frugivores in eastern U. S. forests (Thompson and Willson 1979; Stapanian 1982).
2. Infructescence morphology
    a. Terminal rather than axillary placement of fruits prevents certain groups of birds (e.g., tanagers, emberizid finches) and seed predators from gaining access to fruits (Denslow and Moermond 1985; Levey 1987).
3. Fruit and seed size
    a. Narrow range of fruit sizes in Mediterranean scrublands refects the small size (12–18 g) of avian frugivores; broader range of fruit sizes in the tropics matches the broader range of body sizes (Herrera 1984a, 1987; Wheelwright 1985).
    b. Seed sizes in tropical habitats may in part reflect the relative abundance of fruit "mashers" (tanagers, emberizid finches) and "gulpers" (manakins, trogons, cotingids, etc.); gulpers can disperse larger seeds than mashers (Levey 1987).
4. Fruit pulp composition
    a. In Mediterranean scrublands, watery fruits in summer and lipid-rich fruits in winter match the nutritional needs of resident birds (Herrera 1982).
    b. In temperate U. S. forests, summer fruits produce higher ratios of pulp energy:seed energy to make them competitive with insects for food (Stapanian 1982).
    c. Bitter secondary substances in the arils of large-seeded tropical plants may discourage avian "mashers" (poor dipersers) from eating them (Levey 1987).
5. Fruit color
    a. In bird-dispersed plants in temperate and tropical New World habitats, fruits of many families have converged on the same modal color—black (Wheelwright and Janson 1985).
    b. Conspicuous bicolored fruit displays predominate among plants of small stature in Mediterranean scrublands (Herrera 1987).

TABLE 6-9. (*Continued*)

B. Non-coevolved Traits

1. Phenological traits
   a. Length of fruiting seasons within floras decreases with increasing latitude and is positively correlated with air temperature of the coldest month; climate regulates this trait (Herrera 1984a).
   b. No clear differences in the amplitude of fruiting seasonality exists among plants of different fruit types or seed sizes in an African tropical forest (Gautier-Hion et al. 1985a).
2. Fruit pulp composition
   a. In Mediterranean scrublands, climate (i.e., drought and frost-free periods), rather than birds, selects for watery fruits (Debussche et al. 1987); no influence of frugivores is implicated by (i) the absence of covariation in fruit size, seed size, and pulp:seed mass ratio; (ii) lack of covariation in water, protein, and lipid content; (iii) nearly random cooccurrence of mineral elements across species; and (iv) no overall covariation of fruit design and organic and inorganic pulp constituents (Herrera 1987).
   b. No seasonal trends were apparent in lipids, reducing sugars, energy content/fruit, and quantities of Na, Ca, Mg, and other traits in forests in Illinois and Washington, U. S. A. (Johnson et al. 1985; Piper 1986).

traits come from temperate rather than tropical habitats. The much more diverse nature of tropical floras and faunas, which might promote more "diffuse" kinds of interactions, and the lack of systematic surveys of tropical floras probably lie behind this latitudinal disparity. In view of the endangered status of tropical forests throughout the world, it is imperative that studies of the degree of ecological interdependence and coevolution between tropical angiosperms and their vertebrate seed dispersers be conducted to help determine how "fragile" these systems actually are.

A variety of plant traits appear to be unassociated with fruit–frugivore coevolution (Table 6-9B). In the most-detailed plant survey to date, Herrera (1987) attributed only 3 of 15 patterns (fruit size, seed size and number, and the occurrence of bicolored fruit displays) to selection by avian frugivores in Mediterranean scrublands. Five patterns were of the "null" type, five were attributed to the influence of growth form and/or specific resource and nutrient allocation patterns, and two patterns were of "unknown" origin. From the results of his survey, Herrera (1987, 322) concluded "that historical and phylogenetic constraints on plant distribution and morphology; weak selective pressures from dispersers; generally slower evolutionary rates of plants relative to animals; differential duration of plant and animal taxa; influence of nonmutualistic organisms; spatio-temporally inconsistent selective pressures on plants; and incongruency between the habitat ranges of animal and plant species, all combine to greatly decrease the adaptedness of plant fruiting traits to their current disperser/dispersal environments...."

Compared with other kinds of mutualisms (e.g., certain flowers and their pollinators, ants/acacias, and lichens), the fruiting plant–frugivore interaction is "loose" or "sloppy" (Janzen 1983a). Yet it is pervasive in the tropics and certain other regions, presumably owing to the efficacy, albeit in a highly variable fashion, with which vertebrates disperse the seeds of higher plants. Historically, this mutualism has played a major role in the diversification of angiosperms and certain groups of birds and mammals, including primates (Martin 1984). At present, this interaction still plays an extremely important role in the economy of plants and animals in a variety of habitats. Despite its rather coarse nature, the fruit–frugivore mutualism has played a major role in the evolution of our biosphere in accord with the Biblical injunction, "Be fruitful and multiply." To judge from the title of his autobiography (*The Kindly Fruits of the Earth*), G. E. Hutchinson would approve of this strategy.

ACKNOWLEDGMENTS

I wrote this paper while supported by a Fulbright Fellowship at James Cook University, Queensland, Australia. I thank Professor Rhondda Jones, Head of the Department of Zoology, for arranging my visit and for making facilities and amenities available to me at James Cook. The U. S. National Science Foundation has supported my research on fruits and frugivores in Central America. I thank C. Herrera, H. Howe, and G. Moore for critically reviewing a draft of this paper.

REFERENCES

American Ornithologists Union. 1983. *Check-list of North American birds,* 6th ed. American Ornithologists Union, Washington, D. C.

Appanah, S. 1985. General flowering in the climax rain forests of South-east Asia. *Journal of Tropical Ecology* 1: 225–240.

Baker, H. G. 1972. Seed weight in relation to environmental conditions in California. *Ecology* 53: 997–1010.

Baker, H. G., K. S. Bawa, G. W. Frankie, and P. A. Opler. 1983. Reproductive biology of plants in tropical forests. Pp. 183–215 in F. G. Golley (Ed.) *Tropical rain forest ecosystems.* Elsevier, Amsterdam.

Bronstein, J. L. and K. Hoffmann. 1987. Spatial and temporal variation in frugivory at a Neotropical fig, *Ficus pertusa. Oikos* 49: 261–268.

Chivers, D. J. (Ed.). 1980. *Malayan forest primates.* Plenum Press, New York.

Crepet, W. L. 1984. Advanced (constant) insect pollination mechanisms: Pattern of evolution and implications vis-a-vis angiosperm diversity. *Annals of the Missouri Botanical Garden* 71: 607–630.

Debussche, M., J. Cortez, and I. Rimbault. 1987. Variation in fleshy fruit composition in the Mediterranean region: the importance of ripening season, life-form, fruit type and geographical distribution. *Oikos* 49: 244–252.

Doyle, J. A. 1977. Patterns of evolution in early angiosperms. Pp. 501–546 in A. Hallam (Ed.) *Patterns of evolution*. Elsevier Scientific Publication Company, Amsterdam.

Estrada, A. and T. H. Fleming (Eds.). 1986. *Frugivores and seed dispersal*. W. Junk Publishers, Dordrecht, Netherlands.

Feduccia, A. 1980. *The age of birds*. Harvard University Press, Cambridge, MA.

Fleming, T. H. 1988. *The short-tailed fruit bat. A study in plant–animal interactions*. University of Chicago Press, Chicago, IL.

Foster, R. B. 1982. The seasonal rhythm of fruitfall on Barro Colorado Island. Pp. 151–172 in E. G. Leigh, Jr., A. S. Rand, and D. M. Windsor (Eds.) *The ecology of a tropical forest*. Smithsonian Institution Press, Washington, D. C.

Foster, S. A. 1986. On the adaptive value of large seeds for tropical moist forest trees. A review and synthesis. *Botanical Review* **52**: 260–299.

Foster, S. A. and C. H. Janson. 1985. The relationship between seed size, gap dependence, and successional status of tropical rainforest woody species. *Ecology* **66**: 773–780.

Gardner, A. L. 1977. Feeding habits. Pp. 293–350 in R. J. Baker, J. K. Jones, Jr., and D. C. Carter (Eds.). *Bats of the New World family Phyllostomatidae*, Part 2. Special Publication of the Museum, Texas Tech University, No. 13.

Gautier-Hion, A., J. -M. Duplantier, L. Emmons, F. Feer, P. Heckestweiler, A. Moungazi, R. Quiris, and C. Sourd. 1985a. Coadaptation entre rythmes de fructification et frugivorie en foret tropicale humide du Gabon: mythe ou realite. *Revue Ecologie* **40**: 405–434.

Gautier-Hion, A., J.-M. Duplantier, R. Quiris, F. Feer, C. Sourd, J.-P. Decoux, G. Subost, L. Emmons, C. Erard, P. Heckestweiler, A. Moungazi, C. Roussilhon, and J.-M. Thiollay. 1985b. Fruit characters as a basis of fruit choice and seed dispersal in a tropical forest vertebrate community. *Oecologia* **65**: 324–337.

Gentry, A. H. 1982a. Neotropical floristic diversity: Phytogeographical connections between Central and South America, Pleistocene climatic fluctuations, or an accident of the Andean orogeny? *Annals of the Missouri Botanical Garden* **69**: 557–593.

Gentry, A. H. 1982b. Patterns of Neotropical plant species diversity. Pp. 1–84 in M. K. Hecht, B. Wallace, and G. T. Prance (Eds.) *Evolutionary Biology*, volume 15. Plenum Press, New York.

Gentry, A. H. and L. H. Emmons. 1987. Geographical variation in fertility, phenology, and composition of the understory of Neotropical forests. *Biotropica* **19**: 216–227.

Herrera, C. M. 1981. Are tropical fruits more rewarding to dispersers than temperate ones? *American Naturalist* **118**: 896–907.

Herrera, C. M. 1982. Seasonal variations in the quality of fruits and diffuse coevolution between plants and avian dispersers. *Ecology* **63**: 773–785.

Herrera, C. M. 1984a. A study of avian frugivores, bird-dispersed plants, and their interactions in Mediterranean scrublands. *Ecological Monographs* **54**: 1–23.

Herrera, C. M. 1984b. Adaptation to frugivory of mediterranean avian seed dispersers. *Ecology* **65**: 609–617.

Herrera, C. M. 1985. Determinants of plant-animal coevolution: the case of mutualistic dispersal of seeds by vertebrates. *Oikos* **44**: 132–141.

Herrera, C. M. 1986. Vertebrate-dispersed plants: Why they don't behave the way they should? Pp. 5–18 in A. Estrada and T. H. Fleming (Eds.). *Frugivores and seed dispersal*. W. Junk Publishers, Dordrecht, Netherlands.

Herrera, C. M. 1987. Vertebrate-dispersed plants of the Iberian Peninsula: A study of fruit characteristics. *Ecological Monographs* **57**: 305–331.

Herrera, C. M. 1989. Seed dispersal by animals: A role in angiosperm diversification? *American Naturalist* **133**: 309–322.

Heywood, V. H. 1978. *Flowering plants of the world*. Oxford University Press, Oxford.

Honeycutt, R. L. and V. M. Sarich. 1987. Albumin evolution and subfamilial relationships among New World leaf-nosed bats (family Phyllostomidae). *Journal of Mammalogy* **68**: 508–517.

Houde, P. 1987. Critical evaluation of DNA hybridization studies in avian systematics. *Auk* **104**: 17–32.

Howe, H. F. 1980. Monkey dispersal and waste of a Neotropical fruit. *Ecology* **61**: 944–959.

Howe, H. F. 1984. Constraints on the evolution of mutualisms. *American Naturalist* **123**: 764–777.

Howe, H. F. and G. F. Estabrook. 1977. On intraspecific competition for avian dispersers in tropical trees. *American Naturalist* **111**: 817–832.

Howe, H. F. and J. Smallwood. 1982. Ecology of seed dispersal. *Annual Review of Ecology and Systematics* **13**: 201–228.

Howe, H. F. and G. A. Vande Kerckhove. 1981. Removal of wild nutmeg (*Virola surinamensis*) crops by birds. *Ecology* **62**: 1093–1106.

Howe, H. F. and L. C. Westley. 1986. Ecology of pollination and seed dispersal. Pp. 185–215 in M. J. Crawley (Ed.). *Plant ecology*. Blackwell Scientific, Oxford.

Howe, H. F. and L. C. Westley. 1988. *Ecological relationships of plants and animals*. Oxford University Press, New York.

Janson, C. H. 1983. Adaptation of fruit morphology to dispersal agents in a Neotropical forest. *Science* **219**: 187–189.

Janzen D. H. 1983a. Seed and pollen dispersal by animals: convergence in the ecology of contamination and sloppy harvest. *Biological Journal of the Linnean Society* **20**: 103–113.

Janzen, D. H. 1983b. Dispersal of seeds by vertebrate guts. Pp. 232–262 in D. J. Futuyma and M. Slatkin (Eds.) *Coevolution*. Sinauer Associates, Sunderland, MA.

Johnson, R. A., M. F. Willson, J. N. Thompson, and R. I. Bertin. 1985. Nutritional values of wild fruits and consumption by migrant frugivorous birds. *Ecology* **66**: 819–827.

Jordano, P. 1987a. Patterns of mutualistic interactions in pollination and seed dispersal: Connectance, dependence asymmetries, and coevolution. *American Naturalist* **129**: 657–677.

Jordano, P. 1987b. Avian fruit removal: Effects of fruit variation, crop size, and insect damage. *Ecology* **68**: 1711–1723.

Kochmer, J. P. and S. N. Handel. 1986. Constraints and competition in the evolution of flowering phenology. *Ecological Monographs* **56**: 303–325.

Levey, D. J. 1987. Seed size and fruit-handling techniques of avian frugivores. *American Naturalist* **129**: 471–485.

Marshall, A. G. 1983. Bats, flowers, and fruit: Evolutionary relationships in the Old World. *Biological Journal of the Linnean Society, London* **20**: 115–135.

Marshall, A. G. 1985. Old World phytophagous bats (Megachiroptera) and their food plants: A survey. *Zoological Journal of the Linnean Society* **83**: 351–369.

Martin, R. D. 1984. Body size, brain size and feeding strategies. Pp. 73–103 in D. J. Chivers, B. A. Wood, and A. Bilsborough (Eds.) *Food acquisition and processing in primates*. Plenum Press, New York.

Martin, T. E. 1985. Resource selection by tropical frugivorous birds: Integrating multiple interactions. *Oecologia* **66**: 563–573.

Mather, P. M. 1976. *Computational methods of multivariate analysis in physical geography*. John Wiley and Sons, New York.

McKey, D. 1975. The ecology of coevolved seed dispersal systems. Pp. 159–191 in L. E. Gilbert and P. H. Raven (Eds.) *Coevolution of animals and plants*. University of Texas Press, Austin, Texas.

Moermond, T. C., and J. S. Denslow. 1985. Neotropical avian frugivores: Patterns of behavior, morphology, and nutrition, with consequences for fruit selection. Pp. 865–897 in P. A. Buckley, M. S. Foster, E. S. Morton, R. S. Ridgeley, and F. G. Buckley (Eds.) *Neotropical ornithology*. American Ornithological Union Monograph No. 36.

Muller, J. 1981. Fossil pollen records of extant angiosperms. *Botanical Review* **47**: 1–142.

Murray, D. R. (Ed.) 1986. *Seed dispersal*. Academic Press, Sydney.

Niklas, K. J., B. H. Tiffney, and A. H. Knoll. 1985. Patterns in vascular land plant diversification: An analysis at the species level. Pp. 97–128 in J. W. Valentine (Ed.) *Phanerozoic diversity patterns*. Princeton University Press, Princeton, New Jersey.

Piper, J. K. 1986. Seasonality of fruit characters and seed removal by birds. *Oikos* **46**: 303–310.

Pratt, T. K. and E. W. Stiles. 1983. How long fruit-eating birds stay in the plants where they feed: Implications for seed dispersal. *American Naturalist* **122**: 797–805.

Raven, P. H. and D. I. Axelrod. 1974. Angiosperm biogeography and past continental movements. *Annals of the Missouri Botanical Garden* **61**: 539–673.

Regal, P. J. 1977. Ecology and evolution of flowering plant dominance. *Science* **196**: 622–629.

Salisbury, E. J. 1942. *The reproductive capacity of plants*. Bell, London.

Skeate, S. T. 1987. Interactions between birds and fruits in a northern Florida hammock community. *Ecology* **68**: 297–309.

Smith, J. D. and A. Starrett. 1979. Morphometric analysis of chiropteran wings. Pp. 229–316 in R. J. Baker, J. K. Jones, Jr., and D. C. Carter (Eds.) *Biology of bats of the New World family Phyllostomatidae*, part 3. Special Publication of the Museum, Texas Tech University No. 16.

Snow, D. W. 1971. Evolutionary aspects of fruit-eating by birds. *Ibis* **113**: 194–202.

Snow, D. W. 1981. Tropical frugivorous birds and their food plants: A world survey. *Biotropica* **13**: 1–14.

Stapanian, M. A. 1982. Evolution of fruiting strategies among fleshy-fruited plant species of eastern Kansas. *Ecology* **63**: 1422–1431.

Stebbins, G. L. 1974. *Flowering plants. Evolution above the species level*. Belknap Press, Cambridge, MA.

Stebbins, G. L. 1981. Why are there so many species of flowering plants? *Bioscience* **31**: 573–577.

Stiles, E. W. 1980. Patterns of fruit presentation and seed dispersal in bird-disseminated woody plants in the eastern deciduous forest. *American Naturalist* **116**: 670–688.

Swanepoel, P. and H. H. Genoways. 1979. Morphometrics. Pp. 13–106 in R. J. Baker, J. K. Jones, Jr., and D. C. Carter (Eds.) *Biology of bats of the New World family Phyllostomatidae*, part 3. Special Publication of the Museum, Texas Tech University no. 16.

Terborgh, J. 1983. *Five New World primates*. Princeton University Press, Princeton, NJ.

Terborgh, J. 1986. Community aspects of frugivory in tropical forests. Pp. 371–384 in A. Estrada and T. H. Fleming (Eds.) *Frugivores and seed dispersal*. W. Junk Publishers, Dordrecht, Netherlands.

Thompson, J. N. and M. F. Willson. 1979. Evolution of temperate fruit/bird interactions: Phenological strategies. *Evolution* **33**: 973–982.

Tiffney, B. H. 1984. Seed size, dispersal syndromes and the rise of the angiosperms: Evidence and hypothesis. *Annals of the Missouri Botanical Garden* **71**: 551–576.

Tiffney, B. H. 1986a. Evolution of seed dispersal syndromes according to the fossil record. Pp. 273–305 in D. R. Murray (Ed.) *Seed dispersal*. Academic Press, Sydney.

Tiffney, B. H. 1986b. Fruit and seed dispersal and the evolution of the Hamamelidae. *Annals of the Missouri Botanical Garden* **73**: 394–416.

Traylor, M. A., Jr. and J. W. Fitzpatrick. 1982. A survey of the tyrant flycatchers. *Living Bird* **19**: 7–50.

Van der Pijl, L. 1982. *Principles of dispersal in higher plants*, third edition. Springer-Verlag, Berlin.

Van Tyne, J. and A. J. Berger. 1975. *Fundamentals of ornithology*, second edition. John Wiley and Sons, New York.

Walker, E. P. 1975. *Mammals of the world*, third edition. Johns Hopkins University Press, Baltimore.

Wheelwright, N. T. 1985. Fruit size, gape width, and the diets of fruit-eating birds. *Ecology* **66**: 808–818.

Wheelwright, N. T., W. A. Haber, K. G. Murray, and C. Guindon. 1984. Tropical fruit-eating birds and their food plants: A survey of a Costa Rican lower montane forest. *Biotropica* **16**: 173–192.

Wheelwright, N. T. and C. H. Janson. 1985. Colors of fruit displays of bird-dispersed plants in two tropical forests. *American Naturalist* **126**: 777–799.

Wheelwright, N. T. and G. H. Orians. 1982. Seed dispersal by animals: Contrasts with pollen dispersal, problems of terminology, and constraints on coevolution. *American Naturalist* **119**: 402–413.

Wyles, J. S., J. G. Kunkel, and A. C. Wilson. 1983. Birds, behavior, and anatomical evolution. *Proceedings of the National Academy of Sciences, U.S.A.* **80**: 4394–4397.

APPENDIX 6-1. Families of Fruit-Eating Birds and Mammals

A. Birds

Tinamiformes	Tinamidae
Casuariformes	Casuariidae*
	Dromiceiidae*
Galliformes	Cracidae*
	Phasianidae
Columbiformes	Pteroclidae
	Columbidae
Psittaciformes	Psittacidae
Musophagiformes	Musophagidae
Caprimulgiformes	Steatornithidae*
Coliiformes	Coliidae*
Trogoniformes	Trogonidae*
Coraciiformes	Phoeniculidae
	Bucerotidae*
Piciformes	Ramphastidae*
	Capitonidae
	Picidae
	Jyngidae
Passeriformes	Eurylaimidae
	Cotingidae*
	Pipridae*
	Phytotomidae*
	Tyrannidae
	Oxyruncidae*
	Philepittidae*
	Campephagidae
	Ptilonorhynchidae*
	Paradisaeidae*
	Oriolidae
	Pycnonotidae
	Irenidae*
	Mimidae
	Turdidae
	Zosteropidae
	Bombycillidae*
	Ptilogonatidae*
	Dulidae*
	Callaeidae
	Sturnidae
	Dicaeidae
	Cyclarhidae
	Vireolaniidae
	Vireonidae
	Parulidae
	Tersinidae*
	Thraupidae*

(*continued*)

APPENDIX 6-1. (*Continued*)

	B. Mammals
Marsupialia	Phalangeridae
Dermoptera	Cynocephalidae
Chiroptera	Pteropodidae*
	Phyllostomidae
	Mystacinidae
Primates	Tupaiidae
	Lemuridae
	Indriidae
	Lorisidae
	Cebidae
	Callithricidae
	Cercopithecidae
	Pongidae
Rodentia	Anomaluridae
	Myoxidae
	Platacanthomyidae
	Zapodidae
	Dinomyidae
	Echimyidae
Proboscidae	Elephantidae
Artiodactyla	Tragulidae

[a]Highly frugivorous families are indicated by an asterisk.

7 Variable Composition of Some Tropical Ant–Plant Symbioses

DIANE W. DAVIDSON, ROBIN B. FOSTER, ROY R. SNELLING, and PEDRO W. LOZADA

Thousands of plant species produce food and/or housing to attract ants that protect their foliage, provide nutrient supplements, or disperse their seeds. Opportunistic and casual associations comprise the majority of these ant–plant mutualisms and are common in both temperate and tropical environments (Beattie 1985; McKey 1988). Facultative associations appear not to have elicited significant evolutionary change on the part of ants, nor species-specifity in the relationships of ants or plants (Beattie 1985). In contrast, symbiotic ant–plant mutualisms are largely or wholly tropical in distribution and can exhibit both evolutionary specialization and species-specificity (e.g., Janzen 1966; Schemske 1983; Davidson et al. 1989). More thorough studies of symbiotic ant-plants should then contribute much to our understanding of both the requisites and the geographic distribution of evolutionary specialization in ant–plant relationships.

Despite a relatively low diversity of mutualistic partners, few myrmecophytes from symbiotic mutualisms have monophilic associations with ants. Early tropical naturalists (e.g., Bequaert 1922 and Wheeler 1942) recorded variability in the species composition of most of these relationships, but did not elaborate on its causes. Oligophily, not polyphily, is the rule, with small guilds of often unrelated ant species comprising the majority of associates within particular myrmecophyte taxa (Benson 1985; Huxley 1986; Davidson et al. 1989). Relative abundances of ants in these guilds often vary with habitat and host species.

Monophily and oligophily are not necessarily a consequence of evolutionary specialization on the part of ants or their hosts. Alternatively, as in other mutualisms (Jordano 1987), they may be the outcome of ecological "species sorting," whereby biotic and abiotic environmental factors produce specific and repeatable patterns of association in the absence of evolved specificity. Ecological

Plant-Animal Interactions: Evolutionary Ecology in Tropical and Temperate Regions, Edited by Peter W. Price, Thomas M. Lewinsohn, G. Wilson Fernandes, and Woodruff W. Benson. ISBN 0-471-50937-X © 1991 John Wiley & Sons, Inc.

species sorting is a particularly attractive hypothesis for habitat dependency in ant–plant associations, because variable species composition, by definition, is influenced by the environment. Biotic determinants of species composition may include competition among ants for host plants. Although frequently colonized by more than one foundress, most hosts eventually house only a single ant colony (Davidson et al. 1989). Circumstantial evidence suggests that biotic selection pressures resulting from competition and predation also may have been an important stimulus for evolutionary specialization by ants. Thus, ancestors of many plant ants may have been behaviorally subordinate and/or competitively inferior species that sought out preferred nesting and foraging environments on plants whose growth form or morphology protected them from competing and/or predatory ants (Benson 1985; Davidson et al. 1988 and 1989).

Preliminary studies of several ant–plant symbioses reveal pattern in the organization of relationships between associated ant and plant guilds. Guilds of plant ants are comprised of both generalists and specialists. Ant species with relatively broad tolerances occur across a range of habitats and/or species in the host plant guild, but are most typical of relatively slow-growing hosts. In contrast, narrow specialists are usually confined to habitats and hosts character-ized by rapid plant growth. Thus, niches of specialists are *included* within the niches of generalists, and specialists are confined to favourable resource environments. As in many other natural communities, the included niche pattern may be indicative of specialists outcompeting generalists under favorable conditions, and of generalists being maintained principally by their tolerances for poor conditions (e.g., Colwell and Fuentes 1975).

Below, we outline particular mechanisms of ecological species sorting that may determine the included niche pattern within guilds of plant ants. By presenting this verbal model and some circumstantial evidence, we hope to stimulate explicit experimental tests of our ideas in a range of ant–plant systems.

THE MODEL

Myrmecophytic host plants produce ant resources (housing and food) at rates commensurate with plant growth rates, as determined by developmental and evolutionary responses to available light and nutrients. Associations between particular ant and plant species depend on a match between rates at which host plants produce ant resources and the resource demands of growing ant colonies. Colonies with rapid development and/or costly competitive strategies should be superior competitors for host plants, but may not survive on plants with low rates of resource supply. Conversely, species with slow colony development and poor competitive ability may often occupy slow-growing hosts, but regularly be displaced on fast-growing plants by superior competitors, whose higher energy demands are subsidized there by readily available resources. Niches of the latter species are, thus, included in those of ants with lower energy demands.

The relationship of plant growth rate to ant activity and aggression is more difficult to predict. If high levels of worker activity and aggression are energetically expensive, and if they enhance competitive ability, these traits may be most typical of ants living regularly on fast-growing hosts. However, host-plant ownership often is resolved early in colony establishment, and competitive ability may depend more on colony development rates than on worker activity and aggression. In ant protection mutualisms, worker activity and aggression may be most important in determining the capacities of ants to defend their hosts from herbivory. If herbivory is intense, and ant species differ greatly in the protection afforded their hosts, timid ants that are poor defenders could be restricted to plants in favorable resource environments. There, rapid plant growth may partially compensate for losses to herbivores and allow positive net growth. Whether ants are restricted to fast-growing hosts by their relatively high energy demands or by ineffectual host-plant defense, species specializing on fast-growing plants must be superior colonists and/or competitors for these preferred hosts, in order to be maintained in the guild of plant ants.

Other factors undoubtedly interact with the relative growth rates of plants and ants to influence species associations in ant–plant mutualisms. The internal resources of founding queens may affect their dependence on host-plant resources during the critical stage of colony establishment. Second, at least some tropical myrmecophytes have poorly-developed physical and chemical defenses against herbivory (Janzen 1966; Rehr et al. 1973; Siegler and Ebinger 1987, but see McKey 1984), and seedling ant-plants may be acutely susceptible to depredation by herbivores before they acquire their ants (McKey 1988). Those growing slowly in unfavorable environments may incur greater costs from a fixed rate of herbivory than do fast-growing species in more favorable environments (e.g., Janzen 1974; Coley et al. 1985). As a result, these plants may invest in ant defenses earlier in their development, and may host ant colonies even as comparatively small seedlings with particularly low resource production rates. The predominant associates of such hosts should be ant species with relatively low resource demands and (where herbivory is intense) substantial capacity for antiherbivore defense.

This interpretation of the included niche pattern predicts that host-plant generalists should exhibit increased densities when specialists are removed, but that specialists should not respond to removal of generalists. In the absence of experimental evidence, patterns in species distributions could provide circumstantial support for this hypothesis. Richness and diversity of ant species should be greatest on rapidly growing hosts, where occasional priority of colonization allows even poor colonists and competitors to dominate now and then. However, if competition limits establishment on these favorable hosts, inferior competitors should dominate fast-growing myrmecophytes significantly less often than they colonize these hosts. For these ants, slow-growing myrmecophytes should represent a relatively exclusive resource. Ants characteristic of fast-growing hosts should rarely or never colonize hosts whose intrinsically slow growth rates or unfavorable habitats strongly limit rates of resource supply and compensation for herbivory.

Myrmecophytic *Cecropia*

Ant Associates. Circumstantial support for this model comes from preliminary studies of myrmecophytic *Cecropia* (Cecropiaceae, Berg 1978a), a widespread and important group of early successional and forest light gap trees in Neotropical forests. The genus contains > 100 species, most of which are evolutionarily specialized to house and feed ants (Berg 1978a, b and mss.). Distribution of both myrmecophytic *Cecropia* and their typical ant partners (genus *Azteca*) are strongly habitat-correlated. Ant residents of particular *Cecropia* are thought to depend more on the host tree's habitat than on the tree species itself. Although poorly understood at present, habitat specificity of both plants and *Azteca* ants is repeated throughout the range of *Cecropia* in South and Central America (Harada 1982; Benson 1985; J. Longino 1989; Davidson and Fisher in press). It suggests that coevolution between *Cecropia* and ants has been "diffuse" rather than pairwise and species-specific (*sensu* Janzen 1980; Futuyma and Slatkin 1983).

Myrmecophytic *Cecropia* have long been thought to form associations only with *Azteca* ants. However, specialized ants in other subfamilies are often the most frequent inhabitants of certain shade-tolerant *Cecropia* in western Amazonia (Davidson and Fisher in press). A host species provisionally named *Cecropia tessmannii* may have evolved morphological and developmental traits in response to one of these partners, *Pachychondyla luteola* (Ponerinae). In contrast to the *Cecropia–Azteca* associations, this relationship suggests the possibility of pairwise coevolution.

Mutual Benefits

Superficially, all myrmecophytic *Cecropia* provide similar benefits to symbiotic ants (e.g., Berg 1978a, b; Fig. 7.1). At a species-specific point in sapling development, hollow stems expand and internodes produced thereafter develop with prostomata, or small, soft areas of unvascularized tissue where ant queens cut entrances to internodes. In most species, opened prostomata seal over rapidly, and colony development proceeds in a relatively safe and secluded environment, but without access to external resources. Eclosed workers reopen prostomata and forage for glycogen-rich Müllerian food bodies produced at the bases of petioles on hairy platforms termed trichilia (Rickson 1971, 1973, 1976). Variable across *Cecropia* species are additional food types in the form of spongy parenchyma lining internal stem walls (Bequaert 1922; Benson 1985), lipid-rich pearl bodies on abaxial leaf surfaces (Rickson 1976), and exudates of coccids tended within hollow stems (Bequaert 1922; Janzen 1969).

Ants specialized to *Cecropia* benefit their hosts in at least two ways. Several *Azteca* species attack and kill vines that climb and shade saplings in early successional environments (Janzen 1969; Schupp 1986; Davidson unpublished). *Pachychondyla luteola* prunes only when vines have been occupied recently by enemy ants (Davidson et al. 1988; Davidson mss.). This behavior of attacking and

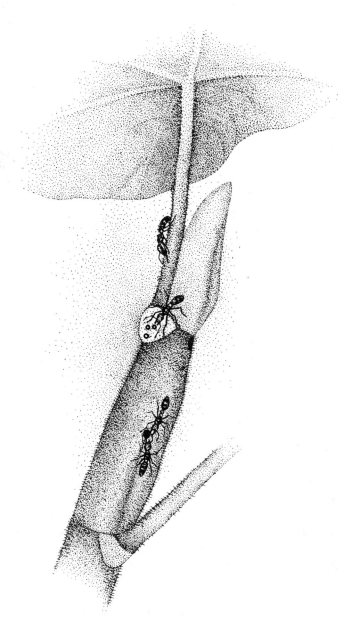

Figure 7-1. Myrmecophytic *C. tessmannii*, showing swollen stems with an open prostoma (entry site), and petiolar bases with white trichilia (sites of Müllerian body production). Ants are workers of *Pachychondyla luteola*. Figure by Ali Partridge based on a photograph by Diane Davidson.

killing vines may have evolved generally as a means of controlling entrance to the host plant by behaviorally dominant, competing and/or predatory ants (David-son et al. 1988). Its importance for both the ant's and the host's well-being may vary with habitat and geography (Janzen 1969 versus Putz 1984). Second, *Azteca* species (Janzen 1969; Schupp 1986) and *P. luteola* (Davidson mss.). attack and repel insect pests of their hosts. At least some *Azteca* are effective in reducing rates of herbivory on these plants (Schupp 1986). Lower rates of herbivory can lead to faster plant growth rates (Coley 1983 for *C. insignis*), and symbiotic ants may also enhance their host's survival and reproductive potential.

Much of the literature on *Cecropia*–ant associations has either neglected the richness of both plant and ant species, or conveyed the impression of uniformity in the benefits that hosts and ants afford one another. Yet these associations are diverse in their species composition, and interspecific differences in both plants and ants suggest that uniformity is unlikely. To illustrate this point, we describe our studies of myrmecophytic *Cecropia* in tropical moist forests (Holdridge 1967) of western Amazonia.

Cecropia in Southeastern Perú

Six myrmecophytic *Cecropia* species are common at Tambopata Nature Reserve and Cocha Biological Station in Madre de Dios, Perú. The species include three pairs of apparent relatives. Morphological distinctions are much more evident between than within species pairs, but additional evidence will be necessary to verify the proposed relationships. *Cecropia engleriana* and unidentified *Cecropia* sp. A may be especially close relatives. Approximately $< 2\%$ of offspring cultivated from seeds of individual parents exhibited the normally very distinctive seedling morphology of the alternative species. A range of intermediate morphologies also occurs within progeny of a single parent. Apparent intermedi-ates between *Cecropia polystachya* Trécul and *Cecropia ficifolia* are found occasionally in natural populations. *Cecropia membranacea* and *C. tessmannii* are unique in numerous traits that set them apart from the other four species. *Cecropia tessmannii* is very closely related to *C. membranacea* (C. C. Berg, personal communication), and we distinguished it from the latter species principally by its stilt roots and fruiting season (mature plants) and by the sizes of prostomata and Müllerian bodies (see below).

Within each pair, species grow on average in different light regimes (Table 7-1). *Cecropia membranacea, C. engleriana,* and *C. polystachya* are restricted to large disturbances along river and stream banks and in human settlements. Hypo-thesized relatives, respectively, *C. tessmannii, Cecropia* sp. A, and *C. ficifolia,* inhabit primary forest and may establish principally in small light gaps. Subtle interspecific differences in traits relevant to ants were habitat-correlated (Table 7-1). Not surprisingly, under controlled greenhouse conditions, shade-tolerant species grew more slowly and with shorter internodes on average than did their paired, light-demanding congeners (Table 7-1). (The exception is *Cecropia* sp. A, whose relatively long internodes are associated with uniquely narrow stems.)

TABLE 7-1. Attributes of Light-Demanding[D] and Shade-Tolerant[T] *Cecropia* **species in Greenhouse and Field Studies**

Cecropia Species	Greenhouse Studies				Field Studies	
	Final Height (cm)	Internode length (cm)	Node with First Trichilia	MB wt. (mg)	Colonization of plants (%)	
					≤ 1 m	≤ 20 IN's
membranacea[D]	255(19)[a]	20.8(19)[b]	16(17)[c]	0.09(10)[d]	17.4(23)[e]	15.6(32)[f]
tessmannii[T]	103(31)[a]	11.0(31)[b]	6(25)[c]	0.16(10)[d]	88.2(17)[e]	90.0(10)[f]
engleriana[D]	350(30)[g]	21.3(30)[h]	20(31)[i]	0.05(10)[j]	42.9(14)	43.8(16)[k]
sp. A[T]	260(29)[g]	16.8(29)[h]	19(31)[i]	0.04(10)[j]	33.3(15)	0.0(10)[k]
polystachya[D]	310(21)[l]	21.4(21)	19(16)[m]	0.06(10)[n]	23.8(21)[o]	29.2(24)
ficifolia[T]	250(17)[l]	23.0(17)	11(17)[m]	0.05(10)[n]	75.0(8)[o]	60.0(5)

Note: Table entries are median values with sample sizes in parentheses. MB = Müllerian body (dry weights of individual bodies, weighed in lots). IN = total internodes. Measurements with identical superscripts differ significantly in Mann-Whitney U Tests (greenhouse data) or Fisher Exact Tests (field data) at two-tailed P values: ≤ 0.00005 (a,b,c,e,f,m); ≤ 0.002 (d,g,h,n); ≤ 0.05 (k,l,o); i = a marginally significant difference at $P = 0.06$. All other comparisons within species pairs are not statistically significant.

Germination took place in a growth chamber with 12 hr of light (30°C) and 12 hr of darkness (24°C). Seeds were germinated in Petri dishes on filter paper moistened with distilled water. Seedlings germinating within a 2-wk period were transferred after 1 wk to sphagnum moss and then (individually, but on the same time schedule) to consecutively larger pot sizes. A standardized soil mix consisted of 2 parts low pH, low sodium, topsoil loam, 6 parts silica sand, 2 parts bark mulch, 2 parts bark chips, 4 parts peat moss, 3 parts vermiculite, and 4 parts perlite. Plants were grown under the natural light regime in Salt Lake City, Utah.

Space and seed limitations prevented us from growing the six focal *Cecropia* simultaneously, and growth parameters of *C. tessmannii* (January 1987–October 1987) were related to those of *C. membranacea* (July 1987–February 1987) indirectly by comparison with *C. engleriana*, which was cultivated during each of the two trials. Data for *C. membranacea* were biased toward smaller plant sizes and internode length by inclusion of several individuals unaccountably stunted early in the trials.

Shorter internodes provide fewer resources to claustral foundresses and their incipient colonies. In addition, the nutritional parenchyma of internal stem walls is highly developed in light-demanding species and barely discernible in shade-tolerant hosts.

Greenhouse studies also reveal that shade-tolerant species produce trichilia with Müllerian bodies at earlier leaf nodes and several weeks sooner than do their light-dependent counterparts (Table 7-1). Differences between *C. engleriana* and *Cecropia* sp. A, (possibly the species that have diverged most recently) are not significant. If interspecific differences in myrmecophytic traits persist when plants are grown in their natural habitats in the field, these differences would signify that shade-tolerant species invest earlier than light-demanding species in their partnerships with ants.

Consequently, rates of resource supply to incipient ant colonies are likely much lower in shade-tolerant than in light-demanding *Cecropia*. Not only are growth rates intrinsically slow, but plants are smaller at the time of colonization (Table 7-1), and growth rates are positively size-dependent (Davidson unpublished). *C. tessmannii* is the extreme example, with the slowest growth rate and the earliest trichilia development. Seedlings as small as 10 cm in height have often been colonized. The comparison of *C. engleriana* and *Cecropia* sp. A is again exceptional, but these species differed by only a single leaf node in the onset of trichilia production.

Cecropia-*ants of Southeastern Perú.* Four specialized *Cecropia*-ants inhabit the six plant species at Cocha Cashu and Tambopata. These ants are *Azteca alfari* and *Azteca xanthochroa* (Subfamily Dolichoderinae), *Camponotus balzani* (Formicinae), and *Pachychondyla luteola* (Ponerinae) (Davidson et al. 1989; Davidson mss.). Each ant species colonizes more than one *Cecropia* species, and all but *P. luteola* can establish successfully on more than one host (Table 7-2). Five of the six *Cecropia* are colonized by more than one ant species and can support established colonies of more than one species, though not simultaneously.

The four ant species share several attributes that evidence specialization to their *Cecropia* hosts: (1) Queens regularly found colonies on *Cecropia* after

TABLE 7-2. Frequencies of Four Specialized *Cecropia* Ants among Inhabitants of Six *Cecropia* spp. from Southeastern Perú, for Larger Plants with Established Colonies (and for Seedlings with Only Foundresses or Incipient Colonies)

Cecropia N(N)[a]	*Azteca alfari*	*Azteca xanthochroa*	*Camponotus balzani*	*Pachychondyla luteola*
Light-demanding species				
1. *membranacea* 53	49.1	20.8	30.2	0.0
(91)	(46.7)	(17.6)	(15.4)	(19.8)
2. *engleriana* 6	83.3	16.7	0.0	0.0
(41)	(9.8)	(75.6)	(14.6)	(0.0)
3. *polystachya* 7	57.1	28.6	14.3	0.0
(64)	(12.5)	(76.6)	(10.9)	(0.0)
Shade-tolerant species				
1. *tessmannii* 31	0.0	0.0	9.7	90.3
(49)	(0.0)	(2.0)	(12.2)	(85.7)
2. sp. A 17	0.0	100.0	0.0	0.0
(43)	(0.0)	(100.0)	(0.0)	(0.0)
3. *ficifolia* 24	0.0	47.8	52.2	0.0
(18)	(0.0)	(83.3)	(16.7)	(0.0)

[a]N = total number of trees sampled per *Cecropia* species (or, for seedlings, total number of foundresses or incipient colonies censused per *Cecropia* species). Along river banks, seasonally inundated lower trunks of *Azteca*-occupied *Cecropia* frequently housed queenless colony fragments of unspecialized ant species (genera *Camponotus*, *Solenopsis*, and *Pachychondyla*) during the dry season. These ants foraged principally off their hosts, were never seen on the host plant's foliage, and rejected Müllerian bodies when these were offered. They may compete little or not at all with the specialized *Cecropia*-ants.

locating prostomata of uncolonized internodes and cutting entrances to stems at these sites. Congeneric species nesting facultatively in *Cecropia* at our study areas do not locate their nest entrances at prostomata. (2) Ants do not leave their host plants for foraging or other activities. (3) Workers of each of these species collect and consume Müllerian bodies. For unknown reasons, these food bodies are not attractive to ants that are generalized foragers on plant surfaces, nor even to those that reside occasionally in *Cecropia* stems (e.g., *Pachychondyla unidentata* and *Camponotus crassus*). (4) Extensive searches of other ant plants at the study sites revealed that these ants did not live regularly on other host genera. (5) With the single exception of *Camponotus balzani*, the ants reject tuna fish and cheese baits that attract foraging by congeneric species not living regularly on *Cecropia*. (6) Again excepting *C. balzani*, the ants attack and prune vines and other vegetation impinging on their host plants.

Queens of the four *Cecropia*-ants differ in ways that may influence their dependence on plant-produced food rewards during colony foundation. Foundresses of *A. alfari* have a proportionately smaller thorax and apparently less flight muscle than those of *A. xanthochroa*. In many claustrally founding ant species, breakdown of muscle tissues used only in mating flights provides resources for production of the first worker broods. Few internal resources for colony development may make *A. alfari* more dependent on parenchymal resources provided within the host-plant stem. Consistent with this hypothesis is the observation that *A. alfari* eggs are markedly smaller than eggs of *A. xanthochroa*, even though worker sizes in the two species are similar. The smaller flight muscles of *A. alfari* might also restrict powers of flight and ability to colonize rare, isolated trees in the primary forest. Queens of *C. balzani* are similar to those of *Azteca* in founding colonies claustrally and, like *A. xanthochroa* females, have relatively well-developed flight muscles. Finally, *P. luteola* queens maintain open prostomata on colonized internodes, and forage regularly on Müllerian bodies during the period of colony establishment (Davidson mss.).

Specialized *Cecropia*-ants also differ in ways that may affect host-plant fitness, though present evidence is anecdotal. For example, although three of the four principal ants are active both diurnally and nocturnally, relatively secretive workers of *C. balzani* forage principally at night. Second, *A. xanthochroa* is more active and aggressive than **A**. *alfari*. Third, colonies of the two *Azteca* species have large numbers of small workers, while those of *Pachychondyla* and *Camponotus* consist of relatively small numbers of large workers. Worker–herbivore ratios are known to influence the protection that ants afford *Cecropia* and other ant-plants (Schupp 1986; Inouye and Taylor 1979). However, the efficacy of protection by a given worker population may vary with plant size, and even individual *Pachycondyla* queens may offer limited protection to tiny seedlings of *C. tessmannii*. Fourth, *C. balzani* is distinctive both in not pruning vegetation around its host, and in tending coccids within host-plant stems.

Finally, McKey (1988) argues that since ant defenses must be acquired, ant-plants are especially susceptible to herbivore damage when they are seedlings and have comparatively few resources. Beetles in the genus *Coelomera* (Chrysomelidae) exact a heavy toll on young *Cecropia* seedlings in southeastern Perú, and in other areas of the Neotropics (Andrade 1984; Schupp 1986; Jolivet 1987). Schupp's data verify the importance of ants as a deterrent to these herbivores. Species-specific rates of colonization and establishment by ants may then have a significant bearing on plant fitness. Growth of incipient colonies appears to be especially slow in *P. luteola*, possibly because the egg-laying rates of foundresses are proportional to the supply of Müllerian bodies (Davidson mss). Moreover, parasitoids attack brood through the open prostomata of stem nests. Colonies may contain as few as 2–4 workers a full year after colonization. Brood sizes in incipient colonies are greatest for the *Azteca* species (*A. alfari* > *A. xanthochroa*). Colony development in these species keeps pace with sapling growth that can exceed 2–3 m per year in *C. membranacea* and *C. polystachya*.

Correspondence to the Model

Ant–Ant Competition. Strong circumstantial evidence suggests that specialized *Cecropia* ants compete for host plants. Multiple stem internodes of *Cecropia* represent spatially partitioned modules permitting redundant colonization of individual plants by multiple foundresses of both specialized and unspecialized plant-nesters. At our study sites in the Manu National Park, approximately 70% of the foundresses or incipient colonies censused on small seedlings occurred on plants with additional ant queens or colonies. Yet trees with established ant colonies and workers foraging externally appeared to be inhabited by a single species of specialized *Cecropia*-ant. Workers of *Azteca alfari* and *A. xanthochroa* are safely distinguished only by the absence or presence (respectively) of microscopic hairs on the tibiae, and we cannot state categorically that trees hosting *Azteca* ants housed just one species. However, all samples collected from trees with established *Azteca* colonies were homogeneous for one or another of the two species when examined in the laboratory. Moreover, we never found established *Azteca* colonies coexisting with those of *Camponotus balzani* or *Pachychondyla luteola* on a single host, nor colonies of the latter two species with one another.

The hostility observed between specialized *Cecropia*-ants also suggests intraspecific and interspecific competition for hosts. Combat regularly ensued between callow workers of different colonies when plant stems were opened and incipient colonies came into contact. Workers from established colonies also attacked introduced queens (Davidson et al. 1989). Host-plant ownership appears to be determined at the seedling or sapling stage. When one of the many incipient colonies of a given host develops sufficiently to leave its private internode, its workers eliminate all other queens and colonies. Territoriality, a distinctive feature of plant ants (Levings and Traniello 1981), assures continued dominance thereafter by numerically superior established colonies.

Specialization on Fast-Growing Plants. The pattern of association between ants and plants strongly suggests that the species dominant on fast-growing hosts may regularly fail to colonize and establish on slow-growing hosts. *Azteca alfari* queens were found only on the three probably unrelated, light-dependent and fast-growing host species, which developed swollen stems and trichilia at relatively large plant sizes (Table 7.2). Present data do not permit us to distinguish whether host-plant restriction is due to habitat-specific searching, active rejection of slow-growing host species, or some other factor. *Azteca alfari* has the large colony sizes and aggressive workers typical of its genus (Carroll 1983), and may require relatively high rates of energy supply. Moreover, if its low activity level in relation to congener *A. xanthochroa* makes it less effective in antiherbivore defense, it may survive only where favorable plant resource environments allow rapid plant growth to compensate for losses to herbivores. Domination of fast-growing hosts by *A. alfari* could be favored by greater allocation to reproductive propagules and/or by the tendency to produce large incipient colonies, nutritionally dependent on host stem parenchyma.

Host-Plant Generalists. Two species appear to be generalists, but to establish more frequently on slow-growing, shade-tolerant species than on fast-growing hosts in more open environments (Table 7.2). *Camponotus balzani* colonizes all but *Cecropia* sp. A, whose uniquely thin stems and tiny internodes may exclude large foundresses of both this species and *Pachychondyla*. *Camponotus* succeeds in establishing on each of the colonized species except *C. engleriana*, for which our sample size is not adequate to characterize the spectrum of ant associates. *Azteca xanthochroa* colonizes and establishes on a similar range of hosts. On shade-tolerant hosts in primary forest, it accounts for less than half of established colonies on *C. ficifolia*, but all of those on *Cecropia* sp. A, which is colonized by no other ants.

Interestingly, the behavior of *A. xanthochroa* in colony founding may differ on light-demanding and shade-tolerant *Cecropia*. At Tambopata, sample sizes were sufficiently large for only four host × habitat combinations to characterize whether queens colonized internodes individually or in groups of multiple foundresses. Frequencies of single- and multiple-queen colonizations did not differ between host species in the same habitat (*C. engleriana* and *C. polystachya* in clearings, and *Cecropia* sp. A and *C. ficifolia* in primary forest). However, individual internodes of the two fast-growing species were more likely to contain multiple foundresses of *A. xanthochroa* than were those of the two slow-growing species (two-tailed Fisher Exact Probability, $P \sim 0.01$, $N = 28$ and 38 internodes with *A. xanthochroa* for the respective host categories). Colony foundation by multiple queens (pleometrosis) appears to be associated with higher survivorship and more rapid production of the first worker brood (Hölldobler and Wilson 1977). Pleometrosis in *A. xanthochroa* may be an effective competitive strategy on fast-growing hosts, but may not be supportable on slow-growing plants with relatively small internodes and little stem parenchyma. Experiments will be necessary to distinguish this possibility from the alternative hypothesis that densities of potential colonists are lower in primary forest.

Pachychondyla–An exception? An extreme host-plant specialist, *Pachychondyla luteola* establishes successfully only on *C. tessmannii*, the host with the slowest intrinsic growth rate and earliest development of myrmecophytic traits (Table 7-1). Superficially then, this species appears to contradict our prediction that ants with relatively low resource demands should be generalists. Closer inspection favors a different interpretation. *P. luteola* may be a recent addition to the guild of *Cecropia*-ants, and its body size and mode of colony establishment may have preadapted it to colonize only two closely related hosts, *Cecropia* sp. nov and *C. membranacea* (Table 7-2).

Closely related *C. tessmannii* and *C. membranacea* share at least three traits that may explain why colonization by *P. luteola* is restricted to these hosts. First, comparatively large prostomata (Fig. 7-2) can accomodate large queens of *Pachychondyla* (and *Camponotus*, body lengths $\sim$ 15 and 12 mm, respectively, in contrast to $\sim$ 7 mm for *Azteca* queens). Second, host-plant stems are covered with long, stiff bristles differing markedly from the short stem hairs of the other four *Cecropia* at our study sites. Stem hairs are thought to facilitate ant movements along plant surfaces and tend to be absent from non-myrmecophytes (Berg 1978 and personal communication; Benson 1985). They are longest on species that regularly house large ants [*Cecropia hispidissima* (vel aff) (Davidson mss.) and this study]. Dense concentrations of long stem hairs on prostomata also appear to prevent relatively small and potentially competing *Azteca* queens from cutting entrances.

Perhaps most importantly, *Pachychondyla* colonize only hosts with unusually large Müllerian bodies (Table 7-1, ≥ 0.09 mg). Unlike *Azteca* and *Camponotus* queens, which found their colonies claustrally, *P. luteola* appears to depend heavily on nutrition from Müllerian bodies during its relatively long period of colony establishment (Davidson mss.). Queens store these food bodies beneath leaf stipules until new internodes are available for colonization. Even after entering stem internodes with their stores of Müllerian bodies, foundresses continue to forage at trichilia through open prostomata. *Pachychondyla* queens consistently reject the small Müllerian bodies of *Cecropia* sp. A and *C. ficifolia* in experimental trials, although workers do collect them. Food bodies on hosts other than *C. tessmannii* and *C. membranacea* may be too small to permit economically profitable foraging by large-bodied queens. Hosts of other large

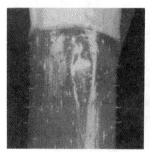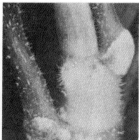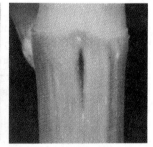

Figure 7-2. Prostoma of *C. tessmannii* (left, note long bristles) in comparison with those of *C. membranacea* (center) and *C. engleriana* (right). The form illustrated at the right is probably most typical for the genus.

Pachychondyla species apparently specialized to *Cecropia* also have distinctively large Müllerian bodies (e.g., *C. hispidissima* of Panama's Caribbean lowlands).

In censuses of both colonizing queens and established colonies, *P. luteola* was the most frequent associate of *C. tessmannii*, which has the slowest intrinsic growth rate of the six host species. In contrast, this ant is represented significantly less often among established colonies than among foundresses colonizing *C. membranacea* (two-tailed Fisher Exact Test, $P \sim 0.00015$), suggesting that it regularly fails to establish on this fast-growing host. Indeed, we have never found workers on *C. membranacea* (Table 7-2). On two occasions, decapitated *P. luteola* queens were discovered on *C. membranacea* seedlings with incipient *Azteca* colonies. Rapidly developing *Azteca* colonies may regularly kill queens of the ponerine ant before their first workers eclose.

Diversity of Symbiotic Associates. The pattern of diversity in ant associates is consistent with predictions of the included niche model. Discounting instances of very low sample size (e.g., established ant colonies on *C. engleriana* in Table 7-2), species richness tends to be greater on fast-growing, light-demanding hosts than on slow-growing, shade-tolerant plants, for both founding queens and established colonies. Shade-tolerant *Cecropia* sp. A and *C. tessmannii* have the narrowest ranges of mutualistic associates. Their typical symbionts colonize other hosts, but are less well represented among established colonies of these alternate hosts.

Other Myrmecophyte Genera

Other guilds of specialized plant ants also appear to have included niches with respect to plant growth rates. However, in contrast to our findings with *Cecropia*-ants, high levels of activity and aggression are usually characteristic of ants associated with fast-growing plants. The explanation for this difference is not yet clear. Relevant distinctions in the *Cecropia* system could include such factors as comparatively intense herbivore pressures, that may restrict the success of relatively timid or inactive ant species (poor plant defenders) on slow-growing plants. The following are examples of the included niche pattern in ants associated with other myrmecophyte taxa or guilds.

In lowland rain forests of Cameroon, myrmecophytic *Leonardoxa africana* (Caesalpiniaceae) can be occupied by either *Petalomyrmex phylax* (Formicinae) or *Cataulacus mckeyi* (Myrmicinae), which interact aggressively with one another (McKey 1984). *Cataulacus* does not protect host plants and is a parasite of the relationship between *L. africana* and *Petalomyrmex*, which reduces herbivory on young leaves with active foliar nectaries. The parasitic species may have superior colonizing ability. It invests more in älate reproductives early in the life history and is favored on shaded plants whose slow growth rates lead to frequent failure of ant colonies. Probably monogynous, colonies of *Cataulacus* are small in comparison to the polygynous colonies of *Petalomyrmex*, and worker densities are lower per swollen internode and on leaf surfaces. McKey's data suggest that the niche of *Petalomyrmex* is included in that of *Cautalacus*, and that *Petalomyrmex* is the superior competitor on relatively fast-growing hosts.

Across the Río Manu from Cocha Cashu, Perú, high terraces with relatively poor soils support myrmecophytic melastomes *Clidemia heterophylla* and *Maieta guianensis* (Davidson et al. 1989). These plants are colonized and sometimes dominated by each of two specialized plant-ants, *Pheidole minutula* (Myrmicinae) and *Crematogaster* cf. *victima*, though individual hosts do not appear to house more than one established colony. The latter ant species is timid and slow-moving, and is eliminated from its hosts by highly aggressive *Pheidole* soliders and workers when plants with different ant species are brought into contact. Restricted to the immediate banks of small streams, *Clidemia* may grow more consistently in favorable light and nutrient regimes than does *Maieta*, which occurs on steep hillsides as well as along streams. Although the relative frequencies of the two ants are indistinguishable on small and large plants of *Maieta*, *Pheidole* occurs significantly more frequently on large than on small individuals of *Clidemia*. The comparison suggests that *Pheidole* may be capable of replacing *Crematogaster* only on relatively fast-growing plants with high rates of resource supply that can support the more aggressive ant. On *Maieta*, *Pheidole* occurs more frequently in streamside plants than on hillside plants, which are occupied principally by *Crematogaster*.

Habitat-correlated distributions are also characteristic of the specialized residents of Costa Rican ant-acacias (Janzen 1983). Both *Pseudomyrmex nigrocinctus* (Janzen's *P. nigrocincta*) and *Pseudomyrmex spinicola* (misidentified by Janzen as *P. ferruginea*) occupy *Acacia collinsii* in moderately shaded dry forests, but only the latter species occurs on *Acacia allenii* in deeply shaded rain forests of the Osa Peninsula. Thus, *P. nigrocinctus*, the more aggressive species, has the narrower niche and may specialize on fast-growing plants. A third and more timid species, *P. flavicornis* (= *P. beltii*), is the typical resident of *A. collinsii* in very open and sunny pastures of Guanacaste. The association of this species with hosts in light-intense environments parallels the pattern for *Azteca* ants on Peruvian *Cecropia*. If the docility of *P. flavicornis* is associated with poor anti-herbivore defense, this species may survive only on plants growing rapidly with high insolation. Alternatively, plant growth rates and productivity of extrafloral nectaries may be water-limited and reduced in relatively open microhabitats within the dry forest. Unlike *P. nigrocinctus* and *P. spinicola*, *P. flavicornis* does not aggressively attack vegetation in the neighborhood of its host (Davidson, personal observation). If clearings around the host plant increase soil moisture by reducing transpiration by neighboring plants (a possibility suggested by J. Lighton, pers. comm.), plants occupied by *P. nigrocinctus* may have higher nectar production rates than those occupied by *P. flavicornis*. Relatively low nectar production rates in hosts of *P. flavicornis* are suggested by the very low worker activity at extrafloral nectaries of these plants.

Finally, the included niche pattern also shows up in the habitat-correlated distributions of ants associated with myrmecophytic epiphytes. In both Asia and South America, individuals and species of epiphytes growing in shaded environments tend to house monogynous ants, while those in open, sunny habitats are tenanted by polygynous species (Davidson 1988; Davidson and Epstein 1989). Relatively unusual in rainforest ants (Wilson 1959), polygyny is, in general, characteristic of species with rapid (Hölldobler and Wilson 1977) or explosive (Fletcher et al. 1980) colony growth and a rich, long-lived resource base. Davidson and Epstein (1989) tested an included niche model for ants of myrmecophytic epiphytes in the Hydnophytinae (Rubiaceae). They predicted that highly aggressive and polygynous *Iridomyrmex cordatus*, the typical occupant of epiphytes in open, sunny habitats (Huxley 1978; Jebb 1985), should be virtually absent from shade-tolerant host plants. In contrast, timid and monogynous *Iridomyrmex scrutator*, the usual associate of shade-tolerant epiphytes, might occasionally occur on light-demanding hosts, in the absence of *I. cordatus*. Data of Jebb (1985) tended to support these hypotheses. Although *I. scrutator* accounted for 11.5% of the ants sampled on three light-demanding *Myrmecodia* species, no ant sample from any of four shade-tolerant host plants (genera *Myrmecodia* and *Anthorhiza*) contained *I. cordatus* ($N = 52$ plants, for both light-demanding and shade-tolerant hosts).

Suggestions for Future Studies

Associations between myrmecophytes and their ants are fascinating for their natural history, but future investigation must probe beyond this superficial level if ant–plant symbioses are to contribute to our understanding of both plant defense systems and the evolution of specialization in mutualism. Ant-protection mutualisms can provide independent tests of certain unifying principles that have emerged from studies of other kinds of plant defenses against herbivores. For example, our demonstration of earlier development of myrmecophytic traits in slow-growing, shade-tolerant hosts parallels the greater investment in toughness and chemical defenses of ecologically similar but nonmyrmecophytic species (e.g., Janzen 1974; Coley et al. 1985). Moreover, we may eventually find that, when herbivory is intense, ants that are poor host-plant defenders are regularly restricted to fast-growing hosts in favorable environments. This pattern would support the suggestion that resource-subsidized rapid plant growth is an effective alternative to active defense against herbivores.

Experimental studies will be necessary to interpret patterns in the species composition of ant–plant symbioses. Correlational approaches cannot discriminate whether these patterns are the outcome of ecological species sorting or of evolutionary specialization. Rigid habitat restrictions of both plants and ants can give the appearance of species-specificity where none exists. The contributions of habitat specificity and species-specificity can be teased apart only by experimental investigations of colonization and establishment of ants on the same host species in different habitats. Manipulations can also indicate whether ecological factors contributing to species sorting include interactive processes such as ant–ant competition for hosts, parasitoid attacks on foundresses and incipient colonies, and interference by ants with host-plant herbivores. Independent experiments should assess not simply *whether* ant associates benefit their hosts, but the *relative* benefits conferred by different ant species. Finally, experiments could also determine the consequences of different levels of antiherbivore protection as a function of habitat-specific differences in the intensity of herbivory and in resources that determine plant growth.

Together, such experiements can supply the data necessary to fully understand the interdependencies among members of ant and plant guilds. Such data would be useful in answering a number of important questions. When is species-specificity the outcome of ecological species sorting, and when it is correlated with evolutionary specializations in morphology, behavior, and other attributes? Does species-specificity in these relationships tend to be symmetric? Is evolutionary specialization contingent on dependence symmetry? Does specialization on one partner occur at the expense of success in forming relationships with other partners? Does specificity promote cladogenesis (a possibility for *C. tessmannii*), and, if so, under what circumstances? Does extreme specialization occur principally in species (e.g., *Pachychondyla luteola*) whose success depends on a refuge from superior competitors and other natural enemies? Are ant-plant symbioses largely or wholly tropical because of more stringent and consistent biotic selection pressures in tropical environments? Attention to such questions will be necessary if the captivating natural history of ant–plant symbioses is ever to contribute to mainstream evolutionary ecology.

ACKNOWLEDGMENTS

Field work was supported by the National Geographic Society Committee for Research and Exploration, the National Science Foundation (RII-8310359), and the John Simon Guggenheim Foundation (all to the senior author). We are grateful to D. Samson for cultivating greenhouse plants and to P. Ward for straightening out inconsistencies in nomenclature of the *Pseudomyrmex* associated with *Acacia*. We also thank C. C. Berg for making provisional identifications of *Cecropia* species. Peru's Ministerio de Agricultura (Dirección General de Forestal y de Fauna) gave permission to work at Tambopata and in the pristine Manu National Park. We thank Director General Marco Romero and Ing. M. Falero for these permissions, and Dr. Gerardo Lamas for logistical assistance. T. M. Aide and P. D. Coley commented on the manuscript.

REFERENCES

Andrade, J. C. 1984. Observaçoes preliminares sobre a eco-etalogia de quatro Coleopteros (Chrysomelidae, Tenebrionidae, Curculionidae) que dependum da embaúba da recreio dos bandeirantes Rio de Janeiro. *Revista Brasileira de Entomologia* **28**: 99–108.

Beattie, A. J. 1985. *The evolutionary ecology of ant–plant mutualisms.* Cambridge University Press, Cambridge.

Benson, W. W. 1985. Amazon ant plants. Pp. 239–266 in G. T. Prance and T. E. Lovejoy (Eds.). *Amazonia.* Pergamon Press, Oxford.

Bequaert, J. 1922. Ants in their diverse relations to the plant world. *Bulletin of the American Museum of Natural History* **45**: 333–621.

Berg, C. C. 1978a. Cecropiaceae, a new family of the Urticales. *Taxon* **27**: 39–44.

Berg, C. C. 1978b. Espécies de *Cecropia* da Amazonia brasileira. *Acta Amazonica* **8**: 149–182.

Carroll, C. R. 1983. *Azteca* (Hormiga Azteca, Azteca Ants, Cecropia Ants). Pp. 691–693 in D. H. Janzen (Ed.) *Costa Rican Natural History*, University of Chicago Press, Chicago, IL.

Coley, P. D. 1983. Intraspecific variation in herbivory on two tropical tree species. *Ecology* 64: 426–433.

Coley, P. D., J. P. Bryant, and F. S. Chapin, III. 1985. Resource availability and plant antiherbivore defense. *Science* 230: 895–899.

Colwell, R. K. and E. R. Fuentes. 1975. Experimental studies of the niche. *Annual Review of Ecology and Systematics* 6: 281–310.

Davidson, D. W. 1988. Ecological studies of Neotropical ant-gardens. *Ecology* 69: 1138–1152.

Davidson, D. W., J. L. Longino, and R. R. Snelling. 1988. Pruning of host plant neighbors by ants: An experimental approach. *Ecology* 69: 801–808.

Davidson, D. W. In manuscript. *Cecropia* with *Pachychondyla*: Possible pairwise coevolution in a symbiotic mutualism.

Davidson, D. W., R. R. Snelling, and J. L. Longino. 1989. Competition among ants for myrmecophytes and the significance of plant trichomes. *Biotropica.* 21: 64–73.

Davidson, D. W. and W. W. Epstein. 1989. Epiphytic associations with ants. In U. Lüttge (Ed.). *Vascular Plants as Epiphytes*. Springer Verlag, Berlin.

Fletcher, D. J. C., M. S. Blum, T. V. Whitt, and N. Temple. 1980. Monogyny and polygyny in the fire ant *Solenopsis invicta*. *Annals of the Entomological Society of America* 73: 658–651.

Futuyma, D. J. and M. Slatkin. 1983. The study of coevolution. Pp. 459–464 in D. J. Futuyma and M. Slatkin (Eds.) Coevolution. Sinauer, Sunderland, MA.

Harada, A. Y. 1982. Contribuição ao conhecimento do gênero *Azteca* Forel, 1978 (Hymenoptera: Formicidae) e aspectos da interação com plantas do gênero *Cecropia* Loefling, 1758. Masters Thesis in Biological Sciences, INPA/FUA, Manaus, 181 pp.

Holdridge, L. R. 1967. Life zone ecology. Tropical Science Center, San Jose, Costa Rica.

Hölldobler, B. and E. O. Wilson. 1977. The number of queens: An important trait in ant evolution. *Naturwissenschaften* 64: 8–15.

Huxley, C. R. 1978. The ant-plants *Myrmecodia* and *Hydnophytum* (Rubiaceae), and the relationships between their morphology, ant-occupants, physiology and ecology. *New Phytologist* 80: 231–268.

Huxley, C. R. 1986. Evolution of benevolent ant–plant relationships. Pp. 257–282 in B. Juniper and Sir R. Southwood (Eds.). *Insects and the Plant Surface*. Edward Arnold, London.

Inouye, D. W. and O. R. Taylor, Jr. 1979. A temperate region plant-ant-seed predator system: Consequences of extrafloral nectar secretion by *Helianthella quinquenervis*. *Ecology* 60: 1–7.

Janzen, D. H. 1966. Coevolution of mutualism between ants and acacias in Central America. *Evolution* 20: 249–275.

Janzen, D. H. 1969. Allelopathy by myrmecophytes: The ant *Azteca* as an allelopathic agent of *Cecroia*. *Ecology* 50: 147–153.

Janzen, D. H. 1974. Tropical blackwater rivers, animals, and mast fruiting by the Dipterocarpaceae. *Biotropica* 6: 69–103.

Janzen, D. H. 1980. When is it coevolution? *Evolution* 34: 611–612.

Janzen, D. H. 1983. *Pseudomyrmex ferruginea* (Hormiga del Cornizuelo, Acacia-Ant).

Pp. 762–765 in D. H. Janzen (Ed.). *Costa Rican Natural History*, University of Chicago, Chicago. IL.

Jebb, M. H. P. 1985. *Taxonomy and tuber morphology of the rubiaceous ant-plants*. Ph.D. Oxford University.

Jolivet, P. 1987. Remarques sur la biocenose des *Cecropia* (Cecropiaceae). Biologie des *Coelomera* chevrolat avec la description d'une nouvelle espece du Bresil (Coleoptera Chrysomelidae Galerucinae). *Bulletin mensuel de la Société Linnéenne de Lyon* **56**: 255–276.

Jordano, P. 1987. Patterns of mutualistic interactions in pollination and seed dispersal: Connectance, dependence asymmetries, and coevolution. *American Naturalist* **129**: 657–677.

Levings, S. C. and J. F. A. Traniello. 1981. Territoriality, nest dispersion and community structure in ants. *Psyche* **88**: 265–320.

Longino, J. T. 1989. Geographic variation and community structure in an ant–plant mutualism: *Azteca* and *Cecropia* in Costa Rica. *Biotropica* **21**: 126–132.

McKey, D. 1984. Interaction of the ant-plant *Leonardoxa africana* (Caesalpiniaceae) with its obligate inhabitants in a rainforest in Cameroon. *Biotropica* **16**: 81–99.

McKey, D. 1988. Promising new directions in the study of ant–plant mutualisms. pp. 335–355 in W. Greuter and B. Zimmer (Eds.). *International Botanical Congress, Berlin 1987–Proceedings*. Koeltz Scientific Books, Königstein.

Putz, F. E. 1984. How trees avoid and shed lianas. *Biotropica* **16**: 19–23.

Rehr, S. S., P. P. Feeny, and D. H. Janzen. 1973. Chemical defence in Central American non-ant-acacias. *Journal of Animal Ecology* **42**: 405–416.

Rickson, F. R. 1971. Glycogen plastids in Müllerian body cells of *Cecropia peltata*, a higher green plant. *Science* **173**: 344–347.

Rickson, F. R. 1973. Review of glycogen plastid differentiation in Müllerian body cells of *Cecropia peltata*. *Annals of the New York Academy of Sciences* **210**: 104–114.

Rickson, F. R. 1976. Anatomical development of the leaf trichilium and Müllerian bodies of *Cecropia peltata*. *American Journal of Botany* **63**: 1266–1271.

Schemske, D. W. 1983. Limits to specialization and coevolution in plant–animal mutualisms. Pp. 67–109 in M. H. Nitecki (Ed.), *Coevolution*. University of Chicago Press, Chicago.

Schupp, E. 1986. *Azteca* protection of *Cecropia*: Ant occupation benefits juvenile trees. *Oecologia* **70**: 379–385.

Siegler, D. S. and J. E. Ebinger. 1987. Cyanogenic glycosides in ant-acacias of Mexico and Central America. *The Southwestern Naturalist* **32**: 499–503.

Wheeler, W. M. 1942. Studies of Neotroical ant-plants and their ants. *Bulletin of the Museum of Comparative Zoology* **90**: 3–262.

Wilson, E. O. 1959. Some ecological characteristics of ants in New Guinea rainforests. *Ecology* **40**: 437–446.

NOTES ADDED IN PROOF

Our *Cecropia tessmannii* should be referred to as *Cecropia* cf. *tessmannii* (C. C. Berg, pers. comm.). Longino's recent revision of the *Azteca alfari* species group (*Los Angeles County Museum, Contributions in Science* 412; 16 pp., Dec. 1989) designate our "*A. alfari*" as *Azteca ovaticeps*.

8 Distribution of Extrafloral Nectaries in the Woody Flora of Tropical Communities in Western Brazil

PAULO S. OLIVEIRA and ARY T. OLIVEIRA-FILHO

Extrafloral nectaries are nectar-secreting glands not directly involved with pollination function, with a widespread occurrence in many plant taxa, particularly among tropical angiosperms. Such glands are extremely variable in structure and may occur on virtually all above-ground plant parts, being especially common on leaves, petioles, stems, stipules, or near the reproductive parts of a plant (cf. Bentley 1977a; Elias 1983; Oliveira and Leitão-Filho 1987). Several analyses of extrafloral nectars have shown that they contain sugars, amino acids, proteins, and vestigial amounts of other organic compounds (Bentley 1977a; Elias 1983). Ants are by far the most frequent visitors of extrafloral nectaries (hereafter EFNs), but a variety of other nectar-feeders (e.g., bees, flies, predatory and parasitic wasps) can also be seen visiting these glands for their secretions.

The controversy on the adaptive significance of EFNs exists since their discovery by early botanists, and is disputed between two groups (cf. Brown 1960). The 'Protectionists' have supported the idea that ant visitors to EFNs protect the plant against herbivores, while the 'Exploitationists' believe that extrafloral nectar secretion has some purely physiological function. Early natural historians used to describe ant–plant relationships with many anatomical, morphological, and behavioral details, and a mutualistic interaction was generally inferred from circumstantial rather than experimental evidence (cf. Wheeler 1942). It was not until Janzen's (1966, 1967) work on the ant x *Acacia* interaction that a continuous sequence of experimental field studies, on a variety of ant–plant associations, provoked an enormous advancement for the protectionist view. Many cases of obligate and facultative ant–plant mutualisms have been experimentally studied since then, from many kinds of environments (reviews in

Plant-Animal Interactions: Evolutionary Ecology in Tropical and Temperate Regions, Edited by Peter W. Price, Thomas M. Lewinsohn, G. Wilson Fernandes, and Woodruff W. Benson.
ISBN 0-471-50937-X © 1991 John Wiley & Sons, Inc.

Beattie 1985; Jolivet 1986; see also D. W. Davidson, Chapter 7). As far as EFN-bearing plants are concerned, several authors have demonstrated that visiting ants can effectively protect the plant against damage from many kinds of herbivores (Bentley 1977a,b; Keeler 1977, 1980a, 1981a; Inouye and Taylor 1979; Schemske 1980; Koptur 1984; Barton 1986; and citations therein).

Provided that the protectionist view is valid, one should expect the distributions of EFNs and ants to be correlated, both on local and global scales. In contrast to the fairly well-documented distribution of ant species (Wheeler 1910, 1942; Kusnezov 1957; Wilson 1971), little is known about the taxonomic distribution and abundance of EFNs in different vegetation types. In tropical habitats (Costa Rica and Jamaica), Bentley (1976) and Keeler (1979a) found that 0–80% of the plants had EFNs, and that this was correlated with ant abundance. In temperate North American areas the mean cover of EFN-bearing plants ranged from 0 to 14% (Keeler 1980b, 1981b). The first study on the abundance of EFNs in a South American environment was that of Oliveira and Leitão-Filho (1987) for the cerrado vegetation of Brazil. These authors found that the abundance of woody plants with EFNs ranged from 7.6 to 20.3% in cerrado areas of São Paulo, SE Brazil.

Plants bearing EFNs are intensively visited by several ant species in the cerrado. A preliminary study with *Qualea grandiflora* Mart. (Vochysiaceae), a typical cerrado plant that bears EFNs on the stem and pedicel, showed that visiting ants can potentially act as antiherbivore agents of this plant (Oliveira et al. 1987). A more thorough study with *Caryocar brasiliense* Camb. (Caryocaraceae) in a cerrado of São Paulo, revealed that 34 ant species collect extrafloral nectar on the plant's sepals (Oliveira 1988). Both day and night the number of *Caryocar* individuals being visited by ants greatly surpassed that of non-nectariferous neighboring plants, resulting in many more live termite-baits attacked by foraging ants on *Caryocar* than on plants lacking EFNs. Moreover, ant-exclusion experiments showed that *Caryocar* plants without ants are significantly more infested by *Eunica* caterpillars (Nymphalidae), *Edessa* sucking bugs (Pentatomidae), *Prodiplosis* bud-destroying flies (Cedidomyiidae) and a stem-galling wasp (Eulophidae) (Oliveira 1988).

Therefore, the possession of EFNs seems to be a relevant antiherbivore tactic within the cerrado community. Comparative data on the taxonomic distribution and abundance of EFNs would help to evaluate the importance of these glands in varying ecological and geographical contexts. The purpose of this chapter is to provide data of this kind for four cerrado areas, a "murundu" field or seasonal marsh, and a gallery forest in Western Brazil. We intend to compare our results with similar studies undertaken in Brazil, as well as in other tropical and temperate habitats.

STUDY SITES AND METHODS

Field work was carried out during 1983–1985 in different vegetation types of West Brazil (Fig. 8-1). The climate is tropical–continental, with high tempera-

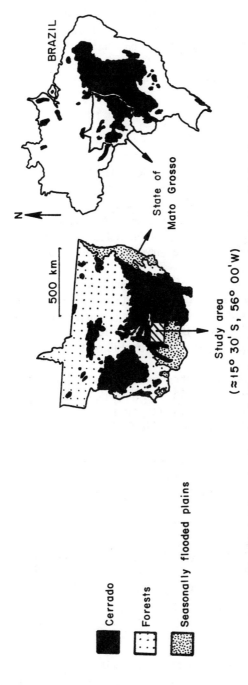

Figure 8-1. Maps showing the distribution of cerrado vegetation in Brazil (after Borgonovi and Chiarini 1965) and of different vegetation types in the State of Mato Grosso (after Oliveira-Filho 1984).

tures throughout the year, and with marked dry and rainy seasons (*Aw* type of Koeppen's system). Mean annual rainfall and temperature are 1421 mm and 25.6°C, respectively (Oliveira-Filho and Martins 1988).

Data on the distribution of EFNs were obtained from floristic surveys conducted in the following vegetation types: cerrado (four sites), murundu field (one site), and gallery forest (one site). The region where these sites are located is shown in Figure 8.1. Only trees and shrubs with a basal trunk diameter of at least 3 cm were included in the surveys. Monocotyledons and treelike ferns were not included in this study.

The Cerrado Vegetation

The cerrados are savanna-like formations which cover nearly 25% of Brazil (Fig. 8-1). They present several intergrading physiognomic forms within their distribution, ranging from forest with more or less merging canopy ("cerradão") to open grassland with scattered shrubs ("campo sujo") (Goodland 1971). In Mato Grosso the cerrados cover approximately 28% of the state's physical area (Fig. 8-1). Four cerrado sites were surveyed in this region, as described below.

Cerrado Site 1. Dense scrub of shrubs and trees (cerrado *sensu stricto* of Goodland 1971) growing on a fine-textured soil with a thick layer of pebbles at the surface. Woody plants were sampled using the point-centered quarter method (Cottam and Curtis 1956), with 140 points distributed in a 4 × 35 systematic grid, and interpoint distances of 10 m (Oliveira-Filho 1988a).

Cerrado Site 2. Forest with open canopy ("cerradão" of Goodland 1971) growing on fine-textured lithosol over phyllite bedrock. The point-centered quarter method was employed, with 60 points distributed in a 5 × 12 systematic grid, and interpoint distances of 10 m (Oliveira-Filho 1988a).

Cerrado Site 3. Occurring on deep arenosols, this cerrado varied physiognomi-cally from cerrado *sensu stricto* to cerradão. It was sampled by 100 contiguous plots of 10 × 10 m, arranged in a transect of 50 paired plots (Oliveira-Filho et al. 1988).

Cerrado Site 4. Open scrub with scattered trees ("campo cerrado" of Goodland 1971) growing on coarse-textured lithosols over sandstone bedrock. The point-centered quarter method was used, with 50 points distributed in two 5 × 5 systematic grids and interpoint distances of 10 m (Oliveira-Filho 1988a).

The Murundu Field

Murundu fields are common landscapes in seasonally flooded plains of West Brazil (Fig. 8-1). They are seasonal marshes with many scattered earthmounds, which are locally known as "murundus." The earthmounds are covered by woody cerrado plants, and their surroundings by a grassy vegetation. Termite nests commonly occur on the top of the earthmounds (detailed descriptions in Diniz de Araujo Neto et al. 1986; Furley 1986). The woody vegetation of 80 earthmounds (0.5–22.0 m diameter) was entirely surveyed in a 5 ha field, comprising 0.3 ha of murundus (Oliveira-Filho 1988a). The physiognomy of the vegetation within earthmounds varied from very closed scrub to dense forest.

The Gallery Forest

Gallery forests are very common next to cerrado areas of West Brazil (Fig. 8-1). They occur in the bottom of valleys, as narrow strips along the streams. Tall trees with closed canopy and many lianas characterize this vegetation type. Woody plants were surveyed with 67 plots of 30 m^2 (0.2 ha) (Oliveira-Filho 1988b).

Plants with Extrafloral Nectaries (EFNs)

All plant species registered in surveys, as well as those found out of the sampled areas, were searched in the field for EFNs. Evidence for the presence of EFNs was obtained *a priori* from lists of species bearing such glands (Schnell et al. 1963; Bhattacharyya and Maheshwari 1971; Bentley 1977a; Elias 1983; Oliveira and Leitão-Filho 1987). Although not used as a criterion, the presence of ants or other nectar-feeders occasionally helped us to detect nectar-secreting structures on a plant. Plant taxonomy follows Cronquist (1981), and voucher specimens are deposited in the herbarium of the Universidade Estadual de Campinas (UEC).

RESULTS

The taxonomic distribution of EFNs within woody species from different vegetation types surveyed in Mato Grosso, as well as the gland location on plants, is summarized in Table 8-1. EFNs were present on 37 plant species belonging to 26 genera and 17 families. The Mimosaceae (6 species), Bignoniaceae (5 species), Vochysiaceae and Chrysobalanaceae (4 species each) were the families most frequently bearing EFNs; these are also among the most common families in the region (Oliveira-Filho and Martins 1986). Twenty-nine EFN-bearing species are typical from cerrado formations (including those found in the murundus), while only eight species seem to be exclusive to the gallery forest (Table 8-1).

Although EFNs were most commonly located on the leaf blade (16 of 37, 43%), glands were also found associated with other plant parts such as petiole, rachis, stem, or calyx (Table 8-1). Nectaries located on vegetative structures (35 of 37, 94%) were more frequent then those found near the bud or flower (5 of 37, 13%) ($X^2 = 22.50$; $P < 0.001$).

Table 8-2 summarizes the data on proportion and abundance of species with EFNs in the woody flora of the surveyed areas. Within the cerrado sites the

TABLE 8-1. Woody Plant Species with Extrafloral Nectaries in Different Vegetation Types in Mato Grosso, W Brazil

Plant Species	Site of nectary[b]	Cerrado Sites[a] (1)	(2)	(3)	(4)	Murundu Field	Gallery Forest[a]
Bignoniaceae							
Cybistax antisyphillitica	LB	X	0	0	0	X	
Jacaranda cuspidifolia	LB	X	X			X	0
Tabebuia aurea	LB	X	0	X		X	
T. caraiba	LB	X	0	X	0	X	0
T. ochraceae	LB	X	X	X	0		
Bombacaceae							
Eriotheca gracilipes	PE	X	X	X	0		0
Caesalpinaceae							
Bauhinia bongardi	IT	0				X	
Caryocaraceae							
Caryocar brasiliense	CA			X	X		0
Chrysobalanaceae							
Hirtella glandulosa	LB	0					X
H. gracilipes	LB	X					X
H. hoehney	LB						X
Licania humilis	LB			X	0		
Combretaceae							
Terminalia brasiliensis	PE	0		X	0		0
T. fagifolia	PE				X		0
T. subsericea	PE	0	X				
Ebenaceae							
Diospyrus coccolobifolia	LB	X	0	X		X	
Euphorbiaceae							
Hieronyma alchorneoides	LB						X
Richeria grandis	PE						X
Lythraceae							
Lafoensia paccari	LB	X	X	X	0		
Malpighiaceae							
Heteropteris byrsonimifolia	LB	0			X	X	
Marcgraviaceae							
Norantea guianensis	LB				X		X
Mimosaceae							
Anadenanthera falcata	RA	0	X				
Inga uruguensis	RA						X
I. heterophylla	RA						X
Mimosa xanthocentra	RA			0	X		
Plathymenia reticulata	ST	0	X	X			X
Stryphnodendron obovatum	RA	X	0	0	0		0
Myrsinaceae							
Rapanea guianensis	LB						X

TABLE 8-1. (*Continued*)

Plant Species	Site of nectary[b]	Cerrado Sites[a]				Murundu Field	Gallery Forest[a]
		(1)	(2)	(3)	(4)		
Ochnaceae							
Ouratea castaneifolia	SL						X
O. hexasperma	SL	X	0	X	0	X	
O. spectabilis	SL	X	0	0	X		
Rubiaceae							
Tocoyena formosa	CA	X	X	X	0	X	
Verbenaceae							
Aegiphila sellowiana	LB						X
Vochysiaceae							
Callisthene fasciculata	ST	0	X				
Qualea grandiflora	ST, PD	X	X	X	0		X
Q. multiflora	ST, PD	0	X	0	0		0
Q. parviflora	ST, PD	X	X	X	X	X	X
Number of species sampled		15	12	14	7	10	14
Number of species found out of the sampled area		9	7	5	12	0	8
Total number of species recorded		24	19	19	19	10	22

[a](X) Species registered within sampled areas; (0) species registered out of the sampled areas.
[b]LB = leaf blade; PE = petiole; IT = intrastipular trichomes; CA = calyx; RA = rachis; SL = stipules; ST = stem; PD = pedicel.

TABLE 8-2. Occurrence and Abundance of Woody Species with Extrafloral Nectaries (EFNs) in Different Vegetation Types, Mato Grosso, Brazil[a]

Vegetation Type	Number of Species Sampled	Percentage of Species with EFNs	Percent Abundance of Plants with EFNs
Cerrado site 1	64	23.4 (15/64)	28.1 (157/559)
Cerrado site 2	47	25.5 (12/47)	31.2 (74/237)
Cerrado site 3	68	20.6 (14/68)	27.6 (421/1524)
Cerrado site 4	30	23.3 (7/30)	21.6 (39/181)
Murundu field	45	22.2 (10/45)	8.3 (72/871)
Gallery forest	85	16.5 (14/85)	14.2 (121/854)

[a]See also Table 8-3 for between-area comparisons.

proportion and abundance of nectary species ranged from 20.6 to 25.5% and from 21.6 to 31.2%, respectively. These values were 22.2% and 8.3% for the murundu area, and 16.5% and 14.2% for the gallery forest (Table 8-2).

Paired comparisons revealed no significant differences between the proportions of nectary species registered in the surveyed areas (Table 8-3). On the other hand, plants with EFNs were significantly less abundant in the murundu than both the cerrado sites and gallery forest. This last area, however, had a

TABLE 8-3. Paired Comparisons (X^2 tests) Between Surveyed Areas, Relative to the Proportion (Lower Left X^2 Values) and Abundance (Upper Right X^2 Values) of Woody Species with Extrafloral Nectaries

| | Cerrado | | | | Murundu Field | Gallery Forest |
	Site 1	Site 2	Site 3	Site 4		
Cerrado site 1	—	$0.65^{ns\,a}$	0.02^{ns}	2.68^{ns}	97.98^d	40.53^d
Cerrado site 2	0.00^{ns}	—	1.14^{ns}	4.40^b	83.83^d	35.61^d
Cerrado site 3	0.03^{ns}	0.16^{ns}	—	2.37^{ns}	125.87^d	55.55^d
Cerrado site 4	0.63^{ns}	0.00^{ns}	0.00^{ns}	—	26.62^d	5.67^c
Murundu field	0.01^{ns}	0.02^{ns}	0.00^{ns}	0.03^{ns}	—	14.53^d
Gallery forest	0.73^{ns}	1.05^{ns}	0.20^{ns}	0.32^{ns}	0.32^{ns}	—

[a] ns = Not significant.
[b] $P < 0.05$.
[c] $P < 0.02$.
[d] $P < 0.001$.
See also Table 8-2.

TABLE 8-4. Comparisons Between EFNs Data (Weighted Averages) from Cerrados of Mato Grosso (Present Study) and São Paulo (Oliveira and Leitão-Filho 1987)[a]

Variable	State of Mato Grosso (4 sites)	State of São Paulo (5 sites)	Significance of Difference
Percentage of woody species with EFNs	$22.95\% \pm 1.38\%$	$18.33\% \pm 2.45\%$	$P < 0.025$
Percent abundance of woody species with EFNs	$27.62\% \pm 1.97\%$	$16.96\% \pm 4.26\%$	$P < 0.05$

[a] One-tailed t tests made after arc $\sin \sqrt{x}$ transformations of proportions (Sokal and Rohlf 1981).

significantly lower abundance of EFNs than the cerrados. All but one (site 2 vs site 4) between-cerrado comparisons revealed no significant differences with respect to the abundance of EFNs (Table 8-3).

Our data from the cerrados of Mato Grosso furnished significantly higher mean values, for both the proportion and abundance of nectary species, than those obtained by Oliveira and Leitão-Filho (1987) from the cerrados of São Paulo in SE Brazil (Table 8-4).

DISCUSSION

This study adds several species and genera to previously published lists of EFN-bearing plants (Schnell et al. 1963; Bentley 1977a; Elias 1983). However, the families reported here as most frequently bearing EFNs (Mimosaceae, Bignoniaceae, Vochysiaceae, and Chrysobalanaceae) are also mentioned as commonly having nectary species in the taxonomic survey of Elias (1983).

Although occurring in a variety of plant locations, the observed tendency of EFNs to concentrate mainly on vegetative parts, especially on the leaf blade, was also detected by Oliveira and Leitão-Filho (1987) for cerrados in SE Brazil and by Keeler (1979b) in a taxonomic survey of a temperate flora (Nebraska, USA). Provided that these glands would be involved in mutualistic association with ants at different habitats, the possible selective pressures (i.e., herbivory) responsible for such a within-plant spatial tendency of EFNs remains unknown and needs further investigation. The fact that EFNs occur widely and with diverse ontogenies, in many orders of angiosperms, strongly suggests that they have evolved independently many times in different plant families (Elias 1983), and that their evolution might have been a relatively "easy" process (Bentley 1977a).

Although the gallery forest did not differ from the cerrados with respect to the proportion of nectary species, several of these plants registered in the forest (6 of 14) are more typical from the adjoining cerrados (cf. Table 8-1). Perhaps this fact could partially explain why EFNs had a lower abundance in the gallery forest than in the neighboring cerrados: typical cerrado species are poorly represented in the forest (Oliveira-Filho 1988b). Nevertheless, the values for proportion and abundance of nectary species within the gallery forest greatly surpassed those from temperate deciduous and riparian forests (Keeler 1979b, 1980b, 1981b). Moreover it is worth noting that the myrmecophyte *Tococa formicaria* Mart. (Melastomataceae) is abundant (3.3% of woody individuals) in the gallery forest (Oliveira-Filho 1988b), indicating that antiherbivore strategies mediated by ant–plant mutualisms can be important in this community.

All nectary species found in the murundus are characteristic of the cerrados, and their proportions did not differ between these two vegetation types (cf. Tables 8-1, 8-3). The lower abundance of EFNs in the murundus than in any of the cerrado sites was due to the nonnectary murundu-specialized tree, *Curatella americana* L., which accounted for 45.5% of the woody individuals sampled in the murundus (cf. Oliveira-Filho 1988a).

Our surveys in the cerrados of Mato Grosso recorded a total of 29 nectary woody species, against 34 found in São Paulo by Oliveira and Leitão-Filho (1987). Seventeen species occurred in both surveyed regions. The lower number of nectary species found in Mato Grosso may be due to a lower sampling effort. Despite this fact, the proportion and abundance of EFN-bearing species was significantly higher for the cerrados of Mato Grosso than for those of São Paulo (cf. Table 8-4).

As a whole the values for proportion and abundance of nectary species in the Brazilian cerrados are greater than those obtained in North American habitats (summarized in Keeler 1981c), supporting Bentley's (1976) and P. D. Coley's (Chapter 2) suggestion that EFNs are more common in tropical than temperate communities. The causes provoking such discrepancies in the geographical distribution of EFNs are still poorly understood. Provided that EFNs, in general, are involved with ant–plant mutualisms (cf. Bentley 1977a), we should expect the distribution of nectary plants to be correlated with the distribution of ants. The cover of plants with EFNs was found to be positively correlated with ant abundance in Costa Rica (Bentley 1976) and Nebraska (Keeler 1980b). Moreover, such a correlation was also detected by Keeler (1979a) along an altitudinal gradient in Jamaica, with both EFNs and ants being more abundant at lower elevations. The number of ant species is known to increase toward the tropics (Kusnezov 1957). Overall predation rate by ants and ant species richness were found to be positively correlated on a latitudinal scale (Jeanne 1979).

Recent studies have shown that ants may play important ecological roles for the woody flora of cerrado vegetation. The arboreal ant community of a cerrado area in Mogi-Guacu (São Paulo) is formed by a stem-nesting subguild of 27 species, and a ground-nesting one of 13 species; 2 species can nest in either substrate (Morais 1980). Plants bearing EFNs represented 41% of the live woody individuals housing ant colonies within $1075\,m^2$ of cerrado in Mogi-Guacu (Morais 1980). Thus the intense foraging by ants on foliage, coupled with herbivore activity, might have represented a strong selective pressure for the evolution of EFNs within the flora of cerrado, resulting in a relatively high proportion and abundance of nectary species within this community. The nectary species *Qualea grandiflora* and *Caryocar brasiliese* are visited, respectively, by 12 and 34 ant species in the cerrado of Itirapina (SP), and both can gain protection from their ant visitors against insect herbivores (Oliveira et al. 1987; Oliveira 1988). These results indicate that EFNs are important promoters of ant activity on cerrado plants, and strengthen the suggestion that ant–plant mutualisms may constitute a relevant antiherbivore tactic in this community. It is worth noting, however, that myrmecophytes (i.e., specialized ant-plants) have not evolved in any cerrado plant taxa. The marked 4- to 5-month dry season (during which most plants have either only old leaves, or no leaves at all) might perhaps explain the absence of specialized ant–plant mutualisms in the Brazilian cerrados (see also Janzen 1966; D. W. Davidson, Chapter 7). Further investigation attempting to correlate EFN abundance with ant abundance, as well as with herbivory pressure, would help to explain why these glands are more common in some cerrado areas than in others, and why they tend to be more abundant in tropical than in temperate communities.

ACKNOWLEDGMENTS

We are grateful to K. S. Brown, Jr., K. H. Keeler, and I. Sazima for helpful suggestions on the manuscript. A. Piedrabuena helped with the statistical analyses. The preparation of the final version of the manuscript was greatly facilitated by the logistic support provided by B. Hölldobler at the MCZ Labs, Harvard University. Financial support to P. S. Oliveira was provided by CNPq (proc. 200512/88.9).

REFERENCES

Barton, A. M. 1986. Spatial variation in the effect of ants on an extrafloral nectary plant. *Ecology* **67**: 495–504.

Beattie, A. J. 1985. *The evolutionary ecology of ant–plant mutalisms.* Cambridge University Press, New York.

Bentley, B. L. 1976. Plants bearing extrafloral nectaries and associated ant community: Interhabitat differences in the reduction of herbivore damage. *Ecology* **57**: 815–820.

Bentley, B. L. 1977a. Extrafloral nectaries and protection by pugnacious bodyguards. *Annual Review of Ecology and Systematics* **8**: 407–427.

Bentley, B. L. 1977b. The protective function of ants visiting the extrafloral nectaries of *Bixa orellana* L. (Bixaceae). *Journal of Ecology* **65**: 27–38.

Bhattacharyya, B. and J. K. Maheshwari. 1971. Studies on extrafloral nectaries of the Leguminales. I. Papilionaceae, with a discussion on the system of the Leguminales. *Proceedings of the Indian National Science Academy* **37**: 11–30.

Borgonovi, M. and J. V. Chiarini. 1965. Cobertura vegetal do estado de São Paulo. I. Levantamento por foto interpretacão das áreas cobertas com cerrado, cerradão e campo em 1962. *Bragantia* **24**: 159–172.

Brown, W. L., Jr. 1960. Ants, acacias and browsing mammals. *Ecology* **41**: 587–592.

Cottam, S. and J. T. Curtis. 1956. The use of distance measures in phytosociological sampling. *Ecology* **37**: 451–460.

Cronquist, A. 1981. An integrated system of classification of flowering plants. Columbia University Press, New York.

Diniz de Araujo Neto, M., P. A. Furley, M. Haridasan, and C. E. Johnson. 1986. The 'Murundus' of the 'cerrado' region of central Brazil. *Journal of Tropical Ecology* **2**: 17–35.

Elias, T. S. 1983. Extrafloral nectaries: Their structure and distribution. Pages 174–203 in B. L. Bentley and T. S. Elias, editors. *The biology of nectaries.* Columbia University Press, New York.

Furley, P. A. 1986. Classification and distribution of murundus in the cerrado of central Brazil. *Journal of Biogeography* **13**: 265–268.

Goodland, R. 1971. A physiognomic analysis of the cerrado vegetation of central Brazil. *Journal of Ecology* **59**: 411–419.

Inouye, D. W. and O. R. Taylor. 1979. A temperate region plant–ant–seed predator system: Consequences of extrafloral nectar secretion by *Helianthella quinquinervis*. *Ecology* **60**: 1–7.

Janzen, D. H. 1966. Coevolution of mutualism between ants and acacias in Central America. *Evolution* **20**: 249–275.

Janzen, D. H. 1967. Interaction of the bull's-horn acacia (*Acacia cornigera* L.) with an ant inhabitant (*Pseudomyrmex ferrugineous* F. Smith) in eastern Mexico. *The University of Kansas Science Bulletin* **47**: 315–558.

Jeanne, R. L. 1979. A latitudinal gradient in rates of ant predation. *Ecology* **60**: 1211–1224.

Jolivet, P. 1986. *Les fourmis et les plantes: un example de coévolution*. Éditions Boubée, Paris.

Keeler, K. H. 1977. The extrafloral nectaries of *Ipomoea carnea* (Convolvulaceae). *American Journal of Botany* **64**: 1182–1188.

Keeler, K. H. 1979a. Distribution of plants with extrafloral nectaries and ants at two elevations in Jamaica. *Biotropica* **11**: 152–154.

Keeler, K. H. 1979b. Species with extrafloral nectaries in a temperate flora (Nebraska). *Prairie Naturalist* **11**: 33–38.

Keeler, K. H. 1980a. The extrafloral nectaries of *Ipomoea leptophylla* (Convolvulaceae). *American Journal of Botany* **67**: 216–222.

Keeler, K. H. 1980b. Distribution of plants with extrafloral nectaries in temperate communities. *American Midland Naturalist* **104**: 274–280.

Keeler, K. H. 1981a. Function of *Mentzelia nuda* (Loasaceae) postfloral nectaries in seed defense. *American Journal of Botany* **68**: 295–299.

Keeler, K. H. 1981b. Cover of plants with extrafloral nectaries at four Northern California sites. *Madroño* **28**: 26–29.

Keeler, K. H. 1981c. A model of selection for facultative nonsymbiotic mutualism. *American Naturalist* **118**: 488–498.

Koptur, S. 1984. Experimental evidence for defense of *Inga* (Mimosoideae) saplings by ants. *Ecology* **65**: 1787–1793.

Kusnezov, N. 1957. Numbers of species of ants in faunae of different latitudes. *Evolution* **11**: 298–299.

Morais, H. C. 1980. Estrutura de uma comunidade de formigas arborícolas em vegetação de campo cerrado. Master's thesis, Universidade Estadual de Campinas, Campinas, Sao Paulo.

Oliveira, P. S. 1988. Sobre a interação de formigas com o pequi do cerrado, *Caryocar brasiliense* Camb. (Caryocaraceae): o significado ecológico de nectários extraflorais. Doctor in Science thesis, Universidade Estadual de Campinas, Campinas, Sao Paulo.

Oliveira, P. S. and H. F. Leitão-Filho. 1987. Extrafloral nectaries: Their taxonomic distribution and abundance in the woody flora of cerrado vegetation in southeast Brazil. *Biotropica* **19**: 140–148.

Oliveira, P. S., A. F. da Silva, and A. B. Martins. 1987. Ant foraging on extrafloral nectaries of *Qualea grandiflora* (Vochysiaceae) in cerrado vegetation: Ants as potential antiherbivore agents. *Oecologia* (Berlin) **74**: 228–230.

Oliveira-Filho, A. T. 1984. Estudo florístico e fitossociológico em um cerrado na Chapada dos Guimarães—Mato Grosso—Uma análise de gradientes. Master's thesis, Universidade Estadual de Campinas, Campinas, São Paulo.

Oliveira-Filho, A. T. 1988a. A vegetação de um campo de monchões—microrrelevos associados a cupins—na região de Cuiabá (MT). Doctor in Science thesis, Universidade Estadual de Campinas, Campinas, Sao Paulo.

Oliveira-Filho, A. T. 1988b. Composição florística e estrutura comunitária da floresta de galeria do Córrego da Paciência (MT). *Acta Botanica Brasilica* (in press).

Oliveira-Filho, A. T. and F. R. Martins. 1986. Distribuição, caracterizacão e composição florística das formações vegetais da região da Salgadeira, na Chapada dos Guimarães (MT). *Revista Brasileira de Botânica* **9**: 207–223.

Oliveira-Filho, A. T., G. J. Shepherd, F. R. Martins, and W. H. Stubblebine. 1988. Environmental factors affecting physiognomic and floristic variations in an area of cerrado in central Brazil. *Journal of Tropical Ecology* (in press).

Schemske, D. W. 1980. The evolutionary significance of extrafloral nectar production by *Costus woodsonii* (Zingiberaceae): An experimental analysis of ant protection. *Journal of Ecology* **68**: 959–967.

Schnell, R., G. Cusset, and M. Quenum. 1963. Contribution à l'étude des glandes extra-florales chez quelques groupes de plantes tropicales. *Revue Generale de Botanique* **70**: 269–341.

Sokal, R. R. and F. J. Rohlf. 1981. *Biometry*. Freeman, San Francisco.

Wheeler, W. M. 1910. *Ants, their structure, development and behavior*. Columbia University Press, New York.

Wheeler, W. M. 1942. Studies of neotropical ant-plants and their ants. *Bulletin of the Museum of Comparative Zoology, Harvard* **90**: 1–262.

Wilson, E. O. 1971. *The insect societies*. Belknap Press, Cambridge.

SECTION 3
Antagonistic Relationships Between Plants and Animals

Adversity was not something dwelt upon by the early travellers in their journals. Belt (1888) admitted to being plagued by mosquitos, Darwin (1860) and Bates (1910) by horseflies, and Wallace (1869) by chiggers, but their real enthusiasm was for the beauty and diversity around them. Spruce (1908, p. 365), travelling in the Amazon region in the 1850s, emphasized the pleasures for insects of eating plants and their products: "Some caterpillars seem to have a decided taste for bitters; and narcotics are rarely objected to; indeed, I should say that most insects are decidedly partial to them, while bees and wasps seem to have a positive pleasure in getting drunk." Wallace was transfixed by the large and captivating butterflies and beetles especially. In Borneo he collected about 24 new species of beetle every day, many of them large wood-boring brentids, longhorns, and weevils (Wallace 1869). For such a keen collector perhaps we can forgive his lack of enthusiasm for caterpillars, for they are sluggish, frequently cryptic, and look terrible when pickled. Belt, however, was an avid gardener and so noticed the insect pests, and the antagonistic relationship between plant and herbivore more acutely. He had cause to study some insect herbivores in detail in his garden; "...the greatest plague of all were the leaf-cutting ants, and I had to wage a continual warfare against them." No doubt such antagonisms focused his interest on ant–plant interactions, predisposing him to the pioneering work on relationships between bull's horn acacias, *Pseudomyrmex* ants, and herbivores. He noted that "...the ants are really kept by the acacias as a standing army, to protect its leaves from the attacks of herbivorous mammals and insects."

These were the beginnings of what is now a thriving area of ecology. The antagonistic relationships between plants and animals enters into most sections of this book in one way or another. Researchers on tropical herbivores still suffer the problems of the early explorer, for feeding insect herbivores are commonly hard to find, although holes in leaves caused by them are not. Therefore, damage caused is more tractable than the biology of the herbivores themselves. However, the painstaking research needed to understand the plants and their herbivores is underway, and two chapters in this section (9 and 10) illustrate the long-term commitment to a more comprehensive understanding of plant–herbivore interactions. This broader view is also developing from studies on temperate systems, for it is becoming clear that although herbivores damage plants, the effect on the plant may not be negative in every case. The continuum of effects is emphasized in Chapter 11. Plants are also good for herbivores in the sense that they provide food, but it is frequently well defended, compromising the herbivore's relish for destruction. The subtle nature of the shifting balance between nutrition and toxicity is captured in Chapter 12. In Chapter 13 the sublety of herbivore impact is cryptically evident when largely nocturnal mammalian herbivores recede as humans encroach on their habitat.

REFERENCES

Bates, H. W. 1910. *The naturalist on the river Amazons.* Murray, London.

Belt, T. 1888. *The naturalist in Nicaragua.* 2nd. ed. Bumpus, London.

Darwin, C. 1860. *The voyage of the Beagle.* Natural History Library, New York.

Spruce, R. 1908. *Notes of a botanist on the Amazon and Andes,* Volume 2. MacMillan, London.

Wallace, A. R. 1869. *The Malay archipelago.* Macmillan, London.

9 Herbivore Fauna of *Piper* (Piperaceae) in a Costa Rican Wet Forest: Diversity, Specificity, and Impact

ROBERT J. MARQUIS

Knowledge of the composition of an herbivore fauna for individual plant species is necessary to understand the selective impact of herbivore species on their host plants. Despite the assertion that herbivores represent a greater selective force in tropical habitats than in temperate regions (Doutt 1960; Baker 1970; Levin 1976; Langenheim 1984), herbivore faunas of tropical plant species are much less studied than their temperate counterparts. For example, there are no published accounts of the complete herbivore fauna for any tropical plant species. Instead, host-usage patterns in tropical systems have been most widely investigated from the viewpoint of the herbivore, defining the host range of a group of closely related herbivores (e.g., Benson et al. 1975; Hopkins 1983; Janzen 1984; Janzen and Waterman 1984; DeVries 1985).

In order to understand the potential of herbivores to select for changes in resistance in their host plants, we need to know the diversity and abundance of individual species that compose an herbivore fauna. Secondly, because faunal composition is not necessarily stable temporally or spatially, it is important to know the factors, both intrinsic (e.g., plant secondary chemistry) and extrinsic (e.g., local abiotic environment) (Janzen 1985), which determine overall faunal diversity and species abundances. Thirdly, extent and timing of damage caused by individual species must be quantified because the selective importance of an individual herbivore will depend on the amount and type of tissue removed (Janzen 1979; Winder and van Emden 1981), the pattern of damage produced within the canopy of a plant (Lowman 1982; Marquis 1988a), and timing of

Plant-Animal Interactions: Evolutionary Ecology in Tropical and Temperate Regions, Edited by Peter W. Price, Thomas M. Lewinsohn, G. Wilson Fernandes, and Woodruff W. Benson. ISBN 0-471-50937-X © 1991 John Wiley & Sons, Inc.

damage relative to plant phenology (Dirzo 1984; Marquis 1988b). Finally, because the predicted evolutionary response to herbivory depends in part on whether or not the herbivore has alternate hosts (Feeny 1976), we need to know the degree of host specificity of the herbivore. An evolutionary change in resistance to a monophagous insect may require a structural modification in defensive compound(s) while an increase in quantity of a particular defense relative to neighbors may suffice against a more polyphagous species (Feeny 1976). Because plant diversity is characteristically high for many tropical communities, delineation of the host range for a given insect species is likely to require years of massive sampling effort, especially for relatively aseasonal habitats where peaks of herbivore abundance do not occur or are not necessarily as predictable as they are in more seasonal tropical forests (e.g., Janzen 1988).

Here I describe the diversity of herbivore faunas and individual species abundance and specificity, as they are presently known for a set of sympatric species of the genus *Piper* (Piperaceae). There are at least 45 species of *Piper* at the relatively aseasonal tropical study site, and these species vary greatly in abundance, growth form (vines, perennial herbs, shrubs, and small trees), and habitat preference. Occurrence of a large number of congeneric species at one site allows comparison of factors which may influence plant–herbivore relationships, while minimizing the potential effect of phylogenetic diversity (Baldwin and Schultz 1988). I first present patterns of host specificity for two sets of *Piper* specialist herbivores: weevils of the subfamilies Baridinae and Rhyncophorinae, and geometrid moths of the subfamily Larentiinae. I then compare these patterns with patterns reported for other tropical herbivore groups. Finally, I describe the complete herbivore faunas of two *Piper* species, *P. arieianum* and *P. holdridgeianum* and compare the impact on their respective hosts. Specifically, I consider what factors are associated with the diversity of herbivore faunas of individual *Piper* species, the relationship between herbivore faunal diversity and impact on the host plant.

STUDY SITE

The study site was the Estación Biológica La Selva (83°59′W, 10°26′N), owned and operated by the Organization for Tropical Studies and located in the Atlantic lowland rain forest of Costa Rica. La Selva is tropical wet forest (Hartshorn 1983). The average rainfall is 4000 mm per year, and never less than 100 mm on average per month, with predictable peaks in rainfall in July and November–December. There is a dry season of variable length and duration from January through May.

NATURAL HISTORY OF WEEVILS

There are 28 species in three genera of the subfamily Baridinae and one species of *Rhodobaenus* of the subfamily Rhynchophorinae, all of the family Curculionidae

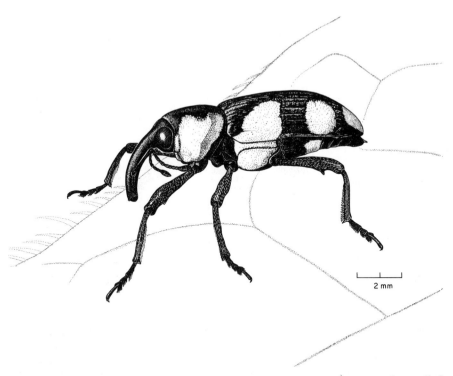

2 mm

Figure 9-1. One of the *Piper* weevils in the subfamily Baridinae, *Ambates cretifer*, studied at La Selva. Drawing by A. Prickett.

(Coleoptera), which feed as adults on leaves of La Selva *Piper* (Fig. 9-1). Members of this group range from Mexico to southern Brazil (unpublished U.S. National Museum records; R. Marquis, personal observation). Host ranges to be discussed are based on adult feeding records; the larvae feed within stems of *Piper* (and perhaps non-*Pipers*), but at this time I know much less of the larval host ranges.

My analysis consists of feeding patterns of only 25 species: I was unable to distinguish in the field individual species of one group of two species (*Ambates* sp. 5 and 6) and one group of three species (*Peridinetus* sp. 3, 9, and 10). These species, along with other species in this group, are members of a set of at least four different mimicry complexes: (1) *Peridinetus cretaceus*, *P.* sp. 7, and *Ambates cretifer* have the same yellow and black aposematic coloration as the baridine *Procholus biplagiatus* (Desb.), which likely feeds only on *Pentagonia donnell-smithii* (Rubiaceae) at La Selva (personal observation); (2) *Peridinetus* sp. 4, 5, and 6 are bird excrement mimics; (3) *Peridinetus* sp. 1, 3, 8, 9, 10, and *P. coccineifrons* are members of a proposed speed mimicry complex (Hespenheide 1973), which is based on mimicry of fast-flying flies; and (4) *Peridinetellus subnudus* is similar in color (black and shining) and shape to members of another complex proposed by Hespenheide (1987). Adult weevils generally feed on maturing leaves; rarely are leaves which have completed expansion attacked (Appendix 9-1). Damage

TABLE 9-1. Feeding Patterns and Behavior, and Preference for Leaf Age, Plant size, and Habitat of the Four Weevil Species which Feed as Adults on *Piper arieianum* at La Selva

Weevil Species	Leaf Age	Habitat	Plant Developmental Stage	Damage Type	Feeding Behavior
Peridinetus sp. 8	1/4–1/2 of full size	Understory	Seedlings	Kidney-shaped holes, 2–5 mm diam	Escape by flight, does not rest on plant
Peridinetus sp. 6	1/2 of full size to full size	Light gaps	Seedlings	Oval holes, 5–10 mm diam	Hides in damage hole, mimics bird excrement, defecates at leaf edge
Ambates sp. 5	1/4–1/2 of full size	Understory and light gaps	Mature plants	Oblong holes, 3–6 mm diam, often at leaf base	Hides in axils of lower leaves and on stems
Peridinetus spp. 3, 9, 10	1/2 of full size to full size	Understory	Mature plants	Contiguous top-surface scraping	Forages from central living hole in which it hides, defecates at hole

patterns, age of leaf attacked, and in some cases habitat and plant size preference are distinct for different weevil species (Table 9-1). For most weevil species, I have observed mating pairs on their known host plants.

NATURAL HISTORY OF GEOMETRID MOTHS

Twenty-one species of three closely related genera of the family Larentiinae feed on leaves of *Piper* at La Selva. This group ranges from Mexico to Argentina (unpublished records of the American Museum of Natural History). There are at least 100 species in the genus *Cambogia* alone, but no systematic work has been conducted since the original descriptions in the late 1800s. At La Selva, larvae are solitary feeders, with one larva per leaf in most species. Almost all species feed on fully mature leaves, often on the oldest leaves of a plant (Appendix 9-2). The larvae attack the lower surface of mature leaves between secondary veins, consuming all but the upper cuticle.

METHODS

Collections, feeding trials, and behavioral observations were made from 1979 to 1988, including all months. Field observations of weevils on plants in the wild were supplemented with feeding trials in the laboratory when actual feeding was not observed in the field. Because both weevils and larvae of geometrid moths make distinctive damage (Table 9-1), it was possible to determine whether all resident weevil and geometrid moth species on *Piper* at La Selva were collected. I make two assumptions in this regard. First, I assume that I have included in my analysis all species of weevils which feed as adults on La Selva *Piper*. In the last three years I have found no new weevil species and documented only two new host records. For the geometrids, I am missing at this time adult specimens of larvae which feed on three *Piper* species (Table 9-2); I have seen these larvae five or fewer times and have not been able to raise them to adults. The second assumption is that the weevil and geometrid species to be discussed feed only on *Piper*. I have neither seen these insects nor their damage on non-*Piper* species, although I have spent at least 750 hr in the field searching for caterpillars on non-*Piper* vegetation. Patterns of host specificity for *Piper*-feeding weevils and geometrids were compared by means of a G-test for goodness of fit (Sokal and Rohlf 1981) with patterns reported in literature for other tropical herbivore groups.

All weevil species are identified by D. R. Whitehead, with voucher specimens deposited in the United States National Museum and the Museo Nacional of Costa Rica. All geometrid species were classified to morphospecies by the author in collaboration with S. Passoa, D. H. Janzen, D. G. Ferguson, and F. Ringe using larval, exuvial, pupal, and adult morphology (including genitalia) and comparison with specimens at the American Museum of Natural History (New York) and the United States National Museum (Wasington, D.C.), where voucher specimens have been placed. Specimens of the remaining herbivores of *P. arieianum* and *P. holdridgeianum* are in the author's personal collection. Voucher specimens of the *Piper* species have been placed in the University of Illinois Herbarium, the British Museum, and the Museo Nacional of Costa Rica.

Quantification of the insect herbivore faunas and the abundance of individual species on *Piper arieianum* at La Selva was through a combination of field observation of numerous plants during all months of the year and regular censuses of marked plants over 1.5 years. For the latter, 200 marked plants (approximately 10,000 leaves total), scattered over a 350-ha area, were censused monthly from April 1981 until October 1982. All censuses were conducted at night, usually between 2000 and 0200 hr. Species of Diptera, Lepidoptera, and Orthoptera were classified as common if they were encountered five or more times during the censuses, and rare if encountered fewer than five times or only during field observations on noncensus plants. The cutoff number was 10 for Coleoptera and Hemiptera (see Appendix 9-3).

Criteria for including an insect species as an herbivore of either *P. arieianum* or *P. holdridgeianum* were: caterpillars found on plant and raised to adulthood on

TABLE 9-2. Number of Weevil and Geometrid Species, and Abundance, Size, Pubescence, and Soil and Habitat Preference of *Piper* Host Species at Estación Biológica La Selva, Costa Rica

Plant Species	Number of Geometrid Species	Number of Weevil Species	Disturbance[a]	Pubescence[b]	Maximum Height (m)	Soil Type[c]	Abundance[d]
P. aduncum L.	0	0	S	2	5	1	C
P. aequale Vahl	0	0	F	1	2	2	R
P. arboreum Aublet	4	2	F	1	4.5	1	R
P. arieianum C. DC.	2	4	F	1	3	3	C
P. augustum Rudge	1[e]	3	F	1	3	3	M
P. auritum H.B.K.	2	6	S	2	4	1	C
P. biolleyi C. DC.	4	1	S	3	3.5	4	R
P. biseriatum C. DC.	1	5	F	3	4	3	M
P. carrilloanum C. DC.	1	4	S	1	3.5	4	M
P. cenocladum C. DC.	2	6	F	1	4	3	C
P. culebranum C. DC.	2	4	S	3	6	1	M
P. concepcionis Trel.	0	2	F	1	2	3	M
P. darienense C. DC.	0	0	F	1	1	4	R
P. decurrens C. DC.	1	4	F	1	3	1	M
P. dolichotrichum Yuncker	2	4	F	3	3	1	C
P. euryphyllum C. DC.	1	1	F	1	5	3	R
P. friedrichsthalii C. DC.	0	0	S	2	2	4	R
P. garagaranum C. DC.	0	1	F	3	1	3	C
P. glabrescens (Miq.) C. DC.	2	2	S	1	2	5	M
P. holdridgeianum W. Burger	1[e]	0	F	1	1.5	3	C
P. imperiale (Miq.) C. DC.	3	3	F	2	5	3	M
P. melanocladum C. DC.	1	2	F	1	2.5	3	M
P. multiplinervium C. DC.	1	3	S	1	5	3	C

P. nudifolium C. DC.	0	1	F	1	1	4	R
P. otophorum C. DC.	0	1	F	3	1	5	R
P. peracuminatum C. DC.	0	3	S	2	5	5	R
P. perbrevicaule Yuncker	0	0	F	2	1	0.5	R
P. phytolaccaefolium Opiz	0	4	S	1	2	1	C
P. reticulatum L.	1	1	S	1	6	5	C
P. sancti-felicis Trel.	8	8	S	2	4	1	C
P. silvivagum C. DC.	2	4	S	1	2	1	M
P. terrabanum C. DC.	1[e]	1	F	1	2	5	R
P. tonduzii C. DC.	0	4	F	3	0.5	4	R
P. urophyllum C. DC.	0	3	F	1	3	4	R
P. urostachyum Hemsley	3	3	S	3	3	1	C
P. virgultorum C. DC.	3	1	F	1	3.5	5	M
P. xanthostachyum C. DC.	2	2	F	1	3	1	M
P. sp. 1	0	3	F	1	5.5	1	C
P. sp. 2	0	6	S	2	3	5	M
P. sp. 3	1	5	S	2	3.5	5	M
P. sp. 4	0	1	F	1	4	4	R
P. sp. 5	2	2	F	3	5.5	3	C
P. sp. 6	0	1	F	2	5	4	R
P. sp. 7	1	0	F	1	2.5	5	R
P. sp. 8	0	0	S	3	1	5	R

[a]S = Secondary growth, light gaps, and open river and stream banks; F = forest understory.

[b]1 = Glabrous, 2 = slightly pubescent, 3 = pubescent to hairy.

[c]1 = All soils, 2 = old alluvium, 3 = all soils except recent alluvium, 4 = recent alluvium, 5 = old and recent alluvium, and/or swamp soils.

[d]R = 5 or fewer populations known, M = intermediate abundance on soils or habitats in which it typically occurs, C = common on soils or in habitats in which it typically occurs.

[e]No adult raised.

185

leaves of the respective host plant (Lepidoptera); adults feeding on foliage or nymphs raised to adulthood on leaves of the respective host plant (Tettigoniidae and Phasmidae); adults and nymphs feeding on plant (Hemiptera).

To determine the range of herbivory which occurs within the genus *Piper* at La Selva, damage was estimated for 10 of the most common species, all understory species except *P. sancti-felicis*, which is almost entirely restricted to light gaps and second growth. Damage was assessed by first collecting leaves which had recently abscised from the plant (minimum 50 leaves per species from 30 or more plants). Because leaves decompose quickly after abscission, damage measured was that which occurred on the plant. Paper tracings of abscised leaves were then made as if they were undamaged; areas of the real leaves and paper tracings were measured with an area meter (LI-Cor Model L1-3000, Lincoln, NE). This measurement is an estimate of the amount of area loss that would occur for an average leaf of the species. Measurement of damage in this way excludes leaves that are entirely consumed or abscise due to feeding by leaf cutting ants, *Atta cephalotes*, and two lepidopteran moth species on La Selva *Piper*. In addition, average standing leaf area missing per plant was compared for *Piper arieianum* and *P. holdridgeianum*. This estimate was made by collecting all the leaves from 25 individuals (*P. arieianum*) or tracing all the leaves while on the plant, for 25 individuals (*P. holdridgeianum*). Tracings were then made of the real leaves in the former species or the field-drawn tracings of damaged leaves in the latter species as if they were undamaged, and the areas of all tracings and real leaves were measured with a leaf area meter.

STATISTICAL ANALYSIS

To examine which plant characteristics might affect the number of insect species associated with a given plant species, I first classified plant species based on plant size (maximum height attained at La Selva), pubescence, relative abundance, soil type(s) on which the species most commonly occurs (Sanchez and Mata 1987), and the openness of habitat in which plants are commonly reproductive (either forest understory or open second growth, light gaps, and/or open river and stream banks) (Table 9-2). I then used multiple regression analysis to determine which of these characteristics were significant predictors of the number of geometrid species, weevil species, or both groups on a given plant species.

RESULTS

There is great variation in host specificity among both weevils and geometrid moths which feed on La Selva *Piper* (Fig. 9-2a, b). Some insect species are restricted to single host plant species while others feed on as many as six (one geometrid species) and 18 (one weevil species) *Piper* species. The geometrid moths are significantly more host-specific than the weevils ($P < 0.005$, G-test); a greater proportion of geometrids are specialists (71% versus 48% of all weevil species feeding on two *Piper* species or less), and a greater proportion of weevils are generalist species (four weevil species with 10 or more hosts and no geometrids with more than six host plants). *Piper*-feeding geometrids fall mid-range in host specificity compared to other tropical herbivore groups, while weevils on La

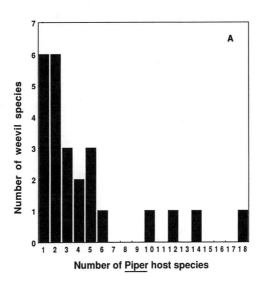

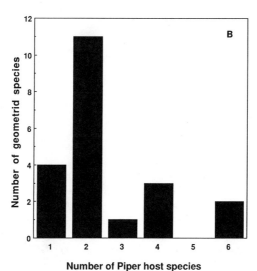

Figure 9-2. Patterns of host specificity for weevils of the subfamilies Baridinae and Rhyncophorinae and moths of the subfamily Larentiinae on *Piper* at La Selva, Costa Rica. (*a*) Number of host *Piper* plant species per adult weevil species. (*b*) Number of host *Piper* plant species per geometrid species.

TABLE 9-3. Distribution Patterns of the Number of Host Species per Insect Species for *Piper* Weevils and Geometrids at La Selva and Patterns Reported for Other Sets of Tropical Herbivores.

Insect Group	Total Insect Species	Number of Plant Species per Herbivore Species				Source
		Mean (SD)	Median	Range		
Guanacaste seed weevils:						
Costa Rica	110	1.4(1.0)	1	1–8	A	Janzen 1980
La Selva butterflies:						
Costa Rica[a]	75	1.7(1.8)	1	1–7	A	DeVries 1985
Corcovado butterflies:						
Costa Rica[a]	49	1.6(1.4)	1	1–6	A	DeVries 1985
Santa Rosa butterflies:						
Costa Rica[a]	50	1.2(0.5)	1	1–3	A	DeVries 1985
Limoncocha Ithomiinae:						
Ecuador	27	1.4(2.7)	1	1–5	A	Drummond 1986
La Selva *Heliconius*	7	2.3(2.6)	1	1–8	A	Smiley 1982
Arima Valley *Heliconius*:						
Trinidad	14	1.8(1.0)	1	1–5	B	Benson 1978
Rincón *Heliconius*:						
Costa Rica	15	1.5(0.7)	1	1–3	B	Benson 1978
La Selva *Piper* Geometridae	20	2.6(1.4)	2	1–6	C	This study
Campinas Ithomiinae:						
Brazil	18	4.3(3.0)	3	1–12	CD	Brown 1987
Caribbean hispine beetles:						
Costa Rica	8	4.6(3.4)	4	1–11	CD	Strong 1977a
La Selva flea beetles	7	3.4(2.1)	2	2–7	CDE	Smiley 1982
Rio de Janeiro *Heliconius*	11	3.2(2.1)	3	1–6	CDE	Benson 1978
Parkia seed weevils:						
South America	10	3.6(2.3)	3	1–7	DE	Hopkins 1983, 1984
La Selva *Piper* leaf weevils	25	4.4(4.5)	3	1–18	E	This study

[a]Includes Nymphalidae, Pieridae, and Papilionidae. Herbivore distributions with same letter not significantly different, $P > 0.05$, G-test.

Selva *Piper* are the least host-specific relative to previously reported patterns of tropical herbivores from a single region (Table 9-3).

Piper species also vary greatly in the number of geometrid moth and weevil species (0–8 species for both groups) which feed on them (Fig. 9-3 *a, b*). The number of herbivore species per host species differs between the two insect groups

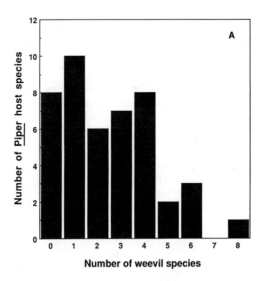

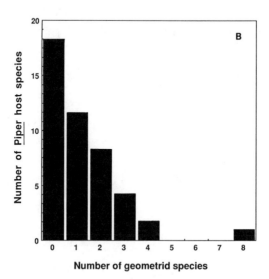

Figure 9-3. Patterns of number of herbivore species, either weevils of the subfamilies Baridinae and Rhyncophorinae or moths of the subfamily Larentiinae, per *Piper* species. (*a*) Number of adult weevil species per host *Piper* species. (*b*) Number of geometrid species per host *Piper* species.

($P < 0.005$, G-test, Table 9-4); twice as many *Piper* species completely escape attack from geometrids than from weevils (Fig. 9-3 *a, b*: 19 versus eight species, respectively).

Plant species abundance and size explained a significant portion of the variance in numbers of geometrid and weevil species, both individually and collectively (Table 9-5). More abundant *Piper* species at La Selva have more associated weevil species, and more weevil and geometrid species together. Plant size (maximum height attained at La Selva) was a significant positive predictor of the number of geometrids associated with a given *Piper* species, and the weevils

TABLE 9-4. Distribution Patterns of the Number of Herbivore Species per Host Plant for Weevils and Geometrids on *Piper* at La Selva and Patterns Reported for Other Sets of Tropical Herbivores[a]

| Host Group | Total Host Species[b] | Number of Herbivore Species per Plant Species | | | Source |
		Mean (SD)	Median	Range	
Santa Rosa woody plants	790	0.6 (0.2)	0	0–9 A	Janzen 1980
Limoncocha Solanaceae	42	1.0 (0.2)	1	0–3 B	Drummond 1986
La Selva *Piper* (geometrids)	45	1.2 (1.4)	1	0–8 C	This study
La Selva *Passiflora* (*Heliconius*)	10	1.6 (1.1)	1	0–3 DE	Smiley 1982
La Selva *Piper* (weevils)	45	2.5 (2.0)	2	0–8 D	This study
La Selva *Passiflora* (flea beetles)	10	2.4 (1.6)	2	0–5 DEFG	Smiley 1982
Campinas Solanaceae	39	1.5 (1.0)	1	0–4 E	Brown 1987
Rincón *Passiflora*	11	1.8 (2.6)	1	1–7 EF	Benson 1978
Arima Valley *Passiflora*	8	3.1 (2.0)	2.5	1–8 EFG	Benson 1978
Rio de Janeiro *Passiflora*	11	3.3 (2.8)	2	1–8 FG	Benson 1978
South American *Parkea*	10	3.6 (1.9)	3.5	1–8 GH	Hopkins 1983, 1984
Caribbean *Heliconia*	11	3.4 (1.1)	3	1–6 H	Strong 1977a

[a]Herbivore distributions with same letter not significantly different, $P > 0.05$, G-test.
[b]Includes host species with no recorded herbivores.

TABLE 9-5. Relationship Between Number of Geometrid Species, or Weevil Species, or Geometrid plus Weevil Species Associated with a Given *Piper* Species and the Size and/or Abundance of that Plant Species at Estación Biológica La Selva, Costa Rica[a]

Dependent Variable	Independent Variables	b	F	P	Model r^2	Model Intercept
Number of geometrid species	Maximum height	0.967	4.22	0.046	0.089	− 0.433
Number of weevil species	Abundance	0.984	8.64	0.005	0.278	− 1.220
	Maximum height	1.125	4.03	0.051		
Number of geometrid and weevil species	Abundance	1.931	7.92	0.007	0.290	− 2.160
	Maximum height	1.342	5.36	0.026		

[a]The regression models which explained the largest portion of the variance but included only significant independent variables are reported.

and geometrids together. In all cases, the variables included in the models explained a low proportion of the total variance in numbers of herbivore species (maximum = 29.0%). None of the remaining plant characteristics tested explained a significant proportion of the variance ($P > 0.20$).

Although the number of insect species that attack a given *Piper* species is associated with the abundance of that plant species, the abundance of the host plant does not appear to influence the abundance of a particular herbivore species. Of the six weevil species (Appendix 9-1) and 10 geometrid species

TABLE 9-6. Herbivore Faunas of *Piper arieianum* and *P. holdridgeianum* at Estación Biológica La Selva

	Piper arieianum		*Piper holdridgeianum*	
Insect Taxon	Stems, Foliage	Reproductive Parts	Stems, Foliage	Reproductive Parts
Orthoptera	19	1	3	0
Lepidoptera	27	0	1	0
Hemiptera	33	1	11	1
Coleoptera	7	4	0	3
Diptera	2	0	1	0
Hymenoptera (*Atta*)	1	0	1	0
Subtotal	89	6	17	4
Total per species	95		21	

(Appendix 9-2) that are rare at La Selva, none is associated with only rare host plant species. Thus, rare herbivore species are not necessarily rare because they feed only on rare hosts.

Table 9-6 gives the distribution of herbivore species by taxa which feed on La Selva *Piper arieianum* and *P. holdridgeianum*, the two host species for which I have the most complete data (see Appendices 9-3 and 9-4 for listings of identified species). I have no information as to possible root-feeders. At this time, 19 species of leaf-chewers on *P. arieianum* (33.9% of total leaf-chewing species) and one species on *P. holdridgeianum* (17%) are known to feed only on their respective hosts (Appendices 9-3 and 9-4). Twenty-one species on *P. arieianum* or 38% (one species on *P. holdridgeianum*) feed on more than one *Piper* species but not outside of the genus. Seven species of leaf-chewers on *P. arieianum* (12.5%) and two species on *P. holdridgeianum* (33.3%) are known relative generalists: they feed on species from at least one family other than Piperaceae. This categorization of host specialization must be considered preliminary until further rearing records are accumulated. Degree of specialization of sap-feeders on these two *Piper* species is not known at this time. Mammals (species unknown) feed on young stems of *P. arieianum* but only very infrequently; no mammalian damage has been observed on *P. holdridgeianum*.

Diversity of herbivore species for *P. arieianum* is four times as great as that for *P. holdridgeianum* (Table 9-6), despite the fact that these two species are approximately the same size and of equal abundance at La Selva. Herbivore species on these two plant species are not equally abundant (Table 9-7, Appendices 9-3 and 9-4). For example, of the 27 species of Lepidoptera which feed on *P. arieanum* at La Selva, only nine species were found on census plants five or more times over 1.5 years (Figure 9-4). If the leaf-chewing species of these two plant species are classified on the basis of their frequency of occurrence, common species of *P. arieianum* herbivores outnumber those of *P. holdridgeianum* by a factor of almost 7 (20 versus 3 species).

Damage produced by their respective herbivore faunas is not related to the number of species which feed on *P. arieianum* and *P. holdridgeianum* (Table 9.7). Even though there are 10 times as many leaf-chewing herbivore species on *P. arieianum* as on *P. holdridgeianum* (seven times as many common species), damage was not significantly greater in *P. arieianum* than *P. holdridgeianum* (ANOVA: $F = 1.01$, $df = 1, 48$, $P > 0.20$).

TABLE 9-7. Relationship between Number of the Herbivore Species and the Damage They Cause[a] for *Piper arieianum* and *P. holdridgeianum* at Estación Biológica La Selva

Plant Species	$\bar{X} \pm SD$ Leaf Area Missing (%)	Total Number Leaf-Chewing Insect Species	Number Common Leaf-Chewing Insect Species
Piper arieianum	16.7	56	20
Piper holdridgeianum	14.8	6	3

[a]Mean leaf area missing ($\pm$ SD), $n = 25$ plants per species, all leaves per plant.

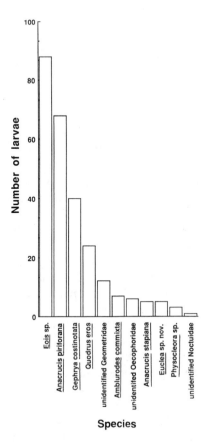

Species

Figure 9-4. Total number of larvae encountered on 200 plants of *Piper arieianum* summed over all census periods from April 1981 to October 1982 (see Methods for further information on sampling method).

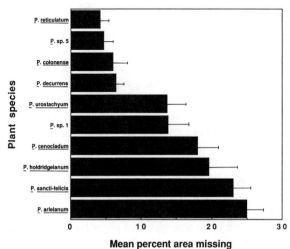

Figure 9-5. Mean percent area missing (± SE) per *Piper* species for a minimum 50 freshly abscissed leaves per species.

Damage levels per species, based on a sample of 50 or more newly abscised leaves per species, varied greatly (Fig. 9-5). Damage ranged from 5.1% for *P. reticulatum* to 25.5% for *P. arieianum*, and varied significantly among this set of 10 species (ANOVA: $F = 2.18$, $df = 9$, 756, $P < 0.05$).

DISCUSSION

Local species richness of the herbivore fauna of *Piper* is positively associated with host plant abundance and size. Similar results were found for two other plant genera at La Selva and their respective herbivore faunas: the number of hispine beetle (Chrysomelidae) species associated with a given *Heliconia* (Heliconiaceae) (Strong 1977a,b) and the number of *Heliconius* butterflies and flea beetle (Chrysomelidae) species on *Passiflora* (Passifloraceae) (Gilbert and Smiley 1978). These same patterns hold for local species richness of herbivores on temperate plant species (Opler 1974; Ward and Lakhani 1977; Strong et al. 1984; but see Karban and Ricklefs 1983). The factors affecting regional herbivore faunal diversity in tropical systems await further investigation; while host plant range, architectural complexity, habitat diversity, and taxonomic isolation (reviewed by Strong et al. 1984) have all been shown to be important correlates of regional faunal diversity of temperate plant species, local and regional diversities are not necessarily related (Gilbert and Smiley 1978; Strong et al. 1984; Cornell 1985).

Evidence from this study suggests that rare herbivore species which feed on La Selva *Piper* are not rare because their host plants are rare. None of the weevil or geometrid species is restricted to rare *Piper* species; all have been recorded on common *Piper* species as well. However, there are at least two other sources of rarity of attack on a given *Piper* species at La Selva which probably are important: (1) rarity due to only infrequent oviposition or colonization on a particular *Piper* by an herbivore which is common on other *Piper* or non-*Piper* species; (2) absolute rarity, i.e., an herbivore which is rare on all potential hosts at the site. An example of the first type of rarity is the nymphalid butterfly *Consul fabius*, which feeds commonly on La Selva *Piper* of open, second growth vegetation, mainly *P. sancti-felicis*. I have encountered *C. fabius* feeding on *P. arieianum* only twice; in both cases the plant was growing on the edge of open second growth vegetation. It is possible to raise *C. fabius* larvae to apparently normal adults on *P. arieianum* (personal observation). Thus, almost complete exclusion of *P. arieianum* from the diet of *C. fabius* appears to be a case of "ecological restriction" of the diet (Smiley 1978): adults do not search in all habitats in which suitable hosts occur. There are also cases of species which are truly rare. For example, I have seen damage, and less commonly larvae, of an unknown species of larentiine geometrid on *P. holdridgeianum* in two of seven years. The larva is distinct from other known *Piper* feeders, and damage has only appeared during December–January. The fact that the abundance of this species fluctuates in the face of what would appear to be an abundant resource (old leaves) suggests that, from the point of view of the host plant, the herbivore population is not constant. Whether in fact population fluctuation actually occurs at sites other than La Selva, or instead the rare appearance of adults at La Selva represents immigration from a nonfluctuating population, is not known at this time (see also Janzen 1987a, b). At least for *Piper* weevils, there is geographic variation in herbivore abundance: two weevil species associated with La Selva *Piper* occur only in certain parts of the La Selva property despite the fact that their host plants are more widely distributed (personal observation).

Diversity of herbivore species associated with *Piper arieianum* (95 species total) is 2.7–12.6 times higher than that reported for the few temperate plant species comparable in size and for which records are available. *Juniperus communis* (Cupressaceae) (a small tree to 10 m) has an average 7.5 associated herbivore species per site ($n = 21$ sites) with a maximum of 15 (Ward and Lakhani 1977). For *Opuntia* (Cactaceae) species of architectural rating 10 or less (*P. arieianum* = 6) mean number of herbivore species per plant species is 8.2 (Moran 1980). For creosote bush, *Larrea tridentata* and *L. cuneifolia* (Zygophyllaceae) (each 0.5–3 m tall) local number of herbivore species is 22 and 19, respectively (Schultz et al. 1977). Scotch broom, *Sarothamnus scoparius* (Leguminosae), has a maximum 35 associated herbivore species at Silwood Park (Waloff and Richards 1977). These values are equal to or less than (except for that of *S. scoparius*) the 20 species associated with *P. holdridgeianum*. Although the number of species of herbivores associated with La Selva *P. arieianum* may be higher than for similar temperate plant species, the distribution pattern of individuals among species is similar. Temperate herbivore faunas are also composed of a few species with many individuals and varying numbers of remaining species which are much less frequently represented (e.g., Lawton 1976; Marquis and Passoa 1989).

Comparison of host specificity of tropical and temperate plant–herbivore systems must take into account the wide range in variability observed among systems within and among tropical habitats (Tables 9-3 and 9-4). Even for the same set of host plants at the same location (*Piper* geometrids and weevils at La Selva), the rules of organization for number of herbivore species on host plants and number of host plants per herbivore species apparently differ. *Piper*-feeding weevils may be more generalized than *Piper*-feeding moths, perhaps because a lesser-flying capability restricts ability of the weevils to find rare hosts. The fact that host-plant abundance is a significant positive predictor of number of weevil species associated with a given *Piper* species, but not so for number of geometrids, is consistent with this hypothesis. In turn, larger *Piper* species may have both more weevils and geometrids than small species because larger species are more apparent (sensu Feeny 1976) to weevils but also of higher quality (greater abundance and predictibility of a particular leaf age class) to geometrids. For *Heliconius* butterflies and flea beetles on La Selva *Passiflora* species, Gilbert and Smiley (1978) suggest that both *Passiflora* abundance and quality (rate of new shoot production) determine the total number of species associated with a given *Passiflora* species. The fact that soil type was not a significant predictor of the number of herbivores associated with a given La Selva *Piper* species suggests that leaf quality factors alone do not determine interspecific choices by *Piper* herbivores. Soil nutrient levels vary greatly over the La Selva property (Sanchez and Mata 1987, Vitousek and Denslow 1986), and these differences are great enough to affect both leaf nutrient content and phenolic chemistry within a plant species (Denslow et al. 1987).

Despite the fact that it is possible to predict the number of herbivore species associated with a plant species based on its overall geographic range and local abundance (Strong et al. 1984), evidence provided here suggests that knowledge of the factors which structure the herbivore community of a given plant species does not necessarily reveal the impact of that community on the plant. Two *Piper* species which differed greatly in herbivore faunal diversity experienced very similar damage levels. Thus, rarity may lead to fewer herbivores, but not necessarily to lower damage (see also Futuyma and Wasserman 1980; Clark and Clark, Chapter 10). The tentative conclusion, as has been suggested previously (Rhoades and Cates 1976, pp. 204–205; Strong et al. 1984, p. 71), is that the number of herbivores and the amount of damage they cause are not necessarily related. This implies that even if tropical herbivore faunas are more diverse than those of temperate plant species, as might be predicted by the pattern of increased species diversity with decreasing latitude, enumeration of the herbivore species per plant species will not allow us to determine if selective impact of herbivores is greater in tropical than in temperate systems (Baker 1970; Levin 1976; Levin and York 1978; Langenheim 1984; and see P. Coley, Chapter 2).

Piper species of La Selva, growing in the same habitat at the same relative abundance, vary greatly both in the number of herbivores associated with them and the amount of damage caused by those herbivores. For *P. arieianum* and *P. holdridgeianum*, even the specificity of the major damaging herbivores differs: more than 95% of the leaf damage to *P. holdrigeianum* is caused by polyphagous tettigoniids, while damage to *P. arieianum* is by both *Piper* specialists and more polyphagous species (in preparation). The conclusion is that information about one particular *Piper* species system does not allow us to generalize to all congenerics. In addition, there is as much variation in damage levels within this set of *Piper* species as there was for a set of 30 La Selva species, representing 27 genera and 18 families (Gonzalez et al. 1985 in Marquis and Braker 1991) (range in mean damage levels per species = 0.5–26.5%; coefficients of variation for the two sets of species were not significantly different, 174.19 versus 180.21, $t = 0.256$, $P > 0.50$, $df = 38$; Sokal and Braumann 1980). This suggests that, contrary to what we would predict *a priori*, individual La Selva *Piper* species are perceived to be as diverse in their suitabilities to La Selva herbivores as are a set of very unrelated plant species. What roles plant abundance, nutritional composition, and physical and chemical defenses have in shaping the herbivore faunas and in turn damage levels of individual plant species await further investigation.

ACKNOWLEDGMENTS

I thank the staff of the Organization for Tropical Studies for logistical support. D. R. Whitehead, S. Passoa, D. H. Janzen, D. G. Ferguson, M. A. Solis, J. F. Gates Clarke, M. Epstein, R. W. Poole, F. Ringe, A. Freeman, D. C. F. Rentz, and

L. H. Rolston together made possible the discussion of patterns through their help with insect identification. Financial support came from NSF grants DEB 81-10197 and BSR-8600207, a Noyes pre-doctoral and Noyes post-doctoral fellowship in association with the Organization for Tropical Studies, and two grants-in-aid from Sigma Xi. C. Kelly, A. Zangerl, M. Berenbaum, M. Cohen, E. Heininger, N. Greig, and J. Schultz provided helpful comments on an earlier draft, and May Berenbaum kindly provided laboratory space during the writing of the paper. I also wish to thank John Lawton for a number of important references. I dedicate this paper to Donald R. Whitehead (1938–1990), who provided the necessary expertise for continuation of this research, as well as much inspiration through his enthusiasm for tropical weevils.

REFERENCES

Baker, H. G. 1970. Evolution in the tropics. *Biotropica* **2**: 101–111.

Baldwin, I. T. and J. C. Schultz. 1988. Phylogeny and the patterns of leaf phenolics in gap- and forest-adapted *Piper* and *Miconia* understory shrubs. *Oecologia* **75**: 105–109.

Benson, W. W. 1978. Resource partitioning in passion vine butterflies. *Evolution* **32**: 493–518.

Benson, W. W., K. S. Brown, and L. E. Gilbert. 1975. Coevolution of plants and herbivores: Passion flower butterflies. *Evolution* **29**: 659–680.

Brown, K. S., Jr. 1987. Chemistry at the Solanaceae/Ithominiinae interface. *Annals of the Missouri Botanical Garden* **74**: 359–397.

Cornell, H. V. 1985. Local and regional richness of cynipine gall wasps on California oaks. *Ecology* **66**: 1247–1260.

Denslow, J. S., P. M. Vitousek, and J. C. Schultz. 1987. Bioassays of nutrient limitation in a tropical rain forest soil. *Oecologia* **74**: 370–376.

DeVries, P. J. 1985. Hostplant records and natural history notes on Costa Rican butterflies (Papilionidae, Pieridae & Nymphalidae). *Journal of Research on the Lepidoptera* **24**: 290–333.

Dirzo, R. 1984. Herbivory: A phytocentric overview. In R. Dirzo and J. Sarukhán (Eds.) *Perspectives on Plant Population Ecology.* Sinauer, Sunderland, MA, pp. 141–165.

Doutt, R. L. 1960. Natural enemies and insect speciation. *Pan-Pacific Entomologist* **36**: 1–14.

Drummond, B. A. III. 1986. Coevolution of Ithomiine butterflies and solanaceous plants. In W. G. D'Arcy (Ed.) *Solanaceae, Biology and Systematics.* Columbia University Press, pp. 307–327.

Feeny, P. 1976. Plant apparency and chemical defense. *Rec. Adv. Phytochem.* **10**: 1–40.

Futuyma, D. J. and S. S. Wasserman. 1980. Resource concentration and herbivory in oak forests. *Science* **210**: 920–922.

Gilbert, L. E. and J. T. Smiley. 1978. Determinants of local diversity in phytophagous insects: Host specialists in tropical environments. In L. A. Mound and N. Waloff (Eds.) Diversity of Insect Faunas. *Symposia of the Royal Entomological Society of London* **9**: 89–105.

Hartshorn, G. S. 1983. *La Selva.* In D. H. Janzen (Ed.) *Costa Rican Natural History.* University of Chicago Press, Chicago Illinois, pp. 136–141.

Hespenheide, H. A. 1973. A novel mimicry complex: Beetles and flies. *Journal of Entomology (A)* **48**: 49–56.

Hespenheide, H. A. 1987. A revision of *Lissoderes* Champion (Coleoptera: Curculionidae: Zygopinae). *The Coleopterist's Bulletin* **41**: 41–55.

Hopkins, M. J. G. 1983. Unusual diversities of seed beetles (Coleoptera: Bruchidae) on *Parkia* (Leguminosae: Mimosoideae) in Brazil. *Biological Journal of the Linnean Society* **19**: 329–338.

Hopkins, M. J. G. 1984. The seed beetles (Bruchidae) of *Parkia* (Leguminosae: Mimosoideae) in Brazil: Strategies of attack. In A. C. Chadwick and S. L. Sutton (Eds.) *Tropical Rain-Forest: The Leeds Symposium*. W. S. Maney & Sons, Leeds. pp. 139–145.

Janzen, D. H. 1979. New horizons in the biology of plant defenses. In G. A. Rosenthal and D. H. Janzen (Eds.) *Herbivores: Their Interactions with Plant Secondary Metabolites*. Academic Press, New York, pp. 331–350.

Janzen, D. H. 1980. Specificity of seed-attacking beetles in a Costa Rican deciduous forest. *Journal of Ecology* **68**: 929–952.

Janzen, D. H. 1984. Two ways to be a tropical big moth: Santa Rosa saturniids and sphingids. *Oxford Surveys in Evolutionary Biology* **1**: 85–140.

Janzen, D. H. 1985. A host plant is more than its chemistry. *Illinois Natural History Survey Bulletin* **33**: 141–174.

Janzen, D. H. 1987a. When, and when not to leave. *Oikos* **49**: 241–243.

Janzen, D. H. 1987b. How moths pass the dry season in a Costa Rican dry forest. *Insect Sci. Applic.* **8**: 489–500.

Janzen, D. H. 1988. Ecological characterization of a Costa Rican dry forest caterpillar fauna. *Biotropica* **20**: 120–135.

Janzen, D. H. and P. G. Waterman. 1984. A seasonal census of phenolics, fibre and alkaloids in foliage of forest trees in Costa Rica: Some factors influencing their distribution and relation to host selection by Sphingidae and Saturniidae. *Biol. J. Linn. Soc.* **21**: 439–454.

Karban, R. and R. E. Ricklefs. 1983. Host characteristics, sampling intensity, and species richness of Lepidopteran larvae on broad-leaved trees in southern Ontario. *Ecology* **64**: 636–641.

Langenheim, J. H. 1984. The roles of plant secondary chemicals in wet tropical ecosystems. In E. Medina, H. A. Mooney, and C. Vazques-Yanes (Eds.) *Physiological Ecology of Plants of the Wet Tropics*. Dr. W. Junk Publishers, The Hague. pp. 189–208.

Lawton, J. H. 1976. The structure of the arthropod community on bracken. *Biol. J. Linnean Soc.* **73**: 187–216.

Levin, D. A. 1976. Alkaloid-bearing plants: An ecogeographic perspective. *Amer. Nat.* **110**: 261–284.

Levin, D. A. and B. M. York. 1978. The toxicity of plant alkaloids: An ecogeographic perspective. *Biochem. Syst. Eco.* **6**: 61–76.

Lowman, M. D. 1982. Effects of different rates and methods of leaf area removal on rain forest seedlings of coachwood (*Ceratopetalum apetalum*). *Aust. J. Bot.* **30**: 477–483.

Marquis, R. J. 1988a. Intra-crown variation in leaf herbivory and seed production in striped maple, *Acer pensylvanicum* L. (Aceraceae). *Oecologia* **77**: 51–55.

Marquis, R. J. 1988b. Phenological variation in the neotropical understory shrub *Piper arieianum*: Causes and consequences. *Ecology* **69**: 1552–1565.

Marquis, R. J. and H. E. Braker. 1991. Plant–herbivore relations at La Selva: Diversity, specificity and impact. In L. McDade, K. S. Bawa, G. S. Hartshorn, and H. E. Hespendheide (Eds.). *La Selva: Ecology and Natural History of a Neotropical Rain Forest*. University of Chicago Press, Chicago, in press.

Marquis, R. J. and S. Passoa. 1989. Seasonal diversity and abundance of the herbivore fauna of striped maple *Acer pensylvanicum* L. (Aceraceae) in western Virginia. *Amer. Midl. Nat.* **122**: 313–320.

Moran, V. C. 1980. Interactions between phytophagous insects and their *Opuntia* hosts. *Ecological Entomology* **5**: 153–164.

Opler, P. A. 1974. Oaks as evolutionary islands for leaf-mining insects. *American Scientist* **62**: 67–73.

Rhoades, D. F. and R. G. Cates. 1976. Toward a general theory of plant antiherbivore chemistry. *Recent Advances in Phytochemistry* **10**: 168–213.

Sanchez, M. F. and R. Mata Ch. 1987. *Estudio detallado de suelos, Estación Biológica "La Selva"*. Organizacíon para Estudios Tropicales, San José, Costa Rica.

Schultz, J. C., D. Otte, and F. Enders. 1977. *Larrea* as a habitat component for desert arthropods. In T. J. Mabry, J. H. Hunzicker, and D. R. Difeo (Eds.). *Creosote Bush: Biology and Chemistry of Larrea in New World Deserts*. Dowden, Huchinson & Ross, Stroudsberg, PA, pp. 176–208.

Smiley, J. T. 1978. Plant chemistry and the evolution of host specificity: New evidence from *Heliconius* and *Passiflora*. *Science* **201**: 745–747.

Smiley, J. T. 1982. The herbivores of *Passiflora*: Comparison of monophyletic and polyphyletic feeding guilds. *Proceedings Fifth International Symposium Insect–Plant Relationships*, Wageningen, pp. 325–330.

Sokal, R. R. and C. A. Braumann. 1980. Significance tests for coefficients of variation and variability profiles. *Syst. Zool.* **29**: 50–66.

Sokal, R. R. and R. J. Rohlf. 1981. *Biometry*. W. H. Freeman, San Francisco, CA.

Strong, D. R., Jr. 1977a. Rolled-leaf hispine beetles (Chrysomelidae) and their Zingiberales host plants in Middle America. *Biotropica* **9**: 156–169.

Strong, D. R., Jr. 1977b. Insect species richness: Hispine beetles of *Heliconia latispatha*. *Ecology* 58: 573–582.

Strong, D. R., J. H. Lawton, and T. R. E. Southwood. 1984. *Insects on plants, community patterns and mechanisms*. Harvard University Press.

Vitousek, P. M. and J. S. Denslow. 1986. Nitrogen and phosphorous availability on treefall gaps of a lowland tropical rain forest. *J. Ecol.* **74**: 1167–1178.

Waloff, N. and O. W. Richards. 1977. The effect of insect fauna on growth, mortality and natality of broom, *Sarothamnus scoparius*. *Journal of Applied Ecology* **14**: 787–798.

Ward, L. K. and K. H. Lakhani. 1977. The conservation of juniper: The fauna of food-plant island sites in southern England. *Journal of Applied Ecology* **49**: 121–135.

Winder, J.A. and H. F. van Emden. 1981. Selection of effective biological control agents from artificial defoliation/insect cage experiments. In E. S. Del Fosse (Ed.) *Proc. Fifth Int. Symp. Biol. Contr. Weeds*, pp. 415–439.

APPENDIX 9-1. Abundance and Leaf Age Preference of Weevils (Curculionidae) on Their *Piper* Host Plants at La Selva

Weevil Species	Host Species	Weevil Abundance[a]	Leaf Age Preference[b]
Ambates cretifer			
Pascoe	*Piper biseriatum*	C	1–2
	Piper cenocladum	C	2
	Piper sp. 2	R	2
	Piper sp. 5	R	2
Ambates solani			
Champion	*Piper auritum*	C	2
	Piper sancti-felicis	C	2
Ambates sp. 1	*Piper carrilloanum*	R	2
	Piper urophyllum	R	2
Ambates sp. 2	*Piper auritum*	C	2
	Piper culebranum	C	2
	Piper decurrens	R	2
	Piper dolichotrichum	R	2
	Piper sancti-felicis	R	2
	Piper tonduzii	C	1
	Piper urostachyum	C	1
	Piper xanthostachyum	C	1
	Piper sp. 2	C	1–2
	Piper sp. 3	C	1–2
Ambates sp. 3	*Piper sancti-felicis*	C	2
	Piper sp. 3	C	2
Ambates sp. 4	*Piper* sp. 1	R	1
Ambates sp. 5	*Piper arieianum*	C	1
	Piper augustum	C	1
Ambates sp. 6	*Piper culebranum*	C	2
	Piper decurrens	C	2
	Piper peracuminatum	C	2
	Piper sp. 4	R	2
	Piper sp. 6	R	2
Ambates sp. 7	*Piper arboreum*	C	1
	Piper biseriatum	C	1
	Piper cenocladum	C	1
	Piper euryphyllum	R	1
	Piper imperiale	C	1
Ambates sp. 8	*Piper sancti-felicis*	C	2
	Piper sp. 2	C	2
	Piper sp. 3	C	2
Ambates sp. 10	*Piper imperiale*	R	2
Ambates sp. 11	*Piper biolleyi*	R	2
	Piper glabrescens	R	1–2
	Piper terrabanum	R	2
	Piper tonduzii	R	1–2
Ambates sp. 12	*Piper cenocladum*	R	2

APPENDIX 9-1. (*Continued*)

Weevil Species	Host Species	Weevil Abundance[a]	Leaf Age Preference[b]
Ambates sp. 13	*Piper biseriatum*	R	2
	Piper cenocladum	C	2
	Piper melanocladum	R	2
Peridinetellus subnudus Champion	*Piper concepcionis*	C	1
	Piper dolichotrichum	C	1
	Piper multiplinervium	C	1
	Piper silvivagum	C	1
	Piper xanthostachyum	C	1
Peridinetus coccineifrons Champion	*Piper carrilloanum*	R	2
	Piper urophyllum	R	2
Peridinetus cretaceus Pascoe	*Piper sancti-felicis*	C	2
	Piper virgultorum	R	2
	Piper sp. 2	R	2
Peridinetus sp. 1	*Piper imperiale*	C	2
Peridinetus sp. 2	*Piper augustum*	C	1
	Piper auritum	R	1
	Piper cenocladum	C	1
	Piper carrilloanum	R	2
	Piper decurrens	R	1–2
	Piper multiplinervium	R	1
	Piper phytolaccaefolium	R	1
	Piper sancti-felicis	C	2
	Piper silvivagum	C	1
	Piper tonduzii	R	1
	Piper urophyllum	R	1
	Piper sp. 1	R	1
	Piper sp. 2	C	1
	Piper sp. 3	C	1
Peridinetus spp. 3, 9, and 10[c]	*Piper arieianum*	C	2
	Piper culebranum	C	2
	Piper concepcionis	C	2
	Piper dolichotrichum	C	2
	Piper garagaranum	C	2
	Piper nudifolium	C	2
	Piper peracuminatum	R	2
	Piper reticulatum	R	2
	Piper sancti-felicis	C	2
	Piper silvivagum	C	2
	Piper urostachyum	C	2
	Piper sp. 2	R	2

(*continued*)

APPENDIX 9-1. (*Continued*)

Weevil Species	Host Species	Weevil Abundance[a]	Leaf Age Preference[b]
Peridinetus sp. 4	*Piper auritum*	C	2
Peridinetus spp. 5 and 6[c]	*Piper arieianum*	C	2
	Piper phytolaccaefolium	C	2
Peridinetus sp. 7	*Piper arboreum*	R	1–2
	Piper auritum	R	2
	Piper biseriatum	C	1–2
	Piper cenocladum	C	1–2
	Piper melanocladum	R	2
	Piper sp. 5	C	2
Peridinetus sp. 8[d]	*Piper arieianum*	C	1
	Piper augustum	R	1
	Piper auritum	R	1
	Piper biseriatum	R	1
	Piper carrilloanum	C	1
	Piper culebranum	C	1
	Piper decurrens	R	1
	Piper dolichotrichum	R	1
	Piper glabrescens	C	1
	Piper multiplinervium	C	1
	Piper otophorum	R	1
	Piper peracuminatum	R	1
	Piper phytolaccaefolium	R	1
	Piper sancti-felicis	C	1
	Piper silvivagum	R	1
	Piper tonduzii	C	1
	Piper urostachyum	C	1
	Piper sp. 3	R	1
Rhodobaenus cf. *nawradii* (Kirsch)	*Piper* sp. 1	C	2

[a]C = common, 20 or more sitings and/or collections; R = rare, 5 or fewer sitings and/or collections.
[b]1 = 1/4–1/2 full size; 2 = 1/2 full size to just fully expanded.
[c]Could not be distinguished in the field.
[d]Feeds exclusively on seedlings of all species.

APPENDIX 9-2. Abundance and Leaf Age Preference of Geometridae (subfamily Larentiinae) on Their *Piper* Hosts at La Selva Costa Rica

Geometrid Species	Note[a]	Host Species	Geometrid Abundance	Leaf Age Preference
Amaurinia sp. 1	Near	*Piper carrilloanum*	R	N
	cassandra Druce	*Piper reticulatum*	R	M–O
		Piper xanthostachyum	R	M–O
		Piper sp. 7	R	M–O
Cambogia sp. 1	Near	*Piper arboreum*	C	M–O
	apyraria Guen.	*Piper biseriatum*	C	M–O
		Piper cenocladum	C	M–O
		Piper euryphyllum	C	M–O
		Piper imperiale	C	M–O
		Piper melanocladum	C	M–O
Cambogia sp. 2	Near	*Piper sancti-felicis*	R	M–O
	cedon Druce	*Piper virgultorum*	R	M–O
Cambogia sp. 3	Near	*Piper dolichotrichum*	C	M–O
	crocina Schaus	*Piper xanthostachyum*	C	M–O
Cambogia sp. 4	Near	*Piper urostachyum*	C	M–O
	crocina Schaus	*Piper dolichotrichum*	R	M–O
Cambogia sp. 5	Near	*Piper auritum*	C	M–O
	mexicaria	*Piper sancti-felicis*	C	M–O
	Walker	*Piper urostachyum*	C	M–O
		Piper virgultorum	C	M–O
Cambogia sp. 6	Near	*Piper biolleyi*	C	M–O
	numida Druce	*Piper culebranum*	C	M–O
Cambogia sp. 7	Near	*Piper sancti-felicis*	C	M–O
	tegularia Guen.	*Piper* sp. 3	R	M–O
Cambogia sp. 8	Near	*Piper multiplinervium*	R	M–O
	tegularia Guen.			
Cambogia sp. 9		*Piper sancti-felicis*	R	M–O
Cambogia sp. 10		*Piper biolleyi*	R	M–O
		Piper culebranum	R	M–O
		Piper decurrens	C	M–O
		Piper sancti-felicis	C	N
		Piper virgultorum	R	M–O
		Piper sp. 5	C	M–O
Cambogia sp. 11	Near	*Piper biolleyi*	R	M–O
	obada Druce	*Piper glabrescens*	R	M–O
Cambogia sp. 12		*Piper biolleyi*	R	M–O
		Piper culebranum	R	M–O
		Piper sancti-felicis	R	M–O
Cambogia sp. 13		*Piper cenocladum*	R	M–O
		Piper imperiale	R	M–O
Cambogia sp. 14	Near	*Piper* sp. 5	R	M–O
	apyraria Guen.			

(*continued*)

APPENDIX 9-2. (*Continued*)

Geometrid Species	Note[a]	Host Species	Geometrid Abundance	Leaf Age Preference
Eois sp. 1	Near	*Piper arieianum*	C	M–O
	binaria Guen.	*Piper auritum*	R	M–O
		Piper imperiale	R	M–O
		Piper urostachyum	R	M–O
Eois sp. 2	Near	*Piper arboreum*	C	M–O
	binaria Guen.			
		Piper glabrescens	R	M–O
Eois sp. 3	Near	*Piper sancti-felicis*	C	M–O
	binaria Guen.	*Piper arieianum*	R	M–O
Eois sp. 4	Near	*Piper arboreum*	R	M–O
	binaria Guen.	*Piper silvivagum*	R	M–O
Eois sp. 5	Near	*Piper arboreum*	R	M–O
	binaria Guen.	*Piper silvivagum*	R	M–O

[a]Based on adult morphology.
[b]R = 3 or fewer adults raised; C = 15 or more adults raised, and/or larvae encountered in the field.
[c]N = new, not fully expanded; M–O = fully expanded, medium to old aged leaves.

APPENDIX 9-3. Insect Herbivore Species of *Piper arieianum* at La Selva, Costa Rica

Insect taxon	Species	Host Range at La Selva[a]	Plant Part[b]	Abundance[c]
Orthoptera				
Eumasticidae	*Homeomastox robertsi* (Descamps)	*Piper* spp., Leguminosae, Rubiaceae, Flacourtiaceae	L	C
Phasmidae	15 unidentified species	?	L	5 spp. C, 1 spp. R
Tettigoniidae	2 unidentified species	*Piper* spp., Solanaceae	L, R	R
Tettigoniidae	*Orophus tessellatus* Saussure	*Piper* spp.	L	R
Tetrigidae	1 unidentified species	*Piper* spp.	L	C
Hemiptera				
Aleyrodidae	1 unidentified species	?	S	R
Aphididae	1 unidentified species	?	S	R
Cercopidae	1 unidentified species	*Piper* spp.	S	C
Cicadellidae	11 unidentified species	?	L	3 spp. C, 8 spp. R
Fulgoroidea	18 unidentified species	?	L,S	8 spp. C, 10 spp. R
Pentatomidae	*Engelmani* Rolston	*Piper* spp.	R	C
Psyllidae	1 unidentified species	?	S	R
Coleoptera				
Curculionidae	*Ambates chaetopus* Champion	*Piper* spp.	R	C
	Ambates sp. 5	*Piper* spp.	L,S,R	C
	Cyrionix sp. 1	*Piper* spp.	R	C
	Cyrionix sp. 2	*Piper* spp.	R	R
	Cyrionix sp. 3	*Piper* spp.	R	R

(continued)

205

APPENDIX 9-3. (*Continued*)

Insect taxon	Species	Host Range at La Selva[a]	Plant Part[b]	Abundance[c]
	Peridinetus sp. 3	*Piper* spp.	L,S	C
	Peridinetus sp. 5	*Piper* spp.	L,S	R
	Peridinetus sp. 6	*Piper* spp.	L,S	C
	Peridinetus sp. 8	*Piper* spp.	L,S	C
	Peridinetus sp. 9	*Piper* spp.	L,S	C
	Peridinetus sp. 10	*Piper* spp.	L,S	R
Lepidoptera				
Apatelodidae	*Tarchon* sp.	*Piper* spp., Solanaceae, Araceae	L	R
Geometridae	*Amblurodes commixta* Warr.	*Piper* spp.	L	R
	Eois sp. 1	*Piper* spp.	L	C
	Eois sp. 3	*Piper* spp.	L	R
	Oxydia occiduata Guen.	?	L	R
	Paragonia planimargo Warr.	?	L	R
	Physocleora sp.	*Piper* spp.	L	C–R
	Species unidentified	*Piper* spp.	L	C–R
Hesperiidae	*Quadrus cerealis* (Stoll)	*Piper* spp.	L	C
Limacodidae	*Euclea* sp. nov.	?	L	C
	Species unidentified	?	L	R
Noctuidae	*Gonodonta* sp.	?	L	R
	Species unidentified	?	L	R
	Species unidentified	*Piper* spp.	L	R
Nymphalidae	*Adelpha heraclea* (Felder)[d]	Verbenaceae	L	R
	Consul fabius (Dbldy.)[d]	*Piper* spp.	L	R
	Zaretis itys (Cr.)[d]	Flacourtiaceae	L	R

Oecophoridae	Species unidentified	?	L	R
Pyralidae	*Gephyra costinotata* Schaus	*Piper* spp.	L	C
	Phostria nr. *humeralis* Guen.	*Piper* spp.	L	R
Tortricidae	*Amorbia* nr. *rectangularis* Meyrick	?	L	R
	Amorbia sp.	*Piper* spp.	L	R
	Anacrucis piriforana (Zeller)	*Piper* spp.	L	C
	Anacrucis stapiana (Felder)	?	L	R
	Anacrucis sp.	?	L	R
	Tortrix nephrodes Walshingham	?	L	R
	Species unidentified	*Piper* spp.	L	R
Diptera				
Cecidomyiidae	Species unidentified	?	L	C
	Species unidentified	?	L	C
Hymenoptera				
Formicidae	*Atta cephalotes* L.	Many families	L	C

[a] ? = No other host species known at this time; *Piper* spp. = feeds on one or more *Piper* species other than *P. arieianum*, with no records from other families.
[b] Plant part fed upon: L = leaves, S = stems, R = reproductive parts.
[c] C = common, 5 or more census records (see text) for Diptera, Lepidoptera, and Orthoptera; 10 or more census records for Coleoptera and Hemiptera. R = rare.
C–R = Abundant during only 3-month period during entire study period.
[d] See DeVries 1985.

APPENDIX 9-4. Insect Herbivores of *Piper holdridgeianum* at La Selva, Costa Rica

Insect Taxon	Species Name	Host Range at La Selva[a]	Plant Part[b]	Abundance[c]
Orthoptera				
Tettigoniidae	*Orophus tessellatus* Saussure	*Piper* spp.	L	C
	Orophus conspersus (Bruner)	*Piper* spp.	L	R
	Philophyllia sp.	*Pothomorphe peltata*		
		Sapindaceae, Solanaceae		
Hemiptera				
Cicadellidae	5 unidentified species	?	L	?
Cercopidae	1 unidentified species	?	S	?
Fulgoroidea	5 unidentified species	?	L,S	?
Pentatomidae	1 unidentified species	?	R	?
Coleoptera				
Curuculionidae	*Ambates chaetopus* Champ.	*Piper* spp.	R	?
	Cyrionix sp. 1	*Piper* spp.	R	?
	Cyrionix sp. 2	*Piper* spp.	R	?
Lepidoptera				
Geometridae	*Cambogia* sp.	?	L	R
Diptera				
Cecidomyiidae	Species unidentified	?	L	R
Hyemenoptera				
Formicidae	*Atta cephalotes* L.	Many families	L	R

[a]? = No other host species known at this time. *Piper* spp. = feeds on one or more *Piper* species other than *P. arieianum*, with no records from other families. [b]Plant part fed upon: L = leaves, S = stems, R = reproductive parts. [c]Abundance based on amount of damage to 25 different plants over 2.5 yr, censused every 2–4 months. C = common, caused 95% or more of total damage damage; R = caused less than 5% of total damage; ? = no information on relative abundance.

10 Herbivores, Herbivory, and Plant Phenology: Patterns and Consequences in a Tropical Rain-Forest Cycad

DAVID B. CLARK and DEBORAH A. CLARK

In any habitat there are four general options for reducing the impact of herbivory. Plants may hide in time or space. Hiding in space could involve becoming too rare to be found by herbivores, or moving into a microhabitat not accessible to them. Hiding in time could involve only producing edible tissue at times when herbivores are not active. Alternatively, plants could produce an ephemeral superabundance of edible tissue, temporarily overwhelming herbivores ("predator satiation", Janzen 1971). A second class of options involves what to do assuming the herbivores do find a plant. Here there are two general solutions: mechanical obstacles and chemical defenses. The relative contributions of these four lines of defense are difficult to quantify. However, by examining plant and herbivore behavior over the life cycle of the plant, it should be possible to determine which defenses are operating at different stages of plant-life history.

In this chapter we will document in detail how one tropical wet forest plant interacts with herbivores. The plant, *Zamia skinneri*, is a cycad attacked by only one major herbivore. The herbivore, larvae of the lycaenid butterfly *Eumaeus minyas*, eat only *Zamia skinneri*. The study is unusual in providing a complete record of leaf production and reproductive output for a natural population over a 7-yr period. We approach this analysis mainly from the plant's point of view, examining the consequences of different patterns of allocation in terms of attack and damage by arthropod herbivores. Overall, our goal is to demonstrate that multiple aspects of the plant's biology can be interpreted as adaptations to reduce the impact of herbivores.

Plant-Animal Interactions: Evolutionary Ecology in Tropical and Temperate Regions, Edited by Peter W. Price, Thomas M. Lewinsohn, G. Wilson Fernandes, and Woodruff W. Benson. ISBN   0-471-50937-X   © 1991 John Wiley & Sons, Inc.

FACTORS INFLUENCING HERBIVORY IN TROPICAL WET FORESTS

The plant we describe is a small shrub in the shaded understory of tropical wet forests. For several reasons this habitat is of unusual interest for studies of herbivory. The terrestrial habitats of the world differ greatly in the ecological and evolutionary relations between arthropod herbivores and the plants they eat. A major ecological axis influencing these relations is climate, which determines the length of plant growing seasons as well as herbivore activity patterns. Tropical wet forests occupy one end of this axis; there is the potential for year-round plant growth and herbivore activity. Tropical wet-forest plants must confront herbivores without a reliable temporal shelter to escape assault or decimate enemies. In addition, these plants face what is probably the greatest diversity of potential herbivores found in any community.

Understory plants face the additional problem of the extreme scarcity of light on the forest floor. Tropical rain-forest understories usually receive only 1–2% of the light that strikes the canopy (Chazdon and Fetcher 1984; Chazdon 1986). In this semiobscurity, plants struggle to maintain a positive carbon balance. The impact of herbivory should be especially important under these light-limited conditions (Coley 1983). As the methods of measuring light environments and plant responses improve, it is becoming increasingly clear that very small-scale events—a few minutes more direct light per day due to a nearby fallen branch, or the loss of one newly formed leaf—can significantly influence the growth, reproduction, and survival of tropical understory plants (Marquis 1984; Clark and Clark 1985, 1987; Oberbauer et al. 1988, 1989; Chazdon 1988).

ZAMIA SKINNERI, A TROPICAL RAIN FOREST CYCAD

Zamia skinneri (Zamiaceae) is a modest understory rosette shrub native to wet forests of Central America (Gomez 1982). The species is the only cycad that occurs in the tropical wet forest of the La Selva Biological Station in Costa Rica, where it has been the focus of a long-term study of patterns of allocation to above-ground growth and reproduction (Clark and Clark 1987, 1988). The species is especially suitable for this type of analysis. Individuals produce relatively few and long-lived leaves and cones (strobili), facilitating the accounting of allocation. Dead leaves usually leave persistent leaf bases, and new leaves are distinctively colored for several months. By counting the total number of leaves, dead-leaf rachises, and new leaves on a plant several times a year, it is possible to determine total production and mortality of leaves and also monitor reproductive output.

Based on our previous studies (Clark and Clark 1987, 1988) we can summarize the basic ecology of *Z. skinneri* as follows. New leaves first become visible as tender croziers. Over the course of approximately 65 days (unpublished data) they expand to 40–180 cm in rachis length and become quite tough. Leaves are produced in discrete episodes (flushes) of 1–12 leaves. New leaves on an individual emerge and harden almost simultaneously, such that the total period when soft new leaves are present is not much longer than the minimum time for one leaf to harden. The median incidence of leaf production is less than one leaf flush per year, and intervals up to several years without leaf production are common. Overlapping or directly consecutive flushes of new leaves do not occur. Leaf production is low, averaging a little over one half per year per plant, and median leaf longevity of expanded leaves exceeds 4 yr (unpublished data, $N = 329$ leaves). Median plant size of individuals in the marked population is 4 leaves (0–39).

Like all cycads, the species is strictly dioecious. On the average, females ($\bar{x} = 13$ leaves, minimum 6) are larger than males ($\bar{x} = 9$ leaves, minimum 3), and both are larger than nonreproductives. Reproduction is rare; only 26% of adult-sized individuals produced cones over a 6-yr period. A substantial commitment to increased leaf area precedes reproduction; leaf number increases 28–41% in individuals which will reproduce 1 yr later. Reproduction also involves a significant time commitment which is greater for females than males; male cones abscise after about 5 months, while female cones remain on the plant for about 18 months. Both reproduction and leaf production are limited by light; small differences in canopy openness are significantly correlated with increased leaf production and reproduction. There is a significant cost of reproduction for both males and females, expressed as a decreased production of new leaves following reproduction. Decreased leaf number on males and females is still evident up to 5 yr after reproduction.

In summary, *Zamia skinneri* is a slow-growing, long-lived denizen of an energy-poor habitat; it makes few reproductive and light-gathering organs, and the production of these organs has demonstrable effects on the plant's growth, reproduction, and survival.

From November 1980 until September 1987 we censused 195 individuals of *Zamia* 3–5 times per year at the La Selva Biological Station (mean rainfall ca. 4000 mm). About three-fourths of the plants were located in primary tropical wet forest, and the rest were in secondary forest. In addition to monitoring leaf production and reproduction (Clark and Clark 1987, 1988), we also recorded presence of eggs and larvae of the specialist herbivore *Eumaeus minyas* or adult languriid beetles (see below), and noted instances of leaf, cone, and stem destruction.

HERBIVORES OF *ZAMIA SKINNERI*

At La Selva there are only two major herbivores known for *Zamia skinneri*: the highly specialized lycaenid butterfly *Eumaeus minyas*, and a group of very similar species of languriid beetles. Together these insects produce nearly all herbivore damage observed on *Z. skinneri*. The impact of *E. minyas* is considerably greater than that of the beetles (see below). An unidentified leaf-miner has also been observed in *Z. skinneri* leaves (DeVries 1983; H. Hespenheide, personal communication), and a distinctive, rarely observed pattern of leaf damage indicates that another minor herbivore of *Z. skinneri* exists at La Selva. Notably absent from the list of *Zamia* herbivores is the abundant and polyphagic leaf-cutter ant, *Atta cephalotes*. Although many of the individuals we studied were located close to *Atta* nests, we observed only one instance of *Atta* cutting *Zamia* leaves. This individual was located directly on top of an *Atta* nest, and was repeatedly defoliated over a 7-yr period. This demonstrates that *Atta's* avoidance of *Zamia* was not due simply to an inability to harvest the tough adult leaves.

Our information on the two major herbivores is based on data from *Zamia* population censuses at intervals of several weeks to months (see methods). We documented the presence and activity of herbivores on obviously damaged plants at each census; total damage to the plant was assessed at the time of census and by comparison with condition at the next or previous census. Our data on the herbivores provide only a first-level analysis of the plant–herbivore interactions, and are subject to several limitations: *Eumaeus* larvae and the beetles are capable of moving among plants, and could be absent during the censusing of a plant they had damaged; when leaves or other plant parts were destroyed, any *Eumaeus* eggs on them would not be documented as having occurred on the plant; when leaf production occurred during intercensus intervals, total loss of the flushed leaves in some cases could have been missed. Because of these factors the results discussed below probably represent a minimum estimate of the impact of herbivores on *Z. skinneri*.

EUMAEUS MINYAS

Butterflies of the genus *Eumaeus* are aposematic specialists on cycads. *E. atala*, which feeds on the Florida cycad *Zamia floridana*, accumulates cycasin (Rothschild et al. 1986), a glucoside of the extremely toxic secondary compound characteristically produced by cycads (methylazoxymethanol). High concentrations of cycasin were found in the larvae, pupae, and adults. *Eumaeus* larvae (*E. atala*, Rothschild et al. 1986; *E. debora*, Ross 1964; *E. minyas*, Ross 1964, DeVries 1976, personal observation) are strikingly colored (bright red and white or yellow) and gregarious.

The only known food plant of *Eumaeus minyas* at La Selva is *Zamia skinneri* (the only cycad present). Most egg clusters are laid on plants with young tender leaves (78%; $N = 67$ observed egg clusters), and the eggs are nearly always laid on the young leaves (84% of clusters on these plants). In 15% of cases, the eggs had been laid on plants with cones (strobili) and tended to be on the cones themselves (7 of 8 observations). Clusters averaged 35 eggs (range = 4–81; $N = 22$). In Mexico, Ross (1964) found *E. minyas* clutches to be much smaller, usually on 3–8 eggs; this difference may be related to the considerably smaller size of the food plant, *Zamia loddigesii* var. *angustifolia*, as compared to *Z. skinneri*.

Although *Eumaeus minyas* larvae can feed on all above-ground parts of *Zamia*, our observations indicate that the smallest instars can only consume tender new leaves or cones. As noted above, most eggs are laid on either new foliage or cones, so most first instar larvae are in direct contact with these plant parts. When young leaves were available on plants damaged by *Eumaeus* ($N = 28$), in no cases were mature leaves destroyed while the young leaves escaped; in all cases of leaf loss some or all of the new leaves were destroyed. When we did see small larvae on mature leaves, they appeared to eat only the margins of the leaflets. The much greater toughness of mature leaves may protect them from the initial instar(s). If partially-toughened new leaves are also unavailable to first instar larvae, there may be a very brief period during leaf development when oviposition can lead to successful larvae. In 54% of cases of egg clusters on plants that were then flushing new leaves ($N = 41$), no leaves were lost. In some of these cases, the eggs may have been laid on new leaves that had already hardened sufficiently to deter the first-instar larvae. Most instances of *Eumaeus* damage to *Z. skinneri* ($N = 51$) include loss of some or all new leaves (57%) or damage to cones (16%). Although mature leaves and the woody stem appear to be safe from the smallest larvae, they can be consumed by later instars; there was damage to these plant parts in 61% of *Eumaeus* attacks ($N = 51$) on *Z. skinneri*. The ability to eat all parts of the cycad contrasts with the observation made by Ross (1964) in Mexico that the larvae of this species can only eat the young foliage of *Zamia loddigesii*, and that eggs are deposited only on tender new leaves. Similar restriction to young leaves has also been reported for *Eumaeus atala* on a Florida *Zamia* (Schwarz 1888, cited in Ross 1964; Rothschild et al. 1986).

Mortality of eggs and larvae appeared to be high. For 53% of egg clusters ($N = 57$), no larvae, pupae, or damage were subsequently observed; we do not know the proportions of failures that occurred prior to or after hatching. We never observed ants associated with eggs, larvae, or pupae, as is common in some lycaenids. Larvae may move from their initial plant to another, perhaps in response to poor food quality or total consumption of appropriate food items on the first plant. Ross (1964) reports that *E. minyas* larvae are often seen crawling on the ground in his study area in Mexico. We know of no such observations at La Selva, although we did observe larvae disappear from one plant and appear on a neighbor on one occasion. In 59% of the cases where larvae or pupae were found on plants ($N = 39$), however, no eggs were noted; although the eggs could have disappeared with consumed plant parts or we could have failed to note them, some of these cases may be due to larval movement between plants. Only two estimates of egg duration were obtained, 6 and 7 days; both observations were incomplete and therefore are minimum estimates. There is substantial mortality in the egg and larval stages; the mean number of larvae observed on plants was 15 ($N = 29$), less than half the mean number of eggs laid. Similarly, in 6 cases where we had counts of both eggs and larvae, mortality up to that point averaged 63% (larval number = 26–91% of egg number). Larval growth rates on *Zamia skinneri* may be slower than those for *Eumaeus minyas* on *Zamia loddigesii*; Ross (1964) reported that the total larval period on the latter species was 15–17 days. In three cases for which we have multiple records of a given group of larvae, the group was present on a plant for at least 17–29 days. Pupae were only observed in 7 cases. They were aggregated, often on the underside of a leaf of a non-*Zamia* individual near the attacked plant ($\bar{X} = 20$ pupae per group; range: 6–51; $N = 5$). In 5 of the 7 observations of pupae, the plant had been completely defoliated (although it is likely we searched harder for pupae around totally defoliated plants). The lack of sequelae of most egg masses, as well as the low number of pupae encountered, together suggest that the success rate of egg deposition by *E. minyas* on *Z. skinneri* is low.

THE BEETLE HERBIVORES

The other class of herbivore we regularly observed on *Zamia skinneri* was a group of species of very similar languriid beetles: *Camptocarpus longicollis*, *Nomotus aenescens*, and *Dasydactylus* sp. A sample of 10 beetles feeding on the same *Z. skinneri* leaf consisted of 8 *C. longicollis* and one each of the other two species (determined by J. M. Kingsolver). Because we did not distinguish these in the field, we will treat them together.

We have observed adults of these beetles eating *Zamia skinneri* foliage and cone tissue, but no other parts of the plants. In 30% of the observations of the beetles on *Zamia* ($N = 33$), the plants had male cones and the beetles were scraping the surfaces of the thick sporophylls; we never saw the beetles cause more than superficial damage to the cones. When the beetles were found on nonreproductive plants, the plants nearly always had new leaves (78% of cases). The beetles often were eating previously damaged (brown, often rotten-looking) leaflets. They were also found in association with a characteristic hollowing of the stout base of the leaf rachis, usually on immature leaves. We have not observed larvae. Females of some languriids are known to deposit their eggs in holes cut into living stems (Crowson 1981), and some larvae are known to be stem borers (Peterson 1951). It is possible that their larvae are responsible for the severely damaged leaf rachises.

There was a notable pattern of association of these beetles with the larvae of *Eumaeus*. When they cooccurred on plants, we often saw them in close proximity, sometimes with the beetles crawling among or on the larvae. In one instance we also found several of the beetles on a dense group of *Eumaeus* pupae, although they did not appear to be feeding on the pupae. Although the beetles rarely cooccurred with *Eumaeus* on plants with cones (only 1 out of 10 cases of beetles on reproductive plants), when on plants with new leaves they cooccurred with *Eumaeus* 67% of the time ($N = 18$). In the few instances when the beetles were found on plants with only mature leaves ($N = 5$), the beetles were always seen in association with *Eumaeus*. When nonreproductive plants were damaged and only beetles were observed on them, the damage was nearly always confined to the new foliage (5 of the 6 cases; 83%). These patterns suggest that the beetles by themselves cannot cause major damage to mature leaves or stems, and that they may be attracted to the odor of *Eumaeus* larvae and/or to the odor of damaged *Zamia* tissue.

THE RELATIVE IMPACTS OF *EUMAEUS MINYAS* AND THE BEETLE HERBIVORES

E. minyas is a much more important herbivore of *Z. skinneri* than are the languriid beetles. Total defoliation occurred in 23% of the *Eumaeus* attacks overall ($N = 80$). Only the butterfly larvae were capable of destroying cones, damaging the stem, and killing whole plants. *Eumaeus* attacks were also much more frequent; this species (eggs, larvae, or pupae) was observed on the plants in twice as many cases of leaf loss than were the beetles (25% vs. 13% of cases; $N = 157$). *Eumaeus* presence was four times more frequent in cases when such damage occurred and either herbivore was observed alone (16% vs. 4% of incidents). Finally, as noted above, the pattern of association between these herbivores suggests that foliage feeding by the beetles, particularly on mature leaves, was facilitated by the *Eumaeus* larvae.

PHENOLOGY OF NEW LEAF PRODUCTION AND HERBIVORE ATTACKS

In the sections that follow we analyze patterns of leaf disappearance. We attribute all loss of new leaves to herbivory, even if we did not observe the leaves being

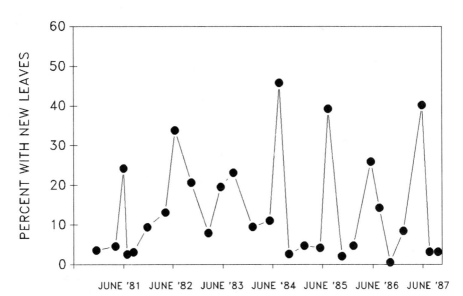

Figure 10-1. The percentage of individuals of *Zamia skinneri* at the La Selva Biological Station, Costa Rica, with new leaves at each census.

eaten. In many cases, the appearance of the demaged plants clearly indicated that the herbivores were responsible (based on numerous actual observations of herbivory). In many other cases, all or most of the new leaves completely disappeared from the center of the crown, with the rest of the plant intact. The only agents we have observed that produce this type of damage are herbivores. Loss of new leaves to pathogens or physiological stress was never observed on these plants.

There was a distinct annual cycle in initiation of new leaves (Fig. 10-1). Production of new leaves showed a pronounced annual peak between May and July, and was considerably lower during the rest of the year (Clark and Clark 1988). However, there was no period when leaf production was absolutely zero; at any time of year there were always at least a few individuals producing new leaves.

The number of *Zamia* with *Eumaeus* eggs or larvae closely paralleled the incidence of new leaf production (Fig. 10-2; $r_s = 0.37$, $P < 0.05$). As with leaf production, there was no period of the year when egg/larval occurrence consistently dropped to zero; eggs or larvae were found on at least one individual in 26 of the 28 censuses. The number of plants that lost new leaves at a census was significantly correlated with the number of plants with eggs or larvae ($r_s = 0.537$, $P < 0.005$, $N = 28$ censuses).

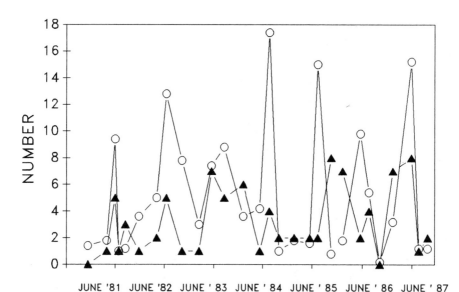

Figure 10-2. The relation between the total number of *Zamia skinneri* flushing new leaves at different censuses (O — O, original data divided by 5 for scaling) and the number with either eggs or larvae of *Eumaeus minyas* (▲ — ▲).

HERBIVORY AND THE MAGNITUDE OF NEW LEAF PRODUCTION

We documented the fate of 674 individual leaf flushes. The average number of leaves flushed was 2.2 (SD = 1.35). Of 1475 new leaves initiated, 21% were lost to herbivores. The number of surviving new leaves per flush increased with the number of leaves in the flush (Fig. 10-3; $r_s = 0.665$, $P < 0.001$). However, the probability of herbivores destroying at least one new leaf on a plant also increased with the number of new leaves flushed. Fifteen percent of the single-leaf flushes resulted in the loss of one leaf; of flushes of $\geqslant 5$ leaves, however, 50%

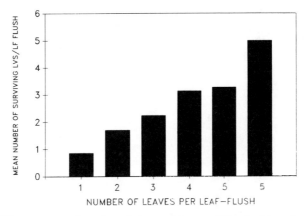

Figure 10-3. The mean number of surviving new leaves of *Zamia skinneri* as a function of the total number of leaves produced per episode of leaf production.

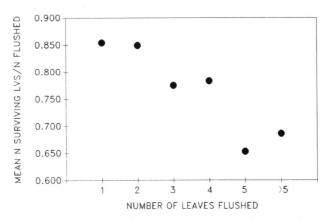

Figure 10-4. The mean number of surviving leaves per new leaf produced, as a function of the total number of new leaves produced per flush.

led to loss of ≥ 1 new leaf ($X^2 = 39.07$, $P < 0.0001$, 5 df, $N = 674$ flushes). The efficiency of new leaf production, expressed as the number of surviving new leaves divided by the total number of new leaves initiated, decreased significantly as the number of leaves per flush increased (Fig. 10-4; Kruskal-Wallis one-way ANOVA, $P < 0.001$).

PATTERNS OF HERBIVORY RELATED TO THE TIMING OF NEW LEAF PRODUCTION

The percentage of plants that produced new leaves and were attacked by herbivores varied from 0 to 56% over the 28 censuses (Fig. 10-5). At times when few plants were producing new leaves, rates of attack were variable; at the annual peak leaf production periods, attack rates are intermediate. The percentage of flushing plants which lost at least one leaf was not significantly correlated with the number of plants flushing at a census ($r_s = -0.228$, $P > 0.10$).

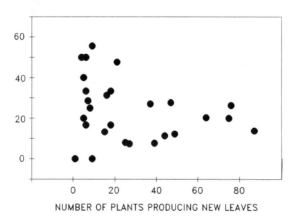

NUMBER OF PLANTS PRODUCING NEW LEAVES

Figure 10-5. The per-individual risk of herbivore attack for plants producing new leaves, as a function of the total number of plants producing new leaves.

RESULTS OF HERBIVORE ATTACKS

Of 726 flushes of new leaves (Table 10-1), 80% resulted in no loss of new leaves. When plants were attacked, frequently only the new leaves were killed. In over a third of the attacks, however, both young and old leaves were consumed. In 6% of the attacks the stem was also damaged.

Both male and female plants were attacked when they had strobili present (Table 10-2). Reproductive females were attacked proportionately more frequently than males and with more severe consequences. Only 22 instances of cone production by females were observed over the 7 yr, and 2 of these cones were lost to *Eumaeus* herbivory. In contrast, the maximum damage from *Eumaeus* attacks on reproductive males was limited to the loss of one or two mature leaves.

TABLE 10-1. Patterns of Herbivore Damage to _Zamia skinneri_ Flushing New Leaves at the La Selva Biological Station, Costa Rica

Damage Type	Percent of Flushes	Number of Flushes
No damage	80	580
Only young leaves lost	13	94
Young and old leaves lost	7	52
		726

TABLE 10-2. Incidence and Effects of Eggs or Larvae of _Eumaeus minyas_ on Reproductive _Zamia skinneri_ Plants at the La Selva Biological Station

Plants with Female Cones (N = 22)[a]

Eggs or larvae observed, cone destroyed, some/all mature leaves lost, $\pm$ some stem damage	9.1%
Eggs laid but no damage to plant, no larvae survived	18.2%
No _Eumaeus_ eggs or larvae observed	72.7%

Plants with Male Cones (N = 100)

Eggs or larvae observed, no cones destroyed, maximum of 1–2 mature leaves lost, no pupae observed	7.0%
No Eumaeus eggs or larvae observed	93.0%

[a]Sample size refers to total number of reproductive episodes observed during the study.

Complete defoliation by _Eumaeus_ and other agents such as litterfall are common occurrences for these long-lived plants; 19% of the study plants experienced one or more periods of leaflessness during this 6.8-yr study. Nevertheless, mortality of totally defoliated individuals was rarely observed. Mortality is not easy to document in _Z. skinneri_; totally defoliated plants may remain leafless up to 2 yr, then resprout. Our operational definition of "dead" therefore was "leafless for more than 2 yr." Over the course of the study 6 of 195 individuals died; 1 of the 6 was definitely killed by larvae. _Eumaeus_ therefore accounted for at least 17% of the _Zamia_ mortality during this study. There was no obvious cause of death for the other individuals which died, and it is possible that _Eumaeus_ killed some of them.

AT THIS POINT IN EVOLUTIONARY TIME

The current relationship between _Zamia skinneri_ and its specialist herbivore poses substantial problems for both the plant and the insect. _Zamia_ loses on the average 1 of 5 of the few new leaves it produces. The simple act of producing new leaves, which is vital for growth and reproduction, opens the door not only to loss of the new leaves but also on occasion to total defoliation, stem damage, and even death. Female reproduction is a rare event, and is made even more difficult by the potential for total destruction of the valuable female cones by herbivores. Finally, the plant has no reliable seasonal shelter from the attacks of *Eumaeus*. Given this set of conditions, what might be predicted of the plant's behavior, and how does the observed behavior accord with the predicted?

For any phenological pattern, there should be selection for individuals to minimize the period that they are at risk. If new leaves are vulnerable, the risk of herbivory could be decreased by shortening maturation time of each leaf and/or decreasing the period over which tender new leaves are present on an individual.

Individual *Zamia* in fact only produce leaves in discrete flushes, thereby minimizing the time over which first-instar larvae have access to young leaves. The time to leaf hardening (ca. 65 days) is still somewhat longer than the generation time for the butterfly (ca. 50 days). Theoretically this could mean that a given *Eumaeus* could begin two generations on one *Zamia* leaf flush. In fact we never observed this, nor do we think it very likely. It is quite possible that 50-day old leaves are already too hard for first-instar larvae. Even if these leaves were accessible, it would require extremely precise egg-laying and mating behavior to fit two egg-layings into this interval. Perhaps most importantly, there are seldom any resources left for a second generation of larvae. We observed 17 cases of larvae on plants which produced new leaves; in 14 instances (82%) the total production of new leaves was consumed.

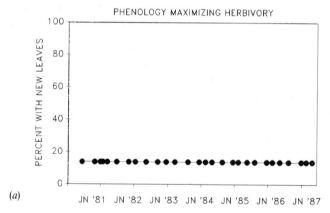

(a)

Figure 10-6. The relation between phenology of new leaf production and herbivory. Figures 10-6*a*–*c* are calculated assuming the same total quantity of new leaves as was actually initiated over this period (as in Fig. 10-6*d*). (*a*) Constant leaf production affords maximum opportunities for the specialist herbivore to encounter suitable larval food resources. (*b*) Highly synchronized, multiannual leaf production minimizes herbivores' chances of finding new leaves, but forces individual plants to wait long periods for new leaves. (*c*) Annual synchrony minimizes vulnerable period while allowing yearly opportunity for leaf production. (*d*) Actual production of new leaves by *Zamia skinneri* during the study.

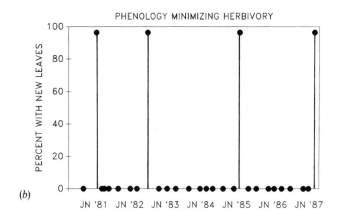

(b)

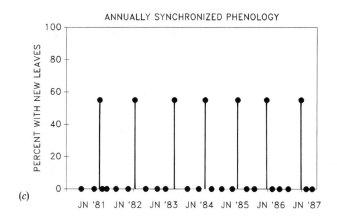

(c)

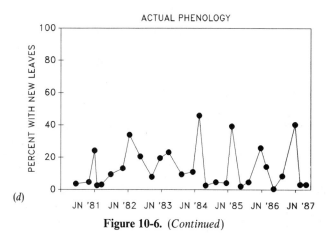

(d)

Figure 10-6. (*Continued*)

When should individual *Zamia* produce their new leaves in order to minimize losses to *Eumaeus*? The answer to this question probably depends on which factors usually limit *Eumaeus* populations. It is quite possible that parasitoids are important in regulating *Eumaeus* abundance (cf. Strong 1984). It is interesting, however, to consider the *Zamia–Eumaeus* system without the parasitoid interaction; what might be the consequences of different phenological patterns for *Zamia* if *Eumaeus* were strictly limited by larval food abundance?

It is easy to predict the phenological pattern that will maximize leaf loss for a *Zamia* population under these conditions (Fig. 10-6a). For a given total production of new leaves, a pattern of random flushing by individuals would lead to a continual steady availability of new leaves. This would afford constant and predictable food resources for larvae hatched at any time of year. Such a pattern of leaf production should lead to a consistently high butterfly population. The ideal temporal behavior for *Zamia*, on the other hand, would be for all individuals to flush synchronously at long, unpredictable intervals (Fig. 10-6b). This predator satiation strategy (Janzen 1971) would ensure that emerging adult butterflies experience maximal difficulties encountering suitable sites for larval development. This may not be a feasible strategy for a tropical wet forest understory plant. Plants in this environment suffer constant leaf attrition due to falling litter (Clark and Clark 1989), and maintaining and increasing leaf area is critical to reproduction in *Zamia* (Clark and Clark 1987). Also, it is not obvious what supraannual environmental cue could serve as the basis for selection for this type of phenology. Given that long periods of zero leaf production impose a cost, and that possibly the only reliable phenological cue is changing day length, one would predict that selection should favor individuals which tend to flush new leaves at one specific time each year (Fig. 10-6c). The observed behavior (Fig. 10-6d) approaches this. While there may well be other reasons for the observed population synchrony in *Zamia* leaf production, the pattern is consistent with the hypothesis that individuals' phenologies have been adjusted to decrease *Eumaeus* herbivory.

Does phenological synchrony among individual *Zamia* in fact decrease a plant's risk of herbivore attack? In ecological time, we cannot demonstrate that it does. The odds of suffering a leaf-destroying attack did not decrease significantly when more plants produced leaves together (Fig. 10.5). However, high rates of attack were never observed during the peak flushing events. It appears that *Eumaeus* currently lacks the capacity to overwhelm the annual peak of *Zamia* leaf production at La Selva. Synchronized leaf production at the individual and population level, coupled with a small window of vulnerability of new leaves to larvae, combine to limit opportunities for population increases of *Eumaeus*.

Another aspect of phenology concerns the magnitude of leaf production: how many leaves should an individual produce in a given episode of new leaf production? A clear prediction emerges from Figure 10-4: to produce a given number of new leaves under the prevailing ecological conditions, the safest course is to produce them in separated flushes of one leaf at a time. Table 10-3 shows the actual behavior of the plants in terms of how many episodes of leaf production were used to produce a given number of leaves. The solid diagonal represents the

TABLE 10-3. The Relation Between the Total Number of New Leaves Flushed by Individual *Zamia skinneri*, and the Number of Episodes of Leaf Production Employed to Produce that Number of Leaves at La Selva Biological Station, Costa Rica[a]

Total New Leaves Produced Per Plant	Number on Episodes of New Leaf Production Per Plant												
	1	2	3	4	5	6	7	8	9	10	11	12	Total
1	2												2
2	2	7											9
3	1	6	4										11
4	1	12	7	2									22
5		2	13	7	2								24
6		4	11	6	4	1							26
7		6	3	6	2	1	1						19
8		1	5	5	1								12
9		1	1	3	5	2							12
10			1	4	2		2	1					10
11			2	3	5	1							11
12			2	1	2								5
13			3		1								4
14					1			1					2
15					1		2	1					4
16					1		1	1					3
17					1	1	1		1				4
18								1	1				2
19					1								1
≥ 20					2	1	2		2	2		1	10
Totals	6	39	52	43	27	11	6	6	2			1	193

[a]The numbers in the table refer to the number of individuals in the sample which exhibited that combination of total leaf production and leaf production episodes. The diagonal solid line indicates the pattern of leaf production which is predicted to minimize herbivore damage to new leaves (see Fig. 10.4 and text).

predicted herbivory-minimizing behavior of single-leaf flushes. Apparently *Zamia* is not minimizing the pre-leaf risk of herbivory in most cases. One reason is related to plant size. Many plants averaged more than one new leaf per year over the 7-yr study, so that the least risk, one-a-year option was not possible. However even among plants which averaged one new leaf or less per year ($N = 113$), only 17% followed the predicted lowest-risk pattern. One cost of the herbivory-minimizing strategy is a net loss of effective leaf area. A plant which flushes 5 leaves at once has the potential for 25 leaf-yr of leaves present over a 5-yr period, compared to only 15 leaf-yr for the plant which flushes 1 leaf a year for 5 yr. Increasing plant size, measured in units of leaves, is critical to individual fitness. Small plants do not reproduce at all, and among reproductives, several traits related to fitness increase with size (Clark and Clark 1987). In this light-limited species, physiological and fitness advantages from adding leaf area as quickly as possible may outweigh the additional risk from flushing new leaves in groups.

CONCLUSION

In terms of the four basic mechanisms for defeating herbivores, we can now say the following about *Zamia skinneri*. Defense in time is clearly of major importance, as evidenced by the pattern of individuals' leaf production in discrete flushes. The population synchrony of flushing remains intriguing, but has not been demonstrated to lower an individual's risk of attack. Further research, including simulations based on field data, are necessary to investigate if the absence of high rates of herbivore attacks during leaf flush peaks is a sufficient explanation for this pulsed phenology.

Defense in space was not studied. Are isolated plants less vulnerable? A mapped population over a large area would be necessary to investigate this. The fact that *Eumaeus* was able to locate flushing *Zamia* even when very few plants were flushing indicates that its prey-locating mechanism are quite sensitive. One would also like to know something about the flight range of adult *Eumaeus*: how large an area can they sample? Defense in space remains an open question.

Mechanical defense and chemical defense are both important for *Zamia*. Assuming *Zamia skinneri* follows the same pattern found by Rothschild et al. (1986) for *Zamia floridana*, mature leaves have lower levels of chemical defence than young leaves. Nevertheless, the only significant herbivores we observed eating mature leaves were large *Eumaeus* larvae. Presumably mature leaves are chemically available to the young larvae, but are mechanically defended by their toughness. Young leaves must rely almost totally on chemical defense. These leaves are very soft and visually conspicuous, and they must be obvious to a wide variety of herbivores. Nevertheless, only one species of herbivore is a major predator on these apparent delicacies.

The interaction we have described in this chapter must approach one end of an axis describing the specificity of plant–herbivore interaction: a monophagic herbivore and a plant attacked mainly by a single herbivore. The patterns we have described may be limited to similarly highly specialized interactions. However, the approach we have taken, that of monitoring the plant's phenological behavior and determining the consequences of different patterns of allocation, appears to us to be generally useful. By carefully observing plants in natural populations in intact habitats, a great deal will be learned about the evolution of plant–herbivore interactions.

ACKNOWLEDGMENTS

We thank Michael Grayum for introducing us to cycads and inspiring this study with his original observations and *Zamia* mapping. Robert Marquis, Deborah Letourneau, and Robert Raguso suggested many improvements on an earlier draft; J. M. Kingsolver identified the languriid beetles. Logistic and financial support were provided by the Organization for Tropical Studies. We are grateful to Sr. Gerardo Vega for his careful assistance in data gathering and data management during a substantial part of this research.

REFERENCES

Chazdon, R. L. 1986. Light variation and carbon gain in rain forest understory palms. *Journal of Ecology* **74**: 995–1012.

Chazdon, R. L. 1988. Sunflecks and their importance to forest understorey plants. *Advances in Ecological Research* **18**: 1–63.

Chazdon, R. L. and N. Fetcher. 1984. Photosynthetic light environments in a lowland tropical rainforest in Costa Rica. *Journal of Ecology* **72**: 553–564.

Clark, D. B. and D. A. Clark. 1985. Seedling dynamics of a tropical tree: Impacts of herbivory and meristem damage. *Ecology* **66**: 1884–1892.

Clark, D. A. and D. B. Clark. 1987. Temporal and environmental patterns of reproduction in *Zamia skinneri*, a tropical rain forest cycad. *Journal of Ecology* **75**: 135–149.

Clark, D. B. and D. A. Clark. 1988. Leaf production and the cost of reproduction in the neotropical rain forest cycad, *Zamia skinneri*. *Journal of Ecology* **76**: 1153–1163.

Clark, D. B. and D. A. Clark. 1989. The role of physical damage in the seedling mortality regime of a neotropical rain forest. *Oikos* **55**: 225–230.

Coley, P. D. 1983. Herbivory and defensive characteristics of tree species in a lowland tropical forest. *Ecological Monographs* **53**: 209–233.

Crowson, R. A. 1981. The biology of the Coleoptera. Academic Press, New York.

DeVries, P. J. 1976. Notes on the behavior of *Eumaeus minyas* (Lepidoptera: Lycaenidae) in Costa Rica. *Brenesia* **8**: 103.

DeVries, P. J. 1983. *Zamia skinneri* and *Z. fairchildiana*. Pages 349–350 in D. H. Janzen (Ed.), *Costa Rican Natural History*. University of Chicago Press, Chicago, IL.

Gomez, P. L. D. 1982. Plantae mesoamericanae novae. II. *Phytologia* **50**: 401–404.

Janzen, D. H. 1971. Seed predation by animals. *Annual Review of Ecology and Systematics* **2**: 465–492.

Marquis, R. J. 1984. Leaf herbivores decrease fitness of a tropical plant. *Science* **226**: 537–539.

Oberbauer, S. F., D. B. Clark, D. A. Clark, and M. Quesada. 1988. Crown light environments of saplings of two species of rain forest emergent trees. *Oecologia* **75**: 207–212.

Oberbauer, S. F., D. A. Clark, D. B. Clark, and M. Quesada. 1989. Comparative analysis of photosynthetic light environments within the crowns of juvenile rain forest trees. *Tree Physiology* **5**: 13–23.

Peterson, A. 1951. *Larvae of insects*. Part II. Ohio State University, Columbus, OH.

Ross, G. N. 1964. Life history studies on Mexican butterflies. III. Nine Rhopalocera from Ocotal Chico, Vera Cruz. *Journal of Research on the Lepidoptera* **3**: 207–229.

Rothschild, M., R. J. Nash, and E. A. Bell. 1986. Cycasin in the endangered butterfly *Eumaeus atala florida*. *Phytochemistry* **25**: 1853–1854.

Strong, D. R. 1984. Exorcising the ghost of competition past: Phytophagous insects. Pages 28–41 in D. R. Strong, D. Simberloff, L. G. Abele, and A. B. Thistle (Eds.), *Ecological Communities, Conceptual Issues and the Evidence*. Princeton University Press, Princeton, NJ.

11 Plant Responses to Herbivory: The Continuum From Negative to Positive and Underlying Physiological Mechanisms

THOMAS G. WHITHAM, JOYCE MASCHINSKI,
KATHERINE C. LARSON and KEN N. PAIGE

Belsky (1987) states that "the negative effect of herbivory on plant growth and fitness approaches a paradigm in ecology." McNaughton (1986) argues that this view is too simplistic and that "plants have the capacity to compensate for herbivory and may, at low levels of herbivory, overcompensate for damage so that fitness may be increased." Although there is a large body of literature supporting the idea that herbivores have detrimental effects on plant growth and reproduction, few studies unquestionably document the deleterious effects on host-plant fitness (Strong et al. 1984; Crawley 1987). In addition, most negative impact studies fall under the same criticisms leveled at studies that purport to demonstrate the beneficial effects of herbivory. In short, the study of plant–herbivore interactions has polarized into camps, one arguing that all direct impacts on plants must be negative and another arguing that the direct effects can be positive.

To unequivocally demonstrate the impact of herbivores on plants, whether negative or positive, requires that grazed plants exhibit a significant change in fitness relative to ungrazed controls. In reality this turns out to be difficult data to obtain, particularly for iteroparous plants with different modes of reproduction. For example, the reproductive efforts from multiple bouts of sexual and asexual reproduction must be accumulated over a lifetime and we must evaluate the ecological and/or evolutionary value of a vegetative ramet versus a sexually derived seed. Furthermore, one must then follow the progeny through reproductive maturity to gain a true measure of fitness. Increases or decreases in biomass alone cannot be used as critical evidence in studies attempting to categorize the

Plant-Animal Interactions: Evolutionary Ecology in Tropical and Temperate Regions, Edited by
Peter W. Price, Thomas M. Lewinsohn, G. Wilson Fernandes, and Woodruff W. Benson.
ISBN 0-471-50937-X © 1991 John Wiley & Sons, Inc.

effects of herbivores on plants. Unfortunately, there are very few studies of under-, equal, or overcompensation in natural systems that have succeeded in measuring the lifetime fitness of grazed and ungrazed controls (but see Crawley and Nachapong 1985).

In recognition of this controversey and the problems involved in measuring herbivore impacts on plant fitness, the main goals of this chapter are to (1) recognize that both direct negative and positive effects can occur, (2) determine which species are most likely to under-, equal, and/or overcompensate, and (3) develop an understanding of the mechanisms that result in some species having a tendency to exhibit one of the three compensatory responses while others are far more plastic and may exhibit all three.

THE NEGATIVE EFFECTS OF HERBIVORY ON PLANTS

Herbivores can have both direct and indirect impacts that negatively affect all aspects of plant growth, reproduction, and status in the community. Here we concentrate on the direct effects of herbivory on plant fitness and the importance of understanding the cumulative impacts of the suite of pests and pathogens that plants are exposed to.

Direct Impacts

Numerous studies show that herbivores can have a dramatic impact on relative fitness. For example, Hendrix (1984) found that when the cow parsnip *Heracleum lanatum* suffered floral herbivory by the parsnip webworm *Depressaria pastinacella*, control plants produced 40% more seeds than damaged plants and seed biomass was 53% greater.

Although partial reductions in relative fitness are common, the complete loss of seed production has also been documented. In marine salt marshes, sea lavender *Limonium vulgare* is commonly fed upon by the aphid *Staticobium staticis*. In 4 to 6 yr of study, the majority of lavender plants in the center of the marshes failed to produce seed owing to aphid infestations (Foster 1984). Experimental removals with insecticides confirmed these observations and demonstrated that with the elimination of aphids, 100% of the inflorescences produced fruits while only 23% of the controls were equally successful.

Some studies show that while high levels of herbivory can be very detrimental to plants, these impacts may also be highly conditional. For example, Lee and Bazzaz (1980) found that when the annual *Abutilon theophrasti* was grown at low densities it could withstand 75% defoliation without significant losses in reproduction. However, when the same species was grown at high densities at identical levels of defoliation, seed production declined by 50%. At low plant densities, high levels of herbivory had no measurable impact while at high plant densities the same levels of herbivory resulted in dramatic impacts. Thus, the negative effects of herbivory were highly conditional and dependent upon other factors such as competitive interactions with other plants.

Striking direct impacts of herbivory have also been documented at the boundaries of plant populations. For example, Randall (1965) observed a conspicuous band of bare sand averaging about 10 m in width that separated coral reefs and beds of sea grasses *Thalassia testudinum* and *Cymodocea manatorum*. These bare zones resulted from heavy grazing by parrotfishes and surgeonfishes that use reefs as refugia from predaceous fishes. When artificial reefs were experimentally added to beds of sea grasses, within a year new bare zones formed. This demonstrated the capacity of herbivores to completely eliminate vegetation and implicated the importance of predators in restricting herbivores to rather narrow foraging zones. A similar pattern of herbivory was experimentally demonstrated by Bartholomew (1970) in which several mammalian herbivores were responsible for the bare zones between coastal sage, *Salvia leucophylla*, and annual grassland. These and other examples concerning hybrid zones at the margins of populations (Drake 1981a, b) suggest intense selection by herbivores may be most important at plant boundaries.

Direct impacts of herbivores on plant architecture can delay sexual maturity, alter sex expression, and have complex interactions with other taxa and trophic levels. For example, Whitham and Mopper (1985) showed that the feeding of moth larva *Dioryctria albovitella* on pinyon pine, *Pinus edulis*, altered tree architecture and made upright open-canopied trees into prostrate close-canopied shrubs (Fig. 11-1). Because female cones are borne on the apically dominant

Figure 11-1. Photographs of two small defoliated pinyon pines, *Pinus edulis*, show how the accumulated herbivory of chronic attack negatively affects tree architecture, growth, and reproduction. The tree on the left suffered shoot mortality from the stem- and cone-boring moth *Dioryctria albovitella* of approximately 30% per year, while the tree on the right suffered < 3%. The right tree produced 59% more trunk wood per year during its 63-yr life span than the smaller 91-yr-old shrublike left tree. The right tree was monecious, while the left tree had lost all female function, becoming a male plant. Adapted from Whitham and Mopper (1985).

terminal shoots, the herbivore-induced change in plant architecture resulted in a loss of terminal shoot production. With the change in architecture, only the lateral male strobli-bearing shoots remained and these normally monoecious trees lost all female function to become male trees. Christensen and Whitham (1991) found that this herbivore-mediated change in pinyon sex expression subsequently had a pronounced impact on the tree's seed dispersal agent, the pinyon jay *Gymnorhinus cyanocephalus*. With a reduced cone crop the birds nearly abandoned these stands and even though some cones were produced, few seeds were dispersed. Craig et al. (1986) observed delayed reproduction due to herbivory; architectural changes in the arroyo willow *Salix lasiolepis,* due to attack by the sawfly *Euura lasiolepis,* indefinitely maintained plants in a juvenile state.

Cumulative Impacts

A critical issue in examining the evolutionary responses of plants to herbivory is whether or not genotypes resistant to one herbivore are indeed resistant to other pests. For example, if resistance has a cost in terms of reduced growth and reproduction as some studies have demonstrated (Leonard 1977; Berenbaum et al. 1986; Coley 1986), then resistance to one pest could come at the expense of increased susceptibility to other very different pests. Since most plants have multiple herbivores and pathogens, studies that emphasize only one "dominant" herbivore may in fact underestimate, overestimate, or totally misrepresent selection pressures thought to be important in plant evolution. Conceivably, the negative impacts on one plant genotype may be canceled out when other herbivores are added to the equation such that all existing plant genotypes have equal fitnesses.

Pests Acting in Concert

Plants attacked by one herbivore may become more susceptible to attack by other pests. For example, Fedde (1973) found that in addition to the damaging effects of the balsam wooly aphid *Adelges piceae* on Fraser fir, *Abies fraseri,* aphid attack facilitated a nearly tenfold increase in attack by another pest. More than 30% of the seeds from aphid-infested trees contained larvae of *Megastigmus specularis,* while only 3.1% of the seeds from uninfested trees were affected by this chalcid.

If attack by one herbivore facilitates attack by another pest, then the entire suite of diverse and often unrelated herbivores must be considered if we are to understand plant responses to herbivory. For example, Drake (1981a, b) examined fruit production and survival of two species of *Eucalyptus* and their natural hybrids in response to very different mortality agents, insects, and fungi (Table 11-1). In all three plant categories, the cumulative impacts of both insects and fungi were far greater than their individual impacts and hybrids were most affected. Thus, while the impacts of a single herbivore may be low and appear

TABLE 11-1. Estimated Number of Fruits per Tree and Fruit Damage by Fungus and Insects

Species	Mean Fruits/Tree	Fruit Damage (%)		
		By Fungus	By Insects	Total
Eucalyptus melanophloia	105,450	4	12	16
Hybrid	10,950	24	33	57
E. crebra	100,500	7	10	17

[a]Adapted from Drake (1981a, b). Hybrids produced not only less fruit but suffered greater damage from two different mortality agents.

insignificant, the cumulative impacts of the suite of herbivores, pathogens, and other agents of mortality may be great.

Pests Acting in Opposition

Just the opposite reaction may occur in which resistance to one pest is strongly associated with susceptibility to another. Using clones of the narrow-leaf cottonwood, *Populus angustifolia*, Whitham (unpub. data) examined resistance to very different herbivores, the gall-producing aphid *Pemphigus betae* and the elk *Cervus elaphus*. In common-garden experiments, the patterns of clone resistance to these very different herbivores were revealing (Table 11-2). With clone #1008, high resistance to aphids (i.e., 98% of the colonizing aphids died), was associated with high susceptibility to elk (i.e., 100% of all branches below 2 m were removed by elk). Just the opposite occurred with two other clones; with clone #1000, high susceptibility to aphids was strongly associated with high susceptibility to elk, and with clone #996 high resistance to aphids was strongly associated with high resistance to elk. With both of these clones there were strong

TABLE 11-2. Resistance of Individual Clones of Narrowleaf Cottonwood, *Populus angustifolia*, to Very Different Herbivores, the Gall Aphid *Pemphigus betae* and the Elk *Cervus elaphus*[a]

Resistance	Narrowleaf Cottonwood Clone Number		
	1000	1008	996
Mean resistance to aphids (aphid mortality)[b]	30% (low)	98% (high)	100% (high)
Mean resistance to elk (branches below 2 m not eaten)[b]	1% (low)	0% (low)	98% (high)

[a]In common garden experiments, clones may be susceptible to both herbivores, resistant to both, or resistant to one but susceptible to the other.
[b]Each data point represents the mean of at least six replicate ramets of each clone and no overlap exists between low and high resistance.

positive associations such that very different herbivores acted in concert. Thus, depending upon host genotype, the cumulative impacts of herbivory may act in concert or act in opposition. Such results strongly argue for field experiments that measure the cumulative impacts of diverse herbivores and pathogens.

If exposure to one herbivore or pathogen induces resistance to subsequent attack by another pest, then the negative effects of any one pest are less clear and the overall impacts may cancel one another. Karban et al. (1987) showed that when cotton seedlings were innoculated with the fungal pathogen *Verticillium dahliae*, spider mite populations, *Tetranychus urticae*, grew more slowly than on control, pathogen-free plants. Similarly, when plants had previously been exposed to mites, they were subsequently less likely to suffer from fungal attack. In such cases where plants are utilized by diverse pests with complex interactions, simple generalizations as to the impact on plant fitness and ultimately plant evolution are probably not valid (see also Fritz et al. 1986; Maddox and Root 1987).

Small but Statistically Significant Impacts

Most researchers have generally accepted the idea that small but significant negative impacts are detrimental to plant fitness, particularly when accumulated over long periods of time. Such generalities become less clear when considering cumulative impacts of diverse herbivores that may act in opposition or in light of neutral and positive examples presented in following sections. An example of small but significant impacts on plant fitness is presented by Sacchi et al. (1988), in which they found that the stem-galling sawfly significantly reduced seed production of the arroyo willow *Salix lasiolepis*. Sixty-nine percent of all plants suffered no measurable losses, 20% suffered a 10% loss, and 11% suffered a 20% or more loss. Because cloning experiments strongly suggest that heavily attacked willows are genetically susceptible (see also Fritz et al. 1986), it appears that sawfly attack can reduce the relative fitness of one genotype relative to another. However, the chance determination of where wind-dispersed seeds land, combined with an extremely low survival rate, increases the probability that plant fitness is independent of herbivory. Thus, the impact of herbivores may be unclear. It becomes critical for future research to identify the importance of small but statistically significant impacts on fitness relative to the other selection pressures that plants experience.

THE POSITIVE EFFECTS OF HERBIVORY ON PLANTS

The underlying assumption of the "herbivore impact" paradigm is that the direct effects of herbivory result in a reduction in plant fitness. In the mid-1970s, however, several authors (Chew 1974; Harris 1974; Dyer 1975; Mattson and Addy 1975; Owen and Wiegert 1976) suggested that herbivory may increase, rather than decrease, the growth and reproductive success of some plant species.

Over the past decade these suggestions have stimulated the publication of more than 40 studies presenting evidence in support of this view (see review by Belsky 1986). Yet, rather than the emergence of a new synthesis, the field has remained polarized (e.g., Belsky 1986, 1987; McNaughton 1986; Verkaar 1986; Crawley 1987). In this section we briefly review the controversy, present results from recent studies supporting the contention that herbivory can benefit plants, and present hypotheses which attempt to explain the origins of overcompensation.

The Controversy

One of the major problems surrounding the controversy hinges on what exactly is meant by "good for plants" (Crawley 1987). The idea that individual plants benefit from the *direct effects* of herbivory is an organismal argument at the level of the population and differs from the *indirect effects* of herbivory at the community and ecosystem levels. At the community and ecosystems levels, there are well-documented studies of indirect effects showing that herbivores benefit some plant species by removing their competitors, altering successional trends, removing litter, and fertilizing plants (McNaughton 1979; Belsky 1986, 1987; Crawley 1987). Belsky (1987) argues that these issues have been confused and the issue of direct effects has erroneously gained validity from the recognized indirect effects.

We believe the major controversial issues are: 1. Do the purported studies of overcompensation critically demonstrate direct positive effects of herbivory on plant fitness? 2. How could the ability to overcompensate arise over time? 3. What are the precise mechanisms that would permit browsed plants to achieve greater fitness than controls? Belsky (1986) has reviewed the first issue and strongly criticized studies purporting to demonstrate overcompensation. She concluded that all of the studies allegedly providing evidence for increased growth and fitness following herbivory were seriously flawed for a variety of reasons and that "no convincing evidence supports the theory that herbivory benefits grazed plants." Although we believe Belsky's review accurately reflects the "intuition and gut feelings" of most plant–herbivore researchers, more recent findings suggest that the issue is far from settled.

Supportive Studies

A typical example of current experiments examining overcompensation is by Cargill and Jefferies (1984). They determined net primary production for both grazed and ungrazed individuals of *Carex subspathacea* and *Puccinellia phryganodes* in La Perouse Bay, a subartic salt marsh east of Manitoba, Canada. Grazing by lesser snow geese was shown to result in a significant increase in net aboveground primary production in both years of the study. In 1979, grazed, mixed swards of *C. subspathacea* and *P. phryganodes* were 35% more productive than ungrazed swards. In 1980, grazing resulted in an 80% increase for both species in comparison to the ungrazed controls. While providing evidence for overcompensation in aboveground biomass in response to herbivory, increased biomass is not necessarily correlated with fitness and it is difficult to compare the value of sexual versus vegetative reproduction.

Studies by Paige and Whitham (1987a) overcome many of the criticisms leveled at impact studies and appear to represent the first clear-cut evidence of overcompensation in a natural system. Scarlet gilia, *Ipomopsis aggregata*, is a red-flowered, biennial herb of western montane regions that usually flowers once and dies. Although lack of seed set can induce cloning and repeated bouts of reproduction, herbivory does not (Paige and Whitham 1987b). Thus, this natural system is well suited for examining the lifetime reproductive success of individual plants that allocate all their reproductive effort to one pathway during one season and then die.

Experimental and observational studies showed that when elk and deer browsed scarlet gilia, there was a pronounced benefit to having been eaten. The

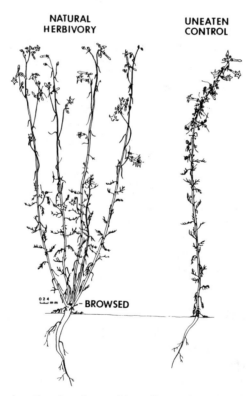

NATURAL
HERBIVORY

UNEATEN
CONTROL

BROWSED

Figure 11-2. Illustration showing the positive effects of herbivory by comparing the architecture and numbers of inflorescences of browsed and uneaten control plants of scarlet gilia, *Ipomopsis aggregata*. See Table 11-3 for comparative data. "Browsed" indicates the position where the original inflorescence was cut by the herbivore. Adapted from Paige and Whitham (1987a).

increase in relative fitness was largely due to an architectural change in the plant (Fig. 11-2); browsed plants on average produced 4 inflorescences, whereas uneaten controls produced only a single infloresence. The increase in inflorescence number resulted in a 2.76-fold increase in flower production and a 3.05-fold increase in fruit production (Table 11-3). Because there were no significant differences in the number of seeds per fruit, differences in seed weight, germination success, or rates of seedling survival, an increase in total fruits produced by grazed plants resulted in increased fitness (Table 11-3). These observations were experimentally replicated by clipping plants to simulate natural herbivory and the performances compared with controls. The resultant experimental manipulations produced patterns that were quantitatively and qualitatively similar.

Cumulative estimates of plant performance demonstrated that browsed plants achieved a 2.4-fold increase in relative fitness over uneaten control plants. By combining the most direct measures of plant fitness (i.e., number of seeds per fruit times the number of fruits per plant times percent germination success times percent seedling survival after 4.5 months), a cumulative estimate of plant performance was obtained. Plants experiencing natural or simulated herbivory

TABLE 11-3. Overcompensation by Scarlet Gilia in Response to Herbivory[a]

Relative Fitness	Natural Herbivory	Simulated Herbivory	Control
Biomass[b]			
Whole plant (g)	10.37 A	7.69 A	4.91 B
Root (g)	3.24 A	2.68 A	1.56 B
Flowering[b]			
Inflorescences per Plant	4.07 A	3.33 A	1.00 B
Flowers/plant	102.3 A	68.9 B	37.0 C
Fruits/plant	57.7 A	44.9 A	18.9 B
Seed[b]			
Seeds/fruit	9.25 A	9.57 A	9.00 A
Seed wt. (mg)	1.18 A	0.95 A	1.14 A
% Germination (on filter paper)	43.0 A	42.0 A	46.0 A
% Survival after 4.5 months (greenhouse)[b]	38.0	48.0	47.0
% Survival after 1.7 years (greenhouse)[b]	19.0	25.0	11.0

[a]Unpublished data and data adapted from Tables 1 and 2 from Paige and Whitham 1987a.
[b]Some plants experienced natural herbivory by deer and elk; others were clipped to simulate herbivory and uneaten plants served as controls. Means with the same letters show no significant difference at the 0.05 level, Student–Newman–multiple-range test.
[c]X^2 tests for the numbers of living and dead after 4.5 months and 1.7 years show no significant differences.

could both expect to produce an average of 87 surviving seedlings, whereas uneaten control plants could expect to produce only 37 surviving seedlings. These results are clearly differentiated from the indirect beneficial effects reported at the ecosystem and community level, and demonstrate direct positive effects on individual fitness.

These results are particularly impressive when considering that even with very high levels of herbivory, overcompensation still resulted. With both the natural herbivory by deer and elk, and the simulated herbivory, approximately 95% of the aboveground biomass or 72% of the plant's total biomass was removed. Importantly, inflorescence regeneration did not come at the expense of the root system. In fact, the root systems of naturally browsed and experimentally clipped plants were 107 and 71%, respectively, greater than the uneaten controls. Thus overcompensation in response to herbivory occurred with both above- and below ground vegetative structures.

It is also interesting to note that the overcompensation response of *Ipomopsis aggregata* to herbivory continues over a wide range of natural field conditions. For example, when plants are growing in close association with pine and grasses to add in the potential negative effects of competitive interactions, browsed plants continue to outperform control plants (Paige, unpublished data).

Although these studies are still inconclusive because true fitness (i.e., individual reproduction and survival of progeny to maturity) were not measured, recent field studies (Paige and Maschinski, unpublished data) indicate that browsed plants are also likely to achieve greater true fitness. Browsed scarlet gilia were transplanted into field plots devoid of vegetation and unbrowsed plants were transplanted into adjacent plots to serve as controls. As expected, browsed individuals produced 2.8 times as many fruits as the controls. Seeds were allowed to naturally disperse and germinate. One year later counts of the number of surviving seedlings were made in each plot. Browsed plants produced 6.6 times more surviving progeny than ungrazed plants under natural field conditions.

Hypotheses for Overcompensation

In light of these results we feel that overcompensation is a real phenomenon and cannot be easily dismissed. We must then address the issue of whether or not overcompensation represents an evolved trait. Four hypotheses are offered which await testing.

Nonadaptation Hypothesis. This hypothesis states that overcompensation is a response in plants that are not yet highly adapted to their environment. For example, apical dominance in scarlet gilia, which is presumably genetically controlled, prevents lateral bud development. When apical dominance is destroyed by herbivores, lateral buds are released, more inflorescences develop, and plant fitness is increased. Thus, the increase in fitness may be the result of a herbivore performing a function that the genetic and/or developmental constraints of the plant cannot. In such instances, the relationship is mutually beneficial, but if plants acquire the genetic ability to perform the function themselves, the relationship should be antagonistic for those genotypes. This hypothesis is similar to one developed by Janzen (1979) and Janzen and Martin (1982) in which they argue that many trees depending upon now extinct large vertebrates for seed dispersal are evolutionary anachronisms. Having lost their dispersal agents these plants failed to change with their environment and still possess dispersal traits that are more appropriate for another time.

Extreme Adaptation Hypothesis. This hypothesis states that overcompensating species have become so well adapted to high levels of herbivory that the absence of herbivory represents an unnatural condition. If a plant species is highly adapted to a nearly certain probability of being browsed, then the responses of experimental control plants protected from herbivory may be outside the normal range of responses and the experimental controls are inappropriate. Adequate controls would require plants from a population which evolved in association with much lower levels of herbivory where plant responses to not being browsed represent the normal range of possibilities. This hypothesis is not likely involved with the studies of *I. aggregata* since in most years 10–40% of the plants escape herbivory.

Bet-Hedging Hypothesis. This hypothesis predicts that plants have evolved intermediate strategies that appear inappropriate over the short-term but are appropriate over the long term. For example, if *I. aggregata* initially produced four inflorescences rather than the normal one, and subsequently suffered herbivory, undercompensation might be the norm. By bet-hedging against the high probability of the first inflorescence being eaten, the plant initiates one flowering stalk rather than four and cuts its initial losses. Although the few plants that escape herbivory produce only a single inflorescence and experience reduced fitness relative to the eaten plants that come back with four, both may do far better than those that initially produce four inflorescenses and are eaten. Since most plants suffer early-season herbivory, the bet-hedging strategy could be favored. Similar arguments could be made with plant responses to abiotic factors such as drought and freezing. The mechanisms underlying the evolution of this bet-hedging response to herbivory may be difficult to disentangle from a mutualistic response.

Mutualism Hypothesis. This hypothesis states that overcompensation represents a mutualism in which herbivory provides a beneficial service to plants. How is this possible? Most mutualistic interactions probably evolved by way of host–parasite, predator–prey, or plant–herbivore interactions (Thompson 1982; Boucher et al. 1982; Law 1985). Yet the theoretical problem in understanding the evolution of a mutualistic interaction from an antagonistic interaction is in determining the ecological conditions that favor selection toward mutualism rather than toward the avoidance of the interaction. Thompson (1982) suggested that a change in outcome from antagonism to mutualism is most likely in interactions that are inevitable within the lifetimes of individuals. If it is unlikely that individuals can avoid being eaten, selection will favor individuals that have traits causing the interaction to have at least less of a negative effect on them. This selection regime sets the stage for the evolution of the interaction toward mutualism. Undoubtedly, grasses have coevolved with their grazers and are adapted to being grazed (Stebbins 1981; Belsky 1987). High levels of silicates, low nutritional quality, sharp awns, and vegetative modes of reproduction probably represent an adaptive syndrome to deal with large grazers. These adaptations, plus the inevitability of the interaction, may have set the stage for the evolution of a mutualistic interaction from a previously anatagonistic one.

Although these hypotheses are not necessarily mutually exclusive and in some cases may be difficult to experimentally examine, they do, however, represent an initial framework for viewing the origins of overcompensation.

THE COMPENSATORY CONTINUUM

As we have seen in the previous two sections, many plants undercompensate in response to herbivory while some have the capacity to overcompensate. We suggest that there is yet another group of plants that are more flexible in their responses to herbivory, such that under one set of circumstances they undercompensate while under other conditions they equally or overcompensate. We suggest that the apparent contradictions in the studies reporting variable impacts of herbivores on plants represent extremes of an intra- and interspecific continuum of compensatory responses (Maschinski and Whitham 1989). Here we examine the patterns of variation in intraspecific compensatory response to herbivory and suggest that this variation can result from both ecological and physiological factors.

Intraspecific Variation in Compensatory Response

An examination of the variation in individual plant responses to herbivory represents a first step in understanding the evolution of compensation. First we will discuss several ecological factors that influence intraspecific variation in compensatory response. These include the single and interactive effects of timing and intensity of herbivory; nutrient, water, and light availability; competition with other plants; plant size, and the type of tissue eaten (Crawley 1983; McNaughton 1983; Dirzo 1984). In general, we would expect that any factor which directly or indirectly influences plant growth will also regulate the degree of compensatory response.

Timing of Herbivory

The timing of herbivory relative to a plant's phenological stage is perhaps the most important external factor affecting compensation. If herbivory occurs at critical times during a plant's development, that is, during the seedling stage or after tissues have matured, mortality or little compensatory growth will occur. Before seedlings have established root systems and photosynthetic machinery, they are extremely vulnerable to herbivory, whereas older plants are less affected by herbivores because of their ability to compensate and/or their increased defenses. For example, in a survey of *Dipteryx panamensis* seedlings, Clark and Clark (1985) demonstrated that seedling longevity was directly related to levels of herbivory within the first month following germination. Further, with complete defoliation of the tropical palm *Astrocaryum mexicanum*, immature plants suffered 82% greater mortality than mature ones (Mendoza et al. 1987).

The interaction of timing and intensity of herbivory also influences the degree to which a plant can compensate. For example, Turnipseed (1972) reported that soybeans could withstand 17% defoliation at any stage of plant development without suffering loss in yield. Higher levels of defoliation (67%) were only detrimental to yield when plants were in an advanced phenological stage (pod set). Smith and Bass (1972) reported similar results for soybeans.

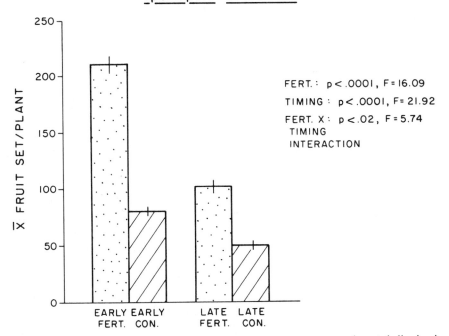

INTERACTION OF TIMING AND FERTILIZER ON COMPENSATION BY Ipomopsis arizonica

FERT.: p < .0001, F = 16.09

TIMING: p < .0001, F = 21.92

FERT. X: p < .02, F = 5.74
TIMING
INTERACTION

Figure 11-3. The ability of *Ipomopsis arizonica* to compensate to experimental clipping is a function of timing, supplemental nutrients, and the interaction of both (see text). Adapted from Maschinski and Whitham (1989).

The compensatory response of semelparous species generally declines as the timing of herbivory comes later in the season. This is largely due to time constraints; plants suffering late season herbivory have little time to recover before the end of the growing season (Crawley 1983). For example, compensatory response of the biennial *Ipomopsis arizonica* drops off markedly when plants are experimentally clipped later in the growing season (Maschinski and Whitham 1989, Fig. 11-3). To determine how timing of herbivory affected fruit set and whether nutrient supplementation could overcome the effects of timing, groups of *I. arizonica* were subjected to two different times of herbivory (May and July) and two different initial fertilization times. The 20 plants clipped in May (early season) set 25% more fruits than plants clipped in July (late season). Further, the effects of timing of herbivory could be overcome by fertilization (Fig. 11-3). Plants clipped in July and supplemented with fertilizer set equal numbers of fruits as unfertilized plants clipped early in the season. Thus, in this system the effects of timing interacted with nutrient availability to determine the level of recovery.

Nutrient Availability

Variation in nutrient availability can produce a gradient of plant compensatory responses. Several studies have demonstrated that nutrient supplementation of clipped or grazed plants promotes overcompensation of fruits (Benner 1988; Maschinski and Whitham 1989) or biomass (Julien and Bourne 1986; Verkaar et al. 1986), while control plants not supplemented with nutrients, equally or undercompensated. Further, for species capable of equal compensation, a gradient of responses can be attributed directly to nutrient levels. For example, when control and clipped (95% above ground biomass removed) *Ipomopsis aggregata formosissima* were subjected to five increasing concentrations of a balanced fertilizer, the degree of compensatory response increased along a nutrient gradient (Maschinski and Whitham, unpublished data). When nutrients were absent or very low in concentration, clipped plants undercompensated, producing 44% less fruits than controls. However, when nutrient levels were high, clipped plants produced 36–41% more fruits than controls.

In general, environments rich in nutrients will be more likely to promote equal or overcompensation in plants, whereas poor nutrient sites will be more likely to result in equal or undercompensation (Belsky 1986, 1987; Maschinski and Whitham 1989; Maschinski and Paige, unpublished data). For example, *Ipomopsis arizonica* can overcompensate from herbivory if grown under high nutrient conditions, whereas under low nutrient conditions there is a higher probability of undercompensation. This same pattern is seen in ragwort, *Senecio jacobaea*; when grown in poor soils it fails to produce any regrowth, while vigorous regrowth occurs in rich soils (Crawley 1983).

We do not expect that nutrients will universally increase plant compensatory response, but rather that the range of response to nutrient supplementation is in part regulated by the genetics of the species and the nutrient regime in which it evolved. Species adapted to low nutrient soils may only compensate from herbivory when nutrients are supplemented because of their inherently slow growth rate and low nutrient reserves (Bryant et al. 1983). Conversely, species adapted to nutrient-rich soils that have high nutrient reserves and rapid growth rates may be less sensitive to nutrient supplementation. For example, *Kyllinga nervosa*, which is adapted to low phosphorus soils, suffers a greater loss of biomass when defoliated under low phosporus levels than when defoliated under high phosphorus levels. In contrast, *Digitaria macroblephara*, which is adapted to high phosphorus soils, has equal responses to defoliation under high and low nutrient regimes (McNaughton and Chapin 1985).

Water Availabilty

Water availability can affect general plant status, growth, and compensatory response. Cox and McEvoy (1983) have shown that the compensatory ability of *Senecio jacobaea* to feeding by cinnabar moth, *Tyria jacobaeae*, was positively correlated with moisture availability. Plants receiving 10 weekly irrigations equally compensated from moth herbivory, whereas plants receiving 0–2 irrigations undercompensated. Studies which report no influence of water include Turnipseed (1972), who demonstrated that while timing and intensity of herbivory affected ultimate yield of soybeans, irrigation had no effect. Irrigation of these soybean plots occurred in late August and early September, 2–3 months after planting. Because no data on natural rainfall are reported, it is impossible to determine whether natural levels of precipitation were adequate for plant growth, causing irrigation to have no effect.

Light Availability

Low light levels can stress plants so that they cannot compensate from herbivory. Understory seedlings of *Omphalea oleifera* grown at high density were capable of equal compensation at four levels of defoliation when growing in full sun, but suffered increasing mortality from herbivory when grown in the shade (Dirzo 1984). At low densities, however, plants growing in sun and shade were equally capable of recovering from herbivory. Apparently, intraspecific competition interacted with low light levels to decrease the compensatory ability of *O. oleifera*.

Influence of Competition. When competition and herbivory are examined simultaneously, two general patterns emerge. First, herbivory can interact with competition to exacerbate the negative effect on yield (Harper 1977; Bentley et al. 1980; Lee and Bazzaz 1980). For example, mortality of *Gutierrezia microcephala* exposed to grasshopper herbivory increased 25% when competitors were present over levels when competitors were absent (Parker and Salzman 1985). Second, competition may have a greater negative effect on yield than does herbivory (Mueggler 1972; Duggan 1985; Fowler and Rausher 1985). For example, competition with grasses significantly decreased reproductive mass of *Aristolochia reticulata*, whereas simulated herbivory had no effect (Fowler and Rausher

1985). In cases where competition does not interact with herbivory to decrease plant fitness, it is likely that competitors do not significantly reduce resources available for compensation in the target species.

Interaction of Factors

The interaction of competition, nutrient availability, and timing of herbivory can also cause a gradient in compensatory responses. Maschinski and Whitham (1989) documented that the compensatory ability of *I. arizonica* was dependent upon these factors (Fig. 11-4). While the majority of naturally grazed and clipped plants equally compensated from herbivory, 4% of the plants studied significantly overcompensated and 13% significantly undercompensated. Overcompensation occurred when plants were grown in the absence of competitors, grazed early in the season, and were supplemented with nutrients (Fertilized and Watered Open Group; Fig. 11-4). Conversely, undercompensation occurred when plants were grown with grass competitors, without supplemental nutrients, and when grazed late in the season (Control Grass Group; Fig. 11-4).

We suggest that external factors of nutrient availability and the timing of herbivory and competition affect plant growth and can lead to probabilistic compensatory responses. The probability of overcompensation increases with increased nutrient availability, decreased plant competition, and early timing of

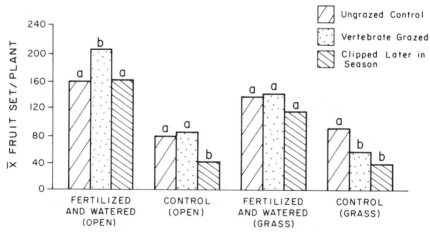

Figure 11-4. The interaction of competition, nutrient availability, the timing of herbivory result in a gradient of compensatory responses that range from overcompensation to undercompensation (see text). Lowercase letters indicate comparisons differing at $P < 0.05$. Adapted from Maschinski and Whitham (1989).

herbivory. Conversely, the probability of undercompensating increases as the timing of herbivory comes later in the growing season, as plant competition increases, and as nutrients become less available. Equal compensation will occur at intermediate levels of these factors or when one or more of the factors is favorable. For example, if a plant experiences competition, but was grazed early in the season and fertilized, it can equally compensate (Fertilized and Watered Grass Group; Fig. 11-4).

Size Effects

When herbivores feed on plants of different sizes within a population, a gradient of compensatory responses can occur. Hendrix (1979) demonstrated that when large *Pastinaca sativa* had their primary umbel destroyed by the parsnip webworm *Depressaria pastinacella*, they were able to replace 100% of the seeds lost to herbivory while small plants were able to replace only 50%.

Type of Tissue Loss

Within a single population of plants, numerous herbivores may selectively feed upon different plant tissues, thereby creating a gradient of compensatory responses from positive to negative (Inouye 1982). For example, when lepidopteran larvae eat the central portion of the basal rosette of the biennial *Jurinea mollis*, plants overcompensate, producing multiple rosettes and up to three times as many seeds as those without multiple rosettes. Conversely, *J. mollis* undercompensates for tephritid fly damage to flower head receptacles (Inouye 1982).

Maschinski and Whitham (unpublished data) have found a similar gradient in the compensatory responses of the biennial *I. arizonica* to varying herbivore attack. In response to meristem damage by tortricid moths, the plants overcompensate, producing 43% more fruits than undamaged plants. In contrast, plants only equally compensate from meristem and fruit damage by the noctuid *Heliothis phloxiphagus*. As the intensity of fruit predation increases, we would expect that the probability of undercompensation would increase.

The examples presented here demonstrate that phenological stage, type of tissue consumed, nutrients, competition, and intensity of herbivory determine the degree of compensation. It is important to note that the range of a species' compensatory response is dictated by morphological and physiological constraints as well as ecological conditions. Consequently, while some species may exhibit the whole gamut of responses to herbivory, others may be far more limited and incapable of any compensation.

THE PHYSIOLOGY OF COMPENSATION

From the previous sections it is clear that plant responses to herbivory vary greatly. For example, both *I. arizonica* and *Pinus edulis* grow in the same habitat and suffer similar herbivore damage (the destruction of apical meristems leading to increased branching), but exhibit very different responses. Impact on *P. edulis* was strongly negative, while on *I. arizonica* impact was usually neutral and with experimental manipulations could span the whole gamut from very negative to very positive. In this section we examine some of the underlying physiological mechanisms that may account for this variation in compensatory response.

We will focus on the plant as a balanced system of sources and sinks. The plant body is divided into regions specialized for growth and reproduction that are nonphotosynthetic, or only marginally so, and regions specialized for net carbon gain. These can be termed sinks and sources, respectively; growing meristems, flowers, and fruits are examples of sinks, while sources are those regions of the plant with high concentrations of labile assimilates, either the photosynthetic organs or storage tissues. Thus, growth of shoot meristems, roots, and reproductive organs is dependent on the rate at which carbon assimilates are supplied by vascular transport from the source. Plant response to herbivore feeding involves the integrated responses of both sources and sinks. Plants grow by reiterating modules, thus increasing the number of parts rather than the size of parts (Harper 1977). Since modules are derived from nonphotosynthetic meristems, clearly a temporary supply of assimilates from source organs is needed to initiate growth. Without a supply of meristems, no increase in number of parts could occur.

We view and discuss two factors as being critical in determining how plants physiologically compensate from herbivore damage: (1) responses by source organs to increase assimilate supply and (2) responses by sink organs to maintain or increase the number of growing points or reproductive organs.

Source Responses to Herbivory

Defoliation reduces the supply of assimilates to roots (Fitter 1986), developing seeds and fruits (Egli et al. 1985; Stephenson 1980), and growing shoots (Marshall and Sagar 1968). Undamaged regions may respond by increasing assimilate supply to sink organs on damaged parts. Increased assimilate supply could result from increased photosynthetic rates in undamaged leaves, translocation of assimilates from undamaged parts to damaged parts, or by mobilization of assimilates from storage tissues.

1. Photosynthetic Responses. Increased photosynthesis in undamaged leaves has been proposed as a major mechanism contributing to compensatory response (McNaughton 1979; Crawley 1983). Although many laboratory studies have found that plants respond to partial defoliation with an increased photosynthetic rate in the remaining leaves (Hodgkinson 1974; Detling et al. 1979; Painter and Detling 1981; Wallace et al. 1984), the importance of increased photosynthesis for compensatory response has not been demonstrated. For example, increased photosynthesis could increase source supply to reproductive sinks remaining on defoliated shoots and to actively growing meristems, but no studies have shown these direct effects.

Few studies have examined compensatory photosynthesis of plants in natural environments, nor have increased photosynthetic rates measured at the single-leaf level been related to increased performance of the whole plant. Nowak and Caldwell (1984) found compensatory photosynthesis in natural field populations of *Agropyron* sp. after artificial defoliation, although increases were generally lower than those found in laboratory studies. The oldest, lower leaves had the largest increase in photosynthetic rate, but the contribution of these leaves to total plant carbon assimilation was small when calculated over the season for the whole plant. The youngest, upper leaves did not significantly increase photosynthetic rate (Fig. 11-5). Furthermore, both the grazing-tolerant *A. desertorum* and

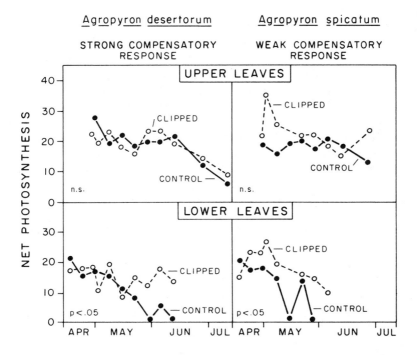

(adapted from Nowak and Caldwell 1984)

Figure 11-5. In their studies of two wheatgrasses exhibiting very different compensatory responses (*A. desertorum*, strong compensation vs. *A. spicatum*, weak compensation), Nowak and Caldwell (1984) demonstrate three major points: (1) the photosynthetic response was not different for the two species, (2) the increased photosynthesis was in the very oldest leaves due to delayed senescence. (3) the oldest leaves made up only 10% of the whole plant biomass. In total these results suggest that compensation through increased photosynthesis following herbivory is not as important as has been suggested in the literature.

the grazing-sensitive *A. spicatum* had very similar photosynthetic responses to defoliation (Fig. 11-5). They concluded that compensatory photosynthesis did not play a major role in recovery from herbivory. The few other field studies have either shown compensatory photosynthetic increases, but have not related it to whole-plant response (Heichel and Turner 1983; Hartnett and Bazzaz 1983) or have found a reduced photosynthetic rate (Larson and Whitham, unpublished data).

2. Vascular Integration. The vascular plumbing between source and sink acts as a constraint on the potential of compensatory photosynthesis to supply assimilates to regrowing shoots or fruits on defoliated branches. Vascular pathways can be highly restricted within a plant, physiologically dividing it into autonomous subunits which function relatively independently of other subunits (Watson and Casper 1984; Pitelka and Ashmun 1985). Clearly sinks on herbivore-damaged parts must have vascular connections to undamaged sources if they are to get assimilate support.

In some species, normally independent modules become physiologically reintegrated following herbivory, and damaged modules draw assimilates from neighbouring modules acting as sources. In many grasses, increased source demand due to defoliation or shading results in the stressed ramet or module acting as a sink for resources from surrounding, undamaged parts (Marshall and Sagar 1965, 1968; Nyahoza et al. 1973; Welker et al. 1985, 1987). In a nongrass species, Hartnett and Bazzaz (1983) found that although ramets of *Solidago* sp. were normally independent, a shaded ramet would draw resources from a connected, unstressed ramet. Schmid et al. (1988) showed that individuals which had vascular connections to sister ramets severed suffered most from defoliation, while connected individuals responded to defoliation with the rapid production of new leaves. Ashmun et al. (1982) also found defoliated ramets of *Clintonia borealis* accumulated resources produced in a sister ramet. Thus, reintegration may be important in the temporary support of sinks (growing meristems or reproductive organs) on parts of the plant with reduced source supply due to herbivory.

In other plants, however, vascular connections between modules appear to be lacking, such that there is no assimilate support of damaged modules by undamaged modules (Ashmun et al. 1982). For example, defoliation of flowering branches of *Catalpa speciosa* results in almost complete abortion of fruit within the branch, despite an abundance of undamaged leaves on surrounding branches (Stephenson 1980). In another example, complete defoliation of one side of *Artemisia tridentata* shrubs resulted in high rates of branch death compared with plants which had the same biomass of foliage removed, but spread over the whole plant. Apparently the completely defoliated branches could not draw resources from the undamaged parts of the plant, whereas partially defoliated branches could maintain growth, although at a reduced rate compared with controls (Cook and Stoddard 1960).

3. Stored Reserves. In contrast to current photosynthates fueling compensatory growth as discussed above, stored reserves may contribute source supply to sinks (Bryant et al. 1983; McNaughton 1979). Gulmon and Mooney (1986) and Meijden et al. (1988) argue that allocation to storage tissue unavailable to herbivores can be an alternative to allocation to defensive chemicals. In an example supporting the role of stored reserves, perennial ryegrass plants were found to allocate about 4% of daily assimilated carbon to storage in stem bases, which was later reallocated to new growth following defoliation (Danckwerts and Gordon 1987). Contribution by storage reserves was not directly measured by Hendrix (1979), but was implied by greater compensatory responses of plants with large basal stem diameters compared with smaller plants.

Other studies have not supported an important role for stored carbon, finding that current photosynthates supported new growth (Davidson and Milthorpe 1966; Ryle and Powell 1975; Richards 1984). It may be that carbon is not the appropriate currency to measure storage reserves. For example, the biennial monocarp *Arctium tomentosum* stores both carbohydrates and nitrogen in the roots during the first year of growth. Reallocation of carbon to the leaves and flowering stalks the following year is relatively small, while nitrogen reallocation is very large (Heilmeier et al. 1986).

Sink Responses to Herbivory

Studies of compensatory mechanisms have tended to concentrate on the responses of source organs to increase assimilate supply following herbivore damage. Since plant growth and reproduction occur by increasing the number of parts rather than the size of parts (Harper 1977), the availability of meristems capable of reiterating modules should play an equally important role in determining compensatory response. For example, Lovett Doust and Eaton (1982) found that pod yield of bean plants was influenced more by the number of modular units produced than assimilate availability. They induced the continued production of modular units by removing pods as they matured, and thus did not increase assimilate supply to remaining pods, but probably instead prevented production of a hormonal signal to end reproductive module production.

1. Reserve Meristems. Maintenance of a reserve bank of dormant meristems or buds, and the ability to activate them, allows plants to rapidly regenerate new modules following damage (Halle' et al. 1978; Maillette 1982, Hardwick 1986). Plants often respond to removal of either apical meristems, developing flowers or fruits by the release of dormant meristems, which results in lateral branching. For example, the grazing-tolerant *Agropyron desertorum* responded to defoliation by increasing the number of growing tillers, while the grazing-sensitive *Agropyron spicatum* did not increase tiller number (Caldwell et al. 1981; Fig. 11-6). Interestingly, both species responded equally in terms of photosynthetic increases following defoliation (Nowak and Caldwell 1984, Fig. 11-6). In this example, meristem responses were more important than photosynthetic rate in determining compensatory response.

Almost all cases of reported overcompensation have been due to an increase in

IMPORTANCE OF MODULAR RESPONSES
TO COMPENSATION BY WHEATGRASSES

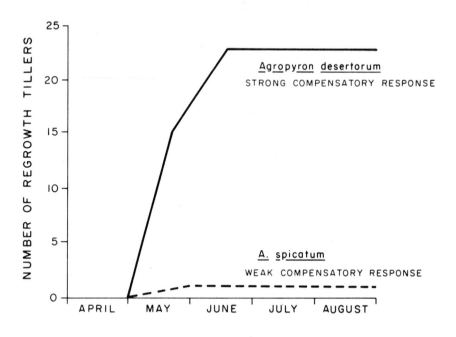

(adapted from Caldwell et al. 1981)

Figure 11-6. In their studies of two wheatgrasses exhibiting very different compensatory responses (*A. desertorum*, strong compensation vs. *A. spicatum*, weak compensation), Caldwell et al. (1981) show the major difference between the two species to be the number of reserve meristems. The strong compensator produces many new tillers while the weak compensator produces few.

the number of modular units when apical dominance and the hormonal suppression of dormant meristems has been removed by herbivores. The response of *I. aggregata* (Paige and Whitham 1987a) resulted from an increase in the number of modules per plant from one to four (Fig. 11-2). Inouye (1982) found a similar result for *Jurinea mollis*, where herbivore damage of the apical meristem increased the number of modules per plant; herbivore damaged plants with four modules produced almost three times more ovules than undamaged plants which consisted of only one module. Overcompensation measured as increased vegetative growth was also due to increase in the number of modules (Jefferies 1984).

   2. Lack of Meristems. Growth of some plants may be limited by the number of meristems available rather than resource supply (Watson 1984). The century plant *Agave desertii* has no reserve meristem population from which new modules can be reiterated following herbivore damage. If a plant has a small portion of its terminal flowering meristem removed by bighorn sheep, no capacity exists to develop another flowering stalk, even though extensive resources still remain in the leaves and basal portion of the flower stalk. Such semelparous plants may survive for 2 yr and then die without flowering (Whitham, personel observation). Similar patterns should be observed for primitive gymnosperms like cycads that have a single apical meristematic cell; with the destruction of this cell, the plant has lost its meristematic regions and capacity to develop new modules.

3. *Predetermined Function.* For many long-lived evergreen species, development of meristems is determinate and almost invariant (Hardwick 1986). For example, the dwarf shoot structure of the lateral meristems of grand fir, *Abies grandis*, is predetermined by position, and maintained even when the terminal leader is removed (Worrall 1984). Thus, in this example, compensatory response is not limited by the number of meristems or modules, but because the growth form of meristems is predetermined such that a lateral meristem cannot function as a terminal meristem.

In combination, meristem limitations have probably been one of the least-appreciated mechanisms of plant compensation to herbivory and should receive much more research attention.

GENERAL PATTERNS OF INTERSPECIFIC RESPONSES

In summary, it seems that several general patterns emerge concerning the abilities of species to compensate from herbivory. The continuum of responses described for some species also exists between species. The level of resources available in the environment and the developmental morphology of the plant determine where on the compensatory continuum a species will fall (Table 11-4).

At one extreme are species which have little potential for compensatory response regardless of ecological conditions. Growth of these species is limited by resource availability to fuel regrowth and by sink responses to increase the meristems necessary to initiate new growth (e.g., *Pinus edulis*, Fig. 11-1). Plants adapted to low-resource environments have inherently slow growth rates, low photosynthetic rates even under ideal conditions, and small below ground

TABLE 11-4. Growing Conditions and Plant Traits Important in Determining a Plant's Probable Compensatory Response

Equal to Overcompensation	Undercompensation
Herbivory early in season	Herbivory Late in season
Abundant water, nutrients, and light	Low water, nutrients, and light
Low competition	High competition
No meristem limitation	Meristem limitation
Fast growth	Slow growth
Integrated plant modules	Nonintegrated modules
Annuals and biennials	Woody perennials

storage reserves (Chapin 1980; Bryant et al. 1983). Bryant et al. (1983) and Coley et al. (1985) have argued that these slow-growing species are adapted for reducing tissue loss to herbivory by heavily investing in constitutive defenses. Opportunistic growth following tissue loss by this group of species may also be limited by an inflexible and determinate growth form (e.g., *Agave desertii*). Strict adherence to genetically determined growth form may prevent species such as conifers from responding to tissue damage (Worrall 1984).

At the other extreme are species with rapid growth rates and strong responses by source organs to increase assimilate supply, coupled with a large population of dormant buds (e.g., *Ipomopis aggregata*, Fig. 11-2). Species growing in resource-rich environments have resources necessary for rapid regrowth of tissues removed by herbivores. They have high growth rates, high photosynthetic rates, and large underground storage reserves (Bryant et al. 1983; but see Paige and Whitham 1987a). These species maintain relatively lower levels of defensive chemicals and suffer much higher rates of herbivory than species growing in low-resource environments (Coley et al. 1985). These species maintain a large stand-by population of dormant buds and respond to abiotic and biotic environmental disturbance by rapidly reiterating new modules (Hardwick 1986; Halle' et al. 1978). We predict that the modules of these species are highly integrated, likely to reiterate lost parts, capable of equal to overcompensation, and may show, a gradient of compensatory responses according to ecological conditions.

Stenseth (1978) suggested a mathematical model which predicts that herbivory is more likely to increase the fitness of short-lived annuals and biennials rather than long-lived perennials. In general, the examples presented here support this view; plants that seem to suffer the most from herbivory are the long-lived woody perennials, while the examples of equal to overcompensation are generally annuals and biennials.

In brief, we believe that the continuum view in part unites opposing schools of thought. While we believe that some groups of plants are predisposed to either being positively or negatively affected by herbivores because of their genetics, life history traits, morphology, and the environment in which they are found, other species are capable of the whole gamut of responses, depending upon local conditions.

SUMMARY

We examined the range of compensatory responses of plants to herbivory and discussed the physiological mechanisms that seem to be involved. Many studies support the paradigm of direct negative impacts of herbivores on plants. At outbreak levels herbivores can dramatically reduce plant growth, alter plant architecture, delay sexual reproduction, eliminate sexual reproduction, and even mediate a sex change in their host plants. Although we know much about the extremes of herbivory, relatively little is known about the impacts of herbivores at low densities, the cumulative impacts of diverse herbivores and pathogens, and the problem of increased resistance to one herbivore coming at the expenses of increased susceptibility to other herbivores.

At the other extreme, numerous studies have suggested that herbivores can have direct positive effects on plant fitness. We reviewed the controversy and developed an example in which *Ipomopsis aggregata* produced 2.8 times more fruits and seeds after having 95% of its aboveground biomass removed by ungulate herbivores. Four hypotheses were developed to account for the origin of overcompensation: (1) nonadaptation hypothesis in which plants are anachronisms; (2) extreme adaptation hypothesis in which absence of herbivory represents an unnatural condition; (3) overcompensation as a form of bet-hedging; and (4) herbivores are plant mutualists.

Having examined the apparent contradiction of under- and overcompensation, we developed the concept of a continuum of compensatory responses. Individual species may be capable of the whole gamut of compensatory responses from negative to positive, depending upon both biotic and abiotic factors. We examined the intraspecific variation in plant responses to the timing of herbivory, nutrient, water and light availability, influence of competitors, and size effects. In general, we predicted that the harsher the conditions, the more severely plants will be impacted by herbivores, while with more benign conditions plants may be able to equally or overcompensate.

We examined potential mechanisms of compensation by viewing the plant as a balanced system of sinks and sources and concluded that the physiological relationships between these structures are crucial in determining the compensatory response. The vascular integration of plant modules has an important effect on compensation. Ramets or shoots that are vascularly integrated and share resources are more likely to compensate from herbivory than those lacking such integration. Equally important is the maintenance of a reserve bank of dormant meristems or buds that allows plants to rapidly regenerate new modules following damage. Plants lacking reserve buds may suffer little tissue loss yet be killed, while plants with reserve buds may suffer extremely high tissue loss but overcompensate from herbivory.

Plants that are most negatively affected by herbivory should be woody perennials with determinate growth, few reserve meristems, nonintegrated modules, and slow growth. Their compensatory abilities should be further reduced by growing under conditions of high competition, low water, low nutrients, low light, and sustaining herbivory late in the growing season. Plants growing under the opposite conditions, often annuals and biennials, should be most likely to equally or overcompensate. While some species exhibit largely fixed responses owing to genetic and physiological limitations, others may be quite flexible. We believe this continuum view in part unites opposing schools of thought.

ACKNOWLEDGMENTS

This research has been supported in part by NSF grants BSR-8705347, BSR-8604983, and BSR-8312758, USDA grants GAM-8700709 and 84-CRCR-1-

1443, Sigma Xi, the Bilby Research Center, and Organized Research of Northern Arizona University. We thank the Utah Power and Light Co., the Wilson Foundation, the Museum of Northern Arizona, and the U. S. Forest Service for their support and facilities. H. Cushman, J. Richards, and F. H. Wagner provided valuable comments on the manuscript.

REFERENCES

Ashmun, J. H., R. J. Thomas, and L. F. Pitelka. 1982. Translocation of photoassimilates between sister ramets in two rhizomatous forest herbs. *Annals of Botany* **49**: 403–416.

Bartholomew, B. 1970. Bare zone between California shrub and grassland communities: the role of animals. *Science* **170**: 1210–1212.

Belsky, A. J. 1986. Does herbivory benefit plants? A review of the evidence. *American Naturalist* **127**: 870–892.

Belsky, A. J. 1987. The effects of grazing: Confounding of ecosystem, community, and organism scales. *American Naturalist* **129**: 777–783.

Benner, B. L. 1988. Effects of apex removal and nutrient supplementation on branching and seed production in *Thlaspi arvense* (Brassicaceae). *American Journal of Botany* **75**: 645–651.

Bentley, S., J. B. Whittaker, and A. J. C. Malloch. 1980. Field experiments on the effects of grazing by a chrysomelid beetle (*Gastrophysa viridula*) on seed production and quality in *Rumex obtusifolius* and *Rumex crispus*. *Journal of Ecology* **68**: 671–674.

Berenbaum, M. R., A. R. Zangerl, and J. K. Nitao. 1986. Constraints on chemical coevolution: Wild parsnips and the parsnip webworm. *Evolution* **40**: 1215–1228.

Boucher, D. H., S. James, and K. H. Keeler. 1982. The ecology of mutualisms. *Annual Review of Ecology and Systematics* **13**: 315–347.

Bryant, J. P., F. S. Chapin, III, and D. R. Klein. 1983. Carbon/nutrient balance of boreal plants in relation to vetebrate herbivory. *Oikos* **40**: 357–368.

Caldwell, M. M., J. H. Richards, D. A. Johnson, R. S. Nowak, and R. S. Dzurec. 1981. Coping with herbivory: Photosynthetic capacity and resource allocation in two semiarid *Agropyron* bunchgrasses. *Oecologia* **50**: 1424.

Cargill, S. M. and R. L. Jefferies. 1984. The effects of grazing by lesser snow geese on the vegetation of a sub-artic saltmarsh. *J. Appl. Ecology* **21**: 669–686.

Chapin, F. S., III. 1980. The mineral nutrition of wild plants. *Annual Review of Ecology and Systematics* **11**: 233–260.

Chew, R. M. 1974. Consumers as regulators of ecosystems: An alternative to energetics. *Ohio J. of Science* **74**: 359–370.

Christensen, K. M., and T. G. Whitham. 1991. Indirect herbivore mediation of avian seed dispersal in pinyon pine. *Ecology* (in press).

Clark, D. B. and D. A. Clark. 1985. Seedling dynamics of a tropical tree: Impacts of herbivory and meristem damage. *Ecology* **66**: 1884–1892.

Coley, P. D. 1986. Costs and benefits of defense by tannins in a neotropical tree. *Oecologia* **70**: 238–241.

Coley, P. D., J. P. Bryant, and F. S. Chapin, III. 1985. Resource availability and plant antiherbivore defense. *Science* **230**: 895–899.

Cook, C. W. and L. A. Stoddart. 1960. Physiological responses of big sagebrush to different types of herbage removal. *Journal of Range Management* **13**: 14–16.

Cox, C. S. and P. B. McEvoy. 1983. Effect of summer moisture stress on the capacity of tansy ragwort (*Senecio jacobaea*) to compensate for defoliation by cinnabar moth (*Tyria jacobaeae*). *Journal of Applied Ecology* **20**: 225–234.

Craig, T. P., P. W. Price, and J. K. Itami. 1986. Resource regulation by a stem-galling sawfly on the arroyo willow. *Ecology* **67**: 419–425.

Crawley, M. J. 1983. *Herbivory*. University of California Press, Berkeley, CA.

Crawley, M. J. 1987. Benevolent herbivores? *Trends in Ecology and Evolution* **2**: 167–168.

Crawley, M. J. and M. Nachapong. 1985. The establishment of seedlings from primary and regrowth seeds of ragwort (*Senecio jacobaea*). *Journal f of Ecology* **73**: 255–261.

Danckwerts, J. E. and A. J. Gordon. 1987. Long-term partitioning, storage and remobiliz-ation of ^{14}C assimilated by *Lolium perenne*. *Annals of Botany* **59**: 55–66.

Davidson, J. L. and F. L. Milthorpe. 1966. The effect of defoliation on the carbon balance in *Dactylis glomerato*. *Annals of Botany* **30**: 185–198.

Detling, J. K., M. I. Dyer, and D. T. Winn. 1979. Net photosynthesis, root respiration, and regrowth of *Bouteloua gracilis* following simulated grazing. *Oecologia* **41**: 127–134.

Dirzo, R. 1984. Herbivory: A phytocentric overview. Pages 141–165 in *Perspectives on plant population ecology*. R. Dirzo and J. Sarukhan (Eds.). Sinauer, Sunderland, MA.

Drake, D. W. 1981a. Reproductive success of two *Eucalyptus* hybrid populations. I. Generalized seed output model and comparison of fruit parameters. *Australian Journal of Botany* **29**: 25–35.

1981b. Reproductive success of two *Eucalyptus* hybrid populations. II. Comparison of predispersal seed parameters. *Australian Journal of Botany* **29**: 37–48.

Dyer, M. I. 1975. The effects of red-winged blackbirds (*Agelaius phoeniceus*) on biomass production of corn grains (*Zea mays*). *J. Appl. Ecology* **12**: 719–726.

Duggan, A. E. 1985. Pre-dispersal seed predation by *Anthocharis cardamines* (Pieridae) in the population dynamics of the perennial *Cardamine pratensis* (Brassicaceae). *Oikos* **44**: 99–106.

Egli, D. B., R. D. Guffy, L. W. Meckel, and J. E. Leggett. 1985. The effect of source–sink alterations on soybean seed growth. *Annals of Botany* **55**: 395–402.

Fedde, G. F. 1973. Impact of the balsam wooly aphid (homoptera: phylloxeridae) on cones and seed produced by infested Fraser fir. *Canadian Entomologist* **105**: 673–680.

Fitter, A. H. 1986. Acquisition and utilization of resources. Pages 375–405 in *Plant Ecology*. M. J. Crawley (Ed.). Blackwell, Oxford.

Foster, W. A. 1984. The distribution of the sea lavender aphid *Staticobium statics* on a marine salt marsh and its effect on host plant fitness. *Oikos* **42**: 97–104.

Fowler, N. L. and M. D. Rausher. 1985. Joint effects of competitors and herbivores on growth and reproduction in *Aristolochia reticulata*. *Ecology* **66**: 1580–1587.

Fritz, R. S., C. F. Sacchi, and P. W. Price. 1986. Competition versus host plant phenotype in species composition: Willow sawflies. *Ecology* **67**: 1608–1618.

Gulmon, S. L. and H. A. Mooney. 1986. Costs of defense and their effects on plant productivity. Pages 681–698 in *On the Economy of Plant Form and Function*. T. J. Givnish (Ed.). Cambridge University Press.

Halle', F., R. A. Oldeman, and P. B. Tomlinson. 1978. *Tropical Trees and Forests: An Architectural Analysis*. Springer-Verlag, New York.

Hardwick, R. C. 1986. Physiological consequences of modular growth in *Plants. Phil. Trans. R. Soc. Lond, B* **313**: 161–173.

Harper, J. L. 1977. *Population biology of plants.* Academic Press, London.

Harris, P. 1974. A possible explanation of plant yield increases following insect damage. *Agro-Ecosystems* **1**: 219–225.

Hartnett, D. C. and F. A. Bazzaz. 1983. Physiological integration among intraclonal ramets in *Solidago canadensis. Ecology* **64**: 779–788.

Heichel, G. H. and N. C. Turner. 1983. CO_2 assimilation of primary and regrowth foliage of red maple (*Acer rubrum* L) and red oak (*Quercus rubra* L): Adaptation to defoliation. *Oecologia* **57**: 14–19.

Heilmeier, H., E. Schulze, and D. M. Whale. 1986. Carbon and nitrogen partitioning in the biennial monocarp *Arctium tomentosum* Mill. *Oecologia* **70**: 466–474.

Hendrix, S. D. 1979. Compensatory reproduction in a biennial herb following insect defoliation. *Oecologia* **42**: 107–118.

Hendrix, S. D. 1984. Reactions of *Heracleum lanatum* to floral herbivory by *Depressaria pastinacella. Ecology* **65**: 191–197.

Hodgkinson, K. C. 1974. Influence of partial defoliation on photosynthesis, photorespiration and transpiration by lucerne leaves of different ages. *Australian Journal of Plant Physiology* **1**: 561–578.

Inouye, D. W. 1982. The consequences of herbivory: A mixed blessing for *Jurinea mollis* (Asteraceae). *Oikos* **39**: 269–290.

Janzen, D. H. 1979. New horizons in the biology of plant defenses. Pages 331–350 in *Herbivores: Their interactions with secondary plant metabolites.* G. E. Rosenthal and D. H. Janzen (Eds.). Academic Press, New York.

Janzen, D. H. and P. S. Martin. 1982. Neotropical anachronisms: the Fruits the gomphotheres left behind. *Science* **215**: 19–27.

Jefferies, R. L. 1984. The phenotype: Its development, physiological constraints and environmental signals. Pages 347–358 in *Perspectives on Plant Population Ecology.* R. Dirzo and J. Sarukhan (Eds.). Sinauar, Sunderland, MA.

Julien, M. H. and A. S. Bourne. 1986. Compensatory branching and changes in nitrogen content in the aquatic weed *Salvinia molesta* in response to disbudding. *Oecologia* **70**: 250–257.

Karban, R., R. Adamchak, and W. C. Schnathorst. 1987. Induced resistance and interspecific competition between spider mites and a vascular wilt fungus. *Science* **235**: 678–680.

Law, R. 1985. Evolution in a mutualistic environment. Pages 145–170 in *Biology of Mutualism: Ecology and Evolution.* D. H. Boucher (Ed.). Oxford University Press, New York.

Lee, T. D. and F. A. Bazzaz. 1980. Effects of defoliation and competition on growth and reproduction in the annual plant *Abutilon theophrasti. Journal of Ecology* **68**: 813–821.

Leonard, K. J. 1977. Selection pressures and plant pathogens. *Ann. N. Y. Acad. Sci.* **287**: 207–222.

Lovett Doust, J. and G. W. Eaton. 1982. Demographic aspects of flower and fruit production in bean plants, *Phaseolus vulgaris* L. *American Journal of Botany* **69**: 1156–1164.

Maddox, G. D. and R. B. Root. 1987. Resistance to 16 diverse species of herbivoreous insects within a population of goldenrod, *Solidago altissima*: genetic variation and heritability. *Oecologia (Berl.)* **72**: 8–14.

Maillette, L. 1982. Structural dynamics of silver birch. I. The fates of buds. *Journal of Applied Ecology* **19**: 203–218.

Marshall, C. and G. R. Sagar. 1965. The influence of defoliation on the distribution of assimilates in *Lolium multiflorum* Lam. *Annals of Botany* **29**: 365–370.

Marshall, C. and G. R. Sagar. 1968. The distribution of assimilates in *Lolium multiflorum* Lam. following differential defoliation. *Annals of Botany* **32**: 715–719.

Maschinski, J. and T. G. Whitham. 1989. The continuum of plant responses to herbivory: The influence of plant association, nutrient availability and timing. *American Naturalist* **134**: 1–19.

Mattson, W. J. and N. D. Addy. 1975. Phytophagous insects as regulators of forest primary production. *Science* **191**: 515–522.

McNaughton, S. J. 1979. Grazing as an optimization process: Grass–ungulate relationships in the Serengeti. *American Naturalist* **113**: 691–703.

McNaughton, S. J. 1983. Compensatory plant growth as a response to herbivory. *Oikos* **40**: 329–336.

McNaughton, S. J. 1986. On plants and herbivores. *American Naturalist* **128**: 765–770.

McNaughton, S. J. and F. S. Chapin, III. 1985. Effects of phosphorus nutrition and defoliation and C_4 graminoids from the Serengeti plains. *Ecology* **66**: 1617–1629.

Meijden, E. van der, W. Marijke and H. J. Verkaar. 1988. Defense and regrowth, alternative plant strategies in the struggle against herbivores. *Oikos* **51**: 355–363.

Mendoza, A., D. Pinero, and J. Sarukhan. 1987. Effects of experimental defoliation on growth, reproduction and survival of *Astrocaryum mexicanum*. *Journal of Ecology* **75**: 545–554.

Mueggler, W. F. 1972. Influence of competition on the response of bluebunch wheatgrass to clipping. *Journal of Range Management* **25**: 88–92.

Nowak, R. S. and M. M. Caldwell. 1984. A test of compensatory photosynthesis in the field: Implications for herbivory tolerance. *Oecologia* **61**: 311–318.

Nyahoza, F. C., C. Marshall, and G. R. Sagar. 1973. The interrelationship between tillers and rhizomes in *Poa pratensis* L.: An autoradiographic study. *Weed Research* **13**: 304–307.

Owen, D. F. and R. G. Wiegert. 1976. Do consumers maximize plant fitness? *Oikos* **27**: 488–492.

Paige, K. N. and T. G. Whitham. 1987a. Overcompensation in response to mammalian herbivory: The advantage of being eaten. *American Naturalist* **129**: 407–416.

Paige, K. N. and T. G. Whitham. 1987b. Flexible life history traits: Shifts by scarlet gilia in response to pollinator abundance. *Ecology* **68**: 1691–1695.

Painter, E. L. and J. K. Detling. 1981. Effects of defoliation on net photosynthesis and regrowth of western wheatgrass. *Journal of Range Management* **43**: 68–71.

Parker, M. A. and A. G. Salzman. 1985. Herbivore exclosure and competitor removal: Effects of juvenile survivorship and growth in the shrub *Gutierrezia microcephala*. *Journal of Ecology* **73**: 903–913.

Pitelka, L. F. and J. W. Ashmun. 1985. Physiology and integration of ramets in clonal plants. Pages 399–435 in *Population Biology and Evolution of Clonal Organisms* . J. Jackson, L. W. Buss, and R. E. Cook (Eds.) Yale University Press, New Haven, CT.

Randall, J. E. 1965. Grazing effect on sea grasses by herbivorous reef fishes in the West Indies. *Ecology* **46**: 255–260.

Richards, J. H. 1984. Root growth response to defoliation in two *Agropyron* bunchgrasses: Field observations with an improved root periscope. *Oecologia* **64**: 21–25.

Ryle, G. J. A. and C. E. Powell. 1975. Defoliation and regrowth in the graminaceous plant: The role of current assimilate. *Annals of Botany* **39**: 297–310.

Sacchi, C. F., P. W. Price, T. P. Craig and J. K. Itami. 1988. Impact of shoot galler attack on sexual reproduction in the arroyo willow. *Ecology* **69**: 2021–2030.

Schmid, B., G. M. Puttick, K. H. Burgess, and F. A. Bazzaz. 1988. Clonal integration and effects of simulated herbivory in old-field perennials. *Oecologia* **75**: 465–471.

Smith, R. H. and M. H. Bass. 1972. Relation of artificial pod removal to soybean yields. *Journal of Economic Entomology* **65**: 606–608.

Stebbins, G. L. 1981. Coevolution of grasses and herbivores. *Annals of Missouri Botanical Garden* **68**: 75–86.

Stenseth, N. C. 1978. Do grazers maximize individual plant fitness? *Oikos* **31**: 299–306.

Stephenson, A. G. 1980. Fruit set, herbivory, fruit reduction, and the fruiting strategy of *Catalpa speciosa* (Bignoniaceae). *Ecology* **61**: 57–64.

Strong, D. N., J. H. Lawton, and Sir Richard Southwood. 1984. *Insects on plants.* Harvard University Press, Cambridge, MA.

Thompson, J. N. 1982. *Interaction and coevolution.* John Wiley, New York.

Turnipseed, S. G. 1972. Response of soybeans to foliage losses in South Carolina. *Journal of Economic Entomology* **65**: 224–229.

Verkaar, H. J. 1986. When does grazing benefit plants? *Trends in Ecology and Evolution* **1**: 168–169.

Verkaar, H. J., E. Van der Meyden, and C. L. Breebaart. 1986. The response of *Cynoglossum officinale* L. and *Verbascum thapsus* L. to defoliation in relation to nitrogen supply. *New Phytologist* **104**: 121–129.

Wallace, L. L., S. J. McNaughton, and M. B. Coughenour. 1984. Compensatory photosynthetic responses of three African graminoids to different fertilization, watering, and clipping regimes. *Botanical Gazette* **145**: 151–156.

Watson, M. A. 1984. Developmental constraints: Effect on population growth and patterns of resource allocation in clonal plant. *American Naturalist* **123**: 411–426.

Watson, M. A. and B. B. Casper. 1984. Morphogenetic constraints on patterns of carbon distribution in plants. *Annual Review of Ecology and Systematics* **15**: 233–258.

Welker, J. M., E. J. Rykiel, D. D. Briske, and J. D. Goeschl. 1985. Carbon import among vegetative tillers within two bunchgrasses: Assessment with carbon-11 labelling. *Oecologia* **67**: 209–212.

Welker, J. M., D. D. Briske, and R. W. Weaver. 1987. Nitrogen-15 partitioning within a three generation tiller sequence of the bunchgrass *Schizachyrium scoparium*: Response to selective defoliation. *Oecologia* **74**: 330–334.

Whitham, T. G. and S. Mopper. 1985. Chronic herbivory: Impacts on architecture and sex expression of pinyon pine. *Science* **228**: 1089–1091.

Worrall, J. 1984. Predetermination of lateral shoot characteristics in grand fir. *Canadian Journal of Botany* **62**: 1309–1315.

12 Plants as Food for Herbivores: The Roles of Nitrogen Fixation and Carbon Dioxide Enrichment

BARBARA L. BENTLEY and NELSON D. JOHNSON

Herbivores eat plants. This apparently simple statement, in fact, represents a very complex relationship. The anatomy and physiology of plants reflect the requirements for nutrition, support, or defense of the plant, not the nutritional needs of herbivores. Yet an herbivore must obtain all the calories, protein, and other nutrients from its host plant if it is to successfully survive and reproduce.

The complexity of these needs becomes apparent in the design of artificial diets for herbivorous insects (Singh 1977). These diets must contain all the appropriate vitamins, minerals, hormone precursors, and, most importantly for the discussion here, suitable forms of carbon and nitrogen. The *ratio* of these components is equally important. For example, animal tissues are usually about 10% nitrogen, yet plant tissues are often as low as 1–2% nitrogen. Thus, herbivores must consume relatively large quantities of plant tissue to obtain sufficient nitrogen for growth and reproduction (Mattson 1980). In fact, if a diet has a very low nitrogen content, herbivorous insects must consume greater quantities of the diet in order to obtain sufficient nitrogen (Slansky and Feeny 1977).

All nitrogen sources are not equivalent, however. In most artificial diets, the nitrogen is present as protein, which can be easily assimilated and used by the insect. In this case, the nitrogen content is a relatively good measure of the nutritional value of the diet. On the other hand, natural vegetation may also contain nitrogeneous compounds which are either less easily metabolized, such as ureides, or may be poisonous to many herbivores, such as most alkaloids (Pelletier 1983). Such nitrogenous compounds often increase in parallel with total nitrogen content (Gershenzon 1984). In this case, the nitrogen concentration may not reflect the nutritional value of the plant tissue.

In this chapter we discuss three aspects of the nutritional value of plants as food for herbivores. First, we describe how the chemical form of nitrogen

Plant-Animal Interactions: Evolutionary Ecology in Tropical and Temperate Regions, Edited by Peter W. Price, Thomas M. Lewinsohn, G. Wilson Fernandes, and Woodruff W. Benson.
ISBN 0-471-50937-X © 1991 John Wiley & Sons, Inc.

interacts with the nitrogen content in an insect diet; second, we present the data from two experiments with lupines (*Lupinus*; Fabaceae) which illustrate how environmental factors can affect the form and concentration of plant nitrogen; and finally we focus in detail on how one such environmental factor, increasing atmospheric CO_2, may alter plant–herbivore relationships.

INTERACTIONS OF DIFFERENT FORMS OF NITROGEN IN THE DIET OF AN HERBIVORE

Although the role of protein in insect diets is relatively well known, we recently established that toxic forms of nitrogen can interact with protein in an insect's diet (Johnson and Bentley 1988). In these studies, we fed southern army worm (*Spodoptera eridania*) high- and low-protein diets that had 0.0–1.2% dry weight sparteine, a quinolizidine alkaloid. Not surprisingly, the protein content of the diet had a major impact on the growth and survivorship of the larvae: larvae grew faster and lived longer on high-protein diets. In addition, alkaloids influence growth and survivorship: larvae fed high sparteine diets had lower survivorship and slower growth rates than those fed sparteine-free diet. Most interesting, however, is the interaction between protein and alkaloids. Larvae fed high-protein diets were able to tolerate higher levels of alkaloids than those fed low-protein plus high-alkaloid diets. These effects are apparent in both growth rates and larval survivorship (Fig. 12-1).

These experiments were designed to mimic the levels of protein and alkaloids in the foliage of lupines, where both protein and alkaloid content can be quite

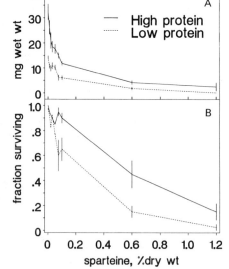

Figure 12-1. Growth (A) and survivorship (B) of *Spodoptera eridania* on artificial diets containing sparteine and two levels of protein. Sample size = ~ 10 per diet; bars represent one standard error. (Reproduced with permission from Johnson and Bentley 1988).

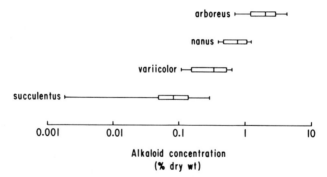

Figure 12-2. Variation in alkaloid concentration in four species of *Lupinus*. The length of the line is the range, length of the bar is the standard error, and cross bar is located at the mean.

variable. High levels of variation occur both among species as well as within species (Fig. 12-2). Although some of the causes of variation are probably related to genetic differences among the plants (especially at the species level), the within-species variation is often a function of environmental factors, including inorganic nitrogen availability, N_2-fixation in root nodules, and damage by herbivores (Johnson et al. 1988).

SOURCES OF VARIATION IN NITROGEN CONTENT: THE ROLE OF N_2-FIXATION AND HERBIVORY.

Like most legumes, plants in the genus *Lupinus* can fix atmospheric nitrogen. In some cases, fixation may be the sole source of nitrogen, though in most natural environments plants have access to some inorganic nitrogen in the soil as well as nitrogen from fixation. Since lupines contain quinolizidine alkaloids, the consequences of N_2-fixation include more than just to provide nitrogen for protein. A healthy, well-nodulated *Lupinus* plant may have as much as 12% of its nitrogen bound in alkaloids. Uninoculated plants growing in nitrogen-poor soils have considerably lower levels of both protein and alkaloids (Johnson et al. 1988). The concentrations of alkaloids and protein are not just a function of the levels of available nitrogen, however. Damage to foliage, such as that by chewing insects, can either increase or decrease alkaloid levels, depending on the intensity of the damage (Johnson et al. 1988, Bentley et al. 1987).

Responses of Plants to Experimental Damage

We documented these effects in a series of experiments in which we damaged lupine (*Lupinus succulentus*) plants that had been provided either with inorganic nitrogen or with N_2-fixing *Bradyrhizobium* (Fig. 12-3). In the first series of experiments, we damaged leaves with a small grid of pierced metal which

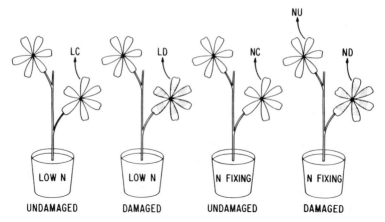

Figure 12-3. Experimental scheme to determine the effects of the interactions between foliage damage and nitrogen source in *Lupinus succulentus*. LC = low nitrate, undamaged control; LD = low nitrate, damaged; NC = nitrogen-fixing, control; ND = nitrogen fixing, damaged; NU = undamaged leaf from a nitrogen fixing, damaged plant.

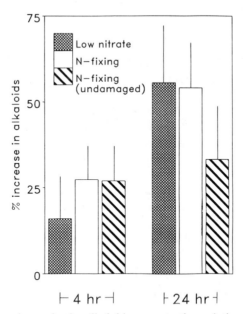

Figure 12-4. Percent change in the alkaloid concentration relative to the undamaged, low-nitrate controls after low levels of damage to foliage of *Lupinus succulentus*. N_2-fixing, $n = 22$; low NO_3^-, $n = 9$. Bars are one standard error. Changes in alkaloids are significant after 24 hr for both the N_2-fixing and low-NO_3^- plants ($p < 0.01$; $p < 0.06$, respectively), and low-NO_3^- and N_2-fixing plants are significantly different from one another after 4 hr ($p < 0.05$). Probabilities were calculated using MANOVAS of individual alkaloid concentrations (Johnson et al. 1988).

produces 36 holes/cm². This level of damage will induce the production of alkaloids, but does not change the photosynthetic surface area. The results of this experiment are illustrated in Figure 12-4. Note that plants fixing nitrogen were able to increase alkaloid levels relative to controls by about 30% within 4 h, whereas alkaloids in the uninoculated plants growing in the low-nitrate medium increased by only 16% after 4 h. Ultimately, however, foliage from all the damaged plants had significantly higher increases in alkaloids relative to undamaged controls (Bentley et al. 1987; Johnson et al. 1988).

The long-term impact of extensive defoliation is quite different. In a second set of experiments, we clipped half of each fully-expanded new leaf from the experimental plants three times between germination and flowering. Both alkaloid levels and growth rates of damaged plants which had access to abundant inorganic nitrogen were not significantly different from the undamaged controls (Fig. 12-5 and 12-6). But N_2-fixing plants had a dramatic drop in both growth and alkaloid concentrations. In fact, N_2-fixing plants were much more similar to the low-nitrate plants than they were to the undamaged N_2-fixing plants. This

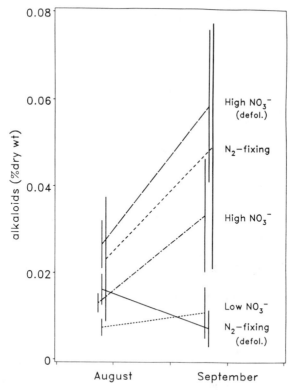

Figure 12-5. Long-term changes in alkaloid concentrations in *Lupinus succulentus* foliage from plants grown under the indicated nitrogen nutrition and defoliation treatments. $N = 10$ plants for all treatments; bars are one standard error.

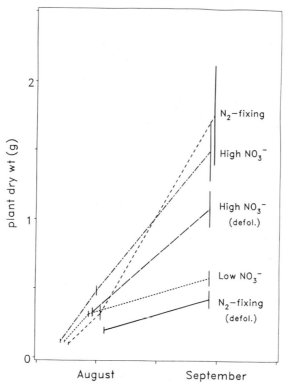

Figure 12-6. Change in dry weight of *Lupinus succulentus* plants following nitrogen and defoliation treatments. N_2-fixing, high NO_3^-, and high NO_3^- (defoliated) groups are not significantly different from each other, but are different from the low NO_3^- and defoliated N_2-fixing plants. $N = 6$ plants at first harvest and 10 plants for all subsequent harvests; bars are one standard error.

response is probably related directly to the availability of photosynthate to fuel the N_2-fixation process. Defoliation removed photosynthetic surface area, and consequently reduced the amount of photosynthate available. In effect, defoliated N_2-fixing plants became nitrogen-stressed (Johnson et al. 1988).

From an herbivore's perspective, these responses are likely to be important. In the short term, feeding on a well-nourished plant can increase the toxicity of the foliage. But over the long term, extensive damage may actually decrease the concentration of toxic compounds in plants with limited access to inorganic nitrogen. On the other hand, poorly-nourished plants, while they may be less toxic, are also less nutritious.

Responses of Plants to Natural Levels of Defoliation

The limitation of the experiments described above is, of course, that they are based on artificial damage of plants grown in the greenhouse. Do plants in their native habitats respond in a similar manner to the actions of natural herbivores? To answer this question, we followed a population of the bush lupine *Lupinus arboreus* growing on the Reserve at the Bodega Marine Laboratory located about 100 km north of San Francisco (Bentley and Johnson, 1990).

Lupinus arboreus is a woody, perennial species of lupine common in the coastal habitats of northern California. At Bodega, it grows in a variety of soil types, including dune sand, weathered granite, and ancient marine deposits. In the spring of 1987, some patches of *L. arboreus* were heavily attacked by the larvae of *Orgyia vetusta*, a tussock moth. This provided a natural experiment from which we could determine the impact of defoliation on alkaloid metabolism and N_2-fixation rates.

In our field study we followed 100 *Lupinus arboreus* plants growing in 10 different areas of the Bodega Reserve. These areas had been selected to represent different soil types and different exposures to salt spray. Four of these sites were infested with *Orgyia* larvae. The remaining sites, even those within 25 m of the infested sites, were free of larvae. The patchiness of the larval distribution is probably a function of the mobility of the insect—female *Orgyia* are flightless and usually lay their eggs on the same plant on which they fed as larvae.

During the season, we collected foliage from each of the plants at all sites on a monthly basis from February through July. This period brackets the time in

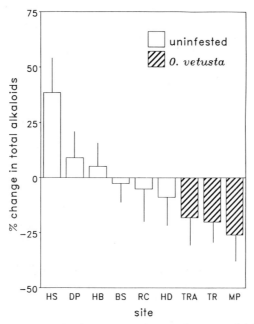

Figure 12-7. Changes in alkaloids in *Lupinus arboreus* after natural defoliation by *Orgyia vetusta* larvae. Values are expressed as percent change between foliage collected in May and that collected in July from the same individual plants. $N = 10$ plants for each site; bars are one standard error.

which *Orgyia* larvae are active. These samples were analyzed for nitrogen concentration and alkaloid concentration and composition. Data for the samples from May (when larvae first hatched) and July (when most larvae had pupated) are shown in Figures 12-7 and 12-8. Alkaloid concentrations in plants which were heavily infested by the larvae showed a significant decrease relative to the uninfested plants—a response similar to our long-term greenhouse experiments. In addition, the amount of nitrogen fixed by the damaged plants also decreased relative to the undamaged plants. Again, these data reflect a response similar to the greenhouse plants subjected to intense defoliation.

In a previous study on *L. arboreus*, Harrison and Karban (1986) found that damage by an early-season herbivore (the woolly bear, *Platyprepia virginalis*) caused a reduction in the growth rate and fecundity of the later-season *Orgyia vetusta*. Our data suggest that this could have been caused by changes in either alkaloids, nitrogen, or both, in the foliage of the host plant. Future studies will confirm the exact cause-and-effect relationships.

In summary, then, the value of plant foliage as food for herbivores is a function, in large part, of both the concentration and composition of the nitrogenous compounds in the host plant. If nitrogen is bound in toxic compounds, then growth or survivorship is reduced. And, if the concentration of nitrogen is low, larvae not only grow more slowly, but are also less tolerant of toxic compounds in their food. Biological activities, such as N_2-fixation and herbivory, can change both the form and the relative concentrations of these compounds. But, as will be seen below, other factors as well may be having a significant effect on the relationships between plants and their herbivores.

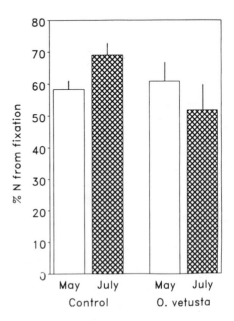

Figure 12-8. Percent of biologically fixed nitrogen in *Lupinus arboreus* before (May) and after (July) defoliation by *Orgyia vetusta* larvae. Fixed nitrogen was determined by the isotope ratio technique developed by Shearer and Kohl (1986). $N = 10$ plants for each site; bars are one standard error.

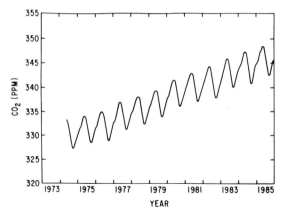

Figure 12-9. Increasing levels of atmospheric CO_2 between 1973 and 1985. Peaks and valleys reflect annual changes of terrestrial photosynthetic rates.

CHANGES IN PLANT NITROGEN CONTENT: THE IMPACT OF INCREASED ATMOSPHERIC CO_2

Until very recently, all studies on the value of plant tissues as food for herbivores were based on the assumption that photosynthetic rates of the host plant were a function of light intensity, water supply, and mineral nutrient availability. These, in turn, would dictate the nitrogen content and other measures of nutritional value of the plant as food. It is becoming increasingly apparent, however, that the concentration of atmospheric carbon dioxide is changing (Fig. 12-9), and that this change is likely to have dramatic impacts on both photosynthetic rates and the composition of plant tissues. In this final section, we would like to review some of the implications of these changes.

Global Effects of CO_2 Enrichment

An increase in atmospheric CO_2 will result in major changes in both the physical and chemical environment of plants. The "greenhouse effect" will cause changes in air and soil temperatures, which ultimately will lead to changes in sea level, in rates of respiration and evapotranspiration, and in the distributions of natural communities (Strain and Cure 1985). Biological systems will respond not only to changes in temperature and water regimes, but also to the direct effects of CO_2. Most importantly, many plants will have increased photosynthetic rates simply because of CO_2 "fertilization." This, in turn, can lead to a wide variety of physiological and ecological responses, as illustrated in Tables 12-1a and 12-1b. Many of the physiological responses have been documented in a number of species, especially those of economic importance (see Acock and Allen, 1985, for a good review). In contrast, very little is known about the ecological responses, especially those of natural systems (Oechel and Strain 1985). Nevertheless, the

TABLE 12-1a. Physiological Features Affected by Increased Atmospheric CO$_2$ (from Strain 1987)

Primary physiological responses
  1. Photosynthesis
  2. Photorespiration
  3. Dark CO$_2$ fixation
  4. Stomatal aperture
Secondary physiological responses
  1. Photosynthate concentration
  2. Photosynthate composition
  3. Photosynthate translocation
  4. Plant water status
      a. Transpiration
      b. Tissue water potential
      c. Water use efficiency
      d. Leaf temperature
  5. Tolerance to gaseous atmospheric pollutants
Tertiary whole plant responses
  1. Growth rate
      a. Weight
      b. Height
      c. Leaf area
      d. Node formation
      e. Nodule formation
      f. Stem diameter
      g. Leaf senescence
  2. Growth form
      a. Height
      b. Branch number
      c. Leaf number
      d. Leaf area
      e. Root/shoot weight
      f. Nodule weight
      g. Stem diameter
      h. Leaf specific weight
      i. Cytological and anatomical changes
  3. Reproduction
      a. Flowering time
      b. Flower size
      c. Flower number
      d. Fruiting time
      e. Fruit size
      f. Fruit number
      g. Seed maturation
      h. Seed germination
      i. Number of seeds per plant

TABLE 12-1b. Ecological Features Affected by Increased Atmospheric CO₂[a] (modified from Strain 1987).

Primary organism interactions
1. Plant–plant
   a. Interference
   b. Competition
   c. Symbiosis
2. Plant–animal
   a. **Herbivory**
   b. Pollination
   c. Dispersal
   d. Shelter
3. Plant–microbe
   a. Disease
   b. Decomposition
   c. Symbioses
      i. Mycorrhizae
      ii. **N₂-fixation**
Secondary population–community interactions
1. Population growth
2. Succession
3. Dominance
4. Changes in gene frequencies
5. Species diversity
6. Geographic distribution
Tertiary ecosystem responses
1. Carbon flux
   a. Productivity
   b. Decomposition
   c. Sequestering
2. Nutrient cycling
3. Water relations
4. Feedback effects (integration of all biotic and abiotic interactions occurring through time)

---

[a]Features emphasized in this paper are in bold type.

data indicate that major changes are likely to occur in photosynthetic rates, which, as noted above, can alter nitrogen fixation and herbivore activity.

Impact of CO₂ Enrichment on Nitrogen Fixation

Possible mechanisms by which N_2-fixation would change with changing CO_2 are presented in Figure 12-10. As CO_2 levels increase, tissue nitrogen levels are likely to decrease either because soil nutrients are depleted (E) or because of a disproportionate increase in carbon in the tissues (A–C). Under these conditions of nitrogen limitation, N_2-fixation is almost always stimulated (Allen and Allen

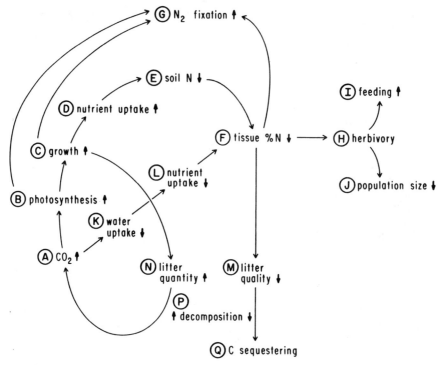

Figure 12-10. Impacts of CO_2 enrichment on the interactions between plants and herbivores.

1981). In addition, it is likely that overall N_2-fixation rates will increase simply because there is more photosynthate to operate the fixation system (B–G).

Although there has been considerable research done on the impact of CO_2 enrichment on soybeans and other legume crops, relatively few studies have focused on the impact on the N_2-fixing system. Of the few studies which do focus on N_2-fixation, the data are both interesting and contradictory. For example, early work by Hardy and Havelka (1975) and Quebedeaux et al. (1975) suggested a close relationship between N_2-fixation rates and increased photosynthate when CO_2 levels were elevated (Fig. 12-10, A, B, G). They concluded that the increased supply of current photosynthate was directly responsible for both the total (g N/plant) and specific (g N/g nodule) N_2-fixation activity of the plants. In contrast, work by Williams et al. (1981, 1982) and Finn and Brun (1982) demonstrated that specific nodule activity did not increase. However, because total nodule weight was greater, total plant fixation was increased under high CO_2 (A–C, G).

In all cases, however, it is apparent that nodule formation and activity is as much a function of soil nitrogen availability as it is of internal carbon availability. Under current CO_2 levels (348 ppm), nodulation and fixation increase with decreasing inorganic nitrogen. In soybeans, for example, plants fertilized with 8.0 mM NH_4NO_3 do not fix nitrogen even when *Rhizobium* inoculum is present in the soil, and yet plants given nutrients solutions containing only 1.0 mM NH_4NO_3 fix at the same rate as plants with no available nitrogen (Williams et al. 1981). The effect of CO_2 enrichment is to change the magnitude of the differences between the treatments: low-nitrate plants have greater changes in fixation rates at elevated CO_2 levels than do the plants with more available nitrogen.

Nevertheless, there is ample evidence that N_2-fixation rates are closely linked to available carbon as well (Anderson et al. 1987; Phillips et al. 1976). Basically, any factor which increases photosynthesis, such as increased light, increased shoot size, or mobilization of starch reserves, will increase N_2-fixation. Even external sources of organic carbon can influence N_2-fixation rates (Shivashankar and Vlassak 1978). Conversely, reduced photosynthate availability will result in decreased N_2-fixation. For example, both prolonged darkness and girdling of the phloem will reduce N_2-fixation rates in experimental plants (Williams et al. 1982; Anderson et al. 1987). As in other studies, the effect of enhanced CO_2 was to reduce the negative impact of the experimental treatment.

The end result is that N_2-fixing plants can maintain higher tissue-nitrogen levels than nonfixers, especially in natural ecosystems. This will be particularly important for plants growing under conditions of CO_2 enrichment, where the nitrogen concentration of most plants decreases with increasing CO_2 levels. Increased CO_2 does *not* always result in a change in the nitrogen concentration in legumes. As will be seen below, this has major implications for foliage-feeding herbivores.

Impact of CO_2 Enrichment on Herbivory

Since the tissue nitrogen content is reduced in most plants grown at elevated CO_2, and since herbivore growth is often a function of the nitrogen content of their food, CO_2 enrichment should affect the quality of the foliage as food for herbivores. Yet, despite the critical role that herbivores play in both agricultural and natural systems, very little work has been done to determine the effect of elevated CO_2 on plant–animal interactions at either the individual or ecosystem level.

The results from the few studies which have been done suggest that CO_2 enrichment may have far-reaching effects. In a series of experiments with the soybean looper on soybean and the army worm on peppermint, for example, Lincoln and his co-workers (1986, 1989) found that larval feeding rates were significantly higher on plants grown in high CO_2. Since the nitrogen concentration was lower in the tissues of the high CO_2 plants, nitrogen intake per bite was lower. Nevertheless, because consumption rates were higher, the *total* amount of nitrogen consumed by the larvae was the same under either treatment (Figure 12-10F, H, I).

Increased CO_2 can also affect the concentration of various secondary metabolites in plant tissues. In foxglove, for example, the concentration of digitoxin is lower on a per gram basis in plants grown under elevated CO_2, even though total yield per unit area is greater because of increased biomass production (Stuhlfauth et al. 1987). In this case, elevated CO_2 would tend to make the plant tissues more palatable.

In an attempt to analyze complex relationships between the level of CO_2 enrichment, and changes in plant chemistry and herbivore activity, Lincoln and Couvet (1989) measured all of these factors in experiments with peppermint and *Spodoptera eridania*. Interestingly, unlike foxglove, the concentration of secondary compounds did not decrease in the peppermint. Nevertheless, tissue nitrogen levels were lower in the high-CO_2 treatments. Thus the effects of elevated CO_2 on the herbivore were all mediated by changes in the levels of leaf nitrogen.

Neither peppermint nor foxglove is likely to be a good model for N_2-fixing species, however. As noted above, increased CO_2 may allow plants to have high fixation rates, which in turn would allow the maintenance of high tissue nitrogen levels. These plants should have adequate nitrogen for protein synthesis. If the defensive compounds of the plant are alkaloids or other nitrogen-based compounds, then nitrogen-fixing species will maintain their ability to defend themselves against most herbivores.

Nevertheless, the consequences of these changes, as summarized in Table 12-2, are most sobering. And unquestionably the consequences will reach far beyond simple plant–herbivore interactions—to interactions at the population, community, and even ecosystem levels. Thus, our current studies of the relationships between plants and their herbivores probably constitute baseline data for an inescapable global experiment on the impact of changing food quality on herbivore success.

TABLE 12-2. Implications of Increased CO_2 on Plant–Herbivore Interactions

1. Biomass production by plants may increase.
   Greater biomass production may increase a plant's "tolerance" of defoliation.
   Herbivore carrying capacity on the host–plant may increase.
2. The decrease in the nitrogen concentration will make plants less nutritious.
   Herbivores must eat more to ingest the same amount of nitrogen.
   Herbivore growth rates and/or fecundity will be lower.
   Herbivore population sizes will decrease (?)
3. The concentration of secondary metabolites will change (per gram dry weight).
   Herbivores may consume a greater biomass of plant tissue.
   Effect of toxins may increase as protein availability decreases.
4. Effects 2 and 3 are less likely to occur in plants with the C_4 photosynthetic pathway or with nitrogen-fixing symbionts.
   C_4 plants are not affected by increased CO_2.
   N_2-fixation increases as photosynthetic rates increase.
      Tissue nitrogen concentrations unchanged.
      Alkaloid concentration change in parallel with nitrogen availability.

REFERENCES

Acock, B. and L. H. Allen. 1985. Crop responses to elevated carbon dioxide concentrations. Pp. 54–97. In B. R. Strain and J. D. Cure (Ed.), Direct Effects of Increasing Carbon Dioxide on Vegetation. US Department of Energy, Washington, DC.

Allen, O. N. and E. K. Allen. 1981. The Leguminosae: A sourcebook of characteristics, uses, and nodulation. University of Wisconsin Press, Madison, WI.

Anderson, M. P., G. H. Heichel, and C. P. Vance. 1987. Nonphotosynthetic CO_2 fixation by alfalfa (*Medicago sativa* L.) roots and nodules. *Plant Physiology* **85**: 283–289.

Bentley, B. L. and N. D. Johnson. 1990. The impact of defoliation by a tussock moth (*Orgyia vetusta*) on a nitrogen-fixing legume (*Lupinus arboreus*). *Bulletin of the Ecological Society of America* **71**: 91.

Bentley, B. L., N. D. Johnson, and L. Rigney. 1987. Short-term induction in leaf tissue alkaloids in *Lupinus* following experimental defoliation. *American Journal of Botany* **74**: 646.

Bleiler, J. A., G. A. Rosenthal, and D. H. Janzen. 1988. Biochemical ecology of canavanine-eating seed predators. *Ecology* **69**: 427–433.

Finn, G. A. and W. A. Brun. 1982. Effect of atmospheric CO_2 enrichment on growth, nonstructural carbohydrate content, and root nodule activity in soybean. *Plant Physiology* **69**: 327–331.

Gershenzon, J. 1984. Changes in the levels of plant secondary metabolites under water and nutrient stree. Pp. 273–320. In B. N. Timmermann, C. Steelink, and F. Loewus (Ed.), *Recent Advances in Phytochemistry: Phytochemical Adaptations to Stree*. Plenum Press, New York.

Hardy, R. W. F. and U. D. Havelka. 1976. Photosynthate as a major factor limiting nitrogen fixation by field-grown legumes with emphasis on soybeans. Pp. 421–439. In P. S. Nutman (Ed.), Symbiotic Nitrogen Fixation in Plants. Cambridge University Press, Cambridge, UK.

Harrison, S. and R. Karban. 1986. Effects on an early-season folivorous moth on the success of a later-season species, mediated by a change in the quality of the shared host, *Lupinus arboreus* Sims. Oecologica (Berl) **69**: 354–359.

Johnson, N. D. and B. L. Bentley. 1988. The effects of dietary protein and lupine alkaloids in the growth of survivorship of *Spodoptera eridania*. Journal of Chemical Ecology: **14**: 1391–1403.

Johnson, N. D., B. Liu, and B. L. Bentley. 1987. The effects of nitrogen fixation, defoliation, and soil NO^- on the growth, alkaloids, and nitrogen levels of *Lupinus succulentus* (Fabaceae). Oecologia (Berl) **74**: 425–431.

Johnson, N. D., L. Rigney, and B. L. Bentley. 1988. Short-term changes in alkaloid levels following leaf damage in lupines with and without symbiotic nitrogen fixation. *Journal of Chemical Ecology* **15**: 2425–2434.

Lincoln, D. E. and D. Couvet, 1989. The effect of carbon supply on allocation to allelochemicals and caterpillar consumption of peppermint. Oecologia **78**: 112–114.

Lincoln, D. E., D. Couvet, and N. Sionit. 1986. Response of insect herbivore to host plants grown in carbon dioxide enriched atmospheres. *Oecologia* (Berl) **69**: 556–560.

Mattson, W. J. 1980. Herbivory in relation to plant nitrogen content. *Ann. Rev. Ecol. Syst.* **11**: 119–161.

Oechel, W. C. and B. R. Strain. 1985. Native species responses to increased atmospheric carbon dioxide concentration. Pp. 118–154 in B. R. Strain and J. D. Cure (Eds.), *Direct Effects of Increasing Carbon Dioxide on Vegetation*. US Department of Energy, Washington, DC.

Pelletier, S. W. 1983. Alkaloids: Chemical and biological perspectives. John Wiley and Sons, New York.

Phillips, D. A., K. D. Newell, S. A. Hassell, and C. E. Felling. 1976. The effects of CO_2 enrichment on root nodule development and symbiotic N_2 reduction in *Pisum sativum* L. *American Journal of Botany* **63**: 356–362.

Quebedeaux, B., U. F. Havelka, K. L. Lival, and R. W. F. Hardy. 1975. Effect of altered pO_2 in the aerial part of soybean on symbiotic N_2 fixation. *Plant Physiology* **56**: 761–764.

Shearer, G. and D. H. Kohl. 1986. N_2-fixation in field settings: Estimations based on natural ^{15}N abundance. *Aust. J. Plant Physiol.* **13**: in press.

Shivashankar, K. and K. Vlassak. 1978. Influence of straw and CO_2 on N_2-fixation and yield of soybeans. *Plant and Soil* **49**: 259–266.

Singh, P. 1977. Artifical diets for insects, mites, and spiders. IFI/Plenum, New York.

Slansky, F. and P. Feeny. 1977. Stabilization of the rate of nitrogen accumulation by larvae of the cabbage butterfly on wild and cultivated food plants. *Ecol. Monogr.* **47**: 209–228.

Strain, B. R. 1987. Direct effects of increasing atmospheric CO_2 on plants and ecosystems. *Trends in Ecology and Evolution.* **2**: 18–21.

Strain, B. R. and J. D. Cure. 1985. Direct Effects of Increasing Carbon Dioxide on Vegetation. U.S. Department of Energy, Washington, DC.

Stuhlfauth, T., K. Klug, and H. P. Fock. 1987. The production of secondary metabolites by *Digitalis lanata* during CO_2 enrichment and water stress. *Phytochemistry* **26**: 2735–2739.

Williams, L. E., D. E. DeJong, and D. A. Phillips. 1981. Carbon and nitrogen limitations on soybean seedling development. *Plant Physiol.* **68**: 1206–1209.

Williams, L. E., T. M. DeJong, and D. A. Phillips. 1982. Effect of changes in shoot carbon-exchange rate of soybean root nodule activity. *Plant Physiology* **69**: 432–436.

13 Altered Patterns of Herbivory and Diversity in the Forest Understory: A Case Study of the Possible Consequences of Contemporary Defaunation

RODOLFO DIRZO and ALVARO MIRANDA

A consistent pattern of the distribution of species on the planet is that biodiversity, in general, decreases from the tropics to the temperate zones (Krebs 1978; Raven and Johnson 1986). Prominent in this pattern is the overwhelming diversity in the tropics of insects—about 50% of which are herbivorous (Strong et al. 1984), of some groups of folivorous mammals (Eisenberg 1981) and flowering plants—the main food resource for herbivores (Strong et al. 1984). This suggests that tropical herbivory—a theme of this volume—should be particularly well represented and varied in the tropics and that its study is of fundamental importance to the understanding of the evolutionary biology of plants and animals. This situation also makes it mandatory to preserve representative portions of the tropics' biotic wealth. Currently, much of our knowledge on tropical biology is being derived from studies in preserved wildlands (see Clark et al. 1987, for an example). However, conservation problems in many tropical and semitropical countries are just formidable. This is so much so that even formally protected areas (e.g., biological stations and national parks) of countries such as Mexico are suffering from anthropogenic disturbances whose consequences are difficult to detect or predict. Also, research in such areas may generate findings which may not be representative of natural-system processes or which may be, at best, anomalous.

The existence of altered processes in tropical preserved wildlands has been suggested, for example, in the case of plant mating systems and their genetic and demographic consequences as a result of reduced population size (Lewin 1989), and in the patterns of tree-fall gaps and their turnover rates as a result of increased

Plant-Animal Interactions: Evolutionary Ecology in Tropical and Temperate Regions, Edited by Peter W. Price, Thomas M. Lewinsohn, G. Wilson Fernandes, and Woodruff W. Benson. ISBN 0-471-50937-X © 1991 John Wiley & Sons, Inc.

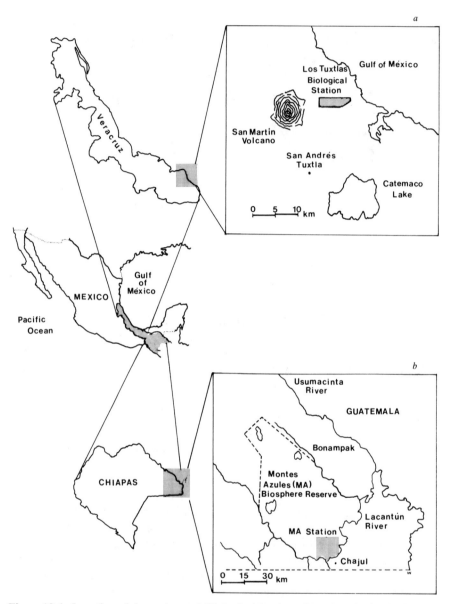

Figure 13-1. Location of the region and Biological Station of Los Tuxtlas in the State of Veracruz (*a*), and of the Montes Azules Biosphere Reserve (Lacandon forest) in the State of Chiapas (*b*).

edge effects (Martinez-Ramos, unpublished data). In this chapter we attempt to integrate our findings on the patterns of herbivory in tropical rain forests of Mexico with our preoccupations for rain forest conservation, with the aim of illustrating (1) how the study of plant–herbivore interactions in some neotropical wildlands may be affected by subtle conservation problems and (2) the ecological (and potentially evolutionary) consequences this may have on the plant–herbivore interface. The conservation problem we address in particular is contemporary defaunation affecting herbivorous vertebrates of a Mexican rain forest.

This study is centered at the Los Tuxtlas tropical research station (Figure 13-1). Los Tuxtlas, is in the State of Veracruz, Southeast Mexico (95° 04′ West long. and 18° 30′ North lat.). It is one of the best-studied areas in the country (see Gòmez-Pompa et al. 1976; Còmez-Pompa and del Amo 1985) and probably in Latin America. Moreover, the area is important for it constitutes the northern-most limit in the distribution of the tropical rain forest in the New World (Dirzo 1987). However, despite its latitudinal position the area is still highly diverse and it combines species of Neotropical origin with Nearctic and endemic species (Andrle 1964; Dirzo 1987; Ibarra-Manriquez and Sinaca 1987; Pérez-Higareda et al. 1987). Yet recent evidence suggests that a number of natural processes under study at the site may be influenced by human activities taking place in the vicinity of the preserved areas. In particular, we submit that the patterns of herbivory on the understory plants may be determined by a process of vertebrate defaunation that has been taking place over the last 20 years or so (see below).

THE ALTERED PATTERNS OF HERBIVORY

The patterns of herbivory in the understory of Los Tuxtlas have been studied in some detail and the main results are summarized here.

Studies with Seedlings in Permanent Quadrats

Continuous observations on individual leaves of seedlings of five species (*Nectandra ambigens, Omphalea oleifera, Pterocarpus rohrii, Faramea occidentalis, Psychotria faxlucens*) in the understory of mature forest showed that the predominant scores of leaf damage were 0 (no damage) and 1–25% of leaf area damaged and that insects (of various orders) were the only herbivores responsible for the damage (Dirzo 1984 and unpublished data). Similar results were obtained with seedlings of *Cecropia obtusifolia* and *Heliocarpus appendiculatus* studied in tree-fall gaps (Nùñez-Farfàn and Dirzo 1988). For these two species yearly averages of the standing levels of damage were close to 15% of leaf area eaten. In all cases there was evidence of insect damage (scars) and no evidence of damage by vertebrates (browsing, pull-out).

In a study directed to assess the rates of pathogen infection on the foliage of 10 understory species in the mature forest (Garcia-Guzmàn 1989), long-term observations on individually marked leaves showed that most of the damage was due to insects and pathogens and none to vertebrates.

Community-Level Surveys of Leaves

For each of 52 species of seedlings, a random sample of 50 leaves was collected and the standing levels of damage were measured to the nearest square millimeter (de la Cruz and Dirzo 1987). The results of this survey (Fig. 13-2) show that 76% ($N = 3394$) of them had levels of damage lower than 10% and 41% were intact or damaged by pathogens. A minimum fraction of the sample (0.5%) bore damage greater than 25% and the overall mean was 10.5%. This information was complemented by an analysis of the types of damage observed (Dirzo unpublished data): 75% of the leaves were damaged by insects and pathogens, 9% were damaged by pathogens alone, 16% were intact, and none of the leaves bore recognizable evidence of damage by vertebrates.

In a more extensive survey using 1103 randomly chosen plants from four 1000-m² plots (Garcia-Guzmàn and Dirzo unpublished data), we found (Table 13-1) that damage by insects and pathogens was the most frequent type of damage, while nearly 40% of the leaves were intact and, again, none of the plants or leaves from such a large sample were damaged by vertebrates.

In summary, the understory vegetation of this forest shows that (1) insects are

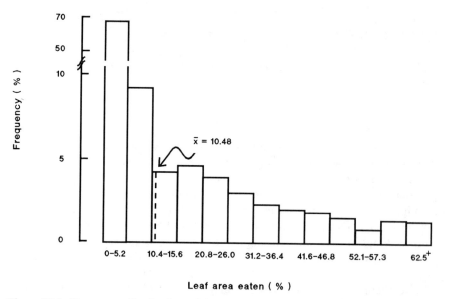

Figure 13-2. Frequency distribution of damage on a sample of 3394 leaves (from 1697 seedlings) from Los Tuxtlas. The broken line indicates the mean leaf area eaten for all sampled leaves. (After de la Cruz and Dirzo 1987).

TABLE 3-1. Types of Damage on the Foliage of Understory
Plants from Los Tuxtlas[a]

Type of Damage	Frequency (%)	
	Plants[b]	Leaves[c]
Insects alone	36.7	15.9
Insects plus pathogens	58.2	43.1
Pathogens alone	1.7	1.4
Intact	3.5	39.6
Vertebrate	0	0

[a]The plants for this survey were randomly chosen from four 1000-m^2
plots. After Garcia-Guzmàn and Dirzo (unpublished).
[b]$N = 1103$.
[c]$N = 8365$.

the main herbivores and, more importantly, (2) there is a consistent absence of damage by vertebrate folivores. The predominant role of insects as foliage consumers has been well established for several tropical sites (e.g., Leigh and Smythe 1978; Janzen 1981b; Dirzo 1987), but this finding has been largely based on the study of the foliage of trees and plants of upper strata and not seedlings or herbs of the forest understory. These insects (together with pathogens and perhaps other invertebrates) should be the exclusive consumers at this level is somewhat surprising, for several folivores, in particular gregarious (e.g., peccaries, coatis) and large (e.g., tapirs, white-tailed deer) ground-dwelling vertebrates should be expected to be important consumers. Absence of evidence of damage by vertebrates, low levels of herbivory, and extremely high-density carpets of seedlings and herbs (see below) are striking and, apparently, recently acquired (G. Pérez-Higareda personal communication) features of the understory of this and other neotropical forests. Are these features anomalous? If so, what are the possible reasons for such a scenario?

POSSIBLE REASONS FOR THE OBSERVED PATTERNS

Three possible reasons could account for the observed patterns: (1) a biogeographical limitation in the distribution of vertebrates, (2) unpalatability of the foliage, and (3) contemporary defaunation. We shall discuss each in turn.

Biogeographical Limitation

Given that Los Tuxtlas is the northernmost limit of the distribution of the tropical rain forest on this continent, it could be expected that this location is out of the range of distribution of some of the typical ground-dwelling neotropical vertebrates, including some folivorous species which, presumably, feed on the plants of the forest understory. However, there is solid evidence in the literature suggesting that Los Tuxtlas is not out of the range of distribution of numerous species relevant to this discussion. Literature accounts include information both at a global level (i.e., the country and the North American region) (Hall 1981; Leopold 1959; Ramirez-Pulido et al. 1983, 1986) and at a more local level (i.e., the State of Veracruz and the specific region of Los Tuxtlas) (Hall and Dalquest 1963; Navarro 1980, 1982). This information and that provided by naturalists and researchers with a long working experience in the area allows us to confidently state that up to some 20 years ago the area maintained sizable populations of ground-dwelling folivorous (or partly folivorous) species such as *Tapirus bairdii, Tayassu pecari, Tayassu tajacu, Mazama americana, Odocoileus virginianus, Dasyprocta mexicana, Agouti paca, Nasua nasua,* as well as arboreal herbivores such as *Ateles geoffroyi, Coendu mexicanus,* and *Potos flavus.* Consequently, we are led to conclude that biogeographic limitation is not responsible for the currently observed patterns of herbivory.

Foliage Unpalatability

An alternative explanation is that chemical characteristics of understory plants, for example high concentrations and diversity of secondary metabolites, low-water contents or other nutritional limitations, render the foliage of these plants highly unacceptable to vertebrate herbivores. We assessed this possibility by means of acceptability experiments in which we offered the foliage of five representative species of seedlings and five representative species of saplings from Los Tuxtlas to four species of folivorous vertebrates (*Tapirus bairdii,* the baird tapir; *Tayassu tajacu,* the collared peccary; *Tayassu pecari,* the white-lipped peccary; and *Odocoileus virginianus,* the white-tailed deer) kept in captivity at the Instituto de Historia Natural in the city of Tuxtla Gutiérrez, Chiapas. We chose the plant species based on their frequency and abundance in randomly established plots. The five species with the highest scores of an index of frequency (number of plots in which a species appeared/total number of plots) × abundance (numbers of individuals) were included in the experiments. Live plants of these species were collected in pots and offered simultaneously to each of the vertebrates. The plants were left in the animals' cages and we observed the herbivore's reactions until feeding was completed. The results of these experiments are given in Table 13-2. It was quite evident that, with one exception (*Omphalea oleifera*), all plant species were readily acceptable to and avidly eaten by the four species of vertebrates. *O. oleifera* was rejected by the tapir and was found moderately acceptable by the two peccaries and the deer. In addition to these short-term responses, we found no evidence that any of the animals used for the experiments experienced any longer-term negative effects associated with their feeding on the experimental plants. Although acceptability trials have a number of limitations (see Dirzo 1980), the results were so clearcut that we conclude that unpalatability is not responsible for the absence of vertebrate damage on these plants in the field.

TABLE 13-2. Acceptability of Foliage of Plants from Los Tuxtlas to Four Species of Vertebrates Maintained in Conditions of Captivity[a]

Plant Species	Baird's Tapir			White-Lipped Peccari			Collared Peccari			White-Tailed Deer		
	RA	PA	R	RA	PA	R	RA	PA	R	RA	PA	R
Seedlings												
Nectandra ambigens	×	—	—	×	—	—	×	—	—	×	—	—
Psychotria faxlucens	×	—	—	×	—	—	×	—	—	×	—	—
Poulsenia armata	×	—	—	×	—	—	×	—	—	×	—	—
Faramea occidentalis	×	—	—	×	—	—	×	—	—	×	—	—
Rinorea guatemalensis	×	—	—	×	—	—	×	—	—	×	—	—
Saplings												
Omphalea oleifera	—	—	×	—	×	—	—	×	—	—	×	—
Chamaedorea tepejilote	×	—	—	×	—	—	×	—	—	×	—	—
Psychotria faxlucens	×	—	—	×	—	—	×	—	—	×	—	—
Faramea occidentalis	×	—	—	×	—	—	×	—	—	×	—	—
Schaueria calycobractea	×	—	—	×	—	—	×	—	—	×	—	—
Overall ($n/10$)	9	0	1	9	1	0	9	1	0	9	1	0

[a]RA = readily eaten, PA = partly eaten, R = rejected.

Contemporary Defaunation

The previous results, together with an evident, ongoing process of contemporary defaunation, suggest that the latter may be responsible, at least in part, for the altered patterns of herbivory we described above. Defaunation in Los Tuxtlas, as in many other forests, has been occurring because of anthropogenic activities such as hunting, illegal trading of animals and their products (e.g., pelts), collection of specimens for pets, and quite importantly, habitat destruction in the immediate vicinity of the preserve. Adequate documentation of this phenomenon is very difficult, partly owing to its illegal nature, the difficulty of studying these animals, and because their absence is not readily evident. In the following section we describe studies attempting to assess the level of defaunation in Los Tuxtlas and to test the hypothesis that contemporary defaunation plays a role in the altered patterns of herbivory and diversity of the forest understory.

THE DEFAUNATION HYPOTHESIS

Clearly, the test of this hypothesis is difficult. As a first approximation we decided to carry out a comparative study, whereby Los Tuxtlas could be compared with a similar site in which the mammalian fauna was adequately represented. Unfortunately, such a site was not to be found within the region of Los Tuxtlas and the comparable area had to be found elsewhere. After an intensive search throughout the remaining distribution of the tropical rain forest in Mexico, we chose the Montes Azules Biosphere Reserve, located in the Lacandon Forest (Selva Lacandona) in the State of Chiapas (Fig. 13-1). This site was chosen for it appeared to be the most promising in terms of its mammalian fauna.

The Selva Lacandona used to be an extensive tract of rain forest covering about 1.3 million ha extending throughout the northeast of the State of Chiapas and into the Petén Zone in Guatemala (Diechtl 1988). Currently, it constitutes the largest tract of tropical rain forest in Mexico, covering an area of approximately 600,000 ha (Diechtl 1988), 300,000 of which have been assigned (as from 1978) to the Montes Azules Biosphere Reserve. This impressive forest is extremely diverse (see Medallin 1986), but its biota is still largely unexplored. Although deforestation has been intensive, the extant large tracts of forest would suggest that most of the faunal elements of tropical rain forests are still well represented.

In this study we included the following aspects: an assessment of the mammalian fauna, analysis of the patterns of herbivory and, as a corollary, an assessment of the structure and diversity of the forest understory. In all cases, the same methodologies were used at both sites.

Comparison of the Mammalian Faunas

To compare both faunas we used a technique of foot-print detection with 100 50×50 cm plots of fine sand randomly positioned along a 500-m transect. The tracks were identified by preparing casts of the foot prints and keying them out with manuals for the identification of animal tracks (Aranda 1981; Murie 1974). Censuses of the 100 plots were done three times during each field session. (There were two field sessions per year, during two years.) This method provided qualitative information regarding the presence–absence of animals. The details and the reliability of this technique are given in Miranda and Dirzo (1990). This survey was complemented with diurnal and nocturnal sightings along the same 500-m transect.

The results of this comparison showed many more species present at Montes Azules than at Los Tuxtlas (15 vs. 9 by the method of tracks and 12 vs. 5 by sightings; Table 13-3). The species list given in Table 13-3 only includes taxa that, by the literature, are expected to occur in Los Tuxtlas. The species detected only

TABLE 13-3. Comparison of the Mammalian Fauna Between Montes Azules and Los Tuxtlas Using Two Methods: Detection of Animal Tracks and Direct Sighting[a]

Species	Montes Azules		Los Tuxtlas	
	Track	Sighting	Track	Sighting
Didelphis sp.	*	*	*	*
Philander opossum	*	*	*	—
Alouatta sp.[b]	—	*	—	*
Ateles geoffroyi[b]	—	*	—	—
Dasypus novemcinctus	*	*	*	—
Sciurus sp.[b]	*	*	*	*
Agouti paca[b]	*	*	*	*
Dasyprocta sp.[b]	*	—	*	*
Nasua nasua[b]	*	*	*	—
Potos flavus[b]	—	*	—	—
Procyon lotor	*	—	*	—
Lutra longicaudis	*	—	—	—
Felis onca	*	—	—	—
Felis pardalis	*	—	*	—
Tapirus bairdii[b]	*	—	—	—
Tayassu tajacu[b]	*	*	—	—
Tayassu pecari[b]	*	*	—	—
Mazama americana[b]	*	*	—	—
Total	15	12	9	5

[a]*, present; —, absent.
[b]Mainly herbivores.

TABLE 13-4. Index of Occurrence of Tracks of a Selected Number of Vertebrate Species from Los Tuxtlas and Montes Azules[a]

Species	Index of Occurrence	
	Los Tuxtlas	Montes Azules
Agouti paca	1.52	3.04
Dasyprocta mexicana	0.30	0.23
Felis onca	0	2.46
Felis pardalis	0.46	1.59
Mazama americana	0	2.07
Tapirus bairdii	0	0.30
Tayassu pecari	0	0.15
Tayassu tajacu	0	0.30
Mean $\pm$ SD	0.29 ± 0.53	1.27 ± 1.17

[a]See text and Miranda and Dirzo (1990) for details on the index used.

in Montes Azules included the tapir, the white-lipped peccary, the collared peccary, and the mazama deer. By their number and biomass (or both) these four species are expected to be important ground-dwelling herbivores. Table 13-3 includes a number of nonherbivorous species whose presence, only in Montes Azules, is indicative of the "health" of the trophic chain in this site and of the converse situation in Los Tuxtlas.

We also calculated a weighted index of occurrence for a number of important species of mammals (Table 13-4). (This index is the product of the frequency with which a species occurs in the plots and the number of recording sessions in which the tracks appeared (for details see Miranda and Dirzo, 1990.) It is clear that the frequency of occurrence of the important herbivores and top predators is substantially larger in Montes Azules (Mann-Whitney's $U = 53$; $P < 0.05$). Moreover, species such as the tapir, the jaguar, the peccaries, and the mazama deer never did appear in the surveyed plots of Los Tuxtlas.

Comparison of the Patterns of Herbivory

The levels of herbivory were determined in permanently marked seedlings (up to 50 cm in height) and saplings (50–150 cm in height). The seedlings were studied in 20-m^2 permanent quadrats and the saplings in ten 10-m^2 quadrats. In both cases, 10 plants were randomly chosen from each individual quadrat, thus totalling 200 seedlings and 100 saplings in each site. The topmost leaves, up to five per seedling and ten per sapling, were individually marked with plastic color rings. At each recording date, all leaves were individually relocated and recorded as intact or eaten with the level of damage and the agent responsible for it noted.

The results of this comparison are summarized in Table 13-5. Although the levels of vertebrate herbivory in Montes Azules would seem to be rather low (overall ca. 29% of the plants and 27% of the leaves were damaged), the comparison with Los Tuxtlas is overwhelming: none of the surveyed plants or leaves were damaged. The implication of this result is that the consistent absence

TABLE 13-5. Damage by Folivorous Vertebrates on Seedlings and Saplings from Both Sites[a]

	Seedlings		Saplings		Overall	
	Plants	Leaves	Plants	Leaves	Plants	Leaves
Montes Azules	58 (29)	289(30.5)	30 (30)	240(24)	88 (29.3)	529(27.2)
Los Tuxtlas	0	0	0	0	0	0

[a]Data are the numbers (percentages) of damaged plants and leaves; pooled values from all quadrats. Original sample sizes for seedling: plants = 200, leaves = 948; saplings: plants = 100, leaves = 1000.

of damage by vertebrates in Los Tuxtlas is, very likely, associated with the marked rarity of this fauna in this site.

Structure and Diversity of the Forest Understory

As a corollary to the comparative study, we included an analysis of the structure and diversity of the understory of the forest from both sites. We surveyed the floristic composition and number of individuals in the same quadrats used for the study of herbivory on seedlings. Seedling density was considerably higher in Los Tuxtlas (Table 13-6, ratio Los Tuxtlas/Montes Azules = 2.3; $P < 0.05$). Nearly 67% of the quadrats from Los Tuxtlas and only 10% from Montes Azules had one dominant species. Likewise, the mean number of different species per quadrat in Los Tuxtlas was about one third that of Montes Azules and, consequently, the mean diversity index is significantly higher in the latter site ($P < 0.001$). These results suggest that the absence of vertebrates may bring about dramatic changes in the structure and diversity of the forest understory.

An obvious alternative explanation for the difference is that Montes Azules, as a forest, has a higher diversity than Los Tuxtlas which, in turn, would determine the differences in the understory. This possibility was explored with a comparative analysis of the diversity of the upper strata of both forests, following the method depicted by Gentry (1982). The survey was carried out in a 0.1-ha plot and included all plants ≥ 1.0 cm dbh.

Both the number of species and the diversity index were higher in Montes Azules than in Los Tuxtlas (Table 13-7; $P < 0.001$ for the index comparison). However, even if Montes Azules is more diverse, the difference between the two forests could still be used to explore the question, by comparing differences in the upperstory communities with the differences between the two understories. For example, while the ratio of the number of species in Montes Azules to Los Tuxtlas is 1.12, the corresponding value for the understory is 2.6 times higher. Likewise, the ratio of Montes Azules to Los Tuxtlas for the diversity index is 1.14 for the upper strata, while that of the understory is 3.46, that is, over three times higher. These relationships suggest that the differences in the upperstory may not be the sole explanation for the differences in the understories.

TABLE 13-6. Structure and Diversity of the Understory from Both Sites[a]

	Los Tuxtlas	Montes Azules
Mean density (range)	52.8 (11–204)	22.6 (10–62)
Number of quadrats with one dominant species[b]	14	2
Mean number of different species[c]	2.3	6.65
Mean diversity index (H')	1.07	3.71

[a]Data from 20 1-m^2 quadrats.
[b]Dominant species = 5 or more conspecific individuals.
[c]Based on 10 randomly chosen plants per quadrat.

TABLE 13-7. Number of Species and Diversity Index for the Trees and Understory of Both Sites

	Number of Species		Diversity Index	
	Trees	Understory	Trees	Understory
Montes Azules (MA)	90	6.65	1.690	3.710
Los Tuxtlas (LT)	81	2.30	1.482	1.071
Ratio MA/LT	1.12	2.89	1.140	3.460

It is instructive to compare the diversity of the upper strata and the understory for each of the two sites. Owing to reasons such as higher species packing and higher turnover rates of individuals and species in the understory, we would expect understories to be more diverse than their corresponding upper strata. Such is clearly the case in Montes Azules, where the diversity index of the understory is 2.2 times higher, whereas in Los Tuxtlas the diversity index of the understory is, surprisingly, even lower than that of the upper strata (Table 13-7). This relationship suggests an anomalous situation in the structure and diversity of the understory of a defaunated forest.

EPILOGUE

Evidence of the role of herbivores as agents that increase vegetational diversity has been documented for agronomic systems (Harper 1969), marine communities (Lubchenco 1978), and grasslands (Harper 1969; Tansley and Adamson 1925), and has been suggested for tropical rain forests (Janzen 1970; Connell 1971). Our results seem to point in this direction. The impact of herbivorous vertebrates on the diversity of this kind of community may be largely augmented because they do not only feed on the foliage, but they also consume and kill large quantities of seeds (Janzen 1981a; Hallwachs 1986).

Our conclusions regarding the role of vertebrate herbivores on the patterns of herbivory and vegetational structure and composition of the understory of Los Tuxtlas is based on correlative findings. Unequivocal conclusions demand experimental, noncircumstantial evidence. In particular, two kinds of experimental manipulation are amenable: (1) reintroduction of vertebrate herbivores to Los Tuxtlas to see if the patterns of herbivory, structure, and diversity of the understory will then resemble those of Montes Azules and (2) the establishment of selective exclosures in Montes Azules to assess the relative importance of different kinds of herbivores (e.g., insects, vertebrates) and to see if, in the absence of grounddwelling herbivores, the understory resembles that of Los Tuxtlas.

If the kind of experimental manipulations we propose here do confirm our findings, they will unequivocally support the hypothesis that herbivorous vertebrates play an important role in the patterns of herbivory in the forest understory and, in the long term, on the structure and dynamics of some tropical rain forests.

ACKNOWLEDGMENTS

We would like to thank the following persons for their assistance with the field work: Santiago Sinaca, Marisela Illescas, Ken Oyama, Alejandro Torres, and Guillermo Ibarra. The manuscript was read and much improved by Victor Jaramillo, Gerardo Ceballos, Cèsar Dominguez, and Lynden Higgins. Our special thanks to Peter W. Price for his encouragement and corrections to the manuscript. This research was partly supported by a CONACyT grant to R.D.

REFERENCES

Andrle, R. F. 1964. *A Biogeographical Investigation of the Sierra of Los Tuxtlas.* Ph.D. thesis, Lousiana State University, Ann Arbor Michigan.

Aranda, J. M. 1981. *Rastros de los Mamiferos Silvestres de México. Manual de Campo.* Instituto Nacional de Investigaciones sobre Recursos Bióticos, Xalapa Veracruz, Mexico.

Clark, D. A., R. Dirzo, and N. Fetcher (Eds.). 1987. *Ecologia y Ecofisiologia de Plantas en los Bosques Mesoamericanos.* Revista de Biologia Tropical 35.

Connell, J. H. 1971. On the role of natural enemies in preventing competitive exclusion in some marine animals and in rain forest trees. In *Dynamics of Populations*, P. J. den Boer and G. R. Gradwell (Eds.). *Proceedings of the Advanced Study Institute in Dynamics of Numbers in Populations, Oosterbeck.* Centre for Agricultural Publishing and Documentation, Wageningen, pp. 298–310.

de la Cruz, M. and R. Dirzo. 1987. A survey of the standing levels of herbivory in seedlings from a mexican rain forest. *Biotropica* 19: 98–106.

Diechtl, S. 1988. *Cae una Estrella. Desarrollo y Destrucción de la Selva Lacandona.* Secretaria de Educación Pública, Mexico.

Dirzo, R. 1980. Experimental studies on slug-plant interactions. I. The acceptibility of thirty plant species to the slug *Agriolimax caruanae. Journal of Ecology* 68: 981–999.

Dirzo, R. 1987. *Propuesta para la Creación de un Parque Ecológico para la Investigación y Educación en Los Tuxtlas, Veracruz.* Secretaria de Desarrollo Urbano y Ecología, Mexico.

Dirzo, R. 1984. Herbivory: A phytocentric overview. In *Perspectives on Plant Population Ecology*, R. Dirzo and J. Sarukhán (Eds.). Sinauer Associates Inc. Publishers, Sunderland, MA, pp. 141–165.

Eisenberg, J. F. 1981. *The Mammalian Radiations. An Analysis of Trends in Evolution, Adaptation, and Behavior*, University of Chicago Press, Chicago, IL.

Garcia-Guzmán, G. 1989. *Estudio sobre Ecologia de Patógenos en el Follaje de plantas en la Selva de Los Tuxtlas.* M.Sc. thesis, Facultad de Ciencias, UNAM, Mexico.

Gentry, A. H. 1982. Patterns of neotropical plant species diversity. *Evolutionary Biology* 15: 1–84.

Gómez-Pompa, A. and S. del Amo, (Eds.). 1985. *Investigaciones Sobre la Regeneración de Selvas Altas en Veracruz, México*. Vol. II. Instituto Nacional de Investigaciones sobre Recursos Bióticos, Editorial Alhambra Mexicana, Mexico.

Gómez-Pompa, A., C. Vázquez-Yanez, S. del Amo, and A. Butanda (Eds.). 1976. *Regeneracion de Selvas. Investigaciones Sobre la Regeneración de Selvas Altas en Veracruz, Mexico*. Compañia Editorial Continental S. A., Mexico.

Hall, R. E. 1981. *The Mammals of North America*. Vols. I and II. John Wiley and Sons, New York.

Hall, R. E. and W. F. Dalquest. 1963. The mammals of Veracruz. *University of Kansas Publications, Museum of Natural History* **14**: 165–362.

Hallwachs, W. 1986. Agoutis (*Dasyprocta punctata*): The inheritors of Guapinol (*Hymenaea courbaril*: Leguminosae). In *Frugivores and Seed Dispersal*, A. Estrada and T. H. Fleming (Eds.). Dr. W. Junk Publishers Dordrech. pp. 285–304.

Harper, J. L. 1969. The role of predation in vegetational diversity. In *Diversity and Stability in Ecological Systems*. Brookhaven Symposia in Biology **22**: 48–62.

Ibarra-Manriquez, G. and S. Sinaca. 1987. *Listados Floristicos de México. VII. Estación de Biologia Tropical Los Tuxtlas, Veracruz*. Instituto de Biologia, UNAM, Mexico.

Janzen, D. H. 1970. Herbivores and the number of tree species in tropical forests. *American Naturalist* **104**: 501–524.

Janzen, D. H. 1981a. Digestive seed predation by a Costa Rica Baird's tapir. *Biotropica* **13**: 59–63.

Janzen, D. H. 1981b. Patterns of herbivory in a tropical deciduous forest. *Biotropica* **13**: 271–282.

Krebs, C. J. 1978. *Ecology: The Experimental Analysis of Distribution and Abundance*. Harper and Row, New York.

Leigh, E. G., Jr. and N. Smythe. 1978. Leaf production, leaf consumption, and the regulation of folivory on Barro Colorado Island. In *The Ecology of Arboreal Folivores*, G. G. Montgomery (Ed.). Smithsonian Institute Press, Washington pp. 33–50.

Leopold, A. S. 1959. *Wildlife of Mexico. The Game Birds and Mammals*. University of California Press, Berkeley, California.

Lewin, R. 1989. How to get plants into the conservationists' ark. *Science* **244**: 32–33.

Lubchenco, J. 1978. Plant species diversity in a marine intertidal community: Importance of herbivore food preference and algal competitive abilities. *American Naturalist* **112**: 23–39.

Medellin, R. 1986. *Murciélagos de Chajul, Chiapas*. B.Sc. thesis, Facultad de Ciencias, UNAM, Mexico.

Miranda, A. and R. Dirzo. 1990. The use of foot prints detection for the comparative analysis of vertebrate faunas. *Journal of Mammalogy*, in press.

Murie, O. J. 1974. *A Field Guide to Animal Tracks*. Houghton Mifflin Company, Boston, MA.

Navarro, D. 1980. *Lista Preliminar de los Mamiferos de las Estación de Biologia Tropical "Los Tuxtlas"*. Publicatión especial del Instituto de Biologia, UNAM, Mexico.

Navarro, D. 1982. *Los Mamiferos de la Estación de Biología "Los Tuxtlas"*. B.Sc. thesis, Facultad de Ciencias, UNAM, Mexico.

Núñez-Farfán, J. and R. Dirzo. 1988. Within-gap spatial heterogeneity and seedling performance in a Mexican tropical forest. *Oikos* **51**: 274–284.

Pérez-Higareda, G., R. C. Vogt, and O. Flores-Villela. 1987. *Lista Anotada de los Anfibios y Reptiles de la Región de Los Tuxtlas, Veracruz*. Instituto de Biologia, UNAM, Mexico.

Ramirez-Pulido, J., M. C. Britton, A. Perdomo, and A. Castro. 1986. *Guía de los Mamíferos de México. Referencias hasta 1983*. Universidad Autónoma metropolitana, Mexico.

Ramírez-Pulido, J., R. López-Wilchis, C. Müdespacher, and I. Lira. 1983. *Lista y Bibliografía Reciente de los Mamíferos de México*. Universidad Autónoma Metropolitana, Mexico.

Raven, P. H. and G. B. Johnson. 1986. *Biology*. Times Mirror/Mosby College Publications, St. Louis Missouri.

Strong, D. R., J. H. Lawton, and T. R. E. Southwood. 1984. *Insects on Plants. Community Patterns and Mechanisms*. Blackwell Scientific Publications, Oxford.

Tansley, A. G. and R. S. Adamson. 1925. Studies on the vegetation of the English chalk. III. The chalk grasslands of the Hampshire–Sussex border. *Journal of Ecology* **13**: 177–223.

SECTION 4
Plant–Butterfly Interactions

The great naturalist-explorers were fascinated by butterflies. While in Brazil, Darwin (1860, p. 32) noted that "The large and brilliantly-coloured Lepidoptera bespeak the zone they inhabit, far more plainly than any other race of animals. I allude only to the butterflies…" Darwin (1860), Belt (1988), and Spruce (1908) were amazed at their mass dispersal, so dense to be like "snowing butterflies." Wallace (1869) used the swallowtails and other families repeatedly in deciphering the biogeography of the Malay archipelago. While on the Amazon, Bates (1862) and Müller (1879) also in Brazil, observed and interpreted the mimicry complexes involving butterflies. Bates (1862, 1910 edition, p. 348) accurately predicted their importance: "The study of butterflies—creatures selected as the types of airiness and frivolity—instead of being despised, will some day be valued as one of the most important branches of Biological science."

Since these early days, the ecology of butterflies has remained central to ecology in general, and a focus for studies on plant–herbivore interactions. Mimicry systems, with chemical defense derived from plants, coevolution of plants and butterflies, behavioral ecology, and chemical communication in animals, as examples, have developed as major themes in ecology. Because butterflies are diurnal, active, and colorful, we have always related to them in these modalities, so important to human experience. They focus our attention and expand our aesthetic senses. For these reasons butterflies will always be a naturalist's delight and a mainstay for the development of ecology.

We need not apologize for this one taxon-specific section of the book, for without butterflies evolutionary ecology, chemical ecology, and plant–herbivore interactions would be the poorer. Brussard (1978) in an overview of ecological genetics, unhesitatingly divided the Animal Kingdom into "*Drosophila*" and "other animals." Likewise, terrestrial animal ecology was long dominated by studies on birds. Among herbivorous animals, however, no other group has been studied in such depth as the butterflies, if one includes knowledge on their population dynamics, host plant choice, behavior, genetics, chemical ecology, and evolutionary ecology. The contributions to this section illustrate richly the breadth of ecological knowledge being developed on butterflies: the toxic Solanaceae and their ithomiine herbivores (Chapter 14); the phytochemical ecology of the plant–herbivore interaction (Chapters 15 and 16); the adaptive radiation of butterflies on plant groups (Chapter 17); and the biodiversity of the genus *Heliconius* (Chapter 18). The butterflies represent the ideal for the scientist with their aesthetic appeal and their aptitude for experiment. We all delight in and benefit from these beauties, for their wonder for us is no less than that for the early naturalist-explorers.

REFERENCES

Bates, H. W. 1862. Contributions to an insect fauna of the Amazon Valley. *Trans. Linn. Soc. London* **23**: 495–566.

Belt, T. 1888. *The naturalist in Nicaragua*. 2nd ed. Bumpus, London.

Brussard, P. F. (Ed.). 1978. *Ecological genetics: The interface*. Springer, New York.

Darwin, C. 1860. *The voyage of the Beagle*. Natural History Library, New York.

Müller, F. 1879. *Ithuna* and *Thyridia*: A remarkable case of mimicry in butterflies. *Trans. Royal Entomol. Soc., London* 1879: xx–xxix.

Spruce, R. 1908. *Notes of a botanist on the Amazon and Andes*. Volume 2, Macmillan, London.

Wallace, A. R. 1869. *The Malay archipelago*. Macmillan, London.

14 Interactions Between Ithomiine Butterflies and Solanaceae: Feeding and Reproductive Strategies

JOÃO VASCONCELLOS NETO

Herbivory has become one of the most important subjects in evolutionary ecology. The mutual impact between plants and herbivores is little known, information being especially scarce for the tropics (Crawley 1983).

The evolution of chemical defenses and host-plant predictability as a result of herbivore selective pressures have been considered by Feeny (1975, 1976) and Rhoades and Cates (1976). Parasitoids and predators are part of the plant defense battery (Price et al. 1980), playing a role in the evolutionary scenario of insect–plant interactions. The mutual impact among plants, herbivores, and parasitoids evolved by several different routes, producing various life strategies inside the community.

Plants in the family Solanaceae and their herbivores constitute a very rich system to study the interactions among three trophic levels in the Neotropical region. There are over 100 genera in the family, with more than 2200 species. Some of the largest genera are *Solanum*, with about 1000 species, *Lycianthes*, with 200 species, *Cestrum*, with 175 species, and *Lycium*, *Physalis*, and *Nicotiana*, all with 75–100 species. The family is distributed around the world, in temperate and tropical regions, especially in South America, which is considered the center of origin, because it is rich in genera and endemic species (Lawrence 1951; Smith and Downs 1966; D'Arcy 1973; D'Arcy personal communication).

Ithomiine butterflies form one of the main groups of herbivores that feed on these plants. They are found from sea level to elevations of 4000 m, in both Central and South America (Fox 1963), extending from southern Mexico to northern Argentina (Brown 1979). These butterflies live inside dense forest in shady and moist places, in pockets which contain many different species (Brown

Plant-Animal Interactions: Evolutionary Ecology in Tropical and Temperate Regions, Edited by Peter W. Price, Thomas M. Lewinsohn, G. Wilson Fernandes, and Woodruff W. Benson.
ISBN   0-471-50937-X   © 1991 John Wiley & Sons, Inc.

and Benson 1974; Brown 1979). Pockets are patches that remain moist during the dry winter, when ithomiine populations concentrate in them.

Although the majority of ithomiine species feed on Solanaceae, a few genera such as *Aeria* and *Tithorea* are associated with Apocynaceae, and one genus, *Hyposcada*, with Gesneriaceae (Drummond 1976, 1986; Haber 1978; Drummond and Brown 1987; Brown 1987; Vasconcellos-Neto 1980, 1986).

Gilbert (1969) indicated these groups of plants and butterflies as appropriate for investigation of the herbivore–host-plant interface.

To understand the ecological relationships between these insects and their host plants, a 4-yr study was carried out in the Sumaré Forestry Station (Horto Florestal de Sumaré), State of São Paulo, Brazil (22°49′ S, 47°17′ W). The vegetation is composed of *Eucalyptus* trees with native vegetation growing in the understory, especially several species of Solanaceae in some patches. The community of ithomiine butterflies was composed of 23 species, nine being residents and the others proceeding from neighboring populations (Brown and Vasconcellos-Neto 1976; Vasconcellos-Neto 1980). The population dynamics of the five most frequent species of resident Ithomiinae and their interaction with Solanaceae were studied in Sumaré.

Additional observations were conducted in two forests, a small one near Campinas (22°50′ S, 47°04′ W) and the other in the Serra do Japi, Jundiaí (23°15′ S, 46°52′ W), both in the State of São Paulo, Brazil.

This chapter will deal with the abundance and distribution patterns of host plants, their phenologies and their impacts on herbivores, giving emphasis to the feeding and reproductive strategies of the latter. At the end a model will be proposed to explain the interactions of three trophic levels: plant–herbivore–parasitoid.

ITHOMIINE POPULATION DYNAMICS

The population dynamics of Ithomiinae and Solanaceae, as well as their interactions, are much influenced by climatic seasonality.

The local climate is warm and moist, with one to two dry months (Nimer 1977), being classified as Cwa according to the Köppen system. The summer is warm and rainy, and in the autumn rain is scattered. The winter is cold and dry (Fig. 14-1).

The butterflies *Mechanitis polymnia casabranca*, *M. lysimnia lysimnia*, and *Hypothyris ninonia daeta* (Fig. 14-2C–E) have similar adult population dynamics. They reproduce and grow during the rainy season (December until March), reducing or stopping reproduction in the dry winter months (July and August), when these mimetic butterflies with the tiger-pattern aggregate in pockets and reach their largest population sizes and local densities (Fig. 14-2). When the first rains come in spring, the butterflies disperse, starting a new reproductive cycle. During the spring and early summer the populations are small, very dispersed, and the rates of attack on host plants are low.

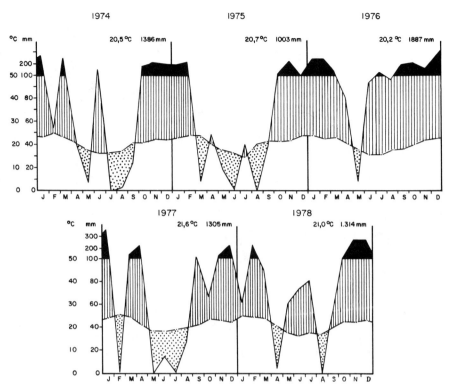

Figure 14-1. Walter and Lieth climatic diagrams for the Horto Florestal de Sumaré, State of São Paulo, Brazil, from 1974 to 1978. The dotted areas on the graphs represent dry periods and black ones very moist periods.

Dircenna dero (Fig. 14-2B) disperses widely, has a small population size, and does not form dry-season aggregations. The population of immatures is as large as those of the three species mentioned above. It reproduces mainly in open areas of the Horto Florestal during the fall and winter. The great difference in population size between *Dircenna* and the first three species of Ithomiinae is not due to available biomass of host plant, but probably to higher mortality caused by harsh microclimatic conditions and the action of natural enemies. Like *Hypothyris ninonia*, this butterfly reproduces mainly in the fall, when eggs and larvae are frequent, but the former species is found mainly in shady and moist habitats inside the forest.

The population of the fifth ithomiine species studied, *Mcclungia salonina* (Fig. 14-2A) is restricted to the moist woods, where *Mechanitis* and *Hypothyris* form their winter aggregations. Its population dynamics differ greatly from the other species, building up during the summer and early fall and reaching greatest density in May. At this time the population falls to a low level that is maintained during the winter and spring (Fig. 14-2) (Vasconcellos-Neto 1980). The cycle of

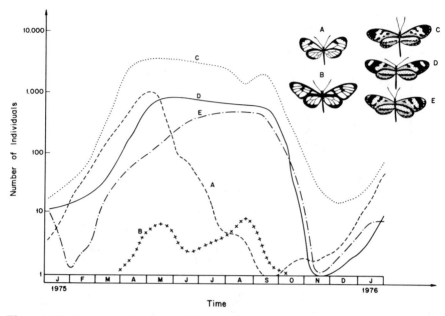

Figure 14-2. Population dynamics of five species of Ithomiinae in the Horto Florestal during 1975. (A) *Mcclungia salonina*; (B) *Dircenna dero*; (C) *Mechanitis polymnia*; (D) *Mechanitis lysimnia* and (E) *Hypothyris ninonia* (modified from Vasconcellos-Neto 1980).

this transparent butterfly is especially affected by egg and larval parasitoids (Monteiro 1981) and very little by climatic conditions and host-plant availability (Vasconcellos-Neto 1980, 1986) (Fig. 14-2).

The populations of the other four species of butterfly varied cyclically in abundance during the 4-yr study. Their cycles were associated with the recurring dry and rainy seasons and with the phenologies and dynamics of their host plants (Vasconcellos-Neto 1980). Predators and parasitoids act on their population sizes, but they are not the main factors molding the reproductive cycles (Vasconcellos-Neto 1980; Monteiro 1981).

HOST-PLANT UTILIZATION

Each species of Solanaceae has a special kind of microhabitat. The evolution of their abundance and distribution patterns was presumably generated by factors such as competition among plants and herbivore pressures. All Solanaceae microhabitats represent a mosaic produced by the combination of light, type of soil, air humidity, and soil water content. *Solanum variabile* and *S. paniculatum* grow typically in open places, but on different types of soil. On the other hand,

Cestrum sendtnerianum grows in shady and moist places, in areas of ithomiine pockets (Fig. 14-3).

The ithomiine butterflies have preferential microhabitats, and adjust their feeding and reproductive strategies to host-plant availabilities in these sites. As an example, *S. paniculatum* is a very common and predictable host for *Dircenna dero* in the thicket, becoming less frequent for *Mechanitis polymnia* in shadier areas and rare for *Hypothyris ninonia* in very shady places (Fig. 14-3). Thus, one Solanaceae species sometimes constitutes either a common resource or a rare one inside a vegetational mosaic.

At the study site there are five perennial species of *Solanum*, which are host plants of *Mechanitis polymnia*, *Hypothyris ninonia*, and *Dircenna dero*. Among these three ithomiine species there are dissociations in reproductive season, microhabitat, and food plant preference. *M. polymnia* shows greater recruitment earlier than the other two (mainly during the summer), while *H. ninonia* and *D. dero* reproduce mainly during the fall and early winter. The latter two species diverge in food preference, *D. dero* laying eggs mainly on *S. variabile* and *S.*

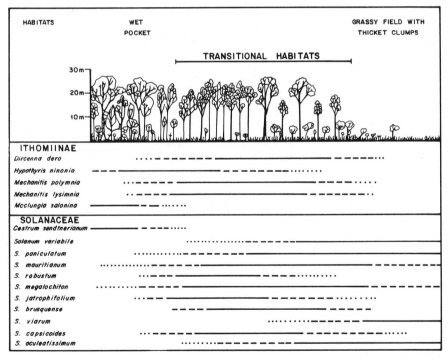

Figure 14-3. Distributions of Solanaceae, and eggs and larvae of five ithomiine species in habitats in the Horto Florestal de Sumaré. This qualitative analysis is represented in a habitat profile. Relative abundance of Solanaceae (below) and of ithomiine eggs and larvae (above) are shown as common (solid lines), uncommon (dashed), or rare (dotted). (Modified from Vasconcellos-Neto, 1986).

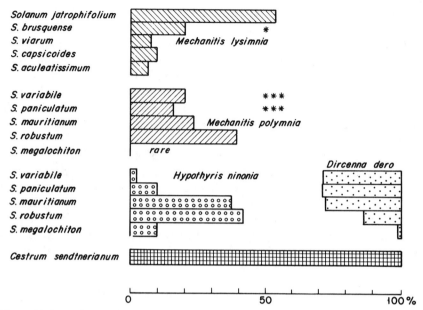

Figure 14-4. Preference for oviposition of five ithomiine species on 11 Solanaceae host plants in the Horto Florestal de Sumaré, State of São Paulo. The histograms represent the percentages of eggs and larvae found on each species of Solanaceae (Modified from Vasconcellos-Neto, 1986). (* rare *Hypothyris ninonia*; *** some times, *Mechanitis lysimnia*). 100% of oviposition by *Mcclungia salonina* occurs on *Cestrum sendtnerianum*.

paniculatum in more open sites (thicket edge), while *H. ninonia* prefers laying eggs on *S. mauritianum* and *S. robustum* in more shady and humid sites (Fig. 14-4). *M. polymnia* lives in transitional habitats, in clearings and forest margins, so that there exists a marked microhabitat dissociation among the three species (Fig. 14-3).

On the other hand *M. lysimnia*, living in the same microhabitat, uses as larval host plants a different group of annual *Solanum* species, and probably one biennial species (*S. brusquense*).

Mcclungia salonina is monophagous, laying eggs on a perennial species, *Cestrum sendtnerianum*. Both the plant and the butterfly occur in moist and shady sites. This butterfly is ecologically distinct in time and space from the other species studied. Thus, these five ithomiine butterflies seem to coexist in part because they diverge in at least three niche dimensions: reproductive season, microhabitat, and host-plant preference (Vasconcellos-Neto 1980).

MUTUAL INTERACTIONS

The Leaves

Ithomiinae and other insects should be important herbivores promoting the evolution of traits of Solanaceae leaves.

Sometimes *Mechanitis lysimnia* lays eggs on the perennial host plants of *M. polymnia*, especially *Solanum variabile*, *S. paniculatum*, and *S. robustum*. These plants have angular lobed leaves similar to those of the usual annual host plants of *M. lysimnia*. This occurs soon after the dry winter, when annual *Solanum* are still almost absent (Vasconcellos-Neto 1980).

A leaf-shape image such as used by other species of butterflies (Rausher 1978; Benson 1978) could be an important cue for *M. lysimnia* to find the host plant. *Passiflora* species have different leaf shapes, sometimes even between young and adult plants. This divergence in appearance in *Passiflora* leaves is interpreted as a defense mechanism against herbivores (Gilbert 1975, Ch. 18 this volume).

Solanum variabile shows a marked polymorphism of leaves, from simple with a normal margin to elongated leaves with a very dentate margin, similar to tomato leaves.

Another species, *Cyphomandra divaricata*, has elongated and dentate leaves when the plant is young but after it attains a certain size, about 1 m, it produces lateral branches with heart-shaped leaves. The adult plant is attacked by *Thyridia psidii*, another ithomiine butterfly, and occasionally by *M. lysimnia* in the Santa Genebra forest (Campinas, SP). At the three communities, Horto Florestal, Santa Genebra forest, and Serra do Japi, the species of Solanaceae show divergence in appearence, which could be interpreted as a mechanism of escape from herbivory by Ithomiinae.

Dircenna dero lays eggs on *S. robustum* and *S. megalochiton*, however, larvae after the third instar were never found feeding on these plants. On the other hand, all *Dircenna* instars were found on *S. paniculatum* and *S. variabile*. The ovipositing female sits on the leaf, moves backwards, bends its abdomen and continues to move back until the tip of the abdomen touches a vein. At this time the butterfly lays a single egg inside the hairs next to the vein. The larvae from the first to third instars feed and rest near the vein, being very cryptic. After this time, from the end of the third instar to the pupa, the larva rolls the leaf, making a shelter which is efficient against desiccation and predation. The "nest" is easily made with leaves of *S. variabile* and *S. paniculatum*. The leaves of *S. robustum* are very large, with a strong and salient vein which has several spines, handicapping the shelter construction. The leaves of *S. megalochiton* are very small, and it would be impossible to hide a fourth or fifth instar larva inside the shelter. Once a shelter made by a *Dircenna* larva with several terminal leaves of *S. megalochiton* was observed.

The annual species of *Solanum* studied have many spines on their leaves and stems. *S. jatrophifolium* especially has a great number of long needles on the stem that obstruct the movement of *Mechanitis lysimnia* larvae from one leaf to another. The gregarious larvae of these butterflies construct a carpet of silk over the spines, facilitating their locomotion. These spines should restrain the access of small rodents and marsupials to the fruits which are dispersed by bats (probably *Carollia perspicillata*).

At Horto Florestal there was an unidentified species of *Solanum* that had a great amount of glandular hairs on the leaves and stem. Only one herbivore, a weevil (probably *Phyrdenus* sp.) was able to walk and feed on the plant without getting stuck. No ithomiine or other herbivore was found feeding on these Solanaceae. Small dead insects were frequently found glued to the leaves. Glandular hairs of Solanaceae could contain secondary toxic compounds, or the insect could die of starvation when immobilized (Harborne 1986).

Size and Plant Phenology

The phenologies of the early stages of Ithomiinae and of the Solanaceae host plants appear to be mutually adjusted. For example, *S. jatrophifolium* germinates and grows in the spring and early summer, when adult population size of *Mechanitis lysimnia* is small (Fig. 14-5). The plants flower and fruit in fall, when the intensity of herbivory is diminishing. Plants germinating later, when *M. lysimnia* is in its reproductive peak, are heavily attacked, do not reproduce, and usually die. Besides the herbivory load, these late-germinating plants must also deal with the harshness of the oncoming dry season. Thus, butterflies as well as

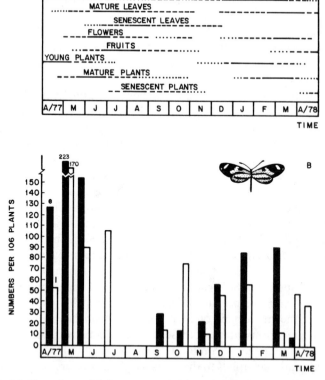

Figure 14-5. (*A*) Phenology of *Solanum jatrophifolium*, with relative abundance shown as common (solid line), uncommon (dashed), or rare (dotted). (*B*) Phenology of immature stages of *Mechanitis lysimnia* (e = eggs; l = larvae).

climatic conditions play roles in the phenology of this annual *Solanum* and probably of the other ones as well.

The host plants of *M. lysimnia* are ephemeral and unpredictable in time and space, because they are short-lived, not being available during the dry winter, and are found scattered inside the forest and woody vegetation. Two evolutionary responses have been selected in this butterfly: adjustment of population dynamics to climatic conditions and host-plant availability, and maximization of reproductive effort.

According to Vasconcellos-Neto and Monteiro (in preparation) there is a positive relationship between the egg clutch size of *Mechanitis lysimnia* and the number of leaves on the host plant (Fig. 14-6). The ovipositing female seems to evaluate the amount of food by inspecting the quantity of leaves and by checking for presence or absence of eggs and possibly caterpillars before deciding to lay eggs and, if laid, how many. These behavioral machanisms should be important in reducing intraspecific competition for the five rare, small, and ephemeral host-plant species of this insect.

The Ithomiinae that lay eggs on perennial *Solanum* reproduce mainly during the summer and fall, when these plants have many young and mature leaves. In general, the *Solanum* studied here sprout, grow, flower, and fruit more intensely in the spring and early summer, when herbivory by ithomiines is low.

During the dry winter, *M. polymnia* interrupts its reproduction, concentrating in pockets. At better times for reproduction there is no limitation of food for larvae, since the host plants are big and perennial. Only on *S. robustum*, the preferential host plant of *M. polymnia*, was there found a positive relationship between the total number of eggs (including several clutches) and the total leaf area. There is no relationship between clutch size and individual leaf area (J. Vasconcellos-Neto and Gisela D. C. M. Silva, unpublished).

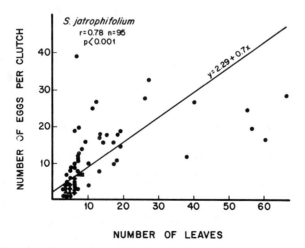

Figure 14-6. Relationship between clutch size of *Mechanitis lysimnia* and the number of leaves per plant of *Solanum jatrophifolium*.

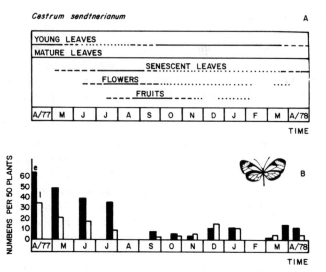

Figure 14-7. (*A*) Phenology of *Cestrum sendtnerianum,* with relative abundance shown as common (solid lines), uncommon (dashed), or rare (dotted). (*B*) Phenology of immature stages of *M. salonina* (e = eggs; l = larvae).

Oviposition in *Mcclungia salonina* is greatest during the summer and fall; its host plant blooms and fruits afterward, during the period of low population density of the butterfly (Figs. 14-2 and 14-7). Although this butterfly is not the only herbivore on this plant, there is a reciprocal adjustment between plant and ithomiine populational dynamics.

INTERACTIONS AMONG THREE TROPHIC LEVELS

Plants have evolved a variety of defense mechanisms against herbivores, such as chemical defenses allied to plant predictability and unpredictability (Feeny 1976; Rhoades and Cates 1976) and nitrogen content associated with chemical and mechanical defenses (McNeill and Southwood 1978; Lawton and McNeill 1979). Natural enemies of the herbivores, together with chemical and mechanical defenses, all make up a plant's defense battery (Lawton and McNeill 1979; Price et al. 1980).

These interactions generate different evolutionary responses within a complex community. The pattern of evolution followed is given by the sum of reciprocal interactions between at least three trophic levels. The observed result is different life-styles, which are often interpreted through their evolutionary mechanisms and consequences rather than the selective causes or forces themselves.

Some emergent questions will be discussed in this perspective, leading to the proposal of a model for the interactions between Solanaceae, herbivores, and parasitoids.

Herbivores

Monophagy vs. Polyphagy. The idea that host ranges are determined on a purely nutritional basis shows important inconsistencies (Dethier 1970). As an alternative suggestion, the role of secondary substances in defense against phytophagous insects has been reviewed by Dethier (1970), Feeny (1975, 1976, Ch. 15), and Rhoades and Cates (1976). Monophagy and polyphagy among insects has also been discussed by Levins and MacArthur (1969), Benson (1978), Cates (1981), Vasconcellos-Neto (1980, 1986), and Futuyma and Peterson (1985).

These days a central question is the principal selective force which pushes herbivores toward specialization [see Special Features, *Ecology* 69(4), 1988]. There are several schools of thinking about host-plant utilization patterns.

Bernays and Graham (1988) argue that natural enemies, especially generalist predators, rather than plant secondary compounds, may be the dominant factor in the evolution of host ranges, as pointed out by Schultz (1988) and Ehrlich and Murphy (1988).

According to Jermy (1984, 1988), host-plant selection is a behavioral phenomenon based on the insect's plant-recognition abilities. Although Futuyma and Peterson (1985) indicated genetic components in the adult choice and larval performance, these facts show how natural selection is acting. These authors are dealing with mechanisms and consequences, not with causes or selective pressure (but see Futuyma, Chapter 19).

The problem is to understand how the factors interact under different ecological conditions in present-day communities, exerting selective pressures on phytophagous insects (Thompson 1988 and Courtney 1988). There are multiple causes acting on the evolution of host-plant range (Barbosa 1988; Rauscher 1988; Thompson 1988; Janzen 1988; Courtney 1988), which perhaps should be thought of as a consequence of diffuse coevolution within complex communities (Fox 1988).

The distribution and abundance patterns of suitable host plants in preferential habitats for oviposition influence the degree of feeding specialization of ithomiine butterflies.

Mechanitis polymnia lays eggs mainly on plants situated in transitional environments of the Horto Florestal and clearings and trails inside the natural forest. In these places, the distance between individuals of one particular host species is often large, which diminishes the probability of a butterfly finding them. Thus, as the distance between host plants as oviposition sites becomes larger, it would be advantageous for a butterfly to reduce it. This would be possible by expanding the host range.

The reasons that led *M. lysimnia* to oligophagy seem to be the rarity and seasonality of the more suitable host plants in its preferential habitats. Benson (1978) and Cates (1981) commented that seasonal fluctuation in abundance of more adequate host plants could lead the herbivore to polyphagy. Sometimes, when populational dispersion of *M. lysimnia* occurred after the dry winter, eggs were found on *Solanum variabile, S. paniculatum,* and *S. robustum,* which are normally not used as hosts. At that time, the usual host plants were rare or absent.

Larvae of both species of *Mechanitis* could complete their development on any of the other's usual host plants in the laboratory. Interspecific competition could play a role in the evolution of their host range, promoting resource partitioning; on the other hand, unpredictability could favor selection for polyphagy.

Dircenna dero lives in more open places, like thickets and transitional habitats (Fig. 14-3); females lay eggs on leaves of five perennial *Solanum* species. However, the caterpillars have highest probability of reaching maturity on *S. paniculatum* and *S. variabile* where they are able to construct a shelter, which protects them against predation and desiccation.

These factors could force this butterfly to use mainly these two *Solanum* species, leading to specialization. These two host plants are more common in thickets than in shady sites, where, however, *D. dero* would probably compete with *Hypothyris ninonia* for its food plants (Vasconcellos-Neto 1980). Smiley (1978) showed that the natural monophagy of *Heliconius melpomene* is not due to host-plant abundance and digestive specialization, but is a result of selective pressures due to predation by ants, parasitism, and competition. Predators could influence host-plant suitability and use (Smiley and Wisdom 1985).

Lawton (1986) reviews the importance of generalist parasitoids in niche differentiation of herbivores and reports examples in which there could be a partitioning of host plants as a result of pressure from different species of parasitoids.

Apanteles concinnus (Braconidae), a common parasite of *Mcclungia salonina* (Monteiro 1981), also attacked *Hypothyris ninonia* and *Dircenna dero* on *S. paniculatum*, and *Mechanitis lysimnia* on *S. jatrophifolium*, whenever both plants and butterflies occurred in the same shady, humid microhabitat as *Mcclungia salonina* and its host plant in Horto Florestal. In another forest (Japi), a species of *Hyposoter* (Ichneumonidae) parasitizes *Epityches eupompe* (Ithomiinae) that feeds on *Athenaea picta*. This parasitoid seems specialized in a special kind of microenvironment—shady and moist sites—where it is common to find *Athenaea*. In the vicinity of this plant, *Mechanitis polymnia* was found feeding on *Solanum mauritianum*; *M. lysimnia* on *Solanum* aff. *acerosum*, and *Pseudoscada erruca* on *Sessea* sp. All these ithomiines were parasitized by *Hyposoter*. Sometimes these species of Solanaceae are distributed along a gradient of ecological conditions, growing mostly in particular habitats. As some parasitoids are specialized in microhabitats, they presumably play a role in the host-plant range of these butterflies.

In a small secondary forest in Campinas, São Paulo State, *H. ninonia* has a narrower food niche, laying eggs on *S. swartzianum* and *S. pabstii*, which occur in higher densities inside the forest in some patches of the environment than do other species of *Solanum* in the habitats of this butterfly in Sumaré. It is clear that host range depends on the ecological conditions of a community.

The unpredictability of host plants in space and time could be selected by herbivory, and favor the selection of certain traits in ithomiine butterflies, such as long life spans, great flight capacity, reproductive diapause (Vasconcellos-Neto 1980), aposematism, and unpalatability (see Brower and Brower 1964; Vasconcellos-Neto and Lewinsohn 1984; Brown 1984, 1987).

These four Ithomiinae are oligophagous, but they show a marked preference in the adult choice for only one of the host plants. *Mechanitis lysimnia* lays eggs principally on *S. jatrophifolium*, although this plant is rarer than its other hosts. If *S. jatrophifolium* becomes abundant for some reason, *M. lysimnia* could specialize on it.

Mcclungia salonina lays eggs only on *Cestrum sendtnerianum*, which occurs in high concentrations in the pocket observed, with an average of 14 plants per transect of 16×1 m. In the small forest of Campinas, where *C. sendtnerianum* is rare, *M. salonina* also fed on *C. laevigatum*, thereby competing with *Pseudoscada erruca*, another ithomiine which uses this species. Monophagy in this case is favored in high concentrations of the proper host plant, while oligophagy is encouraged when host plants are rare. Dethier (1970), Levins and MacArthur (1969), Benson (1978), and Vasconcellos-Neto (1986) argued that monophagous insects are associated with species of host plants that occur in high densities, or are at least concentrated in patches in their habitat.

The pattern of host-plant utilization by Ithomiinae is thus given by the interaction of a constellation of factors (Drummond 1986).

The main factors that would facilitate polyphagy among Ithomiinae are host-plant rarity and seasonality, while monophagy would be promoted by host-plant abundance, or a higher risk of predation and parasitism on some species of plants (Vasconcellos-Neto 1980, 1986).

The distribution, abundance, phenology, and secondary compounds of a host plant and the feeding strategies of the insects are probably coevolved adaptations that result from mutual selective forces, but are not derived exclusively from these interactions.

Ithomiine Reproductive Strategies. The diversity of reproductive strategies is an interesting question to analyze through ecological and evolutionary approaches. This diversity seems to be the result of several factors, such as resource predictability, competition, and enemies, which act in the evolution of strategies that optimize the utilization of available food and the energy allocation for reproductive purposes.

Reproduction includes mating, host location, and the adjustment of the clutch to the available food. This section deals only with the question of clutch size. Early studies on optimization of clutch were developed with birds (e.g., Lack 1947–1949; Cody 1966; Foster 1974). For insect parasitoids, Price (1974) presented the balanced mortality hypothesis, where egg production is adapted to counteract relative environmental harshness. Stamp (1980), Ito et al. (1982), Courtney (1984), and Vasconcellos-Neto (1980, 1986) discussed why some butterfly species cluster their eggs while others lay them singly.

The reproductive strategies of ithomiines should have evolved in response to host-plant abundance and distribution, as well as defenses. When the host plant presents a scattered distribution it seems advantageous to Ithomiinae to deposit eggs in clusters, as Benson et al. (1976) commented for Heliconiinae. Emlen (1973) pointed out that it could be advantageous either to deposit eggs in clusters or to avoid parasites and predators, depending on the pressures exerted by these.

Tinbergen et al. (1967), studying egg predation in different densities by the wild carrion crow, showed that greater spacing among individuals can be interpreted as a defense against predation. If this is true, then when a herbivore's host is scattered this defense is valid; the butterfly could also lay eggs in clusters, because if it is difficult to find the host for most of the butterflies, it is also hard for the parasite or predator.

In a way analogous to Janzen's proposition (1970) for the distribution of tropical trees, the predation and parasitism pressures would "force" the specialist insect to "dilute" its eggs in a dense host-plant population, consequently reducing the detection of its eggs and caterpillars by its natural enemies (Vasconcellos-Neto 1986).

It is expected that different reproductive strategies in phytophagous insects could be selected on the same host plant according to its pattern of abundance in different microhabitats. When the host plant is common, selection should favor solitary eggs on the same plant or on different individuals; when the host is scattered, clusters of eggs should be more advantageous.

Mcclungia salonina, whose plant is abundant in pockets of the Horto Florestal in Sumaré, lays eggs singly; one or two eggs were normally found per plant. Monteiro (1981), studying the parasitoids of these species, found a braconid wasp (*Apanteles concinnus*) to be effective in the control of its population.

Monophagy in this species is related to its host plant's abundance, and the habit of laying solitary eggs may have been selected by pressures of its natural enemies with a low density of attack per plant. Intraspecific competition among larvae could play a role in solitary egg-laying behavior, but not in this case, because the host plant is perennial and has enough biomass to sustain a group of larvae.

The two species of *Mechanitis* deposit their eggs in clusters on distinct groups of *Solanum* species with scattered distribution in transitional habitats of the Horto Florestal. In the small forest that remains in the region and in the forest of the Serra do Japi, distances between host plants are often large, diminishing the probability of being found by the butterflies as well as by natural enemies.

Hypothyris ninonia deposits from one to several eggs on each of *Solanum swartzianum* and *S. pabstii* in shaded places in the small forest in Campinas. These plants are large perennial shrubs that grow inside the forest, where distances between plants are smaller than in clearings in which the species of *Solanum* needing direct sunlight grow. In the Horto Florestal, where the habitats are more illuminated and more open, these two species of *Solanum* are absent. There, *H. ninonia* is oligophagous, laying solitary eggs mainly on plants located in more shady and dense habitats, where the larvae apparently have higher probability of survival. *H. euclea*, a less abundant species, lays eggs in clusters on *S. mauritianum*, which grows very scattered in the forest mainly in small clearings. The predictability of this *Solanum* inside the forest is smaller than that of the host plants of *H. ninonia*.

In open areas, such as edge thicket, where *S. paniculatum* and *S. variabile* occurred concentrated in a few patches, *Dircenna dero* deposits from one to several solitary eggs on the plant, probably as a consequence of pressures by predators and parasitoids. In a little shadier habitats, these plants become rarer, and they are used by *M. polymnia* that clusters its eggs.

Some species of Ithomiinae are known to follow different reproductive tactics in distinct localities. *Mechanitis menapis saturata* and *Hypothyris euclea leucania* deposit eggs in clusters in Costa Rica (Gilbert 1969), while other subspecies (*M. m. mantineus* and *H. e. intermedia*) have been seen to lay eggs singly in Ecuador (Drummond 1976). Comparisons between these two localities might be important to verify the hypothesis that host-plant abundance and natural enemies play roles in the evolution of reproductive strategies. Brown and Benson (1975) mentioned that *Heliconius demeter eratosignis* deposits eggs in clusters in Riozinho, Rondônia, while *H. d. terrasanta* lays eggs singly near TerraSanta, Pará (Brazil), proposing that this difference in reproductive strategies is due to differences in host-plant size in the two localities.

Near Manaus, Amazonas, in the INPA Tropical Silviculture Station, *Hypothyris euclea barii* lays eggs in clusters on *Solanum asperum*, which grows mainly in clearings or places with direct sunlight in the Amazonian tropical forest (Vasconcellos-Neto 1986). In primary forest, populations of this plant are small and very dispersed (see Brown 1980). On the other hand, when roads and clearings are opened in the forest, large populations of *S. asperum* grow in the resulting sunny places. It was verified that when the gregarious larvae of *H. euclea* occur in high densities in the open places, predators such as wasps and ants return constantly to plants searching for food. The larval group is attacked massively and only larvae which separate themselves from the group are more likely to escape predation. If this selective pressure occurs over larger areas and for long periods, laying solitary eggs would be selected (Vasconcellos-Neto 1986).

Young and Moffett (1979) noted that some species of *Solanum* have evolved succulent leaves with dense trichomes that are deterrent to many insects and suggested that other species with thin leaves and without trichomes must have poisonous substances. In the majority of cases, species of Ithomiinae with eggs in clusters are associated with plants with trichomes and spines, while single-egg-laying species are frequently found on thin leaves without trichomes. They suggested that gregarious larvae permit cooperation in feeding behavior, leading to higher efficiency and even exploration of species of *Solanum* with trichomes, which are impenetrable to solitary larvae. However, the solitary larvae of *Hypothyris ninonia* and *Dircenna dero* develop on host plants with trichomes and spines.

Solanaceae that occur in shady and moist habitats normally have leaves with few or no hairs, and those from clearings, edge thicket, and fields have succulent leaves, coriaceous or not, with many hairs or spines. These structures can function as defenses against herbivores, but could also be a primary physiological adaptation against leaf desiccation.

Plant Strategies

Herbivory could lead plants to rarity (Crawley 1983), but, conversely, can abundant plants owe their commonness solely to their antiherbivore chemistry? This could be interpreted as a defense mechanism using a combination of chemical defense and the pressure of predators and parasitoids on herbivores. If the plant is predictable, then so is the herbivore to its natural enemies. As a result of natural selection, herbivores dilute their eggs on plants, reducing the impact of herbivory. In this condition, plant abundance could be understood as a defense mechanism against herbivores if predators and parasitoids are part of its defense battery. On the other hand, a scattered distribution of the host plant functions as a defense against phytophagous insects, being also a result of mutual interaction.

According to Feeny (1975, 1976) and Rhoades and Cates (1975), predictable plants have evolved quantitative chemical defenses which reduce digestibility (like tannins) and are effective against monophagous or specialized herbivores. Unpredictable or unapparent host plants evolve qualitative defenses: several kinds of secondary compounds, like alkaloids, which are effective against generalist insects. The Solanaceae are rich in alkaloids, many of which are highly toxic, and other secondary compounds which defend the plants from microbial infection and herbivores (Harborne 1986; Evans 1986; Lavie 1986; Roddick 1986; Brown 1987).

Although Harborne (1986) described the natural chemical defenses of Solanaceae, nothing is known about their type and diversity in relation to plant predictability and herbivory. If the theory is valid, apparent species of Solanaceae should have less diverse chemical compounds, or quantitative defenses, against monophagous insects, while unpredictable species should have qualitative defenses, a higher spectrum of alkaloids or oils, that protect the plants against generalists.

The many life forms of Solanaceae such as trees, shrubs, herbs and lianas, and perennial or annual plants, should represent strategies which evolved as a result of competition among plants, adaptations to the physical environment, and defense against herbivores caused by phytophagous insects, as well as adaptations to mutualists such as pollinators and seed dispersers.

Solanaceae with more spaced distributions tend to have more extensive periods of flowering and fruiting and produce fewer flowers and fruit each day. The flowers are visited by large bees (bumblebees) and species of Halictidae; and the fruits of several species, mainly *Solanum*, are dispersed by smaller phyllostomid bats, like *Carollia perspicillata*, *Sturnira lillium*, and *S. tildae* (Ueda and Vasconcellos-Neto 1985, unpublished observations).

The production of fewer flowers and fruits is adaptive, permitting outcrossing among scattered plants, and the seeds should have higher probability of dropping in an appropriate habitat patchily distributed in the forest (Humphrey and Bonaccorso 1979). Plants that are more predictable and occur at greater densities, such as *Cestrum*, tend to have bursts of flowering and fruiting. *Solanum* aff. *inaequale* is a common tree with a 12-m height in the Japi forest. It has similar reproductive tactics to those of *Centrum*, being dispersed by *Artibeus lituratus*, a big phyllostomid bat, which searches for concentrated resources. On the other hand, *Carollia* and *Sturnira* seem to have a trap-line behavior when feeding on *Piper* and *Solanum* species (see Heithaus et al. 1975; Heithaus and Fleming 1978; Humphrey and Bonaccorso 1979; Heithaus 1982; Fleming 1982; Sazima and Sazima 1975; Marinho-Filho 1985).

The accumulation of alkaloids in more exposed aerial organs, such as young leaves, flowers, and fruits, may have evolved primarily in response to phytophagous insects. Ripening fruits of *Solanum* and *Lycopersicon* have glycoalkaloids which could be utilized in pigment biosynthesis, suggesting that these compounds may play a role not only in deterrence but also indirectly in the attraction of vectors of seed dispersal (Roddick 1986).

Natural Enemies

Immature stages of ithomiine butterflies are attacked by wasps, ants, bugs (personal observation), and some parasitoids (Monteiro 1981). Although the natural impact among the three trophic levels is little understood, there is some evidence that predictability of plants and herbivores could pressure parasitoids in a similar way, as discussed earlier.

Mcclungia salonina was the species with the highest level of parasitism in the Horto Florestal. Its parasitoid, *Apanteles concinnus*, rarely occurred in other ithomiine larvae (Monteiro 1981). Specialization of this wasp could be favored by plant abundance and host predictability.

The flies *Euexorista brasiliensis* and *Lespesia capitis* occur mainly on *Mechanitis* in Sumaré (Monteiro 1981). *E. brasiliensis* was recorded from pupae of *Danaus erippus*, *D. plexippus* (Danainae), and *Lanomia falcata submaculata* (Hemileucidae) (Guimarães 1977; Dias 1978). These tachinid flies seem to have a wider host range than *A. concinnus*, perhaps related to scattered host insects.

Host range for parasitoids could be affected by many factors, such as secondary compounds incorporated by the larvae from the host plant (Askew 1968; Vinson 1976).

THE MODEL OF THREE TROPHIC INTERACTIONS

Environments are patchy. Factors influencing the fitness of individuals exhibit discontinuities on many scales in time and space. The patterns of these discontinuities produce an environmental patchwork which exerts powerful influences on the distributions of organisms, their interactions, and their adaptations (Wiens 1976).

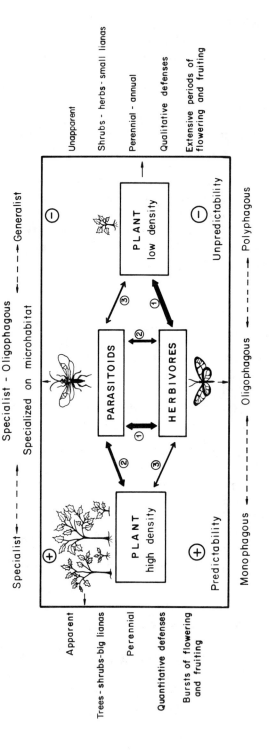

Figure 14-8. Model of interaction between three trophic levels: plant–herbivore–parasitoid. The width of the arrow (1, 2 or 3) indicates the order of importance of the mutual selective pressures. On the left the host plant is predictable (+) and on the right it is unpredictable (−). The feeding and reproduction strategies of herbivores (below) and of parasitoids (above) are shown outside the rectangle, and related to food predictability. Some features of the plant's natural history are mentioned at each side (right and left).

Mutual interactions among the three trophic levels exert selective pressures generating different life-styles. Plants, herbivores, and natural enemies occur in patches of different sizes, from aggregated to scattered, inside communities. Figure 14-8 shows the extremes of host-plant predictability and its consequences on herbivores and their natural enemies (parasitoids). When the host plant is common, the mutual pressure between herbivores and parasitoids is stronger than that between plant and parasitoids, and finally plant and herbivores. As the herbivore is also predictable, parasitism greatly affects its population, and natural selection favors solitary eggs as a defense strategy against enemies. A few solitary eggs on a host plant represents little load, so the mutual impact is weak. Consequently, the importance of mutual interaction between plant and parasitoids comes second, forming part of the plant defense battery, and encouraging host specialization.

On the other hand, rare plants could be interpreted as a defense against herbivory, or a consequence of it. In this case, the main mutual selective pressure is between the plant and its herbivores, with less interactions between herbivores and parasitoids. With the unpredictability of the plant and consequently of the herbivore, parasitoids tend to become specialized on a microhabitat extending their host range. Here the interaction between plant and parasitoid should have the least importance.

The abundance and distribution patterns of plants allied with their chemical and mechanical defenses constitute strategies to escape from herbivory. Other attributes, such as life forms and life cycle including reproductive tactics, also evolved as a result of these interactions. The selective pressures exerted on phytophagous insects by their natural enemies and host-plant predictability will determine the feeding and reproductive strategies of these insects; and this also valid for the natural enemies. The life styles of plant, herbivore, and parasitoids probably evolved principally as a result of mutual selective pressures, but other pressures present in the community should be considered.

ACKNOWLEDGMENTS

I am grateful to Drs. Keith S. Brown Jr, Douglas J. Futuyma, Robert J. Marquis, and Thomas M. Lewinsohn for reading the manuscript and offering useful suggestions, and to Esmeralda Z. Borghi for the drawings. The Fundacão de Apoio à Pesquisa do Estado de São Paulo provided a graduate scholarship during part of the field work.

REFERENCES

D'Arcy, W. G. 1973. Family 170. Solanaceae. *Ann. Missouri Bot. Gard.* **60**: 573–780.

Askew, R. R. 1968. Considerations on speciation in Chalcidoidea (Hymenoptera). *Evolution* **22**: 642–645.

Barbosa, P. 1988. Some thoughts on the evolution of host range. *Ecology* **69**: 912–915.

Benson, W. W. 1978. Resource partitioning in passion vine butterflies. *Evolution* **32**: 493–518.

Benson, W. W., K. S. Brown, Jr., and L. E. Gilbert. 1976. Coevolution of plants and herbivores: Passion flower butterflies. *Evolution* **29**: 659–680.

Bernays, E. and M. Graham. 1988. On the evolution of host specificity in phytophagous arthropods. *Ecology* **69**: 886–892.

Brower, L. P. and J. V. Z. Brower. 1964. Birds, butterflies and plant poisons: A study in ecological chemistry. *Zoologica* (New York) **49**: 137–159.

Brown, K. S., Jr. 1979. Ecologia geográfica e evolução nas florestas Neotropicais. L. D. Thesis, Universidade Estadual de Campinas, SP, Brazil.

Brown, K. S., Jr. 1980. A review of the genus *Hypothyris* Hübner (Nymphalidae), with descriptions of three new subspecies and early stage of *H. daphnis. J. Lep. Soc.* **34**: 152–172.

Brown, K. S., Jr. 1984. Adult-obtained pyrrolizidine alkaloids defend ithomiine butterflies against a spider predator. *Nature* **309**: 707–709.

Brown, K. S., Jr. 1985. Chemical ecology of dehydropyrrolizidine alkaloids in adult Ithomiinae (Lepidoptera, Nymphalidae). *Revta. bras. Biol.* **44**: 435–460.

Brown, K. S., Jr. 1987. Chemistry at the Solanaceae/Ithomiinae interface. *Ann. Missouri Bot. Gard.* **74**: 359–397.

Brown, K. S., Jr. and W. W. Benson. 1974. Adaptive polymorphism associated with multiple Müllerian mimicry in *Heliconius numata* (Lepidoptera, Nymphalidae). *Biotropica* **6**: 205–228.

Brown, K. S., Jr. and W. W. Benson. 1975. The heliconians of Brazil (Lepidoptera: Nymphalidae). Part IV. Aspects of the biology and ecology of *Heliconius demeter*, with description of four new subspecies. *Bull. Allyn. Mus.* (Sarasota) **26**: 1–19.

Brown, K. S., Jr. and J. Vasconcellos-Neto. 1976. Predation on aposematic ithomiinae butterflies by tanagers (*Pipraeidae melanonota*). *Biotropica* **8**: 136–141.

Cates, R. G. 1981. Host plant predictability and the feeding patterns of monophagous, oligophagous, and polyphagous insect herbivores. *Oecologia* **48**: 319–326.

Cody, M. L. 1966. A general theory of clutch size. *Evolution* **20**: 174–184.

Courtney, S. P. 1984. The evolution of batch oviposition by Lepidoptera and other insects. *Amer. Nat.* **123**: 276–281.

Courtney, S. P. 1988. If it's not coevolution, it must be predation? *Ecology* **69**: 910–911.

Crawley, M. J. 1983. *Herbivory.* Blackwell Scientific Publications, Oxford.

Dethier, V. G. 1970. Chemical interactions between plants and insects. In E. Sondheimer, and J. B. Simeone (Eds.). *Chemical ecology.* Academic Press, New York, London, pp. 83–103.

Dias, M. M., Filho. 1978. Parasitismo de *Danaus plexippus erippus* (Lep., Danaidae) por "*Masicera*" *brasiliensis* (Diptera, Tachinidae) e dercrição da larva e pupário do parasita. *Revta. bras. Entomol.* **22**: 5–10.

Drummond, B. A., III. 1976. Comparative ecology and mimetic relationships of ithomiine butterflies in eastern Ecuador. Ph.D. diss., University of Florida, Gainesville.

Drummond, B. A., III. 1986. Coevolution of ithomiine butterflies and solanaceous plants. In W. G. D'Arcy. (Ed.). *Solanaceae: biology and systematics.* Columbia University Press, New York, pp. 307–327.

Drummond, B. A., III. and K. S. Brown, Jr. 1986. Ithomiinae (Lepidoptera: Nymphalidae): summary of known larval food plants. *Ann. Missouri Bot. Gard.* **74**: 341–358.

Ehrlich, P. R. and D. D. Murphy. 1988. Plant chemistry and host range in insect herbivores. *Ecology* **69**: 908–909.

Emlen, J. M. 1973. *Ecology: An evolutionary approach.* Addision-Wesley, Reading, MA.

Evans, W. C. 1986. Hybridization and secondary metabolism in the Solanaceae. In W. G. D'Arcy. (Ed.). *Solanaceae: Biology and systematics.* Columbia University Press, New York. pp. 179–186.

Feeny, P. 1975. Biochemical coevolution between plants and their insect herbivores. In L. E. Gilbert and P. H. Raven (Eds.). *Coevolution between plants and insects.* pp. 3–19. University of Texas Press, Austin, London.

Feeny, P. 1976. Plant apparency and chemical defense. In J. W. Wallace and R. L. Mansell (Eds.), *Biochemical interactions between plants and insects.* Plenum Press, New York, London. pp. 1–40.

Fleming, T. H. 1982. Foraging strategies of plant-visiting bats. In T. H. Kunz (Ed.). *Ecology of bats.* Plenum Press, New York. pp. 287–325.

Foster, M. S. 1974. Rain, feeding behavior and clutch size in tropical birds. *Auk* **91**: 722–726.

Fox, L. R. 1988. Diffuse coevolution within complex communities. *Ecology* **69**: 906–907.

Fox, R. M. 1963. Affinities and distribution of Antillean Ithomiinae. *J. Res. Lep.* **2**: 173–184.

Futuyma, D. J. and S. C. Peterson. 1985. Genetic variation in the use of resources by insects. *Ann. Rev. Entomol.* **30**: 217–238.

Gilbert, L. E. 1969. Some aspects of the ecology and community structure of ithomiid butterflies in Costa Rica. In Advanced Population Biology individual research reports, July–August, pp. 69–93. Organization for Tropical Studies, Ciudad Universitaria San Jose, Costa Rica.

Gilbert, L. E. 1975. Ecological consequences of a coevolved mutualism between butterflies and plants. In L. E. Gilbert and P. R. Raven (Eds.). *Coevolution of animals and plants.* University of Texas Press, Austin, TX.

Guimarães, J. H. 1977. Host–parasite and parasite–host catalogue of South American Tachinidae (Diptera). *Arq. Zool.* (São Paulo) **28**: 1–131.

Haber, W. A. 1978. Evolutionary ecology of tropical mimetic butterflies (Lepidoptera: Ithomiinae). Ph.D. diss., University of Minnesota, Minneapolis.

Harborne, J. B. 1986. Systematic significance of variations in defense chemistry in the Solanaceae. In W. G. D'Arcy (Ed.). *Solanaceae: biology and systematics.* Columbia University Press, New York. pp. 328–344.

Heithaus, E. R. 1982. Coevolution between bats and plants. In T. H. Kunz (Ed.). *Ecology of bats.* Plenum Press, New York. pp. 327–367.

Heithaus, E. R. and T. H. Fleming. 1978. Foraging movement of a frugivorous bat, *Carollia perspicillata* (Phyllostomatidae). *Ecol. Monographs* **48**: 127–143.

Heithaus, E. R., T. H. Fleming, and P. A. Opler. 1975. Foraging patterns and resource utilization in seven species of bats in a seasonal tropical forest. *Ecology* **56**: 841–854.

Humphrey, S. R. and F. J. Bonaccorso. 1979. Population and community ecology. In R. J. Barker, J. K. Jones, Jr., and D. C. Carter (Eds.). *Biology of bats of the New World family Phyllostomidae.* Part III. Spec. Public. Mus. Texas Tech. Univ. v. 16. pp. 1–441.

Ito, Y., Y. Tsubaki, and M. Osada. 1982. Why do *Luehdorfia* butterflies lay eggs in clusters? *Res. Popul. Ecol.* **24**: 375–387.

Janzen, D. H. 1970. Herbivores and the number of tree species in tropical forests. *Amer. Nat.* **104**: 501–528.

Janzen, D. H. 1988. On the broadening of insect–plant research. *Ecology* **69**: 905.

Jermy, T. 1984. Evolution of insect/host plant relationships. *Amer. Nat.* **124**: 609–630.

Jermy, T. 1988. Can predation lead to narrow food specialization in phytophagous insects? *Ecology* **69**: 902–904.

Lack, D. 1947. The significance of clutch size. Parts 1 and 2. *Ibis* **89**: 302–352.

Lack, D. 1948. The significance of clutch size. Part 3. *Ibis* **90**: 25–45.

Lack, D. 1949. Comments on Mr. Skutch's paper on clutch size. *Ibis* **91**: 455–458.

Lavie, D. 1986. The withanolides as a model in plant genetics: Chemistry, biosynthesis and distribution. In W. G. D'Arcy (Ed.). *Solanaceae: biology and systematics.* Columbia University Press, New York.

Lawrence, G. H. M. 1951. *Taxonomy of vascular plants.* MacMillan, New York.

Lawton, J. H. 1986. The effect of parasitoids on phytophagous insect communities. In *Insect parasitoids.* J. Waage and D. Greathead (Eds.). Academic Press, London.

Lawton, J. H. and S. McNeill. 1979. Between the devil and the deep blue sea: On the problem of being a herbivore. In R. M. Anderson, B. D. Turner, and L. R. Taylor, (Eds.). *Population Dynamics.* pp. 223–244. Blackwell, Oxford.

Levins, R. and R. MacArthur. 1969. An hypothesis to explain the incidence of monophagy. *Ecology* **50**: 910–911.

Marinho-Filho, J. S. 1985. Padrões de atividade e utilização de recursos alimentares por seis espécies de morcegos filostomídeos na Serra do Japi, Jundiaí—SP. M.S. thesis, Universidade Estadual de Campinas, SP, Brazil. 78 pp.

McNeill, S. and T. R. E. Southwood. 1978. The role of nitrogen in the development of insect–plant relationships. In J. B. Harborne. *Biochemical aspects of plant and animal coevolution.* pp. 77–98. Academic Press, London.

Monteiro, R. F. 1981. Regulação populacional em Ithomiinae (Lep.: Nymphalidae): Ecologia da interação parasitóide vs. hospedeiro. M.Sc. thesis, Universidade Estadual de Campinas, Campinas.

Nimer, E. 1977. Clima. In *Geografia de região sudeste.* v. 3. pp. 51–89. IBGE. 667 pp.

Price, P. W. 1974. Strategies for egg production. *Evolution* **28**: 76–84.

Price, P. W., C. E. Bouton, P. Gross, B. A. McPheron, J. N. Thompson, and A. E. Weis. 1980. Interactions among three trophic levels: Influence of plants on interactions between insect herbivores and natural enemies. *Ann. Rev. Ecol. Syst.* **11**: 41–65.

Rausher, M. D. 1978. Search image for leaf shape in a butterfly. *Science* **200**: 1071–1073.

Rausher, M. D. 1988. Is coevolution dead? *Ecology* **69**: 898–901.

Rhoades, F. D. and R. G. Cates. 1976. Toward a general theory of plant antiherbivore chemistry. In J. W. Wallace, and R. L. Mansell (Eds.). *Biochemical interaction between plants and insects.* pp. 168–213. Plenum Press, New York, London.

Roddick, J. G. 1986. Steroidal alkaloids of the Solanaceae. In W. G. D'Arcy (Ed.). *Solanaceae: biology and systematics.* Columbia University Press, New York.

Sazima, M. and I. Sazima. 1975. Quiropterofilia em *Lafoensia pacari* St. Hil (Lythraceae), na Serra do Cipó, Minas Gerais. *Cienc. e Cult.* **27**(4): 405–416.

Shultz, J. C. 1988. Many factors influence the evolution of herbivore diets, but plant chemistry is central. *Ecology* **69**: 896–897.

Smiley, J. 1978. Plant chemistry and the evolution of host specificity: New evidence from *Heliconius* and *Passiflora. Science* **201**: 745–747.

Smiley, J. T. and C. W. Wisdom. 1985. Determinants of growth rate on chemically heterogeneous host plants by specialist insects. *Biochemical Systematics and Ecology* **13**: 305–312.

Smith, L. B. and R. Downs. 1966. Solanaceae. In *Flora ilustrada catarinense. Fasc. Sola.* pp. 1–321.

Stamp, N. E. 1980. Egg deposition patterns in butterflies: Why do some species cluster their eggs rather than lay them singly? *Amer. Nat.* **115**: 367–380.

Thompson, J. N. 1988. Coevolution and alternative hypotheses on insect/plant interactions. *Ecology* **69**: 893–895.

Timbergen, N., M. Impekoven, and D. Franck. 1967. An experiment on spacing-out as a defense against predation. *Behaviour* **28**: 307–321.

Uieda, W. and J. Vasconcellos-Neto. 1985. Dispersão de *Solanum* spp. (Solanaceae) por morcegos na região de Manaus-AM. *Revta. bras. Zool.* **2**: 449–458.

Vasconcellos-Neto, J. 1980. Dinâmica de populações de Ithomiinae (Lepidoptera, Nymphalidae) em Sumaré—SP. M.Sc. Thesis, Universidade Estadual de Campinas, SP. 264 pp.

Vasconcellos-Neto, J. 1986. Interactions between Ithomiinae (Lep., Nymphalidae) and Solanaceae. In W. G. D'Arcy (Ed.). *Solanaceae: Biology and systematics.* pp. 366–377.

Vasconcellos-Neto, J. and T. M. Lewinsohn. 1984. Discrimination of unpalatable butterflies by *Nephila clavipes*, a neotropical orb-weaving spider. *Ecol. Ent.* **9**: 337–344.

Vinson, S. B. 1976. Host selection by insect parasitoids. *Ann. Rev. Entomol.* **21**: 109–133.

Weins, J. A. 1976. Population responses to patchy environments. *Ann. Rev. Ecol. Syst.* **7**: 81–120.

Young, A. M. and M. W. Moffett. 1979. Studies on the population biology of the tropical butterfly *Mechanitis isthmia* in Costa Rica. *Amer. Mid. Nat.*, **101**: 309–319.

15 Chemical Constraints on the Evolution of Swallowtail Butterflies

PAUL FEENY

ECOLOGICAL OPPORTUNISM AND HISTORICAL CONSTRAINT

Phytophagous insects have been great opportunists, colonizing virtually all species of land plants and giving rise to at least one quarter of the Earth's eukaryotic organisms (Strong et al. 1984). The number of insect species associated with a particular plant species can be predicted rather well in terms of the plant's geographical range, historical age, and growth form or "architecture"—patterns attributed chiefly to the greater ecological opportunities that larger, older, and more widespread plants have provided for colonization (Strong et al. 1984).

The swallowtail butterflies (family Papilionidae) comprise about 560 species, recently rearranged by Miller (1987a) into seven tribes and 24 genera (Fig. 15-1). In many ways, these insects have been exemplary opportunists. Though primarily associated with tropical latitudes (Slansky 1972; Scriber 1973, 1984), they have colonized all the habitable continents and most terrestrial biomes, from rain forests and hot deserts to arctic and alpine tundra. Their larvae feed, among them, on plants of widely differing growth form, from ephemeral herbs to climax forest trees, and swallowtails have colonized as many as half of the orders of dicotyledonous plants, belonging to all six subclasses (Fig. 15-2).

But in the Papilionidae, as in other phyletic lineages, ecological opportunism has been subject to constraint by past evolutionary history (cf. Cody 1974; Jacob, 1977). Though the swallowtails are reported to feed on plants of 50 families (Scriber 1984), this represents only 17% of the families of dicotyledons (Cronquist 1981). Moreover, disproportionately high percentages of families in the subclasses Magnoliidae (31%) and Asteridae (26%) are reported as larval food plants, while few families in the Caryophyllidae (7%) and Dilleniidae (9%) have apparently been colonized. Only five families, namely the Aristolochiaceae,

Plant-Animal Interactions: Evolutionary Ecology in Tropical and Temperate Regions, Edited by Peter W. Price, Thomas M. Lewinsohn, G. Wilson Fernandes, and Woodruff W. Benson. ISBN 0-471-50937-X © 1991 John Wiley & Sons, Inc.

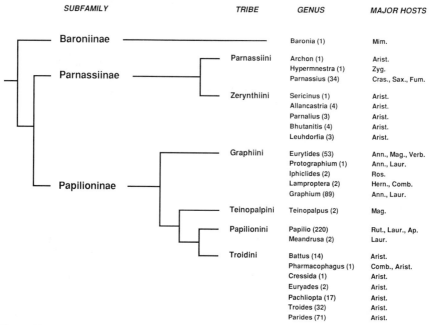

	SUBFAMILY	TRIBE	GENUS	MAJOR HOSTS
	Baroniinae		Baronia (1)	Mim.
	Parnassiinae	Parnassiini	Archon (1)	Arist.
			Hypermnestra (1)	Zyg.
			Parnassius (34)	Cras., Sax., Fum.
		Zerynthiini	Sericinus (1)	Arist.
			Allancastria (4)	Arist.
			Parnalius (3)	Arist.
			Bhutanitis (4)	Arist.
			Leuhdorfia (3)	Arist.
	Papilioninae	Graphiini	Eurytides (53)	Ann., Mag., Verb.
			Protographium (1)	Ann., Laur.
			Iphiclides (2)	Ros.
			Lamproptera (2)	Hern., Comb.
			Graphium (89)	Ann., Laur.
		Teinopalpini	Teinopalpus (2)	Mag.
		Papilionini	Papilio (220)	Rut., Laur., Ap.
			Meandrusa (2)	Laur.
		Troidini	Battus (14)	Arist.
			Pharmacophagus (1)	Comb., Arist.
			Cressida (1)	Arist.
			Euryades (2)	Arist.
			Pachliopta (17)	Arist.
			Troides (32)	Arist.
			Parides (71)	Arist.

Figure 15-1. Cladogram for the tribes of the family Papilionidae, listing genera in each tribe, numbers of species in each genus (in parentheses), and major larval host plants of each genus. Mim. = Mimosaceae, Arist. = Aristolochiaceae, Zyg. = Zygophyllaceae, Cras. = Crassulaceae, Sax. = Saxifragaceae, Fum. = Fumariaceae, Mag. = Magnoliaceae, Verb. = Verbenaceae, Ann. = Annonaceae, Laur. = Lauraceae, Ros. = Rosaceae, Hern. = Hernandiaceae, Comb. = Combretaceae, Rut. = Rutaceae, Ap. = Apiaceae. Data from Miller (1987a, 1987b), Scriber (1984), and Igarashi (1987).

Annonaceae, Lauraceae, Apiaceae, and Rutaceae, account for more than two-thirds of the host records listed by Scriber (1984). Swallowtails are conspicuously absent from many plant families that contain species of great range, age, and individual size (cf. Ehrlich and Raven 1964, p. 601)—families that "should" be colonized according to the ecological rules of the theory of island biogeography. Though the swallowtails have been in existence for at least 48 million years (Miller 1987b), perhaps this has not been long enough for more widespread colonization. Or perhaps swallowtails have been excluded from many plants by particular combinations of ecological factors (cf. Gilbert 1979; Smiley 1985). By no means incompatible with these hypotheses, however, is the likelihood that the extent and direction of swallowtail evolution has been constrained by long-standing relationships to plant chemistry—relationships that favor the coloniz-ation of some plants over others.

That plant chemistry has somehow served as a constraint on the evolution of swallowtails is suggested by the observation that subgroups of the larval food plants are linked with one another by their common content of various classes of

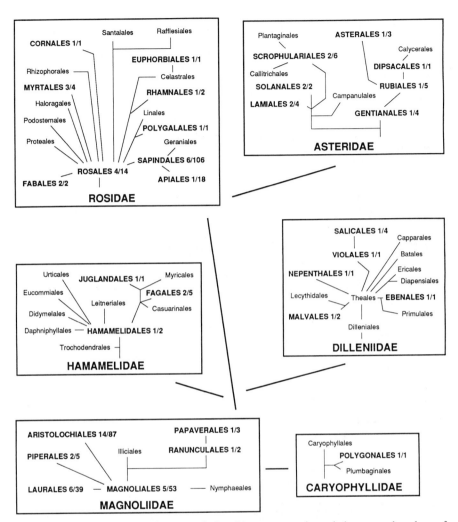

Figure 15-2. Putative evolutionary relationships among the subclasses and orders of dicotyledons (Cronquist 1981), showing orders containing swallowtail host plants (boldface capitals) and the numbers of genera and species of swallowtails reported from each such order. Data from Scriber (1984), Igarashi (1987), and Miller (1987b).

secondary compounds, including coumarins, benzylisoquinoline alkaloids, and essential oils (Ehrlich and Raven 1964; Feeny et al. 1983; Berenbaum 1983). Because the host plants themselves may not be close taxonomic relatives of one another, such patterns cannot be explained simply by parallel cladogenesis (Miller 1987b). Rather, they indicate that adaptation by an insect to particular secondary compounds in its current host–plant species constrains colonization of other plants to species containing similar compounds. The nature of such constraint has been the subject of several hypotheses, which are discussed below.

CHEMICAL CONSTRAINT: A REVIEW OF HYPOTHESES

Behavioral Facilitation

Dethier (1941) and Jermy (1976, 1984) have proposed that the chemical similarities often seen among the host plants of related insects result from preadaptation by the insects in their responses to behavioral cues: colonization of novel host plants by an insect population will be more likely if such hosts contain stimulants and/or lack deterrents that the insects already use as cues for host recognition.

The hypothesis was advanced by Dethier (1941) on the basis of experiments showing that larvae of the black swallowtail *Papilio polyxenes* are attracted by several essential oil components present in their food plants, which belong primarily to the family Apiaceae (Umbelliferae). Since the larvae will also feed on certain plants in the Rutaceae, the primary host family of most *Papilio* species, and since several of the attractant compounds also occur in this family, Dethier suggested that "the transition from one plant family to the other took place because of the presence of identical attractant chemicals in both families" (Dethier 1941, p. 72). Saxena and Prabha (1975) later showed that larvae of *Papilio demoleus* are attracted to essential oil compounds present in their rutaceous hosts, thus providing further support for Dethier's hypothesis.

More recently, attention has shifted to the role of chemistry in oviposition behavior by swallowtails. In most species, the initial larval food plant is determined by the ovipositing female. A female predisposed to lay eggs on an atypical plant may expose dozens or even hundreds of larvae to a novel host, thus increasing the chances for expression of innovative larval feeding behavior. There are numerous reports of female swallowtails laying eggs on host plants typically used by other swallowtail species or even other swallowtail tribes (Feeny et al. 1983).

Both visual and volatile chemical cues influence the decision of female swallowtails to land on potential host plants (Rausher 1978; Saxena and Goyal 1978; Feeny et al. 1983, 1989). Once they have landed on a plant, however, females decide whether or not to oviposit primarily on the basis of stimulants and deterrents perceived by contact chemoreceptors located on their tarsi (Feeny et al. 1983). The contact stimulants are generally polar and females of several swallowtail species can be stimulated to curl their abdomens and lay eggs on filter paper treated with aqueous fractions of host-plant extracts (Nishida 1977; Saxena and Goyal 1978; Ichinosé and Honda 1978; Feeny et al. 1983). If host shifts by insects have been catalyzed by responses to behavioral stimulants, then at least some of the compounds shared by the host plants of related insects should be among those that are used by the insects for host recognition. A preliminary assessment is now possible on the basis of contact stimulants identified recently for three *Papilio* species (tribe Papilionini) and two species in the tribe Troidini.

From methanolic extracts of one of its host plants, *Citrus unshiu*, Nishida and colleagues (Ohsugi et al. 1985; Nishida et al. 1987) have identified several oviposition stimulants for *Papilio xuthus*. Four of the active compounds, identified as vicenin-2, hesperidin, narirutin, and rutin, are flavonoids. These were only weakly active on their own or as a flavonoid mixture. They evoked responses of 100%, however, when mixed with two bases (adenosine and 5-hydroxy-Nω-methyltryptamine), which were likewise almost inactive by themselves (Nishida et al. 1987). T. Ohsugi, R. Nishida, and H. Fukami (personal communication) have recently isolated four further stimulants, namely bufotenine, stachydrine, (−)-synephrine and (+)-*chiro*-inositol. The complete mixture accounts for the activity of *C. unshiu* extracts to *P. xuthus* females.

From the epicarp of *Citrus natsudaidai*, Honda (1986) isolated two flavonoid glycosides, hesperidin and naringin, that stimulated oviposition by *Papilio protenor*, another Rutaceae-feeding swallowtail. The compounds were inactive alone, but were active when combined with a more polar subfraction of the parent extract. The butterflies responded only to flavanones tested, and did not respond to flavones or flavonols or their glycosides (including rutin). More recently, K. Honda (personal communication) has isolated further active compounds, namely L-(−)-proline, L-(−)-stachydrine, (−)-quinic acid, (−)-synephrine, and chlorogenic acid.

Parallel work in my laboratory has revealed that the aqueous phase from extracts of carrot, *Daucus carota*, can be separated into several fractions that stimulate oviposition by *P. polyxenes* females. Two of the stimulants have been identified as *trans*-chlorogenic acid and luteolin 7-O-(6″-O-malonyl)-β-D-glucoside, a labile compound previously unknown from higher plants. These two compounds were inactive alone, but in combination accounted for about 70% of the response to the parent extract (Feeny et al. 1988). Current studies indicate that the females are stimulated by additional compounds, including at least two bases and a sugar or inositol.

The response patterns of three different *Papilio* species, two of them restricted to the Rutaceae and the other primarily to the Apiaceae, are sufficiently similar to provide tentative support for the behavioral-facilitation hypothesis. Though the stimulants identified so far do not belong to classes expected from earlier phytochemical surveys of host plants (Ehrlich and Raven 1964; Berenbaum 1983; Feeny et al. 1983) they indicate an underlying conservatism in oviposition response to mixtures containing similar classes of ingredients (Table 15-1). Superimposed on the basic pattern, however, are apparent differences in specificity to particular compounds within the active classes. These differences may represent adaptations for more accurate recognition of the particular sets of host plants used by each species in the field.

The importance of flavone glycosides as oviposition stimulants for *Papilio* species, incidentally, may help to resolve the minor mysteries of why swallowtails have not colonized the Araliaceae, a family so closely related to the Apiaceae that its lack of colonization by swallowtails puzzled Ehrlich and Raven (1964, p. 600), and why they attack only those umbellifers that belong to the subfamily Apioideae. Both the family Araliaceae and the subfamilies Saniculoideae and Hydrocotyloideae (the other two subfamilies of the Apiaceae) are apparently

TABLE 15-1. Summary of Known Contact Oviposition Stimulants for Swallowtails[a]

Species[b]	Flavonoids	Bases	"Cyclitols"	Other Compounds
Tribe Papilionini				
Papilio xuthus (Rutaceae)	Vicenin-2 Narirutin Hesperidin Rutin	5-Hydroxy-Nω-methyltryptamine Adenosine Synephrine Bufotenine	*chiro*-Inositol	Stachydrine
Papilio protenor (Rutaceae)	Naringin Hesperidin	Synephrine Proline	Quinic acid	Chlorogenic acid Stachydrine
Papilio polyxenes (Apiaceae)	Luteolin 7-O-(6″-O-malonyl)-β-D-glucoside			Chlorogenic acid
Tribe Troidini				
Parides alcinous (Aristolochiaceae)			Sequoyitol	Aristolochic acids
Battus philenor (Aristolochiaceae)			Pinitol	Aristolochic acids

[a]Data from Ohsugi et al. (1985), Honda (1986 and personal communication), Nishida et al. (1987 and personal communication), and Feeny et al. (1983, 1988 and unpublished results).
[b]Major host-plant family listed beneath each species.

devoid of flavones (Crowden et al. 1969; Harborne and Williams 1972; Berenbaum 1983). A major host of Costa Rican populations of *P. polyxenes* is *Spananthe paniculata*, which is currently classified, somewhat uneasily, in the subfamily Hydrocotyloideae (Heywood 1971). From a swallowtail's perspective, however, this species is an apioid umbellifer (Blau and Feeny 1983) and we can predict that it contains flavone glycosides.

Among troidine swallowtails, the major contact oviposition stimulants for *Parides* (= *Atrophaneura*) *alcinous*, isolated from *Aristolochia debilis*, have been identified as sequoyitol (a methyl inositol) and a mixture of aristolochic acids (R. Nishida and H. Fukami, personal communication). From the foliage of *A. macrophylla* we have identified the major stimulants for *Battus philenor* as D-(+)-pinitol (another methyl inositol) and aristolochic acids I and II (D. Papaj, K. Sachdev, P. Feeny, and L. Rosenberry, unpublished results). In both studies, stimulant activity depends on synergism between the inositol and aristolochic acids, none of which were appreciably active alone.

Though belonging to different genera, both troidine species responded to stimulant profiles that are fundamentally similar, again indicating conservatism of oviposition response within tribes (Table 15-1). Though inositols occur widely in plants, and are the one class of stimulants so far known to be shared with the Papilionini, aristolochic acids are restricted to the family Aristolochiaceae (Chen and Zhu 1987). Dependence of troidine swallowtails on these compounds as oviposition cues seems one likely explanation for the rarity with which troidine swallowtails have been reported from other plant families (cf. Miller 1987b).

Metabolic Preadaptation

Though oviposition "mistakes" by swallowtails have been reported frequently, and must therefore be common in nature, shifts to new hosts are unusual. This suggests that behavioral experimentation by adults or larvae, though necessary for initiation of host shifts, is not often sufficient to effect permanent colonization of novel host plants.

Fraenkel (1959) and Ehrlich and Raven (1964) suggested that many secondary compounds evolved as plant defenses and that these have been overcome variously by insects in evolutionary time. Tolerance of the toxic or growth-inhibitory properties of a compound would facilitate colonization of any plants containing it, regardless of botanical affinity, and thus bias host shifts by insects to chemically-similar plants (Ehrlich and Raven 1964). Both Fraenkel (1959) and Ehrlich and Raven (1964) included deterrents as well as toxins in their concept of plant defenses and their scenario thus includes not only metabolic preadaptation but also components of the behavioral-facilitation hypothesis.

To test the idea that plant compounds may have played a role as metabolic barriers in shaping the present-day host distributions of the swallowtails, we have conducted experiments first to determine if plant compounds from nonhosts do in fact represent metabolic (and not merely behavioral) barriers to swallowtails (cf. Bernays and Chapman 1987). Second, we have tested various secondary compounds typical of swallowtail food-plants to assess their potential toxicity to other lepidopterous insects, not associated with the plants in nature. Third, we have tested these compounds with swallowtails to assess their abilities to tolerate not only the compounds typical of their own hosts but also those of hosts of other swallowtails.

In early experiments, we found that larvae of *P. polyxenes* could not survive on carrot or celery (*Apium graveolens*) foliage that had been cultured with doses of allylglucosinolate found typically in cruciferous plants and that the effects could be attributed to toxicity rather than merely to feeding inhibition (Erickson and Feeny 1974; Blau et al. 1978). This result provided support for the hypothesis that nonhosts contain chemical barriers to swallowtails.

Among the many possible chemical barriers in swallowtail host plants themselves, we have so far studied certain furanocoumarins, aristolochic acids, and alkaloids. Berenbaum (1978, 1981a) showed that xanthotoxin, a linear furanocoumarin, is highly toxic to larvae of the southern armyworm *Spodoptera eridania*, but has no ill effects on growth or fecundity of *P. polyxenes*, larvae of which feed typically on plants containing linear furanocoumarins. Angelicin, one of the rarer angular furanocoumarins from certain more advanced genera of the Apiaceae, affected the growth of *P. polyxenes* larvae only slightly, but reduced dramatically the fecundity of female adults (Berenbaum and Feeny 1981). Berenbaum and Neal (1985) showed subsequently that toxicity of xanthotoxin to larvae of the corn earworm *Heliothis zea* was augmented considerably in the presence of myristicin, an inhibitor of insect detoxication enzymes and a compound that often occurs together with furanocoumarins in plants.

Aristolochic acids and several alkaloids, known to occur in swallowtail food plants, proved to be deterrent and/or toxic to larvae of *S. eridania*, *Lymantria dispar*, and *Hyphantria cunea*, all generalist lepidopteran species that do not normally feed on such plants. Of these compounds, aristolochic acids were particularly toxic to all three species (Rausher 1979a; Miller and Feeny 1983). Swallowtail larvae of several species were found to tolerate the alkaloids or aristolochic acids from their own host plants with no ill effects but showed varying degrees of sensitivity to compounds more typical of host plants attacked by other swallowtails (Miller and Feeny 1989). In general, patterns of toxicity of both furanocoumarins and alkaloids to swallowtails are consistent with sequential adaptation to increasingly elaborate structures during evolution (Berenbaum 1983; Miller and Feeny 1983), though the relationships are not clear cut.

Recently, Lindroth et al. (1988) have shown that the phenolic glycosides salicortin and tremulacin from quaking aspen, *Populus tremuloides*, significantly reduce the growth of larvae of the southern subspecies of *Papilio glaucus* (*P. g. glaucus*) but have no such effect on larvae of the northern subspecies *P. g. canadensis*. The effect appears to be due in part to toxicity, rather than simply to feeding inhibition, and suggests that colonization of the Salicaceae by *P. g. canadensis* has required adaptation for overcoming a metabolic barrier.

Futuyma (1983) has argued that the initial shifts by phytophagous insects to new hosts often result only from changes in behavior and that adaptation to specific plant toxins may require little, if any, genetic change. Any such change at the metabolic level is "not immediately necessary for successful change of host but is a fine tuning of adaptation that occurs only after the species has already become specialized on a particular host" (Futuyma 1983, p. 225). Such a scenario may well have characterized many host shifts by swallowtails, especially as some swallowtail larvae are capable of growing on a wider range of plants than is used for oviposition by females of the same species (Stride and Straatman 1962; Wiklund 1975). But the results referred to above, together with field records of swallowtails ovipositing on plants that are toxic to their larvae (e.g., Straatman 1962; Berenbaum 1981b; Papaj 1986), suggest that the metabolic effects of plant compounds have frequently served as constraints on host shifts in this family. Compounds do not need to be acutely toxic to inhibit colonization of novel plants: even a slight increase in the duration of the larval feeding period may lead to a significant increase in losses to predation (cf. Feeny et al. 1985).

Dependence on Plant Chemistry for Insect Defense

Survivorship of the immature stages of several swallowtail species has now been studied in natural habitats. These species include *P. xuthus* (Tsubaki 1973; Hirose et al. 1980; Watanabe 1981), *P. polyxenes* (Blau 1980; Feeny et al. 1985), *Eurytides marcellus* (Damman 1986a), *Luehdorfia puziloi* (Ikeda 1976), *Parides montezuma, Battus philenor,* and *B. polydamas* (Rauscher 1979b). In all cases, predation was a major source of larval mortality. In the case of *P. polyxenes,* it was calculated that a female would have to lay about 100 eggs, on average, to replace herself once in the next generation—the losses being almost entirely attributable to predators and parasitoids (Feeny et al. 1985). Even a slight increase in losses to predation following a shift to a novel host plant could prevent permanent colonization of that plant.

Swallowtails exhibit several traits that are likely to reduce losses to predation. Larvae of all species possess eversible osmeterial glands that release terpenoids or aliphatic acid derivatives when the insects are provoked (e.g., Eisner and Meinwald 1965; Honda 1980; Brower 1984). Inactivation of these glands increased predation of zebra swallowtail, *E. marcellus,* larvae in the field (Damman 1986b). Larvae of several *Aristolochia*-feeding swallowtails sequester aristolochic acids from their food plants—a trait that probably enhances defense of adults and immature stages against predation (Brower and Brower 1964; Rothschild 1972; Brower 1984; Urzua and Priestap 1985; R. Nishida, personal communication; see also Järvi et al. 1981). Though populations of *Battus philenor* (an *Aristolochia*-feeding troidine) in southeast Texas are not immune from predators (Odendaal et al. 1987; M. D. Rauscher, personal communication), the greatest source of larval mortality appears not to be predation but rather failure of the wandering larvae to find new food plants (Rauscher 1979c, 1981; see also Ikeda 1976).

Ehrlich and Raven (1964) have made the reasonable suggestion that dependence on sequestration of plant-derived toxins might constrain colonization by insects to other plants that contain similar compounds. In particular, they argue that the unusual age of certain butterfly–host associations, including that between troidine swallowtails and the Aristolochiaceae, seems to be correlated with plant compounds of unusual toxicity: the compounds have favored long-term survival of the plants and, at the same time, the insects have become "trapped" by specialization and dependence on the compounds for defense.

More generally, the defense traits of swallowtails may preadapt them to tolerate the predators and parasites associated with some plants better than the predator faunae associated with others. When the defenses are dependent on particular classes of plant-derived compounds, as is the case for sequestration of aristolochic acids though apparently not for osmeterial secretions (Honda 1983), selection may favor colonization of chemically-similar plants, while inhibiting colonization of other species.

When plant-derived compounds play no significant role in insect defense, one might expect predation pressure sometimes to favor rather than to inhibit shifts to chemically-unrelated host plants. Such shifts could provide a measure of escape from those parasitoids and predators that use plant compounds as host-finding cues (cf. Read et al. 1970; Elzen et al. 1984; Vinson 1984). Swallowtails suffer significant mortality from parasitoids such as those of the ichneumonid genus *Trogus* (e.g., Watanabe 1981; Feeny et al. 1985; Sperling 1986), various species of which seem to be effective in parasitizing swallowtails on a wide variety of host plants in many kinds of habitats (Heinrich 1962; Mitchell 1979). It is likely that *Trogus* females locate swallowtail larvae by responding, in part, to chemicals originating from the host plant. The apparent inability of most swallowtails to escape from *Trogus*, regardless of host plant or habitat, may indicate that they suffer from the constraint of using some of the same plant compounds for host location as those used by their parasitoids.

CHEMICAL FACILITATION: A STEPWISE MODEL

All three primary hypotheses concerning the nature of chemical constraints on swallowtail evolution seem as reasonable now as they did when first advanced by Dethier (1941), Fraenkel (1959), and Ehrlich and Raven (1964). There is every likelihood that behavioral facilitation, metabolic preadaptation, and dependence on plant compounds for defense have each contributed to the overall pattern of current host-plant relationships. These hypotheses can be consolidated into a stepwise model, indicating the possible roles of plant chemistry as catalysts of colonization (cf. Feeny 1987).

Colonization of a novel host can be considered as a sequence of stages, corresponding to barriers that an insect must overcome (Southwood 1961). Each stage has its own probability of success (Fig. 15-3). First comes behavioral choice: the larvae must choose to feed on the novel plant whether or not the female has chosen to lay eggs on it. Second, the larvae must be able to grow on the novel plant and, third, all stages of growth must be capable of surviving in the face of predation and other sources of mortality. Finally, there must be some likelihood that the surviving progeny will reselect the novel host in the next generation. The overall probability of colonization is the product of all component probabilities (Fig. 15-3).

The combined probabilities of overcoming all the barriers, and thus of colonizing a novel host, may generally be very low. Preadaptation at one or more stages, however, can increase dramatically the chances of overall success. In the case of swallowtails, chemical similarity between current and novel host plants seems to have catalyzed several stages in the colonization sequence (Fig. 15-3). Insofar as the component probabilities for each stage are not independent, of course, the overall probability of successful colonization may be increased. The ability to sequester a plant-derived toxin in the larva, for example, may automatically lead to its storage in the adult. Or there may be positive genetic correlation, due to pleiotropy or linkage disequilibrium, between host-plant preference and host-plant tolerance (cf. Bush 1975; Tavormina 1982; Via 1986; Thompson 1988).

I do not mean to argue that plant chemistry has been the only, or even

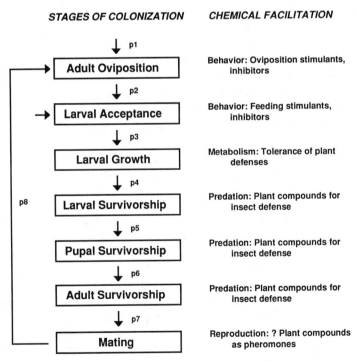

Figure 15-3. Stepwise model for the colonization of a novel host plant by a butterfly, indicating possible facilitation by chemical preadaptation at successive stages. The overall probability of successful colonization is the product of the probabilities (p1, p2, etc.) of completing each step.

dominant, influence on patterns of plant colonization by swallowtails. The probabilities of colonization of a novel host plant will be influenced by the genetic structure of the insect population and by various ecological factors such as similarities of growth form between host and nonhost plants, their relative availabilities in the local habitat, and by whether the insect's mating system is tied to the host plant (cf. Singer 1971; Gilbert 1979; Bush 1975; Futuyma 1983). Other things being equal, however, attempted colonizations are likely to be more frequent on plants of similar chemistry, because of preadaptation at the behavioral level, and a greater proportion of attempted colonizations are likely to be successful on plants of similar chemistry, owing to preadaptation at the metabolic level (Feeny et al. 1983).

ESCAPE FROM CHEMICAL CONSTRAINT: THE CASE OF *PAPILIO GLAUCUS*

One of the most remarkable swallowtails is *Papilio glaucus*, the eastern tiger swallowtail, which is an abundant resident of the temperate forests of central and eastern North America (Tyler 1975) and the most polyphagous species in the family Papilionidae. Though its larvae, like those of its relatives, will feed on the Rutaceae, Lauraceae, and Magnoliaceae, they also feed on trees belonging to at least 10 other families, including several that are rarely if ever reported as hosts for other swallowtails (Scriber 1984) and that display little taxonomic or chemical similarity (Hagen 1986). Evidently, *P. glaucus* has escaped, at least partially, from the chemical constraints that limit the host range of most other swallowtails.

An important prerequisite for this escape seems to have been the ability of *P. glaucus* larvae, unlike those of Lauraceae-feeding relatives, to feed and grow successfully on the mature foliage of trees (Hagen 1986). Most swallowtails attack the young foliage of their host plants and have difficulty growing on the mature leaves; they can tolerate higher levels of toxins that may be present in the young foliage and avoid the toughness and lower nutrient value of the mature leaves (Scriber and Feeny 1979; Rausher 1981; Watanabe 1982). Larvae of *P. glaucus*, by contrast, are relatively insensitive to leaf age (Hagen 1986) and, unlike those of their herb-feeding relative *P. polyxenes*, are quite resistant to the toxic effects of ingested tannins (Steinly and Berenbaum 1985). Though they may grow slowly (Scriber 1975), they exhibit several traits of coloring and behavior that probably offset the greater apparency to predators that slow growth entails. These include widespread dispersion of eggs, crypsis, apparent mimicry of tree snakes, and removal of leaves damaged by feeding (Feeny 1976).

The foliage of temperate-zone trees apparently contains fewer potentially deterrent or toxic compounds than does the foliage of herbaceous plants in similar regions (Futuyma 1976). Moreover, concentrations of many toxins are generally reduced in mature foliage (McKey 1974) and there is convergence among different species of deciduous trees in the quantitative resistance of older leaves (Feeny 1976). Larvae adapted to grow on the mature foliage of one such species are thus more likely to be able to survive on other species in their environment than are larvae in herbaceous habitats. Oviposition "mistakes" by females will not be selected against so rigorously and it will generally be to a larva's advantage to attempt to eat whatever plant it finds itself on. Hagen (1986) found that *P. glaucus* larvae, unlike those of other swallowtails tested, will attempt to feed on the foliage of almost any plant offered to them. Such behavior, in turn, could accelerate the acquisition of further host plants by increasing the probability of adaptation to any toxic barriers that these plants may contain (cf. Lindroth et al. 1988). Hagen (1986) has argued that relaxation of host-choice criteria, combined with larval tolerance of older leaves, has permitted *P. glaucus* to expand its range of host plants in response to ecological opportunities, colonizing a variety of abundant deciduous tree species in temperate-zone forests.

In western North America the larvae of three *Papilio* species feed on overlapping subsets of the plant families attacked by *P. glaucus* further east (Brower 1959a, b; Scriber 1984, 1987). All three species, *P. eurymedon, P. rutulus,* and *P. multicaudatus,* are believed to have been derived quite recently from *P. glaucus* or its immediate ancestor (Brower 1959a, b) and apparently represent an example of the derivation of relatively specialized species from a generalist. Whatever the ecological explanation for the current patterns of host use (cf. Brower 1959b), the derived species are presumably subject to more rigid chemical constraints than is *P. glaucus.* Their host ranges, however, include some plant families not usually attacked by swallowtails, indicating that these constraints are not the same as those possessed by more distant ancestors of *P. glaucus,* which are presumed to have been restricted to the Lauraceae, Magnoliaceae, or Rutaceae (Scriber 1973; Hagen 1986; see also Scriber et al., Chapter 16). This example appears to show how one set of constraints, through a generalist intermediary, can give rise to other sets of constraints—how, in other words, a phyletic lineage can "jump grooves." If *P. glaucus* were not still present to indicate the likely route of past host shifts, the host ranges of the three western species might be more puzzling.

COEVOLUTION AND CHEMICAL CONSTRAINT

There can be little doubt that chemical similarities among swallowtail host plants are best explained in terms of occasional breakthroughs in chemical adaptation, followed by multiple colonization events—the "sequential evolution" postulated by Jermy (1984). Less clear, however, is the role that swallowtails themselves may have played in influencing the patterns of plant chemistry that have helped to shape their own evolution. To what extent, in other words, are swallowtails and their host plants engaged in coevolution (Janzen 1980; Futuyma and Slatkin 1983)? Ehrlich and Raven (1964) suggested that butterflies and their food plants have been engaged in "reciprocal evolutionary relationships" (p. 606). Along with other herbivores, the butterflies have acted as selective pressures bringing about the evolution of chemical resistance in plants.

Such resistance, they argued, led to selective pressures on herbivores to overcome the plant defenses and hence select for yet further diversification of chemical defense.

Berenbaum (1983) has argued in greater detail that papilionine swallowtails, along with other herbivores, pathogens, and fungi, have been responsible for progressive evolution within the phenylpropanoid pathway leading via hydroxy-coumarins to linear and angular furanocoumarins. Her argument is based, first, on biochemical evidence: xanthotoxin (a linear furanocoumarin), but not umbelliferone (a hydroxycoumarin precursor), is toxic to the southern army-worm and, by implication, to many insects not associated with plants containing furanocoumarins. Larvae of *Papilio polyxenes*, however, and presumably those of other *Papilio* species that feed on furanocoumarin-containing plants, can tolerate xanthotoxin and the compound even enhances the growth of *P. polyxenes* larvae (Berenbaum 1981a). Bull et al. (1986) have shown that the enzymes responsible for metabolizing xanthotoxin act much more rapidly in *P. polyxenes* larvae than in those of the fall armyworm *Spodoptera frugiperda*. Yet *P. polyxenes* is relatively sensitive to angelicin (Berenbaum and Feeny 1981), one of the rarer and supposedly more biosynthetically-advanced angular furanocoumarins, at least in part because angular furanocoumarins are metabolized more slowly by *P. polyxenes* larvae than are linear compounds (Ivie et al. 1986). Thus, each new set of plant metabolites seems to confer resistance to herbivore species that have evolved tolerance to the precursors.

Second, swallowtails of the genus *Papilio* have reached their greatest diversity among those sections of the genus that feed on furanocoumarin-containing host plants, notably those in the Rutaceae and Apiaceae. The correlation between furanocoumarin distribution and the host plants of these *Papilio* species is impressive and even accounts for several otherwise-anomalous host plants, such as the genus *Psoralea* in the Leguminosae. Adaptation to furanocoumarins may thus have permitted swallowtails to diversify within a new "adaptive zone" (Ehrlich and Raven 1964), which includes some unrelated plants that happen to contain the same compounds. Third, as evidence that furanocoumarin produc-tion has led to evolutionary success among plants, Berenbaum (1983) shows that the number of species per genus in the Apiaceae increases from those genera with hydroxycoumarins alone to those with linear furanocoumarins and is greatest in those genera containing both linear and angular furanocoumarins.

While Berenbaum's data certainly suggest that adaptation and preadaptation to the presence of furanocoumarins has played a significant role in the radiation of the genus *Papilio*, she emphasizes that the postulated coevolutionary scenario is speculative; many of the proposed steps have yet to be tested experimentally and the generality of existing test results needs to be established (Berenbaum 1983; Thompson 1986). If, for example, furanocoumarins are associated with greater plant success, as her data indicate (though see Thompson 1986), it would be important to demonstrate that this correlation has a causative basis and that reduction of herbivory is at least partially responsible for the selective advantage of furanocoumarin production. Do the plants containing angular furanocoumarins, apparently the most recent biosynthetic advance, suffer less damage from herbivores and pathogens than the plants that lack these compounds?

Since swallowtail host plants share other classes of secondary compounds (Hegnauer 1971; Ehrlich and Raven 1964; Feeny et al. 1983), including essential oils, alkaloids, and flavonoids that are known to affect swallowtails (see discussion above), it would also be important to disentangle effects of these other compounds when establishing the significance of the furanocoumarins. The argument for stepwise coevolution between swallowtails and coumarin-containing plants would be further strengthened by a demonstration of some degree of congruence between the phylogenies of the insects and their host plants, once the systematics of both plants and insects become better known (Miller 1987b). In view of the extensive colonization of novel and sometimes unrelated hosts that may have followed biochemical adaptation at each step, however, exact congruence of cladograms for plants and insects would not be expected. Nobody, for example, is suggesting that swallowtails have coevolved extensively with the legume genus *Psoralea*.

Perhaps the most vigorous recent critic of coevolution (sensu Ehrlich and Raven 1964) is Jermy, who has argued that insects have not appreciably influenced the evolution of chemical resistance in plants (Jermy 1984). As explanations for the occurrence of secondary plant compounds, Jermy (1984) cites the following possible alternative functions of such compounds as listed by Chew and Rodman (1979): regulators for plant growth or biosynthetic activities, storage forms of plant growth regulators, energy or mineral reserves, transport facilitators, waste products, detoxification products of environmental poisons, shields against excessive radiation, and effectors of allelochemic interactions between plants and their plant competitors. However, while some plant secondary compounds undoubtedly serve functions other than defense (see also Seigler and Price 1976), at least as many compounds are known to have adverse effects on herbivores or pathogens (e.g., Hedin 1977; Rosenthal and Janzen 1979; Horsfall and Cowling 1980) or simply to act as deterrents to feeding or oviposition (cf. Bernays and Chapman 1987). The cyanogenic glycosides of acacias are lacking in several species that harbor ants (Rehr et al. 1973), "a complementarity that argues for a defensive role as the adaptive raison d'être of both cyanogenesis and ant attraction" (Futuyma 1983, p. 212).

If the advantages of secondary compounds to plants lay primarily in the realms of internal metabolism or adaptation to the physical environment, one would expect to see a greater tendency toward convergence than is apparent in present-day patterns of plant secondary chemistry (Futuyma 1983). By contrast, one of the most attractive features of the theory of coevolution is the contribution it offers toward explaining the prodigious diversity of secondary compounds in plants. This can be seen from the following scenario, admittedly speculative, for the evolution of a resistance trait in a plant population.

Following the appearance of a novel metabolite, increased deterrence and/or toxicity to insects and other plant enemies may confer an immediate selective advantage on phenotypes in which the novel trait is expressed. As the resistance trait spreads in the plant population and to other populations, three things happen. First, some of the associated insects, through repeated exposure to the novel compound (Berenbaum 1983), evolve tolerance to it, perhaps reinforcing entrainment of groups of specialists. Second, insects that cannot adapt to the novel trait may become extinct (if they have no alternate hosts available) or evolve avoidance of plants containing the compound; the plants thus "shake off" some associated enemies. Third, the novel compound endows the plant with one more barrier against colonization by enemies not yet associated with it; in effect, this reduces the colonization rate of the population by herbivore and pathogen populations in the habitat as a whole. Though not of immediate selective advantage to resistant genotypes, this longer-term consequence of the resistance trait may increase the longevity of the plant population and species in the local flora.

There is a fourth consequence of the appearance in the plant population of a novel compound, namely that the compound may fortuitously increase the probability of colonization by other insects or pathogens—species not previously associated with the plant in question but preadapted to the novel compound by virtue of their own prior evolutionary history. Obviously, the more different in its biological effects is the novel compound from other compounds in the local environment, the less likely is such colonization to occur and (presumably) the greater is the longer-term selective advantage of the novel plant genotype.

Coevolution leads to successive adaptive breakthroughs, many of them chemical in nature, by plants and enemies. Because the fundamental biochemical pathways to natural products in plants are few in number (Whittaker and Feeny 1971), and because many of the enzymes for fine-tuning the structures are quite widespread, different phyletic lineages of plants sporadically evolve similar or identical compounds or combinations of synergistic compounds (cf. Berenbaum 1985) that are already present in other plants. Adaptive breakthroughs by insects and other enemies may thus also preadapt them to colonize other plants that share the same compounds (or that contain other compounds to which the enemies may have become preadapted simultaneously by fortuitous cross resistance; cf. Gould 1983). Such colonization by previously nonassociated enemies would be likely to enhance the long-term selective advantage of chemical uniqueness (Futuyma 1976). Far from being incompatible with coevolution, sequential evolution could be a major driving force for chemical diversification in plants.

In criticizing the theory of coevolution, Jermy (1984) acknowledges many known cases of insects causing substantial damage to wild plants. His criticism highlights, rather, the current dearth of definitive evidence for selective effects of herbivores on the resistance traits of naturally-growing plants. Few appropriate studies have been undertaken, however, so that it seems premature to conclude that such selection does not occur. The ravages of insects, fungi, and pathogens would prevent us from growing many crop plants economically if it were not for the occurrence of heritable resistance traits (Maxwell and Jennings 1980). The work of Berenbaum et al. (1986) does in fact suggest that insect herbivory on wild populations of *Pastinaca sativa* can select for genotypes with differing furanocoumarin content, and Pollard (1986) and Pollard and Briggs (1984) provide evidence that genotypes of plants with high densities of stinging hairs are at a selective advantage over genotypes with fewer hairs in the presence of herbivores. More recently, Simms and Rausher (1989) have provided a direct demonstration of selection by herbivorous insects on the plant *Ipomoea purpurea:* removal of insects in the field was found to eliminate selection on genetic variation influencing damage by corn earworms, *Heliothis zea.*

Since it is sometimes argued (e.g., Jermy 1976) that low levels of herbivore damage to plants are unlikely to lead to significant selection for resistance, it is perhaps worth reemphasizing that damage need not have to be substantial in order to influence the outcome of competition, whether intraspecific or interspecific, between a plant and its neighbors (Harper 1977; Bentley and Whittaker 1979; Fowler and Rausher 1985). It is ultimately this competition that determines which genotypes will produce more offspring than others, and hence whether or not a particular resistance trait may be enhanced in the plant population.

That there has been and continues to be diffuse coevolution (Janzen 1980), involving plants and entire arrays of enemies (including swallowtails), seems probable. Less evident, however, is the extent to which particular swallowtail populations engage in relatively specific "one-on-one" coevolution with individual host-plant populations. Such "true reciprocal coevolution" (Futuyma 1983) would be most obvious, if it occurs at all, where a particular swallowtail population has an unusually large selective impact (cf. Fox 1981). Many swallowtails form a minor part of an extensive insect fauna attacking their host plants and would seem to be unlikely candidates for significant specific coevolution with their hosts. Troidine swallowtails, however, appear to be one of few insect groups that have colonized the Aristolochiaceae and are sometimes sufficiently abundant in relation to their host plants that episodes of relatively specific coevolution are a possibility.

An example is the interaction between the pipevine swallowtail *Battus philenor* and two herbaceous species of *Aristolochia* in the longleaf pine forests of southeast Texas (Rausher 1978). *A. reticulata* is heavily attacked each season by *B. philenor* larvae, which remove about half the annual leaf crop and dramatically reduce annual gain in root weight and hence the number of individuals that flower and the number of seeds produced (Rausher and Feeny 1980). The major source of mortality in the *B. philenor* population seems to be failure of the wandering larvae to find enough host plants to complete their growth (Rausher 1979c, 1981). Both plant and insect thus appear to have a major impact on one another.

A. reticulata plants appear to survive in this habitat because the toughness and low nutrient content of the mature leaves renders them resistant to the wandering larvae (Rausher 1981). The other host-plant species in this habitat,

A. serpentaria, is also attacked by *B. philenor*, but its leaves do not toughen and are vulnerable to larvae at all stages of growth. Its survival depends in part on its narrow, grass-like leaf shape, which reduces the chances of discovery by *B. philenor* females when they are searching visually for the broad-leaved foliage of the more abundant *A. reticulata* (Rausher 1978). It is not known whether the resistance and apparency of the host plants or the behavior and physiology of *B. philenor* have evolved or been modified as a result of coevolution. It is suggestive, however, that in northern Florida, outside the range of *A. reticulata*, the leaves of *A. serpentaria* are markedly broader and resemble closely the leaves of abundant young *Smilax* vines, which *B. philenor* females approach frequently while searching for *A. serpentaria* (P. Feeny, personal observation, Tall Timbers Research Station, May 1977; see also Papaj 1986). It is conceivable that the variation of leaf shape in *A. serpentaria* in the different localities represents an evolutionary response to attack by *B. philenor*.

Cases of specific chemical coevolution between troidine swallowtails and *Aristolochia* species, if they occur at all, are most likely to be revealed in the tropics, where most of the species occur. Though all *Aristolochia* species, as far as is known, contain aristolochic acids (Chen and Zhu 1987), the plants are by no means uniform in their ability to stimulate oviposition or support larval growth. At the same site in eastern Mexico, for example, *Battus philenor* and *Parides montezuma* lay eggs on both *Aristolochia orbicularis* and *A. micrantha*, whereas *Battus polydamas* oviposits only on *A. orbicularis* (Rausher 1979b). Of five wild *Aristolochia* species subject to attack by five troidine species in the vicinity of Campinas, Brazil three species are satisfactory food plants for all five swallowtails, one is somewhat less preferred, and one (*A. brasiliensis*) is quite resistant to all but one *Parides* species (Brown et al. 1981). Another *Parides* species (*P. ascanius*) is restricted to coastal and lowland swamps of the state of Rio de Janeiro, where its only known host plant is *Aristolochia macroura*; 12 other *Aristolochia* species supported little if any larval growth and most failed to elicit any oviposition (Otero and Brown 1986). Detailed studies of such associations might be most interesting.

Lack of congruence between the cladograms available for swallowtails and their host plants argues against long-term specific coevolution, which should lead to parallel cladogenesis (Miller 1987b; Mitter and Brooks 1983). Multiple colonization, however, is not incompatible with diffuse coevolution between the butterflies and their hosts, nor is it incompatible with intermittent bouts of more specific coevolution. The latter can be viewed as eddies of positive feedback within a generally advancing tide of diffuse coevolution between plants and their enemies (cf. Strong et al. 1984, p. 218).

CONCLUSIONS

That the theory of island biogeography (Munroe 1948; MacArthur and Wilson 1967) has been applied with some success to the explanation of the numbers of insect species associated with plants (Strong et al. 1984) can be attributed in part to the freedom of insects to respond to ecological opportunities and thus to obey "ecological rules." As in the swallowtails, however, any particular insect lineage is likely to be subject to its own sets of historical constraints, not necessarily chemical, that restrict the number of host plant taxa that can be colonized according to these rules (cf. Southwood and Kennedy 1983). The success of theories such as that of island biogeography when applied to entire faunas and floras results from the inclusion of many different phyletic lineages, the various idiosyncratic constraints of which are obscured when considered in aggregate (Kennedy and Southwood 1984).

Occasionally, plants seem to have evolved barriers that are qualitatively exceptional in their effectiveness (Southwood 1961). Such seems to be the case in the Aristolochiaceae, in which the aristolochic acids appear to confer resistance that few herbivores or pathogens have yet been able to overcome. The unusual toxicity of aristolochic acids may result not so much from the underlying phenanthrene ring system as from their unique possession of an aromatically-bound nitro group and perhaps, also, by their additional possession of a methylenedioxyphenyl group (Miller and Feeny 1983; cf. Berenbaum 1985). These compounds may account for an herbivore fauna that is apparently smaller than would be expected from the ecological rules of the theory of island biogeography.

ACKNOWLEDGMENTS

I am most grateful to Ian Baldwin, May Berenbaum, Maureen Carter, Vincent Dethier, Robert Hagen, Richard Harrison, Tibor Jermy, John Lawton, James Miller, Mark Rausher, Sir Richard Southwood, Felix Sperling, Sara Via, and Christer Wiklund for comments on parts or all of the manuscript. The research was supported by NSF research grants BSR-8516832 and BSR-8818104, Hatch grant NYC-183413, and a grant from the Cornell Biotechnology Program to P.F. and J. A. Renwick.

REFERENCES

Bentley, S. and J. B. Whittaker. 1979. Effects of grazing by a chrysomelid beetle, *Gastrophysa viridula*, on competition between *Rumex obtusifolius* and *Rumex crispus*. *Journal of Ecology* **67**: 79–90.

Berenbaum, M. 1978. Toxicity of a furanocoumarin to armyworms: A case of biosynthetic escape from insect herbivores. *Science* **201**: 532–534.

Berenbaum, M. 1981a. Effects of linear furanocoumarins on an adapted specialist insect (*Papilio polyxenes*). *Ecological Entomology* **6**: 345–351.

Berenbaum, M. 1981b. An oviposition "mistake" by *Papilio glaucus* (Papilionidae). *Journal of the Lepidopterists' Society* **35**: 75.

Berenbaum, M. 1983. Coumarins and caterpillars: A case for coevolution. *Evolution* 37: 163–179.

Berenbaum, M. 1985. Brementown revisited: Interactions among allelochemicals in plants. *Recent Advances in Phytochemistry* 19: 139–169.

Berenbaum, M. and P. Feeny. 1981. Toxicity of angular furanocoumarins to swallowtail butterflies: Escalation in a coevolutionary arms race? *Science* 212: 927–929.

Berenbaum, M. and J. J. Neal. 1985. Synergism between myristicin and xanthotoxin, a naturally cooccurring plant toxicant. *Journal of Chemical Ecology* 11: 1349–1358.

Berenbaum, M. R., A. R. Zangerl, and J. K. Nitao. 1986. Constraints on chemical coevolution: Wild parsnips and the parsnip webworm. *Evolution* 40: 1215–1228.

Bernays, E. and R. Chapman. 1987. The evolution of deterrent responses in plant-feeding insects. Pages 159–173 in R. F. Chapman, E. A. Bernays, and J. G. Stoffolano, Jr. (Eds.). Perspectives in Chemoreception and Behavior. Springer-Verlag, New York.

Blau, P. A., P. Feeny, L. Contardo, and D. Robson. 1978. Allylglucosinolate and herbivorous caterpillars: A contrast in toxicity and tolerance. *Science* 200: 1296–1298.

Blau, W. S. 1980. The effect of environmental disturbance on a tropical butterfly population. *Ecology* 61: 1005–1012.

Blau, W. S. and P. Feeny. 1983. Divergence in larval responses to food plants between temperate and tropical populations of the black swallowtail butterfly. *Ecological Entomology* 8: 749–757.

Brower, L. P. 1959a. Speciation in butterflies of the *Papilio glaucus* group. I. Morphological relationships and hybridization. *Evolution* 13: 40–63.

Brower, L. P. 1959b. Speciation in butterflies of the *Papilio glaucus* group. II. Ecological relationships and interspecific sexual behavior. *Evolution* 13: 212–228.

Brower, L. P. 1984. Chemical defence in butterflies. Pages 109–134. In R. I. Vane-Wright and P. R. Ackery (Eds.). *The Biology of Butterflies*. Academic Press, London, UK.

Brower, L. P. and J. van Z. Brower. 1964. Birds, butterflies, and plant poisons: A study in ecological chemistry. *Zoologica* 49: 137–159.

Brown, K. S., A. J. Damman, and P. Feeny. 1981. Troidine swallowtails (Lepidoptera: Papilionidae) in southeastern Brazil: Natural history and foodplant relationships. *Journal of Research on the Lepidoptera* 19: 199–226.

Bull, D. L. G. W. Ivie, R. C. Beier, and N. W. Pryor. 1986. *In vitro* metabolism of a linear furanocoumarin (8-methoxypsoralen, xanthotoxin) by mixed-function oxidases of larvae of black swallowtail butterfly and fall armyworm. *Journal of Chemical Ecology* 12: 885–892.

Bush, G. L. 1975. Sympatric speciation in phytophagous parasitic insects. Pages 187–206 in P. W. Price (Ed.). *Evolutionary Strategies of Parasitic Insects and Mites*. Plenum, New York.

Chen, Z. and D. Zhu. 1987. Aristolochia alkaloids. Pages 29–65 in A. Brossi (Ed.). *The Alkaloids: Chemistry and Pharmacology*. Volume 31. Academic Press, New York.

Chew, F. S. and J. E. Rodman. 1979. Plant resources for chemical defense. Pages 271–307 in G. A. Rosenthal and D. H. Janzen (Eds.). *Herbivores*. Academic Press, New York.

Cody, M. L. 1974. Optimization in ecology. *Science* 183: 1156–1164.

Cronquist, A. 1981. An Integrated System of Classification of Flowering Plants. Columbia University Press, New York.

Crowden, R. K., J. B. Harborne, and V. H. Heywood. 1969. Chemosystematics of the Umbelliferae—A General Survey. *Phytochemistry* **8**: 1963–1984.

Damman, A. J. 1986a. *Effects of Seasonal Changes in Leaf Quality and Abundance of Natural Enemies on the Insect Herbivores of Pawpaws.* Ph.D. Dissertation. Cornell University, Ithaca, NY.

Damman, A. J. 1986b. The osmaterial glands of the swallowtail butterfly *Eurytides marcellus* as a defence against natural enemies. *Ecological Entomology* **11**: 261–265.

Dethier, V. G. 1941. Chemical factors determining the choice of food plants by *Papilio* larvae. *American Naturalist* **75**: 61–73.

Ehrlich, P. R. and P. H. Raven. 1964. Butterflies and plants: A study in coevolution. *Evolution* **18**: 586–608.

Eisner, T. and Y. C. Meinwald. 1965. Defensive secretion of a caterpiller (*Papilio*). *Science* **150**: 1733–1735.

Elzen, G. W., H. J. Williams, and S. B. Vinson. 1984. Isolation and identification of cotton synomones mediating searching behavior by parasitoid *Campoletis sonorensis*. *Journal of Chemical Ecology* **10**: 1251–1264.

Erickson, J. M. and P. Feeny. 1974. Sinigrin: A chemical barrier to the black swallowtail butterfly, *Papilio polyxenes*. *Ecology* **55**: 103–111.

Feeny, P. 1976. Plant apparency and chemical defense. *Recent Advances in Phytochemistry* **10**: 1–40.

Feeny, P. P. 1987. The roles of plant chemistry in associations between swallowtail butterflies and their host plants. Pages 353–359 in V. Labeyrie, G. Fabres, and D. Lachaise (Eds.). Insects–Plants. W. Junk Publishers, Dordrecht, The Netherlands.

Feeny, P., W. S. Blau, and P. M. Kareiva. 1985. Larval growth and survivorship of the black swallowtail butterfly in central New York. *Ecological Monographs* **55**: 167–187.

Feeny, P., L. Rosenberry, and M. Carter. 1983. Chemical aspects of oviposition behavior in butterflies. Pages 27–76 in S. Ahmad (Ed.). *Herbivorous Insects: Host-seeking Behavior and Mechanisms.* Academic Press, New York.

Feeny, P., K. Sachdev, L. Rosenberry, and M. Carter. 1988. Luteolin 7-O-(6″-O-malonyl)-β-D-glucoside and *trans*-chlorogenic acid: Oviposition stimulants for the black swallowtail butterfly. *Phytochemistry* **27**: 3439–3448.

Feeny, P., E. Städler, I. Åhman, and M. Carter. 1989. Effects of plant odor on oviposition by the black swallowtail butterfly, *Papilio polyxenes* (Lepidoptera: Papilionidae). *Journal of Insect Behavior* **2**: 803–827.

Fowler, N. L. and M. D. Rausher. 1985. Joint effects of competitors and herbivores on growth and reproduction in *Aristolochia reticulata*. *Ecology* **66**: 1580–1587.

Fox, L. R. 1981. Defense and dynamics in plant–herbivore systems. *American Zoologist* **21**: 853–864.

Fraenkel, G. S. 1959. The raison d'être of secondary plant substances. *Science* **129**: 1466–1470.

Futuyma, D. J. 1976. Food plant specialization and environmental predictability in Lepidoptera. *American Naturalist* **110**: 285–292.

Futuyma, D. J. 1983. Evolutionary interactions among herbivorous insects and plants. Pages 207–231 in D. J. Futuyma and M. Slatkin (Eds.). *Coevolution*. Sinauer, Sunderland, MA.

Futuyma, D. J. and M. Slatkin (Eds.). 1983. *Coevolution*. Sinauer Associates, Sunderland, MA.

Gilbert, L. E. 1979. Development of theory in the analysis of insect–plant interactions. Pages 117–154 in D. J. Horn, R. Mitchell, and G. R. Stairs (Eds.). *Analysis of Ecological Systems*. Ohio State University Press, Columbus, OH.

Gould, F. 1983. Genetics of plant–herbivore systems: Interactions between applied and basic study. Pages 599–653 in R. F. Denno and M. S. McClure (Eds.). *Variable Plants and Herbivores in Natural and Managed Systems*. Academic Press, New York.

Hagen, R. H. 1986. *The Evolution of Host-Plant Use by the Tiger Swallowtail Butterfly, Papilio glaucus*. Ph.D. Dissertation. Cornell University, Ithaca, NY.

Harborne, J. B. and C. A. Williams. 1972. Flavonoid patterns in the fruits of the Umbelliferae. *Phytochemistry* **11**: 1741–1750.

Harper, J. L. 1977. *The Population Biology of Plants*. Academic Press, London, UK.

Hedin, P. A. (Ed.). 1977. *Host plant resistance to pests*. American Chemical Society Symposium Series 62, Washington, D.C., USA.

Hegnauer, R. 1971. Chemical patterns and relationships of Umbelliferae. Pages 267–277 in V. H. Heywood (Ed.). *The Biology and Chemistry of the Umbelliferae*. Academic Press, London, UK.

Heinrich, G. H. 1962. Synopsis of nearctic Ichneumoninae Stenopneusticae with particular reference to the northeastern region (Hymenoptera). Part VII. *Canadian Entomologist Supplement* **29**: 805–886.

Heywood, V. H. (Ed.). 1971. *The Biology and Chemistry of the Umbelliferae*. Academic Press, London, UK.

Hirose, Y., Y. Suzuki, M. Takagi, K. Kiehata, M. Yamasaki, H. Kimoto, M. Yamanaka, M. Iga, and K. Yamaguchi. 1980. Population dynamics of the citrus swallowtail, *Papilio xuthus* Linné (Lepidoptera: Papilionidae): Mechanisms stabilizing its numbers. *Researches in Population Ecology* **21**: 260–285.

Honda, K. 1980. Volatile constituents of larval osmeterial secretions in *Papilio protenor demetrius*. *Journal of Insect Physiology* **26**: 39–45.

Honda, K. 1983. Evidence for *de novo* biosynthesis of osmeterial secretions in young larvae of the swallowtail butterflies (*Papilio*): Deuterium incorporation *in vivo* into sesquiterpene hydrocarbons as revealed by mass spectrometry. *Insect Science Application* **4**: 255–261.

Honda, K. 1986. Flavanone glycosides as oviposition stimulants in a papilionid butterfly, *Papilio protenor*. *Journal of Chemical Ecology* **12**: 1999–2010.

Horsfall, J. G. and E. B. Cowling (Eds.). 1980. *Plant Disease*. Volume 5: How Plants Defend Themselves. Academic Press, New York.

Ichinosé, T. and H. Honda. 1978. Ovipositional behavior of *Papilio protenor demetrius* Cramer and the factors involved in its host plants. *Applied Entomology and Zoology* **13**: 103–114.

Igarashi, S. 1987. On the life history of *Teinopalpus imperialis* Hope in Northern India and its phylogenetic position in the Papilionidae. *Tyô to Ga* **38**: 115–151.

Ikeda, K. 1976. Bioeconomic studies on a population of *Luehdorfia puziloi inexpecta* Sheljuzko (Lepidoptera: Papilionidae). *Japanese Journal of Ecology* **26**: 199–208.

Ivie, G. W., D. L. Bull, R. C. Beier, and N. W. Pryor. 1986. Comparative metabolism of [³H]psoralen and [³H]isopsoralen by black swallowtail (*Papilio polyxenes* Fabr.) caterpillars. *Journal of Chemical Ecology* **12**: 871–884.

Jacob, F. 1977. Evolution as tinkering. *Science* **196**: 1161–1166.

Janzen, D. H. 1980. When is it coevolution? *Evolution* **34**: 611–612.

Järvi, T., B. Sillén-Tullberg, and C. Wiklund. 1981. The cost of being aposematic. An experimental study of predation on larvae of *Papilio machaon* by the great tit *Parus major*. *Oikos* **36**: 267–272.

Jermy, T. 1976. Insect–host-plant relationship—Coevolution or sequential evolution? Pages 109–113 in T. Jermy (Ed.). *The Host-Plant in Relation to Insect Behavior and Reproduction*. Plenum, New York.

Jermy, T. 1984. Evolution of insect/host plant relationships. *American Naturalist* **124**: 609–630.

Kennedy, C. E. J. and T. R. E. Southwood. 1984. The number of species of insects associated with British trees: A re-analysis. *Journal of Animal Ecology* **53**: 455–478.

Lindroth, R. L., J. M. Scriber, and M. T. S. Hsia. 1988. Chemical ecology of the tiger swallowtail: Mediation of host use by phenolic glycosides. *Ecology* **69**: 814–822.

MacArthur, R. H. and E. O. Wilson. 1967. *The Theory of Island Biogeography*. Princeton University Press, Princeton, NJ.

Maxwell, F. G. and P. R. Jennings. 1980. *Breeding Plants Resistant to Insects*. John Wiley & Sons, New York.

McKey, D. 1974. Adaptive patterns in alkaloid physiology. *American Naturalist* **108**: 305–320.

Miller, J. S. 1987a. Phylogenetic studies in the Papilioninae (Lepidoptera: Papilionidae). *Bulletin of the American Museum of Natural History* **186**: 365–512.

Miller, J. S. 1987b. Host-plant relationships in the Papilionidae (Lepidoptera): Parallel cladogenesis or colonization? *Cladistics* **3**: 105–120.

Miller, J. S. and P. Feeny. 1983. Effects of benzylisoquinoline alkaloids on the larvae of polyphagous Lepidoptera. *Oecologica (Berlin)* **58**: 332–339.

Miller, J. S. and P. Feeny. 1989. Interspecific differences among swallowtail larvae (Lepidoptera: Papilionidae) in susceptibility to aristolochic acids and berberine. *Ecological Entomology* **14**: 287–296.

Mitchell, R. T. 1979. A review of swallowtail butterfly (Papilionidae) parasites of the genus *Trogus* (Hymenoptera–Ichneumonidae). *Maryland Entomologist* **1**: 6–7.

Mitter, C. and D. R. Brooks. 1983. Phylogenetic aspects of coevolution. Pages 65–98 in D. J. Futuyma and M. Slatkin (Eds.). *Coevolution*. Sinauer, Sunderland, MA.

Munroe, E. G. 1948. *The Geographical Distribution of Butterflies in the West Indies*. Ph.D. Dissertation. Cornell University, Ithaca, NY.

Nishida, R. 1977. Oviposition stimulants of some Papilionid butterflies contained in their host plants. *Botyu-Kagaku (Scientific Pest Control)* **42**: 133–140.

Nishida, R., T. Ohsugi, S. Kokubo, and H. Fukami. 1987. Oviposition stimulants of a Citrus-feeding swallowtail butterfly, *Papilio xuthus* L. *Experientia* **43**: 342–344.

Odendaal, F. J., M. D. Rausher, B. Benrey, and J. Nunez-Farfan. 1987. Predation by *Anolis* lizards on *Battus philenor* raises questions about butterfly mimicry systems. *Journal of the Lepidopterists' Society* **41**: 141–144.

Ohsugi, T., R. Nishida, and H. Fukami. 1985. Oviposition stimulant of *Papilio xuthus*, a Citrus-feeding swallowtail butterfly. *Agricultural and Biological Chemistry* **49**: 1897–1900.

Otero, L. S. and K. S. Brown, Jr. 1986. Biology and ecology of *Parides ascanius* (Cramer,

1775) (Lep., Papilionidae), a primitive butterfly threatened with extinction. *Atala* **10–12**: 2–16.

Papaj, D. R. 1986. An oviposition "mistake" by *Battus philenor* L. (Papilionidae). *Journal of the Lepidopterists' Society* **40**: 348–349.

Pollard, A. J. 1986. Variation in *Cnidoscolus texanus* in relation to herbivory. *Oecologia* (Berlin) **70**: 411–413.

Pollard, A. J. and D. Briggs. 1984. Genecological studies of *Urtica dioica* L. III. Stinging hairs and plant–herbivore interactions. *New Phytologist* **97**: 507–522.

Rausher, M. D. 1978. Search image for leaf shape in a butterfly. *Science* **200**: 1071–1073.

Rausher, M. D. 1979a. *Coevolution in a Simple Plant–Herbivore System.* Ph.D. Dissertation. Cornell University, Ithaca, NY.

Rausher, M. D. 1979b. Larval habitat suitability and oviposition preference in three related butterflies. *Ecology* **60**: 503–511.

Rausher, M. D. 1979c. Egg recognition: Its advantages to a butterfly. *Animal Behaviour* **27**: 1034–1040.

Rausher, M. D. 1981. Host plant selection by *Battus philenor* butterflies: The roles of predation, nutrition, and plant chemistry. *Ecological Monographs* **51**: 1–20.

Rausher, M. D. and P. Feeny. 1980. Herbivory, plant density and plant reproductive success: The effect of *Battus philenor* on *Aristolochia reticulata. Ecology* **61**: 905–917.

Read, D. P., P. P. Feeny, and R. B. Root. 1970. Habitat selection by the aphid parasite, *Diaeretiella rapae* (Hymenoptera: Braconidae), and hyperparasite, *Charips brassicae* (Hymenoptera: Cynipidae). *Canadian Entomologist* **102**: 1567–1578.

Rehr, S. S., P. P. Feeny, and D. H. Janzen. 1973. Chemical defence in Central American non-ant-acacias. *Journal of Animal Ecology* **42**: 405–416.

Rosenthal, G. A. and D. H. Janzen (Eds.). 1979. *Herbivores: Their Interaction with Secondary Plant Metabolites.* Academic Press, New York.

Rothschild, M. 1972. Secondary plant substances and warning colouration in insects. Pages 59–83 in H. F. van Emden (Ed.). *Insect/Plant Relationships.* Blackwell Scientific Publications, Oxford, UK.

Saxena, K. N. and S. Goyal. 1978. Host-plant relations of the citrus butterfly *Papilio demoleus* L.: Orientational and ovipositional responses. *Entomologia Experimentalis et Applicata* **24**: 1–10.

Saxena, K. N. and S. Prabha. 1975. Relationship between the olfactory sensilla of *Papilio demoleus* L. larvae and their orientation responses to different odours. *Journal of Entomology, Series A* **50**: 119–126.

Scriber, J. M. 1973. Latitudinal gradients in larval feeding specialization of the world Papilionidae (Lepidoptera). *Psyche* **80**: 355–373 (and supplementary table of data).

Scriber, J. M. 1975. *Comparative Nutritional Ecology of Herbivorous Insects: Generalized and Specialized Feeding Strategies in the Papilionidae and Saturniidae* (Lepidoptera). Ph.D. Dissertation. Cornell University, Ithaca, New York.

Scriber, J. M. 1984. Larval foodplant utilization by the world Papilionidae (Lep.): Latitudinal gradients reappraised. *Tokurana Nos. 6/7*: 1–50.

Scriber, J. M. 1987. Population genetics and foodplant use among the North American tree-feeding Papilionidae. Pages 221–230 in V. Labeyrie, G. Fabres, and D. Lachaise (Eds.). *Insects–Plants.* W. Junk Publishers, Dordrecht, The Netherlands.

Scriber, J. M. and P. Feeny. 1979. Growth of herbivorous caterpillars in relation to feeding specialization and to the growth forms of their food plants. *Ecology* **60**: 829–850.

Seigler, D. and P. W. Price. 1976. Secondary compounds in plants: Primary functions. *American Naturalist* **110**: 101–105.

Simms, E. L. and M. D. Rausher. 1989. The evolution of resistance to herbivory in *Ipomoea purpurea*. II. Natural selection by insects and costs of resistance. *Evolution* **43**: 573–585.

Singer, M. C. 1971. Evolution of food-plant preference in the butterfly *Euphydryas editha*. *Evolution* **25**: 383–389.

Slansky, F., Jr. 1972. Latitudinal gradients in species diversity of the New World swallowtail butterflies. *Journal of Research on the Lepidoptera* **11**: 201–217.

Smiley, J. T. 1985. Are chemical barriers necessary for evolution of butterfly–plant associations? *Oecologia* **65**: 580–583.

Southwood, T. R. E. 1961. The evolution of the insect–host tree relationship—A new approach. Proceedings of the 11th. *International Congress of Entomology* (Vienna, 1960). **1**: 651–654.

Southwood, T. R. E. and C. E. J. Kennedy. 1983. Trees as islands. *Oikos* **41**: 359–371.

Sperling, F. A. H. 1986. *Evolution of the Papilio machaon species group in western Canada (Lepidoptera: Papilionidae)*. MS Dissertation. University of Alberta, Edmonton, Alberta, Canada.

Steinly, B. A. and M. Berenbaum. 1985. Histopathological effects of tannins on the midgut epithelium of *Papilio polyxenes* and *Papilio glaucus*. *Entomologia Experimentalis et Applicata* **39**: 3–9.

Straatman, R. 1962. Notes on certain Lepidoptera ovipositing on plants which are toxic to their larvae. *Journal of the Lepidopterists' Society* **16**: 99–103.

Stride, G. O. and R. Straatman. 1962. The host-plant relationship of an Australian swallowtail, *Papilio aegeus*, and its significance in the evolution of host-plant selection. *Proceedings of the Linnean Society of New South Wales* **87**: 69–78.

Strong, D. R., J. H. Lawton, and R. Southwood. 1984. *Insects on Plants*. Blackwell Scientific Publications, Oxford, UK.

Tavormina, S. J. 1982. Sympatric genetic divergence in the leaf-mining insect *Liriomyza brassicae* (Diptera: Agromyzidae). *Evolution* **36**: 523–534.

Thompson, J. N. 1986. Patterns in coevolution. Pages 119–142 in A. R. Stone and D. L. Hawksworth (Eds.). *Coevolution and Systematics*. Clarendon Press, Oxford, UK.

Thompson, J. N. 1988. Evolutionary ecology of the relationship between oviposition preference and performance of offspring in phytophagous insects. *Entomologia Experimentalis et Applicata* **47**: 3–14.

Tsubaki, Y. 1973. The natural mortality and its factors of the immature stages in the population of the swallow-tail butterfly *Papilio xuthus* Linne. *Japanese Journal of Ecology* **23**: 210–217.

Tyler, H. A. 1975. *The Swallowtail Butterflies of North America*. Naturegraph, Healdsburg, CA.

Urzua, A. and H. Priestap. 1985. Aristolochic acids from *Battus polydamas*. *Biochemical Systematics and Ecology* **13**: 169–170.

Via, S. 1986. Genetic covariance between oviposition preference and larval performance in an insect herbivore. *Evolution* **40**: 778–785.

Vinson, S. B. 1984. How parasitoids locate their hosts: A case of insect espionage. Pages 325–348 in T. Lewis (Ed.). *Insect Communication*. Academic Press, London, UK.

Watanabe, M. 1981. Population dynamics of the swallowtail butterfly, *Papilio xuthus* L., in a deforested area. *Researches on Population Ecology* **23**: 74–93.

Watanabe, M. 1982. Leaf structure of *Zanthoxylum ailanthoides* Sieb. et Zucc. (Rutales: Rutaceae) affecting the mortality of a swallowtail butterfly, *Papilio xuthus* L. (Lepidoptera: Papilionidae). *Applied Entomology and Zoology* **17**: 151–159.

Whittaker, R. H. and P. P. Feeny. 1971. Allelochemics: Chemical interactions between species. *Science* **171**: 757–770.

Wiklund, C. 1975. The evolutionary relationship between adult oviposition preferences and larval host plant range in *Papilio machaon* L. *Oecologia (Berlin)* **18**: 185–197.

16 Foodplants and Evolution Within *Papilio glaucus* and *Papilio troilus* Species Groups (Lepidoptera: Papilionidae)

J. MARK SCRIBER, ROBERT C. LEDERHOUSE and
ROBERT H. HAGEN

The concept of chemically-mediated "coevolution" has brought the swallowtail butterfly family, Papilionidae, to the forefront of insect/plant interactions in the last 4 decades (e.g., Dethier 1941, 1954; Forbes, 1958; Munroe and Ehrlich, 1960; Munroe, 1961; Ehrlich and Raven, 1964; Slansky, 1972; Scriber, 1973; Wiklund, 1975; Brown et al. 1980; Berenbaum and Feeny, 1981; Berenbaum, 1983; Feeny et al. 1983; Richard and Guedes, 1983; Thompson 1982, 1988). Most recently, it has been suggested that "stepwise coevolution," "diffuse coevolution," "parallel coevolution," or "sequential coevolution" may be more appropriate terms for the interactions between the Papilionidae and their host plants (Brues, 1920; Jermy, 1976; Berenbaum, 1983; Futuyma, 1983; Mitter and Brooks, 1983; Miller 1987; Thompson 1982, 1986). The Papilionidae are primarily a tropical insect family, with the majority of species and the greatest diversity of lineages found in low latitudes. For example, only 33 of 563 swallowtail species occur in America north of Mexico, and only 11 species are found in Europe (Higgins and Riley 1980; Hodges et al. 1983). Most papilionid species are also specialists, using only one plant family as a larval host (Scriber 1984a). There is little diversity among the plant families that are used as hosts: approximately 80% of the species for which host records exist are restricted to Aristolochiaceae, Annonaceae, Lauraceae, Rutaceae, or Umbelliferae (Scriber 1984a).

Except for Umbelliferae, the major host-plant families for Papilionidae are largely tropical in distribution. It is not surprising therefore that the proportion of specialists among papilionid species is greatest at low latitudes (Scriber 1973, 1984a). The lineages that have been most successful outside the tropics are either

---

Plant-Animal Interactions: Evolutionary Ecology in Tropical and Temperate Regions, Edited by
Peter W. Price, Thomas M. Lewinsohn, G. Wilson Fernandes, and Woodruff W. Benson.
ISBN 0-471-50937-X © 1991 John Wiley & Sons, Inc.

those that have shifted to families of temperate zone plants or those that have broadened their host ranges to include a greater range of plant families. Members of the Holarctic *Papilio machaon* species group, for example, feed on Umbelliferae and Compositae. Considerable evidence indicates that ancestors of the *machaon* group fed on Rutaceae and that the shift to Umbelliferae and Compositae was facilitated by chemical similarities between these plant families (Dethier 1941;

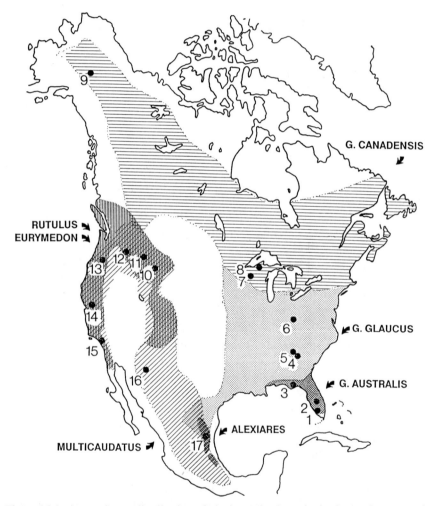

Figure 16-1. Approximate distribution of species and subspecies in the *P. glaucus* species group. The ranges of *P. eurymedon* and *P. rutulus* overlap extensively and are not indicated separately. The subspecies of *P. glaucus* (*australis, glaucus,* and *canadensis*) are shown with allopatric distributions, though *australis* is doubtfully distinct from *glaucus,* and the ranges of *canadensis* and *glaucus* may overlap partially. Numbers indicate primary collecting sites for the electrophoretic study (Hagen and Scriber 1991).

Ehrlich and Raven 1964; Berenbaum 1983; Feeny et al. 1983) and likely to have involved major changes in behavior (Thompson 1988a, b).

Another lineage that has successfully radiated out of the tropics is represented by the North American *Papilio glaucus* species group (Fig. 16-1). The most widely distributed member of this group, *P. glaucus*, is also the most polyphagous species in the Papilionidae. The reported larval host range for *P. glaucus* includes at least

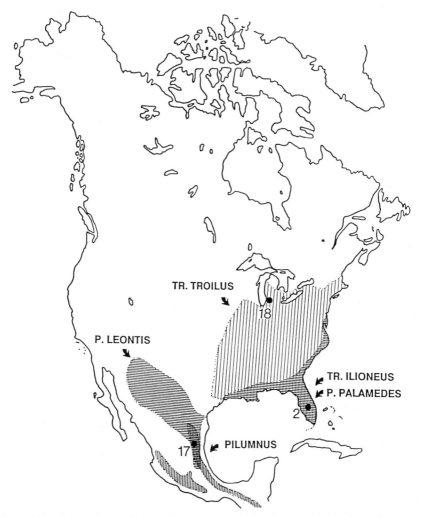

Figure 16-2. Approximate distribution of species and subspecies in the *P. troilus* species group. The ranges of *P. troilus ilioneus* and *P. palamedes palamedes* overlap extensively and are indicated together. *Papilio pilumnus* occupies disjunct areas in the eastern and western cordilleras of central Mexico. The eastern area overlaps partially with the southern end of *P. palamedes leontis'* distribution. Numbers indicate primary collecting sites for the taxa (Hagen and Scriber 1991).

30 genera in 17 plant families (Scudder 1889; Tietz 1972; Scriber 1973, 1984a, 1988). Species in the *glaucus* group are capable of feeding on hosts in plant families used by no other swallowtail butterflies. However, unlike the *machaon* group, there are no clear links between the major swallowtail host plant families and many of the host plants of the *glaucus* group (Brues 1920; Hagen 1986). The *glaucus* group appears to have broken out of the typical swallowtail pattern of sequential coevolution (Hagen 1986; Scriber 1988).

A contrast to the *glaucus* group is provided by the closely related *P. troilus* species group, which is also North American (Fig. 16-2). Unlike the *glaucus* group, *troilus* group species are all specialists on host plants in the Lauraceae. Understanding the historical, behavioral, and physiological bases of differences in host plant use between the *glaucus* and *troilus* species groups may help to illuminate barriers that constrain evolution of host-plant associations within other insect taxa.

In this chapter we discuss hypotheses concerning the evolution of host plant use within the *P. glaucus* and *P. troilus* species groups. Our interpretations are based on comparisons of the biogeography of both species groups and their host plants and upon extensive experimental studies of larval growth and survival on a wide range of plant species. Unfortunately, present knowledge of the phylogeny of *Papilio* is inadequate to test detailed hypotheses about the history of particular host use abilities within the *glaucus* and *troilus* groups (cf. Miller 1987; Mitter and Brooks 1983). However, we are able to draw some tentative conclusions along these lines using data from allozyme electrophoresis of *glaucus* and *troilus* butterflies (Hagen and Scriber 1991).

SYSTEMATICS AND PATTERNS OF HOST PLANT USE IN THE PAPILIONIDAE

Most swallowtail species are dietary specialists as larvae (Scriber 1984a, 1988). Oligophagous feeding behavior prevails in the Papilionidae, with entire lineages (subfamilies and tribes) feeding primarily on a single host-plant family (e.g., the Baroniinae on Leguminosae; the Parnassiinae and the Troidini of the Papilioninae on Aristolochiaceae; most of the Graphiini on the Annonaceae; and most of the Papilionini on the Rutaceae (Igarashi 1984; Scriber 1984a). In spite of close feeding affiliations (see also Forbes 1958), it has been concluded that although plant secondary compounds have influenced the cladogenesis of the Papilionidae, it is less likely that the insects have affected cladogenesis in plants. This has been demonstrated for the Graphiini and their hosts (Miller 1987).

A starting point for discussing the evolution of host plant associations within the *glaucus* and *troilus* species groups is a review of the current state of knowledge on the phylogeny of the Papilionidae. Recent systematic treatments of the Papilionidae generally agree in their interpretation of phylogenetic relationships within the family (Munroe and Ehrlich 1960; Munroe 1961; Hancock 1983; Miller 1987). The largest subfamily, Papilioninae, consists of 4 tribes. The largest of these tribes is the Papilionini with 216 species in 2 genera, *Meandrusa* Moore (one species) and *Papilio* L. (Munroe 1961; Miller 1987). Munroe (1961) was unable to define natural taxa within *Papilio*, and only recognized informal subgeneric groups ("sections" and "species groups"). To date there has been no serious systematic work focused exclusively on the Papilionini: the number of species, their worldwide distribution, and the evident importance of immature stages for character analysis make this a formidable task. It will be several years or more before an adequate phylogeny is available for the tribe as a whole (J. S. Miller, personal communication). For the present, use of Munroe's conservative and informal classification is preferable to alternatives that divide *Papilio* into several genera (Ferris and Brown 1981; Hancock 1983).

Despite the difficulties presented by the Papilionini as a whole, the *glaucus* and *troilus* species groups appear to be monophyletic taxa within it. Rothschild and Jordan (1906), who first described the species groups, noted the cohesiveness of the *glaucus* group in particular. Forbes (1951) recognized the affinities between the *glaucus* and *troilus* groups, which Munroe (1961) subsequently placed together as section III of *Papilio*. Hancock (1983) identified three South American species groups (*scamander, zagreus,* and *homerus* species groups comprising Section V of *Papilio*) as the sister taxon to the *glaucus* and *troilus* groups.

Strong arguments can be made for linking the *glaucus* and *troilus* groups as sister taxa. Characters that appear to be synapomorphies for these include the presence of a medial band in the hind wing (Rothschild and Jordan 1906), the presence of an extra, lateral row of larval proleg crochets (Forbes 1951), and the form of the eyespot on the mature larva. Shared characters of the male genitalia also appear to link the groups (Brower 1959a; J. S. Miller, personal communication). In the *glaucus* and *troilus* species groups the crochet pattern is associated with a method of larval attachment to the host plant that appears to be a derived condition within *Papilio* (for description, see Scudder 1889; R. Hagen, manuscript in preparation).

In spite of their general notice and historical presence in North America (*P. glaucus* was the first insect drawn from North America in the 16th century and the first butterfly described from North America by Linneaus) the relationships of species within the *glaucus* and *troilus* groups have remained poorly known. The *glaucus* species group consists of 5 North American species (Brower 1958a, 1959a): *P. glaucus* L., *P. alexiares* Hoppfer, *P. eurymedon* Lucas, *P. rutulus* Lucus, and *P. multicaudatus* Kirby (Fig. 16-1). Three subspecies have been described within *P. glaucus* (*P. g. canadensis* Rothschild and Jordan, *P. g. australis* Maynard, and *P. g. glaucus*) and two subspecies within *P. alexiares* (*P. a. garcia* Rothschild and Jordan and *P. a. alexiares*). The *troilus* species group consists of 3 species, also exclusively North American: *P. troilus* L., *P. palamedes* Drury, and *P. pilumnus* Boisduval (Fig. 16-2). Two subspecies have been described within *P. troilus* (*ilioneus* Smith and *P.t. troilus*) and within *P. palamedes* (*P.p. leontis* Rothschild and Jordan and *P.p. palamedes*).

Brower (1959a, b) evaluated the status of species within the *glaucus* group. He confirmed the original placement of *P. pilumnus* within the *troilus* group: the superficial resemblance of the adults to those of *P. glaucus* had led Rothschild and Jordan (1906) to include it in the *glaucus* group. On purely phenetic grounds, he suggested that *P. rutulus, P. eurymedon.* and *P. multicaudatus* were more closely related to one another than to *P. glaucus*, while the position of *P. alexiares* was unclear. Brower considered *P. rutulus* to be most similar of these to *P. glaucus*. More recently, Scott has treated *rutulus* and *alexiares* as subspecies of *P. glaucus* along with *P.g. canadensis* (Scott and Shepard 1976; Scott 1986). On the basis of common host-plant use and northern distributions, Scriber (1986a) suggested that *P. rutulus* and *P. glaucus canadensis* may be more closely related than either is to *P. glaucus glaucus*.

The exclusive use of Lauraceae as a larval host by the *troilus*-group species is consistent with retention of an ancestral food plant by this lineage, shared by other clades within the tribes Papilionini and Leptocircini (Munroe and Ehrlich 1960; Hancock 1983; Miller 1987). Specialization on a single plant family is typical for Papilionidae: only 61 (21%) of the 281 species whose hosts are known use more than one family as larval hosts (Scriber 1973, 1984a). Most of these hosts share similar secondary chemistry: only 29 swallowtail species use plant families which produce neither benzylisoquinoline alkaloids nor hydroxycoumarins; Lauraceae produce both (cf. Feeny et al. 1983; Berenbaum 1983; Miller and Feeny 1983; Hagen 1986).

Host shifts between plant families sharing secondary compounds may be relatively easily achieved (all else being equal; e.g., natural enemies), unlike shifts to chemically dissimilar hosts (Dethier 1941, 1954; Ehrlich and Raven 1964; Feeny et al. 1983; Richard and Guedes 1983). The taxonomic distribution of the exceptional papilionid species does not suggest common ancestry distinct from typical species, but rather, independent occurrences of "escape" from specialization on chemically related hosts (Munroe and Ehrlich 1960; Hagen 1986). This type of escape as yet has received little attention in the study of plant–insect interactions.

All of the *glaucus* group species use more than one host-plant family. One of these, Betulaceae, lacks both hydroxycoumarins and benzylisoquinoline alkaloids (Hagen 1986). Larvae of *glaucus* group species are the only *Papilio* species that feed on plants in the Oleaceae, Rosaceae, Salicaceae, Betulaceae, Rhamnaceae, and Tiliaceae (Scriber 1984b; 1986a, b). The most parsimonious hypothesis is that both host shifts (adoption of novel host families) and dietary expansion (evolution of polyphagy) occurred within the *glaucus* group after its ancestor diverged from the *troilus* group ancestor. The relationship between these evolutionary changes is unknown, as are the steps by which they occurred. Comparative studies of *glaucus* and *troilus* group species, offer the best approach toward resolving this uncertainty.

Remington (1951) and Brower (1958a, 1959b) discuss an additional possible scenario for the evolutionary history of the *Papilio glaucus* group. Implicit in this model is the assumption that the common ancestor for the group was polyphagous and that this characteristic is retained by the eastern (allopatric)

P. glaucus species. The more specialized host use of the three western (sympatric) species is assumed to be derived from this ancestral *glaucus*-type polyphagy. It was further presumed that the Pleistocene glaciations aided this reproductive isolation between the eastern and western ancestral components. Subsequent subdivision of foodplant choices (i.e., feeding specialization) may have been reinforced by interspecific competition for escape from predation in the sympatric western populations (Brower 1958b). In contrast, the same predation pressures on tiger swallowtails might favor the use of varied types of foodplants in the east, where only a single species remained (Brower 1958b).

The three western North American species of swallowtails in the *glaucus* group (i.e., *P. rutulus*, *P. eurymedon*, and *P. multicaudatus*) and the Mexican swallowtail *P. alexiares* might be considered to have ecologically subdivided the broad range of *P. glaucus* foodplants since each normally uses only 1–3 mutually exclusive plant families (Brower, 1958a; Scriber, 1988; Scriber et al. 1988). This is consistent with well-known resource-partitioning interpretations (e.g., Benson 1978). Presumably, other reproductive isolating mechanisms evolved in the sympatric western group including different flight periods, different altitudinal (or habitat) preferences in the mountains, and different wing color (yellow or white) patterns (Brower 1959a). The effectiveness of these and other reproductive isolating mechanisms is currently under investigation in our lab.

Experimental Studies of Larval Host Plant Use

Plant species from eight different families were used in no-choice bioassays with newly eclosed first instar larvae from the 10 different *Papilio* taxa of the *glaucus* group and the *troilus* group. *Papilio cresphontes*, which is a sympatric tree-feeder from a different section, was included in the bioassays for comparison.

Field-collected and laboratory-mated females were set up in plastic boxes (10 cm × 20 cm × 27 cm) with a sprig of an appropriate host plant under saturated humidity. The boxes were placed 0.7–1.0 m from continuously lighted 100-watt incandescent bulbs. Eggs were collected at 2-day intervals and larvae were removed as they hatched. Using fine camel-hair brushes, first instar larvae (neonate) were gently placed on fresh leaves of various potential host plants for bioassays of consumption and survival. Leaf moisture was maintained using aquapics, and fresh leaves were provided 3 times per week throughout larval development. Larval survival equaled the percent of first instars set up on a host that successfully molted to the second instar. For some analyses, larvae from different mothers were lumped to yield a single survival value (Tables 16-1 and 16-3). Elsewhere, means were calculated with each mother considered a replicate (Table 16-2).

Neonate Survival: **glaucus** *Group Species.* *Papilio glaucus* (with its three sub-species) is easily the most polyphagous of all of the 563 species of Papilionidae in the world. We have tested larvae on 147 different plant species from 34 different plant families (Hagen 1986; Scriber 1988). No single food plant can suffice

TABLE 16-1. Neonate Survival of 7 Taxa of the Tiger Swallowtail (*glaucus*) Group and *P. cresphontes* on Various Species of 10 Plant Families (1979–1987)[a]

	eurymedon (%)	rutulus (%)	canadensis (%)	alexiares (%)	australis (%)	glaucus (%)	multi-caudatus (%)	cresphontes (%)
Rutaceae								
Ptelea trifoliata (hoptree)	6/6 (100)	8/39 (21)	87/117 (74)	5/8 (63)	11/23 (48)	264/387 (68)	5/7 (71)	78/102 (76)
Citrus limon (grapefruit)	— (?)	— (?)	— (?)	0/30 (0)	50/133 (38)	5/16 (31)	— (?)	2/2 (100)
Lauraceae								
Persea borbonia (redbay)	0/3 (0)	0/8 (0)	0/2 (0)	0/16 (0)	0/212 (0)	0/204 (0)	— (?)	— (?)
Lindera benzoin (spicebush)	0/6 (0)	12/26 (46)	2/175 (1)	1/5 (20)	5/30 (17)	129/549 (24)	— (/)	0/29 (0)
Sassafras albidum (sassafras)	5/6 (83)	25/36 (69)	35/58 (60)	— (?)	9/33 (27)	173/273 (63)	— (?)	0/2 (0)
Cinnamomum camphora (camphortree)	— (?)	4/4 (100)	— (?)	— (?)	84/134 (63)	— (?)	— (?)	— (?)
Magnoliaceae								
Liriodendron tulipifera (tuliptree)	0/17 (0)	1/30 (3)	4/446 (1)	3/7 (43)	233/308 (76)	2216/2818 (79)	— (?)	0/2 (0)
Magnolia virginiana (sweetbay)	0/10 (0)	0/10 (0)	2/85 (2)	5/46 (11)	579/659 (88)	789/531 (52)	— (?)	0/2 (0)
M. acuminata (cucumbertree)	0/8 (0)	1/46 (2)	83/133 (62)	0/2 (0)	15/29 (52)	215/364 (59)	— (?)	0/17 (0)
Rosaceae								
Prunus serotina (black cherry)	90/99 (90)	45/59 (76)	2243/3040 (74)	299/337 (89)	500/641 (78)	4876/6310 (77)	12/15 (80)	0/2 (0)
P. virginiana (choke cherry)	8/8 (100)	14/21 (67)	140/208 (67)	2/2 (100)	15/19 (79)	198/306 (65)	5/7 (71)	— (?)

Oleaceae								
Fraxinus americana (white ash)	2/7 (29)	31/47 (66)	100/128 (78)	3/3 (100)	29/33 (88)	374/442 (79)	5/7 (71)	— (?)
Betulaceae								
Betula papyrifera (paper birch)	5/10 (50)	25/47 (53)	145/192 (76)	2/5 (40)	18/66 (27)	143/474 (30)	— (?)	0/1 (0)
Salicaceae								
Populus tremuloides (quaking aspen)	5/16 (31)	104/148 (70)	346/478 (72)	0/6 (0)	7/137 (5)	177/2188 (8)	— (?)	— (?)
P. balsamifera (balsam poplar)	2/10 (20)	4/6 (67)	121/186 (65)	0/3 (0)	0/2 (0)	9/109 (8)	— (?)	— (?)
P. deltoides (cottonwood)	0/4 (0)	28/41 (68)	28/82 (34)	0/1 (0)	0/15 (0)	2/273 (3)	— (?)	0/2 (0)
P. grandidentata (big-toothed aspen)	0/4 (0)	22/37 (59)	34/88 (39)	1/5 (20)	0/11 (0)	24/343 (7)	— (?)	— (?)
Platanaceae								
Platanus occidentalis (sycamore)	1/3 (33)	50/58 (86)	16/39 (41)	3/6 (50)	4/31 (13)	74/209 (35)	— (?)	0/2 (0)
Rhamnaceae								
Rhamnus spp.	9/19 (47)	0/11 (0)	2/23 (9)	0/2 (0)	0/119 (0)	3/113 (3)	— (?)	— (?)
Tiliaceae								
Tilia americana (basswood)	— (?)	27/58 (47)	52/76 (68)	2/2 (100)	9/18 (50)	66/158 (42)	— (?)	— (?)

[a]Top ratio is survivors/total *n*; percent survival is given in parentheses.

349

throughout the range of *Papilio glaucus* in North America, because no food plant species has a range that large. Therefore, local specialization is a necessity (Scriber 1986b). Unfortunately, little is yet known about the association between oviposition preference and larval performance for these *Papilio* in particular or for insects in general (Futuyma and Peterson 1985; but see Wiklund 1975; Via 1986; Singer et al. 1988; Thompson 1988a, b).

Black cherry, *Prunus serotina* (Rosaceae), is the only plant species tested which was of exceptional suitability for larvae of all 7 taxa of the *P. glaucus* group (Table 16-1). This foodplant serves as our common denominator and primary foodplant for laboratory mass-rearing. In hybrid and backcross studies, it serves as a very handy "control" to assess "heterosis" or "hybrid dysgenesis" (Scriber 1986a). Survival of larvae of all *P. glaucus* taxa was also very good on choke cherry, *Prunus virginiana*, but growth was much slower (Scriber et al. 1982).

The most significant dichotomies in host-use abilities within the *glaucus* group are those for Magnoliaceae and Salicaceae (Table 16-1; Scriber 1982, 1983, 1984b, and 1986a). While *P. g. canadensis*, *P. eurymedon*, and *P. rutulus* survived and grew well on the Salicaceae, they died when fed either sweetbay, *Magnolia virginiana*, or tuliptree, *Liriodendron tulipifera*, of the Magnoliaceae. In contrast, *P. g. australis* and *P. g. glaucus* generally perform excellently on either of the Magnoliaceae hosts, but generally died on the Salicaceae (Table 16-1). The Mexican swallowtail *P. alexiares* and *P. multicaudatus* grew poorly on both of these families (Scriber et al. 1988).

Hosts of any given plant family are not of uniform suitability to *P. glaucus* group species. The Lauraceae species we assayed are a good example. Redbay, *Persea borbonia*, killed all larvae of the *P. glaucus* group species we tested although *P. palamedes* and *P. troilus* developed on this host. However, sassafras, *Sassafras albidum*, was the only host species to approach black cherry in general suitability for all *P. glaucus* group species. Camphor, *Cinnamomum camphora*, and spicebush, *Lindera benzoin*, gave intermediate to low larval survival.

It should also be noted that *P. rutulus* in Sacramento County, California has apparently specialized on sycamore, *Platanus occidentalis*, of the Platanaceae (Brower 1958a). The survival of *P. rutulus* on sycamore was a very good 93%. Much poorer survival was observed for several other *Papilio* taxa on this foodplant family. Similarly, *P. eurymedon* has specialized on the Rhamnaceae in California and in the western United States. Bioassays suggest that *P. eurymedon* was the only one of the 7 *glaucus* group taxa capable of surviving to pupation on *Rhamnus* hosts (see also Brower 1958a).

In contrast to the generalist pattern seen in the *P. glaucus* group species, *P. cresphontes* is a strict plant family specialist (Table 16-1). Larvae survived and grew well on hoptree, *Ptelea trifoliata*, grapefruit, *Citrus limon*, and prickly-ash, *Zanthoxylum americanum*, all Rutaceae. However, *P. cresphontes* larvae failed to develop on any other plant family. This resulted in part from the failure of larvae to initiate feeding on nonrutaceous plants.

Neonate Survival:* troilus *Group Species. Members of the *P. troilus* species group

TABLE 16-2. No-Choice Feeding Bioassays of *P. troilus* Group Species[a]

	palamedes	troilus	pilumnus
Lauraceae			
Persea borbonia (redbay)	77 ± 5	47 ± 13	0
	(30, 562)	(8, 119)	(1, 1)
Lindera benzoin (spicebush)	28 ± 5	86 ± 4	
	(20, 165)	(7, 156)	
Sassafras albidum (sassafras)	78 ± 4	79 ± 6	50 ± 35
	(28, 272)	(10, 161)	(2, 2)
Cinnamomum camphora (camphortree)	52 ± 5	50 ± 8	0
	(28, 285)	(6, 82)	(1, 1)
Persea americana (avocado)	8 ± 5	0 ± 0	
	(9, 80)	(2, 13)	
Rutaceae			
Ptelea trifoliata (hoptree)	0 ± 0	0 ± 0	
	(12, 79)	(7, 63)	
Citrus limon (grapefruit)	0 ± 0	0	
	(14, 131)	(1, 6)	
Magnoliaceae			
Liriodendron tulipifera (tuliptree)	0 ± 0	0 ± 0	
	(12, 117)	(5, 39)	
Magnolia virginiana (sweetbay)	0 ± 0	0 ± 0	
	(19, 194)	(3, 20)	
M. acuminata (cucumbertree)	0 ± 0	0 ± 0	
	(9, 73)	(5, 60)	
Rosaceae			
Prunus serotina (black cherry)	0 ± 0	0 ± 0	
	(12, 80)	(6, 46)	
Oleaceae			
Fraxinus americana (white ash)	0 ± 0	0 ± 0	
	(5, 41)	(4, 33)	
Betulaceae			
Betula papyrifera (paper birch)	0 ± 0	0 ± 0	
	(6, 35)	(4, 35)	
Salicaceae	0 ± 0	0 ± 0	
Populus tremuloides (quaking aspen)	(13, 79)	(4, 35)	
Platanaceae			
Platanus occidentalis (sycamore)	4 ± 2	2 ± 2	
	(14, 114)	(3, 38)	
Tiliaceae			
Tilia americana (basswood)	0 ± 0	0 ± 0	
	(2, 10)	(2, 16)	

[a]Values are mean percents ± one standard error. The numbers of families and individual larvae assayed are given in parentheses.

TABLE 16-3. First Instar Survival of Neonate Larvae of Six *Papilio glaucus* Group Taxa and Their Hybrids (1980–1987)

Geographic Origin	Parents[a]	Quaking Aspen		Tuliptree		Black Cherry	
		n	(%)	n	(%)	n	(%)
I. West and north	P. eurymedon (e)	5/16	(31)	0/17	(0)	90/99	(90)
	P. rutulus (r)	104/148	(70)	1/30	(3)	45/59	(76)
	P.g. canadensis (c)	346/478	(72)	4/446	(1)	2243/3040	(74)
	e × r	12/16	(75)	0/31	(0)	7/10	(70)
	r × e	14/24	(58)	1/16	(6)	37/47	(79)
	e × r	36/52	(69)	0/47	(0)	71/105	(68)
	r × c	1/1	(100)	—	—	1/3	(33)
	e × c	2/2	(100)	0/2	(0)	26/28	(93)
	c × e	2/16	(13)	2/29	(7)	306/341	(90)
	Group I Totals:	**522/753**	**(69.3)**	**8/618**	**(1.3)**	**2826/3732**	**(75.7)**
II. Cross-Pairings (W/N × S/E)	e × m	—	—	—	—	20/27	(74)
	m × e	—	—	—	—	—	—
	e × a	—	—	0/1	(0)	6/8	(75)
	a × e	7/10	(70)	10/12	(83)	5/5	(100)
	e × g	—	—	—	—	—	—
	g × e	14/28	(50)	50/75	(75)	361/418	(86)
	r × m	17/21	(81)	1/18	(6)	19/20	(95)
	m × r	4/5	(80)	1/5	(20)	19/21	(91)
	r × a	—	—	—	—	12/13	(92)

a × r	56/104 (54)	101/131 (77)	136/159 (86)
r × g	—	—	2/3 (67)
g × r	125/190 (66)	152/234 (65)	388/486 (80)
c × m	14/24 (58)	—	29/46 (68)
m × c	—	—	—
c × a	15/24 (63)	21/25 (84)	23/31 (74)
a × c	43/46 (94)	61/92 (66)	83/109 (76)
c × g	223/327 (68)	173/224 (77)	176/260 (68)
g × c	421/690 (61)	664/885 (75)	2550/3192 (80)
Group II Totals:	**922/1448** (63.7)	**1233/1684** (73.2)	**3810/4778** (79.7)
III. South and east			
g × a	10/53 (19)	17/31 (55)	183/298 (61)
a × g	0/8 (0)	9/10 (90)	7/11 (64)
g × m	0/145 (0)	63/74 (85)	558/678 (83)
m × g	—	—	9/9 (100)
a × m	1/2 (50)	14/14 100	40/49 (83)
m × a	—	—	—
P. alexiares (m)	0/6 (0)	3/7 (43)	299/337 (89)
P.g. australis (a)	7/137 (5)	233/308 (76)	500/641 (78)
P.g. glaucus (g)	177/2188 (8)	2216/2818 (79)	4876/6310 (77)
Group III Totals:	**195/2539** (7.7)	**2555/3262** (78.3)	**6472/8333** (77.7)

[a]Female is listed first in hybrid.

used species of the Lauraceae almost exclusively as hosts in no-choice feeding assays (Table 16-2). Larvae rarely even initiated feeding on foliage of other plant families. One exception was sycamore (Platanaceae) on which some larvae did feed and grew. Although all *troilus* group species are Lauraceae specialists, each species differs in its ability to develop on a given host. Larvae of *P. palamedes* survived and grew well on redbay, *Persea borbonia*, but performance on spicebush, *Lindera benzoin*, was very poor. In contrast, *P. troilus* larvae survived best on spicebush, and on average worst on redbay. Sassafras was utilized well by *P. troilus*, *P. palamedes*, and *P. pilumnus*, and is the one lauraceous species that *P. glaucus* group species also use successfully (Scriber 1986a, Table 16-1). To date, we have had too few pure *P. pilumnus* larvae to draw any conclusions about this species. However, we were able to rear one individual to adulthood on sassafras.

Both *P. troilus* and *P. palamedes* showed subspecific–regional differences in host use (Nitao et al. 1991). Larvae of *P. t. ilioneus* from central Florida averaged 76% survival on redbay compared to 19% for Michigan *P. t. troilus*. Redbay may be the only lauraceous host used in the southern half of Florida, but it extends northward only in a narrow band along the Atlantic Ocean and Gulf of Mexico. Some larvae of the Mexican *P. p. leontis* initiated feeding and developed on avocado, *Persea americana*, yet *P. p. palamedes* from Florida starved without even biting avocado. Since the diversity of Lauraceae increases substantially in Mexico, both *P. p. leontis* and *P. pilumnus* undoubtedly use additional hosts that we have not yet identified.

A fundamental difference between larvae of the *troilus* species group and the *glaucus* species group is in the initiation of feeding behavior. In no-choice feeding assays, *troilus* group larvae starved rather than initiate feeding on nonhosts. In contrast, *P. glaucus* larvae initiated feeding on 57 of 59 species from 34 plant families in no-choice assays, although they were able to develop on only 21 of these species (Hagen 1986). Other *glaucus* group larvae showed a similar inclination to initiate feeding on a broad range of woody plant leaves. We suggest the host range in *troilus* group species is restricted primarily by oviposition and feeding behavior and not by digestive physiology. In *glaucus* group species, larvae sample whatever plant they are on and develop on any plant their digestive physiology can handle. Hagen (1986) induced *P. troilus* larvae to initiate feeding on black cherry leaves painted with sassafras extract, although they died shortly afterward. We are conducting similar feeding studies to identify the specific feeding elicitors present in Lauraceae. We hope to determine, if the behavioral hurdle can be cleared, whether the toxicological barrier is also present for *troilus* group species on various *glaucus* group food plants.

Neonate Survival: Interspecific and Intersubspecific Hybrids. Using the technique of hand-pairing in the laboratory, we have hydridized all 10 taxa of the *Papilio glaucus* and *troilus* species groups in virtually every possible combination (Scriber 1986a, 1987, 1988; Scriber and Lederhouse 1988; Scriber et al. 1988). Most of these pairings have resulted in hybrid larvae which have been bioassayed in no-choice tests on key food plants.

The reciprocal inabilities of *P. g. glaucus , P. g. australis*, and *P. alexiares* to use Salicaceae and *P. g. canadensis, P. rutulus*, and *P. eurymedon* to use Magnoliaceae are genetically based. Hybridization of individuals within either of these two groupings will not produce hybrids capable of using both plant families through heterosis (Table 16-3). However, a pairing between individuals of the two groupings will yield offspring which can detoxify both the Salicaceae and Magnoliaceae (Scriber 1986a, 1987; Table 16-3).

Although the Salicaceae and Magnoliaceae may contain toxicological barriers to successful feeding by *troilus* group larvae, we believe that the group may be limited in its larval host range (see Table 16-2) primarily because of behavioral hurdles rather than toxicological barriers. Scriber and Lederhouse (1988) described the feeding behavior of hybrids between *P. glaucus* and *P. pilumnus*, which died on the "common denominator" host for the *glaucus* group, black cherry. Unlike the toxicity of Salicaceae and Magnoliaceae to certain *P. glaucus* taxa in which unadapted individuals feed and produce frass before dying, the *glaucus × pilumnus* larvae starved to death on black cherry. The *glaucus × pilumnus* hybrids fed and survived on spicebush and sassafras. Both parental species have some ability to use these Lauraceae hosts. In addition, the hybrids fed and grew on hoptree (Rutaceae) and tuliptree (Magnoliaceae) on which *P. pilumnus* larvae would not initiate feeding.

Hybrid larvae from *P. glaucus* females paired with Brazilian *P. scamander* males fed and grew to pupation on tuliptree, sweetbay, and cucumbertree of the Magnoliaceae, a host family held in common. In addition, they survived on black cherry (Rosaceae), but not on redbay (Lauraceae). First instar *palamedes × scamander* larvae survived on redbay, camphortree, and sweetbay but not on tuliptree (Scriber et al. 1990).

Bioassays of the larvae of hybrids of *troilus* group species with generalist species indicate that these larvae recognize more plants as potential hosts and can use them once feeding is initiated. It is intriguing that the additional host species were members of the Rutaceae and Magnoliaceae, host families for other *Papilio* groups. We hope to separate initiation of feeding from successful utilization of hosts (i.e., behavioral antixenosis from toxic antibiosis) in part through additional studies of hybrids of *troilus* group species with generalists.

GENETIC DIFFERENTIATION WITHIN THE GLAUCUS AND TROILUS GROUPS

Allozyme electrophoresis is a useful tool for assessing the extent of genetic differentiation among closely related species (Avise 1975; Berlocher 1984). General methods for electrophoresis have been discussed in several sources (Harris and Hopkinson 1978; May et al. 1979; Richardson et al. 1986). An important feature of allozymes is that variation detected by electrophoresis is easly interpreted in terms of allelic variation at discrete genetic loci. Patterns of allozyme variation combined across many loci can be used for estimating phylogenies (Buth 1984; Berlocher 1984). One popular method for combining information across loci is through calculation of genetic distance measures between taxa (Wright 1978; Nei 1972, 1978). Provided that variation at each locus is controlled primarily by genetic drift (or, at least, that selection is uncorrelated across loci), distances can be interpreted as estimates of the average amount of genetic divergence that has occurred between two taxa since they last shared a common ancestor. Assuming further that the rates of increase in genetic divergence do not differ greatly among pairs of taxa being compared, a tree constructed from a matrix of genetic distances can be a good estimate of the true phylogeny (Berlocher 1984; Felsenstein 1984; but see Farris 1981).

A tree for the *glaucus* and *troilus* species groups, using Nei's (1978) genetic distance measure, is shown in Figure 16-3. The genetic distances used in construction of the tree were based on 11 allozyme loci; details are given in Hagen and Scriber (in preparation). Two aspects of this tree are of particular interest in the context of the evolution of host-plant use among the species groups. The first is that the overall amount of divergence among taxa is similar in the two species groups, despite differences in the diversity of host-plant families used by each. If anything, members of the *troilus* group—all Lauraceae specialists—are more divergent from one another (maximum $D = 0.59$) than are the *glaucus* group taxa (maximum $D = 0.38$), which include Salicaceae feeders, Magnoliaceae feeders, and even a unique Rhamnaceae feeder (Table 16-1). At a lower taxonomic level within the *glaucus* group, differentiation in host use is also uncorrelated with

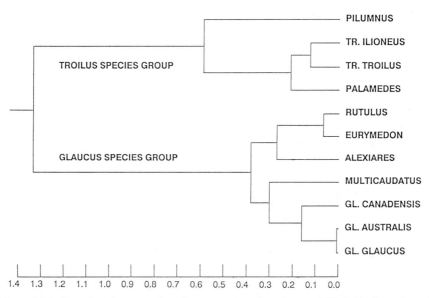

Figure 16-3. Tree for *glaucus* and *troilus* group taxa based upon UPGMA clustering of Nei's (1978) genetic distances, estimated from 11 allozyme loci (Hagen and Scriber 1991). Genetic distances are indicated on the scale.

average genetic divergence. For example, *P. eurymedon* and *P. rutulus* are scarcely differentiated from one another by allozymes, yet *P. eurymedon* larvae are uniquely able to use Rhamnaceae.

The second feature of the tree is that within the *glaucus* group, taxa able to use Salicaceae (*P. eurymedon, P. rutulus,* and *P.g. canadensis*) are not grouped together, distinct from the taxa (*P.g. glaucus, P. alexiares*) capable of using Magnoliaceae. To the extent that the tree may accurately represent the true phylogenetic history of the *glaucus* group, this implies at least two separate origins for the ability to use Salicaceae (see below).

THE RUTACEAE AND LAURACEAE AS HOSTS: ANCESTRAL ADAPTATIONS

The ability of all seven taxa of the *P. glaucus* group to feed and grow well on hoptree (Rutaceae) links the group with other *Papilio* species also able to use the citrus family as host plants (Scriber 1973, 1984b; Hancock 1983; Miller 1987). Munroe (1961), Hancock (1983), and Miller (1987) note that the Rutaceae-feeding habit occurs in Section III, Section IV, and Section II, which consists of a majority of Rutaceae–Umbelliferae feeders. In fact, the Rutaceae are reported as hosts for some of the species in the Graphiini and Troidini tribes as well as the Papilionini (Scriber 1984a). If the reports are accurate, this implies either the Rutaceae were the original hosts for the genus (Hancock 1979) or else convergent colonization resulting from preadaptation to the rutaceous chemicals occurred (Miller 1987). Hancock (1983) believes that, early in the development of the genus, foodplant preferences switched to the Lauraceae and Magnoliaceae from Rutaceae, possibly due to a reduction in the abundance and species diversity of the tropical Rutaceae. If so, then the *troilus* group has lost sensitivity to oviposition and feeding cues of the Rutaceae.

There are presently at least 26 species of Papilionidae known to utilize the Lauraceae as larval foodplants, including the genera *Eurytides, Protographium, Graphium,* and *Papilio* (Scriber 1973, 1984a). As mentioned earlier, the New World has two primary sections of *Papilio* that feed on the Lauraceae: (1) the Neotropical *P. homerus, P. scamander,* and *P. zagreus* groups (Section V) which appear to be polyphagous (Scriber 1984a; Ruszczyk 1986) and (2) the North American *P. glaucus* and *P. troilus* groups (Section III). Both groups have similar green larvae, bearing large mimetic thoracic eyespots, whereas larvae of the Old World Lauraceae-feeding *P. clytia* and *P. agestor* groups (Section I) differ greatly in appearance (Munroe, 1961).

Although all *P. troilus* and *P. palamedes* neonates die on plants of foodplant families other than Lauraceae (the Platanaceae appear to be an exception), death may not be due to toxicity as seems to be the case with the unadapted neonates of other taxa on the Magnoliaceae and Salicaceae. Rather, larvae starve to death because they refuse to initiate feeding on foreign (nonlauraceous) hosts (Fig. 16-4).

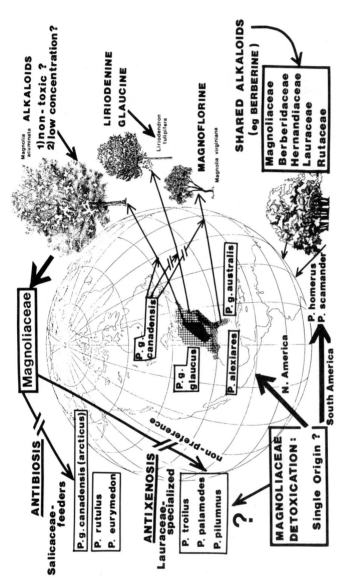

Figure 16-4. Schematic of the ability of various *Papilio* taxa to feed on plants in the Magnoliaceae family. In the southeastern United States *P.g. glaucus*, *P.g. australis*, and *P. alexiares* survive well on all Magnoliaceae species tested (*Magnolia virginiana*, *M. acuminata*, and *Liriodendron tulipifera*). With the exception of *Papilio glaucus canadensis* from the Great Lakes on *M. acuminata*, there is virtually no ability to survive on Magnoliaceae for *P. rutulus*, *P. eurymedon*, or *P.g. canadensis*. The Lauraceae feeding *P. trolius*, *P. palamedes*, and *P. pilumnus* also die on various Magnoliaceae, but unlike the *P. rutulus*, *P. eurymedon*, and *P.g. canadensis* they don't even initiate feeding and starve to death.

Apparently the chemicals in common to the Rutaceae and Lauraceae (e.g., benzylisoquinoline alkaloids, essential oils, and coumarins; Feeny et al. 1983) do not provide sufficient stimulants for larvae we cross-tested (i.e., *P. cresphontes* would not eat spicebush or sassafras of the Lauraceae; *troilus* and *palamedes* would not eat hoptree or prickly-ash of the Rutaceae). *Papilio troilus* has been reported to oviposit on *Aristolochia* (D. West; in Feeny et al. 1983), which is known to contain benzylisoquinoline alkaloids, as do the Rutaceae, Lauraceae, and Magnoliaceae.

The *glaucus* group, in contrast, is able to oviposit and grow on a wide variety of foodplant families in addition to the Rutaceae, Lauraceae, and Magnoliaceae. If Lauraceae feeding was predominant for the proto-*glaucus*, it seems that colonization and host switching has been only partially successful in the *glaucus* group (Scriber et al. 1975; Scriber 1987). Very few of the surviving neonates of *P. glaucus, P.g. australis,* or *P. alexiares* reach pupation on spicebush (Table 16-1), and none survive on redbay, both of the Lauraceae (Scriber 1986b). Natural oviposition by adult *P. glaucus* on Rutaceae (Scriber 1972; Levin and Angleberger 1972) and Lauraceae (Scriber et al. 1975) confirm that these families are still chemosensorily recognized and used as host plants by this polyphagous species in certain locations. More than half a century ago, Forbes (1932) felt that the Rutaceae-feeding behavior was derived from Lauraceae-feeding by "similarity of flavor."

THE ROSACEAE AND OTHER HOST FAMILIES: A GENERAL ADAPTATION IN THE *P. GLAUCUS* GROUP

It is of interest where and when the use of Rosaceae (*Prunus serotina* in particular) may have begun for the *P. glaucus* group. Wild black cherry, as we have seen, is currently an excellent plant for larval growth of all seven taxa, and appears to be a favorite foodplant wherever it occurs in North America. Species of *Prunus* are believed to have evolved from New World tropical ancestors and have been present in North America at least since the early Cretaceous and possibly the Mesozoic (McVaugh, 1951).

The current distribution of wild black cherry (*P. serotina*) in the eastern half of the United States and Mexico north to Arizona conveys the essence of the disjunct geographic distribution patterns observed for numerous plant species shared by the eastern deciduous forest and the montane temperate forest of Mexico (see Martin and Harrell 1957). It is easy to imagine disjunct populations of *Papilio glaucus* with similar ranges, and in fact the distribution map of *P. serotina* from central Mexico to New England very closely describes the range of the *P.g. glaucus* subspecies and also the Mexican *P. alexiares* (Beutelspacher and Howe 1984; Scriber et al. 1988).

The Rosaceae (as well as the Betulaceae and Salicaceae) lack the benzylisoquinoline alkaloids which characterize all *Papilio* foodplant families discussed previously (i.e., the Rutaceae, Lauraceae, Magnoliaceae, Aristolochiaceae,

Annonaceae, and Rhamnaceae; Feeny et al. 1983). No other single class of phytochemicals unites the host plants of the *glaucus* group (Hagen 1986).

Perhaps it was the release from behavioral stimulants for feeding and oviposition (upon which larvae of the *troilus* group appear to rely) which gave the *P. glaucus* group the freedom to become the most polyphagous of all of the world Papilionidae. The willingness of the larvae to initiate feeding on virtually any plant (Hagen 1986; Scriber 1988) and the frequent oviposition "mistakes" by adults (e.g., Brower 1958a; Berenbaum 1981) may have facilitated the expansion of host breadth of the proto-*glaucus* species from ancestral hosts. Adoption of Rosaceae as host plants by the *glaucus* ancestor should be seen not as a shift by a specialist to a new host, but rather as an expansion of the potential host range. The ability to use Rosaceae, Oleaceae, Betulaceae, Platanaceae, and Tiliaceae may have been a consequence of a single evolutionary step. With reduction or elimination of behavioral barriers that enforce specialization, larvae physiologically preadapted for feeding on a few woody plants should have little difficulty growing on a much wider range in response to changes in ecological availability. The addition of hosts from the Rosaceae, Oleaceae, Betulaceae, Platanaceae, and Tiliaceae may reflect the absence of particular toxins in these plants requiring specialized detoxication mechanisms. It certainly required more than just a behavioral shift to add hosts from the Salicaceae and Rhamnaceae, however, since clear toxicity has been described in detail for certain *Papilio* taxa on Salicaceae (Table 16-1; Lindroth et al. 1986).

THE SALICACEAE: AN ADAPTATION TO BOREAL HABITATS

While as widespread in North America during the Cretaceous as were the Lauraceae and Magnoliaceae, the Betulaceae and the Salicaceae (*Populus* in particular) are generally now (in the post-glaciation period) predominantly of north temperate and boreal distribution (Graham 1964; Little 1978). As with the Rosaceae, the Betulaceae and Salicaceae lack benzylisoquinoline alkaloids which are present in the Magnoliaceae, Lauraceae, and Rutaceae. Also like the Rosaceae, the Salicaceae and Betulaceae are particularly rich in phenolic glycosides (Gibbs 1974). Unlike the Rosaceae, the Salicaceae and Betulaceae are poor hosts for the *P.g. glaucus, P.g. australis,* and *P. alexiares* (which survive well on the Magnoliaceae). In contrast, *Papilio rutulus, P. eurymedon,* and *P.g. canadensis* do very well on the Salicaceae and Betulaceae but very poorly on the Magnoliaceae.

The phenolic glycosides appear to constitute the major line of chemical defense in the Salicaceae (Palo 1984), and it has been suggested that the widespread occurrence of salicortin in members of the plant family (Julkunen-Tiito 1985, 1986) may be due to the toxicity of the cyclohexanone saligenin ester (Lindroth et al. 1988a). In fact, of 7 million compounds catalogued by Chemical Abstracts, the 2-hydroxy-3, 4-cyclohexenone group naturally occurs in only 2 compounds: salicortin and tremulacin. We have identified the phytochemicals in

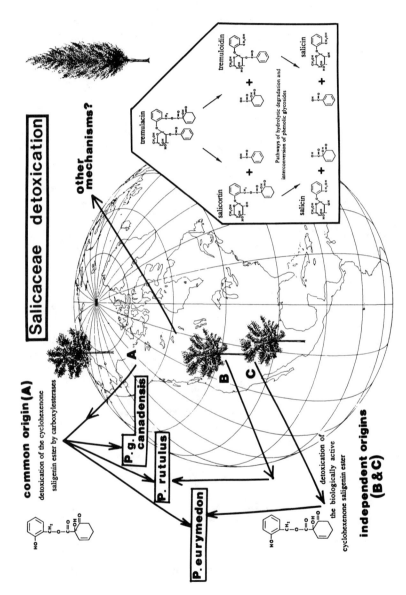

Figure 16-5. Possible origins of detoxication abilities for the active cyclohexanone from Salicaceae. A common origin in Beringia (A), independent origins in the western United States, or alternative detoxication systems to the carboxylesterase enzyme mechanism (Lindroth et al. 1988) are possible. Different detoxication mechanisms in various taxa might indicate either independent phylogenetic paths for the taxa or facile host adaptations in a common lineage.

361

Populus tremuloides (of the Salicaceae) which are toxic to the *P.g. glaucus* subspecies as tremulacin, salicortin, and tremuloidin (Lindroth et al. 1986, 1987; 1988a, b; Lindroth and Pajutee 1987). Lindroth et al. (1987) have shown that tremulacin is twice as active biologically as salicortin, and concluded that the benzoyl ester of tremulacin (as a relatively inexpensive plant metabolite) acts to synergize the toxicity of the cyclohexenone saligenin ester by competitively inhibiting the carboxylesterase enzyme activity. Tremuloidin, though minimally active itself, also contains a benzoyl ester and may act to synergize salicortin and

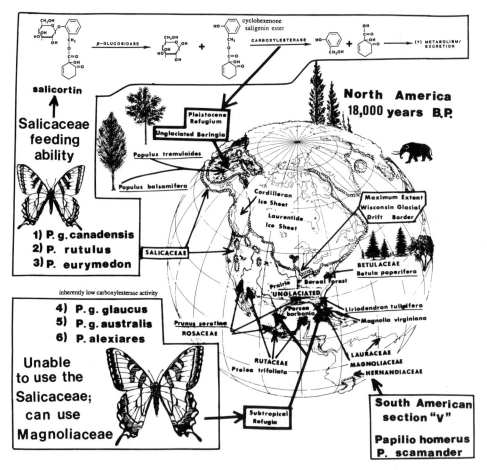

Figure 16-6. Possible Pleistocene landscape in North America depicting the Beringial refuge with *Populus balsamifera* and *P. tremuloides* present for the entire 40,000 years of glaciations. Note also the presumed boreal forest refuges in western North America and eastern North America as well as the subtropical refuge in southeastern North America where the Magnoliaceae and Lauraceae were concentrated. Whether the South American *P. scamander* group on Magnoliaceae and Lauraceae may be from a common lineage as the *P. glaucus* and *P. troilus* is uncertain.

tremulacin (Lindroth et al. 1988b; Fig. 16-5). The mechanism of detoxication by the adapted *P.g. canadensis* is a unique carboxylesterase enzyme (Figs. 16-5 and 16-6). We plan to determine if the same detoxication mechanisms exists fo *P. eurymedon* and *P. rutulus* or whether other mechanisms exist.

BIOGEOGRAPHY AND THE EVOLUTION OF HOST RANGES

Pleistocene glaciations have undoubtedly played a major role in determination of current distributions of North American swallowtails (Remington 1951; Brower 1959a, b; Sperling 1986; Scriber 1988). This effect has been mediated in part through changes in the composition of North American plant communities, as well as through the differential abilities of species to adapt to colder climates or shorter growing seasons (Hagen and Lederhouse 1985; Tauber et al. 1986). However, the cycles of glacial advance and retreat that characterized the Pleistocene were only part of a global cooling that began in the mid-Miocene 15 million years ago and that radically altered North American biomes long before the glaciers (Wolfe 1977, 1978; Axelrod 1966).

For example, the early Miocene forests of North America contained an abundance of Lauraceae, Rutaceae, and Magnoliaceae species, even in Alaska (Wolfe 1977, 1978; Axelrod 1966; Matthews 1979). At present, plants in these families are largely restricted to the southeastern and southwestern parts of the continent. Not surprisingly, *Papilio* species that feed on these plants are similarly restricted in distribution (Opler and Krizek 1984). Scriber (1973) earlier hypothesized that increased breadth of host range in *P. glaucus* at higher latitudes was due to the species either extending its range northward, or remaining in the northern areas as its ancestral hosts were slowly eliminated as a result of climate change. Similarly, Betulaceae- and Salicaceae-feeding populations are currently found primarily from the Rocky Mountains and across the northern United States and Canada where these plants predominate.

Assuming that these differential foodplant utilization abilities or host-plant affiliations were in existence during the glacial maxima (i.e., 18,000 years B.P.), we can attempt to reconstruct geographic histories as well as phylogenetic relationships of the taxonomic groups. Based upon known patterns of food plant affiliations and our data regarding differential abilities for detoxication of plant species by neonate larvae of the various taxa, we propose one scenario for the evolutionary divergence of the North American tiger swallowtail butterflies (see Fig. 16-6). Although *P. alexiares* in Mexico will also be important, the northern *P.g. canadensis* is crucial to the interpretations of this phylogeny and will be described in more detail below.

During the Pleistocene the similarity between the flora of montane Mexico and the southeastern United States was striking (Braun 1955; Martin and Harrell 1957). A general explanation for the floral affinities in these 2 areas is that warm-temperate organisms were driven southward into two major refugia, one in peninsular Florida and one in Mexico by the cold or that a separation of temperate plants occurred because of the arid Pliocene environments (Braun 1955). Exchange of only xero-mesophytic plants was likely during the Pleistocene (Texas is viewed as a barrier to dispersal; Martin and Harrell 1957). One of the plant species able to move across this arid barrier in southern Texas was the wild black cherry *Prunus serotina* (McVaugh 1951 and 1952). This "edge" species did not require a continuous forest corridor, and could make the move across the savannah or prairie-woodland.

Since the Rutaceae, Rosaceae, and Oleaceae can sustain survival and growth of neonate larvae of all taxa tested thus far, it seems feasible that these plant families may have served as a common denominator (see Feeny et al. 1983), and perhaps as hosts, for the ancestor of the *P. glaucus* group. There is a distinct contrast between Magnoliaceae-feeders (*P.g. glaucus* and *P.g. australis*) and Salicaceae-feeders (*P. rutulus, P.g. canadensis,* and *P. eurymedon*). While *P. eurymedon* has specialized on Rhamnaceae, larvae can eat species of Salicaceae successfully. Nothing is known about the ability of *P. multicaudatus* to detoxify and grow on the Salicaceae and Magnoliaceae. Reported foods include the Rutaceae, Oleaceae, and Rosaceae and the current geographic distribution of *P. multicaudatus* is throughout Mexico north through the Rocky Mountains in the western United States. Similarly, *P. alexiares* in Mexico may hold the key to the foodplant phylogeny.

In the eastern half of the United States, *Papilio glaucus glaucus* and *P.g. australis* are abundant. Their choice of and ability to use the Magnoliaceae is unique. The preference for and ability to use quaking aspen and balsam poplar of the Salicaceae by the northern *P.g. canadensis* and the western *P. rutulus* and *P. eurymedon* is coupled with a virtually total inability to use sweetbay and tuliptree of the Magnoliaceae. The reciprocal pattern of abilities is observed for *P.g. glaucus* and *P.g. australis* on the Salicaceae. In other words, fitness (as measured by survival and growth rates) on one host appears to be negatively correlated with fitness on other hosts. Nevertheless, such negative genetic correlations alone are not necessarily the primary driving force in the speciation process. However, if coadapted preference-viability gene complexes exist, they would be more likely to contribute significantly to incipient host races or species (Wiklund 1975; Bush and Diehl 1982; Feeny et al. 1983; Scriber 1983; Rausher 1983; Futuyma and Peterson 1985). The implications in speciation of North American tree-feeding *Papilio* are significant and a likely foodplant use phylogeny with common foods (e.g., Rutaceae, Roseaceae, Oleaceae) can be constructed for additional hypothesis testing (Fig. 16-7).

A shift from the Rutaceae to the Lauraceae and Magnoliaceae for the North American *glaucus* and *troilus* groups (Section III) and also from Rutaceae to the Lauraceae, Magnoliaceae, Berberidaceae, and Hernandiaceae for the *scamander, zagreus,* and *homerus* groups (Section V) may have been facilitated by the numerous shared aporphine alkaloids between all five of these families (p. 253, Gibbs 1974). Some of these aporphine alkaloids are even shared with the Aristolochiaceae (e.g., magnoflorine) or Rhamnaceae and Annonaceae (isochory-dine, Fig. 16-2). Tuliptree, *Liriodendron tulipifera,* is a favorite host plant for *P.g.*

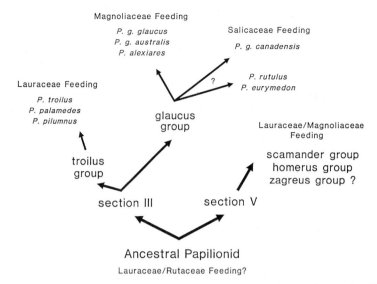

Figure 16-7. Possible sequence for the evolution of host use in *Papilio* Section III. The diagram includes only four key plant families and includes *Papilio* Section V for perspective.

glaucus in the eastern United States and yet it has a relatively diverse array of at least three dozen biologically active secondary compounds (Scriber et al. 1987). Among the benzylisoquinoline alkaloids, it contains glaucine (Fig. 16-2) and liriodenine (Fig. 16-2). A host shift to *Liriodendron tulipifera* would seem to be no small task. In addition to its effects on nonadapted *Papilio*, tuliptree is extremely toxic to or completely avoided by notoriously polyphagous Lepidoptera such as the southern armyworm *Spodoptera eridania*, the gypsy moth *Lymantria dispar*, and the cecropia silk moth *Hyalophora cecropia* (Manuwoto and Scriber 1985).

The feeding abilities of Section III and Section V on Lauraceae and Magnoliaceae may have arisen twice, or once (with subsequent isolation and differentiation of the species groups in North and South America). The Magnoliaceae have remained a southeastern post-Pleistocene glaciation flora. Among the Arcto-tertiary plant communities in southeastern North America (Graham 1964; Axelrod 1958; Wolfe 1978), the bay forests with sweetbay, *Magnolia virginiana*, did not become predominant in the southeast until 5000 B.P. (Watts 1980). This may represent the earliest opportunity for *P.g. australis* to differentiate in association with sweetbay, its favorite (only?) foodplant. Behavioral, physiological, and biochemical adaptations to its hostplant in Florida have been described (Scriber 1986b); however, little allozyme differentiation is observed between *P.g. glaucus* and *P.g. australis* (Hagen and Scriber in preparation; Fig. 16-1). Other recent significant changes in gene frequency in these Florida populations have been observed during only the last 25–30 years (Lederhouse and Scriber 1987).

At the northern end of the continent of North America, we have a different selection of host plants by *Papilio glaucus* than in Florida (Scriber 1984b, 1986b). These insects use only the Salicaceae and Betulaceae. The possibility of an allopatric differentiation of host-use abilities in a northern Beringial refugium has been proposed (Scriber 1988). A vicariance biogeographical approach assumes that barriers divided a once continuous range of a taxon. The Pleistocene glaciations would have provided such a mechanism for *Papilio* especially if the northern refugium (Beringia) is included in the scenario (Braun 1955; Graham 1964; Peterson et al. 1979; Watts 1979, 1980; Colinvaux 1981; Danks 1981; and Murray 1981). The most significant feature about this Beringial refugium for *P.g. canadensis* is that populations of balsam poplar, *Populus balsamifera*, and of quaking aspen, *Populus tremuloides*, are believed to have existed in this area throughout the entire period of glaciation (from 36,000 yr. B.P. to the present post-glacial; Hopkins et al. 1981) although aspen may have been differentiated in the southern Rocky Mts. (Barnes 1975). While "shrub birch" (e.g., *Betula nana, B. glandulosa*, and *B. kenaica*) may have also persisted in this refugium, tree paper birch (*Betula papyrifera*) and spruce were apparently eliminated (absent from 30,000 to 11,500 years ago), with a reintroduction via dispersal from a refugium south of the Laurentide ice sheet only after the glaciation retreat in the Holocene (Hopkins et al. 1981).

Should the ancestral *Papilio* stock have been isolated in this Beringial region (now Alaska) for the period 36,000–9,500 years B.P., it would have been without the common foodplants enjoyed by their relatives remaining south of the Cordilleran and Laurentide ice sheets (e.g., cherry, paper birch, and ash would have been available only as far north as in the spruce forests; Watts 1980). Such isolation in Beringia with basically only the Salicaceae as foodplant options for more than 25,000 years could easily account for the genetic differentiation observed in *"canadensis"* today. Dispersal from this refugium southward and eastward could also explain the striking difference in voltinism currently observed between all of the *P.g. canadensis* (obligate diapause) and *P.g. glaucus* (facultative diapause) and particularly across the narrow blend zone in central Wisconsin (Hagen and Scriber 1989; Rockey et al. 1987a, b) and Michigan. Determination of the extent and direction of gene flow across these zones of parapatry will aid in subsequent interpretations regarding the primary–secondary (sympatric–allopatric) mechanisms of genetic divergence. The "spring brood" of *Papilio glaucus* in the eastern United States is very *"canadensis"*-like in many respects, yet exists in the midst of predominantly *"glaucus"* populations (Scriber 1990). Perhaps the Appalachian "boreal refuge" has harbored *P.g. canadensis* types much as presumed in the "Beringial refuge."

West of the Rocky Mountains in similar montane and boreal forest refugia, an independent origin of Salicaceae feeding occurred in the *P. rutulus–P. eurymedon* lineage. Ecological conditions were probably similar for the ancestors of both *canadensis* and *rutulus–eurymedon* in the western and Beringial (or eastern) refugia, leading to similar selection pressures for new host-use abilities. Whether identical mechanisms of detoxication have evolved in these lineages is a major unresolved question.

ACKNOWLEDGMENTS

This research was supported by the Michigan State University College of Natural Sciences and the Agricultural Experiment Station (Project 8051) and the National Science Foundation (BSR 8306060, BSR 8503464, BSR 8718448), USDA grants #85CRCR-1-1598 and #87-CRCR-1-2851. We are also thankful for support of the Graduate School and College of Agricultural and Life Sciences (Hatch 5134) of the University of Wisconsin. We would particularly like to thank the following people for valuable discussion and/or their assistance in field collections for this study: Matt Ayres, May Berenbaum, Janice Bossart, Keith Brown, Robert Dowell, Mark Evans, Bruce Giebink, Rick Lindroth, Heidi Luebke, Syafrida Manuwoto, Jim Maudsley, James Nitao, Maret Pajutee, Stephen Peterson, David Robacker, Sally Rockey, Frank Slansky, John Thompson, Bill Warfield, and Wayne Wehling.

REFERENCES

Avise, J. C. 1975. Systematic value of electrophoretic data. *Systematic Zool.* **23**: 465–481.

Axelrod, D. I. 1958. Evolution of the Madro-Tertiary geoflora. *Bot. Rev.* **24**: 433–509.

Axelrod, D. I. 1966. Origin of deciduous and evergreen habits in temperate forests. *Evolution* **20**: 1–15.

Barnes, B. V. 1975. Phenotypic variation of trembling aspen in western North America. *Forest Sci.* **21**: 319–328.

Benson, W. W. 1978. Resource partitioning in passion vine butterflies. *Evolution* **32**: 493–518.

Berenbaum, M. 1981. An oviposition "mistake" by *Papilio glaucus* Papilionidae. *J. Lepid. Soc.* **35**: 75.

Berenbaum, M. 1983. Coumarins and caterpillars: A case for coevolution. *Evolution* **37**: 163–179.

Berenbaum, M. and P. Feeny. 1981. Toxicity of angular furanocoumarins to swallowtail butterflies: Escalation in a coevolutionary arms race? *Science* **212**: 927–929.

Berlocher, S. H. 1984. Insect molecular systematics. *Ann. Rev. Entomol.* **29**: 403–433.

Beutelspacher, C. R. and W. H. Howe. 1984. Mariposas de Mexico. *La Prensa Medica Mexicana*, S. A. Mexico 127 pp.

Braun, E. L. 1955. The phytogeography of unglaciated Eastern United States and its interpretation. *Botan. Rev.* **21**: 297–375.

Brower, L. P. 1958a. Larval foodplant specificity in butterflies of the *Papilio glaucus* group. *Lep. News* **12**: 103–114.

Brower, L. P. 1958b. Bird predation and foodplant specificity is closely related procryptic insects. *Amer. Nat.* **92**: 183–187.

Brower, L. P. 1959a. Speciation in butterflies of the *P. glaucus* group. I. Morphological relationships and hybridization. *Evolution* **13**: 40–63.

Brower, L. P. 1959b. Speciation in butterflies of the *Papilio glaucus* group. II. Ecological relationships and interspecific sexual behaviour. *Evolution* **13**: 212–228.

Brown, K. S., A. J. Damman, and P. P. Feeny. 1980. Troidine swallowtails (Lepidoptera: Papilionidae) in southeastern Brazil: Natural history and foodplant relationships. *J. Res. Lepid.* **19**: 199–226.

Brues, C. T. 1920. The selection of food-plants by insects, with special references to lepidopterous larvae. *Amer. Nat.* **54**: 313–332.

Bush, G. L. and S. R. Diehl. 1982. Host shifts, genetic models of sympatric speciation and the origin of parasitic insect species. *Proc. 5th Int. Symp. Insect-Plant Relationships.* pp. 297–305. Wageningen: Pudo.

Buth, D. G. 1984. The application of electrophoretic data in systematic studies. *Ann. Rev. Ecol. & Syst.* **15**: 501–522.

Colinvaux, P. 1981. Historical ecology in Beringia: The south land bridge coast at St. Paul Island. *Quaternary Res.* **16**: 18–36.

Danks, H. V. 1981. The composition, distribution and ecology of arctic insects, with some speculations on the evolution of arctic communities. Pp. 21–23 in (G.G.E. Scudder and J. L. Reveal (Eds.). *Evolution Today (Proc. 2nd Int. Congr. Syst. & Evol. Biol.).*

Dethier, V. G. 1941. Chemical factors determining the choice of foodplants by *Papilio* larvae. *Amer. Nat.* **75**: 61–73.

Dethier, V. G. 1954. Evolution of feeding preferences in phytophagous insects. *Evolution* **8**: 33–54.

Ehrlich, P. R. and P. H. Raven. 1964. Butterflies and plants: A study in coevolution. *Evolution* **18**: 586–608.

Farris, J. S. 1981. Distance data in phylogenetic analysis. In: V. A. Funk and D. R. Brooks (Eds.). *Advances in Cladistics*, Vol. 1 (Proc. 1st Meeting Willi Hennig Soc.). NY Botanic Garden, NY, pp. 3–23.

Feeny, P., L. Rosenberry, and M. Carter. 1983. Chemical aspects of oviposition behavior in butterflies. In: S. Ahmad (Ed.), *Herbivorous insects: Host-seeking behavior and mechanisms.* Academic Press, NY, pp. 27–76.

Felsenstein, J. 1984. Distance methods for inferring phylogenies: A justification. *Evolution* **38**: 16–24.

Ferris, C. D. and F. M. Brown. 1981. *Butterflies of the Rocky Mountains States.* University of Oklahoma Press, Norman.

Forbes, W. T. M. 1932. How old are the Lepidoptera? *Amer. Nat.* **98**: 452–460.

Forbes, W. T. M. 1951. Footnote on Papilio. *Lep. News* **5**: 16.

Forbes, W. T. M. 1958. Caterpillars as botanists. Proc. 10th Int. Congr. *Entomol.* **1**: 313–317.

Futuyma, D. J. 1983. Evolutionary interactions among herbivorous insects and plants. In D. J. Futuyma and M. Slatkin (Eds.), *Coevolution.* Sinauer Assoc., Sunderland, MA, pp. 207–231.

Futuyma, D. J. and S. C. Peterson. 1985. Genetic variation in the use of resources by insects. *Ann. Rev. Entomol.* **30**: 217–238.

Gibbs, R. D. 1974. *Chemotaxonomy of Flowering Plants*. McGill-Queens University Press, Montreal. 4 Volumes.

Graham, A. 1964. Origin and evolution of the biota of southeastern North America: Evidence from the fossil plant record. *Evolution* **18**: 571–585.

Hagen, R. H. 1986. *The evolution of host plant use by the tiger swallowtail butterfly, Papilio glaucus*. Ph.D. Thesis, Cornell University, Ithaca, NY, 293 pp.

Hagen, R. H. and R. C. Lederhouse. 1985. Polymodal emergence of the tiger swallowtail, *Papilio glaucus* (Lepidoptera: Papilionidae): Source of a false second generation in New York State. *Ecol. Entomol.* **10**: 19–28.

Hagen, R. H. and J. M. Scriber. 1989. Sex-linked diapause, color, and allozyme loci in *Papilio glaucus*: Linkage, analysis and significance in a hybrid zone. *Heredity* **80**: 179–185.

Hagen, R. H. and J. M. Scriber. 1991. Systematics of the *Papilio glaucus* and *P. troilus* species groups (Lepidoptera: Papilionidae): inferences from allozymes, Ann. Entomol. Soc. Amer. (in press).

Hancock, D. L. 1979. The systematic position of *Papilio anactus* Macleay (Lepidoptera: Papilionidae) *Australian Ent. Mag.* **6**: 49–53.

Hancock, D. L. 1983. Classification of the Papilionidae (Lepidoptera): A phylogenetic approach *Smithersia* **2**: 1–48.

Harris, H. and D. A. Hopkinson. 1978. *Handbook of enzyme electrophoresis in human genetics*. American Elsevier, New York.

Higgins, L. G. and N. D. Riley. 1980. *A field guide to the butterflies of Britain and Europe*. 4th ed. Collins, London. 384 pp.

Hodges, R. W., T. Dominick, D. R. Davis, D. C. Ferguson, J. G. Franclemont, E. G. Munroe, and J. A. Powell (Eds.). 1983. *Check list of the Lepidoptera of America north of Mexico*. E. W. Classey, London. 284 pp.

Hopkins, D. M., P. A. Smith, and J. V. Matthews. 1981. Dated wood from Alaska and the Yukon: Implications for forest refugia in Beringia. *Quaternary Res.* **15**: 217–249.

Igarashi, S. 1984. The classification of the Papilionidae mainly based on the morphology of their immature stages. *Tyo to Ga* **34**: 41–95.

Jermy, T. 1976. Insect–host-plant relationship: Co-evolution or sequential evolution? Pp. 109–113 in T. Jermy (Ed.). *The Host-plant in Relation to Insect Behavior and Reproduction*. Plenum, NY.

Julkunen-Tiito, R. 1985. Chemotaxonomical screening of phenolic glycosides in northern willow twigs by capillary gas chromatography. *Journal of Chromatography* **324**: 129–139.

Julkunen-Tiito, R. 1986. A chemotaxonomic survey of phenolics in leaves of northern Salicaceae species. *Phytochemistry* **25**: 663–667.

Lederhouse, R. C. and J. M. Scriber. 1987. Increased relative frequency of dark morph females in the tiger swallowtail. *Papilio glaucus* (Lepidoptera: Papilionidae) in south central Florida. *Amer. Midl. Natur.* **118**: 211–213.

Levin, M. P. and M. A. Angleberger. 1972. Observations on foodplant records for *Papilio glaucus* (Papilionidae) *J. Lepid. Soc.* **26**: 177–180.

Lindroth, R. L., J. M. Scriber, and M. T. S. Hsia. 1986. Differential responses of tiger swallowtail subspecies to secondary metabolites from tulip tree and quaking aspen leaves. *Oecologia* **70**: 13–19.

Lindroth, R. L., M. T. S. Hsia, and J. M. Scriber. 1987. Characterization of phenolic glycosides from quaking aspen (*Populus tremuloides*). *Biochem. Systematics and Ecology* **15**: 677–680.

Lindroth, R. L. and M. S. Pajutee. 1987. Chemical analysis of phenolic glycosides: Art, facts, and artifacts. *Oecologia* **74**: 144–148.

Lindroth, R. L., J. M. Scriber, and M. T. S. Hsia. 1988a. Differential food utilization by tiger swallowtail species: Mediation by phenolic glycosides. *Ecology* **69**: 814–822.

Lindroth, R. L., J. M. Scriber, and M. T. S. Hsia. 1988b. Effects of the quaking aspen compounds catechol, salicin, and isoniazid on performance of two subspecies of tiger swallowtails. *Amer. Mid. Natur.* **119**: 1–6.

Little, E. B. 1978. *Atlas of United States Trees.* USDA Forest Service Misc. Publ., Washington, D.C.

Manuwoto, S. and J. M. Scriber. 1985. Antibiosis/antixenosis in tulip tree and quaking aspen leaves against the polyphagous southern armyworm, *Spodoptera eridania Oecologia* **67**: 1–7.

Martin, P. S. and B. E. Harrell. 1957. The pleistocene history of temperate biotas in Mexico and eastern United States. *Ecology* **38**: 468–480.

Matthews, J. M., Jr. 1979. Tertiary and quaternary environments: Historical background for an analysis of the Canadian insect fauna. *Mem. Entomol. Soc. Canada* **108**: 31–86.

May, B., J. E. Wright, and M. Stoneking. 1979. Joint segregation of biochemical lock in Salmonidae: Results from experiments with *Saluelinus* and review of the literature on other species. *J. Fish. Res. Bd. Can.* **36**: 1114–1128.

McVaugh, R. 1951. A revision of the North American black cherries, *Prunus serotina* Ehrh., and relatives. *Brittonia* **7**: 279–315.

McVaugh, R. 1952. Suggested phylogeny of *Prunus serotina* and other wide-ranging phylads in North America. *Brittonia* **7**: 317–346.

Miller, J. S. 1987. Host-plant relationships in the Papilionidae (Lepidoptera): Parallel cladogenesis or colonization? *Cladistics* **3**: 105–120.

Miller, J. S. and P. Feeny. 1983. Effects of benzylisoquinoline alkaloids on the larvae of polyphagous Lepidoptera. *Oecologia* (*Berl.*) **58**: 332–339.

Mitter, C. and D. R. Brooks. 1983. Phylogenetic aspects of coevolution. In: D. J. Futuyma and M. Slatkin (Eds.). Coevolution. Sinauer Assoc., Sunderland, MA, pp. 65–98.

Munroe, E. 1961. The genetic classification of the Papilionidae. *Can. Ent. Suppl.* **17**: 1–51.

Munroe, E. and P. R. Ehrlich. 1960. Harmonization of concepts of higher classification of the Papilionidae. *J. Lepid. Soc.* **14**: 169–175.

Murray, D. F. 1981. The role of arctic refugia in the evolution of the arctic vascularflora— A Beringian perspective. Pp. 11–20 in G. G. E. Scudder and J. L. Reveal (Eds.). *Evolution Today. Proc. 2nd Int. Congr. Syst. & Evol. Biol.*

Nei, M. 1972. Genetic distance between populations. *Amer. Nat.* **106**: 283–292.

Nei, M. 1978. Estimation of average heterozygosity and genetic distance from a small number of individuals. *Genetics* **89**: 583–590.

Nitao, J. K., M. P. Ayres, R. C. Lederhouse, and J. M. Scriber. 1991. Larval adaptation to lauraceous hosts: geographic divergence in the spicebush swallowtail butterfly. *Ecology* (in press).

Opler, P. A. and G. O. Krizek. 1984. *Butterflies east of the Great Plains.* John Hopkins Univ. Press. Baltimore, MD, 294 pp.

Palo, R. T. 1984. Distribution of birch (*Betula* spp.), willow (*Salix* spp.) and poplar (*Populus* spp.) secondary metabolites and their potential role as chemical defense against herbivores. *Journal of Chemical Ecology* **10**: 499–520.

Peterson, G. M., T. Webb, J. E. Kutzbach, T. vander Hammen, T. A. Wijmstra, and F. A. Street. 1979. The continental record of environmental conditions at 18,000 yr B.P.; An initial evaluation. *Quaternary Res.* **12**: 47–82.

Rausher, M. D. 1983. Ecology of host-selection behavior in phytophagous insects. In R. F. Denno and M. S. McClure (Eds.). *Variable plants and herbivores in natural and managed systems.* Academic Press, NY, pp. 223–257.

Remington, C. L. 1951. Geographic subspeciation in the Lepidoptera. I. Introduction. A general outline of subspeciation. *Lep. News* **5**: 17–20.

Richard, D. and M. Guedes. 1983. The Papilionidae (Lepidoptera): Co-evolution with the angiosperms. *Pyton* **23**: 117–126.

Richardson, B. J., P. R. Baverstock, and M. Adams. 1986. Allozyme Electrophoresis. Academic Press, Orlando, FL, 410 pp.

Rockey, S. J., J. H. Hainze, and J. M. Scriber. 1987a. Evidence of sex-linked diapause response in *Papilio glaucus* subspecies and their hybrids. *Physiol. Ecol.* **12**: 181–184.

Rockey, S. J., J. H. Hainze, and J. M. Scriber. 1987b. Diapause in three subspecies of the Eastern tiger swallowtail, *Papilio glaucus* (Lepidoptera: Papilionidae): Evidence of a latitudinal and obligatory diapause. *Amer. Midl. Natur.* **118**: 162–168.

Rothschild, W. and K. Jordan. 1906. A revision of the American papilios. *Novitates Zoologicae* **13**: 411–744.

Ruszczyk, A. 1986. Mortality of *Papilio scamander* (Lep. Papilionidae) pupae in four districts of Porto Alegre (S. Brazil) and the causes of superabundance of some butterflies in urban areas. *Rev. Brasil. Biol.* **46(3)**: 567–579.

Scott, J. A. 1986. *The butterflies of North America. A natural history and field guide.* Stanford Univ. Press. Stanford, CA, 584 pp.

Scott, J. A. and J. H. Shepard. 1976. Simple and computerized discriminant functions for difficult identifications: A rapid nonparametric method. *Pan-Pacific Entomol.* **52**: 23–28.

Scriber, J. M. 1972. Confirmation of a disputed foodplant of *Papilio glaucus* (Papilionidae). *J. Lepid. Soc.* **26**: 235–236.

Scriber, J. M. 1973. Latitudinal gradients in larval feeding specialization of the world Papilionidae (Lepidoptera). *Psyche* **80**: 355–373.

Scriber, J. M. 1982. Foodplants and speciation in the *Papilio glaucus* group. In J. H. Visser and A. K. Minks (Eds.). Proceedings 5th International Symposium on insect–plant relationships. *Pudoc. Wageningen, Netherlands.* pp. 307–314.

Scriber, J. M. 1983. The evolution of feeding specialization, physiological efficiency, and host races in selected Papilionidae and Saturniidae. In R. F. Denno and M. S. McClure (Eds.), *Variable plants and herbivores in natural and managed systems.* Academic Press, NY, pp. 373–412.

Scriber, J. M. 1984a. Larval foodplant utilization by the world Papilionidae (Lep.): latitudinal gradients reappraised. *Tokurana* (Acta Rhopalocerologica) **6/7**: 1–50.

Scriber, J. M. 1984b. Host plant suitability. In W. J. Bell and R. T. Carde (Eds.), *Chemical ecology of insects.* Sinauer Assoc., Sunderland, MA, pp. 159–202.

Scriber, J. M. 1986a. "Allelochemicals and alimentary ecology: Heterosis in a hybrid zone?" Pp. 43–71 in L. B. Brattsten and S. Ahmad (Eds.). *Molecular mechanisms in insect plant interactions.* Plenum Press, NY.

Scriber, J. M. 1986b. Origins of the regional feeding abilities in the tiger swallowtail butterfly: Ecological monophagy and the *Papilio glaucus australis* subspecies in Florida. *Oecologia* **71**: 94–103.

Scriber, J. M. 1987. Population genetics and foodplant use among the North American tree-feeding Papilionidae. Pp. 221–230 in V. Labeyrie, G. Forbes, and D. Lachaise (Eds.). *Proc. 6th Intern. Symp. Insect/Plant Relationships.* W. Junk, Dordrect, Netherlands.

Scriber, J. M. 1988. Tale of the tiger: Biogeography, bionomial classification, and breakfast choices in the *Papilio glaucus* complex of butterflies. In K. C. Spencer (Ed.). *Chemical Mediation of Coevolution,* Academic Press, New York, pp. 241–301.

Scriber, J. M. 1990. Interaction of introgression from *Papilio glaucus Canadensis* and diapause in producing "spring form" eastern tiger swallowtail butterflies, *P. glaucus* (Lepidoptera: Papilionidae). *Great Lakes Entomol.* **23**: 127–138.

Scriber, J. M., M. H. Evans, and R. C. Lederhouse. 1988 [1990]. Hybridization of the Mexican swallowtail *Papilio alexiares garcia* (Lepidoptera: Papilionidae) with other *glaucus* group species and survival of pure and hybrid larvae on potential host plants. *J. Res. Lepid.* **27**: 222–232.

Scriber, J. M., M. T. Hsia, R. Lindroth, and P. Sunarjo. 1987. Allelochemicals as determinants of insect damage across the North American continent: Biotypes and biogeography. Pp. 439–448 in G. Waller (Ed.). *Allelochemicals: Role in Agriculture, Forestry and Ecology.* Proceedings ACS, Washington, D.C.

Scriber, J. M. and R. C. Lederhouse. 1988 [1989]. Handpairing of *Papilio glaucus glaucus* and *Papilio pilumnus* and hybrid survival on various foodplants. *J. Res. Lepid.* **27**: 96–103.

Scriber, J. M., R. C. Lederhouse, and K. Brown. 1990. Hybridization of Brazilian *Papilio* (*Pyrrhosticta*) (Section V) with North American *Papilio* (*Pterourus*) (Section III). *J. Res. Lepid.* (in press).

Scriber, J. M., R. C. Lederhouse, and L. Contardo. 1975. Spicebush, *Lindera benzoin,* a little known foodplant of *Papilio glaucus* (Papilionidae). *J. Lepid. Soc.* **29**: 10–14.

Scriber, J. M., G. L. Lintereur, and M. H. Evans. 1982. Foodplant suitabilities and a new oviposition record for *Papilio glaucus canadensis* (Lepidoptera; Papilionidae) in northern Wisconsin and Michigan. *Great Lakes Entomol.* **15**: 39–46.

Scudder, S. H. 1889. *The butterflies of the eastern United States and Canada.* Vol. II. Published by the author, Cambridge, MA.

Singer, M. C., D. Ng, and C. D. Thomas. 1988. Heritability of oviposition preference and its relationship to offspring performance within a single insect population. *Evolution* **42**: 977–985.

Slansky, F., Jr. 1972. Latitudinal gradients in species diversity of the New World swallowtail butterflies. *J. Res. Lepid.* **11**: 201–218.

Sperling, F. A. H. 1986. *Evolution of the Papilio machaon species group in western Canada.* M. S. Thesis, Univ. of Edmonton, Alberta. 285 pp.

Tauber, M.J., C. A. Tauber, and S. Masaki. 1986. *Seasonal adaptations of insects.* Oxford University Press, NY, 411 pp.

Thompson, J. N. 1982. *Interaction and coevolution.* Wiley, NY.

Thompson, J. N. 1986. Patterns in coevolution. Pp. 119–143 in A. R. Stone and D. J. Hawksworth (Eds.). *Coevolution and systematics.* Oxford University Press, Oxford, UK.

Thompson, J. N. 1988a. Variation in preference in monophagous and oligophagous swallowtail butterlflies. *Evolution* **42**: 118–128.

Thompson, J. N. 1988b. Evolutionary genetics of oviposition preference in swallowtail butterflies. *Evolution* **42**: 1223–1234.

Tietz, H. M. 1972. *An index to the described life histories, early stages, and hosts of the Macrolepidoptera of the continental United States and Canada.* 2 vols. Allyn Museum of Entomology, Sarasota, FL.

Via, S. 1986. Genetic covariance between oviposition preference and larval performance in an insect herbivore. *Evolution.* **39**: 505–522.

Watts, W. A. 1979. Late quaternary vegetation of central Appalachia and the New Jersey coastal plain. *Ecol. Monogr.* **49**: 427–469.

Watts, W. A. 1980. The late quaternary vegetation history of the southeastern United States, *Ann. Rev. Ecol. Syst.* **11**: 387–409.

Wiklund, C. 1975. The evolutionary relationship between oviposition preferences and larval host range in *Papilio machaon L. Oecologia* **18**: 185–197.

Wolfe, J. A. 1977. Paleogene floras from the Gulf of Alaska region. *U.S. Geological Survey Professional Paper* **997**: 1–108.

Wolfe, J. A. 1978. A paleobotanical interpretation of Tertiary climates in the Northern Hemisphere. *Amer. Sci.* **66**: 694–703.

Wright, S. 1978. *Evolution and the genetics of populations.* Vol. 4. Variability within and among natural populations. University of Chicago Press.

17 Aposematic Insects on Toxic Host Plants: Coevolution, Colonization, and Chemical Emancipation

KEITH S. BROWN, Jr., JOSÉ R. TRIGO, RONALDO B. FRANCINI, ANA BEATRIZ BARROS DE MORAIS, and PAULO C. MOTTA

COADAPTATION AND COEVOLUTION

Intense and specific interactions between populations at an ecological interface result in strong mutual selective pressures, leading to adaptations which are evolved and maintained by each side in response to the other. If such "coadaptation" continues at an interface maintained through evolutionary time, with each side reacting sequentially to new adaptations in the other, extensive genetic diversification may occur. The consequences of this process, often called "coevolution" (Fig. 17-1) might be visible today as special adaptations of populations only understandable with reference to the coadaptive interface, in speciation processes giving associations between primitive taxa and between advanced taxa, and in community structure and dynamics.

Since Ehrlich and Raven (1965) introduced the term "coevolution" in this ecological context, examining the relations between a highly diversified group of herbivores (butterflies) and their larval foodplants, it has been extensively used for explanation at the plant–herbivore interface—a + /- interaction in terms of fitness, with one side benefitting at the cost of the other (Table 17-1). The term has also been expanded to explain adaptations at many mutualistic (+ / +), mimetic (+ / + or + / −), predatory and parasitic (+ / −), and competitive (− / −) interfaces, whenever populations are influencing each other's evolution in relatively tight ecological relationships (Table 17-1). Typical adaptations often regarded as results of such mutual selection include specific pollinator attractants

Plant-Animal Interactions: Evolutionary Ecology in Tropical and Temperate Regions, Edited by Peter W. Price, Thomas M. Lewinsohn, G. Wilson Fernandes, and Woodruff W. Benson. ISBN 0-471-50937-X © 1991 John Wiley & Sons, Inc.

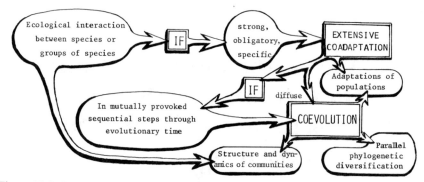

Figure 17-1. Prerequisites, processes, and results of coadaptation and coevolution between populations in strong ecological interaction.

and rewards; warning color patterns; suites of constitutive or inducible defensive chemical (Rhoades 1985) and detoxifying enzymes; special morphological structures such as plant domatia, mycetomes, and arthropod mouthparts; seed size, shape, and hardness; flowering and fruiting phenologies; foraging behavior; and insect life cycles. Speciation and adaptive radiation patterns of specific parasites and their hosts, of reasonably specialized predators and their prey, of pollinators and flowering plants, and of competing animal groups have been attributed to coevolutionary pressures. An attractive and well-documented

TABLE 17-1. Classes of Strong Ecological Interactions and Their Possible Coevolutionary Consequences

Interaction	Effects on Fitness Of Species A/B/C[a]	Coevolutionary Tendencies
Competition	$-/-$	Evolves as far as the minimum adaptation necessary to eliminate the interaction; highly unstable and difficult to observe or follow.
Parasitism	$+/-$	Evolve towards mutualism or evolutionary escape of Species B (harmed by the interaction), giving many adaptations and diversification in the process. Unstable in the long run but easy and fruitful to follow.
Batesian mimicry	$+/-/-$	
Predation	$+/-$	
Mutualism	$+/+$	Evolve to reinforce interactions, giving many adaptations and convergences; evolutionarily stable, trivial, easy to study, but not highly diversifying or edifying.
Müllerian mimicry	$+/+/+$	

[a]The predator which would otherwise attack and perhaps consume the mimic (A), getting sick if it is unpalatable (Müllerian). See Gilbert 1983 for a discussion of these cases.

scheme has been presented by Gilbert (1980) for the structuring of diversified tropical forest communities around coevolved, narrowly defined food webs; coevolution has been implicated in character displacement, niche breadth, and size-ratio characters in competitors (Benson 1978; Roughgarden 1983) as well as some aspects of convergent community structures in widely separated localities with similar climatic regimes (Orians and Paine 1983), including latitudinal and productivity gradients in species diversity.

Inevitably, voices have been raised to restrict this broad use of the term to more closely defined natural processes, such as the definition of Janzen (1980): "Coevolution may be usefully defined as an evolutionary change in a trait of the individuals of a population in response to a trait in the individuals of a second population, followed by an evolutionary response by the second population to the change in the first." A critical and now-classic review book on coevolution (Futuyma and Slatkin 1983, p. 2) pointed out that most alleged examples did not conform to Janzen's definition or to any well-defined one-on-one species interactions, and agreed that the general interactive process of diversification occurring in natural communities could be termed "diffuse coevolution." Futuyma (1983, p. 231) stated that "Genetic changes in pairs of interacting species, each promoted by change in the other, have not yet been documented for plants and insects," and recently Miller (1987) claimed that "There have been no adequately documented cases of parallel cladogenesis between insects and plants." The group examined by Miller (Papilionidae), considered as one of the best examples of coevolution by Ehrlich and Raven (1965), does not show a clear pattern of stepwise diversification with the host-plant families; it was regarded as an additional case of the colonization of already diversified plant groups by radiating and evolving herbivores, which may be simply called "evolutionary host transfer" (Mitter and Brooks 1983) or "sequential evolution" (Jermy 1976, 1984; Futuyma 1983).

CHEMISTRY AND THE COEVOLUTIONARY PROCESS

Almost all studies of coevolution have emphasized the importance of chemical factors (nutrients and coenzymes, secondary metabolites, detoxifying enzymes, and carriers) in the process of reciprocal adaptation. Rather few, however, have sought to identify "hard" chemical aspects of the interaction, which are left to speculation and suggestions for further research, or at best extracted from superficial chemical data in the natural products or pharmacological literature. Recently, the science of chemical ecology, which was born (Brower and Brower 1964), grew (Sondheimer and Simeone 1970), and prospered (Wallace and Mansell 1976) at the plant–herbivore coevolutionary interface, has made much easier the fusion of biological hypotheses with chemical analyses, giving a unified and interdisciplinary experimental research structure. Numerous papers have directly related chemical data to bioassays and evolutionary hypotheses, with the formation of a reasonable body of chemico-evolutionary theory (Feeny 1975;

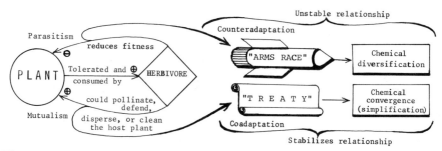

Figure 17-2. Processes, predicted results, effects on chemical diversification, and options for stabilization at the plant–herbivore interface. + = positive influence, − = negative, including effects on energy flow and selection.

Rhoades 1979, 1985; Gottlieb 1982), subject to direct experimental testing, especially at the interface between plants (which produce many ecologically active substances—attractants, deterrents, stimulants, repellents, toxins, hormones, antidigestants, antibiotics) and their herbivores. In light of Table 17-1, this interface may be predicted to produce definable chemical features in relation to different coevolutionary options (Figs. 17-2 and 17-3). For example, many Neotropical specialist herbivores have been observed to efficiently pollinate flowers of their larval host plants (Table 17-2), thereby transforming a potentially disastrous "arms race" into a comfortably stabilized mutualism, in which the plant donates some energy and matter (leaves and toxins) to maintain a fitness-enhancing species alive and nearby. It is notable that many of these plants flower

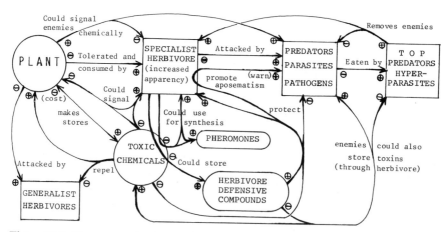

Figure 17-3. Processes, selection, and chemistry in a four-level system based on toxic plants and specialized herbivores. + and − as in Figure 17-2; selection tends to occur at arrowheads. Note the large number of countervalent influences at the middle two levels, leading to delicate optimization processes over evolutionary time.

TABLE 17-2. Some Aposematic Lepidoptera which Apparently Pollinate Their Larval Host Plants (Observed in Brazil and Venezuela)

Group and Species	Host Plant and Family
Pieridae: Pierini	
Ascia monuste	*Brassica oleracea* vars. (Cruciferae), probably also *Capparis* sp. (Capparidaceae) in forests
Nymphalidae: Danainae	
Danaus plexippus, D. gilippus, and *D. eresimus*	*Asclepias curassavica* and other species, *Oxypetalum* spp., probably other Asclepiadaceae
Nymphalidae: Ithomiinae	
Tithorea harmonia, T. tarricina	*Prestonia acutifolia* and other Apocynaceae
Ithomia iphianassa, I. agnosia	*Acnistus arborescens* (Solanaceae)
Nymphalidae: Acraeinae	
Actinote pellenea, A. pyrrha, A. carycina, A. parapheles, A. melanisans, and others	*Eupatorium inulaefolium, E. gaudichaudianum, Vernonia beyrichii, Mikania micrantha,* and many other species in these genera (Compositae)
Actinote surima	*Senecio brasiliensis* (Compositae)
Altinote stratonice	*Liabum hastifolium* (Compositae)
Abananote hylonome	*Verbesina caracasana* (Compositae)
Nymphalidae: Heliconiini	
Heliconius astraea, H. egeria	*Passiflora glandulosa* (Passifloraceae)
Heliconius wallacei, H. burneyi	*Passiflora coccinea* and relatives (Passifloraceae)
Heliconius ethilla	*Passiflora kermesina* (Passifloraceae)[a]
Dioptidae (day-flying moths)	
Scea auriflamma	*Passiflora capsularis* (Passifloraceae)

[a]See Benson et al., 1976.

in response to herbivore attack, suggesting the kind of mutually beneficial, fine-tuned adjustment discussed by Whitham et al. in Chapter 11.

Futuyma and Slatkin (1983, p. 5) remarked that "most of the literature concentrates on the adaptations that may have resulted from coevolution rather than on the process itself." The combination of chemical analysis, bioassay, and coevolutionary theory should, in principle, permit an investigation of the "process itself." An especially well-defined system for the study of this process is that including toxic plants and their speecialist herbivores who have not only breached the host plant's chemical defenses but also, in many cases, have turned them to their own advantage in host-seeking, defense against predators and parasites, and reproduction (Fig. 17-3). In this system, several types of evolutionary responses are possible for coadaptation or counteradaptation of host plants and herbivores, along chemical, behavioral, populational, and ecological dimensions. Thus, detailed study of the optimization of adaptive patterns combining these dimensions, along radiating lines of herbivores and plants, should permit an overview of the coevolutionary process at the interface between them. This chapter will focus on the principal extant cases involving aposematic (warningly colored) Neotropical butterflies and their specific larval hosts, looking at details of the chemistry and ecology of the components of each pair of interacting groups in an attempt to address some or all of the following questions:

(1) How old and stable are the interactions? How have such systems persisted through evolutionary time? Is there evidence for stabilization through transformation of parasitism to mutualism (Fig. 17-2 and Table 17-2)?

(2) How has the "coevolutionary process" acted through history to form the present reality of the interactions? Did older mutual adaptations arise at roughly the same time, or sequentially as a result of reciprocal selective pressures? What other selective agents can be identified? How do these systems fit into coevolutionary theory?

(3) Do these relationships show details of parallel phylogenetic diversification of plants and insects at the interface ("association by descent") or suggest a "sequential evolution" by progressive colonization of previously diversified and defended host plants by insects?

(4) What are the consequence of these associations at the level of Neotropical communities? Are they important in the diversity, complexity, energy flow, importance of mutualism, dynamics, recycling, stability, or other properties of these associations?

(5) What options or syndromes for optimization of structure are represented in these systems? Are there any evolutionary tendencies visible in these syndromes?

(6) Is it possible to predict chemical, ecological, or evolutionary patterns in related systems still not investigated, on the basis of the syndromes visible in the presently known systems?

A PLANT–HERBIVORE INTERFACE: METHODS
FOR CHEMICAL STUDY

Table 17-3 shows the principal groups of Neotropical butterflies (less skippers) with the relative diversity, usual foodplants, and general phylogenetic position of each group. Principal mimetic relationships are indicated for those groups (capitalized) whose adults are considered aposematic, advertising toxic or disagreeable properties. These seven groups contain over 500 species that serve as distasteful models for mimicry rings throughout the Neotropics (and usually in the Old World tropics also, on the same host plants). The main classes of defensive chemicals for the Neotropical host plants of these groups are also indicated. This table defines the system which is examined in this chapter.

Chemical mediation of ecological interactions can be verified by standardized mild extraction of fresh plant or animal material, followed by fractionation to give various acid-base and polarity classes, and bioassay with herbivores or

TABLE 17-3. Phylogenetic Relationships, Foodplants and Their Secondary Chemicals, and Mimetic Interactions of the Principal Groups of Neotropical Butterflies (Skippers not Included)

FAMILY, Subfamily, Tribe (Primitive to Advanced)[a,b]	Approximate Number of Species	Principal Families of Host Plants	Protective Chemicals	Mimetic Relations[c]	
				Müll.	Bates.
PAPILIONIDAE					
Baroniinae	1	Leguminosae (Acacia)			DAN
Papilioninae: TROIDINI[a]	50	ARISTOLOCHIACEAE	Aristolochic acids, Alkaloids, Terpenes, Phenolics	EUM, ITH, HEL	
Graphiini[b]	45	Lauraceae, Annonaceae			TROI, HEL
Papilionini[b]	45	Lauraceae, Rutaceae, Piperaceae, Umbelliferae			TROI, DAN, ITH, ACR, HEL
PIERIDAE: Dismorphiinae	100	Leguminosae			TROI, PIE, DAN, ITH, ACR, HEL
Coliadinae[b]	60	Leguminosae, Bignoniaceae			PIE
PIERINAE (white)[a]	50	CRUCIFERAE, CAPPARIDACEAE	Glucosinolates	—	
Pierinae—other	100	Loranthaceae			TROI, PIE DAN ITH, ACR, HEL
LYCAENIDAE: Riodininae[b]	1300	Many families (often ant mutualists)			PIE, EUM, DAN, ITH, ACR, HEL
Plebejinae	10	Many families; Leguminosae			PIE
THECLINAE: EUMAEUS[a]	5	CYCADACEAE (Zamia)	Pseudocyanogens	TROI	
Other	1300	Many families (ant mutualists)			PIE, EUM
LIBYTHEIDAE	3	Ulmaceae (Celtis)			—

(continued)

TABLE 17-3 (*Continued*)

FAMILY, Subfamily, Tribe (Primitive to Advanced)[a,b]	Approximate Number of Species	Principal Families of Host Plants	Protective Chemicals	Mimetic Relations[c] Müll.	Mimetic Relations[c] Bates.
NYMPHALIDAE: DANAINAE[a]	12	APOCYNACEAE, ASCLEPIADACEAE, MORACEAE, CARICACEAE	Cardenolides, PAs, Saponins, Phenolics	ITH, ACR, HEL	
ITHOMIINAE[a]	300	APOCYNACEAE. SOLANACEAE	Alkaloids, Steroids, Terpenes, Cardenolides, Phenolics	TROI, DAN, ACR, HEL	TROI, PIE, ITH, HEL, ACR
Satyrinae[b]	1000	Gramineae, Cyperaceae, Marantaceae			
(incl. Brassolini)[b]	50	Palmae, Musaceae, Zingiberaceae			—
Morphinae[b]	30	Gramineae, Palmae, Leguminosae, Menispermaceae			
Charaxinae[b]	100	Lauraceae, Myrtaceae, Piperaceae, Euphorbiaceae, Leguminosae, Flacourtiaceae, Ulmaceae			PIE, ITH, ACR, HEL
Apaturinae	30	Ulmaceae			
ACRAEINAE[a]	50	COMPOSITAE (in New World)	PAs, Sesquiterpenes	DAN, ITH, HEL	
Nymphalinae: HELICONIINI[a]	65	PASSIFLORACEAE	Cyanogens, Alkaloids	TROI, DAN, ITH, ACR	
Other tribes	400	Violaceae, Acanthaceae, Compositae, Urticaceae, Sapindaceae, Euphorbiaceae, Verbenaceae, Rubiaceae, Moraceae, Ulmaceae			TROI, PIE, EUM, DAN, ITH, ACR, HEL

[a] Aposematic (distasteful, warningly colored) groups and their host plants are CAPITALIZED, and chemicals are indicated.
[b] These groups are heartily eaten by vertebrate predators orienting by vision (birds and lizards).
[c] Müll. = Müllerian (distasteful) mimics of; Bates. = Batesian (palatable) mimics of; names of mimicked protected groups abbreviated from those capitalized in first column (TROI = Troidini; EUM = *Eumaeus*, etc.).

predators in the laboratory or field to verify isolation of the appropriate biological activity (Fig. 17-4). Active fractions are further purified to give single compounds whenever possible. After these are identified, the extraction scheme can be modified and simplified to permit their direct isolation, and analytical methods developed to help in their quantitative determination in different life

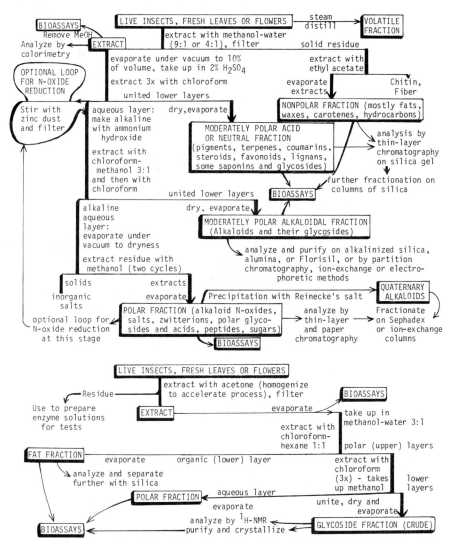

Figure 17-4. Standard rapid fractionation schemes for ecologically active compounds in plants or animals, starting with fresh material and leading to defined chemical fractions for bioassay and isolation of chemical groups or individual compounds (after Brown 1985, 1987, where more details may be found). Upper figure adapted for alkaloid-containing material, including *N*-oxides and quaternary alkaloids. Lower figure adapted for rapid isolation of polar glycoside fractions.

stages, plant or animal parts, or interactions. In this way, chemical and biological experiments are combined to aid in the study of each interface, always seeking rapidity and efficiency in quantitative isolation and microanalysis of active substances.

These basic methods for chemico-ecological research have been more thoroughly discussed in previous publications (Brown 1985, 1987) and widely circulated in meetings of the International Society for Chemical Ecology. They are simple and operational for most species of plants and animals in very rudimentary laboratories (the first extraction can be done in the field with alcohol–water in a film canister), and can be easily adapted to special cases through straightforward application of the principles of separation science (Karger et al. 1973; Miller 1975) and biological assay (Clark et al. 1967; Southwood 1978). They give best results when maintained in a standard form throughout the investigation of a particular interface, and they should be accompanied by a specific chemical test for the active class of compounds, usable on thin-layer plates or paper chromatograms, and adaptable for colorimetry in solution.

We will now review the results of the application of these methods to the system defined above, and discuss their significance in relation to the questions posed about coevolutionary processes in this system. The interactions may be conveniently divided into three chemical syndromes (combinations of adaptations) which can be represented, in order of evolutionary appearance (Table 17-3), as larval sequestration of defensive chemicals from their host plant with passage of these on to adults (Syndrome A), adult sequestration of defensive compounds with rejection of larval host plant toxins (Syndrome B), and *de novo* biosynthesis of defense substances in all stages (Syndrome C).

THE BASIC SYNDROME (A): SIMPLE SEQUESTRATION AND PERILOUS APOSEMATISM

Almost all butterfly larvae feed on plants containing poisonous defensive compounds (Table 17-3), and many are aposematic. Yet very few accumulate the ingested toxins outside the gut, and even fewer pass them on to the adult. Most plant defenses are either directly excreted or metabolized (transformed) by the insects. Consequently, most butterfly larvae and adults are readily eaten by both invertebrate and vertebrate predators (Chai 1986, 1988) (Table 17-3). There is clearly a cost to accumulating toxic compounds in one's tissues, which can probably be avoided in many cases by simply enjoying the "enemy-free space" (Atsatt 1981; Brower 1984) created by a poisonous plant and its sometimes inedible residents; these associations have been observed by us to be surprisingly free of vertebrate predator activity.

The majority of caterpillars and butterflies hide from predators. Even brightly colored adults either have cryptic undersides (thereby disappearing when they land and close their wings) or show disruptive patterns, making them difficult to distinguish in flight against a variegated background (Papageorgis 1975; Brown 1988). Extensive exposure of butterflies to predators is often required, however, by the necessary movement between resource patches, elaborate courtship sequences, and lengthy oviposition procedures. Thus it has been suggested that all such large diurnal insects should have some chemical protection (Rothschild and Marsh 1978; Brower 1984).

One way for a butterfly to acquire chemical defense would be to tolerate and store toxins from a larval foodplant, large quantities of which are processed during larval growth. A final instar caterpillar may consume leaves amounting to many times its final weight, most of this excreted as frass. This would permit accumulation of several percent of adult weight as toxins even if they are present in very low concentrations in the foodplant leaves.

In principle, however, storage of such toxins is difficult, since they are not friendly to metabolic enzymes or animal cell membranes (plants can compartmentalize them more easily). Since an adult butterfly's visual image is dominated by an expanded surface whose coloration and pattern can be controlled by very few genes (Sheppard et al. 1985; Nijhout 1986; Gilbert et al. 1988), defense by cryptic or sudden-flash colors and patterns including deflective marks, large eyespots, false heads and snake mimicry, or by behavioral mechanisms such as fast, erratic flight, may represent a good alternative to chemical defense for adults (Edmunds 1974).

Once a chemical defense option is adopted and works to convince predators to reject a butterfly, this may be predestined to adopt the "aposematic life-style" (Rothschild 1981) whereby it can ward off attack *before* the predator initiates hostilities, reinforcing a negative search-image by flaunting obvious colors, behaviors, sounds or smells, associated with chemical punishment inflicted upon the hapless predator (Brower 1984; Fig. 17-3). Distasteful sprays, exudates, or exoskeleton (including wings) and a rubbery cuticle are useful adjuncts to survive attacks from naive, forgetful, or very hungry predators. Although this flamboyant life-style may permit more security during the long periods of unavoidable exposure of adult butterflies, it is a perilous road to choose, since any advertisement is dangerous and predators can always adapt to or circumvent a prey's defenses (Brown and Vasconcellos-Neto 1976; Calvert et al. 1979; Brown 1988).

Thus, it is not surprising that only a few butterfly groups have in fact been selected for this life-style (Table 17-3). It is also understandable that the more primitive of these groups acquire their primary defenses from the larval foodplants (Syndrome A). Troidine swallowtail adults show compounds of known biological activity (aristolochic acids, benzylisoquinoline alkaloids, terpenes; Fig. 17-5a–c) which are present in their larval foodplants (Von Euw et al. 1968; Rothschild et al. 1970; Urzúa et al. 1983, 1987; Urzúa and Priestap 1985). The adults are common, fly slowly, and bear red or yellow warning marks on their wings and bodies; they are mimicked by large numbers of other swallowtails (usually relished by predators), especially females, and also by many other Pieridae, Satyrinae, Nymphalinae, and day-flying moth groups (Table 17-3; Brown 1988). Adults have a short lifespan, while the likewise aposematic larvae, also mimicked by other presumably more tasty caterpillars, may spend up to 2 months exposed on the *Aristolochia* vines (Brown et al. 1981; Morais and Brown unpublished). The same syndrome occurs in other primitive aposematic butterflies, such as the highly visible white Pierini who store glucosinolates (Fig. 17-5*d*) from larval food (Aplin et al. 1975) and red-and-green *Eumaeus* hairstreaks who inherit pseudocyanogens (Fig. 17-5*e*) from their long-lived and highly aposematic larvae feeding on cycads (Rothschild et al. 1986). The famous Monarch butterfly (a danaine) makes vertebrate predators vomit and then avoid its orange-striped colors, by storing cardiac glycosides (Fig. 17-5*f*) from asclepiadaceous plants eaten by its aposematic larvae (Brower et al. 1967; Brower 1969).

The troidines may have climbed, in addition, a classical coevolutionary "ladder" of reciprocal adaptations with the Aristolochiaceae (Fig. 17-6*a*), where primitive larvae feed only on primitive vines with simple chemistry, while advanced species occupy well-defended derived plants and can detoxify and store the plant toxins. Detailed chemical investigation of this interface is still underway (Morais and Brown unpublished). It can be contrasted with another proposed case of sequential coevolutionary adaptation in a specialized Holarctic group of swallowtails that feed on Umbelliferae, plants which elaborate progressively more toxic coumarins (Fig. 17-6*b*; Berenbaum and Feeny 1981; Berenbaum 1983); the caterpillars are aposematic with good mechanical resistance to being pecked (Järvi et al. 1981), while the adults, which unlike the related troidines are palatable to birds, and wary, fast-flying, and mimetic.

Even these "classical" cases of larval sequestration of defense compounds show variations indicative of the action of optimization processes along other

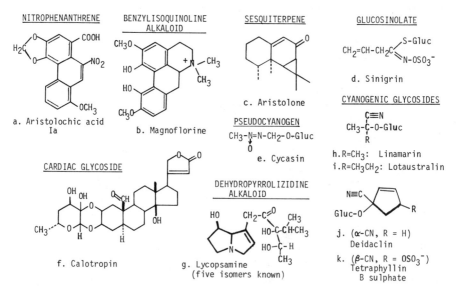

Figure 17-5. Chemical structures of some principal butterfly defensive compounds (representatives of classes or individual compounds).

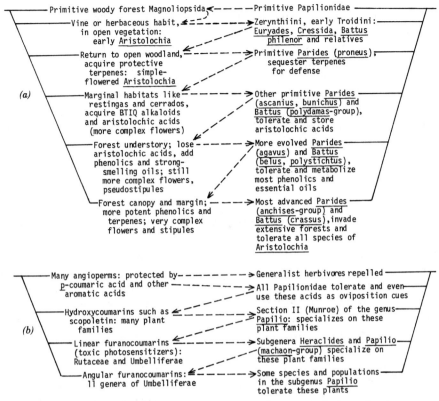

Figure 17-6. Coevolutionary ladders (series of mutually provoked adaptations) presumed for Troidini/Aristolochia (*a*) and proposed for the *Papilio machaon* group on the Umbelliferae (*b*), (after Berenbaum 1983 and Miller 1987).

dimensions within Syndrome A. Miller (1987) has affirmed that swallowtails in general have colonized their host plants, rather than coevolving with them. The chemistry of *Aristolochia* is highly variable, and some troidines mimic other dangerous insects like *Pepsis* wasps and danaine and ithomiine butterflies; the larvae (gregarious in *Battus*, otherwise solitary, defended by a strong-smelling eversible prothoracic gland called the osmeterium) probably biosynthesize many of their own defense compounds (terpenes and lower acids) rather than deriving them from the host plants (Honda 1981). Young *Aristolochia* plants are frequently chewed down to the root and killed by troidine larvae, which must put heavy pressure on these plants to escape by chemical change (Fig. 17-2; Brown et al. 1981; Morais and Brown unpublished). The host-plant flowers have a fetid smell and are pollinated by small flies, not butterflies. Pierini show a fast, erratic flight, inconsistent storage of glucosinolates and cryptic larvae with very rapid development; they are efficient pollinators of some of their host plants. *Eumaeus*, which do not pollinate and often kill their cycad hosts, are locked into these rare primitive plants, and local populations though abundant are highly subject to resource- and climate-provoked extinction (see Clark and Clark, Chapter 10). The adults also defend themselves against predation with pyrazines, probably biosynthesized by the insects themselves (Rothschild et al. 1986). And Monarchs, which may stabilize their foodplant relationship (Fig. 17-2) by pollinating asclepiads, process the cardenolides in their hosts, selectively storing some derivatives and metabolizing others (Brower et al. 1982); they turn out to be a rather exceptional case even within the Danainae (see next section). Thus, even the more primitive "simple" and classic cases offer abundant opportunities to investigate more thoroughly variations in the details of the coevolutionary process; the optimization of conflicting dimensions and selective options has produced solutions not always harmonious with the general syndrome.

AN IDEAL CASE AND A SURPRISING NEW SYNDROME (B)

The Ithomiinae, a sister-group to the Monarch and relatives (Ackery and Vane-Wright 1984), represent in their specific relationship with the Solanaceae (97% of the species; a few primitive groups use Apocynaceae, as do many Danainae) an "ideal case" for studying biochemical coevolution (Table 17-4). They have often been mentioned in the literature as a typical case of Syndrome A (Brower and Brower 1964, p. 154; Young 1972, p. 291; Drummond 1981, p. 63) despite an early failure to detect the commonest *Solanum* toxins (steroidal glycoalkaloids) in adults (Rothschild 1972). Later failures led finally to the unexpected conclusion that, while Solanaceae have one of the broadest suites of protective chemicals of any plant family (up to 12 biogenetic groups of alkaloids, almost all groups of terpenoids, bitter steroids, phenolic and other glycosides, saponins and pungent oils are all common; Brown 1985, 1987), none of these are stored in Ithomiinae adults, and none protect against a common predator of tropical butterflies, the giant orb spider *Nephila clavipes* (Brown 1984, 1985, 1987). Nevertheless, this same spider rejects all field-caught Ithomiinae from its web (Vasconcellos-Neto and Lewinsohn 1984), as it does normally palatable butterflies painted with Ithomiinae extracts, or polar or alkaloidal fractions (Fig. 17-4). The alkaloids (after N-oxide reduction) constituted up to 10–20% of the dry weight of adults, and were composed exclusively of isomeric dehydropyrrolizidine alkaloids, or PAs (Fig. 17-5*g*), traced not to larval foodplants but to flower nectar (Compositae; Eupatorieae) or decomposing foliage (Boraginaceae) heavily visited by adult males (Brown 1984, 1985, 1987). The complete natural flow of these compounds, which are passed over to females during mating and end up on the eggs, was elucidated through microanalysis of many thousands of plant and insect parts along the chain (Fig. 17-7; Brown 1987). It included not only the Ithomiinae but also the Danainae investigated (which store host-plant cardenolides very erratically; Rothschild et al. 1975; Dixon et al. 1978; Rothschild and Marsh 1978; Malcolm and Rothschild 1983), and many species in all three subfamilies of Arctiid moths (Ctenuchinae, Arctiinae, and Pericopinae); all also used these alkaloids as precursors for male sex pheromones, in some cases

TABLE 17-4. Criteria for the Study of Coevolution: The Ithomiinae/Solanaceae Interface, in Theory an Ideal Case

Criterion	Characteristics of the Ithomiinae/Solanaceae System
Diversification	Both groups strongly diversified at tribal, generic, and species levels, as well as chemically, suggesting strong mutual pressure.
Specificity	Highly specific interactions at both generic and specific levels, including in details and in local communities.
Strength	Strong interactions, readily observable in any Neotropical forest system, due to the abundance of both taxa.
Age of the Interaction	The American Ithomiinae line probably moved over from Apocynaceae to Solanaceae at least 15 million years ago.
Geographical consistency	Twenty-one genera in nine tribes show consistent larval host plants from Central America to southern Brazil (Table 17-5).
Present knowledge	Both groups are well-known and well-studied, in terms of systematics, genetics, evolution, biochemistry, behavior, ecological interactions, and biogeography.
Ease of study	Both groups are facile to find and study in the field and both can be maintained with relative ease in the laboratory.
Evolutionary potential	Both groups show appreciable evolutionary plasticity, indicated by cytogenetics (variable karyotypes). Although widely distributed, both groups have rather restricted gene flow between local populations, permitting local adaptations and microevolution.
Chemistry and defense	Both groups are very well defended chemically against most potential enemies, with highly diversified suites of toxic secondary metabolites in the plants.

developing courtship or copulatory structures only after assimilation of PAs (Schneider et al. 1975, 1982; Edgar et al. 1976; Conner et al. 1981; Eisner 1982; Boppré 1986). An abundant and evolutionary stable source of PAs of this single structure (Fig. 17-5g) is encouraged by pollination of PA-producing flowers (Figs. 17-2 and 17-7) or by invading the decomposer trophic level, always abundantly represented in higher-productivity tropical forests where ithomiines and their larval host plants abound.

As would be expected in this scenario, most Solanaceae-feeding Ithomiinae larvae are highly cryptic, and adults and a few more brightly colored larvae that might store toxins are often gregarious (reinforcing the negative search image), and the coevolutionary picture is inverted: the most primitive Ithomiinae leave the Apocynaceae to feed directly on the Solanaceae which many consider most

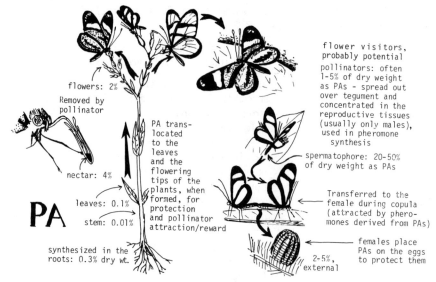

Figure 17-7. Flow of PAs in natural ecosystems, from synthesis in the plants (usually in roots) to dispersion on butterfly and moth eggs (see Brown 1987).

advanced, and the more derived butterflies use foodplants regarded as more primitive, both among and within tribes (Fig. 17-8; Brown 1985, 1987; Drummond 1986; Drummond and Brown 1987). This "anti-coevolutionary process" may be the result of progressive colonization of more complexly defended primitive plants; see Gottlieb (1982, 1984) for a discussion of evolutionary trends toward simplification in plant defensive chemistry. Detailed chemical analysis of host plants (Brown 1984, 1985, 1987) leads to a suggestive picture of preadaptive tolerance for each step in the parasitic colonization (Table 17-5; Brown and Henriques, in press). Adult Ithomiinae do not usually visit the flowers of, or otherwise derive benefit from or offer advantages to, their larval host plants, and apparently obtain very little chemical support from them, simply remaining their annoying and persistent enemies.

This picture is not completely "clean," however, at the transition between

---

Figure 17-8. Estimates of phylogenies for Ithomiinae (left: tribes from top to bottom in order of advancement index for the most primitive member, genera within tribes in order of advancement) and their hostplants (right: tribes and genera in order of advancement from top to bottom), with the principal relationships indicated by lines crossing the center (Brown and Motta in preparation; see also Drummond 1986; Drummond and Brown 1987; Brown 1987). Note the preponderance of associations between primitive insects and advanced plants, and vice versa, giving a strong "X" in the middle of the figure. Some Solanaceae systematists would invert the order of primitive and advanced genera, which would make a less strong "X" in the middle, without giving any clear parallel phylogenesis between the two groups.

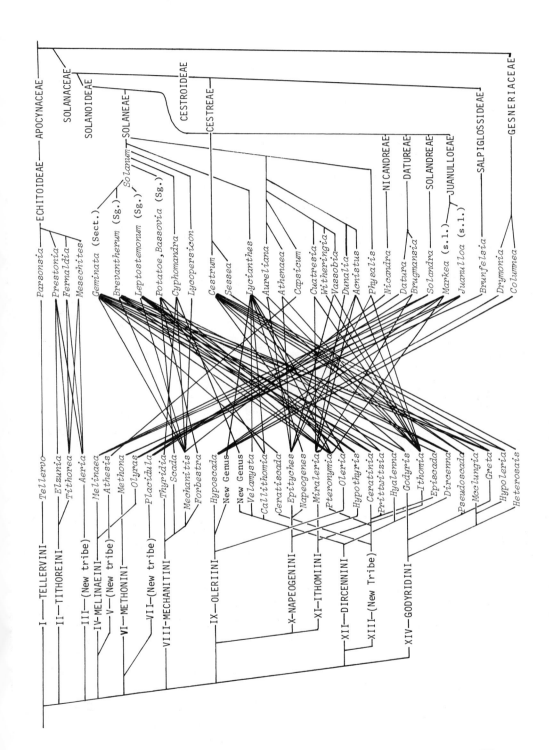

TABLE 17-5. Hypothetical Sequence for Colonization of Progressively More Primitive Solanaceae by Predaptation in Evolving Lineages of Ithomiinae, to Tolerate Classes of Toxic Secondary Chemicals (see Fig. 17-8)

Step	Tribes[a]	Genera of Ithomiinae	Plants Colonized	Substances Tolerated (Classes)
0	I	*Tellervo*	Apocynaceae: Parsonsieae (*Parsonsia,*	Dehydropyrrolizidine alkaloids (PAs)
	II	*Tithorea,*[b] *Elzunia*	*Prestonia, Fernaldia,*	and their *N*-oxides, cardenolides,
	III	*Aeria*[b]	*Mesechites* and others)	simple steroidal alkaloids, terpenes
1a;a'	IV	*Melinaea,*[b] *Olyras,*	Solanaceae: *Markea, Juanulloa,*	Tropane alkaloids and their *N*-oxides,[c]
	IX	*Eutresis, Hyposcada*	*Solandra* and related genera	Phenolic glycosides, and terpenes
1b	V	*Athesis*	Solanaceae: *Capsicum rhomboideum*	Tropane alkaloids, bitter steroids?
2a	VI	*Methona*[b]	Solanaceae: *Brunfelsia*	Brunfelsamidine, other alkaloids,
				coumarin glycosides, and other phenolics
2b;b'	VII	*Placidula*	Solanaceae: *Brugmansia, Datura;*	Tropane alkaloids, bitter steroids,[d]
	XI	*Miraleria, Ithomia*[b]	*Athenaea, Aureliana; Acnistus,*	steroidal glycoalkaloids, saponins,
	X	*Epityches*	*Physalis, Vassobia,* and relatives	terpenes, and phenolics
3a	VIII	*Thyridia*[b]	Solanaceae: *Cyphomandra*	Tropane alkaloids, steroidal glyco-
	XII	New Genus		alkaloids, saponins, terpenes
3b	IX	*Hyposcada* (part)	Gesneriaceae: *Drymonia, Columnea*	Terpenes and phenolics
4a	VIII	*Sais, Scada,*[b] *Forbestra*	Solanaceae: *Lycianthes* and *Solanum,*	Steroidal glycoalkaloids and sapo-
4b	IX, X	*Oleria,*[b] *Napeogenes.*[b]	subgenera *Potatoe* and *Bassovia,*	nins, solamines, possible tropane
4c	XII	*Callithomia,*[b] *Velamysta*	rarely other subgenera	alkaloids, phenolics
5a	VIII	*Mechanitis*[b]	*Solanum* (subgenera *Leptostemonum,*	Steroidal glycoalkaloids, saponins,
5b	IX	Advanced *Oleria,*[b] *Hyalyris,*[b]	*Brevantherum*)	solamines, occasional nicotines
	X	*Hypothyris,*[b] *Garsauritis*		and pungent oils (pyrazines)
5c	XII	*Dircenna,*[b] *Hyalenna*[b]		
	XIII	*Ceratinia,*[b] early *Pteronymia*[b]		
6c	XIII	*Ceratinia,*[b] *Pteronymia*[b]	*Solanum* (section *Geminata*),	Steroidal glycoalkaloids, saponins,
	XIV	*Godyris,*[b] *Greta,*[b] *Hypoleria*[b]	*Cestrum*	pungent oils (pyrazines, terpenes)

[a]Numbers follow the arrangements of the subfamily in Brown (1987) and Brown and Henriques (1990).
[b]Genera with consistent hostplant usage from the northern to southern extremes of their range, usually from Central America to southern Brazil. Others (not in table) with the same consistency include *Episcada* and *Ceratiscada* (Tribe XIII) and *Pseudoscada* and *Heterosais* (Tribe XIV), all in Step 6c.
[c]Tropane alkaloids are close structurally to PAs, possibly interconvertible in some biological systems, and similar in chemical and biological activity.
[d]Withanolides are rather similar structurally to Cardenolides, with similar reactivity and lactone group.

Syndrome A, shown by the Monarch, several other Danainae, and the most primitive Ithomiinae—*Tellervo* (Edgar 1982; Orr, Trigo, Motta, and Brown in preparation) and *Tithorea* (Edgar 1982, Trigo 1988), feeding on Apocynaceae from which they derive PAs—and the chemically less dependent Ithomiinae who seek out protective compounds after hatching from the pupa (Syndrome B). In the middle are several groups of Ithomiinae (Fig. 17-9) that show indefinite processes which may be due to sequestration, transformation and/or excretion of the host plant compounds, with varying intensity of search for and accumulation of PAs as adults (Trigo and Brown 1988). Chemical analysis of these groups is proceeding and should help in understanding details of the process of evolution from Syndrome A to B. The plants, after having diversified chemically (perhaps partly in response to herbivore pressure; Fig. 17-2), can clearly no longer be trusted to provide the same protective chemicals to their enemies. The herbivores must thereby colonize new, modified, or still-unselected hosts, using the well-known insect potential to seek them out or to develop tolerance just as fast as the plants develop resistance. They will of course continue to use plant toxins whenever possible, including through chemical modification, but must cope with the instability of the $+/-$ interface through evolutionary time (Fig. 17-2). If they can discover alternatives through mutualistic relationships or independent synthesis of defense compounds, these should be positively selected.

It is also interesting that the Apocynaceae and Solanaceae-feeding Ithomiinae

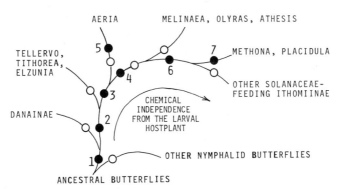

Figure 17-9. A variety of chemico-evolutionary patterns at the transition between Syndromes A and B: primitive Ithomiinae (Tribes I through VIII) and Danainae. Derived traits (solid circles in each branching) are (after Trigo and Brown 1988): (1) Cardenolide and PA-dependent for defense and pheromones, probably sequestering them from ancestral larval host plants (pre-Apocynaceae) as suggested by Edgar (1975). (2) Incorporate PAs but not cardenolides from larval hosts. (3) Do not incorporate PAs from larval host plants, since they are no longer present there; depend on PAs from adult nectar sources. (4) Switch to Solanaceous larval foodplants. (5) Transform non-PA chemicals from larval host plants into PAs. (6) Larvae lose flexible tubercles and danaiform striped pattern, indicating different chemistry, behavior, and foodplant interaction. (7) Usually store negligible amounts of PAs in adults, may transform host-plant chemicals directly into protective chemicals similar to those derived from PAs.

as well as the Asclepiadaceae-feeding Danainae still retain their ability to sequester PAs when these are painted on the larval foodplant leaves (Rothschild and Edgar 1978; Trigo and Motta 1986). This supports Edgar's proposal (1975) that PA uptake from ancient foodplants, for both defensive and reproductive purposes, is very old in this lineage. Ancestral butterflies may have fed upon primitive Apocynaceae which contained both PAs and cardenolides; the Solanaceae pulled away from this lineage and these herbivores by elaborating other alkaloids and losing those on which the insects depended, forcing the Ithomiinae to seek their defensive and pheromone-precursor substances (obviously of fundamental importance to their fitness) in adult food (Trigo and Motta, in preparation).

THE ADVANCED SYNDROME (C): CHEMICAL INDEPENDENCE FROM PLANTS

The Heliconiini accumulate large amounts of the cyanogenic glycoside linamarin (Fig. 17-5h) accompanied by smaller amounts of its homologue lotaustralin (Fig. 17-5i), both presumably useful in defense against vertebrate predators (Nahrstedt and Davis 1981, 1983; Davis and Nahrstedt 1985; Nahrstedt 1985). While these cyanogens are present in some Passifloraceae, the hostplants usually include other structures with cyclopentene rings and sulphate moieties (Fig. 17-5j,k) that are more toxic to heliconiine larvae and adults. Labelled dietary precursors such as ^{14}C-valine are efficiently converted into linamarin by various stages of Haliconiini (Nahrstedt and Davis 1983); it seems probable that these insects biosynthesize their own defense compounds *de novo*, and indeed an elegant coevolutionary ladder has been proposed *at the population level* to permit heliconians to neutralize the effects of progressively more potent host-plant cyanogens (Spencer 1988). In extreme cases the heliconians may even reject linamarin itself when encountered in host-plant tissue, since the larva's maceration would lead to its hydrolysis with the liberation of intolerable amounts of cyanide; the larvae therefore make β-glucosidases which inhibit the enzymes in the plant, and apparently excrete the whole bundle with the attached cyanogens. Chemical emancipation is thus carried to an extreme with the catabolism of plant components which are being simultaneously biosynthesized in other parts of the same caterpillar, for its protection. Probably the larvae cannot distinguish dietary linamarin, the friendly cyanogen, from other deadly ones the plant might slip into the platter, before the latter are hydrolyzed in the gut and start poisoning the insect.

Above the population level, an attractive picture of coevolution through diverse radiating lines has been advanced for the Heliconiini–Passifloraceae interface (Benson et al., 1976). This has been justifiably criticized by Mitter and Brooks (1983), however, and in fact it lacks rigorous phylogenies on both sides, as well as objectivity in definition of the relationship. In theory, a scheme of multiple radiations can be shown to correlate adequately with almost any other such scheme, even though the relationship may in fact be random. Better evidence for parallel phylogenesis in these two groups awaits a more complete analysis of chemical and other characters in specialized species-pair interactions, which are now well enough known to permit this detailed examination.

An even more unusual case exists in the American Acraeinae, close relatives of the Heliconiini (Table 17-3), whose African cousins often feed on Passifloraceae and other cyanogenic plants, and contain linamarin, probably biosynthesized, not sequestered (Nahrstedt and Davis 1983). Perhaps as a result of competitive pressure from the American heliconians (Ackery 1988), New World Acraeinae abandoned cyanogenic plants and use almost exclusively Compositae (Eupatorieae, Senecioneae, and Heliantheae) as larval foodplants; the adults also spend most of their short lives (1–2 weeks) feeding on flowers of these same plants (Table 17-2). Both leaves and nectar of most of these host plants are rich in PAs, the same substances so avidly sought by Danainae and Ithomiinae, more primitive members of the Nymphalid line (Table 17-3). The Acraeinae summarily reject these PAs, and instead like the heliconians accumulate large amounts of linamarin (Fig. 17-5*h*), apparently biosynthesized in all stages (no lotaustralin has been detected). Microanalysis for PAs indicates low concentrations in the gut of larvae and adults, none in eggs, pupae, or pre-pupae; very occasionally individuals seem to store higher concentrations of PAs but most are PA-free, and are not usually rejected by *Nephila* spiders (Brown and Francini 1988). These animals are highly gregarious and aposematic in all four stages, make no effort to conceal themselves, and are rarely taken by any predators—through their very own biosynthetic efforts. They may collaborate with the foodplants by pollinating their flowers (which often appear after a major leaf flush has been devastated by acraeine larvae) or helping disperse their seeds (usually windblown). With their unpredictable massive attacks, however, they cannot be expected to (and do not) show coherent patterns of parallel phylogeny with the composite hosts.

SUMMARY: PROGRESSIVE CHEMICAL EMANCIPATION

The overall picture of the aposematic butterfly–toxic host-plant interface in the Neotropics is thus a story of evolution away from chemical dependence on the host, often accompanied by correlative population parameters such as gregariousness and life-history variations (Table 17-6). Especially when the inherently antagonistic and unstable plant–herbivore relationship cannot be stabilized by mutualistic short-circuits (Fig. 17-2; Benson et al. 1976), it would seem that in an environment of reciprocal distrust, with the plant ever seeking new ways to effectively repel or poison the herbivore and the latter continually perfecting its methods of detoxifying any surprise offered by the plant, a "coevolutionarily stable strategy" might be the collection of defense substances from less treacherous associates (Syndrome B) and finally *de novo* synthesis (Syndrome C) (Brown and Francini 1988). The interface more resembles a "cold war," with inexpensive advantageous adaptations developed in the secret backrooms of the genome and sprung upon the opponent in time to neutralize any new

TABLE 17-6. Summary of Syndromes A, B, and C, Aposematic Butterfly Groups, and Relevant Population Parameters

Syndrome	Source of Toxins	Options	Group	Toxin concentration (% dry weight)	Gregarious? Larva	Gregarious? Adult	Life Span[a] Larva	Life Span[a] Adult	Aposematic? Larva	Aposematic? Adult
A	Larval host plant	Sequester, transform, excrete	Troidini							
			Battus[b]	0.1	++	±	long	short	+	+
			Parides[b]	0.1	−	−	long	short	++	++
			Pierinae	0.2	±	±	short	short	−	+
			Eumaeus[c]	1.0	++	+	long	short	++	+
B	Adult food	Sequester, transform, excrete	Danainae[d]	CG 0.1 PA 1.5	−	−	short	long	++	+
			Ithomiinae							
			Primitive[d]	1.0	−	+	short	long	+	+
			Advanced	3.0	±	+	short	long	−	+
C	Endogenous	Biosynthesize	Acraeinae	3.0	++	++	long	short	+	+
			Heliconiini	3.0	±	±	short	long	+	+

[a]"Short" is usually 10–20 days, while "long" is usually from 60 to 180 days.

[b]Battus and Parides larvae probably both sequester and biosynthesize components of the osmeterial secretion.

[c]Eumaeus may contain pyrazines produced from endogenous biosynthesis.

[d]Danainae participate of both Syndromes A and B, as do some primitive Ithomiinae, obtaining some of their defensive compounds from the larval foodplant (CG, eventually PA) and another part through adult feeding (PA).

development from that side, rather than an "arms race" involving a balance of expensive, potentially offensive adaptations, never really used. Intensification of the mutual pressure could easily lead to the dangers of extinction or breaking off of relations between the two parts (as seen in Syndromes B and C), or else a pollination "treaty" (Fig. 17-2) or other mutual aid agreement. The loan of toxic chemicals by the plant's biosynthetic banks to help defend the enemy against its own enemies would surely not be recommended by strategic defense considerations, which would prefer to use this energy to stimulate the enemy's enemies (Fig. 17-3).

Considering the questions raised in the early part of this chapter, it may be noted that the results of phylogenetic and chemical analysis with bioassay suggest that:

(1) The parasitic interactions presently observed between these aposematic insects and their toxic host plants are indeed very old, with a very ample (usually cosmopoliton, at least hemispheric) geographical distribution. They have persisted through a combination of coadaptation, transformation to mutualism, tolerance, and discovery of alternative new food and drug sources.

(2) The "coevolutionary process" has been carried out in many ways, including that of promoting its own termination. Many other selective agents have also acted, giving adaptations whose age, origin, and function are difficult to determine or relate to stepwise coevolution at the primary interface between the insect and its larval host plant.

(3) Primitive groups may follow an "association by descent" model, but most interactions seem to have developed through progressive colonization of already diversified host plants by radiating preadapted herbivores.

(4) *Diversification* processes in Neotropical communities probably include some one-on-one or diffuse coevolutionary components, but are probably not dominated by these mechanisms. Other simpler, less deterministic, or even nonbiotic factors may leave more options to individual lineages to follow a variable or optimized mix of selective paths. The importance of *mutualism*, as well as diverse factors of community *structure* and *dynamics*, are probably strongly tied into coevolutionary processes (Gilbert, 1980).

(5) Three main syndromes of chemical defense can be discerned in the evolutionary advancement within this system, showing additional subdivisions provided by population biology and life-history options: larval sequestration (A), adult sequestration (B), and *de novo* biosynthesis (C).

(6) While these syndromes may be used to suggest patterns to be expected in other closely related groups (nonaposematic butterflies or other insect taxa), the complexity of evolutionary options and dimensions may make prediction as risky today as it was before these chemical results were available. It is still advisable to collect chemical data in each case before speculating in print.

Further work underway in Campinas, in addition to the lines mentioned above, includes studies with vertebrate and hymenopteran predators, volatile or minor components with enhanced ecological activity, and details of phylogenies and host-plant relationships at the species level. These should help refine, expand, and in some cases profoundly change the emergent patterns discussed here, which nevertheless will continue to reflect the complexity and fragility of the interface between toxic plants and their associated specialist herbivores.

ACKNOWLEDGMENTS

This manuscript was stimulated and criticized at various stages, in part or in its entirety, by Drs. Paul Feeny, Miriam Rothschild, Lincoln Brower, Lawrence Gilbert, Woodruff Benson, Adolf Nahrstedt, John Edgar, Jacques Pasteels, Paul Ehrlich, Daniel Janzen, George Sheppard, Paulo S. Oliveira, and João Vasconcellos-Neto. Peter Price and Thomas Lewinsohn gave special help in its development and refinement, and permitted its presentation to the Symposium in Campinas in March 1988. Hermógenes F. Leitão Filho aided greatly in the identification of plants; Boyce A. Drummond III, James D. Miller, and T. Eisner shared manuscripts and reprints, and Silvana Aparecida Henriques performed innumerable chemical analyses and fractionations with unvarying precision and good will. We are grateful to all these persons and the many others who have helped our growth, work, and comprehension in the field of chemical ecology.

REFERENCES

Ackery, P. R. 1988. Hostplants and classification: a review of Nymphalid butterflies. *Biol. J. Linn. Soc.* **33**: 95–203.

Ackery, P. R. and R. I. Vane-Wright. 1984. *Milkweed butterflies: Their cladistics and biology.* British Museum (Natural History), London, ix + 425 pp.

Aplin, R. T., R. D'Arcy Ward, and M. Rothschild. 1975. Examination of the large white and small white butterflies (*Pieris* spp.) for the presence of mustard oils and mustard oil glycosides. *J. Ent. (A), London* **50**: 73–78.

Atsatt, P. R. 1981. Lycaenid butterflies and ants: Selection for enemy-free space. *Amer. Natur.*, **118**: 638–654.

Benson, W. W. 1978. Resource partitioning in passion vine butterflies. *Evolution,* **32**: 493–518.

Benson, W. W., K. S. Brown Jr., and L. E. Gilbert. 1976. Coevolution of plants and herbivores: Passion flower butterflies. *Evolution* **29**: 659–680.

Berenbaum, M. R. 1983. Coumarins and caterpillars: A case for coevolution. *Evolution,* **37**: 163–179.

Berenbaum, M. R. and P. Feeny. 1981. Toxicity of angular furanocoumarins to swallowtail butterflies: Escalation in the coevolutionary arms race. *Science* **212**: 927–929.

Boppré, M. 1986. Insects pharmacophagously utilizing defensive plant chemicals (pyrrolizidine alkaloids). *Naturwissenschaften* **73**: 17–26.

Brower, L. P. 1969. Ecological chemistry. *Scient. Amer.* **220**: 22–29.

Brower, L. P. 1984. Chemical defence in butterflies. In Vane-Wright, R. I. and P. R. Ackery (Eds.), *The Biology of Butterflies.* Academic Press, London, pp. 109–134.

Brower, L. P. and J. v. Z. Brower. 1964. Birds, butterflies and plant poisons: A study in ecological chemistry. *Zoologica (N.Y.)* **49**: 137–159.

Brower, L. P., J. v. Z. Brower, and J. M. Corvino. 1967. Plant poisons in a terrestrial food chain. *Proc. Nat. Acad. Sci. USA.* **57**: 893–898.

Brower, L. P., J. N. Sieber, C. J. Nelson, S. P. Lynch, and P. M. Tuskes. 1982. Plant-determined variation in the cardenolide content, thin-layer chromatography profiles, and emetic potency of Monarch butterflies, *Danaus plexippus* reared on the milkweed, *Asclepias eriocarpa* in California. *J. Chem. Ecol.* **8**: 579–633.

Brown Jr., K. S. 1984. Adult-obtained pyrrolizidine alkaloids defend ithomiine butterflies against a spider predator. *Nature* **309**: 707–709.

Brown Jr., K. S. 1985. Chemical ecology of dehydropyrrolizidine alkaloids in adult Ithomiinae (Lepidoptera, Nymphalidae). *Rev. bras. Biol.* **44**: 435–460.

Brown Jr., K. S. 1987. Chemistry at the Solanaceae/Ithomiinae interface. *Ann. Miss. Bot. Garden* **74**: 359–397.

Brown Jr., K. S. 1988. Mimicry, aposematism and crypsis in Neotropical Lepidoptera: The importance of dual signals. *Bull. Soc. Zool. France* **113**: 83–101.

Brown Jr., K. S., A. J. Damman, and P. Feeny. 1981. Trodine swallowtails (Lepidoptera: Papilionidae) in southeastern Brazil: Natural history and foodplant relationships. *J. Res. Lepid.* **19**: 199–226.

Brown Jr., K. S. and R. B. Francini. 1988. Cianogênese em Acraeinae americanos (Lepidoptera: Nymphalidae) e padrões de defesa química em Lepidópteros aposemáticos. *Resumos XV Congresso Brasileiro de Zoologia, Curitiba, Paraná*, p. 234.

Brown Jr., K. S. and S. A. Henriques. 1990. Chemistry, coevolution and colonization of Solanaceae leaves by ithomiine butterflies. In Lester, R. N., M. Nee, and N. Estrada (Eds.), Academic Press, London. In press.

Brown Jr., K. S. and J. Vasconcellos-Neto. 1976. Predation on aposematic ithomiine butterflies by tanagers (*Pipraeida melanonota*). *Biotropica* **8**: 136–141.

Calvert, W. H., L. E. Hedrick, and L. P. Brower. 1979. Mortality of the Monarch butterfly (*Danaus plexippus* L.) due to avian predation at five overwintering sites in Mexico. *Science* **204**: 847–851.

Chai, P. 1986. Field observations and feeding experiments on the responses of rufous-tailed jacamars (*Galbula ruficauda*) to free-flying butterflies in a tropical rain forest. *Biol. J. Linn. Soc.* **29**: 161–189.

Chai, P. 1988. Wing color of free-flying neotropical butterflies as a signal learned by a specialist avian predator. *Biotropica* **20**: 20–30.

Clark, L. R., P. W. Geier, R. D. Hughes, and R. F. Morris. 1967. *The Ecology of Insect Populations in Theory and Practice*. Methuen & Co., London, xiii + 232 pp.

Conner, W. E., T. Eisner, R. K. Van der Meer, A. Guerrero, and J. Meinwald. 1981. Precopulatory sexual interaction in an Arctiid moth (*Utetheisa ornatrix*): Role of a pheromone derived from dietary alkaloids. *Behav. Ecol. Sociobiol.* **9**: 227–235.

Davis, R. H. and A. Nahrstedt. 1985. Cyanogenesis in insects. In Kerkut, G. A. and L. I. Gilbert (Eds.), *Comprehensive Insect Physiology, Biochemistry and Pharmacology*, Vol. 11, Pergamon Press, Oxford, pp. 635–654.

Dixon, C. A., J. M. Erickson, D. N. Kellett, and M. Rothschild. 1978. Some adaptations between *Danaus plexippus* and its food plant, with notes on *Danaus chrysippus* and *Euploea core* (Insecta: Lepidoptera). *J. Zool. London* **185**: 437–467.

Drummond III, B. A. 1981. Ecological chemistry, animal behavior, and plant systematics. *Solanaceae Newsletter* **2**: 59–66.

Drummond III, B. A. 1986. Coevolution of ithomiine butterflies and solanaceous plants. In D'Arcy, W. G. (Ed.), *Solanaceae, Biology and Systematics.* Columbia University Press, N.Y., pp. 307–327.

Drummond III, B. A. and K. S. Brown, Jr. 1987. Ithomiinae (Lepidoptera: Nymphalidae): Summary of known larval foodplants. *Ann. Miss. Bot. Garden* **74**: 341–358.

Edgar, J. A. 1975. Danainae (Lep.) and 1,2-dehydropyrrolizidine alkaloid-containing plants—with references to observations made in the New Hebrides. *Phil. Trans. R. Soc. London B* **272**: 467–476.

Edgar, J. A. 1982. Pyrrolizidine alkaloids sequestered by Solomon Island danaine butterflies. The feeding preferences of the Danainae and Ithomiinae. *J. Zool. London* **196**: 385–399.

Edgar, J. A., C. C. J. Culvenor, and T. E. Pliske. 1976. Isolation of a lactone, structurally related to the esterifying acids of pyrrolizidine alkaloids, from the costal fringes of male Ithomiinae. *J. Chem. Ecol.* **2**: 263–270.

Edmunds, J. 1974. Defence in Animals. Longmans, Harlow, xvii + 357 pp.

Ehrlich, P. R. and P. H. Raven. 1965. Butterflies and plants: A study in coevolution. *Evolution* **18**: 586–608.

Eisner, T. 1982. For love of nature: Exploration and discovery at biological field stations. *Bioscience* **32**: 321–326.

Feeny, P. 1975. Biochemical evolution between plants and their insect herbivores. In Gilbert, L. E. and P. H. Raven (Eds.), *Coevolution of plants and animals,* University of Texas Press, Austin, pp. 3–19.

Futuyma, D. J. 1983. Evolutionary interactions among herbivorous insects and plants. In Futuyma, D. J. and M. Slatkin (Eds.), *Coevolution.* Sinauer, Sunderland, MA, pp. 207–231.

Futuyma, D. J. and M. Slatkin (Eds.). 1983. *Coevolution.* Sinauer, Sunderland, MA, x + 555 pp.

Gilbert, L. E. 1980. Food web organization and the conservation of neotropical diversity. In Soulé, M. E. and B. A. Wilcox (Eds.), *Conservation Biology.* Sinauer, Sunderland, MA, pp. 11–33.

Gilbert, L. E. 1983. Coevolution and mimicry. In Futuyma, D. J. and M. Slatkin (Eds.), *Coevolution.* Sinauer, Sunderland, MA, pp. 263–281.

Gilbert, L. E., H. S. Forrest, T. D. Schultz, and D. J. Harvey. 1988. Correlations of ultrastructure and pigmentation suggest how genes control development of wing scales of *Heliconius* butterflies. *J. Res. Lepid.* **26**: 141–160.

Gottlieb, O. R. 1982. *Micromolecular evolution, Systematics and Ecology—An essay into a novel botanical discipline.* Springer-Verlag, Berlin, xi + 170 pp.

Gottlieb, O. R. 1984. Phytochemistry and the evolution of angiosperms. *An. Acad. bras. Ciênc.* **56**: 43–50.

Honda, K. 1981. Larval osmeterial secretions of the swallowtails (*Papilio*). *J. Chem. Ecol.* **7**: 1089–1113.

Janzen, D. H. 1980. When is it coevolution? *Evolution* **34**: 611–612.

Järvi, T., B. Sillen-Tullberg, and C. Wiklund. 1981. The cost of being aposematic. An experimental study of predation on larvae of *Papilio machaon* by the great tit, *Parus major.* *Oikos* **36**: 267–272.

Jermy, T. 1976. Insect–host plant relationships—Co-evolution or sequential evolution? *Symp. Biol. Hung.* **16**: 109–113.

Jermy, T. 1984. Evolution of insect/host plant relationships. *Amer. Natur.* **124**: 609–630.

Karger, B. L., L. R. Snyder, and C. Horvath. 1973. *An introduction to separation science.* Wiley, N.Y., xxi + 586 pp.

Malcolm, S. and M. Rothschild. 1983. A danaid Müllerian mimic, *Euploea core amymone* (Cramer) lacking cardenolides in the pupal and adult stages. *Biol. J. Linn. Soc.* **19**: 27–33.

Miller, J. M. 1975. *Separation methods in chemical analysis.* Wiley-Interscience, N.Y., x + 309 pp.

Miller, J. S. 1987. Host-plant relationships in the Papilionidae (Lepidoptera): Parallel cladogenesis or colonization? *Cladistics* **3**: 105–120.

Mitter, C. and D. R. Brooks. 1983. Phylogenetic aspects of coevolution. In Futuyma, D. J. and M. Slatkin (Eds.). *Coevolution.* Sinauer, Sunderland, MA. pp. 65–98.

Nahrstedt, A. 1985. Cyanogenic compounds as protecting agents for organisms. *Plant Syst. and Evol.* **150**: 35–47.

Nahrstedt, A. and R. H. Davis. 1981. The occurrence of the cyanoglycosides, linamarin and lotaustralin in *Acraea* and *Heliconius* butterflies. *Comp. Biochem. Physiol.* **68B**: 575–577.

Nahrstedt, A. and R. H. Davis. 1983. Occurrence, variation and biosynthesis of the cyanogenic glucosides linamarin and lotaustralin in species of the Heliconiini (Insecta: Lepidoptera). *Comp. Biochem. Physiol.* **75B**: 65–73.

Nijhout, H. F. 1986. Pattern and pattern diversity on lepidopteran wings. *Bioscience* **36**: 527–553.

Orians, G. H. and R. T. Paine. 1983. Convergent evolution at the community level. In Futuyma, D. J. and M. Slatkin (Eds.), *Coevolution.* Sinauer, Sunderland, MA. pp. 431–458.

Papageorgis, C. 1975. Mimicry in Neotropical butterflies. *Amer. Scient.* **63**: 522–532.

Rhoades, D. F. 1979. Evolution of plant chemical defense against herbivores. In Rosenthal, G. A. and D. H. Janzen (Eds.), *Herbivores: Their interaction with secondary plant metabolites.* Academic Press, N.Y., pp. 3–54.

Rhoades, D. F. 1985. Offensive–defensive interactions between herbivores and plants: Their relevance in herbivore population dynamics and ecological theory. *Amer. Natur.* **125**: 205–238.

Rothschild, M. 1972. Secondary plant substances and warning coloration in insects. In van Emden, H. F. (Ed.), Insect/Plant relationships. *Royal Ent. Soc. London, Symp. No. 8,* Blackwells, Oxford, pp. 59–83.

Rothschild, M. 1981. Mimicry, butterflies and plants. *Symb. Bot. Upsal.* **22**: 82–99.

Rothschild, M. and J. A. Edgar. 1978. Pyrrolizidine alkaloids from *Senecio vulgaris* sequestered and stored by *Danaus plexippus. J. Zool. London* **186**: 347–349.

Rothschild, M. and N. Marsh. 1978. Some peculiar aspects of Danaid/plant relationships. *Ent. Exp. & Appl.* **24**: 437–450.

Rothschild, M., T. Reichstein, J. von Euw, R. Aplin, and R. R. M. Harman. 1970. Toxic Lepidoptera. *Toxicon* **8**: 293–299.

Rothschild, M., J. von Euw, T. Reichstein, D. A. S. Smith, and J. Pierre. 1975. Cardenolide storage in *Danaus chrysippus* (L.) with additional notes on *D. plexippus* (L.). *Proc. Roy. Entomol. Soc. London B,* **190**: 1–31.

Rothschild, M., R. J. Nash, and E. A. Bell. 1986. Cycasin in the endangered butterfly *Eumaeus atala florida*. *Phytochemistry* **25**: 1853–1854.

Roughgarden, J. 1983. Coevolution between competitors. In Futuyma, D. J. and M. Slatkin (Eds.), *Coevolution*. Sinauer, Sunderland, MA, pp. 383–403.

Schneider, D., M. Boppré, H. Schneider, W. R. Thompson, C. J. Boriack, R. L. Petty, and J. Meinwald. 1975. A pheromone precursor and its uptake in male *Danaus* butterflies. *J. Comp. Physiol.* **97**: 245–256.

Schneider, D., J. Zweig, S. B., Horsley, T. W. Bell, J. Meinwald, K. Hansen, and E. W. Diehl. 1982. Scent organ development in *Creatonotes* moths: Regulation by pyrrolizidine alkaloids. *Science* **215**: 1264–1265.

Sheppard, P. M., J. R. G. Turner, K. S. Brown Jr., W. W. Benson, and M. C. Singer. 1985. Genetics and the evolution of muellerian mimicry in *Heliconius* butterflies. *Phil. Trans. R. Soc. London B* **308**: 433–613.

Sondheimer, E. and J. B. Simeone (Eds.). 1970. *Chemical Ecology*. Academic Press, NY, xv + 336 pp.

Southwood, T. R. E. 1978. *Ecological methods—with particular reference to the study of insect populations*. Chapman & Hall, London, 524 pp.

Spencer, K. C. 1988. Chemical mediation of coevolution in the *Passiflora-Heliconius* interaction. In Spencer, K. C. (Ed.), *Chemical mediation of coevolution*. Academic, NY, pp. 167–240.

Trigo, J. R. 1988. Ecologia química na interação Ithomiinae (Lepidoptera: Nymphalidae)/Echitoideae (Angiospermae: Apocynaceae). Master's thesis, Universidade Estadual de Campinas, xi + 178 pp.

Trigo, J. R. and K. S. Brown Jr. 1988. Ithomiinae: Evolutionary patterns in the obtaining of PAs (dehydropyrrolizidine alkaloids) from plants. *An. Simp. Intern. Ecol. Evol. Herb. Tropicais*, Campinas, SP, p. 63.

Trigo, J. R. and P. C. Motta. 1986. Implicações evolutivas da assimilação de alcalóides deidropirrolizidínicos (PAs) por larvas de Ithomiinae (Lepidoptera: Nymphalidae). *Resumos VII Reunião sobre Evolução, Sistemática e Ecologia Micromoleculares (RESEM)*, Rio de Janeiro.

Urzúa, A. and H. Priestap. 1985. Aristolochic acids from *Battus polydamas* (Lepidoptera: Papilionidae). *Biochem. Syst. Ecol.* **13**: 169–170.

Urzúa, A., R. Rodríguez, and B. Cassels. 1987. Fate of ingested aristolochic acids in *Battus archidamas*. *Biochem. Syst. Ecol.* **15**: 687–689.

Urzúa, A., G. Salgado, B. K. Casels, and G. Eckhardt. 1983. Aristolochic acids in *Aristolochia chilensis* and the *Aristolochia* feeder *Battus archidamas* (Lepidoptera). *Coll. Czech. Chem. Commun.* **48**: 1513–1519.

Vasconcellos-Neto, J. and T. M. Lewinsohn. 1984. Discrimination and release of unpalatable butterflies by *Nephila clavipes*, a neotropical orb-weaving spider. *Ecol. Entom.* **9**: 337–344.

Von Euw, J., T. Reichstein, and M. Rothschild. 1968. Aristolochic acid in the swallowtail butterfly *Pachlioptera aristolochiae*. *Israel J. Chem.* **6**: 659–670.

Wallace, J. and R. Mansell (Eds.). 1976. *Biochemical interactions between plants and insects*. Recent Advances in Phytochemistry, 10, Plenum, NY, xii + 425 pp.

Young, A. M. 1972. On the life cycle and natural history of *Hymenitis nero* (Lepidoptera: Ithomiinae) in Costa Rica. *Psyche* **79**: 284–294.

18 Biodiversity of a Central American *Heliconius* Community: Pattern, Process, and Problems

LAWRENCE E. GILBERT

In this chapter, I summarize evidence that two evolutionary innovations, pupal mating and pollen feeding, have opened new adaptive zones within the butterfly tribe Heliconiini. I suggest ways that these innovations alter the ecological and evolutionary interactions between *Heliconius*, its adult and larval resource plants, as well as coevolution among heliconiines. For example, experiments indicate that host-associated killing of teneral butterflies by pupal-mating males generates selection for restricted larval diet in many *Heliconius*.

Insights into these historical processes emerge from ecological, physiological, behavioral, genetic, systematic, and natural historical studies of contemporary communities. In turn, the historical scenario provides an important basis for understanding local patterns of biodiversity in this system. My chapter is a celebration of neotropical diversity, and an admission of relative ignorance, meant to encourage others to work on this and similar systems of speciose herbivore genera from both evolutionary and ecological perspectives.

HELICONIUS AS A REPRESENTATIVE OF TROPICAL BIODIVERSITY

Like most ecologists, I am drawn to the problem of tropical species diversity in an almost aesthetic way. Yet I center my research primarily on a single genus of phytophagous insects, *Heliconius*, at one rain-forest location in Costa Rica. I am interested in the mechanisms by which properties of this community arise from the properties, traits, and interactions of individuals which compose the populations of each coexisting species. This is a reductionist and natural historical approach to understanding emergent properties of the community. We

Plant-Animal Interactions: Evolutionary Ecology in Tropical and Temperate Regions, Edited by Peter W. Price, Thomas M. Lewinsohn, G. Wilson Fernandes, and Woodruff W. Benson. ISBN 0-471-50937-X © 1991 John Wiley & Sons, Inc.

need to know a few representative ecological and evolutionary subsystems in great depth in order to develop better generalities about how diversity evolves and persists. This is the ultimate goal that motivates my approach to the Heliconiini and associated organisms.

I have four reasons for optimism that the results of studies on *Heliconius* will be relevant to developing generalities about tropical biodiversity:

1. Phytophagous insects, including *Heliconius*, constitute the majority of all animal species both globally and in the tropics.

2. A large, but undetermined component of the increase in species diversity along the temperate–tropical gradient results from the progressive expansion of species richness within the generic level, a pattern reflected by *Heliconius*.

3. The *Heliconius* community can be conveniently studied as component individuals and populations. Thus, one has the opportunity to understand patterns at the community level in terms of individual behavior, population dynamics, and interactions of individuals from different species.

4. A large fraction of biodiversity in the tropics consists of low density or rare species. Although conspicuous, *Heliconius* species are typically low-density organisms in natural habitats, and therefore may provide insights concerning the persistence of rare species.

In this chapter I discuss some aspects of the evolution and ecology of diversity in heliconiine butterflies primarily from the perspective of the research of my colleagues, students, and myself in Central America. It should therefore be kept in mind that the account may be limited by my lack of experience with genera of Heliconiini and species groups of *Heliconius* restricted to parts of South America. Moreover, since my primary intent is to stimulate interest in the *Heliconius* system, I have emphasized interesting patterns and phenomena in need of further work rather than a systematic review of all the literature on *Heliconius* and associated organisms.

THE EVOLUTION OF HELICONIINE DIVERSITY: A SCENARIO

A close relative of New World heliconiines is the Old World genus *Cethosia*. It, like typical members of the tribe, uses plants of the family Passifloraceae as larval food. Species of *Cethosia* possess wing patterns which can be easily seen to fit the standard nymphaline ground plan (Nijhout 1986). In the New World, where heliconiines and Passifloraceae diversify dramatically, several genera possess elements of this basic nymphaline pattern (e.g., *Agraulis, Dione*) and/or lack distinctive mimetic patterns (e.g., *Philaethria, Podotricha, Dryadula, Dryas*). These genera share a conspicuous trait with *Cethosia*: the wing venation of the hindwing fails to form a closed discal cell. In the remaining four and most diverse heliconiine genera the discal cell is closed and departure from the basic nymphaline ground-plan is greater, with the appearance of several novel mimetic patterns.

The "open-cell" heliconiines are generally fast-flying and wide-ranging as individuals. Their highly vagile populations are associated with open sunny habitats where they visit unspecialized butterfly-pollinated flowers with short corollas and large floral displays (e.g., *Lantana*). These genera are relatively edible (Chai 1990) and appear to rely on fast flight to avoid predation. In contrast, "closed-cell" genera such as *Heliconius* and *Laparus* are relatively unpalatable, aposematic, slow-flying, and pollen-feeding.

Brown's (1981) summary of the biology and phylogeny of the heliconiines, which places the "open-celled" groups as ancestral and "closed-cell" genera as derived, is supported by recent molecular evidence (Lee et al. 1991). Brown's work provides the foundation for the scenario of adaptive radiation and life-history evolution presented in Figure 18-1. Below, I use this scenario to describe

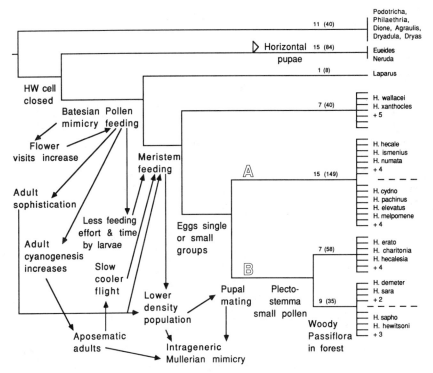

Figure 18-1. Hypothesized relationships of heliconiines based on Brown (1981), illustrating major adaptive zones for *Heliconius*. The cascade of life-history innovations set in motion by pollen feeding is indicated by arrows and is discussed in the text. Numbers on horizontal lines indicate number of species, followed by number of races in parentheses. Horizontal pupae on the line leading to *Eueides* and *Neruda* are hypothesized to originate after the origin of pupal mating in *Heliconius* (open triangle).

important events in the evolution of heliconiines and show how each event may have catalyzed further adaptive radiation.

Evolution of Wing Pattern Mimicry

In the New World, "open-celled" heliconiines are represented by 6 genera and 11 species, while the more derived "closed-celled" members of the tribe belong to 4 genera and 54 species, a sevenfold increase in intrageneric diversity. I hypothesize that the first step toward elaborating the modern themes of heliconiine diversification was the evolution of Batesian mimicry in ancestors of the "closed-cell" genera. Mimicry of such highly distasteful groups as ithomiine butterflies (see Brower and Brower 1964; Chai 1986; Brown 1987) would have allowed two behavioral changes. These in turn were critical for subsequent diversification in the tribe. First, Batesian mimicry, by decreasing risk of predation, would have allowed more conspicuous and leisurely flower visitation. Second, mimicry could have permitted expansion of microhabitat range to include successional stages utilized by model species and occupied by unexploited Passifloraceae.

Pollen Feeding as an Adaptive Zone

The innovation of pollen feeding in an ancestor of present-day *Laparus* and *Heliconius* (Gilbert 1972) is apparently unique among the butterflies. The first pollen-feeding heliconiine must have used a range of butterfly flowers opportunistically, as pollen-feeding *Heliconius* still do when deprived of favored flowers. Pollen gathering is slow, and requires an individual to spend much more time on a flower than does the acquisition of nectar. The evolution of mimicry may have been a prerequisite for that of pollen collecting by assuring the relatively predator-free time on flowers needed for removing pollen.

Pollen-feeding genera include the monotypic *Laparus* and *Heliconius* with 38 species. *Heliconius* not only accounts for over 50% of New World heliconiine species, but embodies an even greater fraction of heliconiine biodiversity in terms of life history variety among species (see Brown 1981), and genetic differentiation (i.e., racial variety) within species (see Fig. 18-1 and Turner 1981).

The evolution of pollen feeding brought about a dramatic increase in the availability of free amino acids to the adult (Gilbert 1972), and set the stage for three major evolutionary trends which culminate in modern *Heliconius*. These trends, described in detail below, are (1) the increase of adult cyanogenesis, (2) the shift of reproductive effort from larval to adult feeding, and (3) the increase in adult behavioral sophistication.

Pollen Feeding and Cyanogenesis

Heliconius contain stored cyanogenic glycosides in both larval and adult tissues. Because such compounds characterize the passifloraceous larval hosts, it was originally assumed (e.g., Brower and Brower 1964) that these were sequestered during larval feeding and passed on to the adults. However, it is now known that *Heliconius*, both as larvae and as adults, manufacture cyanogenic glycosides from amino acid precursors (Nahrstedt and Davis 1983). Because the butterfly's system of cyanogenesis can be used to counter that of the plant (Spencer 1988), it seems likely that this capability of protoheliconiines fostered the initial shift onto Passifloraceae.

Pollen feeding is strongly correlated with unpalatability of heliconiines to a natural predator, the jacamar (see Table 18-1 and Chai 1990). This finding suggests that the evolution of pollen feeding enhanced the capacity for cyanogenic glycoside production by adult *Heliconius* and therefore promoted distastefulness and aposematism. Thus the scenario I suggest begins with Batesian mimicry, leads to pollen feeding, increased distastefulness, aposematism, and finally the intrageneric Mullerian mimicry characteristic of modern *Heliconius* (Turner 1981; Gilbert 1983; and see Fig. 18-1).

A shift to unpalatability and warning color for predator avoidance is correlated with shifts in flight physiology. Unpalatable butterflies fly more slowly and at cooler temperatures (Srygley and Chai 1990). This enables them to invade cool microhabitats and extend their flight periods into overcast conditions or cooler morning and evening hours unavailable to related butterflies. Such a capacity for foraging in cool temperatures would have given early pollen-feeding

TABLE 18-1. Feeding Response of Rufous-Tailed Jacamars to Free-Flying Heliconiines at Sirena[a]

	Heliconiine Species	Total	Eaten	% Eaten	Rank
	Philaethria dido	12	12	1.00	1
	Dryas julia	24	22	0.92	2
	Dryadula phaetusa	16	14	0.88	3
	Agraulis vanillae	12	9	0.75	4
P[b]	*Heliconius charitonia*	24	15	0.63	5
	Eueides isabella	8	4	0.50	6
	Eueides lybia	33	13	0.39	7
	Eueides aliphera	36	14	0.39	8
	Dione juno	44	9	0.20	9
P	*Heliconius melpomene*	20	2	0.10	10
P	*Heliconius ismenius*	42	2	0.05	11
P	*Laparus doris*	35	0	0.00	12
P	*Heliconius cydno*	2	0	0.00	13
P	*Heliconius pachinus*	22	0	0.00	14
P	*Heliconius erato*	32	0	0.00	15
P	*Heliconius hecalesia*	3	0	0.00	16
P	*Heliconius hecale*	30	0	0.00	17
P	*Heliconius sara*	6	0	0.00	18
P	*Heliconius hewitsoni*	19	0	0.00	19

[a]Data extracted from Chai (1990). [b]P = pollen feeding. Pollen feeders are different from nonpollen feeders. Mann-Whitney U test $p < 0.001$

heliconiines access to floral resources previously exploited and dominated by hummingbirds or bees.

The evolution of pollen feeding in *Heliconius* affected the evolution of the cucurbit genera *Gurania* and *Psiguria*. The details of this interaction are yet to be fully explored. Some species of the Guraniinae have retained their association with hummingbirds and evolved resistance to visitation by heliconiines, while others have become specialized for pollination by these insects. Resistance in *Gurania* is achieved by spikes of the calyx that hinder butterfly access to male flowers, or by corolla tubes that fail to open and are forcible by hummingbirds but not by *Heliconius* (see Condon and Gilbert 1990).

Adaptation to pollination by *Heliconius* involved the evolution of flowers more accessible to butterflies and less rewarding to hummingbirds. Examples from both *Gurania* and *Psiguria* include species with smaller flowers, less nectar production, and presentation of floral rewards later in the day (Condon and Gilbert 1990; Gilbert 1975). Indeed, *Psiguria* may well represent a genus of the Guraniinae which arose and diversified because of the existence of the pollen-feeding heliconiines. The odd species from several other plant families (Rubiaceae, Boraginaceae, Polygalaceae) appears to have specialized for *Heliconius* pollination, but these cases do not appear comparable to the development of *Psiguria* and *Gurania* as diverse, geographically widespread, coevolved mutualists of *Heliconius*.

Pollen Feeding and Life History Evolution

The second major trend set in motion by pollen feeding is the shift of collecting resources for reproduction from the larval to the adult stage of the life history. Even if distasteful to birds, larval lepidoptera are highly vulnerable to invertebrate predators and parasitoids. In contrast, an aposematic, mobile adult should have relatively higher daily survivorship than a larva. Pollen feeding is not only correlated with greater unpalatability but provides nutrition for adult somatic maintenance and extended reproductive longevity (Dunlap-Pianka et al. 1977) and nuptial gifts (Boggs and Gilbert 1979).

Two theoretical predictions have been made about the impact on life-history traits that pollen feeding should have. First, Boggs (1981) predicted that butterfly species with lower expectation of obtaining high-quality nitrogenous resources in the adult stage should allocate relatively more of larval resources to future reproduction than species with high expectations of acquiring such resources. In terms of *Heliconius* biology, this is a prediction that species with more access to pollen should carry forward fewer nutritional reserves into the adult stage. Boggs (1981) estimated the proportion of larval-gathered reserves from the relative nitrogen contents of abdomen and thorax in teneral adults of three species that differed in their access to pollen. As predicted, the most intensive pollen feeder, *H. cydno*, had the lowest proportion of abdominal nitrogen; *H. charitonia*, an opportunistic pollen gatherer, ranked second in the trait, and *Dryas julia*, which does not feed on pollen at all, ranked third.

The second and partly overlapping prediction (Gilbert 1972) is that high-quality adult nutrition would allow reduction of the larval stage because fewer resources would be collected for reproduction and less time would be spent in that vulnerable stage. Singer (1981) made a related prediction by stating that improvement in the quality of life of the adult insect relative to that of the larva should reduce time spent and resources gathered as a last-instar larva and increase resource gathering by adults. The quality of life of pollen-feeding adult *Heliconius* is high in both resource availability and predator protection. Larvae of *Heliconius* species should therefore develop rapidly, when compared to those of less-aposematic, nonpollen-feeding relatives of similar size. This prediction has not been tested for the heliconiines generally. However, preliminary comparisons of *Philaethria dido* and *H. hecale* growing on *P. vitifolia* support the prediction (R. Srygley, personal communication). In this way, time of exposure to predation on larvae is reduced, as is the leaf biomass required to support larval development, thus allowing the evolution of specialization on small, high-quality parcels of host such as seedlings and expanding meristems (see below).

Pollen Feeding and Behavioral Sophistication

The third major trend of heliconiine evolution which is correlated with pollen feeding is the increase in behavioral sophistication (Gilbert 1975). Since the fraction of resources for reproduction collected by the adult stage increased to the 80% level known for modern *Heliconius* (Dunlap-Pianka et al. 1977), and since that fraction of reproductive effort is due to pollen and nectar foraging over an extended reproductive life, any behavioral changes which improved foraging ability would have been more strongly favored by natural selection than would be the case in short-lived species less dependent on adult foraging. The well-developed eyes, visual orientation to landmarks, location memory, faithfulness to roosts and resource locations, and other traits of *Heliconius* have been discussed in regard to the pollen-feeding adaptive zone (Gilbert 1975). However, many such morphological and behavioral details deserve more rigorous, comparative study within the Heliconiini among pollen-feeding and nonpollen-feeding taxa. In one recent study, Sivinski (1989) has determined that *H.* charitonia has 2.5–7 times the mushroom body volume to brain volume ratio of other nymphalines studied, including heliconiines *Dryas* and *Agraulis*. The mushroom body in insects appears to be connected with memory although its role in arthropods generally is debated (see Sivinski (1989) for a review).

The population-level manifestation of landmark learning and orientation by individual foragers of *Heliconius* is the highly sedentary nature of local populations (Turner 1971; Ehrlich and Gilbert 1973; but see Mallet 1986a). Such population structure emerges from the fact that each individual occupies a home range based in large part on a network of pollen plants such as *Psiguria* which flower over extended periods (many months). Combining mark–release–recapture with techniques for indirectly measuring within-day movement among floral resources by single individuals, it is possible to show that resident

Heliconius often circulate through their entire local home range within a given day (Murawski and Gilbert 1986), a behavior more characteristic of small birds and mammals than of butterflies. Furthermore, although *Heliconius* learn to avoid flower patches where they have been caught and handled, they do not abandon their home range, and remain at risk for capture in the area (Mallet et al. 1987) during the remainder of their 2–6 month residency. One reason for home range stability is the faithfulness of individuals to nocturnal roosts (Waller and Gilbert 1982; Mallet 1986b).

These sedentary, long-lived insects that feed on predictable adult resources can survive and remain resident through long periods when the less predictable larval resources are scarce or absent. Most of our observations on this phenomenon involve *H. hewitsoni* and its host, *P. pittieri*, at Sirena Station, Corcovado National Park, Costa Rica. We observed the same marked females investigating certain plants for days, even weeks, without oviposition when no new shoots were being produced (see Gilbert 1982a, b; Longino 1984 for discussion of *H. hewitsoni* and *P. pittieri*). These insects retain the capacity to reproduce after extended periods without hosts: one captive *H. hewitsoni* survived for 2 months before new shoots became available, at which point she laid a cluster of viable eggs (L. E. Gilbert, personal observation). These observations suggest a relationship between pollen feeding and ability of *Heliconius* to exist at low population density (see below), since part of the resilience of *Heliconius* populations is this capacity to forage while waiting for new growth scattered in space and time.

To develop a simple model of the role of pollen feeding in shaping the biology of *Heliconius*, I have liberally extrapolated from existing data and have made assumptions which need testing. Almost all facets of this discussion would profit from better empirical support. For example, the findings that adult cyanogenesis is linked to both pollen consumption and distastefulness are based on laboratory studies that need to be tested in the field. Students of living *Heliconius* marvel at the sophistication and plasticity of behavior displayed by marked individuals, yet rigorous experimental studies of learning, memory, or other aspects of behavior are lacking.

Evolution of Resource Use Patterns and its Consequences for Dynamic Interactions Between Heliconiines and Passiflora

The closest living relative to *Heliconius* is, according to Brown, the monotypic genus *Laparus*. *Laparus doris* is the only pollen-feeding species known outside *Heliconius*. Brown's (1981) scheme of heliconiine relationships implies a single origin of pollen feeding. However, this innovation did not immediately result in a burst of speciation, as it is *Heliconius*, not *Laparus*, that is highly speciose. I suggest that a shift from oviposition on old leaves to use of young shoots and seedlings was the important catalyst for further diversification.

L. doris females deposit rafts of eggs on mature leaves of large forest vines such as *Passiflora ambigua* (Benson et al. 1976). Modern representatives of the presumed early radiation of *Heliconius*, including *H. wallacei*, *H. xanthocles*, *H.*

burneyi, and four other species, resemble *Laparus* in that all cluster eggs on large forest vines, but place them on meristems rather than on older leaves (Benson et al. 1976; Mallet and Jackson 1980). The subsequent and major radiation of *Heliconius* was apparently based on the use of smaller plants or plant parts by the deposition of one or a few eggs. (This category includes about 24–25 species of *Heliconius* in the so-called advanced radiations A and B. See Benson et al. 1976 and Fig. 18-1).

Species of radiation A tend to associate with host plants in the *Granadilla* subgenus of *Passiflora* (Benson et al. 1976) and to associate with high-quality pollen sources in the forest such as *Psiguria*. One part of radiation B, including *H. erato*, *H. charitonia*, and related species, utilize the *Plectostemma* subgenus of *Passiflora*, a group characterized by nonwoody, short-lived plants of successional habitats. Smiley (1985a) suggested that the exploitation of this group of *Passiflora* entailed a shift in habitat away from the forest and in pollen-feeding behavior toward more efficient exploitation of less optimal pollen sources characteristic of successional habitats.

Boggs et al. (1981) and Murawski (1986) provide evidence that radiation A tends to collect larger pollen grains such as those of *Psiguria*, whereas radiation B tends to collect small pollen grains such as those of *Lantana*. Such differences reflect more than habitat choice since *H. erato* from Group B outcompetes comimic and sympatric *H. melpomene* (Group A) for small pollen in greenhouse experiments (Boggs et al. 1981) and *H. pachinus* (Group A) consistently collects a greater fraction of *Psiguria* pollen than comimic and sympatric *H. hewitsoni* (Group B) (Murawski 1986).

Together with the trait of pupal mating shared by all Group B species (see below), patterns of pollen feeding support the contention of Benson et al. (1976) that another part of radiation B secondarily returned to woody forest *Passiflora* such as the subgenus *Astrophea* and to the habit of placing large egg clusters on new shoots (see Fig. 18-1). Some members of this group (e.g., *H. hewitsoni* and *H. sapho*) have evolved highly refined oviposition behavior tightly adjusted to the phenology and pattern of host-shoot production (Gilbert 1982a, b; Longino 1984, 1986). The temporal spacing of new shoots on these plants may be as long as 84 days (Longino 1984).

The different feeding behaviors I have described are correlated with the population dynamic interactions of heliconiines and their hosts. Nonpollen-feeding species that lay large clusters on old leaves rapidly defoliate their hosts and reach high abundance in captivity, but in the field do so only rarely, principally when their hosts reach high density in early succession or in agricultural settings. In more stable habitats, heliconiines are less abundant than in heavily disturbed successional areas, and we find a predominance of sedentary pollen-feeding insects that lay single eggs on young shoots or seedlings. These larvae often destroy the shoots or kill the seedlings, but they cannot perform well on old leaves. Hence resources are often limiting to each larva, and it is not surprising that many of these species produce hatching larvae that are aggressive and cannibalistic toward conspecifics. Both in enclosures and in nature, such

Heliconius (e.g., *H. erato*, *H. pachinus*) are limited in part by the one shoot–one mature larva outcome of such aggression. In captivity, higher butterfly densities and greater plant damage are characteristic of pollen-feeding *Heliconius* species such as *H. sara* and *H. hewitsoni* (Longino 1984), which place egg clusters on young shoots, resulting in gregarious, nonaggressive larvae that move onto older leaves when their shoot has been demolished.

The foregoing examples suggest that the population dynamic interactions between heliconiines and their hosts may depend both on the egg-clustering behavior and on the age of leaves eaten. Experiments are needed to elucidate the reasons for these different dynamic interactions between hosts and the various feeding guilds of heliconiines. However, it seems clear that the persistent attack on small plants and new growth promoted by the sedentary behavior of pollen-feeding *Heliconius* makes these insects a significant force in the evolution of *Passiflora*.

The Pupal-Mating Syndrome and Community Evolution

Almost half the *Heliconius* species (42%, $N = 16$) possess the unusual habit of mating during female eclosion (Group B in Fig. 18-1). Males search for pupae, sit on them, usually in groups of 2–4, and compete to mate when eclosion occurs (see Gilbert 1975). It seems likely pupal-mating behavior evolved in the context of low-density populations wherein the discovery of an eclosed virgin female was a very low-probability event for a male searching at random. Under such conditions, the ability to locate prepupae or pupae and to learn routes should have altered male searching tactics to focus in the areas where newly hatching individuals were likely to appear. At some point, an ancestor of this group increased the chances of success by waiting for emergence on the pupa itself. Some contemporary species, like *H. hewitsoni* and *H. charitonia*, do not wait, but pry into the pupa and actually mate with the female as she drops from the pupal case (Fig. 18-2).

Initial molecular studies indicate a single origin of pupal mating (Lee et al. 1990). Moreover, the pupal-mating syndrome is associated with at least two other traits or phenomena in a manner that suggests an important role of this innovation in the phylogenetic expansion of *Heliconius* as well as in the packing of *Heliconius* species into local habitats.

First, pupal mating appears to have enhanced the possibilities for intrageneric mimicry (Fig. 18-1). The remarkable Müllerian mimicry within *Heliconius* (Turner 1981) is expressed locally between pairs of species that share the same microhabitats (Smiley and Gilbert, manuscript). In almost all cases, each mimetic species pair consists of a pupal-mating (Group B) and a nonpupal-mating (Group A) species (Gilbert 1983, 1984). At our study site in Corcovado Park, Costa Rica, three very different wing patterns are "occupied" by three such pairs of species. It appears that the strikingly different mating tactics of these groups allow phenotypically identical species to occupy the same microhabitats. Such species may have complex evolutionary relationships (Templeton and Gilbert 1985).

Figure 18-2. *H. hewitsoni* male 958 has inserted 2–3 of terminal abdominal segments between pupal skin and teneral female abdomen. Male 1652 (below and out of focus) has just mated with a female whose wings are still expanding.

Second, this mating tactic may influence host-plant specialization in an unexpected way. Pupal-mating species may displace other heliconiines from their hosts by interference competition. Males of these species sit on, mate with, and disrupt eclosion of other *Heliconius* species of both mating types (Gilbert 1984). I have experimentally estimated the extent of this disruption caused by *H. hewitsoni* at Sirena, Corcovado National Park, Costa Rica. At various times over a 4-yr period I placed pupae of *other species* near the host plant of *H. hewitsoni*. Whenever prepupae or pupae of *H. hewitsoni* were on or near the plant, male search was intense and survival of heterospecific pupae was reduced. Teneral individuals of nonpupal-mating species (*D. julia*, *H. hecale*, *H. ismenius*, *H. melpomene*, and *H. pachinus*) suffered almost 100% mortality from attack by male *H. hewitsoni*. Pupal-mating species (*H. sara*, *H. charitonia*, and *H. erato*) fared better. Males often emerged unscathed, but surviving females were mated and sterilized. I am now expanding this study to see whether other pupal-mating species are exerting a similar negative impact around their hosts, and if so, under what conditions. If these preliminary observations of *H. hewitsoni* represent a general phenomenon, pupal mating may be the basis for some of the unexplained host-plant and habitat segregation observed in local *Heliconius* communities (Gilbert 1984 and see below). Aggressive pupal-mating males may prevent any other heliconiine species from evolving preference for their host plants.

Two heliconiine genera (*Eueides* and *Neruda*) that share the forest habitat with Group B species have evolved a novel pupal attitude: their pupae project horizontally instead of hanging vertically (Fig. 18-1). It would be interesting to know whether this morphological change has protected them from attack by visually searching pupal-mating males, allowing their coexistence with pupal-mating species on the same host species. Did this trait arise after pupal mating evolved in Group B (Fig. 18-1)?

EVOLUTIONARY RESPONSES OF PASSIFLORA TO HELICONIINE ATTACK

A complete scenario of heliconiine evolution would include a parallel scenario of evolution of the larval hosts and an analysis of coevolution between insect and plant. Although *Passiflora* phylogeny is not well understood, it is worth pointing to probable ways that interactions with *Heliconius* have promoted the evolution of these plants. Here I briefly review this topic as a logical connection between the historical account of *Heliconius* evolution and the final discussion of ecological interactions in diverse communities.

Although there are now many known examples in which variation among individual plants influences herbivore attack (reviewed by Karban, 1991), we have as yet only one clear demonstration of herbivorous insects currently imposing selection on their hosts (Simms and Rausher 1989). Simms and Rausher used an annual plant, in which fitness is more readily measured and hence selection more readily detected than in long-lived plants such as passion vines. In consequence, direct estimation of current selection pressures on hosts may not be feasible in the *Passiflora–Heliconius* system (Billington et al., manuscript). In any case, such estimates are not necessarily pertinent tests of the published hypotheses that *past* selection caused by *Heliconius* has been responsible for the evolution of host traits such as egg mimicry (Williams and Gilbert 1981) and diversity of both leaf shape (Gilbert 1975, 1982a) and secondary chemistry (Spencer 1988). Below, I discuss the role of *Heliconius* in evolution of each of these three traits.

1. *Egg Mimicry.* The production by *Passiflora* of structures that mimic eggs of *Heliconius* has been shown experimentally to be a functional adaptation that reduces attack by these butterflies in a greenhouse environment (Williams and Gilbert 1981). This finding, coupled with the striking accuracy of the mimicry and its multiplicity of phylogenetic and developmental origin (Gilbert 1982a) constitute strong circumstantial evidence that it has evolved in response to attack by *Heliconius*. As far as is known at present, the mimicry is variable among rather than within populations. Hence it has not been feasible to ask whether naturally-existing intrapopulation variation of mimicry influences either herbivore attack or plant fitness. However, the potential exists for artificially creating variation for the trait by hybridization. Such engineered plant populations could be exposed

to natural populations of *Heliconius* to test the strength of selection for egg mimicry.

2. *Plant Chemical Diversity.* Ehrlich and Raven (1965) suggested that selection by herbivorous insects could lead to evolutionary diversification of plant secondary chemicals. Spencer (1988) points out that *Passiflora* is by far the most diverse group of plants in terms of the variety of cyanogenic glycosides within and among species. He attributes this pattern to the selective impact of *Heliconius* which use their own cyanogenic glycoside system to counter that of the plant. He points out two ways that the β-glucosidase of the butterfly might inhibit cyanide release. First, by complexing with the plant glycoside, it may compete with the plant's own enzyme and prevent its action. Second, the butterfly enzyme may bind to and inactivate the plant's glucosidase, again preventing hydrolysis of the plant cyanogenic compounds. This is the type of system in which step–counter step coevolution is possible and further work on the system should reveal whether this process has been or is occurring. If the insects are currently applying selection for chemical diversity, we may predict that the chemical diversity of *Passiflora* within habitats should be higher than expected from random assembly of the overall diversity within the relevant subgenera (or other appropriate taxonomic level).

Dolinger et al. (1973) suggested that insect attack may entail selection for chemical diversity within attacked plant populations, and that this might render heavily-attacked populations more diverse than those with historically lower rates of attack. Since Dolinger et al. did not demonstrate this effect in the plants they studied (Stermitz et al. 1990), this remains an interesting, untested idea. Now that the relevant chemistry is beginning to be understood (Spencer 1988), the *Heliconius–Passiflora* system would seem to be appropriate for testing hypotheses about correlations among insect and plant chemical traits within interacting populations, as well as among populations and species.

3. *Diversity of Leaf Shape.* Just as a plant may escape herbivory by possessing unusual chemical traits, so it may also escape by unusual physical appearance that hinders recognition by visually searching insects. This argument was used (Gilbert 1975, 1982a) to suggest that *Heliconius* have been responsible for the extreme diversity of leaf shapes within the genus *Passiflora*, and in particular among different species in the same habitat. Like the analogous hypothesis about chemical diversity, this hypothesis could be tested by comparing local leaf-shaped diversities from those that would be produced by random association of the relevant taxonomic groups.

Proper elucidation of the evolutionary relationships of *Passiflora* and *Heliconius* awaits improved systematic treatment of both genera, and a much deeper understanding of chemical aspects of their interaction. As I will discuss in the final section, the diversity of *Passiflora* locally is a major factor in the diversity of *Heliconius* communities. It will be interesting to know the extent to which these insects have promoted the diversification of their larval resources in evolutionary time.

THE ECOLOGY OF HELICONIUS DIVERSITY

Today, diverse assemblages of heliconiines can be found in forests across the neotropics. In a given rain-forest locality as many as 20 species might be encountered (e.g., Fig. 18-3) 15–17 of which could be breeding residents, typically including about 10 species of *Heliconius*. Much of the research of our group has been focused on looking for pattern in the comparative behavior and ecology of individuals and populations of coexisting *Heliconius* in hopes of developing hypotheses on the maintenance of local diversity in this system.

Two studies have addressed community patterns in heliconiines through comparing assemblages of butterflies and host plants in different regions (Benson 1978; Gilbert and Smiley 1978). The latter authors summarized several major factors which appear to determine local heliconiine diversity.

1. *Regional Fauna.* The number of species living in a local site cannot exceed the numbers which occur in the surrounding region.

2. *Host Diversity.* Numbers of heliconiine species correlate with numbers of species of Passifloraceae.

3. *Variety and Diversity of Successional Stages.* Natural disturbance helps maintain a greater equitability and variety of larval and adult host plants and makes available combinations of microhabitat and host species upon which some heliconiine species specialize.

Figure 18-3. The heliconiine community in Corcovado Park, Costa Rica. Column 1: *Philaethria dido, Dryas julia, Dryadula phaetusa, Heliconius ismenius, H. hecalesia, H. hecale.* Column 2: *Dione juno, Agraulis vanillae, Laparus doris, Heliconius charitonia, H. melpomene, H. erato.* Column 3: *Eueides isabella, E. lineta, E. lybia, E. aliphera, Heliconius sara, H. hewitsoni, H. pachinus* (see also Tables 18-1 and 18-2).

Figure 18-4. *Passiflora* species in Corcovado Park, Costa Rica (see also Table 18-2). Top row: *Passiflora vitifolia, P. ambigua, P. pittieri,* (the long leaf overlapping into rows 2 and 3) *P. costaricensis, P. menispermifolia.* Second row: *Passiflora lobata,* (*P. pittieri*), *P. pseudo-oerstedi.* Third row: *Passiflora auriculata, P. talamancensis,* (*P. pittieri*), *P. lancearia, P. quadrangularis.* Bottom row: *Passiflora coriacea.*

TABLE 18-2. Relationship Between Adult Abundance During 1981 and Availability of Host Passiflora in a Trailside Census

Heliconius	Number caught (1981)	Abundance Rank	Host Availability Rank	Hosts[a]
hewitsoni	762	4	7	a
pachinus	873	3	3	a, c, d, e, h–k
erato	729	5	4	c–e
melpomene	249	6	8	h
hecalasia[b]	2	10	10	g
hecale	1009	1	1	k
ismenius	921	2	2	i
sara[b]	5	9	9	f
charitonia	34	8	5	b
doris	219	7	6	j

[a] *Passiflora:* (a) *pittieri,* (b) *lobata,* (c) *talamancensis,* (d) *coriacea,* (e) *costaricensis,* (f) *auriculata* (not naturally resident within 5 km of Sirena but grown there in garden), (g) *lancearia* (not resident at or near Sirena), (h) *menispermifolia,* (i) *quadrangularis,* (j) *ambigua,* (k) *vitifolia.* $N = 9$ (exclude *hecalesia*); $P < 0.05$; Kendall Tau $= 0.667$.

[b] Not natural breeding resident during the study.

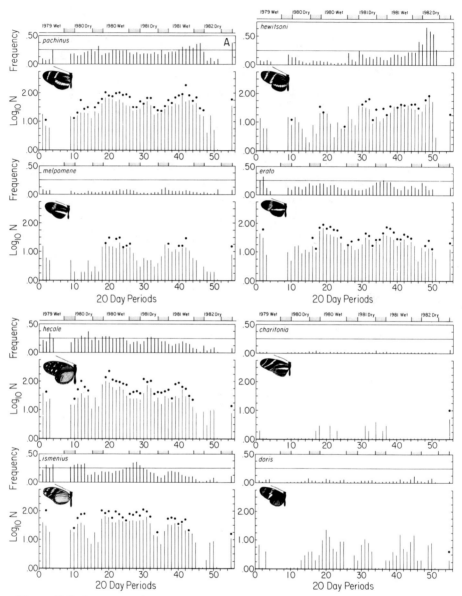

Figure 18-5. Summary of population dynamics and relative abundance data for *Heliconius* populations at Sirena, Corcovado Park, Costa Rica. The *X* axis represents the duration of the study, 1100 days between July 1979 and June 1982. Estimates were made at 20-day intervals. Panels summarize the relative abundance and population estimates of the eight resident species.

The lower portion of each panel shows first the Log$_{10}$ of the total of individuals seen or known alive during a period (vertical line) and the Manly–Parr estimate of total numbers

4. *Climatic predictability*. With similar numbers of host species, less predictable sites had fewer heliconiines and these had broader host range than in more predictable sites (Benson 1978).

Our understanding of the maintenance of local diversity has been assisted by our long-term studies of heliconiine and *Passiflora* communities at Sirena in Corcovado National Park, Costa Rica (Figs. 18-3 and 18-4). Relative abundances of the *Heliconius* and their larval resources have been estimated from censuses of *Passiflora* and associated *Heliconius* larvae (Mallet 1984) and of *Heliconius* adults (Table 18-2 and Fig. 18-5). In the data presented here (Table 18-2), the rank abundance of each *Heliconius* was determined from the total numbers of different individuals of each species captured over one year in the mark–release–recapture program. The host availability rank for a *Heliconius* was based on an index which combined counts of leaves of all hosts used by a *Heliconius* within the microhabitats where that species searches for host plant and larvae are found. Thus, although *H. pachinus* uses eight of the local *Passiflora* (Table 18-2), its host availability rank is only three since it restricts its oviposition search to forest habitats.

Since the species diversity of *Heliconius* across sites correlates with numbers of host species (Gilbert and Smiley 1978), it is important to ask how such a relationship arises in terms of local community interactions. On the one hand, there could be a tight chemical restriction to host plant, with each *Heliconius* species occurring in proportion to the abundance of its host. On the other hand, *Heliconius* could be generalists on *Passiflora* chemically, but specialists on microhabitats, the diversity of which somehow might govern *Passiflora* diversity. Examination of Table 18-2 shows that both of these hypotheses are partly true. Most species are host specialists (e.g., *H. melpomene*, *H. hecale*, *H. hewitsoni*, *H. ismenius*), but at least one (*H. pachinus*) is a habitat specialist and host generalist. However, of the Sirena species which appear as specialized on a single host based on the brief census (Table 18-2), only two, *H. melpomene* and *H. hewitsoni*, restrict oviposition to a single species of the available *Passiflora* both in the field and in captivity (see below).

The Persistence of Rare Species

Relative to temperate zone insect populations (Varley et al. 1974), all *Heliconius* species exist at extremely low densities. Rarely do the most common species reach

---

(solid circle). Since males and females differ in recapture rates, separate estimates were carried out for the sexes, then summed.

The upper part of each of these same panels shows the relative abundance of species, obtained by dividing the number of different individuals actually caught in a period by the total of all *Heliconius* caught in the period.

densities greater than five individuals per hectare and some species, such as *H. melpomene*, persist at densities of one individual per hectare and below. These are densities comparable to those of neotropical birds and mammals (e.g., several studies cited in Leigh et al. 1982).

Mark–release–recapture data for the component populations of this community at Sirena are shown in Figure 18-5. The upper portion of each panel shows the frequency of a given species in the *Heliconius* community while the lower shows the estimated numbers (dot) or minimum number known to be alive (top of bar) in a 30-ha area at 20-day intervals for approximately 2 yr.

During our study, species that were initially rare in the Sirena *Heliconius* community tended to remain rare, while abundant species also preserved their status (Fig. 18-5). For example, *H. pachinus* maintained its position at 20–25% of the total *Heliconius* caught at Sirena over most of this 2-yr period, while *H. melpomene* was consistently at a level of 2–6% of the total. Likewise *H. hecale*, whose host is the very common vine, *P. vitifolia*, was typically abundant, whereas *H. charitonia* was always rare in spite of a relatively abundant host, *P. lobata* (see Table 18-2). These basic relationships held over much of the 1980s at Sirena (L. E. Gilbert, personal observation and unpublished data).

Because much of overall species diversity in the tropics and elsewhere consists of rare species, it is important to ask first, why certain species tend to be rare and second, by what means such low-density species persist in local communities. Below, I discuss the biology of *H. melpomene* and *H. charitonia* in the context of their persistent rarity at Sirena.

H. charitonia can be a common species in midelevation montane forest having high rates of disturbance. It is common in lowland seasonal tropical forest and is the only *Heliconius* occurring outside the northern fringes of neotropical habitats. However, at lowland wet forest sites such as Sirena, where the *Heliconius* community is highly diverse, *H. charitonia* is always rare, sporadic, and confined to early successional areas, even though its host(s) may be abundant. Since no other heliconiine uses *P. lobata* at Sirena, larval competition is probably not a factor that causes its rarity. Moreover, first instar larvae placed on *P. lobata* at Sirena survived to pupation at rates typical of other species on their hosts (Gilbert, unpublished data). *H. charitonia* is normally pollen feeding and will use *Psiguria* in greenhouses. However, at Sirena it fails to effectively exploit high-quality pollen sources such as *Psiguria* as evidenced by a long-term study of pollination biology of that plant (Murawski 1986). This suggests that *H. charitonia* is excluded from this resource by competition from other *Heliconius* species in the *Heliconius*-rich lowland forests.

Further evidence that the principal factors accounting for the rareness of *H. charitonia* at Sirena involve adult rather than larval biology comes from experimental pulses of the adult populations of *H. charitonia* and *H. melpomene*. In these experiments (Gilbert, unpublished data), I released laboratory-reared adults into the study site at Sirena during on-going mark–release–recapture studies. Recapture and/or resightings of individually marked insects provided information on survival and/or residency. The relatively rapid decline of *H.*

charitonia (Fig. 18-6) is evidence that adult dispersal or mortality is higher for it than for *H. melpomene*, a species apparently limited by larval resources.

It appears that factors limiting adult residency reduce the opportunity of *H. charitonia* to locate larval hosts and successfully breed at Sirena. Natural source areas for *H. charitonia* near Sirena are large zones of secondary succession where

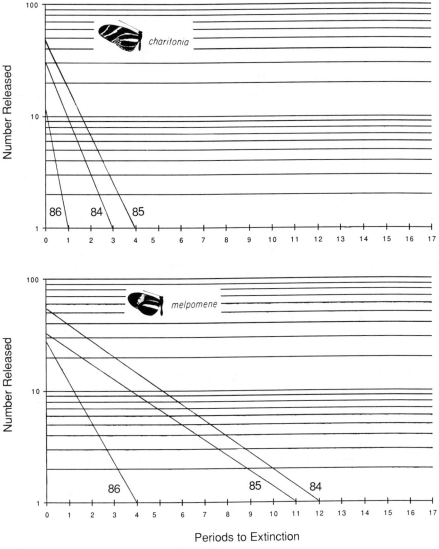

Figure 18-6. Trajectories of residency of pulse-released *H. charitonia* and *H. melpomene* at Sirena in December 1984–1986. Periods are 7 days. *Y* axis indicates numbers released; *X* axis indicates time until released cohort is no longer detected by ongoing mark–release–recapture program.

landslides in steep terrain continually provide disturbance. Without adult dispersal from these sources, *H. charitonia* would probably disappear at Sirena.

While Sirena was a sink, not a source for *H. charitonia* during the 1980s, the reverse held true for *H. melpomene*. Although in relatively low density at Sirena, *H. melpomene* is even rarer nearby with the nearest breeding population occurring some 3–4 km away. In contrast to *H. charitonia*, *H. melpomene* appears to be a permanent resident whose abundance is limited by larval rather than adult resources. Shoot occupation by *H. melpomene* eggs or larvae on its host, *P. menispermifolia*, was > 50% in Mallet's (1984) survey and has been typically in that range. Since not all shoots are likely to be suitable due to such factors as predaceous ants (Smiley 1985b), the level of shoot occupation observed must represent near saturation of suitable host material. At times, virtually every known *P. menispermifolia* will possess at least one shoot occupied by *H. melpomene* eggs or larvae. With respect to adult feeding, *H. melpomene* is able to exploit *Psiguria* and occurs on that resource approximately in proportion to its frequency in the source community.

In contrast to *H. charitonia*, the high level of shoot occupancy shown by *H. melpomene* at its current rarity shows that this butterfly is prevented from ever becoming one of the most abundant species at Sirena by scarcity of its host. Indeed, according to Table 18-2, the rank of abundance of *H. melpomene* (6) at Sirena is higher than expected from its host availability rank (8) given a perfect rank correlation. In contrast, the rank of abundance of *H. charitonia* (8) is below expectation, given its host availability rank (5). These two departures from perfect rank correlation are not sufficient to destroy the overall significance of *Heliconius* abundance versus host availability at Sirena ($p < 0.05$ by Kendall rank test). If restriction to a rare larval host limits *H. melpomene*, what prevents the evolution of broader diet?

The Causes of Host Specialization in Heliconius: The Ecological Monophagy Hypothesis

The hypothesis of "ecological monophagy" describes the restriction of an insect to a single host species for reasons other than physiological adaptation to plant chemical defenses. This was first proposed by Gilbert (1979) based on Smiley's (1978) study of *H. melpomene* at La Selva, Costa Rica. Smiley found that larvae of this monophagous species performed just as well on several other local *Passiflora* as on their own host species, and suggested that natural selection associated with patterns of ant predation on the different passion vines was responsible for the restriction of *H. melpomene* to a single host species.

Further evidence to support the hypothesis of ecological monophagy stems from studies of *H. melpomene* at Sirena, where it is also monophagous, though not on the same host species it uses at La Selva. In field experiments larval instars 1–3 were placed on different species of *Passiflora*. They grew well on four *Passiflora* species, including both their normal host, *P. menispermifolia* (Smiley and Wisdom

1985) and *P. pittieri*, among others. I have already described experiments in which adults emerging from *H. melpomene* pupae that had been placed on *P. pittieri* were killed by pupal-mating male *H. hewitsoni*. This interference competition constitutes an ecological source of selection against use of *P. pittieri* by *H. melpomene*. Since *P. pittieri* is much more abundant than the sole host of *H. melpomene* at Sirena (*P. menispermifolia*) and since *H. melpomene* currently exists at the upper limit of abundance set by host availability, the ecological monophagy hypothesis once again seems to be a plausible explanation of the restricted diet of this butterfly (cf. Bernays and Graham 1988). An explanation based on natural selection rather than phylogenetic constraint is required in this case (e.g., Gilbert, 1984), because *H. melpomene* is known to occasionally exchange genes with the closely related species *H. cydno* and *H. pachinus* (L. E. Gilbert, personal observation), both of which are diet generalists which coexist sympatrically with *H. melpomene*. Therefore, a source of genetic variation in host range is not a constraint in evolutionary time.

H. melpomene is not the only *Heliconius* that does not exploit all potential host species on which larvae are able to feed. Indeed, the phenomenon of ecological monophagy may be quite general. Below, I list evidence from two other species at Sirena.

1. Apart from *H. melpomene*, *H. hewitsoni* is the only species at Sirena that refuses to oviposit on more than one host species either in the field or in captivity. It is a strict specialist on young shoots of *P. pittieri*. Mated females may be enclosed for many months in the presence of other *Passiflora* species, even old growth of *P. pittieri*, without depositing eggs. Only *P. pittieri* new shoots elicit oviposition. Yet I routinely rear larvae on *Plectostemma* group *Passiflora* such as *P. biflora*. The extreme specialization of *H. hewitsoni* to *P. pittieri* probably reflects adaptation to the phenology of shoot production of that plant, in that egg clusters must be laid during a narrow phase of shoot development (Gilbert 1982a; Longino 1984).

2. *H. sara* is restricted to the rare *P. auriculata* and therefore is rare at Sirena. However, in South America this species uses the *Astrophea* subgenus, represented at Corcovado by *P. pittieri* (Benson et al. 1976). In greenhouse culture, I have observed *H. sara* ovipositing on *P. pittieri* in the presence of abundant and suitable *P. auriculata*. Once again, the defence of this plant by pupal-mating male *H. hewitsoni* may prevent *H. sara* from evolutionary incorporation of *P. pittieri* into its diet at Sirena. I have also observed a natural transfer of *H. sara* larvae from *P. auriculata* to an adjacent *P. coriacea* in an outdoor garden at Sirena. *P. coriacea* turns out to be a suitable host for *H. sara*, which apparently never receives eggs in the field or in the greenhouse. It is however, a minor host for *H. erato* which is both cannibalistic as larvae and possesses pupal-mating adults.

We need to know much more about the ecology and evolution of local host use patterns in *Heliconius* since these largely define how many ways there are to be a *Heliconius*, yet coexist within a stable community. In particular, we need more knowledge concerning the chemical aspects of host specialization by *Heliconius* as well as detailed studies on the nature of ecological differences between coexisting *Passiflora* species.

CONCLUSION

The example of *Heliconius* amply demonstrates how easily the study of tropical herbivorous insects can potentially divert us from the elegance of simple generalities to the joy of natural history for its own sake. Yet a closer look provides evidence that within large assemblages of congeneric herbivores, most readily found in rain-forests, one finds the comparative tools for addressing many fundamental questions on the ecology and evolution of biological diversity. Furthermore, some fraction of the diversity in such systems may be understood only in light of certain details of natural history unique to particular taxa and therefore resistant to explanations at high levels of generality.

ACKNOWLEDGMENTS

I am grateful to the Costa Rican Park Service and to a succession of directors at Sirena Station for allowing long-term research in Corcovado Park. Many past students and associates have helped gather data, have provided unpublished manuscripts, reprints, and discussions, or have helped in other important ways. Thanks to W. Benson, H. Billington, C. Boggs, S. Boinski, S. Bramblett, K. Brown, R. Canet, P. Chai, M. Condon, P. DeVries, C. Duckett, N. Greig, K. Kernan, M. Linares, J. Longino, J. Mallet, R. Moore, D. Murawski, P. Phillips, C. Reis, A. Simpson de Gamboa, J. Smiley, K. Spencer, R. Srygley, and C. Thomas. Support for *Heliconius* research at Sirena has been supported by NSF BSR-8315399, NSF DEB 790633, and World Wildlife Fund grants to L.E.G. as well as NSF, National Geographic, and WWF grants to associated students. My sincere thanks to M. Singer and J. Lanza for critical reviews of the early drafts of this chapter. Special thanks to M. Singer and S. Bramblett for major efforts in editing and typing, respectively.

REFERENCES

Benson, W. W. 1978. Resource partitioning in passion vine butterflies. *Evolution* **32**: 493–518.

Benson, W. W., K. S. Brown, Jr., and L. E. Gilbert. 1976. Coevolution of plants and herbivores: Passion flower butterflies. *Evolution* **29**: 659–680.

Bernays, E. and M. Graham. 1988. On the evolution of host specificity in phytophagous arthropods. *Ecology* **69**: 886–892.

Boggs, C. L. 1981. Nutritional and life-history determinants of resource allocation in holometabolous insects. *American Naturalist* **117**: 692–709.

Boggs, C. L. and L. E. Gilbert. 1979. Male contribution to egg production in butterflies: Evidence for transfer of nutrients at mating. *Science* **206**: 83–84.

Boggs, C. L., J. T. Smiley, and L. E. Gilbert. 1981. Patterns of pollen exploitation by *Heliconius* butterflies. *Oecologia* **48**: 284–289.

Brower, L. P. and J. V. Z. Brower. 1964. Birds, butterflies and plant poisons: A study in ecological chemistry. *Zoologica* **49**: 137–159.

Brown, K. S. Jr. 1981. The biology of *Heliconius* and related genera. *Annual Review of Entomology* **26**: 427–56.

Brown, K. S. Jr. 1987. Chemistry at the Solanaceae/Ithomiinae interface. *Annals of Missouri Botanical Gardens* **74**: 359–397.

Chai, P. 1986. Field observations and feeding experiments on the responses of rufous-tailed jacamars (*Galbula ruficauda*) to freeflying butterflies in a tropical rainforest. *Biological Journal of Linnean Society* **29**: 161–189.

Chai, P. 1990. Relationships between visual characteristics of rainforest butterflies and responses of a specialized insectivorous bird. Pages 31–60 in M. Wicksten (compiler). Adaptive Coloration in Invertebrates. Texas A & M University Sea Grant College Program, Galveston. Symposium Sponsored by the American Society of Zoologists.

Condon, M. A. and Gilbert, L. E. 1990. Reproductive biology and natural history of neotropical vines *Gurania* and *Psiguria*. Pages 150–166 in D. Bates, R. W. Robinson, and C. Jeffrey (Eds.). *Biology and Utilization of the Cucurbitaceae*. Cornell University Press, Ithaca, NY.

Dolinger, P. M., P. R. Ehrlich, W. L. Fitch, and D. E. Breedlove. 1973. Alkaloid and predation patterns in Colorado lupine populations. *Oecologia* **13**: 191–204.

Dunlap-Pianka, H. L., C. L. Boggs, and L. E. Gilbert. 1977. Ovarian dynamics in heliconiine butterflies: Programmed senescence versus eternal youth. *Science* **197**: 487–490.

Ehrlich, P. R. and L. E. Gilbert. 1973. Population structure and dynamics of the tropical butterfly *Heliconius ethilla*. *Biotropica* **5**: 69–82.

Ehrlich, P. R. and P. H. Raven. 1965. Butterflies and plants: A study in coevolution. *Evolution* **18**: 586–608.

Gilbert, L. E. 1971. Butterfly–plant coevolution: Has *Passiflora adenopoda* won the selectional race? *Science* **172**: 585–586.

Gilbert, L. E. 1972. Pollen feeding and the reproductive biology of *Heliconius* butterflies. *Proc. Nat. Acad. Sci. USA* **69**: 1403–1407.

Gilbert, L. E. 1975. Ecological consequences of a coevolved mutualism between butterflies and plants. Pages 210–240 in L. E. Gilbert and P. R. Raven (Eds.). *Coevolution of Animals and Plants*. International Congress of Systematics and Evolutionary Biology, Boulder, CO. University of Texas Press, Austin, TX.

Gilbert, L. E. 1979. Development of theory in the analysis of insect–plant interactions. Pages 117–154 in D. Horn, R. Mitchell, and G. Stairs (Eds.) *Analysis of Ecological Systems*. Ohio State Press, Columbus, OH.

Gilbert, L. E. 1982a. The coevolution of a butterfly and a vine. *Scientific American* **247**: 110–121.

Gilbert, L. E. 1982b. Oviposition by two *Heliconius* species: Comments on a paper by Dr. A. Young. *New York Entomological Society* **90**: 115–116.

Gilbert, L. E. 1983. Coevolution and mimicry. Pages 263–281 in D. Futuyma and M. Slatkin (Eds). *Coevolution*. Sinauer, Sunderland, MA.

Gilbert, L. E. 1984. The biology of butterfly communities. Pages 41–54 in R. Vane-Wright and P. Ackery (Eds.). *The Biology of Butterflies*, XI Symposium of the Royal Entomological Society of London. Academic Press, New York.

Gilbert, L. E. and J. T. Smiley. 1978. Determinants of local diversity in phytophagous insects. Pages 89–104 in L. A. Mound and N. Waloff (Eds.). *Diversity of Insect Faunas*. Blackwell Scientific, London, England. Symposium of the Royal Entomological Society of London.

Karban, R. 1991. Plant variation: Its effects on populations of herbivorous insects. In R. S. Fritz and E. L. Simms (Eds.). *Ecology and Evolution of Plant Resistance*. John Wiley, New York, in press.

Lee, C. S., B. A. McCool, J. L. Moore, D. M. Hills, and L. E. Gilbert. 1991. *Phylogenetic study of heliconiine butterflies using restriction analysis of their ribosomal RNA genes*. Submitted to Zoological Journal of the Linnean Society.

Leigh, E. G. Jr., A. S. Rand, and D. M. Windsor. 1982. *The Ecology of a Tropical Forest: Seasonal rhythms and long-term changes*. Smithsonian Institution Press, Washington, 468 pp.

Longino, J. T. 1984. *Shoots, parasitoids, and ants as forces in the population dynamics of Heliconius hewitsoni in Costa Rica*. Ph.D. Dissertation. University of Texas, Austin, TX.

Longino, J. T. 1986. A negative correlation between growth and rainfall in a tropical liana. *Biotropica* **18**: 195–200.

Mallet, J. 1984. *Population structure and evolution of Heliconius butterflies*. Ph.D. Dissertation. University of Texas, Austin, TX.

Mallet, J. 1986a. Gregarious roosting and home range in *Heliconius* butterflies. *National Geographic Research* **2**: 198–215.

Mallet, J. 1986b. Dispersal and gene flow in a butterfly with home range behavior: *Heliconius erato* (Lepidoptera: Nymphalidae). *Oecologia* **68**: 210–217.

Mallet, J. L. B. and D. A. Jackson. 1980. The ecology and social behaviour of the Neotropical butterfly *Heliconius xanthocles* Bates in Colombia. *Zoological Journal of the Linnean Society* **70**: 1–13.

Mallet, J., J. T. Longino, D. Murawski, A. Murawski, and A. Simpson de Gamboa. 1987. Handling effects in *Heliconius*: Where do all the butterflies go? *Journal of Animal Ecology* **56**: 377–386.

Murawski, D. A. 1986. *Pollination ecology of a Costa Rican population of Psiguria warscewiczii in relation to foraging behavior of Heliconius butterflies*. Ph.D. Dissertation. University of Texas, Austin, TX.

Murawski, D. A. and L. E. Gilbert. 1986. Pollen flow in *Psiguria warscewiczii*: A comparison of *Heliconius* butterflies and hummingbirds. *Oecologia* **68**: 161–167.

Nahrstedt, A. and R. H. Davis. 1983. Occurrence, variation and biosynthesis of the cyanogenic glucosides linamarin and lotaustralin in the species of the Heliconiini (Insecta: Lepidoptera). *Comparative Biochemistry and Physiology* **75**: 65–73.

Nijhout, H. F. 1986. Pattern and pattern diversity of lepidopteran wings. *Bioscience*. **36**: 527–533.

Simms, E. L. and M. D. Rausher. 1989. The evolution of resistance to herbivory in *Ipomoea purpurea*. II. Natural selection by insects and costs of resistance. *Evolution* **43**: 573–787.

Singer, M. C. 1981. Sexual selection for small size in male butterflies. *American Naturalist* **119**: 440–443.

Sivinski, J. 1989. Mushroom body development in nymphalid butterflies: A correlate of learning. *Journal of Insect Behavior* **2**: 277.

Smiley, J. 1978. Plant chemistry and the evolution of host specificity: New evidence for *Heliconius* and *Passiflora*. *Science* **201**: 745–747.

Smiley, J. 1985a. Are chemical barriers necessary for evolution of butterfly–plant associations? *Oecologia* **65**: 580–583.

Smiley, J. T. 1985b. *Heliconius* caterpillar mortality during establishment on plants with and without attending ants. *Ecology* **66**: 845–849.

Smiley, J. T. and C. S. Wisdom. 1985. Determinants of growth rate on chemically heterogeneous host plants by specialist insects. *Biochemical Systematics and Ecology* **13**: 305–312.

Spencer, K. C. 1988. Chemical mediation of coevolution in the *Passiflora–Heliconius* interaction. Pages 167–239 in K. C. Spencer (Ed.). *Chemical Mediation of Coevolution*. Academic Press, New York.

Srygley, R. B. and P. Chai. 1990. Predation and the elevation of thoracic temperature in brightly-colored, neotropical butterflies. *American Naturalist* **135**: 766–787.

Stermitz, F. R., G. N. Belofsky, D. Ng, and M. C. Singer. 1990. Quinolizidine alkaloids obtained by *Pedicularis semibarbata* (Scrophulariaceae) from *Lupinus fulcratus* (Leguminosae) fail to influence the specialist herbivore *Euphydryas editha* (Lepidoptera). *Journal of Chemical Ecology* **15**: 2521–2530.

Templeton, A. R. and L. E. Gilbert. 1985. Population genetics and the coevolution of mutualism. Pages 128–144 in D. H. Boucher (Ed.). *The Biology of Mutualism*. Croom Helm, London, England.

Turner, J. R. G. 1971. Experiments on the demography of tropical butterflies. II. Longevity and home-range behavior in *Heliconius erato*. *Biotropica* **3**: 21–31.

Turner, J. R. G. 1981. Adaptation and evolution in *Heliconius*: A defense of neoDarwinism. *Annual Review of Ecology and Systematics* **12**: 99–121.

Varley, G. C., G. R. Gradwell, and M. P. Hassell. 1974. *Insect Population Ecology: An analytical approach*. University of California Press, Berkeley, CA 212 pp.

Waller, D. and L. E. Gilbert. 1982. Roost recruitment and resource utilization: Observations on a *Heliconius charitonia* L. Roost in Mexico (Nymphalidae). *Journal of the Lepidopterists' Society* **36**: 178–184.

Williams, K. S. and L. E. Gilbert. 1981. Insects as selective agents on plant vegetative morphology: Egg mimicry reduces egg laying by butterflies. *Science* **212**: 467–469.

SECTION 5
Specificity in Plant Utilization

"What a noble subject would be that of a monograph of a group of beings peculiar to one region but offering different species in each province of it—tracing the laws which connect together the modifications of forms and colour with the *local* circumstances of a province or station—tracing as far as possible the actual *affiliation* of the species." Thus wrote Bates (1856) in the upper Amazon to Wallace. Of course, Wallace was heavily engaged at this time in a biogeographic view of nature, and major works appeared soon after (Wallace 1869, 1876). But concentrating on the large scale precluded the study of the detailed associational patterns between plants and their herbivores, closer to what Bates may have had in mind. In fact, Wallace (1876) did not discuss the plants on which the insects fed in his pioneering study. However, the subject of insect herbivore specificity was already advanced beyond the general level of the field by Walsh's (1864) studies on biotypes.

The subject of specificity of herbivores is still a central theme in plant–herbivore interactions, as evidenced by a special feature in the journal *Ecology* (e.g., Bernays and Graham, 1988). The contributions in this section extend our knowledge on the evolution of specificity in insect herbivores (Chapter 19) and flower mites (Chapter 20), and the development of geographic and host races (Chapter 21). The large ungulates relate to the green world in a very different way from the majority of insects on plants, but even here there is heretofore unsuspected distributional and nutritional fine-tuning (Chapter 22). This section dovetails with several others in the book, for the subject of specificity is relevant to many other studies. It has served as a focus in plant–animal relationships now for over 120 years, and the fascinating problems posed by specificity will fuel the field for decades to come.

REFERENCES

Bates, H. W. 1856. Letter to A. R. Wallace. In J. Marchant. 1916. *Alfred Russel Wallace: Letters and reminiscences.* Harper, New York, pp. 52–53.

Bernays, E. A. and M. Graham. 1988. On the evolution of host specificity in phytophagous arthropods. *Ecology* **69**: 886–892.

Wallace, A. R. 1869. *The Malay archipelago.* Macmillan, London.

Wallace, A. R. 1876. *The geographical distribution of animals.* Harper, New York.

Walsh, B. 1864. On phytophagic varieties and phytophagic species. *Proc. Entomol. Soc. Philadelphia* **3**: 403–430.

19 Evolution of Host Specificity in Herbivorous Insects: Genetic, Ecological, and Phylogenetic Aspects

DOUGLAS J. FUTUYMA

Species of herbivorous insects vary greatly in their breadth of diet. Some, such as the fall webworm *Hyphantria cunea* (Arctiidae), have been recorded from hundreds of species of plants in numerous families; others, such as the bruchid beetle *Caryedes brasiliensis*, are believed to feed on only a single plant species (Janzen 1978). The majority of phytophagous species lie toward the specialized end of the spectrum, typically feeding on only some of the species in a single genus, or a group of closely related genera, of plants. This observation poses two large, challenging questions: What is the selective advantage of host specificity? What has determined which plants have been adopted during the evolution of an insect lineage?

Accounting for the origin of host specificity in insects is a particularly dramatic instance of the more general problem of the evolution of ecological specialization (Futuyma and Moreno 1988). It is specialization rather than generalization that presents a puzzle. In every generation of a host-specialized species, some individuals must fail to find suitable hosts (see, e.g., Dethier 1959), so there must exist at least some selection for a broader diet. What countervailing forces of selection, then, favor specialization?

It is important to recognize that the selection that maintains host specificity may be different from the selection that operated in the origin of a specialized from a more generalized diet. By its behavior, specifically its choice of host plant, an insect species may largely determine the selection pressures to which it is subject. Thus the host-selection behavior of an insect species affects the evolution of features such as its capacities for digestion and detoxification, its phenology, its defenses against predators and parasites, and in some cases its likelihood of

Plant-Animal Interactions: Evolutionary Ecology in Tropical and Temperate Regions, Edited by Peter W. Price, Thomas M. Lewinsohn, G. Wilson Fernandes, and Woodruff W. Benson. ISBN 0-471-50937-X © 1991 John Wiley & Sons, Inc.

finding mates (see Colwell, Chapter 20). A long history of association with a particular host may therefore bring about the successive evolution of numerous features that in aggregate confine it to that host, but these features may differ greatly from those responsible for the original evolution of its specialized habit. To contrast different insect species, then, and to demonstrate that each survives or grows better on its own host than on the others', may shed little light on the selective advantages responsible for the origin of specialization. Finally, the great number of specialized species tells us little about the advantages of the specialized compared to the generalized habit, because numerous species share a specialized host association simply by virtue of common ancestry. The proportion of specialized species in a clade will be affected by the relative rates of speciation and extinction of specialized versus generalized species (Vrba 1980), and asymmetry in the rate of transition between specialized and generalized habits will likewise affect the proportion of specialists. But these macroevolutionary phenomena do not bear on the origin of specialization, for which we must look to genetic changes within populations.

FACTORS INFLUENCING THE ORIGIN OF SPECIALIZATION

When a novel plant is sufficiently similar to an insect's normal hosts, it may be added to the diet without evolutionary change, as we know from experience with crop pests and other cases in which an insect immediately adopts a newly introduced plant (Strong et al. 1984; Thomas et al. 1987). In many instances, however, a switch to a new host, or addition of a new host to the diet, requires genetic variation in behavioral traits and often in morphological and physiological traits that affect "performance" (growth, survival, fecundity). At a behavioral level, at least some insect genotypes must feed, and often must oviposit as well. Because the origin of any change in host affiliation occurs by selection among genotypes within populations, it is in population and ecological genetic studies of genetic variation in and selection on these characters that the advantages of host specificity may best be sought. By identifying factors that affect polymorphism at loci that affect host utilization, it is possible to shed some light on factors that affect the evolution of diet breadth.

To illustrate this point, I shall summarize our observations on the fall cankerworm *Alsophila pometaria* (Lepidoptera: Geometridae) in New York. In this univoltine species, winged males and wingless females emerge from underground pupation sites in late autumn and early winter (November and December), when the females, which do not feed, lay eggs on the bark of leafless trees. Larvae emerge at the time of budbreak (late April–early May), may disperse by ballooning, and feed on the foliage of numerous deciduous trees (especially *Quercus* spp., *Carya* spp., *Acer rubrum*, and *Prunus serotina* on Long Island, New York). After several weeks they drop from the canopy and remain under ground until autumn. Because larval survival and growth and pupal weight (which is highly correlated with subsequent fecundity) depend on completing development before foliage is fully mature (Mitter et al. 1979; Futuyma and Saks 1981), synchrony of egg hatch with budbreak is an important correlate of fitness.

Although about 10% of the *Alsophila* population on Long Island consist of normally sexual females and males (Harshman and Futuyma 1985a), electrophoretic studies show that the majority of the population on Long Island and elsewhere consists of parthenogenetically reproducing females that transmit the maternal genotype (even if heterozygous) intact to most or all of their all-female offspring (Mitter and Futuyma 1977). Parthenogenetic females do not lay viable eggs unless they mate, however, and a small fraction of the brood of such females show evidence of recombination and inclusion of paternal alleles (Harshman and Futuyma 1985b). Moreover, even in the relatively few progeny tests we have performed, we have found some instances of asexual reproduction in daughters of sexual females (Harshman and Futuyma 1985b). We believe, therefore, that new asexual genotypes arise at a reasonably high frequency from the sexual population and perhaps from other asexual genotypes, and that this may account for the high frequency of individually rare asexual genotypes in Long Island populations. We have no indication that the mode of reproduction is a facultative response to host plant, density, or other environmental factors.

Most of our studies have focused on two common genotypes, identified by their profile at four polymorphic enzyme loci. Genotype *A* has a high frequency in stands dominated by red maple (*Acer rubrum*), and genotype *B* is the predominant genotype in the oak-dominated stands (especially *Quercus coccinea, Q. alba*, and *Q. ilicifolia*) that cover much of Long Island. Abrupt changes in the frequencies of these genotypes occur at narrow ecotones between stands of maple and oak, both on Long Island and at Princeton, New Jersey (Mitter et al. 1979), although we have failed to document genotype frequency differences between samples taken from neighboring oak and maple trees within a mixed stand. The association between genotype and vegetation type led us to suspect a role for natural selection, and we have investigated several factors. (Other possible factors, such as differential susceptibility to predators or parasites associated with different vegetation types, have not been examined.)

First, the genotypes differ in phenology. Adults of genotype *A* emerge earlier in the winter than those of *B*; the significance of this pattern is obscure, except that eggs laid earlier in the winter, or exposed experimentally to greater insolation during the winter, hatch earlier in the spring (Schneider 1980). Moreover, eggs of *A* hatch earlier even if the eggs of the two genotypes are refrigerated, and so prevented from developing, until both are placed outdoors at the same time. We postulated (Mitter et al. 1979) that this phenological difference may contribute to the genotype–host association, because *Acer rubrum* (the host of *A*) foliage develops earlier than *Quercus* (the host of *B*) foliage. Although the correlation consists of only two points, it is in the direction expected if selection favors genotypes that hatch as synchronously with budbreak as possible because of the advantage of feeding on immature foliage.

Second, we found that although newly hatched, "naive" larvae of the two genotypes do not differ in behavioral responses to foliage of *Quercus*, their response to *Acer* differs: a much greater proportion of *B* than of *A* larvae ballooned off *Acer* foliage almost immediately after they were placed on it (Futuyma et al. 1984). Acceptance of *Acer* would be advantageous in an *Acer*-dominated canopy. The genotypes therefore differ in their threshold of acceptance of *Acer*, but a difference in specificity is not the same as a difference in preference (Singer 1986). When third-instar larvae that had been reared on *Prunus serotina* (Rosaceae) were tested for 28 hr in dishes with three leaf discs of *Acer rubrum* and three of *Quercus coccinea*, more larvae of genotype *B* (17 of 21) consumed more *Quercus* than *Acer* ($p < 0.05$ by sign test), whereas larvae of genotype *A* showed no overall preference (13 of 24 consumed more *Quercus*).

Third, because genotypes might suffer differential mortality during the summer in the relatively moist soil of typical *Acer*-dominated sites compared to the drier, well-drained soil of typical *Quercus*-dominated sites, we attempted to test for differential tolerance to dessication. Prepupae were placed individually in 20-ml glass vials, 3/4 filled with either a 2:1 (by volume) mixture of vermiculite and sand (moist treatment) or a 1:2 mixture (dry treatment), into which they burrowed and pupated. The loosely capped vials were placed in an incubator, the photoperiod (16–8 hr light) and temperature (25–8°C) of which were changed weekly to approximate current weekly ambient conditions. The medium was moistened with 2 ml of water once a week for the moist treatment and once every 2 weeks for the dry treatment. Emergence began on a natural schedule in early November and continued until the end of December. Vials in which emergence did not occur were checked for unemerged survivors (there were none). The survival of both genotypes in both treatments was high (Table19-1), so the dessication stress may have been insufficient to test for an effect. Nevertheless, survival of genotype *A* was higher under moist than dry conditions, whereas the treatments did not significantly affect survival of genotype *B* (although the direction of the difference is opposite from that in genotype *A*). A *G* test for interaction indicates that the effect of treatment on survival depended on genotype ($G = 6.833$, $df = p < 0.01$). This experiment does not test for differential susceptibility to other factors that might affect survival in the soil, such as pathogens or predators.

Fourth, we have measured conversion efficiencies of both these genotypes on both *Acer* and *Quercus* foliage (Futuyma et al. 1984). AD (approximate digestibility) of both foliage types was higher for genotype *A* than *B*; there was no

TABLE 19-1. Survival Through Pupation of *Alsophila* Genotypes Under Moist and Dry Conditions

Genotype	Treatment	Survived	Died	% Survival
A (progeny of 10 ♀ ♀)	Moist	65	3	95.6
	Dry	61	11	84.7
B (progeny of 8 ♀ ♀)	Moist	29	7	80.6
	Dry	37	3	92.5

genotype X host interaction. ECI (efficiency of conversion of ingested food) did not differ on *Acer*; on *Quercus*, genotype A had a higher efficiency than B. ECD (efficiency of conversion of digested food) showed significant genotype X host interaction, but unexpectedly, the mean value for each genotype was higher on the other genotype's usual host.

Although I am not sure what to make of the last result, it is safe to say that these genotypes provide no support (at least on the basis of conversion efficiencies) for the common supposition (e.g., Dethier 1954, Janzen 1978) that host differences in secondary chemistry favor specialization because of a trade-off in the ability to detoxify (or otherwise handle) different kinds of compounds. [*Acer* differs from *Quercus* in several chemical respects, including its possession of saponins and at least one alkaloid, and it harbors a very different insect fauna (Futuyma and Gould 1979).] Because these two asexual genotypes, inasmuch as they are common, may be unrepresentative of genetic variation in a population (especially in a sexual population, for which we would like to draw inferences from this study), we sought further for the negative genetic correlations that the trade-off hypothesis should lead us to expect (Futuyma and Philippi 1987). On the assumption that many of the rare asexual genotypes of *Alsophila* have arisen recently from the sexual population and so may represent a nearly random sample of the genotypes that may arise by recombination, we divided the progeny of rare asexual genotypes among four species of host plants (*Quercus coccinea; Q. alba; Castanea dentata*, like *Quercus* a member of the Fagaceae; and *Acer rubrum* [Aceraceae]). Nine genotypes were reared in the laboratory and 10 in the field. Using live weight late in the larval period as a criterion of performance, we found no negative correlations among genotypes for any pair of host plants; numerous positive correlations were found instead. Most of the correlations became nonsignificant when individual values were corrected for mean genotypic differences (as measured on the superior host, *Q. coccinea*); of nine correlations, one (*Castanea* vs. *Acer*) was positive and one (*Castanea* vs. *Q. alba*) was negative.

On the whole, then, our work with *Alsophila* has pointed to genotypic differences in phenology, host preference, and possibly host-associated habitat differences (dessication stress in the soil) as factors that may select for different genotypes on different host plants—a differentiation in genotype frequency that may be viewed as a model of the incipient evolution of a specialized host association. We have been unable to find in physiological responses to the foliage of different species evidence of the trade-off predicted by the hypothesis that superior ability to tolerate the secondary chemistry of one plant species is bought at the cost of the ability to tolerate other species' compounds. A similar failure to find strong trade-offs has characterized most of the other studies that sought such effects by measuring or inferring genetic correlations (Gould 1979; Hare and Kennedy 1986; James et al. 1988; Rausher 1984; Via 1984).

The hypothesis that specialized host association evolves *ab initio* because of the greater physiological efficiency of specialists requires that negative correlations exist among genotypes within populations. The greater efficiency of specialists than of generalists that is sometimes evident from interspecific comparisons (e.g., Scriber 1979) may be a secondary consequence of a history of specialized host affiliation, as discussed earlier in this paper. If negative genetic correlations in physiological responses to different plants generally prove as elusive in the future as they have so far, explanations of the origin of host specificity based on selective factors extrinsic to the physiology of the insect–plant interaction may take precedence over those based on plant chemistry. Such factors may include avoidance of predators or competitors (cf., Bernays and Graham 1988 and associated commentaries), or preference based on features of the habitat (e.g., Williams 1983). In a similar vein, Colwell (1985, 1986) has proposed that although host-specific flower mites may have evolved special adaptations to their hosts as a consequence of the association, the origin of the specificity of association may have had another cause, namely the advantage of using a specific plant as a rendezvous for mating. Simply the high local abundance of one or another plant, which in some cases seems to have influenced the evolution of the degree of host specificity (Wiklund 1981, Rowell-Rahier 1984) may in some circumstances lead to the evolution of a specialized host affiliation (Futuyma 1983; Futuyma and Moreno 1988).

This conclusion would not deny a role for plant secondary compounds in the evolution of phytophagous insects. A host-specific insect must recognize its host somehow, and many plant compounds clearly are used as stimuli; closely related insects are likely to feed on chemically similar plants because the insects have inherited similar stimulus–response patterns from their common ancestor. Moreover, many plant compounds clearly have an antibiotic, defensive effect, constitute barriers to colonization by nonadapted insects, and evoke the evolution of physiological adaptation in the insects that use them. To question the role of secondary compounds in the origin of host specificity means only that if physiological trade-offs are not general, we may doubt whether incorporation of a new host into the diet requires loss of association with a former host.

CAN HOST ASSOCIATIONS BE PREDICTED?

Even from the present taxonomy of phytophagous insects, which generally lacks explicit phylogenetic information, interesting variation in the history of host affiliation is evident. In some taxa, host association is phylogenetically conservative; for example, within the Papilionidae (swallowtail butterflies), the large pantropical and subtropical tribe Troidini is restricted to Aristolochiaceae. Ehrlich and Raven (1964) cite other such instances among butterflies. In some groups, most species have similar host associations, but a few depart radically; for example, although most satyrine butterflies feed on monocots, a few species of *Euptychia* feed on *Selaginella* (a lycopod) and mosses (Singer et al. 1971, DeVries 1987). In some taxa, each species is specialized, but closely related species are found on very different hosts: for instance, species of *Pyrrhalta*, subgenus *Tricholochmaea* (Coleoptera: Chrysomelidae) feed on such plants as *Salix* (Salicaceae), *Spiraea* (Rosaceae), and *Vaccinium* (Ericaceae) (Wilcox 1965).

Patterns such as these invite comparative investigations of factors governing host specificity. Conservative patterns, as in the Troidini, suggest that stabilizing selection favors specialization for a particular kind of plant, perhaps because of specialized physiological or morphological traits shaped by a long history of association. In other groups, such as *Tricholochmaea*, there may exist selection for "specialization per se" (Futuyma and Moreno 1988), as a consequence not of physiological factors intrinsic to the insect–plant relationship, but of ecological factors such as a competition, predation, or mate-finding.

In all these cases, we may ask why some plants are utilized rather than others: what has determined which plants have been adopted? Our answer will depend on our beliefs about genetic variation. If we believe that variation is seldom limiting, citing as evidence responses to artificial selection on almost any trait chosen (Lewontin 1974), or adaptation by hundreds of insect species to a great array of historically unprecedented toxins, that is, insecticides (Georghiou and Taylor 1977), then we will be inclined to believe that most populations at most times have harbored the genetic variation necessary for adaptation to any of a great variety of plants. The actual hosts adopted would then be a consequence of a largely unknowable history of ecological selection pressures. On the other hand, species are not indefinitely adaptable: most species, after all, are extinct. We may therefore suppose that limitations on genetic variation exist, that a species is more prone to vary in some directions than others, and that availability of genetic variation has made the adoption of those hosts actually used more likely than host shifts that from an ecological point of view seem equally plausible, but which have not been realized during evolution.

In asking whether most populations at most times have had sufficient genetic variation both for realized host transitions and those not realized, or had genetic variation predisposing them toward realized transitions only, the possibility arises that genetic characterization of species might in some degree predict phylogenetic history—a possible link between the largely disjunct halves of evolutionary biology that treat of evolutionary mechanisms and evolutionary history (Futuyma 1988). For since "most times" include the present as a random point, we may address the question by determining whether contemporary populations of a species that retains an ancestral host association have genetic variation that would enable adaptation to a "derived host" that has actually been adopted by derived species in the clade, but not to plants that have not been adopted in the history of the clade. That is, suppose insect taxa A, B, and C feed on plants a, b, and c respectively, and that phylogenetic analysis shows that A and B are sister taxa, with C a distant outgroup. By mapping host use onto the phylogeny (which is based on characters independent of the host associations), it may be possible to say that association with b is derived from association with a. The hypothesis that the history of host associations has been guided by availability of genetic variation predicts that species A is more likely to display genetic variation for traits necessary to colonize plant b than plant c. It is also interesting to know if there exists asymmetry in the amount of genetic variation in the responses of species A and B to each other's host, for this may cast light on the relative likelihood of reversal versus origination *de novo* of host affiliations. For instance, it might be expected that a derived species (*B*) will carry more "vestigial" genetic variation for responses to the ancestral host (*a*) than the "ancestral" species harbors for a derived host (*b*). In passing, I note that information of this kind has implications for agriculture, if we wish to predict which insects could rapidly evolve to become pests of a newly introduced species or resistant cultivar of a crop plant. I am aware of only six studies that have looked for genetic variation in insects' responses to plants on which the species does not ordinarily feed (Gould 1979; Guthrie et al. 1982; Pathak and Heinrichs 1982; Painter et al. 1931; James et al. 1988; Thompson 1988).

PRELIMINARY STUDIES OF *OPHRAELLA*

To illustrate the directions such a study might take, I shall describe some preliminary studies that as yet lack genetic content, but which suggest patterns that will become the object of study at the genetic level.

Ophraella Wilcox (Fig. 19-1), a monophyletic North American genus of leaf beetles (Chrysomelidae, Galerucinae), contains 13 species, according to LeSage (1986). The hosts, on which all the life stages are found, fall into four tribes of Asteraceae (Compositae) (Table 19-2). All the species have specialized host associations, typically occurring on only certain species within a single plant genus. The exceptions are *O. pilosa*, which has been found on the closely related genera *Aster* and *Solidago*, and *O. communa*, which is associated primarily with *Ambrosia* species but has also been found in various parts of its range on *Iva, Xanthium* (both, like *Ambrosia*, in subtribe Ambrosiinae, tribe Heliantheae), and *Helianthus* (subtribe Helianthinae, tribe Heliantheae).

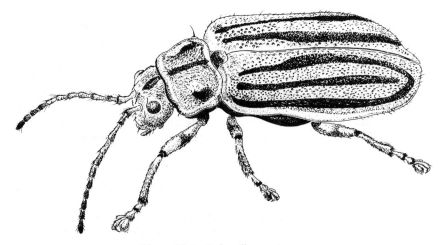

Figure 19-1. *Ophraella sexvittata.*

TABLE 19-2. Distribution and Known Hosts of *Ophraella* Species

Species	Distribution	Hosts[a]
americana	E. of Rocky Mts., s. Canada-FL	*Solidago* sp. (A)
pilosa	s. Canada and n.U.S., coast to coast; MO[c]	*Aster* (A) species, incl. *A. urophyllus*[c], *A. laevis*[c]; *Solidago bicolor*[c] (A)
notata	E.N.A., Great Lakes to Gulf Coast	*Eupatorium* species (E): *perfoliatum*[c], (*hyssopifolium*[c]), (*maculatum*[c])
conferta[b]	E.N.A., s. Canada to N.C.	*Solidago* species (A): *altissima*[c] complex, *rugosa*, *juncea*[c]
sexvittata[b]	s.e. U.S.	*Solidago* species (A): *altissima*[c], *leavenworthii*[c]
macrovittata[b]	s.e. U.S.	*Solidago altissima*[c] (A)
cribrata	s. Canada, U.S. except SW	*Solidago* species (A): *juncea*[c], (*altissima*[c]), (*pinetorum*[c]), (*nemoralis*), (*bicolor*[c])
communa	s. Canada to s. Mexico, coast to coast	*Ambrosia* species (AM): *artemisiifolia*[c], *psilostachya* (in west); *Iva axillaris* (AM) (in Calif.); *Xanthium* (AM) (in Calif.); (*Helianthus* (H)).
notulata	Atlantic and Gulf coastal regions	*Iva frutescens*[c] (AM)
nuda	s.e. Alberta	*Iva axillaris*[c] (AM)
bilineata	Great Plains of S. Canada, n.U.S.	*Chrysopsis villosa*[c] (A)
arctica	n. Canada (n. N.W.T., n. Man.)	*Solidago multiradiata*[c] (A)
californiana	California	*Artemisia Douglasiana* (AN)

[a] Host tribe after each name indicated as A: Astereae; AM: Heliantheae, Ambrosiinae; E: Eupatorieae; H: Heliantheae, Helianthinae; AN: Anthemideae. Host records from the literature are in LeSage (1986) and references therein, and Goeden and Ricker (1985); several records are based on personal communication from R. Goeden, W. Palmer, and J. Sullivan. Host names in parentheses indicate few records; may be uncommonly or incidentally occupied.
[b] *conferta* and *sexvittata* may be geographic variants, and *macrovittata* a color morph of *sexvittata*.
[c] Personal record (often confirming literature records).

Adoption of a new host by *Ophraella* requires that the plant elicit oviposition and both larval and adult feeding, and that it sustain larval growth. That at least some of these potential barriers are independent of each other is indicated by tests of New York *O. communa* and *O. notulata*. These closely related species occur sympatrically, associated with *Ambrosia artemisiifolia* and *Iva frutescens*, respectively. I have never found a specimen of either on the other's host, despite

TABLE 19-3. Percent Survival of Larvae of *Ophraella communa* and *O. notulata*

O. communa (Host: *Ambrosia*) Ambrosia				Iva		O. notulata (Host: *Iva*) Ambrosia		
day	%	n	day	%	n	day	%	n
Lab								
13	60.4	619	12	4.4	722			
			20	2.4	722			
			8	8.8	68			
Greenhouse								
33	19.8	263				33	21.7	1076
Field								
			28	19.6	255	22	41.0	432
			47	6.3	255			

extensive collecting; free-ranging adults therefore must differ in host preference. Table 19-3 presents results of rearing hatchlings of *O. communa* on both *Ambrosia* and *Iva*, and of *O. notulata* on *Ambrosia*, in the laboratory (in a constant-temperature incubator, with groups of larvae provided fresh leaves every 2 days) and in the greenhouse and the field (enclosed in mesh sleeves on intact plants, with larvae introduced as egg clusters). Hatchlings of both species leave abundant feeding damage on the host plant of their congener, but survival of *O. communa* is far lower on *Iva* than on *Ambrosia*. It is likely, then, that *Iva* foliage is toxic (or perhaps nutritionally inadequate) to *communa* larvae. At least in the field, *O. notulata* appears to have higher survival on *Ambrosia* than *O. communa* does on *Iva*; by 21 days of age, 41% had survived, and 46% of these had attained pupation (a rare occurrence for *communa* on *Iva*). Many *O. notulata* reared on *Ambrosia* became normal-looking adults. In the greenhouse, egg clutches from 22 wild-caught female *O. notulata* were split and placed on two *Ambrosia* plants; the proportion of offspring that survived to 33 days varied significantly among families (heterogeneity $X^2 = 187.87$, $df = 21$, $p < 0.001$), suggesting (but not at all conclusively demonstrating) that there may exist genetic variation for features affecting survival on the congener's host.

Adults of *O. communa* and *O. notulata* feed readily and abundantly (Fig. 19-2*A*, *B*), and can survive for at least 3 weeks on each other's host in the laboratory. Thus the host preference that must typify free-ranging animals can be overridden by confinement without choice of plant (this is by no means true of all *Ophraella* species confronted with congeners' hosts: see below). Oviposition responses to *Iva* by *O. communa* were tested by maintaining laboratory-reared individuals on *Ambrosia* until they had mated and deposited two clutches of eggs; 49 females were then placed on *Iva* leaves (renewed every 2 days) in petri dishes in an 18/6 L:D, 23°/19° incubator. Almost all laid eggs on *Iva* within the first four days; I interpret these to have been eggs matured while feeding on Ambrosia (the normal interclutch interval is 1–3 days). Thereafter, only 3 of 42 survivors laid eggs on day 9, 1 of 42 on day 12, 1 of 41 on day 16, 1 of 35 on day 19, and 1 of 35 on day 20, when the test was terminated. In all, 6 of 49 females laid eggs that had been matured while feeding on *Iva*; those that had not oviposited proved, upon dissection, to have no ripening eggs in the ovarioles. Of 9 females returned to *Ambrosia* from *Iva*, 2 died within 3 days, but 6 laid eggs within 1 to 6 days and had ripe eggs when dissected. Of 26 females kept on *Iva* but exposed to the odor of *Ambrosia* or to a water slurry of crushed *Ambrosia* painted on *Iva* leaves, only one oviposited within 8 days. These results suggest that the alien plant may affect both vitellogenesis and the behavioral component of oviposition (Johansson 1964; Labeyrie 1968; Pouzat 1976), and that individual beetles vary in these respects.

On the basis of electrophoretic and morphological data (Futuyma and McCafferty, in press), *O. notulata* and *O. communa* appear to be closely related. By mapping the host associations onto the phylogeny we have inferred, it appears probable that the association of *O. communa* with *Ambrosia* may be ancestral to *notulata*'s association with *Iva*, although this conclusion cannot be strongly maintained. Whatever the direction of the host shift, the responses of these species to each other's host imply that the transition required evolutionary change in adult attraction to the plant (host preference), oviposition response, and physiological properties affecting larval survival. The greater survival of *O. notulata* larvae on *Ambrosia* than of *O. communa* on *Iva* is interesting in that it may imply retention of adaptation to *Ambrosia* in *O. notulata*, and partial preadaptation to *Iva* in *O. communa*, if indeed the direction of the host shift has been as postulated.

If we suppose that the first step in the evolution of a new host association consists of colonization of a novel plant by dispersing beetles that are unable to find their normal hosts, a positive feeding response by the adult is the first, necessary (but not sufficient), step. In 1987 I tested such responses in trials designed as an exploration for possible patterns. As the availability of beetles and plants provided opportunity, I tested wild-caught adult beetles individually in 5-in plastic Petri dishes in which were placed moistened filter paper and a leaf of a congener's host plant. The tests were carried out in an incubator at 18/6 L:D, 23/19 C, and about 80% R.H. The various tests were not run concurrently, so they cannot be considered controlled experiments. Every 24 hr for 1–3 days, consumption was scored and leaves were renewed before the beetles were moved to another test plant or were returned to their normal host. No beetles were deprived of the normal host for more than 3 days. Adults feed voraciously, and attack the normal host immediately if starved for 24 hr. Leaf consumption is presented (Figure 19-2) as the number of ocular reticle units of area (8.2 units = 1 mm^2). Note that the figures vary in scale and in plant species tested.

At one extreme of the range of responses, *O. bilineata* did not feed (except for trace feeding) on any novel plant (it fed voraciously on its natural host, but feeding was not quantified). *O. conferta* and *O. cribrata* fed on only one novel plant, in the same genus as their natural hosts. At the other extreme, *O. arctica* and several populations of *O. communa* collected from different hosts accepted some plants both in the same tribe as their natural hosts and in different tribes. For most species, consumption of plants in the same tribe as the natural host ranked higher (by Kruskal–Wallis test) than of plants in other tribes (Table 19-4). It is quite possible that this pattern reflects chemical stimuli shared by related plants (Ehrlich and Raven 1964).

On the basis of morphology, LeSage (1986) places these species of *Ophraella* in four groups: (1) *notata*; (2) *pilosa, cribrata*; (3) *conferta*; (4) *communa, arctica, bilineata, notulata, nuda*. My unpublished electrophoretic and morphological analyses strongly affirm the unity of group 4, but indicate that *notata, conferta,* and *cribrata* form a monophyletic group. *O. pilosa*, morphologically and electrophoretically isolated from the others and sharing plesiomorphic characters with an outgroup (*Monoxia*), appears to be the sister group of the rest of the genus. In general, positive responses to the hosts of other species were elicited by hosts of other members of the same species group (as indicated by the electrophoretic data), but there are some exceptions to this pattern (e.g., *O. notata* and *O. communa* displayed some response to each other's host). Overall, however, beetles that accept plants other than their usual hosts appear to have a lower threshold of response to the hosts of apparently close than of apparently distant relatives, as expected if there were a certain degree of behavioral "preadaptation" for host transitions that have been realized during the evolution of the group.

It is interesting that acceptance of plants unrelated to the natural host was documented primarily in the "communa group" (4), and that the "unrelated" plants accepted by these species are the natural hosts (or, in the case of *Artemisia vulgaris*, closely related to the natural host) of other members of the same group.

Figure 19-2. (A–E). Twenty-four hour consumption by individual adult *Ophraella* beetles of various plants, measured as leaf area consumed (number of reticule units, see text). All beetles were collected in the field on their natural hosts, except for Californian *O. communa*, reared for several generations in the laboratory on *Ambrosia psilostachya* and then on *A. artemisiifolia*. Test plants, arrayed along the abcissa, vary among panels; the natural host of a population is in uppercase letters. Numbers on the abscissa refer to scores taken after 24 (1), 48 (2), or 72 (3) hours after removal from natural host; individuals were transferred to fresh foliage of the test plant on successive days (curved arrows) or to different plant species on successive days. Each point represents an individual beetle; large circles indicate zero consumption by a number of individuals indicated by an adjacent number in parentheses. "x" indicates a mean of *n* beetles confined together in a few instances. All the plants are natural hosts of an *Ophraella species* except *Xanthium strumarium*, recorded only as an incidental host of *O. communa* in California, and *Artemisia vulgaris*, a European species not utilized by any *Ophraella*, but to which *A. Douglasiana*, the host of *O. californiana*, was originally referred. (*A*) Upper panel: *O. communa* from Long Island, NY; lower panel: *O. communa* from San Diego Co., CA (*ex Ambrosia*), *O. communa* from Inyo Co., CA (*ex Iva*), and *O.* cf. *communa* from Davis Mts., TX. (*B*) Upper: *O. notulata* from Long Island, NY; lower: *O. nuda* from Orion, Alberta. (*C*) Upper: *O. arctica* from Inuvik, N.W.T.; lower: *O. bilineata* from Cascade, MT. (*D*) Upper: *O. notata* from Ithaca, NY; lower: *O. pilosa* from Ithaca, NY. (*E*) Upper: *O. cribrata* from long Island, NY; lower: *O. conferta* from Ithaca, NY.

19-2A

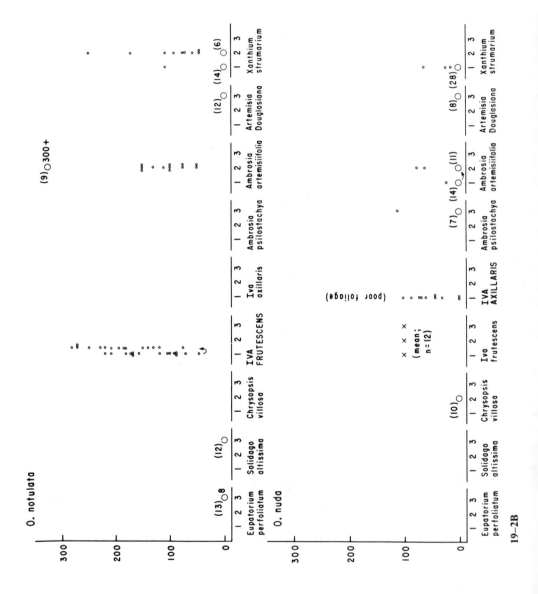

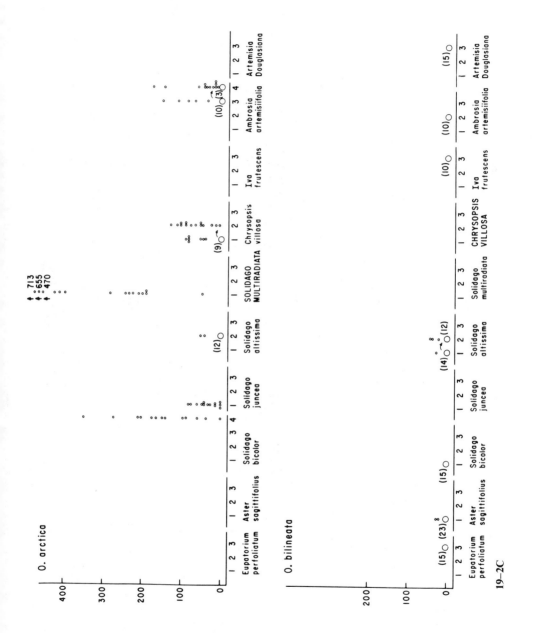

O. arctica

O. bilineata

19—2C

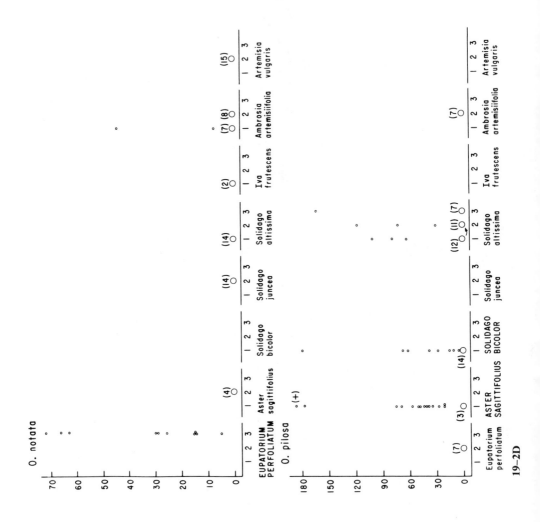

19-2D

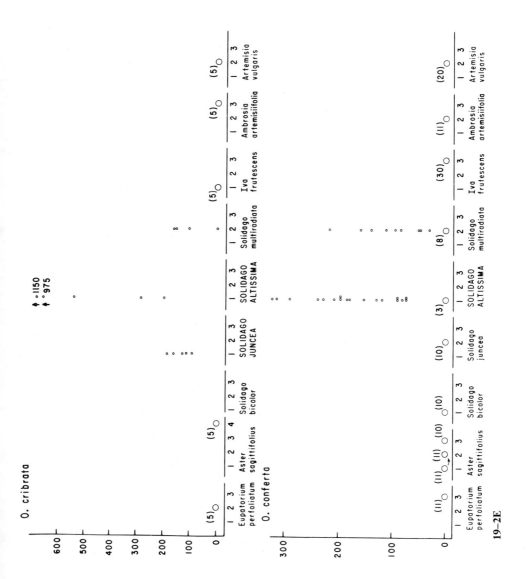

19-2E

447

TABLE 19-4. Summary of Feeding Responses of Adult *Ophraella* to Hosts of Congeners[a]

Species	Responses to Plants in Same Tribe as Host	Responses to Plants in Different Tribe from Host	Responses to Plants in Different Tribe, but Host to Related *Ophraella* Species[b]	Comparison of Responses to Plants in Same vs. Different Tribes from Host[c]
arctica	+[d],+,+,+	+[d]	+	$H = 0.371, p > 0.50$
bilineata	0,0	+,+[d],0,0,0,0	0,0,0	$H = 5.259, p < 0.025$
communa (NY)	+,+	+,+,+	+,+	$H = 23.771, p < 0.001$
communa (CA)	+	+,0,0	+,+	NT[b]
communa (TX)	+[d]	+,0	+,0	NT[b]
conferta	+,0,0,0	0,0,0	—	$H = 14.444, p < 0.001$
cribrata	+,0	0,0,0	—	$H = 4.508, p < 0.05$
notata	—	+[d],0,0,0,0	0,0,0	NT[b]
notulata	+,+	0,0	—	$H = 21.639, p < 0.001$
nuda	+[d],+[d],+,+	0,0	0	$H = 2.189, p > 0.01$
pilosa	+	0,0	—	$H = 1.051, p > 0.10$

[a]See Figure 19-2. +: feeding recorded; 0: no feeding; —: comparison not made; NT: no test possible.

[b]"Related species" are those in the same species group as determined by preliminary phylogenetic analysis (group 1: *notata, conferta, cribrata*; group 2: *arctica, bilineata, communa, notulata, nuda*; group 3: *pilosa*). For purposes of this analysis, *communa* (TX) is treated as a different species from other *communa*.

[c]Kruskal-Wallis H, with data consisting of individual feeding scores on two groups of plants, those in same vs. different tribe as natural host of the species.

[d]Positive feeding response by only a few of the individuals tested (see Fig. 19-2).

For example, the Texan population referred to *O. communa* in Figure 19-2 (to be designated a new species by Futuyma, in press) feeds on *Artemisia*, but accepted *Ambrosia*; conversely, populations of *O. communa* taken from Ambrosiinae fed on *Artemisia* at least slightly. Electrophoretic data indicate that the *Artemisia*-associated species is very closely related to, *O. communa*.

These tests do not address the question of whether genetic variation within *Ophraella* species has directed the history of host shifts; tests of this hypothesis are yet to come. However, instances in which a few individuals consumed considerable quantities of a plant while the majority rejected it (e.g., *O. notata* on *Ambrosia*, *O. pilosa* on *Solidago altissima*) raise the possibility that some of the variation has a genetic basis. Moreover, in three instances in which consumption by individual beetles was measured on two successive days, there was significant variation among individuals and substantial repeatability (Falconer 1981): for *O. pilosa* on *S. altissima* the repeatability (r) was 0.76, and p (by F test) < 0.001 ($n = 15$); for *O. arctica* on *Ambrosia artemisiifolia*, $r = 0.66$, $p < 0.01$ ($n = 15$); for *O. communa* on *Iva frutescens*, $r = 0.29$, $p < 0.05$ ($n = 19$). Repeatability does not necessarily imply a genetic basis for the variation, but the large differences in phenotypic variation among different beetle–plant combinations suggest that if the heritability of the observed responses were even roughly constant across combinations, there would exist differences in levels of genetic variance in response threshold. For example, none of 74 *O. communa* consumed any *Solidago altissima* whatever, whereas the majority accepted *Iva*, suggesting that the genetic potential for colonization of the latter may be greater than of the former. Preliminary phylogenetic analyses indicate that *Ambrosia*-feeding *O. communa* is closely related to *Iva*-feeding *O. notulata*, whereas *Solidago altissima* is host to species in a different clade within *Ophraella*. I expect, then, that genetic analysis will reveal more genetic variation for response to the host of its close relative than of distant relatives.

If this expectation holds, there will be some hope that the genetic characterization of insect species may provide some basis for predicting phylogenetic patterns of host utilization. The only basis for such prediction that has been developed to date is entirely correlational: discerning instances in which host transitions have occurred between plants that share common chemical features (Ehrlich and Raven 1964). In few instances is there direct evidence that a shared compound is indeed the cue that has facilitated the host shift, and there are numerous instances in which any chemical commonality among the hosts of closely related insects is obscure. Such cases challenge us to ask whether the evolutionary events can be understood or predicted at all. A genetic approach to problems in evolutionary history may offer some hope.

SUMMARY

Among the numerous questions raised by the narrow host specificity that is so prevalent among phytophagous insects are these: What selective advantages account for specialization? What factors have determined which plants have been adopted during the evolution of a clade of insects?

The first question requires that we distinguish between selection that maintains a particular host association and the selective factors responsible for the origin of host specificity from a more polyphagous habit. A long history of association with a particular group of plants undoubtedly brings about the progressive evolution of numerous special adaptations that increasingly commit an insect lineage to that host association. Thus stabilizing selection is undoubtedly responsible for the evolutionarily conservative host associations observed in higher taxa consisting of species that feed on related plants. During the origin of specialization, however, many of these constraints may not yet have evolved.

Explaining the origin of host specificity requires data on the selective advantages and disadvantages of relatively polyphagous and oligophagous genotypes within populations. Research on host-associated genotypes in the geometrid moth *Alsophila pometaria* is reviewed, and several selective factors are described that favor genetic differentiation in forests that differ in species composition. These include correspondence between the phenology of genotypes and that of the prevalent hosts, genotypic differences in threshold for acceptance of locally abundant plants, and perhaps adaptation to ecological factors (specifically, soil conditions) correlated with the distribution of the hosts. As in a considerable number of other studies, we have found no evidence in favor of the widely entertained hypothesis that the advantage of specialization lies in a trade-off in physiological capacities to grow on food plants that differ in secondary chemistry. Thus at least in the incipient stages of specialization, ecological factors extrinsic to the physiology of the insect–plant interaction may impose the selection that initiates the evolution of a specialized host association. I pose the hypothesis that ecological factors of this kind are of major importance, imposing selection for "specialization *per se*" rather than stabilizing selection for a phylogenetically conservative host association, in those insect taxa in which closely related species are variously associated with unrelated, chemically dissimilar plants.

The question, what has determined which plants have been adopted during the history of an insect clade, may best be approached from a phylogenetic perspective, whereby one first attempts to document what host shifts have occurred, before attempting to find in them an explicable pattern. Given a provisional history of host shifts, we may ask if the species are characterized by patterns of genetic variation that would imply that realized host shifts were more likely than shifts not realized. If genetic constraints have guided the history of host affiliation, rather than a largely unknowable history of ecological selection pressures, the phylogenetic pattern may prove predictable in some degree from genetic data on extant species. The very first stages in such a research program are described for the chrysomelid beetle genus *Ophraella*. For a pair of species that represent what are believed to be ancestral and derived host affiliations, several characters are described that would require genetic change for the historically realized host shift to have been accomplished. Although genetic data are not yet available, the feeding responses of various *Ophraella* species to the host plants of their congeners describe several patterns. In some instances, all individuals tested responded positively to a test plant; in other cases, none did, leading to the hypothesis that in such instances the threshold of response may be so high as to be passed by few genotypes. The majority of positive responses were to plants related to the species' natural host, suggesting that chemical commonality of plants affects the likelihood of host shifts. Positive responses to plants unrelated to the species' natural host were restricted primarily to hosts of closely related species of *Ophraella*. If patterns of genetic variation correspond to average responses, it is likely that the phylogenetic history of host shifts will be at least dimly reflected in the genetic potentialities of extant species.

ACKNOWLEDGMENTS

I am grateful to the several collaborators on the *Alsophila* study, whose names appear in the bibliography. For assistance with the study of *Ophraella*, I am grateful to Laurent LeSage for enthusiastically sharing his knowledge of the genus, to Richard Goeden for sending stocks of *O. communa* from California, to Gabriel Moreno and Craig Longtine for field assistance, and to Cynthia D'Urso for heroic help with the experiments. I appreciate suggestions by Robert Colwell and John Thompson that improved the manuscript. The *Alsophila* and *Ophraella* studies were supported by National Science Foundation grants BSR8306000 and BSR8516316 respectively. This is Contribution No. 735 in Ecology and Evolution from the State University of New York at Stony Brook.

REFERENCES

Bernays, E. and M. Graham. 1988. On the evolution of host specificity in phytophagous arthropods. *Ecology* **69**: 886–892.

Colwell, R. K. 1985. Community biology and sexual selection: Lessons from hummingbird flower mites. In J. Diamond and T. J. Case (Eds.), *Community Ecology*, pp. 406–424. Harper and Row, New York.

Colwell, R. K. 1986. Population structure and sexual selection for host fidelity in the speciation of hummingbird flower mites. In S. Karlin and E. Nevo (Eds.), *Evolutionary Processes and Theory*, pp. 475–495. Academic press, New York.

Dethier, V. G. 1954. Evolution of feeding preferences in phytophagous insects. *Evolution* **8**: 33–54.

Dethier, V. G. 1959. Food-plant distribution and density and larval dispersal as factors affecting insect populations. *Can Entomol.* **88**: 581–596.

DeVries, P. J. 1987. *The Butterflies of Costa Rica and Their Natural History*. Princeton University Press, Princeton, NJ.

Ehrlich, P. R. and P. H. Raven. 1964. Butterflies and plants: A study in coevolution. *Evolution* **18**: 586–608.

Futuyma, D. J. 1983. Selective factors in the evolution of host choice by insects. In S. Ahmad (Ed.), *Herbivorous Insects: Host-seeking Behavior and Mechanisms*, pp. 227–244. Academic Press, New York.

Futuyma, D. J. 1988. *Sturm und Drang* and the evolutionary synthesis. *Evolution* **42**: 217–226.

Futuyma, D. J., R. P. Cort, and I. van Noordwijk. 1984. Adaptation to host plants in the fall cankerworm (*Alsophila pometaria*) and its bearing on the evolution of host affiliation in phytophagous insects. *Amer. Natur.* **123**: 287–296.

Futuyma, D. J. and F. Gould. 1979. Associations of plants and insects in a deciduous forest. *Ecol. Monogr.* **49**: 33–50.

Futuyma, D. J. and S. S. McCafferty. 1991. Phylogeny and the evolution of host plant associations in the leaf beetle genus *Ophraella* (Coleoptera, Chrysomelidae). *Evolution* (in press).

Futuyma, D. J. and T. E. Philippi. 1987. Genetic variation and covariation in responses to host plants by *Alsophila pometaria* (Lepidoptera: Geometridae). *Evolution* **41**: 269–279.

Futuyma, D. J. and M. Saks. 1981. The effect of variation in host plant on the growth of an oligophagous insect, *Malacosoma americanum*, and its polyphagous relative, *Malacosoma disstria*. *Ent. Exp. Appl.* **30**: 163–168.

Georghiou, G. P. and C. E. Taylor. 1977. Pesticide resistance as an evolutionary phenomenon. *Proc. XV Internat. Entomol. Congr.*: 759–785.

Gould, F. 1979. Rapid host range evolution in a population of the phytophagous mite *Tetranychus urticae* Koch. *Evolution* **33**: 791–802.

Guthrie, W. D., J. L. Jarvis, G. L. Reed, and M. L. Lodholz. 1982. Plant damage and survival of European corn borer cultures reared for sixteen generations on resistant and susceptible genotypes of corn. *J. Econ. Entomol.* **75**: 134–136.

Hare, J. G. and G. G. Kennedy. 1986. Genetic variation in plant–insect associations: Survival of *Leptinotarsa decemlineata* populations on *Solanum carolinense*. *Evolution* **40**: 1031–1043.

Harshman, L. G., and D. J. Futuyma. 1985a. Variation in population sex ratio and mating success of asexual lineages of *Alsophila pometaria* (Lepidoptera: Geometridae). *Ann. Ent. Soc. Amer.* **78**: 456–458.

Harshman, L. G. and D. J. Futuyma. 1985b. The origin and distribution of clonal diversity in *Alsophila pometaria* (Lepidoptera: Geometridae). *Evolution* **39**: 315–324.

James, A. C., J. Jakubczak, M. P. Riley, and J. Jaenike. 1988. On the causes of monophagy in *Drosophila quinaria*. *Evolution* **42**: 626–630.

Janzen, D. H. 1978. The ecology and evolutionary biology of seed chemistry as relates to seed predation. In J. B. Harborne (Ed.), *Biochemical Aspects of Plant and Animal Coevolution*, pp. 163–206. Academic Press, New York.

Johansson, A. S. 1964. Feeding and nutrition in reproductive processes in insects. In K. C. Highnam (Ed.), *Insect Reproduction*, pp. 43–55. Royal Entomol. Soc., London.

Labeyrie, V. 1968. Longévité et capacité reproductrice de lignées d' *Acanthoscelides obtectus* sélectionnées en fonction de la rèponse aux stimuli de ponte. *CR Soc. Biol.* **162**: 2203–2206.

LeSage, L. 1986. A taxonomic monograph of the Nearctic galerucine genus *Ophraella* Wilcox (Coleoptera: Chrysomelidae). *Mem. Entomol. Soc. Canada No. 133.*

Lewontin, R. C. 1974. *The Genetic Basis of Evolutionary Change.* Columbia University Press, New York.

Mitter, C. and D. J. Futuyma. 1977. Parthenogenesis in the fall cankerworm, *Alsophila pometaria* (Lepidoptera: Geometridae). *Ent. Exp. Appl.* **21**: 192–198.

Mitter, C., D. J. Futuyma, J. C. Schneider, and J. D. Hare. 1979. Genetic variation and host plant relations in a parthenogenetic moth. *Evolution* **33**: 777–790.

Painter, R. H., S. C. Salmon, and J. H. Parker. 1931. Resistance of varieties of winter wheat to Hessian fly *Phytophaga destructor* (Say). *Kans., Agric. Exp. Sta. Bull.* **27**: 1–58.

Pathak, P. K. and E. A. Heinrichs. 1982. Selection of populations of *Nilaparvata lugens* by exposure to resistant rice varieties. *Environ. Entomol.* **11**: 85–90.

Pouzat, J. 1976. Le comportement de ponte de la Bruche du Haricot en présence d'extrait de plante-hôte. Mise en évidence d'interactions gustatives et tactiles. *CR Acad. Sci.* **282**: 1971–1974.

Rausher, M. D. 1984. Tradeoffs in performance on different hosts: Evidence from within- and between-site variation in the beetle *Deloyala guttata*. *Evolution* **38**: 582–595.

Rowell-Rahier, M. 1984. The food-plant preferences of *Phratora vitellinae* (Coleoptera: Chrysomelidae). B. A laboratory comparison of geographically isolated populations and experiments on conditioning. *Oecologia* **64**: 375–380.

Schneider, J. C. 1980. The role of parthenogenesis and female aptery in microgeographic, ecological adaptation in the fall cankerworm, *Alsophila pometaria* Harris (Lepidoptera: Geometridae). *Ecology* **61**: 1082–1090.

Scriber, J. M. 1979. The effects of sequentially switching food plants upon biomass and nitrogen utilization by polyphagous and stenophagous *Papilio* larvae. *Ent. Exp. Appl.* **25**: 203–215.

Singer, M. C. 1986. The definition and measurement of oviposition preference in plant-feeding insects. In J. H. Miller and T. A. Miller (Eds.), *Insect–Plant Relations*, pp. 65–94. Springer-Verlag, New York.

Singer, M. C., P. R. Ehrlich, and L. E. Gilbert. 1971. Butterfly feeding on a lycopsid. *Science* **172**: 1341–1342.

Strong, D. R., J. H. Lawton, and R. Southwood. 1984. *Insects on Plants: Community Patterns and Mechanisms*. Harvard University Press, Cambridge, MA.

Thomas, C. D., D. Ng, M. C. Singer, J. L. B. Mallet, C. Parmesan, and H. L. Billington. 1987. Incorporation of a European weed into the diet of a North American herbivore. *Evolution* **41**: 892–901.

Thompson, J. N. 1988. Variation in preference and specificity in monophagous and oligophagous swallowtail butterflies. *Evolution* **42**: 118–128.

Via, S. 1984. The quantitative genetics of polyphagy in an insect herbivore. II. Genetic correlations in larval performance within and among hosts. *Evolution* **38**: 896–905.

Vrba, E. S. 1980. Evolution, species and fossils: how does life evolve? *S. Afr. J. Sci.* **76**: 61–84.

Wiklund, C. 1981. Generalist vs. specialist oviposition behaviour in *Papilio machaon* (Lepidoptera) and functional aspects on the hierarchy of oviposition preferences. *Oikos* **56**: 163–170.

Wilcox, J. A. 1965. A synopsis of the North American Galerucinae (Coleoptera: Chrysomelidae). *New York State Mus. Sci. Serv. Bull.* No. 400. Albany, New York.

Williams, K. S. 1983. The coevolution of *Euphydryas chalcedona* butterflies and their larval host plants. III. Oviposition behavior and host plant quality. *Oecologia* **56**: 336–340.

NOTES ADDED IN PROOF

(1) Using principal components analysis, J. Jaenike (*Ann. Rev. Ecol. Syst.*, in press) has reanalyzed the data of Futuyma and Philippi (1987) on *Alsophila*, and has found evidence for a trade-off between growth on *Acer* and species of Fagaceae. (2) Since this manuscript was revised in 1989, I have described the *Artemisia*-associated Texan population of "*Ophraella communa*" as a new species, and have synonymized *O. sexvittata* and *O. macrovittata* (Futuyma 1990, *J. New York Entomol. Soc.* 98: 163–186). An estimate of the phylogeny of *Ophraella*, with inferences on the history of host shifts in the genus, is presented by Futuyma and McCafferty (in press).

20 Host Plant Discrimination: Experiments with Hummingbird Flower Mites

AMY JO HEYNEMAN, ROBERT K. COLWELL, SHAHID
NAEEM, DAVID S. DOBKIN, and BERNARD HALLET

Hummingbirds and the flowers they visit are associated throughout most of their range with nectarivorous mites of two closely related genera, *Rhinoseius* (Fig. 20-1) and *Proctolaelaps* (Gamasida: Ascidae). These "hummingbird flower mites," of which there may be some 200 species (Colwell 1979), feed on nectar and pollen, mate, and lay eggs inside or near the corollas of ornithophilous flowers (Colwell 1973, 1983, 1985, 1986a; Colwell and Naeem 1979; Dobkin 1984, 1985, 1987). For dispersal between plants the mites travel on the bills of hummingbird visitors.

The diversity of species involved in this hummingbird–flower–mite complex decreases with increasing latitude, elevation, and insularity. Thus, a few hectares of New World lowland tropical rainforest may contain a dozen or more sympatric species of hummingbird flower mites and as many kinds of hummingbirds, feeding on over 20 plant species, whereas at higher latitudes and elevations and on Caribbean islands remote from the South American mainland the system is much simpler (Colwell 1986a). In California, for example, a single mite species, *Rhinoseius epoecus*, is carried between a few species of a single native plant genus (*Castilleja*) by three hummingbird species (Colwell and Naeem 1979).

Each species of hummingbird flower mite depends on one or very few host plant species that, in a given region, together provide a reliable, year-round supply of flowers. In all known cases, hummingbird flower mite species are extremely faithful to their host plants. Based on some 25,000 specimens of more than 20 species examined, only about one in 200 hummingbird flower mites is found on a host plant other than its usual species. Many species of hummingbird flower mites are monophagous—the majority, in tropical lowland forests (Colwell 1986a). Moreover, in most cases each mite species *monopolizes* the host

Plant-Animal Interactions: Evolutionary Ecology in Tropical and Temperate Regions, Edited by
Peter W. Price, Thomas M. Lewinsohn, G. Wilson Fernandes, and Woodruff W. Benson.
ISBN 0-471-50937-X © 1991 John Wiley & Sons, Inc.

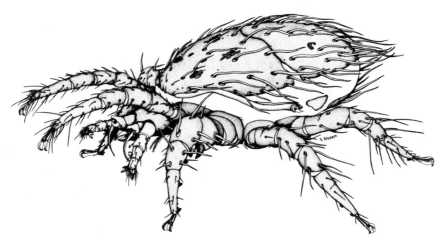

Figure 20-1. A hummingbird flower mite, a male *Rhinoseius hirsutus* from Trinidad, W.I. This species inhabits the flowers of the plant *Cephaelis muscosa* (Rubiaceae) and travels between inflorescences primarily on the hummingbird *Amazilia tobaci*. The mite is less than a half millimeter in length, about the size of the period at the end of this sentence. Drawing and copyright by Shahid Naeem.

species in its repertoire. Only rarely do two mite species share the use of the same host species in the same locality. In all 6 known cases of "host-sharing," the pair of mite species involved are noncongeners—a species of *Rhinoseius* and a species of *Proctolaelaps* (Colwell 1973, 1986a).

Overlap on hummingbirds, however, is common. Based on samples taken from hummingbirds (by aspiration), the mites evidently travel between plants of their own host species on any hummingbird that visits their host flowers. Thus, mite species that inhabit short-corolla flowers tend to travel together on short-billed birds, and those that inhabit long-corolla flowers travel on long-billed birds. An individual hummingbird that visits the flowers of several plant species in the period of a few hours may carry mites from all of them—four or five species of mites at a time. Thus the mixing of mite species on birds contrasts sharply with their largely nonoverlapping distribution among host plants. Given the lack of overlap of mite species in host plant use (or complete overlap, in the case of host-sharing species pairs) and the high and variable cooccurrence of mite species on birds, it is apparent that each individual mite disembarks only at flowers of its own usual host species, among all the plants visited by its hummingbird carriers.

At each locality we have studied, *all* hummingbird flower mites are conspicuously absent from the flowers of one or more potential host-plant species. These "miteless" hummingbird-pollinated species would seem to be suitable hosts in that, like actual host species, they provide the protection of tubular flowers, abundant nectar, and access via hummingbird visitors known to carry mites. Nonetheless, intensive searching reveals virtually no inhabitants in these "miteless" plants (fewer than one mite in 200 flowers at our Volcan Colima site, for example).

Hummingbird flower mites, then, are capable of both discriminating between potential host species and miteless plant species, and of distinguishing among potential host plants, disembarking only at flowers of their own host. Their decisions are probably based on olfactory information (Colwell 1979, 1986a); as mites ride in the birds' nares, they are continuously exposed to a flux of chemical information about the immediate environment of the bird's bill. Although bird-pollinated flowers typically have no fragrance detectable to the human nose and probably none detectable to the hummingbird nose either [there is no evidence that hummingbirds can smell (Tyrrell and Tyrrell 1985)], many secondary plant compounds are nevertheless volatile and could well serve as olfactory cues for mites. An ability to detect minute quantities of specific compounds at long and short range has been demonstrated in a mite closely related to hummingbird flower mites, *Proctolaelaps nauphoetae*, whose choice of cockroach host is based on a chemical unique to that insect (Egan 1976, Egan et al. 1975). This mite species senses compounds with specialized sensory setae located on the first pair of legs. Other mites have also been shown experimentally to detect host plant odors (Rodriguez and Rodriguez 1987) and prey odors (McMurtry and Rodriguez 1987).

We conducted a series of preference experiments (in the sense of Singer 1986) to investigate the ability of hummingbird flower mites to discriminate among floral tissues and nectars. These tests (1) examine the ability of mites to distinguish acceptable (host) from unacceptable (miteless and alien host) plants, (2) explore preferences between host plant species used in different seasons by the same mite species, and (3) test for possible preferences between unoccupied host plant flowers and host flowers inhabited by conspecific mites. The experiments also aim to determine the basis on which mites make their choices—whether by attraction to favored compounds or by avoidance of nectars that are repellent, nutritionally inadequate, or produced by plants unacceptable for ecological or behavioral reasons. Many of the questions posed by our studies and perhaps some of the answers they provide are equally appropriate to other groups of host-specific arthropods, including many herbivorous insects.

STUDY SITES AND METHODS

Study Sites

Preference experiments were carried out at four field sites. The California study site, in hard chaparral and open, grassy foothills, is located in Strawberry Canyon, Alameda County, at about 100 m elevation, on property owned by the University of California at Berkeley. Here, a single mite species, *Rhinoseius epoecus* (Colwell and Naeem 1979), seasonally inhabits several native species of Indian Paintbrush (Scrophulariaceae: *Castilleja*) and is carried between plants by resident and migrant Anna Hummingbirds (*Calypte anna*), and by migrant Rufous and Allen Hummingbirds (*Selasphorus rufus* and *S. sasin*) (Colwell 1986a).

Rhinoseius epoecus is also the sole species of hummingbird flower mite at the Volcan Colima study site, 170 km south of Guadalajara, Mexico. The upper slopes of this active volcano, at elevations between about 2000 and 2500 m, provide a primary wintering site for migrant *Selasphorus rufus*. The site also supports a high concentration of hummingbirds of 20 other species. In winter, when preference experiments were conducted, a profusion of hummingbird pollinated shrubs, herbs, and vines flower in light gaps in the pine–oak forest (Des Granges 1977, 1978). Of the more than 20 species of plants visited by hummingbirds, only two commonly support *Rhinoseius* mites: *Castilleja integrifolia*, a congener of the California host species, and *Lobelia laxiflora* (Lobeliaceae).

One of our lowland tropical sites is La Selva Biological Station in the Sarapiqui region of Costa Rica, owned and operated by the Organization for Tropical Studies. Here, in premontane wet forest, 8–10 common hummingbird species visit more than 50 plant species (Stiles 1975, 1978 provides details), and transport 8–10 mite species. During late summer, when preference experiments were conducted, the most abundant host plants in flower are 9 species of *Heliconia*. Hummingbird flower mites are known to inhabit at least 20 other plant species at this site as well (unpublished data).

The second lowland tropical site is the Arima Valley in the Northern Range in Trinidad, West Indies, a land-bridge island separated from Venezuela by a narrow strait. The fauna and flora are essentially a "mainland" biota, similar in general composition to that of La Selva. The two sites share a number of hummingbird species, and many of the mite host plants of the Arima Valley are congeners of those at La Selva. At least one mite species, *Proctolaelaps kirmsei*, occurs at both sites in the same host plant (*Hamelia patens*). Snow and Snow (1972), Feinsinger and Swarm (1982), and Feinsinger et al. (1982, 1985) provide data on hummingbird visitation to plants of the Arima Valley, and report flowering phenologies for many of these species.

Experimental Methods

Each preference experiment was conducted with a series of as many as 60 individual mites, each tested separately in its own apparatus—a transparent acrylic block with a central chamber into which three passages lead, each fitted with a removable glass capillary tube (Fig. 20-2). [Naeem's design for these chambers was inspired by preference chambers used by Egan (1975)]. A capillary tube containing the mite to be tested was inserted in the leg of the "T." The two lateral capillary tubes, with test substances at their capped tips, constituted the two options available to the mite. The distance to test substances from the central chamber (70 mm) was on the order of the length of a typical hummingbird bill— the distance that these mites run in a matter of seconds between a bird's nares and a flower when they disembark.

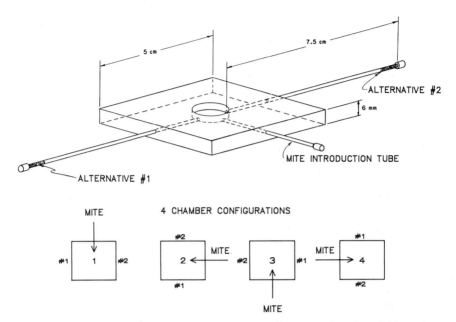

Figure 20-2. Apparatus for testing nectar preferences of hummingbird flower mites. Three glass capillaries, one (the leg of the "T") containing a single mite and the other two (the arms of the "T") containing test solutions at their distal tips, are inserted into holes drilled into a clear acrylic block. These tubes, capped at their distal ends, open into a central chamber covered with a circular glass cover slip. From the central chamber the mite has access to both alternatives. The location of the two test solutions was reversed in half the chambers and equal numbers of chambers faced in each of the four cardinal directions, to control for any tendency of the mites to turn right or left or to orient to external stimuli.

In each experiment (or replication), we set up a series of 10–60 of these chambers simultaneously, each with a single mite and an identical pair of test substances. The location of the two test solutions was reversed in half the chambers to control for any tendency of the mites to turn in a particular direction. In addition, equal numbers of chambers faced in each of the four cardinal directions, to control for any tendency of the mites to orient in a particular compass direction or toward a particular stimulus, such as light. Chambers were shielded from direct sunlight, but were exposed to ambient field temperatures and indirect light levels.

To set up an experiment, we collected fresh flowers containing mites and placed them in a small plastic bag. Whenever possible we selected flowers with mites that appeared ready to disperse—actively gathering at the entrance to a flower. Nectar was collected with 5 or 10- microliter capillary tubes from fresh flowers not inhabited by mites. These flowers had previously been bagged with tulle fabric to prevent the introduction of mites, removal of nectar, admixture of pollen, or other contamination by hummingbird feeding. In the case of flowers with long or curved corollas, the distal part of the flower was cut off with a razor blade, below the anthers and above the nectar line, to minimize nectar contamination that might be caused by scraping the inner corolla with the collecting capillary or by accidental introduction of pollen. Nectar from flowers found to contain mites or other arthropods was not used. In a sterile tube, we pooled nectar from a number of flowers to promote uniformity among chambers within an experiment.

If an artificial sugar solution was required as a test solution, aqueous solutions were generally prepared with equal weights of the three sugars most abundant in hummingbird-pollinated flower nectars—sucrose, glucose, and fructose (Baker 1975; Baker and Baker 1983). In a few early experiments, sucrose alone was used to make the test solutions. The total sugar concentration of the test solution matched, on a weight-to-weight basis (see Bolten et al. 1979), the concentration of the natural nectar to be used as the other alternative for a given experiment. One tip of each capillary tube to be used in a preference experiment was filled with approximately 6 mm (about 1 μl) of pooled nectar or matched sugar solution. That tip was then capped to minimize changes in concentration, by either dehydration or hydration.

To minimize separation from moisture and food, mites were removed from their flowers only after all tubes containing test solutions had been prepared. Adults were extracted from collected flowers on the tip of a #0000 artist's paint brush, and coaxed individually into capillary tubes that were then capped temporarily at both ends. The sex ratio of mites used in experiments was similar to the sex ratio of mites in flowers, which is in general female-biased (Colwell 1981; Wilson and Colwell 1981). When all test solution tubes and mite tubes had been prepared, the tubes were quickly inserted into the chambers (Fig. 20-2). Approximately 3–4 hours usually elapsed from the time flowers were collected to the time the tubes were inserted into chambers.

Although capable of traversing the length of a capillary tube in less than 15 seconds, the mites varied considerably in rate of movement from the introduction tube to one of the two lateral tubes containing test solutions. Like herbivorous insects in similar preference chambers (Lewis and van Emden 1986; Mulkern 1969), hummingbird flower mites spend a large proportion of time resting, generally near the test solution on which they had most recently fed. The presence of a mite in a capillary tube was therefore taken as an indication of preference for the solution at the tip of that tube. In chambers that contained a female mite, any eggs laid near the meniscus of a test solution were also taken to indicate preference for the solution. Mites in the entrance tube or in the central chamber were not counted.

Mites occasionally moved between the two lateral tubes containing test solutions. Therefore we recorded the position of the mite and positions of eggs, if any had been laid, every half hour during the day, and every two to three hours at night. Each experiment was continued for approximately 24 hours, ample time to allow the mites to demonstrate a preference. After 24 hours, fungal hyphae often became evident in some nectars, which appeared to induce mites to relocate.

Moreover, any volatile compounds present in one nectar would be expected eventually to diffuse into the other, confounding the results. In the next section, we discuss our attempt to resolve these complications statistically.

To minimize contamination from previous experiments capillary tubes and tube caps were generally used only once, then discarded. Chambers were boiled 2 minutes in a distilled water–Microwash solution, rinsed twice in boiling water, shaken dry, then submerged and agitated for 1 minute in 50% ethanol, rinsed in distilled water, excess moisture shaken off, and finally dried thoroughly at low temperature in an oven or vacuum chamber. In a few cases when supplies ran low, tubes and tube caps were washed in the same manner and reused.

Studying behavior by means of such preference experiments is highly labor-intensive, and yields relatively small sample sizes. The largest tests involved 60 chambers—two simultaneous tests of 30 chambers each. Because not all mites were near one of the two test substances when preferences were recorded (some individuals were occasionally in the central chamber, and a few never left the entrance tube), actual sample sizes were substantially smaller. Results, however, were generally sufficiently clear-cut that relatively small samples yielded statistically significant results. Because mites and flower nectars were generally abundant and easily collected in the study areas selected, sample sizes could be increased when necessary by repeating experiments on successive days.

Statistical Methods

Each time the positions of mites and the number and distribution of eggs in the test chambers were recorded, the number of points in favor of each of the two alternative test solutions was totaled. The presence of a mite in a test solution tube was counted as one point in favor of that alternative. Similarly, if eggs had been laid, their presence (in any number) near a test solution was counted as a single point in favor of that alternative. The distribution of mites is assumed to be, in principle, independent of the distribution of egg masses; indeed, females were not infrequently recorded in one arm of the "T" after laying eggs in the other. Because mite activity varied throughout the day, and mites did not always remain in the lateral tubes (mites also had continuous access to the entrance tube and to the central chamber), sample size varied from reading to reading over the course of an experiment.

The position of each mite over the course of an experiment reveals its direction of preference, which in general remained stable throughout the 24 hours that each experiment was continued. Because successive readings of the same mites are not independent, however, we chose to use a single set of observations at an intermediate time as the estimate of preference. To maximize the likelihood that adequate time for mites to make a choice had elapsed, but degradation or homogenization of test solutions was not yet causing mites to relocate, the measure of mite preference was taken to be the total number of points in favor of each alternative at the observation time that this distribution was most asymmetrical. The statistical significance of the maximally asymmetrical distribution, compared to a binomial expectation, may be somewhat inflated for a single experiment (because it may be an outlier), but the *direction* of asymmetry is clearly unbiased. Therefore, arithmetically summed data from several independent tests of the same hypothesis should yield no inflation if the null hypothesis in fact is true and sample sizes are commensurate. As a check for consistency, this index of preference was compared with time averages for position; in every case, the two measures were in accord with regard to direction of preference.

Each test, then, ultimately produces two numbers: (1) the number of mites plus the number of sets of eggs at the first alternative, and (2) the number of mites plus the number of sets of eggs at the second, at the time of maximum differentiation. If the mites do not prefer one alternative over the other, the expectation is that both mites and egg masses will be approximately evenly divided between the two alternatives. Thus, a binomial distribution, with an expectation of 0.5, can be used to approximate the probability that the experimental outcome is due to chance alone.

In several cases experiments were repeated with identical protocol (the same mite species presented with the same alternatives at the same field site, usually on successive days), and the two independent sets of data were combined for statistical analysis. To test more general hypothesis, statistical analysis was conducted on data pooled from results of several experiments that tested the same hypothesis for several different mite species or host plant species.

In certain cases, differences in preference between experiments, rather than within-experiment significance of results, must be evaluated. For example, when two mite species are each tested for preference between their own host nectar and that of the other species, the data from these two related experiments can be cast as a 2×2 contingency table. Although the number of mites utilized in the experiment from each of the two host species is determined by the experimenter, at any time only a subset of the mites are scored at either of the two alternatives. The marginal totals of the 2×2 table are thereby limited, but not determined by the experimenter. Thus the G-test of independence, with Williams' correction for a 2×2 table (Sokal and Rohlf 1981, Chapter 17.4), is more appropriate than Fisher's exact test or chi-squared to determine the statistical significance of difference in preferences between two sets of mites.

EXPERIMENTS AND INTERPRETATION

Results from the preference experiments we conducted are presented in the text below and in Figures 20-3–20-11. The main conclusions from these presentations are summarized in Figure 20-11, which may also serve as a preview of the experiments to be reported in detail below.

Moisture Requirements—Vapor Density Gradient

In a brief series of preliminary experiments, we gave mites a choice between water as one alternative and an empty capillary tube as the other. Of the total score of 14 (mites plus egg masses—4 for *Proctolaelaps certator* and 3 for *Rhinoseius trinitatis* from Trinidad, and 7 for *R. epoecus* from California), 12 points indicated a preference for water ($P = 0.006$ by the binomial test). We may presume that water at the extreme tip of a capillary tube in these experiments produced a humidity gradient, with vapor concentrations highest near the water. Because they lack the protective coating of a sclerotized or waxy cuticle (Krantz 1978), hummingbird flower mites are extremely vulnerable to desiccation (Dobkin 1985). These experiments showed that mites are capable of negotiating the pathways of the choice chamber and of following vapor density gradients to the moisture they require. Insects also respond to vapor density gradients produced by wet cotton wool or plant leaves, either positively or negatively, according to their degree of desiccation (e.g., Saxena 1978).

All other tests reported here involved choices between flower nectars or between nectar and equivalently concentrated simple sucrose or sucrose–glucose–fructose solutions. Although floral nectar of hummingbird-pollinated species generally contains a diversity of nutritional elements, including free amino acids, proteins, and lipids, the three most abundant constituents generally present are the sugars sucrose, glucose, and fructose (Baker 1975, 1977, 1978; Baker and Baker 1982, 1983; Heyneman 1985). These nectar–nectar and nectar–sugar choices offered to the mites therefore present negligible concentration differences. Consequently, vapor density gradients should have had little or no influence on the remaining test results.

Miteless Plant Species

When mites at each of the four field sites were offered nectar from their own host and from local miteless species known to be visited by the hummingbird carriers of the test mites, preferences were consistently exhibited for host plant nectar (Fig. 20-3). At the behavioral level, the avoidance of miteless species suggested by these experiments and by the field distribution of hummingbird flower mites may indicate that the mites are (1) repelled by some substance or substances present in miteless nectars, (2) attracted by compounds present in host nectars, or (3) a combination of the two phenomena. If the avoidance of miteless nectars is adaptive, it suggests (1) that miteless nectars are actually toxic to mites (and perhaps to other arthropods, but not to hummingbirds), (2) that nectars from miteless species do not adequately meet the nutritional requirements of the mites, or (3) that miteless species are palatable and nutritionally adequate, but are avoided for ecological or behavioral reasons.

A series of experiments was designed to answer the behavioral question—whether the absence of mites in apparently suitable plant species is primarily a consequence of active avoidance of these plants by hummingbird flower mites or of attraction to their own host plants, or a combination of these two responses. To test for attraction to the nectar of host species (Fig. 20-4), experiments were conducted at all four sites using four mite species. Each mite was offered nectar of the host species from which it was collected and a sucrose or sucrose–glucose–

HOST NECTAR vs. MITELESS NECTAR

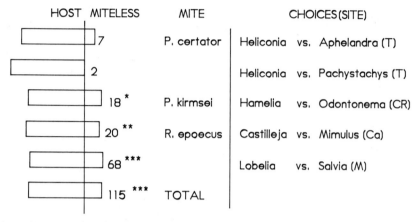

HOST MITELESS	MITE	CHOICES (SITE)
7	P. certator	Heliconia vs. Aphelandra (T)
2		Heliconia vs. Pachystachys (T)
18 *	P. kirmsei	Hamelia vs. Odontonema (CR)
20 **	R. epoecus	Castilleja vs. Mimulus (Ca)
68 ***		Lobelia vs. Salvia (M)
115 ***	TOTAL	

Figure 20-3. Results of preference experiments for hummingbird flower mites presented with a choice between nectar of their usual host species and nectar of a nearby species, also visited by hummingbirds, that never supports hummingbird flower mites (a "miteless" species). For each test, the same hummingbirds visit the host plant and the miteless plant. In this figure and in Figures 20-4 through 20-10, bars represent the proportion of individuals that showed a preference for each alternative. Numbers give the sample size for each emperiment and for the combined total. Asterisks indicate the significance of binomial tests: $*P < 0.05$, $**P < 0.01$, $***P < 0.001$. Abbreviations for sites: CR = Costa Rica (La Selva Biological Station), Ca = California (Strawberry Canyon), M = Mexico (Volcan Colima), and T = Trinidad (Arima Valley). Details of the results are as follows (the first plant listed is the host species; the number of mites and egg masses indicating a preference for each alternative is shown in parentheses): *Proctolaelaps cerator*: *Heliconia wagneriana* (6), *Aphelandra incerta* (1); *P. certator*: *Heliconia wagneriana* (2), *Pachystachys coccinea* (0); *Proctolaelaps kirmsei*: *Hamelia patens* (14), *Odontonema* sp. (4); *Rhinoseius epoecus*: *Castilleja foliolosa* (16), *Mimulus* sp. (4); *R. epoecus*: *Lobelia laxiflora* (50), *Salvia iodantha* (18); totals: host nectar (88), miteless nectar (27).

fructose solution of equivalent concentration. Overall, mites significantly preferred host-plant nectar over sugar solution. (The sole significant exception was the unaccountable preference of *Proctolaelaps kirmsei* for sugar solution.) Because no noxious compounds exist in the simple sugar solutions, the tests imply attraction to host nectar rather than repulsion from the alternate choice.

To test for avoidance of miteless plant species (Fig. 20-5), mites at the four study sites were offered nectar from one of six miteless species representing three plant families, against an alternative sucrose glucose–fructose solution matched in concentration with the miteless nectar. In four of the six experiments mites preferred sugar solutions over miteless nectars. This preference is statistically significant in two cases. In two other cases nectar appeared to be preferred over sugar solutions, but neither of these outcomes was significant. Thus, the only

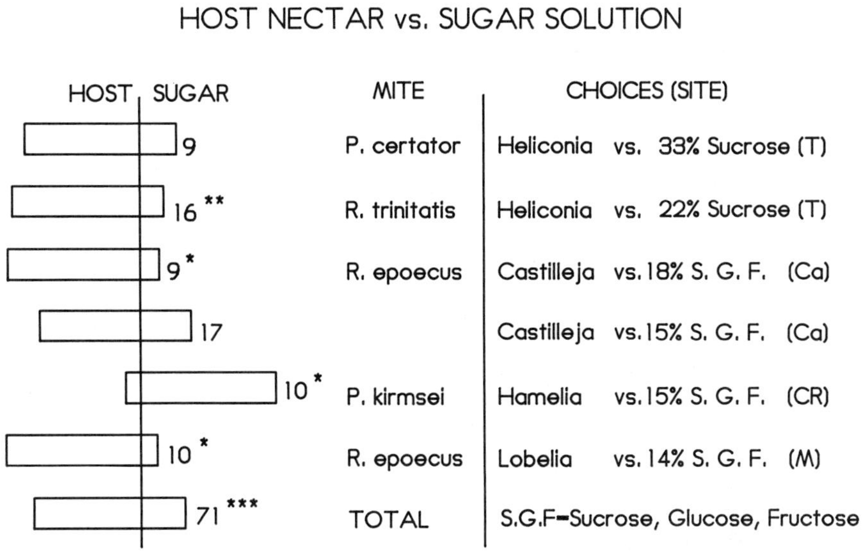

HOST NECTAR vs. SUGAR SOLUTION

HOST	SUGAR		MITE	CHOICES (SITE)
	9		P. certator	Heliconia vs. 33% Sucrose (T)
	16 **		R. trinitatis	Heliconia vs. 22% Sucrose (T)
	9 *		R. epoecus	Castilleja vs. 18% S. G. F. (Ca)
	17			Castilleja vs. 15% S. G. F. (Ca)
	10 *		P. kirmsei	Hamelia vs. 15% S. G. F. (CR)
	10 *		R. epoecus	Lobelia vs. 14% S. G. F. (M)
	71 ***		TOTAL	S.G.F—Sucrose, Glucose, Fructose

Figure 20-4. Results of preference experiments for mites presented with a choice between nectar of their usual host species and a sugar solution of the same concentration. See the caption for Figure 20-3 for information on figure conventions. Details of the results are as follows: *Proctolaelaps certator: Heliconia wagneriana* (7) 33% sucrose (2); *Rhinoseius trinitatis: Heliconia trinitatis* (14), 22% sucrose (2); *R. epoecus: Castilleja foliolosa* (8), 18% S.G.F. (sucrose–glucose–fructose) (1); *R. epoeous: Castilleja foliolosa* (11), 15% S.G.F. (6); *P. kirmsei: Hamelia patens* (1), 15% S.G.F. (9); *R. epoecus: Lobelia laxiflora* (9), 14% S.G.F. (1); totals: host nectar (50), sugar solution (21).

statistically significant results indicate that hummingbird flower mites tend to prefer simple sugar solutions to nectar from miteless species, even though the latter is nutritionally superior to sugar solutions (Heyneman 1985).

Together, these two sets of preference experiments suggest that avoidance of miteless species involves a marked attraction to host nectar, in concert with a relatively weak and variable repulsion from miteless nectar.

Familiar Alien Host Plants

Especially in biologically rich tropical communities, the hummingbird species in a local assemblage typically vary considerably in bill length and curvature, and each species tends to feed on plant species whose flowers more or less match the morphology of its bill (Feinsinger and Colwell 1978; Feinsinger et al. 1982, 1985). Thus species of hummingbird flower mite that inhabit different short-corolla flowers in the same area, for example, tend to be transported together on short-billed birds that visit their host-plant species. Likewise, a long-billed bird species typically carries mites from several long-corolla host-plant species that it visits regularly (unpublished data). As a consequence, a given mite species is regularly

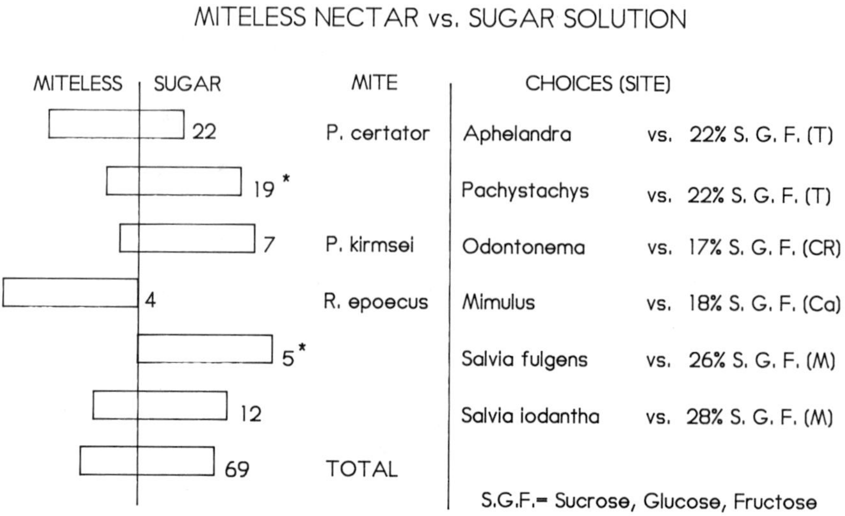

MITELESS NECTAR vs. SUGAR SOLUTION

MITELESS	SUGAR	MITE	CHOICES (SITE)		
	22	P. certator	Aphelandra	vs.	22% S. G. F. (T)
	19 *		Pachystachys	vs.	22% S. G. F. (T)
	7	P. kirmsei	Odontonema	vs.	17% S. G. F. (CR)
4		R. epoecus	Mimulus	vs.	18% S. G. F. (Ca)
	5 *		Salvia fulgens	vs.	26% S. G. F. (M)
	12		Salvia iodantha	vs.	28% S. G. F. (M)
	69	TOTAL			

S.G.F.= Sucrose, Glucose, Fructose

Figure 20-5. Results of preference experiments for mites presented with a choice between nectar of a plant species that is visited by hummingbirds but that never supports hummingbird flower mites (a "miteless" species) and a sugar solution of the same concentration. For each test, the same hummingbirds visit the host plant (listed below) and the miteless plant. See the caption for Figure 20-3 for information on figure conventions. Details of the results are as follows: *Proctolaelaps certator* from *Heliconia wagneriana*: *Aphelandra incerta* (15), 22% S.G.F. (sucrose–glucose–fructose) (7); *P. certator* from *H. wagneriana*: *Pachystachys coccinea* (4), 22% S.G.F. (15); *P. kirmsei* from *Hamelia patens*: *Odontonema* sp. (1), 17% S.G.F. (6); *Rhinoseius epoecus* from *Castilleja foliolosa*: *Mimulus* sp. (4), 18% S.G.F. (0); *R. epoecus* from *Lobelia laxiflora*: *Salvia fulgens* (0), 26% S.G.F. (5); *R. epoecus* from *L. laxiflora*: *Salvia iodantha* (4), 28% S.G.F. (8); totals: miteless nectar (28), sugar solution (41).

exposed to only a subset of the hummingbird-pollinated plant species in a locality, in addition to its own host-plant species.

From the point of view of a particular mite species (call it the "focal species"), we call the normal host of another hummingbird flower mite species in the same locality an "alien host." If an alien host is frequently visited by the usual hummingbird carriers of the focal mite species, we refer to that host as a "familiar alien host." An alien host rarely or never visited by usual carriers of the focal species are designated "unfamiliar alien hosts." [The easiest way to make this distinction in practice is to assess which mites are found together on hummingbirds and which are not. Foraging observations and pollen sampling from birds (Feinsinger et al. 1985) provide concordant information.]

Figure 20-6 reports the results of experiments conducted to test the ability of mites to distinguish their own host plant species from familiar alien hosts. At the lowland tropical site in Trinidad, where some 18 species of hummingbird flower mites coexist, two pairs of mite species were selected for these experiments. Sibling species *Proctolaelaps contentiosus* and *P. certator*, the first pair, are both carried

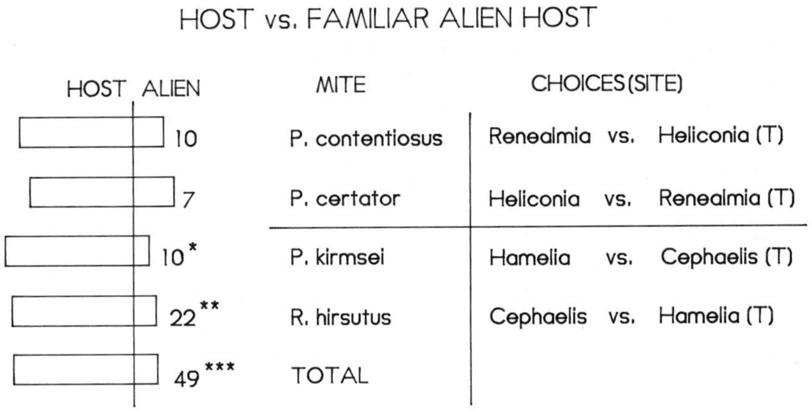

HOST vs. FAMILIAR ALIEN HOST

HOST ALIEN	MITE	CHOICES (SITE)	
10	P. contentiosus	Renealmia vs.	Heliconia (T)
7	P. certator	Heliconia vs.	Renealmia (T)
10 *	P. kirmsei	Hamelia vs.	Cephaelis (T)
22 **	R. hirsutus	Cephaelis vs.	Hamelia (T)
49 ***	TOTAL		

Figure 20-6. Results of preference experiments for mites presented with a choice between nectar of their usual host species and nectar of a "familiar alien host" (the host of another mite species that shares use of the same hummingbirds for transport). The two mite species above the horizontal line are both carried by the hummingbird *Glaucis hirsuta*, the two below by the hummingbird *A. tobaci*. Note that the same alternatives were presented to both members of each pair of mite species; a more powerful statistical test of these results is described in the text. See the caption for Figure 20-3 for information on figure conventions. Details of the results are as follows (the first plant listed is the usual host species): *Proctolaelaps contentiosus*: *Renealmia exaltata* (8), *Heliconia wagneriana* (2); *P. certator*: *Heliconia wagneriana* (5), *Renealmia exaltata* (2); *P. kirmsei*: *Hamelia patens* (9), *Cephaelis muscosa* (1); *Rhinoseius hirsutus*: *Cephaelis muscosa* (19), *Hamelia patens* (3); totals: host nectar (41), familiar alien host nectar (8).

chiefly by the hummingbird *Glaucis hirsuta*. *P. kirmsei* and *Rhinoseius hirsutus*, the second pair, frequently share transport on *Amazilia tobaci* (unpublished data). Within each pair, each mite species was offered nectar from the host plants of both species as alternatives, as detailed in the figure.

The results presented in Figure 20-6 can be analyzed in two different ways. Results of binomial tests for individual experiments and for all four experiments combined appear in the figure as for earlier experiments. Taken together, they show a consistent and significant preference for the nectar of each mite's usual host-plant species over a familiar alien host. The second approach (see Statistical Methods) sets up a 2×2 contingency table for each pair of mite species (sources of mites as rows, sources of nectar as columns). In the case of the first pair, the G-test for independence indicates a difference in preference significant at $P < 0.05$, whereas the second pair yields $P < 0.001$. Mites clearly prefer nectar of their own host plant over the nectar of familiar alien hosts.

Unfamiliar Alien Host Plants

In experiments carried out in Trinidad and Costa Rica, hummingbird flower mites were offered nectar from their own host and from unfamiliar alien hosts

(Fig. 20-7). In each case, the mite species tested is carried by different species of hummingbirds than those that carry the mite normally affiliated with the alien host nectar presented as an alternative (based on unpublished data on the occurrence of mites on birds). With one exception, these experiments revealed no significant preference between host and unfamiliar alien host. *Proctolaelaps jurgatus* in Trinidad, for example, whose principal hummingbird carrier is the short-billed *Chlorestes notatus*, expressed no clear preference between the nectar of its sole host species, the short-corolla *Isertia parviflora*, and the long-corolla alien host *Centropogon cornutus*. (*C. cornutus* is the host plant of the mite *P. glaucis*, whose principal carrier is the long-billed hummingbird *Glaucis hirsuta*.)

In the one exceptional case, *Proctolaelaps kirmsei* in Costa Rica did prefer its short-corolla host plant *Hamelia patens* over the long-corolla alien host *Aphelandra storkei*, the host plant of undescribed *Proctolaelaps* species "C," although the latter showed no significant preference given the same choices. (Consequently, a G-test on the contingency table for this pair of mites was also significant, $P < 0.02$.) Of the dozens of hummingbird-pollinated Acanthaceae we have examined, only *Aphelandra storkei* and *A. golfodulcensis* support humming-bird flower mites. It is tempting to suggest that *Proctolaelaps* species "C" from *A. storkei* has evolved some special means of coping with the nectar of its host,

HOST vs. UNFAMILIAR ALIEN HOST

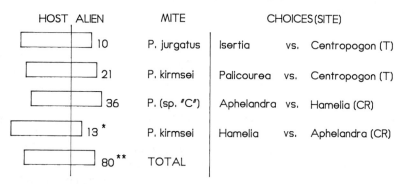

Figure 20-7. Results of preference experiments for mites presented with a choice between nectar of their usual host species and nectar of an "unfamiliar alien host" (the host of another mite species that rarely or never shares use of the same hummingbirds for transport). Note that some alternatives were presented to both members of the second pair of mite species; a more powerful statistical test for this pair is described in the text. See the caption for Figure 20-3 for information on figure conventions. Details of the results are as follows (the first plant listed is the usual host species): *Proctolaelaps jurgatus: Isertia parviflora* (7), *Centropogon cornutus* (3); *P. kirmsei: Palicourea crocea* (13), *Centropogon cornutus* (8); *P.* sp. "C" (undescribed): *Aphelandra storkei* (20), *Hamelia patens* (16); *P. kirmsei: Hamelia patens* (11), *Aphelandra storkei* (2); totals: host nectar (51), unfamiliar alien host nectar (29).

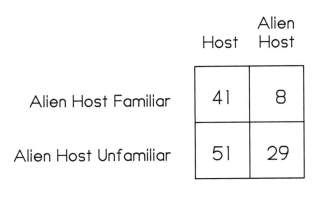

Figure 20-8. Contingency table (G-test) comparing the preference of hummingbird flower mites for their own host nectar vs. nectar of the host of another mite species in the same assemblage (alien host), when the alien host plant species is familiar vs. unfamiliar. The tendency to avoid alien hosts is much stronger when the alien host is familiar.

whereas *P. kirmsei* retains a general ancestral aversion to the nectar of Acanthaceae.

In general, however, hummingbird flower mites apparently do not differentiate between nectar of their own host plant and nectar of those host species not normally visited by their hummingbird carriers as accurately as they discriminate between more familiar choices. Figure 20-8 shows this distinction dramatically in the form of a contingency table, analyzed by the G-test.

Alternate Host Plants

Hummingbird flower mite species whose host plants flower reliably all year are commonly monophagous. Inhabitants of plant species that flower only part of the year must include in their repertoire one or more additional host species with complementary but partially overlapping flowering seasons. At least some individuals in these "sequential specialist" species (Colwell 1973, 1986a) must shift host affiliation seasonally between their asynchronously flowering hosts. To investigate the ability of these mites to recognize and to distinguish between alternate host species, we conducted the series of preference experiments presented in Figure 20-9.

In Trinidad, *Proctolaelaps kirmsei* is a sequential specialist, primarily dependent on two plant species with largely nonoverlapping flowering seasons—*Hamelia patens* and *Palicourea crocea* (both Rubiaceae)—to provide a year round resource base (Colwell 1986a). Preference experiments were carried out during the transition from dry to wet season (in March) when *Palicourea* completes its flowering season and *Hamelia* begins to bloom. The relatively few *Hamelia* flowers available were already heavily occupied by mites. Mites collected from

SEQUENTIAL SPECIALISTS: CURRENT vs. ALTERNATE HOST

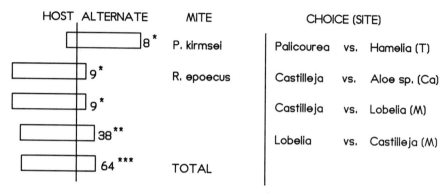

HOST ALTERNATE   MITE   CHOICE (SITE)

8 *   P. kirmsei   Palicourea   vs.   Hamelia (T)

9 *   R. epoecus   Castilleja   vs.   Aloe sp. (Ca)

9 *   Castilleja   vs.   Lobelia (M)

38 **   Lobelia   vs.   Castilleja (M)

64 *** TOTAL

Figure 20-9. Results of preference experiments for sequential specialist mites (which shift hosts seasonally) presented with a choice between nectar of their current host species and nectar of their alternate host species at a time when both are in flower. Note that same alternatives were presented to *R. epoecus* from each of its Mexican hosts in the last pair of experiments; a more powerful statistical test for this pair is described in the text. See the caption for Figure 20-3 for information on figure conventions. Details of the results are as follows (the first plant listed is the current host species): *Proctolaelaps kirmsei*: *Palicourea crocea* (1), *Hamelia patens* (7); *Rhinoseius epoecus*: *Castilleja foliolosa* (8), *Aloe* sp. (1); *R. epoecus*: *Castilleja integrifolia* (8), *Lobelia laxiflora* (1); *R. epoecus*: *Lobelia laxiflora* (29), *Castilleja integrifolia* (9); totals: current host nectar (46), alternate host nectar (18).

Palicourea showed a significant preference for *Hamelia* nectar over the nectar of their current host (Fig. 20-9). *Proctolaelaps kirmsei* may prefer *Hamelia patens* regardless of which host the mites matured in, which would result in an efficient transfer from *Palicourea* to *Hamelia* and a less efficient but nonetheless necessary switch in the opposite direction during the transition from wet to dry season. To establish whether mite preference for *Hamelia* remains stable once *Palicourea* resumes flowering, it will be necessary to carry out preference tests during this second transition period.

In California, *Rhinoseius epoecus* also relies on two asynchronously flowering host species (Colwell and Naeem 1979). This mite inhabits widespread native species of the genus *Castilleja* from early spring through late fall and may survive the winter in a few protected coastal areas in winter-flowering *C. latifolia*. During the winter, in urban gardens, the mites shift to species of *Aloe* and *Kniphofia*—introduced from Africa. [In Africa, plants of these two genera are pollinated by sunbirds (Nectariniidae) and occupied by sunbird flower mites, a distinct lineage of *Proctolaelaps*, which in turn inhabit flowers of cultivated *Hamelia* introduced from tropical America (unpublished data)]. In contrast with *Proctolaelaps kirmsei* in Trinidad, *R. epoecus* in California preferred nectar from its current host *Castilleja foliolosa* to the nectar of *Aloe* flowers, which were just coming into bloom (Fig. 20-9). Again, the reverse experiment has yet to be performed. It would perhaps not be surprising if the native host were preferred in all seasons.

In Mexico *R. epoecus* also utilizes two asynchronously flowering host species. Preference experiments were conducted in December during a three-month period when flowering in *Castilleja integrifolia* was declining and *Lobelia laxiflora* flowering was well underway. *Rhinoseius epoecus* collected from each host species were offered a choice between nectar from their current host and from the alternate host. *Rhinoseius epoecus* from both plants showed significant preferences for nectar of their current host species (Fig. 20-9). Results from this last pair of experiments can be cast in a contingency table to provide an even stronger test—sources of mites as rows, sources of nectar as columns. The G-test yields $P \ll 0.001$.

This last result implies a relatively inefficient transfer of mites between the two host species in both directions. Indeed, the newly-flowering *Lobelia*, though in full flower, was inhabited by relatively few mites. Field experiments in which mites are introduced into unoccupied alien hosts, coupled with preference experiments before and after introduction, have shown that a significant (though incomplete) shift preference may occur by exposure to a new host species in as little time as a single generation (unpublished data). Thus, once mites have moved to the new host, as its flowers become more abundant and those of the old host decline, relatively few reverse transfers may take place. We do not mean to suggest that specific preferences or seasonal shifts in preference have no evolved component, only that learning may consolidate host shifts in sequential specialists (Colwell 1986b).

Host Flowers with Mites vs. Virgin Host Flowers

We scrupulously avoided using nectar from flowers occupied by mites in previous experiments to prevent influencing experimental mites with possible chemical cues from former flower inhabitants. Some of these experiments (Fig. 20-3 and 20-4) showed clearly that mites are attracted to virgin nectar of their own host species. Is this attraction amplified or diminished by the prior presence of conspecifics? A final set of experiments examined this question.

In the experiments outlined in Figure 20-10, we presented mites with a choice between the nectar of previously inhabited and previously uninhabited (virgin) flowers from their own host species. Experiments with two mite species at three localities indicate a consistent and significant preference for nectar from flowers that have not been occupied.

In uncrowded artificial habitats, hummingbird flower mites are highly gregarious (unpublished data of Colwell and Naeem) and in many cases the distribution of mites among individual flowers on a plant in nature shows significant aggregation (Fig. 4 in Colwell 1973; Dobkin, unpublished data). Certainly males, and probably females as well, must mate many times to realize their maximum lifetime reproductive output (Colwell 1986b), suggesting an advantage to aggregation. Within occupied inflorescences, however, density-dependent population growth can be demonstrated (unpublished data of Colwell and Naeem for *Proctolaelaps kirmsei* on *Hamelia patens*)—suggesting an

NECTAR FROM HOST FLOWERS WITH MITES
vs. VIRGIN HOST NECTAR

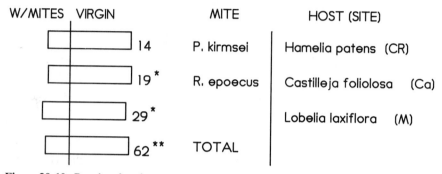

W/MITES	VIRGIN		MITE	HOST (SITE)
		14	P. kirmsei	Hamelia patens (CR)
		19*	R. epoecus	Castilleja foliolosa (Ca)
		29*		Lobelia laxiflora (M)
		62**	TOTAL	

Figure 20-10. Results of preference experiments for mites presented with a choice between (1) nectar of their current host species taken from flowers occupied by conspecific mites and (2) virgin nectar from the current host species. See the caption for Figure 20-3 for information on figure conventions. Details of the results are as follows: *Proctolaelaps kirmsei* from *Hamelia patens*: with mites (4), virgin (10); *Rhinoseius epoecus* from *Castilleja foliolosa*: with mites (5), virgin (14); *R. epoecus* from *Lobelia laxiflora*: with mites (9), virgin (20); totals: with mites (18), virgin (44).

advantage to seeking unoccupied inflorescences, particularly for gravid females. Indeed, females are more likely to disperse than males, in most species of hummingbird flower mites (Wilson and Colwell 1981), and most of the mites involved in preference experiments were gravid females (as indicated by the frequency of egg-laying in preference chambers). Too few males were used in the experiments of Figure 20-10 to assess any sexual difference in preference for virgin vs. occupied nectar. It would be worthwhile to repeat these experiments with that objective.

DISCUSSION

The experiments summarized in Figure 20-11 demonstrate that hummingbird flower mites express preferences on the basis of nectar alone that are consistent with their host affiliations in the field. We do not pretend that the conditions of the preference chambers (a simultaneous choice between alternatives in enclosed, still air) reproduce the conditions under which actual choices must be made in nature (alternatives sequentially presented in the rapidly moving air of a hummingbird's respiratory passages as it moves from plant to plant). Moreover, the design of the test chamber does not permit the clear separation of olfactory from gustatory cues. Nonetheless, the consistency of the experimental results with field distributions argues for a key role for nectar chemistry in host preference and

SUMMARY OF PREFERENCE EXPERIMENTS

Miteless plants
    Host   ◄◄◄◄   Miteless species
    Host   ◄◄◄◄ ►   Sugar solution
    Miteless species   ◄◄ ►►►►   Sugar solution

Alien host plants
    Host   ◄◄◄◄   Familiar alien
    Host   ◄◄◄ =   Unfamiliar alien

Alternate host plants
    Current host   ◄◄◄◄   Alternate host

Occupied host plants
    Virgin host   ◄◄◄   Occupied host

◄ $P < .05$ ◄ $.05 < P < .25$ $= P > .25$

Figure 20-11. Summary of results from Figures 20-3 through 20-10, for experiments on the nectar preferences of hummingbird flower mites. The arrowheads indicate the direction and significance (see figure footnote) for the results of each of one of the separate experiments represented by data bars in Figures 20-3–20-10.

discrimination by hummingbird flower mites. In the following sections, we attempt to infer the causes of host fidelity and patterns of host affiliation in light of these experiments, under the assumption that they adequately represent patterns of host discrimination in nature.

Attraction and Avoidance

Attraction to host nectar appears to play a primary role in host-plant discrimination by hummingbird flower mites. Our experiments show that mites significantly prefer their own host nectar to the nectar of miteless species (Fig. 20-3) as well as to the nectar of alien host species (Fig. 20-6 and 22-7). But they also prefer their own host nectar to sugar solutions (Fig. 20-4). Because sugar solutions are free of potentially repellent, noxious, or toxic compounds, preference expressed for host nectar in the experiments with sugar as an alternative reveal the attractive nature of host nectar.

    We studied two kinds of "nonhosts" (from the point of view of a focal species of hummingbird flower mite)—miteless species and alien hosts. The tendency to prefer sugar solution to miteless nectar suggests active avoidance of miteless nectar (Fig. 20-5). Likewise, the significantly greater preference for host nectar over alien nectar when the latter comes from a familiar alien host plant, rather than from an unfamiliar one (Fig. 20-8, based on Fig. 20-6 and 20-7), suggests that alien nectars tend to be avoided if they are "familiar." Tests with the same mite species and alien hosts used in Fig. 20-6 and 20-7 versus matched sugar solutions would be useful to confirm this tendency toward avoidance of familiar hosts, and to assess whether unfamiliar hosts also provoke some degree of avoidance. Moreover, the biological meaning of "familiar" is still obscure. Is individual experience required for the specific recognition of those alien species most frequently encountered, is this recognition a genetically programmed response, or are both genes and experience involved (Jaenike 1982; Papaj and Rausher 1983; Menzel 1985; Papaj and Prokopy 1986)? Regardless of the mechanism, this result implies that the rare mite that mounts the bill of an atypical hummingbird visitor to its host plant (for example, a long-billed bird visiting a short-corolla host plant) may be a likely candidate for chance colonization of an unfamiliar alien host plant. If that plant happens to be suitable, but unoccupied by any other species of hummingbird flower mite, the conditions may be right for genetic isolation and perhaps speciation, if the event is rare enough (Bush and Diehl 1982; Zwölfer and Bush 1984; Colwell 1986a, 1986b; Feder et al. 1988).

Avoidance of Miteless Species as a Response to Repellent, Noxious, or Toxic Substances

The widespread phenomenon of "miteless" hummingbird-pollinated plants suggests that repellence may be an adaptive response by these plant species to the consumption of nectar by nonpollinators. Baker and Baker (1975, 1978), Baker (1978), Janzen (1977), and Rhoades and Bergdahl (1981) have suggested that flower nectar may contain substances that deter nonpollinating organisms. The nectar of flowers representing many pollination systems contain potential allelochemicals, including phenolic compounds, alkaloids, and nonprotein amino acids (Baker and Baker 1975; Baker 1977, 1978).

In hummingbird-pollinated flowers, such potentially repellent, noxious, or toxic substances presumably would be aimed at deterring arthropods, possibly including hummingbird flower mites, but would have to be relatively innocuous to the vertebrate pollinators—as suggested by Feinsinger and Swarm (1978). To test the hypothesis that such compounds protect nectar from nonpollinators, several sets of preference experiments have been conducted in which flower nectar was offered to ants in a diversity of habitats in Costa Rica (Haber et al. 1981; Guerrant and Fiedler 1981), in Trinidad (Feinsinger and Swarm 1978), and in Brazil (Schubart and Anderson 1978). Some 40 plant species were tested by these authors, including flowers pollinated by large bees, small bees, butterflies, small moths, sphingid moths, hummingbirds, bats, and generalist pollinators. Nectars accepted by ants included several from host plants of hummingbird flower mites (e.g., *Costus spiralis*, *Hamelia patens*, *Heliconia psittacorum*, *H. wagneriana*, *Renealmia exaltata*), as well as nectars from miteless species (e.g., *Aphelandra pilosa* and two species of *Justicia*—both are genera of Acanthaceae).

Overall, these ant preference experiments failed to demonstrate the presence of effectively repellent substances in flower nectars. Guerrant and Fiedler (1981) conclude "... no reliable evidence shows that potentially repellent compounds in floral nectar serve to defend the nectar," and Haber et al. (1981) state that "... the natural occurrence of potentially toxic compounds in floral nectars often does not inhibit nectar consumption by ants." But ants are only one group among the many nectar thieves that plants might find it advantageous to deter (Baker and Baker 1975, 1978). That ants appear to consume test nectars relatively indiscriminately does not necessarily mean that phenolics, alkaloids, and nonprotein amino acids are not effective protection against other arthropods. In the case of miteless hummingbird-pollinated flowers, for example, Heyneman (1985) showed that miteless species have fewer thrips and other flower-inhabiting arthropods than host species of hummingbird flower mites.

Hummingbird flower mites may well be more physiologically sensitive than ants to deterrents in nectar. In addition, because flower nectar generally constitutes a small proportion of ant diets (Carroll and Janzen 1973), the small amounts of associated noxious or toxic substances imbibed may be tolerated (Freeland and Janzen 1974). In contrast, an individual hummingbird flower mite, which feeds exclusively on nectar from one, or at most two or three plant species in its lifetime, cannot minimize its intake of potentially noxious or toxic compounds by mixing its diet.

In detailed chemical analyses of floral nectars from plant species used (hosts) and avoided (miteless species) by hummingbird flower mites, Heyneman (1985) examined all potential deterrents commonly found in hummingbird-pollinated flowers (Baker and Baker 1975; Baker 1978), including alkaloids, phenolic compounds, and nonprotein amino acids. A statistical comparison of host and miteless nectars indicated significant and consistent differences in the phenolic compounds present. Studies of spider mites feeding on strawberry foliage and apple leaves have shown that certain phenolic compounds deter mites (Dabrowski and Rodriguez 1972) and decrease their fecundity (Dabrowski and Bielak 1978). These findings suggest that phenolics may play a role in the avoidance of miteless species by hummingbird flower mites both in the field and in the preference experiments we have presented here.

Volatile compounds in the corolla or other plant tissues, either in addition to or independent of deterrents in the nectar itself (Kerner von Marilaun 1878; Guerrant and Fiedler 1981), may also repel hummingbird flower mites from miteless plant species in nature. Mints (Labiatae), for example, known for their volatile oils, are commonly available to hummingbird flower mites, but are consistently avoided by them (Heyneman 1985). Van der Pijl (1955) demonstrated the presence of such compounds in several Indonesian *Myrmecodia* (Rubiaceae) species pollinated by carpenter bees. He found that flower petals placed in ant pathways were avoided, while leaves placed as controls were not. To test for deterrent compounds in flower petals, including ornithophilous species, Guerrant and Fiedler (1981) offered sugar solutions with and without macerated corolla tissue to ants. Petal tissue from 9 of 17 species tested significantly reduced palatability, 2 marginally reduced it, and 6 did not reduce palatability significantly. Corolla tissue of two hosts of hummigbird flower mites, *Hamelia patens* and *Erythrina fusca*, proved to be significantly deterrent to ants. Other mite host species (*Heliconia wagneriana* and *Calathia lutea*) and miteless species (*Stachytarpheta jamaicensis*) were in the group with corolla tissue not deterrent to ants. In short, floral tissue as well as nectar may be involved in deterring mites from using the flowers of some hummingbird-pollinated plants, but further study is needed.

Nutritional Factors in Host Affiliation

Because the qualitative nutritional requirements of most insects have been shown to be fairly similar (House 1965), and because the approximately 25 nutrients they require are generally omnipresent in plant tissues (Fraenkel 1953), some have argued that nutritional differences among plants are not likely causes for the host specificity of phytophagous insects (Beck and Reese 1976; Fraenkel 1959, 1969; Futuyma 1983; Bernays and Graham 1988). Similarly, floral nectars from plants with a common pollinator type do not differ substantially in their primary nutritional elements (Baker and Baker 1975, 1982; Baker 1977; Heyneman 1985).

In contrast with the remarkable uniformity of their qualitative nutritional requirements, however, insects show considerable diversity in their quantitative requirements—that is, the proportions of essential nutrients required (Auclair 1969; House 1969; Hsiao 1972). For example, varying the ratios of a few constituents in nutritionally complete synthetic diets reduced the efficiency of food conversion by phytophagous larval *Celerio euphorbiae* to a greater degree than uniformly decreasing the concentrations of all nutrients present (House 1969). Thus, relatively slight imbalances in the ratios of nutrients such as amino acids can greatly alter the quality of insect diets (House 1969; Auclair 1969; Beck and Reese 1976).

In addition, some insects can assess the relative nutritional quality of diets offered (e.g., Auclair 1965, 1967, 1969; House 1967, 1969; Hsiao 1969, 1972; Schoonhoven 1969, 1972; Miller and Strickler 1984; Thompson 1988). For example, aphids (Auclair 1965, 1967, 1969) offered diets varying in concentrations of nutritional constituents such as sucrose, free amino acids, and pH values, selected diets on which their rates of growth and reproduction were highest. Small variations in the balance of nutrients can be deleterious to mites as well as to insects (Slansky and Rodriguez 1987). Nutritional differences among host plants (Dabrowski and Bielak 1978; Gerson and Aronowitz 1980) and among artificial diets (Van der Geest et al. 1983) affected mite feeding efficiency, fecundity, and longevity. On the other hand, numerous studies have found little or no correlation between host preference and subsequent growth and survival of insects, corrected for consumption rate (Wiklund 1975; Miller and Strickler 1984; Futuyma and Moreno 1988; Thompson 1988).

Nevertheless, nutritional differences may provide an adaptive explanation for the avoidance of miteless species by hummingbird flower mites. Indeed, imbalances in the ratios of free amino acids optimal to mites could effectively eliminate them from particular plant species without affecting the nutritional quality of the nectar for hummingbird pollinators (or ants), which obtain essentially all nonsugar nutrients from arthropod prey. Significant differences do in fact exist in several nutritional constituents, including ratios of essential amino acids, between host and miteless plant species (Heyneman 1985).

In addition to the avoidance of miteless species, patterns of affiliation of mite species with acceptable host-plant species could also involve nutritional factors. We have shown clearly that mites discriminate among host species, preferring their own host to the host of other mites (Fig. 20-6 and 20-7). Although this pattern has several possible causes (Colwell 1986a and below), in principle each host species might differ critically in the balance of nutritional factors, and each mite species might be specially adapted to the nectar of its host—unable to survive, or at least suffer decreased reproductive fitness, on the nectar of alien hosts. While difficult to test rigorously, strong evidence against this hypothesis arises from the success of experimental "transplants" of mites into (unoccupied) alien hosts (Colwell 1973, 1986a) and from the broad taxonomic scope of the host repertoire of polyphagous and sequential specialist hummingbird flower mites (Colwell 1986a). Nutritional differences may play some role, but cannot alone account for the rigid restriction and fidelity of mites to their hosts.

Ecological and Behavioral Factors in Host Affiliation

Recent studies and reviews of plant-arthropod interactions have revealed the broad significance of ecological and behavioral influences on host-plant affiliation and specialization in insects and mites (Levins and MacArthur 1969; Gilbert 1975; Gilbert and Smiley 1978; Rowell 1978; Smiley 1978; Fox and Morrow 1981; Futuyma and Moreno 1988; Futuyma 1990). Ecological constraints and behavioral factors undoubtedly play a major role in structuring hummingbird flower mite communities as well.

Miteless plant species appear morphologically and phenologically well-suited to mite habitation. Ecological factors influencing avoidance by mites, however, cannot be ruled out—although they remain obscure. Likewise, the avoidance of alien hosts by hummingbird flower mites may be adaptive for ecological or behavioral reasons, rather than strictly physiological factors. Many studies show that insects consistently avoid some edible potential host plants (Singer 1971; Gilbert and Singer 1975; Wiklund 1975; Smiley 1978; Miller and Strickler 1984; Futuyma and Moreno 1988; Thompson 1988), and ecological causes have sometimes been established. For tropical *Heliconius* butterflies, for example, finding and utilizing host plants effectively, while minimizing predation, may require a high degree of host specialization (Gilbert and Smiley 1978).

Similarly, effective use of host-plant species by hummingbird flower mites sometimes requires special behavioral and morphological adaptations (Colwell 1986a). *Proctolaelaps certator*, for example, must be able to walk on water to traverse the water-filled bracts that surround its *Heliconia wagneriana* host flowers. Other species of hummingbird flower mite would drown (and do, in field tests). *Rhinoseius trinitatis* must lay eggs outside its *Heliconia trinidatis* host flowers and leave the flowers before they fall off the plant at a precise time each day (Dobkin 1987 and unpublished data). Certain other hummingbird flower mites would remain in the flowers with their eggs, and be lost. Clearly, special adaptations, or lack of them, for coping with the morphological and phenological idiosyncrasies of particular host plants limit host range. These special adaptations, however, are more likely to be the consequence, rather than the cause of host specialization (see Bernays and Graham 1988; Futuyma 1989).

Several lines of evidence point to a minor role, at most, for interspecific competition as an adaptive explanation for the avoidance of alien hosts by hummingbird flower mites (Colwell 1986a), although in some cases aggression from a resident species may deter colonization by another (Colwell 1973).

Because hummingbird flower mites mate in their host flowers, and both sexes mate many times (Colwell 1986b), selection arising from differential mating success alone (a form of sexual selection) may narrow or focus host range and intensify host fidelity (Colwell 1986a, b). In other words, individual mites that stray from their usual host species will leave fewer descendants than those that stick with the usual host. Futuyma and Moreno (1988) refer to this "sexual rendezvous" principle for host fidelity as an example of selection for "specialization *per se*," which appears to be important in other species of arthropods that mate on their host plants (Zwölfer 1974, 1984; Gilbert 1979; Rosenzweig 1979; Zwölfer and Bush 1984).

In the case of hummingbird flower mites, the limitation of host-sharing to *non-congeners* provides indirect support for the sexual rendezvous hypothesis (Colwell 1986a). Closely related species (congeners) that share a host are more likely to waste time, energy, or gametes on mistaken matings than are more distantly related species (non-congeners). Selection against host-sharing should thus be stronger for congeners than for non-congeners if selection for host fidelity arises through differential mating success. [Zwölfer (1974) has conclusively demonstrated the critical role of host plants in mate-finding and mate recognition for trypetid flies of thistles, which likewise coexist only in non-congeneric assemblages]. In contrast, if host-sharing were based on special adaptation to noxious, toxic, nutritional, morphological, or phenological features of the host, congeners (which are likely to share adaptations by common descent) would be more likely than non-congeners to share hosts, assuming interspecific competition is weak.

Our discovery, in this study, of a difference in degree of avoidance of familiar vs. unfamiliar alien hosts (Figs. 28-6–20-8) is also consistent with a role for mating success in the host fidelity of hummingbird flower mites. The intensity of selection against mistaken host choices, arising from differential mating success, should scale with the frequency of mistaken choices, which will be greater for alien hosts often encountered than for those rarely encountered.

SUMMARY

Hummingbird flower mites breed and feed in hummingbird-pollinated flowers and travel between plants on the birds. Although several mite species often ride together on the same bird, flower censuses show that they disembark only at their own host-plant species, and that congeneric mite species never share hosts. In addition, all hummingbird flower mites avoid a subset of the plant species made available to them by their hummingbird carriers, even though the flowers of these "miteless" plant species offer conditions and resources seemingly suitable for hummingbird flower mites.

To investigate the behavioral mechanisms and to illuminate the causes of patterns of host affiliation and host fidelity in these mites, we conducted a series of preference experiments. Individual mites were given a choice of nectars (or sugar solution controls) in closed chambers, and were scored for their preferences according to their feeding and oviposition behavior in the chambers.

The results of these experiments show that mites discriminate among the nectars of different plant species primarily through attraction to the nectar of their usual host, but also by active avoidance of the nectar of hosts of other hummingbird flower mites ("alien" hosts) and the nectar of miteless species. Alien hosts frequently encountered by a particular mite species while travelling on hummingbirds ("familiar" alien hosts), are avoided more effectively than those that are less frequently encountered ("unfamiliar" alien hosts). This tendency to accept unfamiliar hosts may provide occasional opportunities to colonize unoccupied but suitable host species, and may thus play a role in speciation of these mites.

Mite species that shift hosts seasonally generally prefer nectar of their current host to nectar of their alternate host. The prior presence of conspecific mites appears to diminish the attractiveness of host nectar, rather than enhancing it.

Miteless plant species may produce repellent, noxious, or toxic substances in their nectar, or may be nutritionally inadequate for hummingbird flower mites— as well as for flower-feeding insects, which also tend to avoid these plant species. Indeed, significant differences in nutrients and in potentially repellent phenolic compounds do exist between the nectar of miteless and host-plant species. In contrast, toxic substances and nutritional differences are not as likely to be important in the avoidance of alien hosts.

In nature, the adaptive value to mites of discrimination among hummingbird-pollinated plants is probably complex, perhaps involving the nutritional adequacy of nectar, the presence of toxic substances in some nectars, and special adaptations to the phenology or morphology of hosts. Several lines of evidence suggest that host fidelity is amplified and patterns of host affiliation are influenced by selection for the use of host plants as cues for finding mates.

ACKNOWLEDGMENTS

This paper is based on a chapter from a doctoral dissertation by A. J. Heyneman (1985), who took the lead in planning, executing, and analyzing the results of the experiments, made the important discovery that mites react differently to familiar and unfamiliar host nectar, did the chemical analysis of nectar, and drafted much of the text that appears here. Colwell provided grant support, prepared this paper, devised the data figures—and wrote the Acknowledgments. Colwell and Dobkin participated in planning and carrying out the field work and many of the experiments. Naeem designed the preference chambers, worked out the alpha taxonomy of the Trinidad mites, drew the figures, and participated extensively in planning and experimentation. Hallet provided able assistance with many of the experiments in Mexico and California and helpful comments on data analysis and on the manuscript. We are greatful to Carol Baird and Lloyd Goldwasser for their help with some of the experiments, to Herbert and Irene Baker for advice and assistance with nectar chemistry, and to Peter Feinsinger for providing mite specimens from Trinidad hummingbirds. We thank Herbert Baker, Beth Braker, Roy Caldwell, and Douglas Futuyma for critical reading of the manuscript, and participants in the Campinas Symposium for their useful comments. Our work was supported by the U.S. National Science Foundation (DEB78–12038 and BSR86–04929).

REFERENCES

Auclair, J. L. 1965. Feeding and nutrition of the pea aphid, *Acyrthosiphon pisum* (Homoptera: Aphidae), on chemically defined diets of various pH and nutrient levels. *Annals of the Entomological Society of America* **58**: 855–875.

Auclair, J. L. 1967. Effects of pH and sucrose on rearing the cotton aphid, *Aphis gossypii*, on a germ-free and holidic diet. *Journal of Insect Physiology* **13**: 431–446.

Auclair, J. L. 1969. Nutrition of plant-sucking insects on chemically defined diets. *Entomologia Experimentalis et Applicata* **12**: 623–641.

Baker, H. G. 1975. Sugar concentrations in nectars from hummingbird flowers. *Biotropica* **7**: 37–41.

Baker, H. G. 1977. Non-sugar chemical constituents of nectar. *Apidologie* **8**: 349–356.

Baker, H. G. 1978. Chemical aspects of the pollination biology of woody plants in the tropics. Pages 57–82 in P. B. Tomlinson and M. H. Zimmerman (Eds.). *Tropical Trees as Living Systems*. Cambridge University Press, New York.

Baker, H. G. and I. Baker. 1975. Studies of nectar constitution and pollinator–plant coevolution. Pages 100–140 in L. E. Gilbert and P. H. Raven (Eds.). *Coevolution of Animals and Plants*. University of Texas Press, Austin, Texas.

Baker, H. G. and I. Baker. 1978. Ants and flowers. *Biotropica* **10**: 80.

Baker, H. G. and I. Baker. 1982. Chemical constituents of nectar in relation to pollination mechanisms and phylogeny. Pages 131–171 in M. H. Nitecki (Ed.). *Biochemical Aspects of Evolutionary Biology*. University of Chicago Press, Chicago, IL.

Baker, H. G. and I. Baker. 1983. Floral nectar sugar constituents in relation to pollinator type. Pages 117–141 in C. E. Jones and R. J. Little (Eds.). *Handbook of experimental pollination biology*. Van Nostrand-Reinhold, New York.

Beck, S. D. and J. C. Reese. 1976. Insect–plant interactions: Nutrition and metabolism. *Recent Advances in Phytochemistry* **10**: 41–92.

Bernays, E. A. and M. Graham. 1988. On the evolution of host specificity in phytophagous arthropods. *Ecology* **69**: 886–892.

Bolten, A. B., P. Feinsinger, H. G. Baker, and I. Baker. 1979. On the calculation of sugar concentration in flower nectar. *Oecologia* **41**: 301–304.

Bush, G. L. and S. R. Diehl. 1982. Genetic models of sympatric speciation and the origin of parasitic insect species. Pages 297–306 in J. H. Visser and A. K. Minks (Eds.). *Fifth international symposium on insect–plant relationships*. PUDOC, Wageningen.

Carroll, C. R. and D. H. Janzen. 1973. Ecology of foraging by ants. *Annual Review of Ecology and Systematics* **4**: 231–258.

Colwell R. K. 1973. Competition and coexistence in a simple tropical community. *American Naturalist* **107**:737–760.

Colwell, R. K. 1979. The geographical ecology of hummingbird flower mites in relation to their host plants and carriers. *Recent Advances in Acarology* **2**: 461–468.

Colwell, R. K. 1981. Group selection is implicated in the evolution of female-biased sex ratios. *Nature* **290**: 401–404.

Colwell, R. K. 1983. *Rhinoseius colwelli* (Acaro floral del colibri, totolate floral de colibri, hummingbird flower mite). Page 619 (fig.) and pages 767–768 in D. H. Janzen (Ed.). *Costa rican natural history*. University of Chicago Press, Chicago, IL.

Colwell, R. K. 1985. Stowaways on the hummingbird express. *Natural History* **94**: 56–63.

Colwell, R. K. 1986a. Community biology and sexual selection: Lessons from hummingbird flower mites. Pages 406–424 in T. J. Case and J. Diamond (Eds.). *Ecological communities*. Harper and Row, New York.

Colwell, R. K. 1986b. Population structure and sexual selection for host fidelity in the speciation of hummingbird flower mites. Pages 475–495 in S. Karlin and E. Nevo (Eds.). *Evolutionary processes and theory*. Academic Press, New York.

Colwell, R. K. and S. Naeem. 1979. The first known species of hummingbird flower mite north of Mexico: *Rhinoseius epoecus* n. sp. (Mesostigmata: Ascidae). *Annals of the Entomological Society of America* **72**: 485–491.

Dabrowski, Z. T. and B. Bielak. 1978. Effect of some plant chemical compounds on the behavior and reproduction of spider mites (Acarina: Tetranychidae). *Entomology Experimental and Applied* **24**: 317–326.

Dabrowski, Z. T. and J. G. Rodriguez. 1972. Gustatory response of *Tetranychus urticae* Koch to phenolic compounds of strawberry foliage. *Zesz. probl. Postep. Nauk roln.* **127**: 69–78.

Des Granges, J.-L. 1977. Interactions among resident and migrant hummingbirds in Mexico. Ph.D. thesis, Dept. of Biology, McGill University, Montreal.

Des Granges, J.-L. 1978. Organization of a tropical nectar feeding bird guild in a variable environment. *Living Bird* **17**: 199–236.

Dobkin, D. S. 1984. Flowering patterns of long-lived *Heliconia* inflorescences: Implications for visiting and resident nectarivores. *Oecologia* **64**: 245–254.

Dobkin, D. S. 1985. Heterogeneity of tropical floral microclimates and the response of hummingbird flower mites. *Ecology* **66**: 536–543.

Dobkin, D. S. 1987. Synchronous flower abscission in plants pollinated by hermit hummingbirds and the evolution of one-day flowers. *Biotropica* **19**: 90–93.

Egan, M. E. 1975. Small animal preference testing: Design and technique. *Annals of the Entomological Society of America* **68**: 386–387.

Egan, M. E. 1976. The chemosensory bases of host discrimination in a parasitic mite. *Journal of Comparative Physiology* **109**: 69–89.

Egan, M. E., R. H. Barth, and F. E. Hanson. 1975. Chemically-mediated host selection in a parasitic mite. *Nature* **257**: 788–790.

Feder, J. L., C. A. Chilcote, and G. L. Bush. 1988. Genetic differentiation between host races of the apple maggot fly *Rhagoletis pomonella*. *Nature* **336**: 61–64.

Feinsinger, P. and R. K. Colwell. 1978. Community organization among neotropical nectar-feeding birds. *American Zoologist* **18**: 779–795.

Feinsinger, P. and L. A. Swarm. 1978. How common are ant-repellent nectars? *Biotropica* **10**: 238–239.

Feinsinger, P. and L. A. Swarm. 1982. "Ecological release," seasonal variation in food supply, and the hummingbird *Amazilia tobaci* on Trinidad and Tobago. *Ecology* **63**: 1574–1587.

Feinsinger, P., L. A. Swarm, and J. A. Wolfe. 1985. Nectar-feeding birds on Trinidad and Tobago: Comparison of diverse and depauperate guilds. *Ecological Monographs* **55**: 1–28.

Feinsinger, P., J. A. Wolfe, and L. A. Swarm. 1982. Island ecology: Reduced hummingbird diversity and the pollination biology of plants, Trinidad and Tobago, West Indies. *Ecology* **63**: 494–506.

Fox, L. R. and P. A. Morrow. 1981. Specialization: Species property or local phenomenon? *Science* **211**: 887–893.

Fraenkel, G. 1953. The nutritional value of green plants for insects. *Transactions of the IXth International Congress of Entomology, Amsterdam 1951*, **volume 2**: 90–100.

Fraenkel, G. 1959. The raison d'etre of secondary plant substances. *Science* **121**: 1466–1470.

Fraenkel, G. 1969. Evaluation of our thoughts on secondary plant substances. *Entomologia Experimentalis et Applicata* **12**: 473–486.

Freeland, W. J. and D. H. Janzen. 1974. Strategies in herbivory by mammals: The role of plant secondary compounds. *American Naturalist* **108**: 269–289.

Futuyma, D. J. 1983. Evolutionary interactions among herbivorous insects and plants. Pages 207–231 in D. J. Futuyma and M. Slatkin (Eds.). *Coevolution*. Sinauer Associates, Sunderland, MA.

Futuyma, D. J. 1990. Evolution of host specificity in herbivorous insects: Genetic, ecological, and phylogenetic aspects. Pp. 431–453, this volume.

Futuyma, D. J. and G. Moreno. 1988. The evolution of ecological specialization. *Annual Review of Ecology and Systematics* **19**: 207–233.

Gerson, U. and A. Aronowitz. 1980. Feeding of the carmine spider mite on seven host plant species. *Entomologia Experimentalis et Applicata* **28**: 109–115.

Gilbert, L. E. 1975. Ecological consequences of a coevolved mutualism between butterflies and plants. Pages 210–240 in L. E. Gilbert and P. H. Raven (Eds.). *Coevolution of Animals and Plants*. University of Texas Press, Austin, TX.

Gilbert, L. E. 1979. Development of theory in the analysis of insect–plant interactions. Pages 117–154 in D. Horn, R. Mitchell, and G. Stairs (Eds.). *Analysis of Ecological Systems*. Ohio State University Press, Columbus, OH.

Gilbert, L. E. and M. C. Singer. 1975. Butterfly ecology. *Annual Review of Ecology and Systematics* **6**: 365–397.

Gilbert, L. E. and J. H. Smiley. 1978. Determinants of local diversity in phytophagous insects: Host specialists in tropical environments. Pages 89–105 in L. A. Mound and N. Waloff (Eds.) *Diversity of insect faunas*. Royal Entomological Society of London, Symposium 9.

Guerrant, E. O. and P. L. Fiedler. 1981. Flower defenses against nectar-pilferage by ants. *Biotropica 13* (Reproductive Botany supplement): 25–33.

Haber, W. A., G. W. Frankie, H. G. Baker, I. Baker, and S. Koptur. 1981. Ants like flower nectar. *Biotropica* **13**: 211–214.

Heyneman, A. J. 1985. Selective use of floral nectars by hummingbirds and hummingbird flower mites. Ph.D. Dissertation, University of California, Berkeley.

House, H. L. 1965. Insect nutrition. Pages 769–813 in M. Rockstein (Ed.). *The physiology of insecta*, Vol. 2. Academic Press, New York.

House, H. L. 1967. The role of nutritional factors in food selection and preference as related to larval nutrition of an insect, *Pseudosarcophaga affinis* (Diptera: Sarcophagidae), on synthetic diets. *Canadian Entomologist* **99**: 1310–1321.

House, H. L. 1969. Effects of different proportions of nutrients on insects. *Entomologia Experimentalis et Applicata* **12**: 651–669.

Hsiao, T. H. 1969. Chemical basis of host selection and plant resistance in oligophagous insects. *Entomologia Experimentalis et Applicata* **12**: 777–788.

Hsiao, T. H. 1972. Chemical feeding requirements of oligophagous insects. Pages 225–240 in J. G. Rodriguez (Ed.). *Insect and mite nutrition.* North Holland Publishing Company, Amsterdam.

Jaenike, J. 1982. Environmental modification of oviposition behavior in *Drosophila. American Naturalist* **119**: 784–802.

Janzen, D. H. 1977. Why don't ants visit flowers? *Biotropica* **9**: 252.

Kerner von Marilaun, A. 1878. *Flowers and Their Unbidden Guests.* Translated into English and edited by W. Ogle. C. Kegan Paul and Company, London.

Krantz, G. W. 1978. *A Manual of Acarology.* Oregon State University Book Stores, Inc., Corvallis, OR.

Levins, R. and R. MacArthur. 1969. An hypothesis to explain the incidence of monophagy. *Ecology* **50**: 910–911.

Lewis, A. C. and H. F. van Emden. 1986. Assays for insect feeding. Pages 95–120 in J. R. Miller and T. A. Miller (Eds.). *Insect–plant interactions.* Springer-Verlag, New York.

McMurtry, J. A. and J. G. Rodriguez. 1987. Nutritional ecology of phytoseid mites. Pages 609–644 in F. Slansky, Jr. and J. R. Rodriguez (Eds.). *Nutritional ecology of insects, mites, spiders, and related invertebrates.* John Wiley and Sons, New York.

Menzel, R. 1985. Learning in honey bees in an ecological and behavioral context. Pages 55–74 in B. Holldobler and M. Lindauer (Eds.). *Experimental behavioral ecology and sociobiology.* Sinauer Associates, Sunderland, MA.

Miller, J. R. and K. L. Strickler. 1984. Finding and accepting host plants. Pages 127–157 in W. J. Bell and R. Cardé (Eds.). *Chemical ecology of insects.* Sinauer Associates, Sunderland, MA.

Mulkern, G. B. 1969. Behavioral influences on food selection in grasshoppers (Orthoptera: Acrididae). *Entomologia Experimentalis et Applicata* **12**: 509–523.

Papaj, D. R. and R. J. Prokopy. 1986. Phytochemical basis of learning in *Rhagoletis pomonella* and other herbivorous insects. *Journal of Chemical Ecology* **12**: 1125–1143.

Papaj, D. R. and M. D. Rausher. 1983. Individual variation in host location by phytophagous insects. Pp. 77–124 in S. Ahmad (Ed.). *Herbivorous insects: Host-seeking behavior and mechanisms.* Academic Press, New York.

Rhoades, D. F. and J. C. Bergdahl. 1981. Adaptive significance of toxic nectar. *American Naturalist* **117**: 798–803.

Rodriguez, J. G. and L. D. Rodriguez. 1987. Nutritional ecology of phytophagous mites. Pages 177–208 in F. Slansky, Jr. and J. R. Rodriguez (Eds.). *Nutritional ecology of insects, mites, spiders, and related invertebrates.* John Wiley and Sons, New York.

Rosenzweig, M. L. 1979. Three probable evolutionary causes for habitat selection. Pages 49–60 in G. P. Patil and M. L. Rosenzweig (Eds.). *Contemporary quantitative ecology and related ecometrics.* International Cooperative Publishing House, Fairland, MD.

Rowell, C. H. F. 1978. Food plant specificity in neotropical rain-forest acridids. *Entomologia Experimentalis et Applicata* **24**: 451–462.

Saxena, K. N. 1978. Role of certain environmental factors in determining the efficiency of host plant selection by an insect. *Entomologia Experimentalis et Applicata* **24**: 666–678.

Schoonhoven, L. M. 1969. Gustation and foodplant selection in some lepidopterous larvae. *Entomologia Experimentalis et Applicata* **12**: 555–564.

Schoonhoven, L. M. 1972. Some aspects of host selection and feeding in phytophagous insects. Pages 557–566 in J. G. Rodriguez (Ed.). *Insect and mite nutrition.* North Holland Publishing Company, Amsterdam.

Schubart, H. O. R. and A. B. Anderson. 1978. Why don't ants visit flowers? A reply to D. H. Janzen. *Biotropica* **10**: 58–61.

Singer, M. C. 1971. Evolution of food-plant preferences in the butterfly *Euphydryas editha. Evolution* **35**: 383–389.

Singer, M. C. 1986. The definition and measurement of oviposition preference in plant-feeding insects. Pages 65–94 in J. R. Miller and T. A. Miller (Eds.). *Insect-plant interactions.* Springer-Verlag, New York.

Slansky, F., Jr. and J. R. Rodriguez (Eds.). 1987. *Nutritional ecology of insects, mites, spiders, and related invertebrates.* John Wiley and Sons, New York.

Smiley, J. 1978. Plant chemistry and the evolution of host specificity: New evidence from *Heliconius* and *Passiflora. Science* **201**: 745–747.

Snow, B. K. and D. W. Snow. 1972. Feeding niches of hummingbirds in a Trinidad valley. *Journal of Animal Ecology* **41**: 471–485.

Sokal, R. R. and F. J. Rohlf. 1981. *Biometry.* 2nd ed. 859 pp. W. H. Freeman and Co., San Francisco.

Stiles, F. G. 1975. Ecology, flowering phenology, and hummingbird pollination of some Costa Rican *Heliconia* species. *Ecology* **56**: 285–301.

Stiles, F. G. 1978. Temporal organization of flowering among the hummingbird foodplants of a tropical wet forest. *Biotropica* **10**: 194–210.

Thompson, J. N. 1988. Evolutionary ecology of the relationship between oviposition preference and performance of offspring in phytophagous insects. *Entomologia Experimentalis et Applicata* **47**: 3–14.

Tyrrell, R. A. and E. Q. Tyrrell. 1985. *Hummingbirds: Their life and behavior.* Crown Publishers, New York.

Van der Geest, L. P. S., T. C. Bosse, and A. Veerman. 1983. Development of a meridic diet for the two-spotted spider mite *Tetranychus urticae. Entomologia Experimentalis et Applicata* **33**: 297–302.

Van der Pijl, L. 1955. Some remarks on myrmecophytes. *Phytomorphology* **5**: 190–200.

Wiklund, C. 1975. The evolutionary relationship between adult oviposition preferences and larval host plant range in *Papilio machaon L. Oecologia* **18**: 185–197.

Wilson, D. S. and R. K. Colwell. 1981. Evolution of sex ratio in structured demes. *Evolution* **35**: 882–897.

Zwölfer, H. 1974. Das Treffpunkt-Prinzip als Kommunikationsstrategie und Isolationsmechanismus bei Bohrfliegen (Diptera: Trypetidae). *Entomologia Germanica* **1**: 11–20.

Zwölfer, H. and G. L. Bush. 1984. Sympatrische und parapatrische Artbildung. *Z. zool. Systematik Evolutionsforschung* **22**: 211–233.

21 Biotypes and the Evolution of Niches in Phytophagous Insects on Cardueae Hosts

HELMUT ZWÖLFER and MARIA ROMSTÖCK-VÖLKL

Biological differences between herbivore populations which suggest the occurrence of an infraspecific variation in host relationships are often found in studies of insect–plant systems, particularly when they are extended from single areas to larger regions. Diehl and Bush (1984) have proposed to use the term "biotypes" if a general category is needed for such cases. They define biotypes as conspecific populations which differ in some biological trait. If the genetic basis and the evolutionary status of the biological differences are ascertained, biotypes can be classified as (a) cases of nongenetic polyphenisms, (b) cases of polymorphic or polygenic variation within populations, (c) geographic races, or (d) sympatric host races. A detailed genetic analysis may demonstrate that within complexes of biotypes single taxa have acquired complete reproductive isolation and are therefore host-associated sibling species.

In this contribution we discuss mainly cases c and d. Obviously, geographic races and sympatric or parapatric host races differ much in their mechanisms of reproductive isolation. But in this contribution we do not raise the question how gene pools are separated. We are interested in microevolutionary changes in host relationships, a common trait of geographic races and host races. To lay emphasis on this character, we use the category of "biotypes" and not the terms host race, geographic race, or subspecies. As biotypes can be the origin of the evolution of siblings, there occur cases where the borderline between biotypes and sibling species is not distinct. This is particularly the case in clusters of incipient species, which are probably not rare among herbivorous insects. Some of the complexes of biotypes discussed in the following may indeed include single siblings.

Biotypes of phytophagous insects usually are cumbersome for the taxonomist but they may offer excellent opportunities to study the evolution of ecological niches, of host ranges, and insect–plant relationships. Moreover, biotypes have

Plant-Animal Interactions: Evolutionary Ecology in Tropical and Temperate Regions, Edited by Peter W. Price, Thomas M. Lewinsohn, G. Wilson Fernandes, and Woodruff W. Benson.
ISBN 0-471-50937-X © 1991 John Wiley & Sons, Inc.

important implications for the biological control of insect pests and weeds as well as for pest management.

Diehl and Bush (1984) review papers dealing with biotypes and Strong (1979) discusses studies of the geographic variation in insect–host plant relationships. Informative analyses of complexes of biotypes in herbivorous insects are available for species of the tephritid genus *Rhagoletis* (Bush 1969), the complex of the membracid *Enchenopa binotata* Say (Wood and Guttman 1981) and species of the ermine moth genus *Yponomeuta* (Menken 1981). Examples of detailed studies of infraspecific geographic variation of host selection are those by Wiklund (1975, 1982) on the European swallowtail, *Papilio machaon* L. and by Scriber (1982, 1983 and this volume) on the Nearctic *Papilio* spp. In our contribution we shall deal with examples taken from the insect fauna of European thistles which comprises several complexes of biotypes in various stages of evolutionary divergence. They can serve as models of the evolution of herbivore–plant systems. Before we discuss biotypes of the thistle fauna in detail, a short outline of this herbivore fauna will be given.

THE CYNAROIDEAE AND THEIR FAUNA

Thistles, that is, members of the subfamily Cynaroideae, from a well-characterized group of the family Asteraceae. They are divided into the tribes Echinopeae (2 genera, 150 species), Carlineae (12 genera, 90 species) and Cardueae (60 genera, 2500 species) with the subtribes Centaureinae and Carduinae. The Palearctic herbivorous insect fauna of thistles is dominated by the members of a number of specialized genera (e.g., the curculionid genera *Larinus, Rhinocyllus, Bangasternus, Lixus (subg. Lixochelus)*, the cynipid genus *Isocolus*, the tephritid genera *Urophora, Xyphosia, Orellia, Terellia, Chaetorellia, Chaetostomella*) and a number of Microlepidoptera (e.g., the genera *Metzneria* and *Pterolonche*) which have used the Cynaroideae as radiation platforms and which are exclusively or almost exclusively associated with them (Zwölfer 1965, 1988). All organs of the Cynaroideae are exploited by herbivorous insects. Herbivorous guilds attacking the flower heads are particularly diversified. That insect communities associated with Palearctic Cardueae hosts are structured by evolutionary processes is shown by distinct patterns of niche partitioning among thistle aphids (Völkl 1990), Mediterranean stem miners in Cardueae hosts (Zwölfer and Brandl 1986), or the guild of insects feeding in Cardueae flower heads (Zwölfer 1988).

Geographic gradients in the richness and diversity of the specialized herbivore taxa associated with the Cynaroideae reflect the evolutionary history of this subfamily (Zwölfer 1988). Figure 21-1 shows that the Palearctic thistle fauna is particularly rich in the Mediterranean and western Asia, the evolutionary center of the host (Jäger 1987). Members of many phytophagous taxa occur over large parts of the Palearctic region. This contrasts with the situation in North America, where in spite of the relatively large number of 90 native *Cirsium* spp. the fauna of

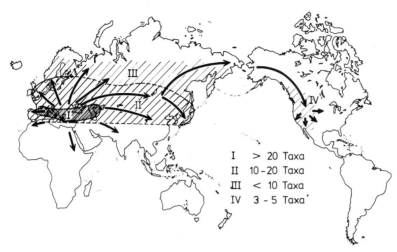

Figure 21-1. Evolutionary center and directions of spreading of the Cardueae (arrows) and diversity of their specialized herbivorous fauna (decreasing species richness from the Mediterranean (zone I) to the Nearctic (zone IV). (Based on data from Small 1919; Jäger 1987; Zwölfer 1988. Genera listed in Appendix 21-1).

specialized herbivorous insects is very poor (Appendix 21-1). Only two taxa of Palearctic thistle insects (the tephritid genera *Chaetostomella* and *Orellia*) were able to follow the Cardueae genus *Cirsium* via Beringia to the Nearctic (Steck 1981) and only very few autochthonous Nearctic insect taxa were able to evolve the necessary adaptations to exploit the Nearctic *Cirsium* spp. in a specialized way. Compared to the Palearctic *Cirsium* flora, the Nearctic species are clearly underused resources (Goeden and Ricker 1987a, b).

EVIDENCE FOR BIOTYPES IN THE FAUNA OF EUROPEAN THISTLES

Table 21-1 summarizes what at present is known about the occurrence of biotypes or presumed biotypes among the European Cardueae. There is little doubt that additional cases will be found, particularly if the guilds associated with roots and stems are investigated in more detail. The degree of evidence that the observed variation in host relationships is infraspecific and genetically determined varies greatly in the species listed in Table 21-1. In *Xyphosia miliaria* the existence of a biotype specialized on *C. heterophyllum* can so far only be deducted from the geographic pattern of host use. In the other cases given in Table 21-1 there are additional criteria which allow the conclusion that biotypes are involved. One of these is the behavior in choice experiments designed to assess the host preference in feeding or oviposition tests (biotypes of *Larinus sturnus* and *Urophora solstitialis* (Fig. 21-2), or of *Rhinocyllus conicus* (Zwölfer and Harris

TABLE 21-1. A List of Biotypes Found on European Cardueae

Species	Hosts	Area[a]	Evidence[b]	Ref.[c]
Larinus	*Arctium* spp.	Pannonic part	abce	1, 2
sturnus		of A, Caucasus		
(Curculionidae)	*Cirsium*	European Alps	ae	3
	spinosissimum			
	Centaurea spp.	Alps, Jura	abcef	1, 2, 4
		mountains		
	Carduus spp.	sw D, se F, Gr	abce	1, 2
L. jaceae	*Centaurea*	SP, w F, w D	aef	1, 2, 5
	Carduus spp.	A, D, CH, F	aef	1, 2, 5
Rhinocyllus	*Silybum*	s F, I, GR	abfg	6, 7, 8
conicus	*Carduus*	s F, I	abfg	6, 7, 8
(Curculionidae)	*pycnocephalus*			
	Carduus nutans	s Europe	abfg	6, 7, 8
Xyphosia	*Cirsium*	n Scandinavia	a	9
miliaria	*heterophyllum*			
(Tephritidae)	*Cirsium* spp.	CH, A, F, D, GB	a	10
Tephritis	*Cirsium*	(Fig. 21-4)	abef	11
conura	*heterophyllum*			
(Tephritidae)	*Cirsium*	(Fig. 21-4)	aef	11
	spinosissimum			
	Cirsium	(Fig. 21-4)	aef	11
	erisithales			
	C. palustre +	(Fig. 21-4)	abef	11
	heterophyllum			
	C. oleraceum +	(Fig. 21-4)	abef	11
	acaule			
Urophora	*Cirsium arvense*	Europe	abcdef	12
cardui	*Cirsium*	GR	abcde	12
(Tephritidae)	*creticum*			
Urophora	*Carduus nutans*	Europe	abde	10, 13
solstitialis	*Carduus*	European Alps	ade	13
	defloratus			
	Carduus crispus	Denmark, n D	abde	13

[a]Area: n = northern, sw = southwestern, se = southeastern, A = Austria, CH = Switzerland, D = West Germany, F = France, GB = Great Britain, GR = Greece, I = Italy, SP = Spain.

[b]Evidence: a = evaluation of field surveys; b = choice tests (adult feeding, oviposition, rate of visits to test plants) under laboratory conditions; c = mating tests; d = production of hybrids; e = biometric analyses; f = analysis of isoenzymes by starch gel electrophoresis; g = evaluation of biocontrol projects (transfer of biotypes from Europe to North America).

[c]References: 1. Zwölfer (1977); 2. Zwölfer (1979); 3. Zwölfer (1975); 4. Zwölfer and Herbst (1990); 5. Herbst (unpublished); 6. Zwölfer and Harris (1984); 7. Goeden (1978); 8. Goeden et al. (1985); 9. Romstöck (1984); 10. Zwölfer (1965); 11. (references in section on *T. conura* in this chapter); 12. Brandl, Zwölfer (unpublished); 13. Möller (unpublished).

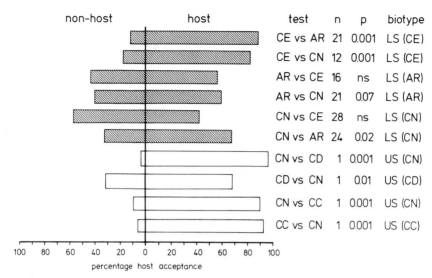

Figure 21-2. Summary of two-choice host selection tests with biotypes of *Larinus sturnus* (Zwölfer 1979) and *Urophora solstitialis* (Möller, unpublished). Hatched columns = *L. sturnus* (= LS), biotypes from *Centaurea scabiosa* (CE), *Arctium* (AR), *Carduus nutans* (CN). Test: percentage of adult feeding on the host plant of the biotype ("host") compared to a simultaneously offered host plant of another *L. sturnus* biotype ("non-host"). n = Number of replicates (each with a group of 4–5 adults). p = Level of significance. Open columns = *U. solstitialis* (= US), biotypes from *Carduus nutans* (CN), *C. defloratus* (CD), *C. crispus* (CC.) Test: percentage of larvae found on the host of the biotype ("host") and the host of another *U. solstitialis* biotype ("non-host") in two-choice oviposition experiments.

TABLE 21-2. Oviposition of *Tephritis conura* in Laboratory Tests with Different *Cirsium* spp.

Origin of the Biotypes	Plant Species Tested[a]		
	C. heterophyllum	*C. palustre*	*C. oleraceum*
C. heterophyllum (Continental Europe)	h	—	—
C. heterophyllum (Scotland)	h	+	—
C. palustre (Scotland)	+	h	—
C. oleraceum	—	—	h

[a] h = original host plant. + = normal oviposition in laboratory tests. − = no or only very slight oviposition in laboratory tests.

1984). For *Tephritis conura* where host plants from the specific "rendez-vous" sites, host acceptance was measured by assessing the frequency of visits to individual plant species in multiple choice tests and by counting oviposition events on different hosts. The results available to date (Romstöck and Arnold 1987; Romstöck unpublished) are summarized in Table 21-2.

In the case of *R. conicus* the existence of biotypes was detected by Goeden (1978), who evaluated results of biological control projects and found that weevils originating from *Carduus pycnocephalus* in Italy showed a persistent ovipositional preference for this host and weevils from *Silybum* favored this host after transfer to California.

In *Larinus sturnus, L. jaceae,* and *Tephritis conura* (Zwölfer 1974) populations associated with different host taxa show statistically significant differences in body size and in characters bearing a functional relationship to the dimensions of specific oviposition sites (i.e., Cardueae flower heads) such as the length of the rostre of *Larinus* females (Fig. 21-3) or the ovipositor of tephritids. Table 21-3 shows that in *T. conura* different biotypes differ in the length of the oviscapt. In Central Europe, the monophagous biotype of *T. conura* from *C. heterophyllum* has a longer oviscapt [confidence interval ($p = 0.05$), $n = 66$, 1.070–1.120 mm]

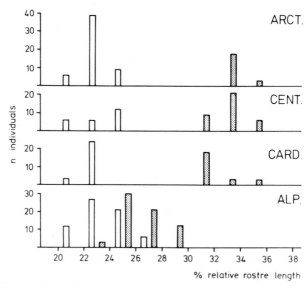

Figure 21-3. Proportions of rostre length (RL) as a percentage of total body length in males (open columns) and females (shaded columns) of four biotypes of *Larinus sturnus*. [Populations from *Arctium* (ARCT), *Centaurea* (CENT), *Carduus* (CARD), and the Alpine thistle *Cirsium spinosissimum* (ALP)]. Note that the sexual dimorphism in rostre length is less pronounced in the Alpine biotype where the particular structure of the oviposition site does not require an elongated female rostre.

TABLE 21-3. Length of Oviscapts in four Biotypes of *T. conura*[a]

Host Plant, Length of Oviscapt (mean) Sample Size	C. oleraceum	C. erisithales	C. heterophyllum
C. oleraceum (Swiss Jura) 0.98 mm (53)	—		
C. erisithales (Alps) 1.03 mm (21)	ns	—	
C. heterophyllum (Alps) 1.10 mm (20)	0.001	0.042	—
C. spinosissimum (Alps) 1.26 mm (57)	0.001	0.001	0.001

[a]Comparison of the oviscapt length in biotypes of *Tephritis conura* associated with four different *Cirsium* spp. The figures give the levels of significance (median test).

than the oligophagous British biotype which exploits the two sympatric host-plant species *C. palustre* and *C. heterophyllum* [confidence interval ($p = 0.05$), $n = 185$, 0.922–0.942 mm] (Romstöck and Arnold 1987).

A difficult problem is the discrimination between biotypes (i.e., conspecific populations) and siblings. In complexes of biotypes the degree of reproductive isolation may vary and certain populations may have reached the status of species (e.g., *Tephritis conura* attacking *Cirsium heterophyllum* in continental Europe, M. Komma, personal communication). To find out whether individuals belonging to different biotypes still recognize each other as potential mating partners "mating tests" were carried out between members of different biotypes of *Larinus sturnus* and *Urophora cardui*. However, as interspecific mating has been induced in the tephritid genera *Urophora* and *Chaetorellia* (Zwölfer 1974), mating tests alone are not a very convincing argument as concerns the conspecificity of tephritid populations. Therefore in laboratory experiments we attempted to hybridize a number of biotypes. Viable hybrid generations were produced by crossing the biotypes of *Urophora cardui* on *Cirsium arvense* and *C. creticum* (Zwölfer, unpublished: larvae of the hybrid generation produce normal galls on *C. arvense*), and by crossing biotypes of *Urophora solstitialis* (Möller, Bayreuth, unpublished).

A further criterion for genetic differences between biotypes is provided by allozyme studies. So far data are available for *Tephritis conura*: Seitz and Komma 1984; Komma 1984, 1986; *Rhinocyllus conicus*: Goeden et al. 1985; Unruh and Goeden 1987; *Larinus sturnus*: Herbst 1986.

COMMENTS ON SINGLE SPECIES

Larinus spp.

The large weevil genus *Larinus* is almost exclusively restricted to Cynaroideae hosts whose flower heads are exploited by the larvae. Most genera of the Cardueae, Carlineae, and Echinopeae have their specialized *Larinus* species and there is some evidence of a parallel cladogenesis between the tribes and subtribes of the Cynaroideae and the subgenera of *Larinus* (Zwölfer and Herbst 1990). *Larinus sturnus* Schall. breeds in the heads of the Carduinae genera *Carduus, Cirsium,* and *Arctium* and the Centaureinae genus *Centaurea* and has thus a broader host range than any of the other *Larinus* spp. investigated by one of us (Zwölfer et al. 1972). There are at least four biotypes of *L. sturnus*: at altitudes between 1700 and 2000 m of the European Alps there exists a morphologically distinct (Fig. 21-3) form on *Cirsium spinosissimum* (Zwölfer 1975). It is parapatric with a widely distributed (Alps, Swiss, and German Jura) biotype on *Centaurea scabiosa* which in choice tests shows a distinct feeding preference (adults) for their host over *Carduus nutans* and *Arctium* sp. (Fig. 21-2). In spite of the fact that in western and central Europe we made an extensive survey of *Arctium*, *L. sturnus* populations on this widely distributed plant genus have only been found in the Pannonic region of Austria (Zwölfer 1979) and in the Caucasus (P. Harris, Regina, Canada, personal communication). In eastern Austria this biotype is parapatric with that on *C. scabiosa*. In choice tests the biotype from *Arctium* fed on both plant species readily but slightly preferred *Arctium* over *Carduus nutans* (Fig. 21-2). *L. sturnus* populations on *C. nutans* and closely related host species were only found in few parts of our observation area (southwestern Germany, Greece, southeastern France). In choice tests they preferred their host over *Arctium* but not over *C. scabiosa* (Fig. 21-2). As has been discussed by Zwölfer (1977 and 1979), the pattern of biotypes of *L. sturnus* is complementary to that of *L. jaceae* F., a closely related species with an almost identical host range (*C. scabiosa, Carduus* spp., *Cirsium* spp.). With the exception of two localities (Kavalla, Greece; Avignon, France) we have never found both *Larinus* spp. exploiting the same host population. Where the two *Larinus* spp. occur in sympatry they show an ecological character displacement, that is, they are represented by biotypes specialized on different host genera.

Rhinocyllus conicus

Taxonomically *Rhinocyllus* is a near relative of *Larinus*. It has a similar biology but differs in details of the oviposition, larval feeding, and pupation behavior and in its complex of parasitoids (Zwölfer and Harris 1984). The larvae of *R. conicus* live in the flower heads of the closely related Carduinae genera *Carduus, Cirsium,* and *Silybum*. Interest was focused on the occurrence of biotypes of *R. conicus*, as this weevil has been widely used as a control agent against *Carduus* spp. and *Silybum marianum*. As discussed in detail by Goeden et al. (1985), follow-up studies made with introduced *R. conicus* populations in North America, oviposition tests, and gel electrophoretic studies confirm the existence of three biotypes in western and central Europe. They are associated with *Carduus nutans, Carduus pycnocephalus,* and *Silybum marianum.* At least the latter two biotypes are sympatric or parapatric at certain localities of southern France. The existence of a fourth biotype which under field conditions almost exclusively breeds in *Cirsium vulgare* (Spain, western France) is suggested by the geographic distribution pattern of field records of *R. conicus* from this thistle (Zwölfer and Preiß 1983). As a study by Klein (1986) showed that West German *R. conicus* populations are oligophagous and exploit several allochronously flowering *Carduus* and *Cirsium* spp. in a predictable sequence, the observations by Zwölfer and Preiß (1983) have to be reexamined. It is possible that the association with *C. vulgare* is primarily due to a different phenology (longer preoviposition period after hibernation) and not to a direct preference for *C. vulgare.* Whatever mechanism is involved, *R. conicus* populations in the Atlantic parts of Europe seem to be particularly adapted to exploit *C. vulgare,* a host plant which in this region is more abundant and predictable as host than the other potential thistle host species of *R. conicus.*

Xyphosia miliaria

The tephritid breeds in the flower heads of *Cirsium* spp. and occasionally of *Carduus nutans.* In central and western Europe and in southern Scandinavia the species attacks mainly *Cirsium palustre* and *C. arvense* and has not been found on *C. heterophyllum,* a thistle which has been extensively surveyed by us (Romstöck 1987). In northern Scandinavia (Lapponia), where *C. heterophyllum* is the only available Cardueae species and a fairly abundant resource, it serves as a regular host (Romstöck 1984). We have not as yet carried out laboratory experiments. The available field observations suggest that the north Scandinavian *X. miliaria* populations constitute a biotype which evolved as an adaptation to the particular resource situation at the northern distribution border of the Cardueae.

Tephritis conura

Our survey (Fig. 21-4) showed that larvae of the tephritid fly *Tephritis conura* feed in flower heads of at least seven "thistle" species in the genus *Cirsium* (tribe Cardueae) (Romstöck 1986). All these host plants are closely related and belong to the section *Cirsium* whithin the genus *Cirsium.* Hybridization between sympatric host-plant species is common. *T. conura* is univoltine and overwinters in the adult stage in still unknown hibernation sites, from where the host populations are recolonized in spring. Eggs are laid in batches in flower buds of an early developmental stage (floret length: 1.0–2.0 mm). The oviposition period lasts for about 4 weeks. Adults die after having finished egg laying. Larvae feed within the florets and the receptacle of the head causing an increased growth of callus tissue and vascular bundles. Emergence of the new generation starts in late

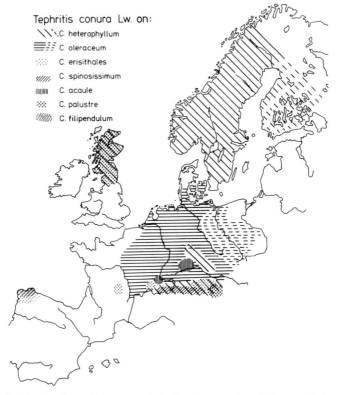

Tephritis conura Lw. on:

\\\\.C. heterophyllum
≡≡∷ C. oleraceum
∷∷ C. erisithales
////// C. spinosissimum
‖‖‖‖ C. acaule
∷∷ C. palustre
▨▨ C. filipendulum

Figure 21-4. Distribution of biotypes of *Tephritis conura* on different *Cirsium* spp. The Spanish populations of *T. conura* on *Cirsium filipendulum* have not yet been studied in detail.

summer. Life table studies (Romstöck 1987) show that the key factor for changes in population numbers is the mortality between adult emergence and the recolonization of the host populations in the following year (winter mortality and losses due to the failure to localize suitable hosts in the proper developmental stage).

Therefore, two features are essential for this species:

1. High population density: dispersal losses of the adults can only be compensated by a high frequency of herbivore–host contacts during the short oviposition period and by high larval survival rates.

2. A precise synchronization of the oviposition phase with bud development of the host populations: the modifications of the flower head tissues necessary for the larval development of *T. conura* (induction of callus growth and increase of vascular bundles of the receptacle) is only possible as long as young, meristematic tissue is available.

In the region covered by our survey (Fig. 21-4) the total distribution area of *T. conura* is by far larger than the distribution areas of the single host species. By

TABLE 21-4. Potential and Actually Used Host Plants of *Tephritis conura* in Different Regions of Europe[a]

	C. heterophyllum	C. palustre	C. oleraceum	C. acaule	C. spinosissimum	C. erisithales	C. filipendulum
Central Europe							
Lowlands		nu	h	h			
Uplands	h	nu					
Central Alps	h				h		
Massif Central						h	
NW-Spain							h
Great Britain	h	h		nu			
Scandinavia	h	nu					

[a] h = Used as host plant. nu = Present in regions where *T. conura* occurs, but not used as host.

forming biotypes adapted to available *Cirsium* species, that is, by "resource tracking" *T. conura* was able to extend its geographical range to the northern limit and to the altitudinal limit of the distribution of Cardueae. In the most areas the number of actually used host-plant species is smaller than the number of available potential host species (Table 21-4). Some *Cirsium* species which belong to the total host range of *T. conura* remain locally or regionally unattacked, even if they are abundant in the area.

Which selection factors were involved in the formation of biotypes in *T. conura*? We shall discuss three ecological constraints which may have led to the evolution of particular biotypes in *T. conura*:

a. It can be assumed that the synchronization of the egg laying period with bud formation plays an important role: The developmental stage of buds in which oviposition takes place is essential for larval survival. The examples in Fig. 21-5 show phenological patterns as they occur in our observation area in northern Bavaria. Bud formation in *C. acaule* and *C. oleraceum* is synchronous. Even with different habitat preferences of the two host plants, spatial contact zones exist in most populations. Under these temporal and spatial conditions the same biotype of *T. conura* uses both *C. oleraceum* and *C. acaule* as hosts. Also in Great Britain, the synchronous bud formation in the sympatric host species *C. palustre* and *C. heterophyllum* (Romstöck and Arnold 1987) was a prerequisite for the adaptation of a single *T. conura* biotype to two different hosts. On the other side, even in the contact zone of *C. heterophyllum* and *C. oleraceum* the maxima of the period of bud formation are separated by an interval of about 2–3 weeks (Romstöck 1987). Females which have colonized *C. heterophyllum* stands have finished their oviposition and die off when *C. oleraceum* starts to form buds. Even in spatial contact zones of both plants the probability of encountering buds of

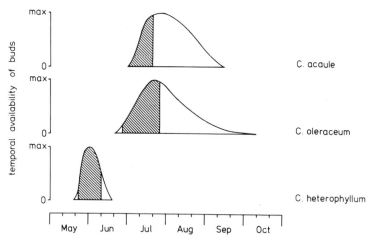

Figure 21-5. The phenological pattern of bud formation in three host plants of *Tephritis conura*. Shaded area = oviposition period of *Tephritis conura*.

both hosts during the period of egg-laying activity is minimal. These differences in host-plant phenologies should exert diverging selection pressures separating the two biotypes. This assumption is supported by host acceptance tests in the laboratory and by allozym studies (Seitz and Komma 1984).

b. Another factor is the availability and abundance of the host plants. Our investigations of the population dynamics of *T. conura* (Romstöck 1987) show that it needs high larval densities to compensate the heavy losses during the annual dispersal phases. High larval densities can only be reached with a high rate of contacts between females and buds in the appropriate developmental stage. In regions where one of the potential host-plant species grows in high densities, specialization on this plant facilitates host finding, mate finding, and synchronization. An example is provided by the contrast in use of *C. hetero-phyllum* and *C. palustre* by *T. conura*: In wide parts of Central Europe where *C. heterophyllum* is common and occurs in large groups, a monophagous biotype evolved which is specialized on this host. The sympatric *C. palustre* with synchronous bud development but a more scattered distribution is not used under field conditions and is largely refused in laboratory tests. In northern Great Britain densities of *C. heterophyllum* are much lower than on the continent and the plant is distributed in isolated patches. Here the *T. conura* biotype is adapted to exploit the widespread *C. palustre* together with *C. heterophyllum*. In this way it has an increased efficiency of host finding and can thus achieve sufficiently high population densities for survival (Romstöck and Arnold 1987).

c. A comparative mortality analysis of the populations of *T. conura* in Great Britain suggests that also a differential predation pressure influences the host relationships of this tephritid. In Great Britain the larvae which exploit the large flower heads of *C. heterophyllum* suffer only little (average = 5%) by the attack of Pteromalid parasitoids whereas larvae in the small heads of *C. palustre* are exposed to a much higher parasitization (up to 50%; Romstöck and Arnold 1987). In this way the relative rarity of *C. heterophyllum* is compensated by the fact that it offers an almost "enemy-free space" which secures high survival rates of the larvae, whereas the value of the abundant *C. palustre* is reduced, as its small flower heads facilitate the access for parasitoids. Exploiting both hosts can be considered as a way of bet hedging of the English biotype of *T. conura*. It balances the disadvantages and the advantages of two resources.

Urophora spp.

This tephritid genus is almost exclusively associated with Cardueae hosts. All *Urophora* spp. are gall makers. With the exception of *U. cardui* (L.) which forms stem galls, the larvae of *Urophora* develop in galls made in the receptacle or in achenes of the flower head, where also pupation takes place. Over most of Europe *U. cardui* is monophagous on *C. arvense*. In Greece, where *C. arvense* also occurs, we did not find it on its usual host but on *C. creticum*, a thistle which occupies similar habitats (river banks, borders of lakes) to those populations of *C. arvense*, on which *U. cardui* builds up particularly stable populations (Zwölfer 1982). It is not as yet clear whether these two types of *U. cardui* populations are geographically separated or whether they are parapatric. A biometric study (Dr. R. Brandl, Bayreuth, personal communication) demonstrated significant differences between *U. cardui* populations from the two *Cirsium* spp. Under laboratory conditions populations from both host plants can be induced to produce viable galls on *C. arvense* as well as on *C. creticum*. The populations associated in the field with the two *Cirsium* spp. interbreed under laboratory conditions and produce viable hybrid offspring (Brandl, Zwölfer, unpublished). We interpret the situation in *U. cardui* as the occurrence of two biotypes which evolved to exploit different resource situations.

The tephritid *Urophora solstitialis* F. restricts its host range to *Carduus* species. A detailed study (Möller, in preparation) which involved extended field surveys, host selection tests (Fig. 21-2), and biometric analyses yielded good evidence that there occur at least three biotypes with different host associations in *U. solstitialis*: the distribution area of populations associated with *Carduus nutans* and marginally with some other *Carduus* spp. covers a large part of Europe. We have records from the Atlantic coast of France to the Black Sea and from southern England to southern Italy. Populations associated with *Carduus crispus* occur in southern Scandinavia and Denmark. Populations on *Carduus defloratus* and *C. personata* are restricted to regions of the European Alps and parts of the Swiss Jura where they can exist in parapatry with the biotype on *C. nutans*. Under laboratory conditions all biotypes can be hybridized. The females of the hybrid of the *C. crispus* biotype (origin Denmark) and the *C. nutans* biotype (origin: northern Bavaria) showed the same host-selection behavior as the biotype living on *C. crispus* (Fig. 21-2), that is, the distinct ovipositional preference for *C. crispus* over *C. nutans* is a dominant genetic trait of the biotype associated with *C. crispus* (Möller, in preparation).

DISCUSSION

The ability to develop biotypes adapted to exploit particular constellations of resources [e.g., the races of *Papilio machaon* L. (Wiklund 1975), the biotypes in the fruit fly genus *Rhagoletis* (Bush 1969), in the small ermine moth (genus *Yponomeuta*, Menken 1981), or in the complex of the membracid *Enchenopa binotata* (Wood and Guttman 1981)] is a widely occurring strategy in herbivores. Fox and Morrow (1981) emphasize that diet breadth can be a flexible attribute in herbivorous insects and the local feeding specialization responds to features of the particular habitats. In our study of the insect fauna of European Cardueae we found by field observations, laboratory tests, the evaluation of biometric data, and enzyme patterns several cases of infraspecific variation in the host relationships which conform with biotypes in the sense of Diehl and Bush (1984). In the following we examine the genetic differentiation of these biotypes and factors which have been involved in their formation. The last sections deal with biotypes as cases of incipient speciation and as templets for the evolution of ecological niches and guilds.

The Genetic Differentiation of Biotypes

For some of the biotypes of insects exploiting Cardueae hosts, isoenzyme patterns have been analyzed. These studies showed genetic differences between conspecific biotypes and assessed genetic distances (Nei 1972) of some of them. They range from 0.047 to 0.086 in *Tephritis conura* (Seitz and Komma 1984); from below 0.1 to 0.3 in *Rhinocyllus conicus* (Unruh and Goeden 1987); from 0.035 to 0.110 (arithmetic mean = 0.071) in *Larinus sturnus* (Herbst 1986); and are 0.033 in *Urophora solstitialis* (R. Brandl, Bayreuth, personal communication). If compared to the data found for *Drosophila* (Ayala et al. 1974), these genetic distances are above the average values given for conspecific local populations ($D = 0.031$) and with one exception they are distinctly lower than the values for *Drosophila* subspecies ($D = 0.230$). [The exception is the *R. conicus* biotype from *Silybum* ($D = 0.3$), which according to Unruh and Goeden (1987) "may warrant subspecific or even species status"]. The available data on the population genetics of biotypes suggest that in most cases the genetic differentiation is slight and that gene flow between biotypes is possible. It therefore can be assumed that the majority of biotypes of Cardueae-associated insects still participate at a common gene pool and that their genetic differentiation may be reversible.

Factors Involved in the Evolution of Biotypes

Biotypes are produced by microevolutionary processes, which will occur in parapatry or even sympatry, as a shift to new hosts usually requires a situation where both the original and the new host are present. A comparative study of biotypes among herbivorous insects on thistles suggests that the following processes (which may operate in various combinations) promoted the evolution of biotypes in thistle insects: (a) resource tracking; (b) host plant abundance; (c) temporal availability of hosts; (d) adaptation to structural particularities of the host; (e) enemy pressure; (f) interspecific competition.

a. Examples of "resource tracking" in thistle insects occur at the latitudinal or altitudinal border of the distribution of Cynaroideae: in northern Scandinavia biotypes of *T. conura* and *X. miliaria* evolved an association with *Cirsium heterophyllum*, the representative of Cynaroideae with the northermost distribution border. In the European Alps special biotypes of *T. conura* and *L. sturnus* became adapted to the "Alpine thistle" *Cirsium spinosissimum*, a host, which reaches altitudes where no other thistle species can survive.

b. The abundance and predictability of hosts is an important factor for the evolution of biotypes, as demonstrated by Wiklund (1982) in his studies on biotypes of *Papilio machaon* L. Shifts between "monophagic strategies" and "oligophagic strategies" (Wiklund 1982) did also occur in *T. conura*. This shows a comparison of the continental biotype of *T. conura*, which is strictly monophagous on *C. heterophyllum* (in Scandinavia and certain European mountains a relatively abundant food source) and the oligophagous Scottish biotype, which exploits the common *Cirsium palustre* together with the relatively rare *C.*

heterophyllum (Romstöck and Arnold 1987). The geographical distribution of biotypes of *Urophora* spp. (biotypes of *U. cardui* on *Cirsium creticum* and of *U. solstitialis* on *Carduus defloratus* and *C. crispus*), of *Rhinocyllus* (biotypes on *Silybum* and *C. pycnocephalus*), and of *Larinus* (biotype on *Centaurea scabiosa*) also conforms with the pattern of resource abundance.

c. The temporal availability or unavailability of hosts is a factor which explains why the oligophagous biotype of *T. conura* associated with *Cirsium oleraceum* attacks also *C. acaule*, whereas the continental biotype on *C. heterophyllum* is monophagous: The relatively short oviposition period and the necessity of a precise synchronization with the development of the host enforces monophagy if the flower periods of locally available potential hosts do not overlap (Fig. 21-5).

d. All the biotypes discussed in our contribution belong to herbivorous species which oviposit into the flower heads of Cardueae. The single Cardueae genera and species possess flower heads with a broad range of size differences and a wide spectrum of structural particularities. Where biotypes are specialized on single-host species, they show morphological adaptations in structures connected with oviposition. Examples are the different sizes of the oviscapt in biotypes of *T. conura* and *U. solstitialis* and of the female rostre of biotypes of *Larinus sturnus* (Fig. 21-2), which is used to make oviposition holes in the capitula of the host species.

e. The Scottish biotype of *T. conura* uses an abundant host where it is exposed to high parasitism (*Cirsium palustre*) together with a relatively rare host where parasitism is low (*C. heterophyllum*). In this case the two host species have different structures (small versus large flower heads) which provide different degrees of protection against parasitoids. We interpret the oligophagous strategy of the Scottish biotype as a result of the opposing forces of differential enemy pressure and different resource concentration (Romstöck and Arnold 1987).

f. As has been suggested and discussed in detail by Zwölfer (1979), interspecific competition between the ecologically almost homologous weevils *Larinus sturnus* and *L. jaceae* seems to be the basis of the character displacement found in these species. Where both species occur in sympatry, they are usually represented by biotypes exploiting different host species. Experiments where both *Larinus* biotypes were forced to breed in the same host-plant population and where the mortality of *L. sturnus* larvae was increased due to the presence of *L. jaceae* larvae (Zwölfer 1979) show that interspecific competition is a mortality factor if both species use the same host plant.

Biotypes as "Ecological Templets" for speciation processes

We assume that it would be wrong to consider biotypes per se as cases of incipient speciation. The available observations suggest that the ability to form biotypes is a flexible attribute of certain herbivorous insect species (Fox and

Morrow 1981) and that niche shifts in biotypes remain at least partially reversible. If, however, situations develop which provide a sufficiently strong reproductive isolation and enough time for increased genetic differentiation, speciation can occur and the niche shifts of biotypes can become irreversible. In such cases the niche shift of biotypes will act as a templet which determines the way in which the niches of incipient species evolve. Indeed, the taxonomic relationships of many herbivorous insect species suggest that an ecological differentiation preceded the complete reproductive isolation and speciation (Zwölfer and Bush 1984).

If the biotypes can serve as "ecological templets" and form a first step in speciation processes, one has to postulate a close correspondence between the feeding specialization of biotypes and the ecological niches of clusters of related species. This correspondence does exist in the specialized herbivorous fauna of Cardueae. The different biotypes of *T. conura* colonized host species belonging to the same section of a genus, those of *X. miliaria*, *U. cardui* and *U. solstitialis* specialized within the limits of host genera, those of *R. conicus* became adapted to hosts belonging to different genera of the same subtribe (Carduinae), and the biotypes of *Larinus* adopted host genera belonging to the same tribe (Cardueae). This is in accordance with the pattern found in the typical genera of the thistle fauna (e.g., *Larinus*, *Bangasternus*, *Rhinocyllus*, *Urophora*, *Terellia*, *Chaetorellia*, *Isocolus*). These taxa have all host ranges which are essentially restricted to the same host genus, subtribe, or tribe (Zwölfer 1988). Thus, the small-scale shifts of the biotypes conform to the general pattern of host ranges found in the majority of herbivorous genera which exploit Cynaroideae hosts in a specialized way: their adaptive radiation was restricted to or largely concentrated on a single host subfamily or single host tribes and subtribes.

The small-scale steps of host transfer and the narrow host pattern of thistle insects forms a marked contrast with the patterns found for instance in many groups of butterflies, where even host races evolved by shifts to taxonomically unrelated but phytochemically often similar host plant families (e.g., by races in selected Papilionidae and Saturniidae studied by Scriber 1983).

Biotypes, Species Richness, and Guild Formation

A striking feature of the herbivorous insect fauna of the Cynaroideae are the geographic gradients in species richness (Fig. 21-1) and ecological diversity of Cardueae-associated herbivores. Our investigations on the phytophagous fauna of Cardueae suggest that host shifts of biotypes are an important element in the microevolution and that they can be a first step in the radiation process of specialized herbivore taxa. A high rate of host shifts requires a pool of preadapted herbivore species as well as high local diversities of potential hosts, that is, of Cynaroideae genera. This situation was realized in west Palearctica (Small 1919) since the Miocene–Pliocene period, but it did not exist in the Nearctic region (Zwölfer 1988). The biogeographic pattern of the herbivore fauna of the Cynaroideae leads to the conclusion that a regional richness in potential host taxa plays an analogous role for radiation processes of specialized herbivores as archipelagoes for the radiation of island faunas.

It is noteworthy that almost all cases of biotypes found so far in our study belong to the guilds of endophytic insects feeding within Cardueae flower heads. In most Cardueae host species this guild contains members with three different trophic strategies (Zwölfer 1987): (a) early attacking and usually gregarious gall formers or callus feeders; (b) achene and receptacle feeders; (c) polyphagous species which can switch from herbivory to predation. Wherever host shifts in biotypes took place, the original trophic strategy is maintained. Most of the studied biotypes are adapted to the first strategy (*Tephritis, Urophora, Rhinocyllus, Xyphosia*), but *Larinus sturnus* is an achene and receptacle feeder. No biotypes were found among the polyphagous subguild. In all cases where we observed biotypes, they are adjusted to the guild structure of the particular host plant species. Our findings show that in Cardueae-associated insects the evolution of biotypes complements herbivorous guilds.

ACKNOWLEDGMENTS

We are indepted to Mrs Marion Preiß and Gabi Lutschinger, who made the drawings and to Dr. R. Brandl (Bayreuth) for reading the manuscript and for valuable comments. The financial support by the Deutsche Forschungsgemeinschaft (SFB 137) is gratefully acknowledged.

REFERENCES

Ayala, F. J., M. L. Tracey, D. Hedgecock, and R. C. Richmond. 1974. Genetic differentiation during the speciation process in *Drosophila*. *Evolution* **28**: 576–592.

Bush, G. L. 1969. Sympatric host race formation and speciation in frugivorous flies of the genus *Rhagoletis* (Diptera: Tephritidae). *Evolution* **23**: 237–251.

Diehl, S. L. and G. L. Bush. 1984. An evolutionary and applied perspective of insect biotypes. *Ann. Rev. Entomol.* **29**: 471–504.

Dittrich, M. 1977. Cynareae- systematic review. In *The Biology and Chemistry of the Compositae*, V. H. Heywood, J. B. Harborne, and B. L. Turner (Ed.). pp. 999–1015. London: Academic Press.

Fox, L. R. and P. A. Morrow. 1981. Specialization: Species property or local phenomenon? *Science* **211**: 887–893.

Goeden, R. D. 1978. Initial analysis of *Rhinocyllus conicus* (Froelich) (Col.: Curculionidae) as an introduced natural enemy of milk thistle, *Silybum marianum* (L.) (Gaertner), and Italian thistle, *Carduus pycnocephalus* L., in southern California. In *The biology and chemistry of the Compositae*, V. H. Heywood, J. B. Harborne, and B. L. Turner (eds.). pp. 39–50. Academic Press, London.

Goeden, R. D., D. W. Ricker, and B. A. Hawkins. 1985. Ethological and genetic differences among three biotypes of *Rhinocyllus conicus* (Col. Curculionidae) introduced into North America for the biological control of asteraceous thistles. *Proc. 6th Symp. Biol. Control Weeds*, Vancouver, Canada, 1984, E. S. Delfosse (Ed.), pp. 181–189. Ottawa: Agric. Canada.

Goeden, R. D. and D. W. Ricker. 1987a. Phytophagous insect faunas of the native thistles, *C. brevistylum, Cirsium congdonii, Cirsium occidentale,* and *Cirsium tioganum,* in Southern California. *Ann. Entomol. Soc. Am.* **80**: 152–160.

Goeden, R. D. and D. W. Ricker. 1987b. Phytophagous insect faunas of native *Cirsium* thistles, *C. mohavense, C. neomexicanum,* and *C. nidulum,* in the Mojave Desert of Southern California. *Ann. Entomol. Soc. Am.* **80**: 161–175.

Herbst, J. 1986. *Biosystematische Untersuchungen an Arten der Gattung Larinus Germ. (Col. Curculionidae).* Diploma thesis, University of Bayreuth. 72 pp.

Jäger, E. J. 1987. Arealkarten der Asteraceen Tribus als Grundlage der ökogeographischen Sippencharakteristik. *Bot. Jahrb. Syst.* **108**: 481–497.

Klein, M. 1986. Anpassungen von *Rinocyllus conicus* Froel. Curculionidae an allochrone Wirtspflanzensituationen. *Verh. Dtsch. Zool. Ges.* **79**: 175–176.

Komma, M. 1984. Wirtsrassenbildung bei der Bohrfliege *Tephritis conura* (Dipt.: Tephritidae). *Verh. Dtsch. Zool. Ges.* **79**: 177–178.

Komma, M. 1986. Mikroevolution bei der Bohrfliege *Tephritis Conura* (Dipt.: Tephritidae): Einnischung bei der Wirtswahl und in der Zeit. *Verh. Dtsch. Zool. Ges* **79**: 177–178.

Menken, S. B. J. 1981. Host races and sympatric speciation in small ermine moths, Yponomeutidae. *Entomol. Exp. Appl.* **30**: 280–291.

Nei, M. 1972. Genetic distance between populations. *Am. Nat.* **106**: 283–292.

Romstöck, M. 1984. Zur geographischen Variabilität des mit *Cirsium heterophyllum* Blütenköpfen assoziierten Phytophagenkomplexes. *Verh. 10. Int. Symp. Entomofaun. Mitteleur.,* Budapest 1983, pp. 123–127.

Romstöck, M. 1986. Mikroevolution bei *Tephritis conura* (Dipt., Tephritidae): biogeographische und populationsökologische Aspekte. *Verh. Dtsch. Zool. Ges.* **79**: 186.

Romstöck, M. 1987. *Tephritis conura Loew (Diptera: Tephritidae) und Cirsium heterophyllum (L.) Hill (Cardueae): Struktur und Funktionsanalyse eines ökologischen Kleinsystems.* PhD thesis. Univ. Bayreuth, West-Germany 147 pp.

Romstöck, M. and H. Arnold. 1987. Populationsökologie und Wirtswahl bei *Tephritis conura* Loew-Biotypen (Dipt.: Tephritidae). *Zool. Anz.* **219**: 83–102.

Scriber, J. M. 1982. Foodplants and speciation in the *Papilio glaucus* group. *5th Proc. Int. Symp. Insect–Plant Relationships,* J. H. Visser and A. K. Minks (Eds.), pp. 307–314, Wageningen.

Scriber, J. M. 1983. The evolution of feeding specialization, physiological efficiency and host races in selected Papilionidae and Saturniidae. In *Impact of Variable Host Quality on Herbivorous Insects,* R. F. Denno and M. S. McClure (Eds.), pp. 373–412, Academic Press, New York.

Scriber, J. M., R. C. Lederhouse and R. H. Hagen. 1990. Foodplants and evolution within *Papilio glaucus* and *Papilio troilus* species groups (Lepidoptera: Papilionidae). In *Plant-Animal Interactions: Ecology in Tropical and Temperate Regions.* P. W. Price, T. M. Lewinsohn, G. W. Fernandes and W. W. Benson (eds.). pp. 341–373, Wiley, New York.

Seitz, A. and M. Komma. 1984. Genetic polymorphism and its ecological background in Tephritid populations (Dipt. Tephritidae). In *Population Biology and Evolution,* K. Wöhrmann and V. Loeschke (Eds.), pp. 143–158. Heidelberg: Springer.

Small, J. 1919. *The origin and development of the Compositae.* London: Addison-Wesley. 4 pp.

Steck, G. J. 1981. North American Terelliinae (Diptera: Tephritidae): *Biochemical systematics and evolution of larval feeding niches and adult life histories.* PhD Thesis. University of Texas, Austin, TX, 250 pp.

Strong, D. R. 1979. Biogeographic dynamics of insect–host plant communities. *Ann. Rev. Entomol.* **24**: 89–119.

Unruh, T. R. and R. D. Goeden, 1987. Electrophoresis helps to identify which race of the introduced weevil, *Rhinocyllus conicus* (Coleoptera: Curculionidae) has transferred to two native southern California thistles. *Environ. Entomol.* **16**: 979–983.

Völkl, W. 1990. Resource partitioning in a guild of aphid species associated with Creeping Thistle (*Cirsium arvense*). *Ent. exp. appl.* in press.

Wiklund, C. 1975. The evolutionary relationship between adult oviposition preference and larval hostplant range in *Papilio machaon*. *Oecologia* **18**: 185–197.

Wiklund, C. 1982. Generalist versus specialist utilization of host plants among butterflies. *Proc. 5th int. Symp. Insect-Plant Relationships*, Wageningen, pp. 181–191.

Wood, T. K. and S. I. Guttman. 1981. The role of host plants in the speciation of treehoppers: An example from the *Enchenopa binotata* complex. In *Insect Life History Patterns*, R. F. Denno and H. Dingle (Eds.), pp. 39–54, Springer, Heidelberg.

Zwölfer, H. 1965. Observations on the distribution and ecology of *Altica carduorum* Guer. (Coleopt., Chrysomelidae.)—*Commonwealth Inst. Biol. Control, Techn. Bull.* **5**: 121–141.

Zwölfer, H. 1974. Innerrartliche Kommunikationssysteme bei Bohrfliegen. *Biologie in unserer Zeit*, **5**: 146–153.

Zwölfer, H. 1975. Vergleichende Untersuchungen an alpinen und nichtalpinen Populationen von *Larinus sturnus* Schall. (Col.: Curculionidae): Diversität und Produktivität im ökologischen Grenzbereich. *Verh. Ges. Ökologie* (Erlangen, 1974): 47–53.

Zwölfer, H. 1977. Der Informationswert faunistischer Daten für populationsökologische Untersuchungen: Das Verteilungsmuster der Wirtsrassen von *Larinus sturnus* Schall. und *L. jaceae* F. (Col.: Curculionidae). *Verh. 6. Int. Symposium Entomofaunistik in Mitteleuropa* (Lunz, 1975) (Ed. H. Malicky): 209–219. The Hague, Junk.

Zwölfer, H. 1979. Strategies and counterstrategies in insect population systems competing for space and food in flower heads and plant galls. Symp. Population Ecology (Mainz, 1978). *Fortschr. Zool.* **25**: 331–353.

Zwölfer, H. 1982. Das Verbreitungsareal der Bohrfliege *Urophora cardui* L. (Dipt.: Trypetidae) als Hinweis auf die ursprünglichen Habitate der Ackerdistel (*Cirsium arvense* (L.) Scop.). *Verh. Dtsch Ges.* **75**: 298.

Zwölfer, H. 1987. Species, richness, species packing, and evolution in insect plant systems. *Ecol. Stud.* **61**: 301–319.

Zwölfer, H. 1988. Evolutionary and ecological relationships of the insect fauna of thistles. *Ann. Rev. Entomol.* **33**: 103–229.

Zwölfer, H. and R. Brandl. 1986. Ökologische Einnischung und morphologische Ähnlichkeitsgrenzen bei Distelinsketen. *Verh. Dtsch. Zool. Ges.* **79**: 195–196.

Zwölfer, H. and G. L. Bush. 1984. Sympatrische und parapatrische Artbildung. *Z. f. zool. Systematik u. Evolutionsforschung* **22**: 211–233.

Zwölfer, H., K. E. Frick, and L. A. Andres. 1972. A study of the host plant relationships of European members of the genus *Larinus* (Col.: Curculionidae). *Techn. Bull. Commonw. Inst. Biol. Control* **97**: 36–62.

Zwölfer, H. and P. Harris. 1984. Biology and host specificity of *Rhinocyllus conicus* Froel.

(Col.: Curculionidae), a successful agent for biocontrol of the thistle, *Carduus nutans* L.—*Z. ang. Ent.* **97**: 36–62.

Zwölfer, H. and J. Herbst. 1990. Präadaptation. Wirtskreiserweiterung und Parallel-Cladogenese in der Evolution phytophager Insekten. *Osche-Festschrift, Z. f. zool. Systematik u. Evol. Forschung.* in press.

Zwölfer, H. and M. Preiß. 1983. Host selection and oviposition behaviour in West-European ecotypes of *Rhinocyllus conicus* Froel. (Col.: Curculionidae).—*Z. ang. Ent.* **95**: 113–122.

APPENDIX 21-1. Insect Genera Associated with the Flower Heads of Cardueae Hosts in the Paleactic and Nearctic[a]

Palearctic	
*Lasioderma**	(Col.: Anobiidae)
*Bruchidius**	(Col.: Bruchidae)
Psylliodes	(Col.: Chrysomelidae)
Eustenopus	(Col.: Curculionidae)
*Larinus**	(Col.: Curculionidae)
*Rhinocyllus**	(Col.: Curculionidae)
Bangasternus	(Col.: Curculionidae)
*Metzneria**	(Lep.: Gelechiidae)
*Homoeosoma**	(Lep.: Pyralidae)
Euxanthis	(Lep.: Cochylidae)
*Cochylis**	(Lep.: Cochylidae)
*Eucosma**	(Lep.: Tortricidae)
*Epiblema**	(Lep.: Tortricidae)
*Lobesia**	(Lep.: Tortricidae)
*Pyroderces**	(Lep.: Momphidae)
*Porphyrinia**	(Lep.: Noctuidae)
*Tingis**	(Heteropt.: Tingitidae)
Isocolus	(Hym.: Cynipidae)
*Urophora**	(Dipt.: Tephritidae)
Ceriocera	(Dipt.: Tephritidae)
*Terellia**	(Dipt.: Tephritidae)
*Orellia**	(Dipt.: Tephritidae)
*Chaetostomella**	(Dipt.: Tephritidae)
Chaetorellia	(Dipt.: Tephritidae)
*Xyphosia**	(Dipt.: Tephritidae)
*Tephritis**	(Dipt.: Tephritidae)
*Acanthiophilus**	(Dipt.: Tephritidae)
Nearctic	
*Platyptilia**	(Lep.: Pterophoridae)
*Rotruda**	(Lep.: Pyralidae)
*Chaetostomella**![b]	(Dip.: Tephritidae)
*Orellia**!	(Dip.: Tephritidae)
*Paracantha**	(Dip.: Tephritidae)

[a]Genera attacking heads of *Cirsium* are marked with an asterisk.
[b]! = Immigrants from the Palearctic.

22 Evolutionary Ecology of Large Tropical Herbivores

SAMUEL J. McNAUGHTON

Large hoofed mammals, the ungulates, have been preeminent herbivores for tens of millions of years in the vast expanses of tropical subhumid to arid climates occupied by woodland, savanna-grassland, and shrubland. Unlike many conclusions in evolutionary ecology, which are derived from short term observations of contemporary phenomena, excellent fossil records of large herbivores and less extensive, but no less revealing, plant fossils provide a firm contextual framework for interpreting the current evolutionary ecology of large mammalian herbivores and the plants they feed upon.

The ungulates are an expansively defined group of two mammalian orders encompassing Earth's principal large terrestrial herbivores. A marked reduction in the functional digits was characteristic during ungulate evolution, leading to ungulatigrade pedalism on massively enlarged digitary nails. All the nonpolar continents except Australia supported populations of ungulates throughout much of the Tertiary (65–7 million yrs BP) and the spectacular adaptive radiations of these animals beginning in the Eocene, about 50 million years ago, is among the evolutionary patterns best documented in the fossil record. Principal features of those radiations were (1) reductions of functional digits and elongation of the legs leading to increasingly cursorial limbs, (2) modifications of skull structure and dentition associated with dietary transitions, and (3) evolution of the digestive tract, also diet-related (evolutionary discussions here are based largely on Scott 1937; Simpson 1945; Romer 1966, 1968; Radinsky 1969; Olson 1971; Maglio and Cooke 1978; Eisenberg 1981). In addition to the true ungulates, subungulates, principally members of the Proboscidea, were important components of the large herbivore fauna that evolved during the Cenozoic Era. Widespread extinctions in the Pleistocene attenuated the radiations, leaving South America with only camelids, cervids, and tapirs, and North America with a limited array of bovids and cervids. Massive 19th century hunting in North America and Eurasia reduced the abundance of ungulates in those regions to tiny remnants of the populations present less than a century and a half ago. Much of

Plant-Animal Interactions: Evolutionary Ecology in Tropical and Temperate Regions, Edited by Peter W. Price, Thomas M. Lewinsohn, G. Wilson Fernandes, and Woodruff W. Benson. ISBN 0-471-50937-X © 1991 John Wiley & Sons, Inc.

the extant abundance and species diversity of large herbivores is in Africa, which contains approximately 90 species; about 75 species are in the single family Bovidae, ranging in size from the 3.5-kg royal antelope (*Neotragus pygmaeus*) to the African buffalo (*Syncerus caffer*), with males up to 850 kg (Leuthold 1977).

Since this book is about tropical herbivores, it is appropriate to recall that the Eocene was characterized by floras similar to present tropical rain forests extending within the Arctic Circle (Wolfe 1978). Thus, although much of the large herbivore radiation took place beyond the contemporary tropics, it occurred in tropical climates nonetheless.

ORIGINS

The ungulate–subungulate adaptive framework arose in the late Cretaceous, about 75 million years ago, with the appearance of condylarths (extinct order Condylarthra), the first proto-ungulates. Their generalized dentition indicates that they descended from earlier insectivores and were forest dwellers feeding largely on soft, dicotyledon leaves. They radiated extensively during the Paleocene (65–54mBP) in South America, North America, and Eurasia, reaching a size comparable to a modern pony, and gradually dwindled to extinction in the Oligocene (ca. 30 m BP).

The Proboscidea, subungulates encompassing Earth's largest contemporary herbivores, appeared slightly later than the condylarths, from presently unde-fined ancestors. They radiated extensively in Eurasia, Africa, and the New World through the Miocene and Pliocene, and then underwent substantial extinctions leading to the two existing species of elephants (Aguirre 1969). In addition to the spectacular evolution of the upper nose, palate, and lip into an extended proboscis and of the incisors into tusks, the cheek teeth evolved into massive, replaceable structures. The worn front teeth are developmentally replaced by new teeth moving forward from the rear. This suggests that the proboscidean diet became exceedingly abrasive as they evolved. The other subungulate order, the Hyracoidea, is limited to small-bodied species occupying restricted habitats in Africa and SW Asia.

First appearing about 60 m BP in the Paleocene was one of two major evolutionary groups accounting for most of Earth's large, herbivorous mammals, the Perissodactyla, or odd-toed ungulates, and by the early Eocene four of the five known superfamilies are recognizable (Romer 1966). The other major large herbivores, the even-toed Artiodactyla, appeared early in the Eocene, about 50 m BP. Both are presumed to be derived from the condylarths. Perissodactyls were dominant from 54 to 38 m BP (the Eocene Epoch), but declined during rapid radiation of artiodactyls in the Oligocene (38–26 m BP).

EVOLUTIONARY PATTERNS

The general trend of locomotory organ evolution in both ungulate orders was a reduction of digits, restriction of surface contact to a hardened corneous layer of the epidermis (the hoof), and modification and fusion of bones leading to longer, cursorial legs. The mesaxonic reduction led to a foot axis passing through the middle digit, with the others greatly reduced. This led to the Perissodactyla, currently represented by horses, tapirs, and rhinoceroses. The paraxonic reduction led to an axis through the third and fourth digits. This led to the Artiodactyla, among the most diverse of mammalian orders, including pigs, camels, and ruminants.

Evolution of the limbs in both ungulate orders accomplished increased mobility. Elongation of medial digit bones (metapodials) to produce a third effective leg segment was a common evolutionary trend. Diagnostic of artiodactyls is the astragalus, a modified ankle bone capable of vertical rotation that increases propulsive thrust of the stride. All of the trends in locomotory organs suggests natural selection for moving rapidly over long distances on hard, tissue-erosive surfaces.

Dental trends were toward enlargement and reinforcement, accompanied by convolution of the grinding surface. An increase in molar height from low-crowned, brachyodont, to high-crowned, hypsodont, structure was associated with major expansions of Earth's grasslands during the Miocene, beginning 26 m BP. The grinding surfaces of teeth became progressively ridged and convoluted. Dental trends and modifications of skull structure associated with musculature and reinforcement indicate natural selection by a diet changing from soft, easily masticated plant tissues to hard, abrasive, mastication-resistant tissues.

Contemporaneous with the appearance and radiation of the ungulates and subungulates was the appearance and radiation of grasses, members of the plant family Poaceae. Unambiguous grass pollen first appears in Paleocene deposits (Clayton 1981). One of the distinctive anatomical features of grasses is the accumulation of characteristic opaline silica bodies, phytoliths, in epidermal cells. The shapes of phytoliths are taxonomically diagnostic (Prat 1932) and among the earliest known grass fossils are phytoliths from Eocene deposits in Patagonia, South America (Frenguelli 1930; Bertoldi de Pomar 1972). These fossils indicate that grasses originated and began to radiate at the same time that the major large herbivore orders were developing. At least one modern grass tribe, the Oryzae, can be identified in epidermal fossils of late Eocene age (Litke 1968). It seems likely that the earliest grass-like plants were, therefore, contemporaneous with, and occupied habitats similar to, the condylarths. Just as the condylarths were tropical forest inhabitants, so grasses are believed to have originated in wet tropical forest glades and margins (Clayton 1981). By the Oligocene, around 30 m BP, the expansion of arid, open habitats, radiation of the Poaceae into those habitats, and the limb specialization and hypsodonty evolving in ungulates indicate that coevolution of grasses and large, mammalian herbivores was well underway (Clayton 1981; Stebbins 1981).

There can be little doubt, I believe, that phytoliths in grasses, combined with the progressive, continuous evolution of hypsodonty in the Perissodactyla, Artiodactyla, and Proboscidea, as well as the unique replacement of the front teeth of the latter as they wear away by teeth growing from the rear, provide the strongest existing fossil evidence for coevolution of herbivores and plants

(McNaughton et al. 1985). A dietary shift within large mammal lines from browsing on less abrasive dicot leaves to grazing on grasses, and continuing progressive evolution of grasses into more and more abrasive forms, were the undoubted driving forces in the evolution of ungulate and subungulate dentition. Limb evolution was certainly associated with avoiding the large predators that were simultaneously evolving, but it also undoubtedly facilitated exploiting a food resource, forage grasses, whose growth is unpredictable in space and time due to the sporadic precipitation of the grassland climate (McNaughton 1979; McNaughton et al. 1985).

Finally, the evolution of rumen and cecal fermentation by artiodactyls and perissodactyls, respectively, was instrumental to their ascendancy as herbivores (Langer 1974; Janis 1976). Much of the potential energy of plant tissues is completely inaccessible to herbivores because no multicellular animal is capable of synthesizing cellulase. However, development of cecal and, later, rumen fermentation by a symbiotic microbial gut fauna greatly expanded the range of foods that the large herbivores could exploit effectively.

DIET, MORPHOLOGY, DIGESTION, AND SIZE

Contemporary large mammalian herbivores are commonly classified as browsers that feed on the leaves and shoots of dicotyledenous trees and shrubs, grazers that feed on the leaves and stems of monocotyledens, principally grasses, and mixed feeders that consume varying proportions of dicot shoots, dicot seeds, and grasses. Mixed feeders are commonly grazers during the grass growing season and browsers during the dormant season.

These broad feeding classes are distinguished by distinctive mouth and dental morphologies (Bell 1969; Owen-Smith 1982). Grazers have broad muzzles and a flattened mandibular synarthrosis forcing the incisors to project outward from the jaw (Fig. 22.1). Although the broad muzzle leads to nonselective feeding within the grass sward, the projecting incisors allow close cropping of low growing, highly nutritious forages in grazing lawns and leads to the evolution of dwarfed, prostrate grass ecotypes where grazing is intense (McNaughton 1984). Most of the variation in grazer mouth morphometrics is in muzzle width, with narrower muzzles allowing greater selectivity, but with little variation in incisor insertion angle. Browsers, in contrast, have narrower muzzles and more erect incisors. This combination of skull and dental traits allows a browser to forage selectively, fitting its muzzle into narrow passages in arborescent vegetation to nip off tender, nutritious shoots. Mixed feeders are intermediate, as expected, but somewhat more similar to browsers than grazers. Most of the morphometric variation within both browsers and mixed feeders is in incisor insertion angle, with muzzle width varying less markedly.

In addition, the incisors of grazers tend to be uniform in size, while browsers and mixed feeders have small lateral and large central incisors, facilitating selective nipping off of shoots in a highly heterogeneous dicotyledenous canopy

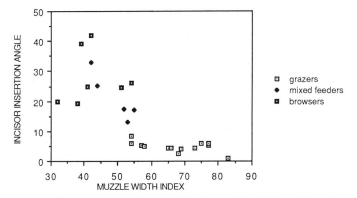

Figure 22-1. Relationship between standardized muzzle width and angle of incisor insertion for grazers (dotted square), browsers (square), and mixed feeders (diamond). After Bell (1969) and Owen-Smith (1982).

(Janis and Ehrhardt 1988). In both recent and Miocene faunas, pure grazers such as horses (Equidae, Perissodactyla) and alcelaphine antelopes (Alcelaphini, Artiodactyla) have decidedly hypsodont dentition while such browsers as giraffes (Giraffidae) and mixed-feeding bovids are more brachyodont (Janis 1984). Thus, major features of the evolutionary ecology of contemporary large herbivores, food sources, feeding selectivity, and methods of feeding, can be directly related to the features of fossil large mammalian herbivores.

One of Earth's great evolutionary revolutions, which had major consequences for ecosystem organization and, later, human beings, was the Eocene to Miocene modification of the artiodactylian digestive tract into the rumen, a complicated storage and fermentation chamber, the reticulum and omasum, all believed derived from the esophagus, and the abomasum, which corresponds to the nonruminant stomach. This structurally complicated digestive system accompanies two functions that are particularly important ecologically. First, the rumen serves as a storage vessel so that ruminants can ingest food rapidly with minimal initial mastication, to regurgitate it later for more thorough particle reduction. This allows them to consume forage rapidly in habitats where herbivores are predator-prone, while they can retire to less hazardous habitats to ruminate (Young 1962). Second, microbial fermentation of cellulose and other carbohydrates in the rumen releases volatile fatty acids serving as the ruminant's energy source while the microbes themselves are digested further down the digestive tract to serve as protein sources. This evolutionary innovation, combined with hypsodont dentition allowing the processing of silicaceous graminoids and limb evolution increasing mobility, made the low-quality, abrasive, spatially and temporally unpredictable primary productivity of grasslands readily available to ruminants (McNaughton et al. 1985) and they became Earth's major large herbivores.

However, because of the relationships among body size, metabolic expenditure, gut capacity, and digestive efficiency, rumen fermentation is most advantageous over body sizes between about 10 and 1000 kg (Demment and Van Soest 1985). Mammalian gut capacity scales as a linear function of body mass while metabolic requirement, of course, scales as a fractional exponent of mass. Therefore, the ratio of metabolic requirement to gut capacity declines with body size. In addition, retention time increases with gut capacity and this is the principal animal trait influencing digestibility of a food source. The rumen is a morphological and symbiotic adaptation that both increases retention time and renders less digestible dietary components, such as plant cell walls, more digestible. Small mammals are constrained to eating a high quality diet of readily digestible foods by a high metabolic rate to gut capacity ratio; they have evolved hindgut fermentation and often produce biphasic feces, some of which are reingested. Very large mammals (i.e., elephants), have low ratios of metabolic rate to gut capacity and very long retention times. But because of their high total metabolic requirements, they must consume low-quality diets.

Ruminants evolved to occupy the broad midrange of body sizes. Their digestive tract evolved to selectively delay passage rate, producing more efficacious digestion than is possible in a nonruminant of similar body size. They are capable of efficiently utilizing the fiber of lower-quality, grass diets. Demment and Van Soest (1985) argue that the first, small, proto-ruminants evolved foregut fermentation to detoxify dietary components and that this facilitated further evolutionary complication of the digestive tract to allow utilization of lower-quality forage as grasslands expanded during drier climatic periods.

BEHAVIORAL ECOLOGY OF LARGE TROPICAL HERBIVORES

Behavioral studies have been a major focus of research on mammalian herbivores since the inception of research in the tropics (Talbot 1965; Leuthold 1977; Gosling 1986). It is impossible to consider that massive body of research here beyond one set of generalizations particularly relevant to the animals' status as herbivores.

Many aspects of ungulate social organization are closely related to the patterns of feeding ecology and body size considered in the immediately preceding section of this chapter, and these have been best characterized for the ruminants (Jarman 1974). Small species, with a body weight of 20 kg or less, are solitary or pair-bonded and territorial. They select a highly diverse diet of high-quality browse, and utilize a cryptic antipredator behavior. A typical species is Kirk's dik-dik (*Madoqua kirki*). Species in the range of body mass from 15 to 100 kg are selective feeders on grass or browse, occur in small female groups with single males in a territory, and exhibit cryptic or short flight antipredator behavior. A typical species is the oribi (*Ourebia ourebi*). Mixed feeders commonly range in body size from 20 to 200 kg; they occur in herds of several to a hundred individuals with territorial, breeding males and bachelor male herds. They feed selectively on mixtures of browse (principally in the dry season) and grass (principally in the wet season). Erratic flight is a common antipredator behavior.

Impala (*Aepyceros melampus*) are typical of this ecological-behavioral class of ungulates. Herbivores with a body size of 75–250 kg are grazers, selective primarily for grass growth stage and leaf blade, depending upon availability. They commonly are highly gregarious, occurring in herds of scores to thousands, and are often migratory over the course of a year, occupying distinct wet and dry season ranges. Flight is frequently a response to predators, though defense against smaller predators is also common. The blue wildebeest (*Connochaetes taurinus*) is typical. The fifth socioecological class of large animals between 200 and 700 kg mass feed unselectively on low-quality forage, commonly though not exclusively, grass. They form moderate-sized female and breeding bull herds in stable home ranges. Antipredator behavior can range from flight to defense. Buffalo are typical.

Although not considered by Jarman in his codification of behavior, the largest existing land mammal, the African elephant (*Loxodonta africana*), is (Wilson 1975, 491) "distinguished by one of the most advanced social organizations." From the standpoint of the evolutionary ecology of tropical herbivores, the most significant aspect of this social organization is its hierarchical nature with matriarchs playing a pivotal role in herd behavior (Douglas-Hamilton 1972). Efficient exploitation of resources within the home range of these herbivores may be regulated to a significant extent by learned behavior in which the matriarchs serve as a repository of long-term experience.

These behavioral studies indicate that a whole suite of traits in large tropical herbivores is related to their feeding ecology. In addition to the anatomical and physiological evolution revealed by the fossil record, there was an accompanying divergence of social organization from small, solitary, forest-dwelling animals selectively feeding on rare but highly nutritious foliage components to large, gregarious, plains dwelling herbivores feeding unselectively on common but less nutritious grass. Differences in breeding and antipredator behavior accompanied these evolutionary trends from condylarths to artiodactyls.

FEEDING ECOLOGY IN A HIERARCHICAL ENVIRONMENT

Large herbivores exist in spatial and temporal environments that are functionally allied and manifestly structured in a hierarchical fashion (Senft et al. 1987; McNaughton 1989). At the lowest spatial and temporal scales are individual bites of food, with volumes on a scale of cubic centimeters selected on time scales of seconds (Fig. 22-2). It has been estimated that a large herbivore consumes 10^7 bites per year (Chacon et al. 1976); assuming 6 hours a day feeding, this means more than one bite per second on average. At the other spatiotemporal extreme, migratory ungulates move over ranges of thousands of square kilometers in the course of a year. Even animals occupying restricted home ranges can encounter environmental heterogeneity comparable to the migratory ungulates as they move from hilltops in the peak of the wet season to lowland flood plains in the dry season (McNaughton 1983, 1985).

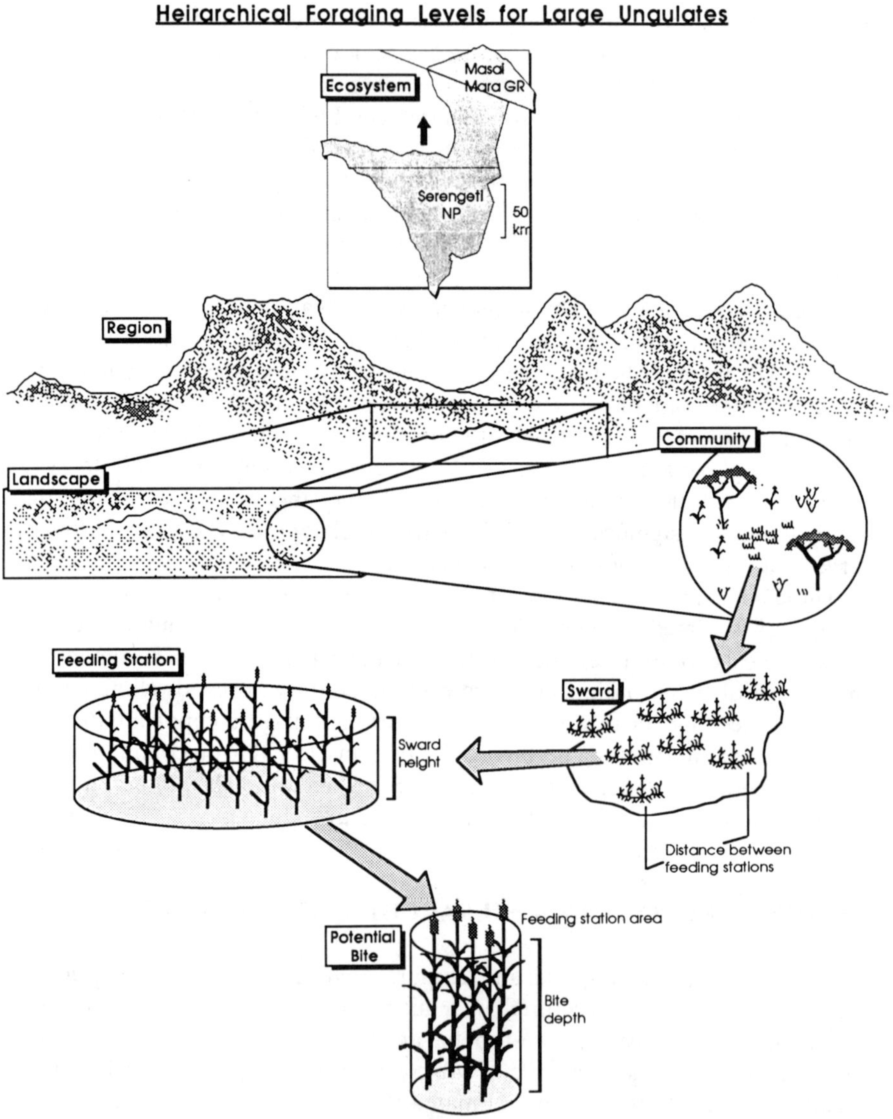

Figure 22-2. Hierarchical spatial scales over which large ungulate feeding ecology has evolved from the smallest scale of individual bites at an operational time scale of seconds to geographic ranges at time scales of months. After Senft et al. (1987), and McNaughton (1989).

There is a huge accumulated literature indicating that ungulates are selective foragers at all hierarchical levels, from landscapes within geographic regions to bites within swards (e.g., Talbot and Talbot 1962; Stewart and Stewart 1970; Jarman 1971, 1972; Field 1972; Sinclair 1977; Jarman and Sinclair 1979; Maddock 1979). The criteria of selection are as complex as the levels and temporal scales at which decisions are required (Senft et al. 1987). Among the factors associated with forage selection in a general order of decreasing spatiotemporal scale, but not necessarily functional importance, are: geographical location, proximity to water, tree-grass balance, grass stature, grassland species composition, species identity, forage leaf–stem balance, protein content, and digestibility. Many of these plant properties are, of course, strongly intercorrelated. For example, a grass sward with a high leaf–stem ratio will also generally be low in stature and high in protein and digestibility. So, while most of the foraging decisions by invertebrate herbivores may be encapsulated in the ovipositioning choice of females, the foraging behavior of large mammals is a complex, highly reticulate process operating at many levels, some inherited and some learned, some environmental and some innate.

GRASS MINERAL HETEROGENEITY AND LARGE MAMMAL EVOLUTIONARY ECOLOGY

Studies of forage quality in natural ecosystems have focused largely on energy and protein contents and digestive yields (Robbins 1983). In the tropics, however, forages are so commonly deficient in mineral elements that complete mineral supplementation has been advocated as a standard husbandry practice for domestic livestock (McDowell et al. 1983) and wildlife in Africa can suffer pathological mineral deficiencies (Jarrett et al. 1964; Wilson and Hirst 1977). It is likely that chronic deficiency depressing fecundity and growth may be more important than evident, acute pathology (Franzmann et al. 1975). But there is no natural ecosystem for which a systematic investigation of potential mineral deficiencies has been determined. I began such a survey of the Serengeti National Park, Tanzania, in 1986. The study is designed to reflect the foraging hierarchy applicable to large ungulates, from the level of the entire geographic region to individual bites.

A first objective of those studies was to determine whether mineral heterogeneity could explain the distributional heterogeneity of resident, nonmigratory grazers in the ecosystem (McNaughton 1988). One of the characteristic features of the distribution of large tropical herbivores is that they are very heterogeneously distributed, and have been so since first observed by European explorers (McNaughton and Georgiadis 1986). Resident animals often occur in multispecific clumps, separated from other clumps by largely unoccupied areas of varying extent. The favored explanations of those distribution patterns have been (Gosling 1986): (a) foraging facilitation among species with different foraging habits, and (b) protection from predators. Neither of those hypotheses is totally persuasive (McNaughton 1988). If foraging facilitation were an overriding factor, the herds should move consistently with localized showers to take advantage of sporadic bursts of primary productivity, but these resident herds typically occupy temporally stable home ranges, as discussed earlier in this chapter. If predation were a pivotal factor, sporadic, unpredictable spatial movements would further reduce predation but, again, the temporally stable nature of the distribution patterns argues against that hypothesis.

My comparison of resident herd distribution centers and adjacent, sparsely occupied areas showed that forage mineral content was a strong discriminator between grasslands supporting resident herds and nearby locations lacking dense animal concentrations (McNaughton 1988). Based on animal requirements and forage contents, Mg, Na, and P were minerals particularly likely to influence animal distribution patterns in the Serengeti. However, soil mineral contents were incapable of predicting animal distributions related to forage mineral contents, suggesting that ecological processes overriding underlying geological and edaphic heterogeneities were influential. These results provide preliminary evidence that a heretofore overlooked ecological factor, plant mineral con-

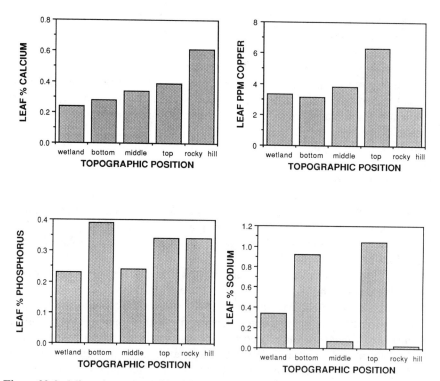

Figure 22-3. Mineral content of leaf blades of actively growing grasses along a topographic gradient from lowland wetlands to rocky hills in the western Serengeti National Park, Tanzania.

centrations, are a major determinant of the distribution and abundance of large tropical herbivores.

As an example of the forage nutritional heterogeneity that grazers confront, data on grass leaf blade mineral contents along a soil catena in the western Serengeti reveal a wide variety of patterns, from no variation to extreme variability (Fig. 22-3). Phosphorus was uniform across the catena ($F_{4,16} = 2.32$, $P = $ n.s.), Ca increased gradually from wetlands to upper hills and then increased dramatically on rocky hills ($F_{4,16} = 10.28$, $P < 0.001$), Cu was low everywhere except on upper hill slopes ($F_{4,16} = 5.53$, $P < 0.01$), and Na was highly variable, being highest on lower slopes and hilltops ($F_{4,16} = 8.38$, $P < 0.001$). Therefore, a grazer seeking to balance its diet for these essential elements would have to mix forages from different microsites to achieve that balance.

Do these patterns matter to ungulates; can they detect and respond to them? Certainly, there is unequivocal evidence that they can respond to dietary Na and, probably, P (Denton 1984). However, it also is generally established that herbivorous mammals respond to dietary deficiency by exploratory behavior, widening the variety of foods consumed, even if specific hungers are not involved (Rozin 1976; Nudds 1980). So it is likely that incipient nutritional deficiency would lead to tasting behavior tending to spread intake over a broader range of grasses than would be the case if a single species, or grassland, met all dietary requirements.

CONCLUSION: LARGE TROPICAL HERBIVORES IN COMPLEX ENVIRONMENTS

Grasslands are seemingly so unvaried or uninteresting as to provoke tedium, boredom even; they are dull places (McNaughton 1989). The American Great Plains were characterized as the Great American Desert by early geographers and were regarded as uninhabitable right up to the time of human settlement, in spite of their abundant herbivores, productive grasslands, and fertile soils (Webb 1931). So, also, it is colloquially familiar to hear large grazers referred to as "dumb" or "brutes" or the two combined. Therefore, both grasslands and large herbivores have a historically pejorative connotation. However, I believe this chapter documents that both grasslands and large herbivores are enormously more complex, structurally and functionality, than either their appearance or common wisdom concede.

The evolution of complex limbs, dentition, skulls, and digestive tracts in large tropical herbivores suggests that they were subject to natural selection by an intricate environment. My recent studies of their distributions in relation to forage mineral concentrations document a degree of distributional and nutritional fine-tuning previously unsuspected. Taken together with their diverse, complicated social systems, more or less manifestly related to foraging modes, the evidence indicates that the evolutionary ecology of large tropical herbivores is among the most elaborate known.

ACKNOWLEDGMENTS

My research is supported by the NSF Ecosystem Studies Program. Mineral analyses were by M. M. McNaughton, sample collection was aided by F. F. Banyikwa and N. J. Georgiadis, and M. Oesterheld assisted with figure preparation.

REFERENCES

Aguirre, E. 1969. Evolutionary history of the elephant. *Science* **164**: 1366–1376.

Bell, R. H. V. 1969. The use of the herb layer by grazing ungulates in the Serengeti National Park, Tanzania. *Ph. D. Diss., U. Manchester.*

Bertolidi de Pomar, H. 1972. Opalo organogenico en sedimentos superficiales de la llanura santafesina. *Ameghiniana* **9**: 265–279.

Chacon, E., T. H. Stobbs, and R. L. Saldland. 1976. Estimation of herbage consumption by grazing cattle using measurements of eating behaviour. *J. Brit. Grassland Soc.* **31**: 81–87.

Clayton, W. D. 1981. Evolution and distribution of grasses. *Ann. Missouri Bot. Gard.* **68**: 5–14.

Demment, M. W. and P. J. Van Soest. 1985. A nutritional explanation for body-size patterns of ruminant and non-ruminant herbivores. *Am. Nat.* **125**: 641–672.

Denton, D. 1984. *The Hunger for Salt.* Springer-Verlag, Berlin.

Douglas-Hamiton, I. 1972. On the ecology and behaviour of the African elephant: The elephants of Lake Manyara. *Ph. D. Diss., Oxford U.*

Eisenberg, J. F. 1981. *The Mammalian Radiations.* University of Chicago Press, Chicago, IL.

Field, C. R. 1972. The food habits of wild ungulates in Uganda by analyses of stomach contents. *E. Afr. Wildl. J.* **10**: 17–42.

Franzmann, A. W., J. L. Oldemeyer, and A. Flynn. 1975. Minerals and moose. *Proc. Am. Moose Workshop and Conf.* **11**: 114–140.

Frenguelli, J. 1930. Particulas de silice organizada en el loess y en los limos pampeanos. Selulas siliceas de Gramineas. *Anales Soc. Cien. Santa Fe* **1**: 1–47.

Gosling, L. M. 1986. The evolution of mating strategies in male antelopes. Pp. 244–281 in D. I. Rubenstein and R. W. Wrangham (Eds.). *Ecological Aspects of Social Evolution.* Princeton University Press, Princeton, NJ.

Janis, C. 1976. The evolutionary strategy of the Equidae and the origins of rumen and cecal digestion. *Evolution* **30**: 757–774.

Janis, C. M. 1984. The use of fossil ungulate communities as indicators of climate and environment. Pp. 85–104 in P. Brenchley (Ed.). *Fossils and Climate.* Wiley, New York.

Janis, C. M. and D. Ehrhardt. 1988. Correlation of relative muzzle width and relative incisor width with dietary preference in ungulates. *Zool. J. Linnean Soc.* **92**: 267–284.

Jarman, P. J. 1971. Diets of large mammals in the woodlands around Lake Kariba, Rhodesia. *Oecologia (Berl.)* **8**: 157–178.

Jarman, P. J. 1972. Seasonal distribution of large mammal populations in the unflooded middle Zambesi Valley. *J. Appl. Ecol.* **9**: 283–299.

Jarman, P. J. 1974. The social organization of antelope in relation to their ecology. *Behaviour* **48**: 215–266.

Jarman, P. J. and A. R. E. Sinclair. 1979. Feeding strategy and the pattern of resource partitioning in ungulates. Pp. 130–163 in A. R. E. Sinclair and M. Norton-Griffiths (Eds.). *Serengeti. Dynamics of an Ecosystem.* University of Chicago Press, Chicago, IL.

Jarrett, W. H. F., J. W. Jennings, M. Murray, and A.m. Harthoorn. 1964. Muscular dystrophy in wild Hunter's antelopes. *E. Afr. Wildl. J.* **2**: 158–159.

Langer, P. 1974. Stomach evolution in the Artiodactyla. *Mammalia* **38**: 295–314.

Leuthold, W. 1977. *African Ungulates.* Springer-Verlag, Berlin.

Litke, R. 1968. Uber den Nachweis tertiarer Graminen. *Monatsber. Deutsch. Akad. Wiss. Berlin* **10**: 462–471.

Maddock, L. 1979. The "migration" and grazing succession. Pp. 104–129 in A. R. E. Sinclair and M. Norton-Griffiths (Eds.). *Serengeti. Dynamics of an Ecosystem.* University of Chicago Press, Chicago.

Maglio, V. J. and H. B. S. Cooke. 1978. *Evolution of African Mammals.* Harvard University, Cambridge, MA.

McDowell, L. R., J. H. Conrad, and G. L. Ellis. 1983. Mineral Deficiencies and Imbalances and Their Diagnosis. *Symp. Herbivor Nutr. in Sub-trop. and Trop., Pretoria, S. Afr.*

McNaughton, S. J. 1979. Grassland–herbivore dynamics. Pp. 46–81 in A. R. E. Sinclair and M. Norton-Griffiths (Eds.). *Serengeti. Dynamics of an Ecosystem.* University of Chicago Press, Chicago, IL.

McNaughton, S. J. 1983. Serengeti grassland ecology: The role of composite environmental factors and contingency in community organization. *Ecol. Monogr.* **53**: 291–320.

McNaughton, S. J. 1984. Grazing lawns: Animals in herbs, plant form, and coevolution. *Am. Nat.* **124**: 863–886.

McNaughton, S. J. 1985. Ecology of a grazing ecosystem: The Serengeti. *Ecol. Monogr.* **55**: 259–294.

McNaughton, S. J. 1988. Mineral nutrition and spatial concentrations of African ungulates. *Nature* **334**: 343–345.

McNaughton, S. J. 1989. Interactions of plants of the field layer with large herbivores, p. 15–29 in G. M. O. Maloiy, and P. A. Jewell (Eds.). *Symp. Zool. Soc. Lond.* Zool. Soc. Lond., London.

McNaughton, S. J. and N. J. Georgiadis. 1986. Ecology of African grazing and browsing mammals. *Ann. Rev. Ecol. Syst.* **17**: 39–65.

McNaughton, S. J., J. L. Tarrants, M. M. McNaughton, and R. H. Davis. 1985. Silica as a defense against herbivory and growth promotor in African grasses. *Ecology* **66**: 528–535.

Nudds, T. D. 1980. Forage preference: Theoretical considerations of diet selection by deer. *J. Wildl. Manage.* **44**: 735–740.

Olson, E. C. 1971. Vertebrate Paleozoology. Wiley-Interscience, New York.

Owen-Smith, N. 1982. Factors influencing the consumption of plant products by large herbivores. Pp. 359–404 in B. J. Huntley and B. H. Walker (Eds.). *Ecology of Tropical Savannas.* Springer-Verlag, Berlin.

Prat, H. 1932. L'epiderme des Graminees. Etude anatomique et systematique. *Ann. Sci. Nat. Bot. ser. 10,* **14**: 117–324.

Radinsky, L. B. 1969. The early evolution of the Perissodactyla. *Evolution* **23**: 308–328.

Robbins, C. T. 1983. *Wildlife Feeding and Nutrition*. Academic Press, New York.

Romer, A. 1966. *Vertebrate Paleontology, 3rd ed.* University of Chicago Press, Chicago, IL.

Romer, A. 1968. *Notes and Comments on Vertebrate Paleontology*. University of Chicago Press, Chicago, IL.

Rozin, P. 1976. The selection of foods by rats, humans, and other animals. *Adv. Study Behav.* **6**: 21–76.

Scott, W. B. 1937. *A History of Land Mammals in the Western Hemisphere*. Macmillan, New York.

Senft, R. L., M. B. Coughenour, D. W. Bailey, L. R. Rittenhouse, O. E. Sala, and D. W. Swift. 1987. Large herbivore foraging and ecological hierarchies. *BioScience* **37**: 789–799.

Simpson, G. G. 1945. *The Principles of Classification and a Classification of Mammals*. Amer. Mus. Nat. Hist. Bull. 85, New York.

Sinclair, A. R. E. 1977. *The African Buffalo*. University of Chicago Press, Chicago, IL.

Stebbins, G. L. 1981. Coevolution of grasses and herbivores. *Ann. Missouri Bot. Gard.* **68**: 75–86.

Stewart, D. R. M. and J. E. Stewart. 1970. Food preference data of some East African wild ungulates. *Zool. Afr.* **5**: 115–129.

Talbot, L. M. 1965. A survey of past and present wildlife research in East Africa. *E. Afr. Wildl. J.* **3**: 61–85.

Talbot, L. M. and M. H. Talbot. 1962. Food preference of some East African wild ungulates. *E. Afr. Agric. For. J.* **27**: 131–138.

Webb, W. P. 1931. *The Great Plains*. Grosset and Dunlap, New York.

Wilson, D. E. and S. M. Hirst. 1977. Ecology and factors regulating roan and sable antelope populations in South Africa. *Wildl. Monogr. No. 54.*

Wilson, E. O. 1975. *Sociobiology. The New Synthesis*. Belknap Press, Harvard University Press, Cambridge, MA.

Wolfe, J. A. 1978. A paleobotanical interpretation of Tertiary climates in the Northern Hemisphere. *Am. Sci.* **66**: 694–703.

Young, J. Z. 1962. *The Life of Vertebrates, 2nd ed.* Oxford University Press, Oxford.

SECTION 6
Community Patterns in Natural and Agricultural Systems

"The limits of faunae are not always geographical in the ordinary sense, inasmuch as a genus or family of plants, or even animals, may support an insect fauna quite as peculiar to it as that of most countries or islands." This emphatic statement came from Cockerell (1882), an entomologist engaged in recording herbivorous insects on native and introduced plants in Jamaica. For a long time, few have developed his proposition at the herbivore community level. However, 100 years later, Cockerell's insect faunae have turned into a central theme of community ecology.

In few other subjects is the discrepancy of temperate and tropical studies so evident. The need for adequate data on insect assemblages on tropical plants is enormous. In this section, two chapters present information on such assemblages. In Chapter 23, insect associates of a selected plant taxon are surveyed on local and regional scales; Chapter 24 reports patterns of insects found on plants of a particular local community. A direct comparison with better-known temperate herbivorous assemblages cannot yet be attempted, but possibilities for such comparisons are manifest.

We thus have come full circle to the subject of our first section. Does the greater tropical diversity of many herbivorous taxa signify that tropical plants support more diverse assemblages, or does it only reflect the greater diversity of the plants themselves? This question, and its underlying processes, is closely tied to the herbivores' trophic specificity, the subject of the preceding section. A lot of effort is likely to be expended on these issues in the next years and, hopefully, improve our understanding of the essential features of tropical versus temperate communities.

Associations of herbivorous insects and their food plants are important not only in themselves, but for their practical consequences also. These are the subject of the final three chapters, which emphasize, in turn, adaptations to introduced crops (Chapter 25); the responses of insect herbivores to different cropping schemes (Chapter 26); and the effects of weeding and chemical control on insect assemblages (Chapter 27).

As early as 1760 the Jesuit João Daniel, imprisoned in Portugal, wrote: "In the Amazon lands the agricultural uses of other countries have no value, one must instead search for new methods of beneficing them." Temperate cultivating practices have failed often enough in tropical regions. The search for better cultural practices will certainly benefit from a more thorough understanding of plant–animal interactions within these settings.

REFERENCES

Cockerell, T. D. A. 1882. Notes on plant faunae. Insect Life 5: 117–121.

Daniel, J. 1757–1776(?). Tesouro descoberto no rio Amazonas. Bibliotecal Nacional, Rio de Janeiro, 1976.

23 Insects in Flower Heads of Asteraceae in Southeast Brazil: A Case Study on Tropical Species Richness

THOMAS M. LEWINSOHN

Phytophagous insect assemblages on plants have become a standard subject of community ecology, and yet their present standing in community studies and theory is somewhat equivocal. On the positive side, since Southwood (1960, 1961) explored the number of species associated with Hawaiian and British trees and attempted to account for them, a number of studies have exposed similar empirical correlations in a variety of plant and insect taxon combinations. Most important among these are correlations between plant area, or range, and number of associated species. Such species-area relations came to be even taken for granted; thus "...the most general expression for community geography is the species/area relationship (which) is to ecological biogeography as allometric relationships are to morphology" so that "species richnesses need special explanation only when they deviate from the species/area relationship appropriate for the region and organisms in question" (Strong 1979, p. 91).

True enough, a species-area relationship has been found in a number of insect–plant systems, and other factors not directly related to plant range seemingly affect phytophage species richness as well (Strong 1979; Strong et al. 1984; Lawton 1983; Kennedy and Southwood 1984). Nonetheless, the subject is far from settled. Questions have been raised on the generality of the relationships; the validity of data and analyses employed to evince species-area and connected relationships; on the explanatory power of the theory and hypotheses proposed, and the criteria to test them.

One particular problem concerns the spatial scale of the described relationships. In some cases, local assemblages showed no correlation with the geographical range of their host plants (Futuyma and Gould 1979; Karban and

Plant-Animal Interactions: Evolutionary Ecology in Tropical and Temperate Regions, Edited by Peter W. Price, Thomas M. Lewinsohn, G. Wilson Fernandes, and Woodruff W. Benson. ISBN 0-471-50937-X © 1991 John Wiley & Sons, Inc.

Ricklefs 1983). More recently, the connection between species richness at the local community level and at the regional scale has emerged as a key component of more complete explanations of the dynamics and maintenance of phytophagous insect assemblages (Cornell 1985a; Ricklefs 1987; Compton et al. 1989).

In this chapter, I will present the first results of a study on phytophagous insect richness in flower heads of Asteraceae in Southeast Brazil. This is one of the first studies on the subject in a tropical region, and also one of the few studies designed to obtain information toward this end, instead of making use of data gathered for other purposes.

The main challenge for such a study is the lack of previous information and background—taxonomic, bionomic, and biogeographic. This is a serious problem, and one may well ask whether, with these restrictions, it is at all feasible to obtain useful results. One must look for particular ways of overcoming these constraints and so, by necessity, one of the components of this chapter is strictly methodological. Other than that, I intend to address three questions:

1. Are there clear correlates—in particular, plant range–with total and local insect richness on different hosts?
2. What relation is there between local and total richness in this system?
3. Are local and/or regional assemblages richer in species than their temperate counterparts?

THEORETICAL EXPLANATIONS FOR THE SIZE OF PHYTOPHAGE ASSEMBLAGES

Explanations for the size of insect faunas associated with various plants have been reviewed several times (Strong 1979; Strong et al. 1984; Kennedy and Southwood 1984; Stevens 1986; Tahvanainen and Niemelä 1987). In this section I will give a brief outline of the relevant theory and try to clarify some terminological and conceptual ambiguities.

In the present study and all the ones which will be referred to, the number of species per plant or sampling unit is the only indicator of community diversity used. Species richness is a simple diversity measure (Pielou 1975) and, in many cases, the only one available. In what follows, "diversity" and "richness' will be used interchangeably and are equivalent.

"Local" richness is a vague term, but it implies a spatial scale congruent with interactive processes within communities. Whittaker (1960) named local diversity alpha diversity, and defined beta diversity as the diversity along a gradient; nowadays it means any kind of among-habitat, or among-site, diversity. Gamma diversity is the total diversity in the gradient, or the region considered, and is a combination of alpha and beta diversities. Cornell (1985a) calls it regional richness, when referring to phytophagous assemblages on host plants, thus stressing that often a study will only deal with a part of a plant species' range.

Total, or regional, richness

Most hypotheses for species richness set out to explain a species-area relationship which is either established or assumed. The following hypotheses are usually invoked (see also Strong et al. 1984):

1. Habitat diversity (Williams 1943): plants with larger ranges will occur in more kinds of habitat, each with its own set of potential or successful colonists.

2. Encounter frequency (Southwood 1961; Southwood and Kennedy 1983): larger plant ranges gives insects more opportunity of contact, and so increase chances of successful colonization (whatever the genetical, developmental, or behavioral mechanisms involved).

3. Area per se (Connor and McCoy 1979; called Equilibrium hypothesis in McGuinness 1984): the larger a plant's range, the larger the insect populations that live off it, and, consequently, the smaller their risk of extinction.

4. The equilibrium theory of Island Biogeography: its application to phytophagous insect communities was first proposed by Janzen (1968). Its full version combines the preceding two (Strong et al. 1984) in a dynamic equilibrium between colonization—immigration rates—which follow encounter-frequency rules—and extinction rates, established by the size of the area per se. Either the immigration or the extinction rate alone can establish the pattern, as long as a dynamic equilibrium is attained. There are several restrictions to the application and tests of this theory to phytophagous assemblages (Rey et al. 1981; Southwood and Kennedy 1983).

5. Random placement (also called passive sampling, Connor and McCoy 1979): this was put forward as a null hypothesis against which alternative explanations should be tested: "species number is controlled by passive sampling from the species pool, larger areas receiving effectively larger samples than smaller ones ... the correlation between species number and area is viewed solely as a sampling phenomenon, rather than the result of biological processes..." (Connor and McCoy 1979, p. 793). This hypothesis may serve for incidental species, such as the "tourists" in large samples obtained by fogging canopies (Southwood et al. 1983), but it would hardly apply to effective phytophages, unless the mechanics of colonizing a host were as simple as settling on a subtidal boulder. Random placement has sometimes been confused with the sampling effect proper, which follows.

6. The collecting effect: if sampling effort concentrates disproportionately on widespread plants, they will tend to have longer lists of recorded species even if they are not really richer in species. Passive sampling by the plants would produce a species-area pattern which is quite real, even if devoid of biological meaning; but a pattern derived from the collecting effect would be solely an artifact due to sampling bias. This, among other reasons, led Kuris et al. (1980) to question the use of species lists compiled for other purposes to detect species-area effects (but see Lawton et al. 1981; Southwood and Kennedy 1983).

Other factors which may promote phytophage richness patterns are not directly related to species-area effects. They include:

1. Structural complexity of plants (Lawton 1983): can be taken as a modality of habitat diversity, inasmuch as morphologically more complex organisms provide a larger and more heterogeneous set of microhabitats for phytophages (Strong et al. 1984; Lewinsohn 1986).

2. Taxonomic isolation: also a comprehensive factor which may refer to chemical distinctness from other plants, to the number of sympatric relatives, or to other features (Strong et al. 1984). Isolation has been treated as a form of distance from a colonization source (i.e., other plants) which affects immigration rates, and thus incorporated in the island equilibrium hypothesis (Janzen 1968).

One cannot expect to validate one of the area-related explanations over the others through analyses based entirely on overall insect record lists, no matter how reliable. Although several studies have investigated effects of various factors on total phytophage richness on different plants (Neuvonen and Niemelä 1981; Kennedy and Southwood 1984; Leather 1986), they do not—and cannot—differentiate, for instance, between habitat heterogeneity and area-per-se effects, nor can they assess the role of passive sampling or the collecting effect on the estimates of richness (Stevens 1986).

Other theoretical explanations for species richness require information on the size and composition of local phytophagous faunas.

The Relation Between Alpha and Beta Diversity

Stevens (1986) proposed a test to discriminate between the hypotheses of area-per-se, habitat diversity, and collecting effort (which he called passive sampling, confusing it with the random-placement hypothesis). A significant correlation of local richness with plant range is required to validate the area-per-se hypothesis. Under the habitat-heterogeneity hypothesis, additional samples in different localities should add more phytophage species than if the same sampling effort is expended in the same site; under the collecting-effort hypothesis, species accumulation should be equal for both alternatives.

As tempting as this scheme is, it does not cover all possibilities of combining within- and between-site richness generated by these hypotheses: for instance, if the same species have different abundance ranks in different localities, species will accumulate more quickly if new localities are added than in the same locality, even without any habitat-heterogeneity effect.

Cornell (1985a, b), with a quite distinct approach, proposed the following alternative hypotheses (offering a combination of both as a third possibility):

1. Local limitation: if a local assemblage is limited by intracommunity interactions, especially competition, its size should be largely independent of total richness, except in very depauperate faunas; on similar host species of different

total richness, mean local richness should converage to a common asymptote dictated by community dynamics. This is equivalent to the community saturation hypothesis (Lawton 1982; Strong et al. 1984).

2. Pool enrichment: if there is no upper limit to local community size, local richness should be proportional to total richness on the host species.

Although a species-area relationship is commonly assumed, it is not obligatory for either hypothesis. Under local limitation, plants with large faunas may be more widespread, in which case there is a significant geographic turnover of species between sites; or not, in which case total richness is controlled by area-unrelated factors (for instance, taxonomic isolation). The pool enrichment hypothesis, too, does not entail the existence of a species-area relation: it only states that plants with large local assemblages also have large total faunas, whether or not area contributes to determine these sizes.

Insect Richness on Tropical Plants

Is there a general well-established species-area relationship between tropical phytophagous insect faunas and their hosts? Strong (1979), in his review, refers to only three tropical examples: one is Southwood's (1960) pioneering treatment of insect faunas of Hawaiian trees; the other two are literature surveys of insects associated with cacao (Strong 1974) and sugar cane (Strong et al. 1977) in different world regions. Except for another study with five additional tropical crops (McCoy and Rey 1983), no further investigation of tropical species-area relations has been reported (Strong et al. 1984; Tahvanainen and Niemelä 1987).

The biota of the Hawaiian islands has a number of unique features, due to their extreme oceanic isolation and to their different age (Cox et al. 1976). The other studies deal with cash crops planted extensively at high densities, which moreover are of more or less recent introduction in most of the areas surveyed. Both factors have been shown to have strong influence on insect richness (Strong 1979; Strong et al. 1984; Janzen 1986). Thus, the existing data set on species-area relationships for tropical plants and insects is very limited and inadequate for general inferences.

The most obvious reason for the scarceness of studies of overall species richness is the dearth of lists of insect species associated with tropical plants, other than some crops. Even the few existing lists are of little use for a number of reasons:

1. Unreliable identification of the plants, the insects, or both.

2. Restricted utility because of unclear taxonomic and/or geographical limitations of the study.

3. Incomplete surveys, with no standardization of sampling effort on different plants/areas, and often without any indication of sampling effort at all.

Studies on local richness in tropical areas are almost as scarce as those on regional richness. In Costa Rica, Strong (1977) examined the richness of hispine beetles on *Heliconia* species, and Gilbert and Smiley (1978) the local diversity of heliconiine butterflies and of their *Passiflora* hosts. Other studies provide related information on phytophagous communities, but are not applicable to this study (Fernandes and Price 1988; Fernandes and Price, Chapter 5; Marquis, Chapter 9; Cytrynowicz, Chapter 24). Seed predators have been studied intensively on some hosts, especially leguminous trees (e.g., Janzen 1980; Lewinsohn 1980; Hopkins 1983). Endophagous seed predators have specific attributes related to their unique resource; to my knowledge, no one has searched for a seed-predator richness–plant area relationship at a local or regional level.

The need for additional studies on tropical species-area patterns is thus manifest. However, practical considerations will inevitably impose limits, geographical or taxonomical (insects, plants, or both) to any investigation.

THE STUDIED SYSTEM

Asteraceae are the largest family of flowering plants; they have worldwide distribution and are often among the foremost families in local or regional floras. Faunas of phytophagous insects on Asteraceae are correspondingly varied. Many Asteraceae are weeds of serious economic status and have been widely screened for potential biological control agents. Furthermore, insects on Asteraceae provide convenient microsystems for laboratory and field studies on the dynamics and evolution of phytophagous insects and their hosts (e.g., Zwölfer and Romstöck, Chapter 21, Futuyma, Chapter 19).

Unlike North American and Eurasian Asteraceae, there are few studies of phytophagous insects on neotropical Asteraceae, although in this region too the family is among the largest and most common. A few species of weeds which are serious invaders in other continents have been surveyed in their native Central and South American range for potential controllers (Bennett 1963; Cruttwell 1974; Cock 1982).

Methods and Study Sites

The survey concentrated on phytophages in flower heads, from which adult insects were reared. Insects were recorded from individual plants, although not from each flower head. A local sample of a plant species consisted of 1–4 series of 20 plants, in most cases totalling 40 or 60 plants.

Because the design of the study called for the largest possible number of samples, it was unfeasible to dissect flower heads systematically. Therefore, immatures and their activity patterns within the flower heads were not recorded, and species richness may have been underestimated in samples in which no adult of a given species ecloded.

The data set analyzed to date comprises 5115 individual plants in 266 sampling series, collected from 1985 to 1987. In all, 70 asteraceous species were collected; in 36 species, 60 plants or more were sampled and 28 species were sampled at two or more localities.

Sampling localities are shown in Figure 23-1. They encompass a wide range of geographical and habitat conditions. Lowland sites are on the coastal plain where plants were collected in dunes, scrubs, and forest and sand "restingas." In upland areas, above 600 m, plants were collected mostly in roadsides, old pastures, and also along trails and clearings within semideciduous or second-growth forest in low mountain ranges, up to about 1100 m.

Most highland localities are in the Mantiqueira, the foremost mountain range in southeast Brazil; samples were collected up to 2300 m, mostly in meadows, on open hillsides, or among rocky outcrops. Sites below about 1200 m in this range

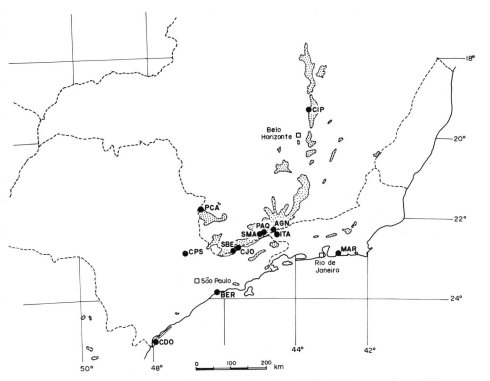

Figure 23-1. Main collection localities in southeast Brazil. Hatched areas, above 1200 m, indicate the main mountain ranges. Coastal localities, less than 100 m elevation: CDO—Ilha do Cardoso; BER—Bertioga; MAR—Maricá. São Paulo plateau: CPS—Campinas (600 m); Serra do Japi and Serra das Cabras (not shown) are close to Campinas (both 800–1100 m). Mantiqueira range: PAQ—Passa Quatro (900–1300 m); ITA—Itatiaia (700–1000 m); AGN—Agulhas Negras (1600–2300 m); SMA—Sertão dos Martins (1500–2000 m); CJO—Campos do Jordão (1500–1900 m); PCA—Poços de Caldas (1000–1300 m). SBE—São Bento do Sapucaí (800 m). Espinhaço range: CIP—Serra do Cipó (1000–1350 m).

were treated as different localities and classed as upland rather than highland. One mountain site, the Serra do Cipó, belongs to the Espinhaço range. It is occupied mostly by natural fields and meadows, with prominent rocky outcrops; physiognomically it is somewhat like the higher parts of the Mantiqueira, but floristically the two ranges are entirely distinct: they have almost no montane, and few widespread, Asteraceae in common.

Variables

Plant Species and Tribes. Tribes in Asteraceae are distinctive and "natural" subdivisions (Heywood et al. 1977). They also have quite distinct geographical distributions and, conversely, each continental flora has a different combination of major tribes (Jäger 1987). Within continents, tribes also have particular distributions which reflect their history as well as ecological preferences.

Preceding studies have concentrated on faunas of certain tribes or species groups; thus, Zwölfer (1988 and references) has worked on the Cynaroideae (formerly a single tribe) in Europe, and in the western USA, major studies have concerned Heliantheae (Goeden and Ricker 1986 and references) and Cardueae (Goeden and Ricker 1987 and references).

Table 23-1 outlines the taxonomic grouping of the Asteraceae included in this study. Major tribes in Brazil were sampled in rough proportion to their sizes; tribes not listed are absent in Brazil or represented by very few, usually introduced, species (Barroso 1986). Whenever possible, "tribe" was entered as a

TABLE 23-1. Number of Species and Genera of the Tribes of Asteraceae Included in the Present Study[a]

Tribe	Number of Species (Genera)					Representative Sampled Genera
	World		Brazil		Sampled	
Astereae	2500	(135)	169	(17)	7	(3)	*Baccharis, Conyza, Erigeron*
Eupatorieae	2000	(150)	336	(25)	22	(4)	*Eupatorium, Ageratum, Trichogonia, Mikania*
Heliantheae	4000	(210)	271	(42)	6	(5)	*Aspilia, Bidens*
Inuleae	2100	(180)	57	(21)	3	(2)	*Pluchea*
Mutisieae	1000	(89)	116	(28)	4	(3)	*Chaptalia, Dasyphyllum*
Senecioneae	3000	(100)	83	(7)	11	(5)	*Senecio, Erechtites*
Tageteae	250	(18)	15	(3)	1	(1)	*Porophyllum*
Vernonieae	1400	(70)	336	(30)	16	(5)	*Vernonia, Centratherum*
Total					70	(28)	
Total including other tribes	23000	(1300)	1400	(180)			

[a]Genera are in parentheses. Sizes of tribes are rounded and based on Barroso (1986), Jäger (1987) and H. F. Leitão Filho (unpublished). Some of the more extensively sampled genera of each tribe are also listed.

categorical variable in analyses, since insect association patterns may well diverge among tribes.

Taxonomic Isolation and Flower Head Size. The taxonomic unit considered in different studies may range from genus to order; as a rule, only species within the studied region are counted. As Table 23-1 shows, there are substantial differences between worldwide and regional (i.e., in Brazil) sizes of tribes, and this applies to genera as well. Several large and worldwide genera, as for instance *Erigeron*, have few species in Brazil. For taxonomic isolation, I used species numbers in Brazil, either per genus or per tribe; tribe size in general gave a slightly better fit and was more commonly used.

A further variable included in the study was the size of flower heads, also a feature of each host species; this was expressed as average dry weight, and no intraspecific variation was taken into account. Flower head size corresponds to the amount of resources individually packaged, from the standpoint of endophagous species. Dry weight is a rather crude measure because it sums protective and supporting structures, such as bracts, to flowers, ovaries, and seeds. Zwölfer (1987) used the diameter as a measure of flower head size.

Geographic Range. Only two studies on phytophagous species richness on Asteraceae have considered plant range. For Heliantheae in California, Goeden and Ricker (1986) summed the areas in all counties from which the plant species were recorded; likewise, for European Cynareae, Zwölfer (1987) employed the summed area of countries in whose national floras the plants appeared. Both are only rough approximations, since plants may vary widely in their spread within these units.

For Brazilian Asteraceae even such approximate information is unavailable. In the present study, plant species records were gathered in publications and from specimens in major Brazilian herbaria; additional information was offered by collectors and taxonomists. Rather than fabricate a pseudocontinuous measure of geographical distribution, I grouped plant species in three classes according to their distribution in SE Brazil, varying from highly endemic to ubiquous. In regression and path analyses, these classes were further reduced to a binary variable, to avoid problems with ordinal variables in these statistical procedures.

Plants were also grouped according to their elevational range in three zones, from coast to highland. Most species of narrow range are confined to highland areas, which reflects the high endemicity of many Asteraceae in neotropical mountains (cf. Smith and Cleef 1988). Conversely, most species that occur in coastal areas extend also into higher elevations.

Sampling Effort. An important component for analyses of incomplete surveys, sampling effort was expressed in three alternative ways: number of sampled localities; number of sampled individuals; and dry weight of collected flower heads. Although correlated, these three convey different kinds of information. Individuals and dry weight are measures of gross sampling, either for local collections or for all collections of a host species; the former is more useful for further field work because plants are readily counted, while the latter is more adequate for interspecific comparisons because sample size is reduced to a common resource unit. In analyses I generally employed sample weight, but in practice it made little difference to results which one was used (which was somewhat unexpected).

Number of localities, on the other hand, is a measure of the spatial extent of sampling, and it is dependent on the geographical spread and frequency of the plants. Lawton and Schröder (1978) even employed it as a measure of host-plant range for their analysis of Zwölfer's (1965) data. Although this is empirically convenient, it adds to the difficulty of separating the effect of plant range, which is a feature of the species, from the effect of collecting effort, which is variable and sampling-dependent.

The Flower Head Insects and Their Host Ranges

Table 23-2 outlines the main phytophagous groups found in flower heads of Brazilian Asteraceae to date. A detailed account of the composition, geography, and host relationships of these taxa will be presented elsewhere.

Three families of Diptera are outstanding components of this fauna, in species diversity, abundance, and frequency in individual plants. They are also found in more plant species, with host records in 33–41 of the 70 sampled species. Lepidoptera contribute substantially to overall phytophagous richness, but occur less often and in lesser numbers than the three major dipteran families. On the other hand, host ranges of four of the families are rather large, which reflects that most of the common species are quite polyphagous.

Tephritids comprise almost half of the 86 species in Table 23-2. The largest genus by far is *Tomoplagia* with 13 recognized species, all associated with Vernonieae. Most other tephritids too are associated with hosts within a single tribe. The 35 hosts recorded for tephritids belong to all sampled tribes, but with one isolated rearing only from Senecioneae; this is quite remarkable since *Senecio* species host several common tephritids in Europe (Cameron 1935) and, especially, in North America (Wasbauer 1972).

Most agromyzids are in the genus *Melanagromyza*; the recognized species are very similar and belong to the *minimoides* complex, which points to a monophyletic origin; however, their hosts span all the sampled tribes except for Inuleae. Although Asteraceae are the main host group for this genus—at least in North America—the flower and seed-feeding habit is comparatively minor (Spencer and Steyskal 1986), and the *minimoides* group may represent an instance of a shift to a novel plant organ preceding radiation and speciation on new hosts; this is one of the pathways to speciation invoked by Zwölfer (1982) for insects on European Cynaroideae.

The number of cecidomyiid species is certainly underestimated, because of the near impossibility in recognizing species from adult specimens only. In fact, in the most common genera, *Asphondylia* and *Neolasioptera*, neotropical species

TABLE 23-2. Major Phytophagous Insect Taxa Occurring in Flower Heads of Asteraceae in Southeast Brazil[a]

Herbivore Family	Number of Species	Abundance Rank	Frequency Rank	Number of Host Plants	Important Genera
DIPTERA					
Tephritidae	38	3	1	35	Tomoplagia, Xanthaciura, Trupanea, Urophora
Agromyzidae	14	2	3	33	Melanagromyza, Liriomyza
Cecidomyiidae	5[b]	1	2	41	Asphondylia, Neolasioptera, Dasineura
Drosophilidae	1	8	10	2	Drosophila
LEPIDOPTERA					
Cochylidae	7	6	6	26	Saphenista, Cochylis
Tortricidae	5	10	8	5	Clarkeulia
Gelechiidae	3	9	7	13	Recurvaria
Pterophoridae	6	5	4	27	Adaina, unidentified
Pyralidae	4	4	5	25	Phycitodes, Unadilla
COLEOPTERA[c]					
Apionidae	3	7	9	3	Apion

[a] For each family the number of species, its rank in number of reared adults (abundance) and in frequency of occurrence in samples, among the listed taxa, is given. Number of host species recorded for the family, and representative genera, are also shown.
[b] Adult morphospecies only; underestimates (see text).
[c] Excludes Curculionidae (see text).

535

described from Asteraceae are diagnosed only by larval characters. Zwölfer (1987) has excluded this family from his analyses on insect richness in European Cynareae because of the difficulty in separating some phytophagous and predacious larvae.

Another problem concerns the curculionids, which are important components of flower head faunas elsewhere (Zwölfer 1988). In Brazil, too, they were occasionally very abundant, but were excluded from the present analyses because in bulk samples one cannot separate true endophages from the majority of other species, of which only the adults feed on pollen or petals.

The taxa shown in Table 23-2 are all endophagous in the flower heads. Ectophagous larvae and nymphs are in general more mobile among flower heads, and they will often feed on protective structures rather than the flowers proper. Contrary to most endophages, often they are also able to transfer and feed on other parts of the plant. Because of the distinct character of their host association, many analyses were carried out for endophagous species separately. Analyses which refer to phytophages include, in addition, the more important ectophagous taxa: Lepidoptera (Geometridae) and Heteroptera (especially Lygaeidae, Miridae, and Rhopalidae).

Host ranges varied from a single to several plant species belonging to different tribes. Strict monophagy is exemplified by the tephritid *Tomoplagia argentiniensis*, found exclusively on *Vernonia scorpioides* in five localities extending from the coast to the lower reaches of the Mantiqueira range. Many insect species appeared in a single sample, in which case one cannot dissociate the restricted host range from their rarity. At the other extreme, wide polyphagy is shown for instance by the pterophorid *Adaina bipunctata*, reared from 17 hosts, and the agromyzid *Melanagromyza* cf. *erechtitidis*, with 16 host plants, although the latter may prove to be a species complex (Lewinsohn 1988). In both cases, hosts belong to three tribes, while, by contrast, the five known hosts of the cochylid *Cochylis securifera* belong each to a different tribe.

Although host ranges of different phytophages graded continuously between these extremes, one can recognize distinct specialist and generalist groups. Specialists are either monophagous or associated with a variable number of species within a genus or a genus group of the same tribe. Generalists occur on hosts of different tribes, although they often are preferentially associated with one or two tribes. Disregarding Cecidomyiidae for the reasons given in the preceding section, the majority of Diptera is specialized; the reverse holds for Lepidoptera, in ectophagous and endophagous families alike.

SPECIES RICHNESS

Local Richness

Since the number of plants sampled locally for each host varied from 20 to 80, both intra- and interspecific comparisons required a reduction to a common basis; this was accomplished through rarefaction, a method devised for this purpose (Sanders 1968; Hurlbert 1971). Rarefaction is superior to the comparison of synthetic diversity indices or of parameters from a fitted distribution because it does not depend on the choice and assumptions of a particular theoretical species distribution.

Rarefaction commonly employs the relative abundance of species within a single large sample to calculate the expected number of species in any sample of smaller size; thus different samples can be compared at any common stipulated size, which usually is given in number of individuals: in our case, the collection of insects obtained from a local plant sample. However, I preferred to use the absolute frequencies of phytophage species on individual plants, instead of their abundances. This produces rarefied frequency curves, which estimate the number of phytophage species expected in the flower heads of a given number of plants: this is easier to grasp and more useful than the number of species expected in a given number of insects. This procedure assumes that species are independently distributed among plant individuals. Rarefaction curves of species frequencies were calculated according to the formula of Shinozaki (1963) as corrected by Kobayashi (1974, 1979).

Figure 23-2 shows rarefied local frequency curves obtained for some plant species. A local sample of 40–60 plants was sufficient to estimate local phytophage richness in most cases, but in some, such as *Ageratum fastigiatum* in Itatiaia (Fig. 23-2*d*), the low frequency of all encountered insects precluded the curve from levelling off.

Samples of the same plant in different localities in general agreed well, suggesting that local richness varies much less within than among host species; see, for instance, *Ageratum conyzoides* (Fig. 23-2*c*) and *Vernonia scorpioides* (Fig. 23-2*e*). *Vernonia petiolaris* (Fig. 23-2*f*, pe) is an outstanding case, since in two localities of comparable altitude and plant abundance, it produced 1 and 9 insect species; no other plant approached such a large intersite difference.

A comparison of local richness among elevational zones was carried out to search for a possible large-scale habitat effect. This could not be done on pooled plant samples, because regions differ strongly in floristic composition. Instead, I examined all available intraspecific comparisons between regions (Table 23-3). There is obviously no trend for higher local species richness in any zone. This can also be seen, for instance, in the samples of *Ageratum conyzoides* and *Vernonia scorpioides* (Fig. 23-2*c, e*): their local richness curves from coastal, upland, and highland localities are completely intermingled.

There is no clear effect of plant range on the size of local faunas. Figure 23-2 also illustrates this point: each pair of graphs shows a species of larger distribution in the upper row, compared to one or more congeneric species of more restricted distribution, below. In *Senecio* there are larger local faunas in the widespread *S. brasiliensis*, compared to the three species restricted to high elevations in the Mantiqueira range. For *Ageratum* the reverse holds; the cosmopolitan *A. conyzoides* clearly has a poorer flower head fauna than *A. fastigiatum* which, although often locally abundant, is restricted to the upland

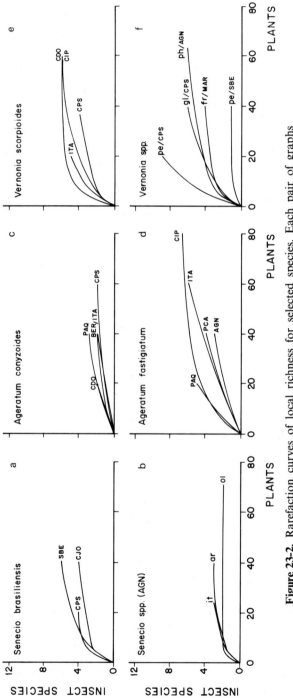

Figure 23-2. Rarefaction curves of local richness for selected species. Each pair of graphs compares congeneric species of different geographic areas. Species in the upper graphs have larger areas, lower ones have smaller ranges within the studied region. Locality codes for individual curves as in Figure 23-1. *Senecio* species are: *argyrotrichus* (ar), *itatiaiae* (it), *oleosus* (ol). *Vernonia* species are: *fruticulosa* (fr), *glabrata* (gl), *petiolaris* (pe), *phaeoneura* (ph).

TABLE 23-3. Pairwise Intraspecific Comparisons of Local Richness Among Elevational Regions[a]

Comparison		Difference in Local Richness			Binomial Probability
A	B	A > B	B > A	A = B	
Lowland	Upland	18	12	5	0.36
Upland	Highland	12	12	0	1
Lowland	Highland	2	2	0	1

[a] The 63 comparisons are from 19 plant species. Two-tailed binomial probabilities are given for the difference between A > B and B > A, with a null hypothesis of equal probability (Zar 1984).

plateau and mountains in SE Brazil. In *Vernonia*, local richness of the widespread *V. scorpioides* is roughly equivalent to its congeners of more restricted range, such as *V. phaeoneura* (Fig. 23-2*f*, ph), a species highly endemic to a part of the Agulhas Negras region at about 1700–2000 m.

Total Species Richness

Number of Localities. Plants sampled in more localities have larger flower head faunas, both of endophages and of all phytophages (Fig. 23-3); in a linear regression model, with each additional locality the total fauna increases by about two endophagous species, or about three phytophagous species. These species-locality regressions are untransformed, since exponential and power regressions did not fit the observations better than a simple linear one; this is often the case in species-area regressions, although they are routinely transformed (Connor and McCoy 1979).

Does this relationship, by itself, have any biological meaning, or is it essentially reproducing a sampling effect; and, is it possible to separate one from the other without additional information? Lawton and Schröder (1978) tried to overcome this difficulty by showing a better fit of the insect species–plant locality regressions in a power (log–log) model, than with an exponential (mono–log) model, which led them to conclude that "the most widespread plants appear to have more species of insects on them not just because they were sampled more intensively, but because they are more widespread" (Lawton and Schröder 1978, p. 60). Their reasoning depends on a combination of premises, especially on the validity of island equilibrium models for phytophagous insect faunas, and so cannot be accepted without corroboration of these assumptions.

Area. Plant geographic range, by itself, is also significantly correlated with total number of observed insect species. Rank correlation of plant area is highly significant both with total endophagous richness (Spearman's $rho = 0.52$, $df = 68$, $p < 0.001$) and with total phytophages ($rho = 0.53$, $df = 68$, $p < 0.001$).

This measure of plant range, contrary to the number of localities, is sampling-independent. Nonetheless, observed total richness, too, is dependent on sampling

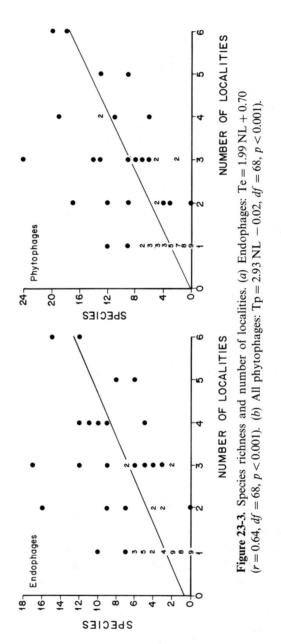

Figure 23-3. Species richness and number of localities. (*a*) Endophages: Te = 1.99 NL + 0.70 (*r* = 0.64, *df* = 68, *p* < 0.001). (*b*) All phytophages: Tp = 2.93 NL − 0.02, *df* = 68, *p* < 0.001).

effort unless the total fauna has been recorded, so that further samples cannot add new species. Thus, the richness-area effect may conceivably still be expressing a sampling phenomenon of no particular biological relevance.

Joint Analyses: Area, Sampling Range and Sampling Effort. To dissect the relationship between species richness and statistically causal variables—plant area and sampling intensity (either as bulk volume or as geographical extent)—two other approaches are employable. First, one may discount the effect of sampling before testing for differences between plants of different area (Karban and Ricklefs 1983). This was applied in an analysis of covariance among area categories, with plant species as observations and the number of localities as covariate. There was no significant heterogeneity between slopes, so that mean richness of area classes could be compared after adjusting for localities (Huitema 1983). Plants of limited range had smaller mean faunas than more widespread species, but the differences were not significant (F among area classes: endophages $= 0.79$, $p = 0.47$; phytophages $= 0.51$, $p = 0.6$, $df = 2/58$). By this procedure, therefore, there is no evidence that plant area has a direct effect on insect richness.

Alternatively, one can inspect the effect of these variables in a multivariate analysis. Path analysis seems especially appropriate, since it allows setting up a model of causal relationships between variables and then evaluating its fit to the correlations observed between them; it is thus more flexible than multiple regression (Li 1977; Schemske and Horvitz 1988). Cornell (1986) and Zwölfer (1987) have applied path analysis to investigate phytophagous insect richness.

In path analysis, the simple correlation coefficient r between a causal and a response variable is decomposed into direct and indirect effects, and a spurious fraction. The direct effect (or path coefficient p) corresponds to the standardized partial regression coefficient in a multiple regression, and represents the effect of that independent variable on the response variable, with other independent variables held constant. The indirect effect is the sum of all effects via chains of successive causal variables, according to the chosen model, excluding nondirectional correlations between causal variables, which are unanalyzable (Schemske and Horvitz 1988). Correlation coefficients are determined only by the observations and are model-independent; path coefficients are determined by the choice of the set of the causal variables that are supposed to affect each response variable, but are independent of the connections among causal variables. All other components—indirect, spurious, or unanalyzable—are entirely model-dependent and thus apply only to the chosen model. For this reason, the model should be chosen according to independent knowledge or assumptions about the causal relationships among variables. Tests of significance are often employed in path analysis, but they are biased against direct, as compared to indirect, effects (Kim and Kohout 1975) and so will not be employed here.

The multiple coefficient of determination R^2 can also be partitioned into components which indicate the contribution of each causal variable to the determination of the response variable. This partitioning, too, is model-dependent, and also affects the total value of R^2, which represents the fraction of

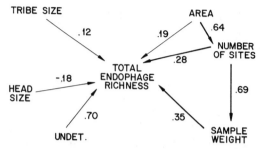

Figure 23-4. Path analysis model for total endophage richness in all sampled species. The arrows indicate direction of influence from causal to response variables, and numbers are path coefficients. Variables unconnected by arrows are completely independent for the model. The undertermined factor is the complement of the multiple determination coefficient R^2.

the variation of the response variable explained by the causal variables in the model.

Figure 23-4 shows a simple model of the relationship among host area, number of sampling localities, total sample weight, and total observed endophage richness; two other causal variables, flower head size and tribe size, are also included. The corresponding partitioning of observed correlation and determination coefficients is shown in Table 23-4. The model accords to the following assumptions: first, plant area has a direct effect on number of sampling localities, which affects total sample weight, but no direct effect of area on sample weight is expected; the inverse paths are meaningless (for instance, the number of localities in which a plant is sampled cannot affect its geographical area). Second,

TABLE 23-4. Path Analysis of Total Endophage Richness According to the Model in Figure 23-4[a]

Variable	Coefficients				Determination	
	Correlation (r)	Direct (p)	Indirect (i)	Effect (e)	Effect (e × p)	Total (r × p)
Area	0.542	0.188	0.355	0.542	0.102	0.102
Localities	0.644	0.283	0.242	0.525	0.148	0.182
Sample weight	0.640	0.351	0	0.351	0.123	0.225
Flower head size	− 0.179	− 0.179	0	− 0.179	0.032	0.032
Tribe size	0.122	0.122	0	0.122	0.015	0.015
R^2					0.420	0.556

[a]The simple correlation coefficient (r) is broken down into direct (p, or path coefficient) and indirect (i) components, whose sum is the total effect (e). The difference (r−e) is the spurious residue of the total correlation, according to the model. The multiple determination coefficient (R^2) is also broken down into components for each causal variable; the effect column (e × p) shows components of determination which are meaningful in the model. N = 70 species.

flowerhead size and tribe size may directly affect species richness, but there is no reason to assume any relationship between them, or with the other variables, so they enter the model as isolated causes.

Despite the fairly high correlation of plant area with species richness ($r = 0.54$), its path coefficient (0.19) is clearly lower than the one for number of sampling localities (0.28), and lower still than the direct effect of sample weight (0.35). Nonetheless, the total effect of area barely surpasses that of number of localities, and is much stronger than that of sample weight. This results mainly from two features of this particular model: the indirect effect of area by way of the other two variables, which accounts for two-thirds of its total effect, and the elimination of almost half of the correlation between sample weight and richness (0.29 out of 0.64) as spurious. The breakdown of the multiple determination coefficient R^2 reflects these patterns: out of a total 56% of variation in total endophage richness determined by the variables in the model, 42% are due to meaningful effects according to the model, with only slight differences among these three variables. The remaining causal variables, flower head size and tribe size, have little effect on total richness; together, they account for less than 5% of the observed variation.

In these observations there is a clear sampling effect, expressed by the path coefficient of sample weight. This variable is a "pure" measure of sampling effort, unaffected by the geographical extent of the samples; its magnitude relative to the alternative variables may indicate insufficient sampling and, in consequence, one could expect this component to diminish with further sampling. Not so with the number of localities; the importance of the geographical spread of sampling is shown both by its fairly high path coefficient and by its foremost position among the meaningful (i.e., effective) components of determination. Plant area, in this analysis, plays a minor determining role, and most of its effect on total endophage richness is indirect.

The model shown in Figure 23-4 is quite robust. Modifying relations among causal variables, or substituting their equivalents (genus instead of tribe size, or number of plants instead of weight) produces no important change. The model for total phytophage richness is similar to the one presented, except for the number of localities, whose path coefficient is increased to 0.37, compared to 0.28 for endophages. This suggests that proportionally more ectophagous than endophagous species are added with sampling in further sites or, in other words, that beta diversity is relatively more important in ectophage richness.

Local and Total Richness: Alternative Models

In this section I explore alternate models for interaction of local and total species richness. Local richness was set through rarefaction at a standard size of 40 plants. For these analyses, then, local richness of a plant species is the mean of all rarefied estimates of local richness in samples with 40 or more individuals, and only the 44 species with at least one local sample of this size were included.

Only interspecific patterns will be explored, although intraspecific variation can also be of interest.

The Source Pool Model. There is a strong relationship between local and total richness. Simple linear regressions show this:

$$L = 1.19 + 0.49\,Te \quad (N = 44, r = 0.79) \qquad \text{(total endophages)}$$

$$L = 1.72 + 0.32\,Tp \quad (N = 44, r = 0.69) \qquad \text{(total phytophages)}$$

in which L is mean local richness, and Te and Tp, respectively, total endophages and phytophages.

This relationship, as before, may be partially attributable to limited sampling, especially for plants sampled at a single locality, where total and local richness are almost equal (but not quite: local richness may be smaller than total observed richness because of rarefaction). Total richness becomes increasingly independent from local richness with the addition of further sampling localities. As a simple check for bias due to this interdependence, the above regressions were repeated for the 25 species sampled at two or more localities. The exclusion of plants sampled at a single site had almost no effect on regression slopes, although intercepts were lower; and the overall fit, as indicated by the correlation coefficients ($r = 0.83$ and 0.72, respectively) was actually improved.

Such regressions imply that total richness is an independent causal variable for local richness; in other terms, total richness is the size of a species pool from which local faunas are drawn. The inverse assumption, that local richness drives total richness, is also tenable, and will be examined.

A path model for local richness is presented in Figure 23-5 and Table 23-5. This model is simplified because estimates of local richness were set at a standard number of individuals, and so they already are corrected for sampling intensity. Number of localities has also been excluded because a priori there is no reason to suppose an effect on local richness, and this is supported by the low correlation ($r = 0.08$) between them. The eventual effect of the number of localities may be partially expressed by way of plant area. Total endophage richness is potentially dependent on the other three causal variables—area, flower head size and tribe size—and, as in the preceding model, these are deemed independent among themselves.

The strongest correlation of local richness by far is with total richness. In fact,

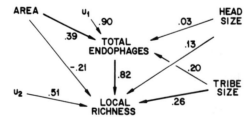

Figure 23-5. Path analysis model for local richness, with total richness among its causal variables. Note that the strong indirect effect of area on local richness is annulled by the negative path between them. U_1 and U_2 are the undertermined fractions of the variation in total and local richness, respectively.

TABLE 23-5. Path Analysis of Mean Local Richness According to the Model in Figure 23-5[a]

Variable	Coefficients				Determination	
	Correlation (r)	Direct (p)	Indirect (i)	Effect (e)	Effect ($e \times p$)	Total ($r \times p$)
Area	0.114	−0.212	0.320	0.108	−0.023	−0.024
Flower head size	0.136	0.134	0.021	0.154	0.021	0.018
Tribe size	0.405	0.259	0.159	0.419	0.109	0.105
Total endophages	0.786	0.815	0.000	0.815	0.664	0.641
R^2					0.771	0.740

[a]See Table 23-4 for explanation. Note that the total multiple determination coefficient is smaller than the one for effect, because the spurious correlation residue (not shown) detracts from the total determination. $N = 44$ species.

since there is no indirect effect and there is a negative spurious component, the path coefficient of total richness even exceeds its correlation coefficient. Total richness accounts for about 85% of the fairly high determination ($R^2 = 0.77$) attained by the model.

Tribe size accounts for the remaining 15% of the variance determined in this model. Most of this effect is exerted directly on the local fauna, as shown by the path coefficient, which is 1.6 times the indirect effect by way of total richness. Thus, tribe (or genus) size has a definite influence on local, but not on total, richness.

Plant area, in this model, has no part in determining local richness, as indicated by its low coefficients of correlation (0.11) and determination (− 0.02); a relatively strong indirect coefficient through total richness (0.32) is cancelled out by a negative path coefficient.

Local Richness as a Contributing Cause of Total Richness. Local richness, instead of being derived from total richness, may be taken to drive it. This assumption leads to a different model than the preceding one, in which the two variables change places; path analysis does not permit reciprocal causation between variables (Li 1977) and, anyway, the biological premises are quite distinct for the two models, which provide alternative descriptions of the local–total faunal dynamics.

A model for total endophage richness according to these assumptions is presented in Figure 23-6 and Table 23-6. Plant parameter variables—tribe and flower head sizes, and plant area—have direct and indirect paths to total richness, while sampling variables—number of localities and sample weight— have no effect by way of local richness, for the reasons given in the foregoing model.

This path model differs from the first one for total endophage richness (Fig. 23-4 and Table 23-4) in the addition of local richness and its structural consequences.

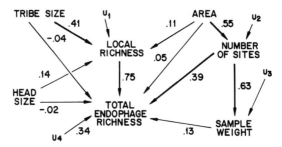

Figure 23-6. Path analysis for total endophage richness, including mean local richness among its causal variables. Compare with Figure 23-4, for a model without local richness, and with Figure 23-5, for the model in which total richness is a cause, not a response, variable to local richness. Except for tribe size, other variables are little affected in their relation to the richness measures.

TABLE 23-6. Path Analysis of Total Endophage Richness According to the Model in Figure 23-6[a]

| Variable | Coefficients | | | | Determination | |
	Correlation (r)	Direct (p)	Indirect (i)	Effect (e)	Effect (e × p)	Total (r × p)
Area	0.393	0.046	0.348	0.393	0.018	0.018
Localities	0.564	0.386	0.084	0.469	0.181	0.218
Sample weight	0.506	0.133	0	0.133	0.018	0.067
Flower head size	0.026	− 0.023	0.102	0.078	− 0.002	− 0.001
Tribe size	0.195	− 0.038	0.303	0.265	− 0.010	− 0.008
Local richness	0.786	0.749	0	0.749	0.561	0.589
R^2					0.766	0.883

[a]For explanation see Table 23-4. $N = 44$ species.

Some changes are caused by the reduction from all 70 plant species used for the first model, to the 44 species included in the second. This accounts for the differences between models in correlation coefficients of the same variables (Tables 23-4 and 23-6): although most variables in common have different absolute values, relative ones are close enough.

With the inclusion of local richness, it assumes most of the variation in total richness accounted for by the model. While there is no obligatory reason for the local–total path to be similar in the converse models, in fact, total and local richness have very similar effects on each other in either direction (compare Tables 23-5 and 23-6), in correlation, path and determination coefficients.

The roles of other variables are substantially affected by the addition of local richness. Sample weight becomes less important, so that in this model the effect of bulk sampling intensity on total richness is secondary. By contrast, the number of sampling localities still contributes about 21% to total determination, almost entirely through a strong direct path ($p = 0.39$) to total richness. Plant area, too, has an expressive total effect ($e = 0.39$), but in this case it is mostly indirect (0.35), with a very weak direct path. The dimensions of the indirect routes (Fig. 23-6) show that, in agreement with the preceding analyses, the main effect of plant area on total richness is exerted by way of number of localities, so that, in the end, it only reinforces the role of the spatial extent of sampling for total richness.

In keeping with the preceding models, tribe size has an insignificant direct effect on total richness. A strong link by way of local richness leads to a total effect coefficient of 0.27 but, with a negative and almost null path coefficient, taxon size can be discounted from the main determinants of total richness.

As before, changing certain variables for their counterparts: tribe size by genus size, or endophage by phytophage richness, entails no substantial change in the overall structure.

Contrasting the Models and Their Consequences. There is no formal test for determining a preferred causal direction between local and total richness, based solely on the confrontation between models. Once a path direction is chosen, from independent knowledge or premises about the mechanics of the system, the corresponding model can only serve for exploring the consequent network of influences among variables, and be judged by its fit to observed values.

The first total–local richness model treats the total fauna as a source pool from which local faunas are drawn, in combinations which may be more or less ordered, according to other determining processes. Hence, the emphasis of this model is on an overall limitation or exhaustion of potential colonists, with little effect from local interactions. Cornell (1985a, b, 1986) and Compton et al. (1989) employed this scheme in their analyses of local vs. regional richness. Stevens (1986) also based his proposition of alternative hypotheses on such a relationship.

In the second model, the emphasis is on local communities. A causal path from local to total diversity suggests that the incorporation of phytophagous species into a plant's fauna is governed by local dynamics, which include interactions among phytophages as well as with the host plant. If this is so, total richness will only express the summation of local faunas over the host species' range. The contribution of beta diversity to total richness will depend on among-site heterogeneity as well as the total extent of the plants' distribution. Zwölfer (1987) proposed a local-driven model for total insect richness on European Cynaroideae, without explicating the reasons for this choice; but they probably conform to the foregoing outline.

The confrontation of the two alternative total–local models is especially useful in showing that, whatever the theoretical scheme proposed, the connections of each causal variable to total and local richness are mostly unaffected by the causal relation between the two. Any difference can readily be explained by changes in the included variables. Thus, the minor direct influence of plant area on local richness is constant among models, while its apparent direct effect on total richness, in the first but not the second model, only reflects the inclusion of number of localities in the latter. Under either theoretical premise, plant area affects total richness mostly by way of the number of localities: its addition to beta diversity. No substantial effect on local richness is evinced in either case. The effect of flower head size is also unchanged between models: it has a minor but constant effect on local richness, and none on total richness.

Tribe size is the only factor to show a different effect: in the first model it has almost equal paths to total and local richness, while in the second one all its effect on total richness is indirect, through a path which in the total source pool model is deemed meaningless.

SPECIES RICHNESS OF TROPICAL AND TEMPERATE FLOWER-HEAD FAUNAS

Table 23-7 shows mean species numbers recorded on all sampled plants, as well as their mean local richness; the latter is not standardized, to permit direct comparison with insect richness in other regions.

Although differences in mean total richness among tribes are quite large, so is within-tribe variation. Total species richness, adjusted for number of localities, was not significantly different among tribes (Ancova, $F = 1.12$, $df = 7/61$, $p = 0.36$). Local richness was not significantly different among tribes either (excluding Mutisieae and Tageteae; single-factor Anova, $F = 1.27$, $df = 5/59$, $p = 0.29$). Although in path analyses tribe size affected local richness, this influence did not emerge as an overall significant difference between tribes, though with further sampling such a difference may be evinced.

Table 23-8 shows sizes of some phytophagous flower-head faunas surveyed in temperate areas. Other published insect surveys give insufficient information on

TABLE 23-7. Mean Recorded Species Richness of Endophages on Hosts of Different Tribes in SE Brazil[a]

Tribe	Number of Species	Total Richness			Local Richness		
		Mean	SD	Range	Mean	SD	Range
Astereae	7	7.29	6.45	2–17	3.64	2.15	2–7.7
Eupatorieae	22	5.41	4.46	0–16	3.84	3.23	0–12
Heliantheae	6	4.00	1.67	2–6	3.04	1.52	2–5
Inuleae	3	1.67	2.08	0–4	1.00	1.00	0–2
Mutisieae	4	2.00	2.16	0–5	1.75	2.21	0–5
Tageteae	1	9.00	—	—	3.50	—	—
Senecioneae	11	3.09	2.51	0–7	2.27	1.52	0–4.7
Vernonieae	16	4.00	3.93	0–12	2.76	2.05	0–5
ALL TRIBES	70	4.49	4.11	0–17	3.02	2.41	0–12

[a] All sampled plants are included ($N = 70$). Means are uncorrected for sampling intensity, to compare with Table 23-8. SD are sample standard deviations.

TABLE 23-8. Species Richness of Endophagous Insects in Flower Heads of Temperate Asteraceae Species

Region–Taxon	Number of Species	Total Richness			Local Richness		
		Mean	SD	Range	Mean	SD	Range
Europe							
Carlineae[a]	5	5.20	2.59	2–8	1.40	0.25	1–1.6
Echinopeae[a]	2	3.50	3.53	1–6	1.50	2.12	0.2–1.3
Centaureinae[a]	31	9.13	6.06	0–22	2.42	1.30	0–5.5
Carduinae[a]	40	7.50	4.20	0–17	1.99	0.81	0–4.0
All tribes	78	7.90	5.04	0–22	2.09	1.06	0–5.5
California							
Carduinae[b]	9	3.00	1.22	1–5	1.01	0.73	0.3–2.2
Heliantheae[c]	10	2.50	1.18	0–4	—	—	—

[a]From Zwölfer, unpublished; summarized in Zwölfer (1985).
[b]Goeden and Ricker (1987) and three preceding papers. Insects were selected according to information in species lists, with the same criteria used for Brazilian faunas. Introduced plants were excluded.
[c]Goeden and Ricker (1986) and six preceding papers; comments as above.

sampling localities, effort, or functional association of the insects, or are concerned with a single plant; thus, even for temperate regions few data sets are suitable for the intended comparison.

Heliantheae in SE Brazil averaged four species of total observed endophages (Table 23-7) against 2.5 in California (Table 23-8), but the difference does not attain significance (Mann-Whitney $U = 15$, $p < 0.10$). Since sampling intensity in Goeden and Ricker's study was much higher than mine, if anything the difference should increase with further sampling. It must be noted that all the plants examined in California are in the Ambrosiinae (and 9 of the 10 species in *Ambrosia*), while the Brazilian plants belong to other subtribes, and this may account for taxonomical differences among their faunas: for instance, agromyzids, among the most common endophages in the Brazilian Heliantheae sampled, are entirely absent from the Ambrosiinae in California, while on *Bidens pilosa* in Florida they are also common (Needham 1948). Differences in size of faunas may therefore be connected with their distinct composition, and so with the history of the assemblages; hence one can hardly attribute the larger faunas in Brazil to their tropical setting alone. Unfortunately, Goeden and Ricker (1986) did not give information of species richness in local samples, so that size of local assemblages cannot be compared.

European and Brazilian flower-head faunas, although having no plant tribes in common, can be contrasted both for total and local endophage richness, keeping in mind that the European Cynaroideae were more extensively and thoroughly sampled (Zwölfer 1965, 1985, 1987). If one considers mean richness on tribes, as shown in Tables 23-7 and 23-8, there is no difference in observed total richness between continents (Mann-Whitney $U = 9$, $p = 0.23$), but this gives undue influence to small and less-sampled tribes. A comparison of plant species (ignoring tribes) gives significantly larger sizes for faunas on European than on Brazilian Asteraceae (Anova, $F = 20.1$, $df = 1/146$, $p < 0.001$). This is strongly related to the extent of sampling; if species richnesses are adjusted to number of sampling localities, the difference becomes nonsignificant (Ancova, $F = 1.63$, $p = 0.20$, $df = 1/145$): adjusted means are 5.8 endophages/host species for Brazil and 6.7 for Europe.

As to local richness, the picture is different. Almost all Brazilian tribes have larger average local assemblages than the European ones, although a nonparametric comparison of tribe means between continents is not significant (Mann-Whitney $U = 4$, $p = 0.08$). On average, Brazilian Asteraceae have 3.0 local endophage species, compared to 2.1 on European Asteraceae. An analysis of covariance shows no effect of sampling extent, as expected, so that adjusted means are only slightly different from those.

Taken together, these comparisons suggest that local assemblages are richer in species in tropical than in temperature regions. Total faunas, on the other hand, are not yet well defined: if, as it seems, the geographical extent of sampling is everywhere an important determining factor of total observed faunas, then these, too, will tend eventually to be larger in tropical regions, which are comparatively undersampled. If that is so, then neotropical faunas will be richer than palearctic ones at all scales.

If, on the other hand, the patterns of the present data sets are reinforced by additional sampling, one is left to explain why local assemblages should, on average, be larger on neotropical Asteraceae, while total flower-head faunas are not. It is tempting to advance suppositions of different mechanisms at each scale, so that community dynamics would set sizes of local faunas, while total assemblages would be due to other factors. However, I do not think such explanations are called for, because I expect that additional sampling will confirm that local and total faunas are coherent, at least with regard to size.

Apart from size, the path model in Figure 23-6 shows some striking similarities with the corresponding model obtained by Zwölfer (1987, Fig. 4). In both, local richness is affected by taxon size, but not by other variables, notably by plant area; and total richness is mainly determined by local richness and by number of sampling localities. Flower head size has no significant effect on either local or total richness; and other variables, experimented in his model but not included in the present study, are also trivial for species richness. Whatever differences may exist between the temperate and tropical insect faunas in Asteraceae, their underlying structure and connections appear to be very similar.

CONCLUSIONS

Equilibrium, Time, and History

There does not seem to be much sense in attempting further tests of the hypotheses summarized in the introduction, as they stand. To a greater or lesser extent, they are quite compatible, and so very hard to test unambiguously; or else they require tests that are unfeasible in practice. A similar point has been made by McGuinness (1984) with reference to the difficulty of associating the same hypotheses to mathematical or statistical distribution functions, and to discriminate among the expected species-area functions.

All the proposed hypotheses depend, more or less directly, on a premise of equilibrium, since phytophagous faunas can only reflect a causal variable if they tend to stabilize at different sizes according to that effect. If the assemblages on each host change at different rates and are determined to different degrees by a common factor, one cannot expect to find a close correspondence between species richness and that factor.

Presupposing an equilibrium state in contemporary tropical phytophagous faunas is doubly hazardous. First one must suppose that such equilibria exist in general; and if they do, one must evaluate the extent to which they are presently attainable. Although steady sizes may be reached rapidly in ecological time (Strong 1974, 1979), tropical regions are being subject to sudden and large-scale changes with some unpredictable consequences: time of plant residence, a robust contributing factor for insect faunas of British trees (Kennedy and Southwood 1984) is virtually worthless under such conditions.

As an illustration, forest cover of the state of São Paulo has been reduced from 80% (a conservative estimate) to about 5% in the last two centuries, and this has certainly affected the plants and possibly their flower head associates. Thus, *Erechtites valerianaefolia* is listed as a noxious weed in the region (Leitão Filho et al. 1975), but I had some trouble even in finding it in sufficient numbers for sampling. This plant requires rich soils, so it is (or rather, was) especially common in recently cleared forest areas, which may explain why a so-called weed nowadays seems to be turning into an occasional and scarce species! Conversely, previously fugitive species may be turning rapidly into common and abundant species in man-made settings; I suppose that *Porophyllum ruderale* qualifies for this category. Both plants support rather rich faunas of similar size (*E. valerianaefolia*: 8 species total, 4.6 local average; *P. ruderale*: 9 and 3.5, respectively) but I wonder whether this is a temporary coincidence, before one loses and the other gains further associates.

The particularities of the plants' and insects' past histories are being assigned a progressively larger role in the explanation of faunal associations in different regions. The preexisting matrix of related plants and potential colonizers may account for seemingly extraordinary rich or impoverished faunas, which are not explained by local deterministic processes (Lawton 1982; Ricklefs 1987; Tahvanainen and Niemelä 1987).

Faunas on thistles (Carduinae) in California are clearly depauperate compared to their European counterparts which, both regionally and locally, are about twice as rich as the former. They also differ in the proportion of endophages to ectophages, and the proportion of specialists to generalists (both are higher in Europe) as well as the taxonomic composition of the faunas (Goeden and Ricker 1987; Zwölfer 1988). Zwölfer (1987, 1988) and Goeden and Ricker (1987) argue that, to a large extent, the difference is due to the more recent origin of the North American Cynaroideae; few insects from the rich Palearctic cynaroid fauna tracked their hosts along the Bering connection, and few Nearctic phytophages were able to switch to these plants. If they are correct, then the history of the plant group combined with the preexisting faunal matrix would account for a better part of the striking difference between continents, without recourse to any contemporary climatical, biogeographical, or ecological feature; indirectly, this would also diminish the relevance of a global temperate–tropical contrast.

The preceding discussion seems to delineate a picture of fundamentally noninteractive phytophagous assemblages. This may be valid for some cases, but by no means for all. Competition between cooccurring insects in asteraceous flower heads has been established several times (Zwölfer 1979; Neuenschwander 1984; Crawley and Pattrasudhi 1988) and Zwölfer (1985) adduced a set of "assembly rules" from observed guild patterns. Together, these studies evince the nonrandom combination of locally cooccurring insects due to interactive processes. Flower head assemblages thus show distinct patterns and influences at different spatial scales, and a dual set of explanations may be required to elucidate them.

Alpha and Beta Diversities

Joint investigation of local and total richness has proved especially rewarding, but, unfortunately, few studies provide the necessary information. For North American cynipid gallers on oaks, Cornell (1985a) concluded that among-site (beta) accumulation makes a fixed contribution to total richness, while local (alpha) diversity increases with total (gamma) richness. In his view, total richness is a cause of local richness; he is less concerned with the eventual reverse influence from local assemblages to total richness, and so beta diversity is also a minor determinant of total richness (Cornell 1985a, b, 1986). Similar conclusions were reached by Compton et al. (1989) for bracken-feeding insects.

The present study, on the contrary, suggests that a large share of the variation in total richness can be attributed to among-site accumulation, without excluding a significant local–total relation. A similar conclusion was reached by Zwölfer (1987) in the European Cynareae, and by Stevens (1986) in North American woodborers.

It remains to be ascertained why there is such a significant geographical turnover of phytophages. Although assembly rules point at a competitive or otherwise interactive structuring of local communities (Zwölfer 1979, 1985), the strong correlation of local and total faunas indicates that there is no internal limitation of local size. A causal path from local to total richness would account for this, but then, one would be left to explain why there is a larger among-site accumulation on species with locally larger faunas (which Cornell did not find in cynipids).

If local dynamics are secondary, insects may be limited by habitat restrictions and this has been verified in several cases (Strong et al. 1984). However, I believe that many local absences reflect the ephemeral and unstable nature of their resource base, with large and often unexplained variations in the insects' populations (Crawley and Pattrasudhi 1988), so that local extinctions may be relatively frequent. Composition of local communities would then be alterable by rather frequent extinctions and recolonizations from the total insect matrix. Taken together, they would form a regional mosaic of more or less unpredictable and possibly nonequilibrial local assemblages.

Polyphagy

Despite the assertion that host area is the most common and often also the strongest correlate of phytophage richness (Strong 1979; Kennedy and Southwood 1984), the generality and strength of this connection have been questioned for some time. In some cases no significant relation of plant range to species richness was found, particularly of local assemblages (Futuyma and Gould 1979; Karban and Ricklefs 1983); in others, although range was a significant factor, it contributed little to explain the overall variation in species richness (Lawton and Price 1979; Claridge and Wilson 1981).

Instead of examining the supposed exceptions to the species-area rule (Strong 1979), the evidence thus seems to induce a converse question: when does plant range give a good estimate of insect richness? From a survey of the literature, Tahvanainen and Niemelä (1987) concluded that area explained more of the variation in phytophage richness on trees and shrubs, and on plants with poor or little diversified chemical defenses; at the same time, a better fit was obtained with polyphagous and/or ectophagous insects, especially when wider taxonomical assemblages were considered, as already noted by Kennedy and Southwood (1984).

Others have alluded to the connection of insect polyphagy and species-area relations. With regard to Asteraceae, Lawton and Schröder (1978) and Zwölfer (1982) concluded by different approaches that specialist richness is less better predicted by plant range than generalist richness. Goeden and Ricker (1986) have found a significant species-area relation for the total faunas on Heliantheae ($r = 0.79$, $p < 0.01$, $df = 10$), but this can be shown to be produced by ectophagous species; for all species on flower heads, the relation is almost equal ($r = 0.81$, $p < 0.001$) but for endophages alone it is nonsignificant ($r = 0.37$, $p > 0.05$); since ectophagy and polyphagy are correlated, this too agrees with the above pattern. Furthermore, contingency analyses show a similar trend in some polyphagous taxa associated with Brazilian Asteraceae (Lewinsohn 1988).

Taxonomic Isolation

Taxonomic isolation, too, can be brought into an integrated scheme. At the evolutionary level, larger plant taxa may promote the appearance of new associates by facilitating radiation onto other species and the subsequent formation of host races (Zwölfer and Romstöck, Chapter 21); this is a variation of the exposure-frequency effect (Janzen 1968; Southwood and Kennedy 1983). At a more immediate level, larger plant taxa tend to have more species in local communities, with two possible consequences for oligophages and particularly for polyphages: first, the local set of related species may effectively form a resource pool so that, in combination, they exceed an acceptability threshold according to optimal foraging rules (MacArthur 1972), or more simply, together they support larger populations—an area-per-se effect. Second, cooccurring host plants may have divergent phenologies and tolerances, so that together they facilitate bridging unfavorable periods.

Zwölfer (1987) and I have found a direct path of host taxon size with local, but not with total, phytophage richness. This may signify, for these insects at least, that the immediate ecological processes play a greater role in establishing a taxon size/insect richness relation than the longer-term evolutionary responses, although they are complementary rather than alternative. Further evidence is adduced by the more frequent occurrence of polyphages on widespread plants, despite the host records showing no preferential association with them (Lewinsohn 1988).

A Joint Model

In retrospect, several correlations have been at least tentatively established in different insect–plant systems:

1. Plant range is better correlated with numbers of generalist than of specialist species.
2. Plant range has little direct effect on local richness, so that species-area relations are generated mostly by beta-diversity.
3. Local abundance and range size are correlated (Brown 1984), so that widespread plants should on average be locally more abundant.
4. Plant local abundance influences the number of associated species, especially of polyphages, through a resource concentration effect (Root 1973; but see Karban and Ricklefs 1983, Futuyma and Gould 1979).

From the foregoing, and extending the conclusions of Tahvanainen and Niemelä (1987), I submit that the species-area relation may be strongest for combinations of polyphagous (or, by correlation, ectophagous) insects with widespread plants.

Although each has been often considered by itself, there is certainly a connection between the size of insect assemblages on different plants, on the one hand, and host range of the insects, on the other. The beta component of plant faunas, and the "host turnover" of locally specialized generalists (Fox and Morrow 1981), are equivalent insofar as both are among-site additions to their respective total ranges. To attain a better understanding of the structure and shaping of insect-plant assemblages, it may be necessary to analyze jointly the plants' insect spectra and the insects' host spectra, and their respective rates of accumulation at different spatial scales; and temporal scales as well, but that may be too much to expect.

As a final practical comment, I am convinced that one need not have an exhaustive survey of a plant's fauna to address the questions posed in this chapter. If local collections are large enough to obtain estimates of local richness independent of sampling intensity; and if enough localities are surveyed to produce reliable estimates of beta accumulation rates, one has enough elements to assess the build-up of total richness without knowing its actual size. This is especially pertinent to studies of tropical communities, due to the difficulties in obtaining complete and large-scale surveys.

ACKNOWLEDGMENTS

I thank the many friends who helped me on field trips and with the laborious rearing and mounting of insects. Several specialists helped with the identifications, especially Victor O. Becker, Ângelo P. Prado, and Hermógenes F. Leitão Filho. I thank Howard Cornell, John Lawton, and Richard Goeden for reading and commenting on the manuscript and Rogério P. Martins, Peter W. Price, Woodruff W. Benson, and many others for stimulating discussion of these ideas. Special thanks are due to Helmut Zwölfer for unrestricted access to his ideas, experience, and unpublished data. Esmeralda Z. Borghi drew the illustrations.

Financial support for the field work was provided by FAPESP (grant 363-4/85). Part of this work was presented to the Universidade Estadual de Campinas in partial fulfillment of requirements for the DSc degree.

REFERENCES

Barroso, G. M. 1986. Sistemática de Angiospermas do Brasil. Vol. 3. Imprensa Universitária, Univ. Fed. Viçosa, Viçosa.

Bennett, F. D. 1963. Final report on surveys of the insects attacking *Baccharis* spp. in the S. E. United States of America and Brazil, 1960–63. *Commonw. Inst. Biol. Contr. Report* **27**: 1–27.

Brown, J. H. 1984. On the relationship between abundance and distribution of species. *Am. Nat.* **124**: 255–279.

Cameron, E. 1935. A study of the natural control of ragwort (*Senecio jacobaea* L.). *J. Ecol.* **23**: 265–322.

Claridge, M. F. and M. R. Wilson, 1981. Host plant associations, diversity and species-area relationships of mesophyll-feeding leafhoppers of trees and shrubs in Britain. *Ecol. Entomol.* **6**: 217–238.

Cock, M. J. W. 1982. Potential biological control agents for *Mikania micrantha* HBK from the neotropical region. *Trop. Pest Manag.* **28**: 242–254.

Compton, S. G., J. H. Lawton, and V. K. Rashbrook. 1989. Regional diversity, local community structure and vacant niches: The herbivorous arthropods of bracken in South Africa. *Ecol. Entomol.* **14**: 365–373.

Conner, E. F. and E. D. McCoy. 1979. The statistics and biology of the species-area relationship. *Am. Nat.* **113**: 791–833.

Cornell, H. V. 1985a. Local and regional species richness of cynipine gall wasps on California oaks. *Ecology* **66**: 1247–1260.

Cornell, H. V. 1985b. Species assemblages of cynipid gall wasps are not saturated. *Am. Nat.* **126**: 565–569.

Cornell, H. V. 1986. Oak species attributes and host size influence cynipine wasp species richness. *Ecology* **67**: 1582–1592.

Cox, C. B., I. N. Healy, and P. D. Moore. 1976. Biogeography—An ecological and evolutionary approach. 2nd. ed. Blackwell, Oxford.

Crawley, M. J. and R. Pattrasudhi. 1988. Interspecific competition between insect herbivores: Asymmetric competition between cinnabar moth and the ragwort seed-head fly. *Ecol. Entomol.* **13**: 243–249.

Cruttwell, R. E. 1974. Insects and mites attacking *Eupatorium odoratum* in the Neotropics 4. An annotated list of the insects and mites recorded from *Eupatorium odoratum* L., with a key to the types of damage found in Trinidad. *Tech. Bull. Commonw. Inst. Biol. Control* **17**: 87–125.

Fernandes, G. W. and P. W. Price. 1988. Biogeographical gradients in galling species richness—tests of hypotheses. *Oecologia* **76**: 161–167.

Fox, L. R. and P. A. Morrow 1981. Specialization: Species property or local phenomenon? *Science* **211**: 887–893.

Futuyma, D. J. and F. Gould. 1979. Associations of plants and insects in a deciduous forest. *Ecol. Monogr.* **49**: 33–50.

Gilbert, L. E. and J. T. Smiley. 1978. Determinants of local diversity in phytophagous insects: Host specialists in tropical environments. pp. 89–104 in Mound, L. A. and N. Waloff (Ed.). Diversity of Insect Faunas. Blackwell, Oxford.

Goeden, R. D. and D. W. Ricker. 1986. Phytophagous insect fauna of the desert shrub *Hymenoclea salsola* in Southern California. *Ann. Entomol. Soc. Am.* **79**: 39–47.

Goeden, R. D. and D. W. Ricker. 1987. Phytophagous insect faunas of native *Cirsium* thistles, *C. mohavense, C. neomexicanum,* and *C. nidulum,* in Southern California. *Ann. Entomol. Soc. Am.* **80**: 161–175.

Heywood, V. H., J. B. Harborne, and B. L. Turner. 1977. An overture to the Compositae. Pp. 1–20 in Heywood, V. H., J. B. Harborne, and B. L. Turner (Ed.). The Biology and Chemistry of the Compositae. Vol. I. Academic Press, London.

Hopkins, M. J. G. 1983. Unusual diversities of seed beetles (Coleoptera: Bruchidae) on *Parkia* (Leguminosae: Mimosoideae) in Brazil. *Biol. J. Linn. Soc.* **19**: 329–338.

Huitema, B. E. 1983. The Analysis of Covariance and Alternatives. John Wiley, New York.

Hurlbert, S. H. 1971. The nonconcept of species diversity: A critique and alternative parameters. *Ecology* **52**: 577–586.

Jäger, E. J. 1987. Arealkarten der Asteraceen-Tribus als Grundlage der oekogeographischen Sippencharakteristik. *Bot. Jahrb. Syst.* **108**: 481–497.

Janzen, D. H. 1968. Host plants as islands in evolutionary and contemporary time. *Am. Nat.* **102**: 592–595.

Janzen, D. H. 1980. Specificity of seed-attacking beetles in a Costa Rican deciduous forest. *J. Ecol.* **68**: 929–952.

Janzen, D. H. 1986. Does a hectare of cropland equal a hectare of wild host plant? *Am. Nat.* **128**: 147–149.

Karban, R. and R. E. Ricklefs. 1983. Host characteristics, sampling intensity, and species richness of Lepidoptera larvae on broad-leaved trees in southern Ontario. *Ecology* **64**: 636–641.

Kennedy, C. E. J. and T. R. E. Southwood. 1984. The number of species of insects associated with British trees: A re-analysis. *J. Anim. Ecol.* **53**: 455–478.

Kim, J. O. and F. J. Kohout. 1975. Special topics in general linear models. Pp. 368–397 in Nie, N. H., C. H. Hull, J. G. Jenkins, K. Steinbrenner, and D. H. Bent (Ed.). Statistical Package for the Social Sciences. 2nd ed. McGraw-Hill, New York.

Kobayashi, S. 1974. The species-area relation. I. A model for discrete sampling. *Res. Pop. Ecol.* **15**: 223–237.

Kobayashi, S. 1979. Species-area curves. Pp. 349–368 in Ord, J. K., G. P. Patil, and C. Taillie (Ed.). Statistical Distributions in Ecological Work. International Coop. Publ. House, Fairland, MD.

Kuris, A. M., A. R. Blaustein, and J. J. Alió. 1980. Hosts as islands. *Am. Nat.* **116**: 570–586.

Lawton, J. H. 1982. Vacant niches and unsaturated communities: A comparison of bracken herbivores at sites on two continents. *J. Anim. Ecol.* **51**: 573–595.

Lawton, J. H. 1983. Plant architecture and the diversity of phytophagous insects. *Ann. Rev. Entomol.* **28**: 23–39.

Lawton, J. H., H. Cornell, W. Dritschilo, and S. D. Hendrix. 1981. Species as islands: Comments on a paper by Kuris et al. *Am. Nat.* **117**: 623–627.

Lawton, J. H. and P. W. Price. 1979. Species richness of parasites on hosts: Agromyzid flies on British Umbelliferae. *J. Anim. Ecol.* **48**: 619–637.

Lawton, J. H. and D. Schröder. 1978. Some observations on the structure of phytophagous insect communities: The implications for biological control. Proc. 4th int. Symp. Biol. Control of Weeds, Gainesville, FL, pp. 57–73.

Leather, S. R. 1986. Insect species richness of the British Rosaceae: The importance of host range, plant architecture, age of establishment, taxonomic isolation and species-area relationships. *J. Anim. Ecol.* **55**: 841–860.

Leitão Filho, H. F., C. Aranha, and O. Bacchi 1975. Plantas invasoras de culturas no Estado de São Paulo. Vol. II. Hucitec, S. Paulo.

Lewinsohn, T. M. 1980. Predação de sementes em *Hymenaea* (Leguminosae: Caesalpinioideae): aspectos ecológicos e evolutivos. MSc Thesis, Unicamp, Campinas.

Lewinsohn, T. M. 1986. Plantas hospedeiras, heterogeneidade ambiental e diversidade de insectos fitófagos. Pp. 39–72 in Tundisi, J. G. (Ed.). *Perspectivas de Ecologia aplicada.* Acad. Ciencias do Estado de São Paulo, São Paulo.

Lewinsohn, T. M. 1988. Composição e tamanho de faunas associadas a capitulos de Compostas. DSc Thesis, Unicamp, Campinas.

Li, C. C. 1977. *Path Analysis—A Primer.* 2nd. ed. Boxwood Press, Pacific Grove, CA.

MacArthur, R. H. 1972. *Geographical Ecology.* Harper and Row, New York.

McCoy, E. D. and J. R. Rey 1983. The biogeography of herbivorous arthropods: Species accrual on tropical crops. *Ecol. Entomol.* **8**: 305–313.

McGuinness, K. A. 1984. Equations and explanations in the study of species-area curves. *Biol. Rev.* **59**: 423–440.

Needham, J. G. 1948. Ecological notes on the insect population of the flowerheads of *Bidens pilosa*. *Ecol. Monogr.* **18**: 431–446.

Neuenschwander, P. 1984. Observations on the biology of two species of fruit flies (Diptera: Tephritidae) and their competition with a moth larva. *Israel J. Entomol.* **18**: 95–97.

Neuvonen, S. and P. Niemelä. 1981. Species richness of Macrolepidoptera on Finnish deciduous trees and shrubs. *Oecologia* **51**: 364–370.

Pielou, E. C. 1975. *Ecological Diversity*. Wiley, New York.

Rey, J. R., E. D. McCoy, and D. R. Strong, Jr. 1981. Herbivore pests, habitat islands, and the species-area relation. *Am. Nat.* **117**: 611–622.

Ricklefs, R. E. 1987. Community diversity: Relative roles of local and regional processes. *Science* **235**: 167–171.

Root, R. B. 1973. Organization of a plant–arthropod association in simple and diverse habitats: The fauna of collards (*Brassica oleracea*). *Ecol. Monog.* **43**: 95–124.

Sanders, H. L. 1968. Marine benthic diversity: A comparative study. *Am. Nat.* **102**: 243–282.

Schemske, D. W. and C. C. Horvitz. 1988. Plant–animal interactions and fruit production in a neotropical herb: Path analysis. *Ecology* **69**: 1128–1137.

Shinozaki, K. 1963. Note on the species-area curve. *Proc. 10th Annual Meeting, Ecol. Soc. Japan*, Tokyo, 5.

Smith, J. M. B. and A. M. Cleef. 1988. Composition and origin of the world's tropicalpine floras. *J. Biogeogr.* **15**: 631–645.

Southwood, T. R. E. 1960. The abundance of the Hawaiian trees and the number of their associated insect species. *Proc. Hawaiian Entomol. Soc.* **17**: 299–303.

Southwood, T. R. E. 1961. The number of species of insect associated with various trees. *J. Anim. Ecol.* **30**: 1–8.

Southwood, T. R. E. and C. E. J. Kennedy. 1983. Trees as Islands. *Oikos* **41**: 359–371.

Spencer, K. A. and G. C. Steyskal. 1986. Manual of the Agromyzidae (Diptera) of the United States. USDA, Washington.

Stevens, G. C. 1986. Dissection of the species-area relationship among wood-boring insects and their host plants. *Am. Nat.* **128**: 35–46.

Strong, D. R., Jr. 1974. Rapid asymptotic species accumulation in phytophagous insect communities: The pests of cacao. *Science* **185**: 1064–1066.

Strong, D. R., Jr. 1977. Insect species richness: Hispine beetles of *Heliconia lathispatha*. Ecology **58**: 573–582.

Strong, D. R., Jr. 1979. Biogeographic dynamics of insect–host plant communities. *Ann. Rev. Entomol.* **24**: 89–119.

Strong, D. R., Jr., J. H. Lawton, and T. R. E. Southwood. 1984. Insects on Plants: *Community Patterns and Mechanisms*, Blackwell, Oxford.

Strong, D. R., Jr., E. D. McCoy, and J. R. Rey. 1977. Time and the number of herbivore species: The pests of sugarcane. *Ecology* **58**: 167–175.

Tahvanainen, J. and P. Niemelä. 1987. Biogeographical and evolutionary aspects of insect herbivory. *Ann. Zool. Fennici* **24**: 239–247.

Wasbauer, M. W. 1972. An annotated host catalog of the fruit flies of America north of Mexico (Diptera: Tephritidae). *Calif. Dep. Agric. Bur. Entomol. Occas. Pap.* **19**: 1–172.

Whittaker, R. H. 1960. Vegetation of the Siskiyou Mountains, Oregon and California. *Ecol. Monogr.* **30**: 279–338.

Williams, C. B. 1943. Area and number of species. *Nature* **152**: 264–267.

Zar, J. H. 1984. *Biostatistical Analysis*. 2nd ed. Prentice-Hall, Englewood Cliffs, NJ.

Zwölfer, H. 1965. Preliminary list of phytophagous insects attacking wild Cynareae (Compositae) in Europe. *Tech. Bull. Commonw. Inst. Biol. Contr.* **6**: 81–154.

Zwölfer, H. 1979. Strategies and counterstrategies in insect population systems competing for space and food in flower heads and plant galls. *Forschr. Zool.* **25**: 331–353.

Zwölfer, H. 1982. Patterns and driving forces in the evolution of plant–insect systems. Proc. 5th int. Symp. Insect–Plant Relationships, Wageningen, pp. 287–296.

Zwölfer, H. 1985. Insects and thistle heads: Resource utilization and guild structure. *Proc. VI Int. Symp. Biol. Contr. Weeds, Vancouver*, 407–416.

Zwölfer, H. 1987. Species richness, species packing, and evolution in insect–plant systems. *Ecological Studies* **61**: 301–319.

Zwölfer, H. 1988. Evolutionary and ecological relationships of the insect fauna of thistles. *Ann. Rev. Entomol.* **33**: 103–122.

24 Resource Size and Predictability, and Local Herbivore Richness in a Subtropical Brazilian Cerrado Community

MICHAEL CYTRYNOWICZ

Should we expect different plants in a given locality to be utilized by equal numbers and kinds of herbivore species? The answer is certainly not, for plants vary in a number of ways, and herbivore "loads" are influenced by a multitude of factors.

The problem is determining the most relevant factors, their individual contributions, and their interactions. This is no easy task, for relevant attributes may reside in plant structure, biochemistry, phenology, life-history, and population bilogy.

Additionally, plants may vary in terms of biotic and abiotic peculiarities of microhabitats occupied (Root 1973; Saxena 1978; Bach, 1980; Gilbert 1980; McCoy, 1984; Price 1984), such as site humidity and insolation, position within the ant mosaic, and amount of enemy-free space for herbivores to cite a few.

Some plants may be more readily colonized by "foreign" herbivores than others, by virtue of a variety of characteristics both of new hosts and of colonizers (McCoy and Rey 1983; Southwood and Kennedy 1983).

There may be historic differences between plant species. Some may have been present for a long time in the community, others may have been exposed for a much shorter period of time to local herbivores. Some plants may have been involved in mutualistic interactions, or in evolutionary arms races with certain herbivores, and, of course, each plant population has a unique evolutionary history, and is composed of individuals which are not identical to one another, with respect to several of the items listed above.

Thus, in order to explain the numbers and kinds of herbivores associated with different plants in a community, we should view plants as highly variegated

Plant-Animal Interactions: Evolutionary Ecology in Tropical and Temperate Regions, Edited by Peter W. Price, Thomas M. Lewinsohn, G. Wilson Fernandes, and Woodruff W. Benson. ISBN 0-471-50937-X © 1991 John Wiley & Sons, Inc.

resources, with both intrinsic and extrinsic dynamic attributes, immersed in an environment that is also extraordinarily variegated (Kareiva 1986).

PLANT ARCHITECTURE

Several of these attributes form what John Lawton (1983) has termed "Plant Architecture," involving plant structural diversity, the variety of parts a plant has, and its size. Several researchers have found correlations between plant size and structural diversity, and the number of herbivore species using that plant as a host (Lawton and Schröder 1977; Moran 1980; Cornell 1986; Cytrynowicz 1988; for other references, Strong et al. 1984).

Understanding the influence of plant structural diversity on herbivore species richness is apparently easy: different herbivores specialize on different plant parts, and a structurally more complex plant can acquire a larger set of preadapted colonists, than a structurally simple one (Price 1984; Strong et al. 1984).

However, when trying to explain the influence of another component of plant architecture, size per se, we run into some difficulties. How can we isolate the effect of size per se from other architecture components? More importantly, how much is the commonly observed plant size × herbivore richness correlation influenced by interspecific herbivore competition?

Regional vs. Local Studies

If we still do not know the importance of interspecific competition, it is in part because herbivore richness comparisons have been mostly made on a geographical not local level, and have been mainly based on faunal lists and on plant data obtained in the literature, rather than work in the field.

In the last decades, herbivore richness studies focused on what we might term "macro" patterns. Southwood, in his classical papers (1960, 1961) correlated herbivore richness with the presence of the plant taxon along the ages in the British Isles, or with regional abundance in Hawaii. At that time, the theoretical questions most hotly debated were how fast did introduced plants acquire a set of herbivores, and whether this acquisition was asymptotic or not. The debate was sparked by Strong (1974a, b), who reinterpreted Southwood's data as evidence that herbivore accumulation must be fast, in "ecological time," with herbivores drawn from the local source pool, and limited by present plant range size.

Details about local herbivore richness, however, were almost nonexistent, as stressed by Gilbert and Smiley (1978), particularly for nontemperate regions, and evidence on the role interspecific competition plays in determining herbivore richness should come mainly from these studies. There were few studies trying to establish a link between regional and local herbivore richness, among them a classical paper in which Opler (1974) presented some evidence that widely distributed plants (oaks) also have locally rich herbivore faunas (leaf-miners). Later on, this link would be intensively examined by researchers, notably Cornell (1985). See also Lewinsohn (Chapter 23) on the fauna associated with composite flower heads.

Plants as Islands

In the last two decades, ecological theory suggested that larger plants should have larger herbivore faunas. Plants were viewed as islands, both in ecological and evolutionary time (Janzen 1968), as the real and virtual islands described by MacArthur and Wilson (1967) in the *Theory of Island Biogeography*. At a given locality, larger islands (i.e., more abundant plants, and/or with larger individuals) would have larger equilibrated faunas. Larger plant-islands would be found by a greater number of colonists, and would "lose" fewer residents than smaller islands, because larger islands could support larger herbivore populations. At this point interspecific competition was usually invoked, and, if it was not obvious, nevertheless it could be automatic (Janzen, 1973), or a kind of "small intensity times a long time equals large effects" competition.

Recently, the appropriateness of viewing plants as MacArthur–Wilson islands in evolutionary time has been questioned (Southwood and Kennedy 1983). Moreover, several authors questioned the importance of interspecific competition itself, in structuring insect × plant communities (Lawton and Strong 1981; Lawton 1984; Strong 1984). After a great deal of ghost exorcizing (Connell 1980) and a plethora of null models (Colwell and Winkler 1984), the relevance of interspecific competition is still questioned.

Size per se

Size per se clearly explains part of the variability in the number of herbivore species associated with structurally similar plants (Moran 1980; Neuvonen and Niemela 1981). Its influence is difficult to understand, however, if we rule out interspecific competition between herbivores and its consequences (niche separation, stratification) as a cause.

A parallel between size per se, and what Feeny (1976) has termed "conspicuousness" of plants and their parts was drawn by Lawton (1983). In evolutionary time (and involving also plant abundance and distribution), this conspicuousness results in a greater exposure of certain plants to colonization by herbivores, as proposed by Southwood in his "Exposure Frequency Hypothesis" (see Southwood and Kennedy 1983).

Strong et al. (1984) commented on the difficulties in deciding how much size per se influences the greater species richness of some plants via the conspicuousness, "larger targets" hypothesis, or via the "larger, less prone to extinction, herbivore populations," but they pointed out that more studies like Moran's (1980) on cacti, dissecting the components of size, would be useful.

The Need for Local Studies

In summary, part of the difficulties in explaining the influence of size per se on herbivore richness would be resolved if size could be adequately measured. Also, herbivore species richness studies have usually been based on data obtained from faunal lists, or on samplings involving a small subset of the plant or herbivore species present in a given community, which can slightly to severely bias generalizations (Southwood et al. 1982; Karban and Ricklefs 1983).

In 1981 I set out to study the influence of plant attributes, size included, on local herbivore richness of Cerrado plants in Campininha, southeastern Brazil. A local study was almost obligatory, for Neotropical faunas, with a few exceptions (sites in Costa Rica or Panama) are poorly known, and there are no reliable faunal lists or plant distribution maps. Aware of the need for taxonomically comprehensive data, I chose to sample plant species from several families, and all herbivorous arthropod taxa found on their foliage. Also, I tried to keep sampling effort fairly constant, to avoid a possible "entomologist effect."

My results were, in a way, puzzling. The climate at Campininha seems to be fairly unpredictable, with a rainy summer, and a cold, often very dry winter with irregular frosts. There is a huge amount of phenological variation, both between and within plant species. The herbivores do not seem to be organized in clearly defined component communities, and potentially polyphagous herbivore species seem to be dominant. Nevertheless, I found a very strong plant size vs. herbivore richness association.

In the next pages I will present my Campininha results in more detail, and I will suggest how the size × richness correlation might occur even in the absence of competition, in a community which might be described as nonequilibrium, sensu Wiens (1984).

THE CERRADO

Cerrado is one of the most extensive biomes in Brazil, covering about a quarter of its territory. It is most exuberant in Central Brazil (Mato Grosso, Goiás, Minas Gerais), but extends up to the Northeast and the Amazon, and south (about 25°S) to the State of Paraná. General descriptions of Cerrado can be found in Eiten (1972) and Goodland and Ferri (1979).

It is frequently difficult, for many Cerrado species, to tell whether they are trees or shrubs (Heringer et al. 1977; Goodland and Ferri 1979), given the amount of site-to-site intraspecific variation in size, trunk branching, or canopy form.

Many Cerrado species exhibit very tortuous trunks, and thick bark (fine drawings can be found in Ferri 1969), possibly an adaptation against fire (Goodland and Ferri 1979). Leaves are commonly leathery and tough; some species have rigid leaves, which break with a sharp, cracking noise when crumpled. Leaf pubescence is rather common.

The Study Site

Campininha is a 300-ha patch of Savanna-like Cerrado, surrounded by plantations and stretches of gallery forest, located in the State of São Paulo, Southeastern Brazil, 22° 18′S and 47° 11′W, slightly above the Tropic of Capricorn. Thus, Campininha lies close to the southernmost distribution limit of the Cerrado. Its subtropical climate is seasonal, Cwa-type, with rainy summers and dry winters (Eiten 1963), which can be fairly severe, with frosts, and temperatures sometimes below 0°C. Winter can be very dry: in 1981 and 1982, for example, winter precipitation totals of only 5 and 10 mm were recorded.

The Plants and the Sampling Scheme

The floristic composition of the Cerrado at Campininha is well known. Gibbs et al. (1983) list 107 woody species and supply vegetation profiles for Campininha, one of them of the area where I carried out my sampling (Fig. 1, Quadrat 15 of these authors). Campininha, like other Cerrado areas, is rich in species from cosmopolitan families, such as Fabaceae and Asteraceae, as well as more endemic families such as Malpighiaceae and Melastomataceae.

During the winter, many plant species shed their leaves, others dry out because of the frost (Fig. 24-1). The aspect and color of the vegetation suggest severe drought, though some authors (Goodland and Ferri 1979, for references) affirm that most Cerrado plants, with deep root systems that reach into the water table, have sufficient water for their vital functions. During the dry period, many leaf-shedding species produce flowers.

From April 1981 to February 1982, I regularly sampled arthropods on the foliage of 357 marked plants, from 31 species (10–12 plants per species), and 19 plant families (Table 24-1), located within 15, 75-m² plots. These plant species

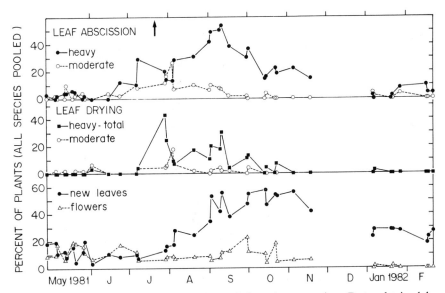

Figure 24-1. General phenological patterns of Cerrado vegetation. Data obtained by pooling the 31 species sampled. Points are percentage of plants, in a given plot and date, with leaf abscission, leaf drying, new leaves, or flowers. The arrow points to July 21, 1981, when a severe frost (air temperature reaching −2.6°C) occurred.

TABLE 24-1. Campininha Plant Species Sampled

Species	Family	Number of Plants Species Sampled	Code
Acosmium dasycarpum (Vog.) Yak.	Leguminosae	12	Ad
Acosmium subelegans (Mohl) Yak.	Leguminosae	12	As
Aspidosperma tomentosum Mart.	Apocynaceae	11	At
Bauhinia holophylla Steud.	Leguminosae	11	Bh
Byrsonima coccolobifolia (Spr.) Kunth	Malpighiaceae	11	Bc
Byrsonima intermedia Adr. Juss	Malpighiaceae	11	Bi
Casearia sylvestris Sw.	Flacourtiaceae	12	Cs
Didymopanax macrocarpum (C. & S.) Seem	Araliaceae	11	Dm
Didymopanax vinosum (C. & S.) March	Araliaceae	11	Dv
Diospyros hispida DC.	Ebenaceae	10	Dh
Erythroxylum suberosum St. Hil.	Erythroxylaceae	12	Es
Erythroxylum tortuosum Mart.	Erythroxylaceae	11	Et
Gochnatia barrosii Cabr.	Compositae	12	Gb
Gochnatia pulchra Cabr.	Compositae	11	Gp
Leandra involucrata DC.	Melastomataceae	11	Li
Myrcia albotomentosa	Myrtaceae	11	Mt
Miconia albicans (Sw.) Triana	Melastomataceae	11	Ma
Neea theifera Oerst.	Nyctaginaceae	11	Nt
Ouratea spectabilis (Mart.) Engl.	Ochnaceae	11	Os
Palicourea rigida H. B. K.	Rubiaceae	12	Pr
Piptocarpha rotundifolia (Less.) Baker	Compositae	10	Pf
Rapanea guianensis Aubl.	Myrsinaceae	11	Rg
Roupala montana Aubl.	Proteaceae	23	Rm
Rudgea viburnioides (Cham.) Benth.	Rubiaceae	11	Rv
Serjania erecta Radlk.	Sapindaceae	11	Se
Styrax ferrugineus Nees & Mart.	Styracaceae	11	Sf
Tabebuia caraiba (Mart.) Bur.	Bignoniaceae	11	Tc
Tabebuia ochracea (Cham.) Standl.	Bignoniaceae	12	To
Tocoyena formosa (C. & S.) K. Schum.	Rubiaceae	11	Tf
Vernonia rubriramea Mart.	Compositae	11	Vr
Xylopia aromatica (Lam.) Mart.	Annonaceae	11	Xa

were chosen because they figured among the most common woody plants at Campininha (W. H. Stubblebine and H. F. Leitão Filho, personal communication).

Each individual plant was visited 4–5 times; at each visit, the plant had its phenological state, number of leaves, and degree of leaf damage annotated. Other plant information was obtained in the field, as well as from the UNICAMP Herbarium (HUEC, Campinas, São Paulo), and from Ferri (1969), Heringer (1971), Rodrigues (1971), Heringer et al. (1977), Goodland and Ferri (1979), Gibbs et al. (1983), Lopes (1984), and H. F. Leitão Filho (pers. comm.),

F. W. Martins (pers. comm.), W. H. Stubblebine (pers. comm.), and W. Mantovani (pers. comm.). The data thus obtained related to plant species phenology and susceptibility to frost, dimensions and growth form, foliar characteristics, local abundance (Campininha), regional distribution (in Brazil), phytosociology, and taxonomic isolation (local and regional levels).

For a period of up to 5 min, the visited plant was carefully searched for arthropods. All arthropods had their activities, position on the plant, and development stage annotated, were manually collected or sampled, and brought to the lab for rearing and identification. More details about the sampling procedure can be found in Cytrynowicz (1988).

The 5 min alotted per plant were adequate for the majority of species; examination of small plants (e.g., *Palicourea rigida* with 4 or 5 leaves) usually took no more than 45 or 60 sec. Large plants, however (e.g., *Rapanea guianensis* with several thousand leaves) had their "momentary" faunas certainly underestimated.

Not unexpectedly, there were pronounced phenological differences among the 31 plant species (Fig. 24-2), two plant groups being prominent: species which grow continually (e.g., *Didymopanax vinosum*, Fig. 24-3a), and species which shed their leaves during the winter (e.g., *Acosmium subelegans*, Fig. 24-3b). Additionally, several species were severely affected by frost—as indicated by the drying

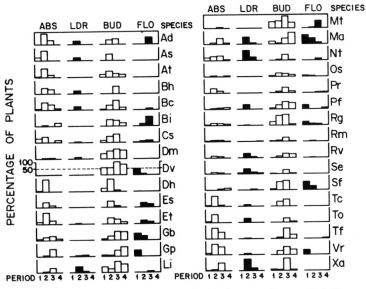

Figure 24-2. Phenological variation among the 31 plant species sampled. Bars represent percentage of plants of a given species, in a given period, with ABS = leaf abscission; LDR = leaf drying; BUD = new leaves, and FLO = flowers. For species of plants see Table 24-1. Periods: (1) April 29 to July 8, 1981; (2) July 29 to September 16; (3) September 30 to November 18; (4) January 6 to February 22, 1982.

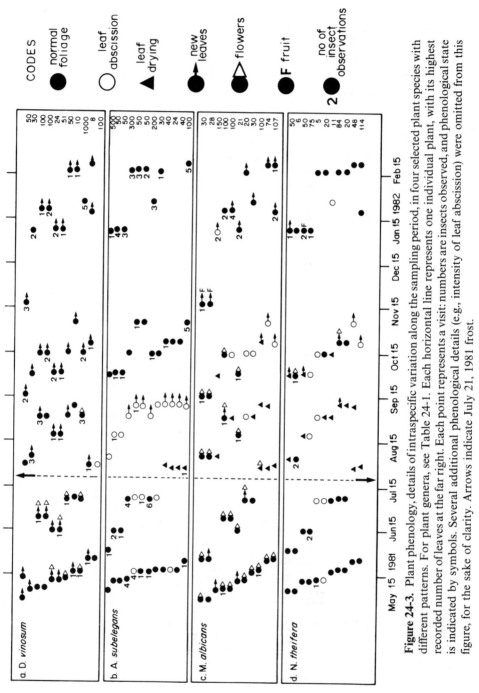

Figure 24-3. Plant phenology, details of intraspecific variation along the sampling period, in four selected plant species with different patterns. For plant genera, see Table 24-1. Each horizontal line represents one individual plant, with its highest recorded number of leaves at the far right. Each point represents a visit: numbers are insects observed, and phenological state is indicated by symbols. Several additional phenological details (e.g., intensity of leaf abscission) were omitted from this figure, for the sake of clarity. Arrows indicate July 21, 1981 frost.

out of leaves, sometimes entire portions of a plant, followed by the (tardy) production of new leaves at the base of the plant.

I also found marked intraspecific variation in phenology, mainly on deciduous species or those affected by frost (e.g., *Miconia albicans* and *Neea theifera*, Fig. 24-3c, d).

The susceptibility of Cerrado plants to frost-induced drying of leaves is analyzed by Silberbauer-Gottsberger et al. (1977), who argue that frost may be an important factor delimiting the geographic distribution of Cerrado species (but see Gibbs et al. 1983).

Though some species had very similar phenologies, this was not the rule for all pairs of congeners, not for plants belonging to the same family, and I did not find a striking correspondence between phenological similarity and taxonomic affinity.

The Herbivores

I recorded slightly less than a thousand arthropod observations (an arthropod species found on a given plant/date), in about 180 genera and 260 species, from 1722 plant visits, an average of 0.56 arthropod observations per visit. Nearly all observations were of insects, most of them herbivores.

Homopterans comprised about 45% of the observations, followed by Coleopterans, Lepidopterans, and Hemipterans (25, 13, and 10%, respectively); the three most commonly observed insect families, Cicadellidae, Membracidae, and Chrysomelidae, comprised about 25% of the observations, the remaining 75% including 59 families from 7 arthropod orders. The most commonly observed (and also most abundant) insect was the highly generalist aphid (*Toxoptera aurantii* (recorded from at least 41 plant visits).

INSECTS ON PLANTS

The average number of insect observations per visit (grouping all plant species) remained fairly stable through the autumn and winter, rose in spring, and was maximal in summer, roughly mirroring the increases in precipitation, daily temperature, and presence of new foliage (Fig. 24-4). The seasonal representation of different families was heterogeneous. Homopterans were the only insects common from late winter to early spring, a dry period with increasing daily temperatures, and with many of the plants still leafless. After the spring rains, other insect orders start to be increasingly better represented, notably Coleoptera.

I rarely observed insects, especially leaf-chewers, actually feeding on the plants (night feeding may be common in the Cerrado). Confirmation of feeding was possible for lepidopterans, but impracticable for most of the other groups, so I classified the insects alternatively as "associated" with the plant when they were found feeding, or when they occurred in great quantity or repeatedly on it

TABLE 24-2 Number of Herbivore Species on the Foliage of Sampled Campininha Plant Species

Plant Species	Herbivore Order[a]					Species[b]	No. of Visits[c]	No. of Species per Visit
	Col.	Hem.	Hom.	Lep.	Other Orders			
A. dasycarpum	5	2	5	3	2	17 (24)	52	0.33 (0.46)
A. subelegans	5	1	10	8	1	25 (64)	57	0.44 (1.12)
A. tomentosum	4	3	8	0	1	16 (30)	54	0.30 (0.56)
B. holophylla	5	1	6	2	0	14 (26)	51	0.27 (0.55)
B. cocolobifolia	3	1	7	3	0	14 (29)	52	0.27 (0.56)
B. intermedia	5	0	7	4	0	16 (27)	55	0.29 (0.49)
C. sylvestris	3	2	13	1	2	21 (37)	58	0.36 (0.64)
D. macrocarpum	4	1	7	0	1	13 (27)	54	0.24 (0.50)
D. vinosum	4	2	10	0	2	18 (46)	51	0.35 (0.90)
D. hispida	5	2	7	3	1	18 (43)	48	0.38 (0.90)
E. suberosum	2	6	8	2	1	19 (37)	58	0.33 (0.64)
E. tortuosum	4	2	10	2	2	20 (42)	53	0.79 (0.38)
G. barrosii	4	3	7	2	3	19 (39)	59	0.32 (0.66)
G. pulchra	2	1	9	0	0	12 (19)	55	0.22 (0.35)

L. involucrata	2	3	6	2	1	14 (20)	49	0.29 (0.41)
M. albotomentosa	5	2	10	2	1	20 (58)	55	0.36 (1.05)
M. albicans	2	3	6	1	1	13 (20)	53	0.25 (0.38)
N. theifera	0	2	3	1	0	6 (10)	55	0.11 (0.18)
O. spectabilis	7	2	9	3	2	23 (47)	55	0.42 (0.85)
P. rigida	1	1	8	2	2	14 (20)	60	0.23 (0.33)
P. rotundifolia	3	0	4	2	1	10 (17)	49	0.20 (0.35)
R. guianensis	5	1	10	0	1	17 (43)	54	0.31 (0.80)
R. montana	4	4	17	0	2	27 (66)	111	0.24 (0.59)
R. viburnioides	0	0	5	0	0	5 (9)	53	0.09 (0.17)
S. erecta	0	1	3	1	0	5 (10)	54	0.09 (0.19)
S. ferrugineus	3	3	14	1	0	21 (46)	51	0.41 (0.90)
T. caraiba	6	1	5	1	1	14 (27)	55	0.25 (0.49)
T. ochracea	4	5	5	0	0	14 (18)	55	0.25 (0.33)
T. formosa	1	3	6	3	0	13 (25)	53	0.25 (0.47)
V. rubriramea	1	0	6	3	0	10 (14)	50	0.20 (0.28)
X. aromatica	2	2	8	1	0	13 (16)	52	0.25 (0.31)

[a]Col: Coleoptera; Hem.: Hemiptera; Hom.: Homoptera; Lep.: Lepidoptera. Only "associated" and "potentially associated" herbivore species.

[b]Number of insect observations within parentheses. One observation is an insect species on a given plant–date combination.

[c]One visit = 1 plant sampled on a given date.

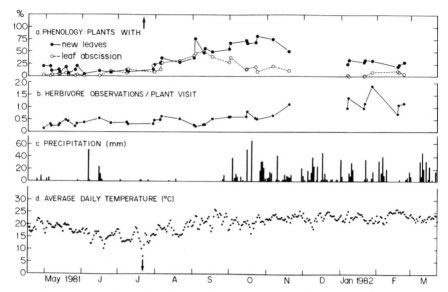

Figure 24-4. Progression, along the sampling period, in precipitation and average daily temperatures, in plant phenology, and in the number of herbivore observations per plant visit (all plant species pooled). Phenological data as in Figure 24-1. Herbivore data: values are total number of herbivore observations in a given plot and date, divided by number of plants sampled in that plot. Precipitation and temperature data kindly provided by CTH/DAAE, Secretaria de Obras e Meio Ambiente do Estado de São Paulo, meteorological station "D4–100" in Campininha (22°18's 47°11'W, alt. 600 m). Arrows indicate July 21, 1981 frost.

(especially as immatures); as "potentially associated," when the insect was herbivorous but did not fall in the cases above, or as "nonassociated", nonherbivorous arthropods (which of course might be associated with the plant in another way). Table 24-2 lists the number of arthropod species recorded from the 31 plants species.

Faunal Uniqueness and Faunal Similarity

About 25% of the insects, mostly homopterans, were observed on more than one plant species. On the average, more than half of the insect fauna observed on a given plant species was not unique to it (Table 24-3), the degree of faunal uniqueness being unrelated to faunal size on a given plant.

Despite the low faunal uniqueness observed, no two plant species had closely similar insect faunas. Among the 465 possible pairs of species, 200 had no insect species in common, 158 had just one, 85 had two, 17 had three, and only 5 plant pairs had 4 insect species in common. On the average, a given plant species "shared" less than one insect species with another plant (last column, Table 24-3).

These apparently contradictory findings—low faunal uniqueness and low

TABLE 24-3. Faunal Uniqueness and Degree of Herbivore Species Sharing

Plant Species	Total[a]	Positively Identified[b]	Not Exclusive[c] (n)	Not Exclusive[c] (%)	Shared Pairwise[d] with other Plant Species
A. dasycarpum	17	12	9	0.75	0.93
A. subelegans	25	22	10	0.45	1.23
A. tomentosum	16	15	8	0.53	0.80
B. holophylla	14	12	8	0.67	0.73
B. coccolobifolia	14	11	9	0.82	1.47
B. intermedia	16	11	6	0.55	1.03
C. sylvestris	21	13	7	0.54	0.97
D. macrocarpum	13	11	9	0.82	1.37
D. vinosum	18	13	10	0.77	1.03
D. hispida	18	15	8	0.53	0.63
E. suberosum	19	16	8	0.50	0.97
E. tortuosum	20	17	8	0.47	0.83
G. barrosi	19	12	7	0.58	0.63
G. pulchra	12	8	5	0.62	0.53
L. involucrata	14	11	5	0.45	0.73
M. albotomentosa	20	18	9	0.50	1.23
M. albicans	13	10	7	0.70	0.87
N. theifera	6	4	1	0.25	0.07
O. spectabilis	23	18	9	0.50	1.33
P. rigida	14	11	6	0.55	0.73
P. rotundifolia	10	5	0	0.00	0.00
R. guianensis	17	12	5	0.42	0.80
R. montana	27	23	14	0.61	1.60
R. viburnioides	5	2	1	0.50	0.77
S. erecta	5	4	2	0.50	0.20
S. ferrugineus	21	17	10	0.59	1.10
T. caraiba	14	12	6	0.50	0.70
T. ochracea	14	11	8	0.73	0.97
T. formosa	13	12	5	0.42	1.27
V. rubriramea	10	7	4	0.57	0.33
X. aromatica	13	10	7	0.70	0.73

[a]"Associated" plus "potentially associated" herbivore species (see text).
[b]Positively identified: herbivore species whose presence of absence in other plants could be confirmed.
[c]Not exclusive: also found on other plant species, whether as "associated" or as "potentially associated" to these species.
[d]Average number of herbivore species a given plant species "shares" with another plant species. Example: *Didymopanax vinosum* has no insect species in common with 10 other plant species; shares 1 insect species with 10 plant species; 2 insect species with 9 plant species, and 3 insect species with just a single plant species—*Ouratea spectabilis* (*Hylax* sp. 1, Anobiidae; *Toxoptera aurantii*, Aphididae, ?*Polyxenus* sp., Polyxenidae). *D. vinosum* shares an average of $(10 \times 0) + (10 \times 1) + (9 \times 2) + (1 \times 3)/30 = 1.03$ insect species with another plant species.

faunal similarity, can thus be exemplified: of the 18 insect species found on *Didymopanax vinosum*, 13 could be "positively identified" (that is, their absence–presence on other plant species could be confirmed, column "Positively identified" in Table 24-3). Out of these 13 "PI" species, 10 were also present on other plants (column "Not exclusive" in Table 24-3). Thus, 77% of the "positively identified" insects on *D. vinosum* occurred on other plant species (and 23% were unique to it). However, pairwise faunal resemblance between *D. vinosum* and any given plant was very low (on the average, 1.03 insect species, last column, Table 24-3). *D. vinosum* "shared" insects with 19 other plants, usually a different insect with each plant, or a single, highly polyphagous insect (e.g., the aphid *T. aurantii*) with a number of plants.

In some cases, a given insect species was shared by, and unique to, congeneric plant species, or seemed to be closely associated to a plant taxon. However, by far the majority of cases of coocurrence involved (a) abundant, apparently highly polyphagous insects, found over a variety of plants, and (b) rare to common insects occurring on two or more adjacent (usually unrelated) plants, within a given plot and on a given date, in a mosaic-like fashion in space-time.

At the plant family level, and considering insect species or genera, faunal resemblance was low (very low within the Fabaceae, Rubiaceae, or Melastomataceae). The three legumes, for example had in common only the cicadellid genus *Idiocerus*, a genus found over several other plants.

Thus, so far, the herbivorous entomofauna at Campininha does not seem to be highly structured in terms of component communities (i.e., insect taxa particularly common over a given plant taxon). Rather, it is evocative of the loosely structured temperate forest system reported by Futuyma and Gould (1979).

Insect Species Richness and Plant Attributes

Among the enormous list of possible plant variables, I selected for correlation tests and for multiple-regression analysis a few fairly independent, biologically relevant attributes, such as leaf numbers, toughness, and hairyness, plant taxonomic isolation and abundance at Campininha, frequency of leaf shedding, and of frost-induced leaf-drying.

The average size of plant species (as measured by leaf number) and frequency of leaf drying strongly correlate with the number of herbivore species (Table 24-4). For "associated" and "potentially associated" insect species combined, number of leaves explains 36% (0.597^2) of the variation in insect species richness; number of leaves explains 39% (0.627^2) of the variation in stem or leaf-sucking insect species.

The numbers of "associated" or "potentially associated" insect species were subjected to separate multiple regressions, as were the numbers of chewing or sucking insect species (Table 24-5). Number of leaves and frequency of leaf drying were, overall, the best predictors of insect species richness among the 31 plant species. Plants with tougher (usually glabrous) leaves have fewer stem or leaf-

TABLE 24-4. Correlation Coefficients Between Herbivore Species Richness and Selected Plant Species Attributes

Plant Variables[a]	Herbivore Species				
	Associated	Potentially Associated	A + PA	Leaf–Stem Sucking	Leaf Chewing
LEAVES	0.501[b]	0.405[b]	0.597[b]	0.627[b]	0.289 ns
TOUGH	−0.284 nt[c]	−0.152 nt	−0.284 nt	−0.415 nt	−0.023 nt
HAIRS	−0.304 nt	0.178 nt	−0.067 nt	0.037 nt	−0.138 nt
ISOLN	0.024 ns[d]	−0.084 ns	−0.043 ns	−0.302 ns	0.232 ns
ABUND	0.223 ns	0.289 ns	0.342 ns	0.289 ns	0.235 ns
LFABS	0.111 ns	−0.070 ns	0.021 ns	−0.127 ns	0.157 ns
DRY	−0.404[e]	−0.413[e]	−0.541[b]	−0.431[e]	−0.398[e]

[a]Leaves: Average, for all plants sampled per species, of the highest number of leaves recorded for each plant during the sampling period. Tough: Leaf toughness, 0–1 (1 = tough) variable. Hairs: Leaf hairyness, 0–1 (1 = hairy) variable. Isoln: $1n + 1$ transform of number of confamilials in Campininha. Abund: $1n + 1$ of local abundance, data obtained from Lopes (1984). Lfabs: Leaf abscission. Percentage of the visits each species had strong to complete leaf abscission, arc-sin $(p + 0.5)$ transformation. Dry: Leaf drying. Percentage of the visits leaves were found to be very dry to completely dry, arc-sin $(p + 0.5)$ transformation.
[b]$P < 0.01$.
[c]nt = 0–1 variables, coefficients cannot be tested.
[d]ns = Nonsignificant.
[e]$0.05 > p > 0.01$.

sucking insect species than do plants with softer (often hairy) leaves, all else being equal.

At the plant intraspecific level, a detailed analysis of insect richness was made difficult by the sampling scheme adopted, for the 11 or 12 conspecific plants were no examined on the same dates, and were localized in different plots (so that seasonal and site peculiarities could obscure other plant differences). To find out whether a plant size × herbivore richness trend emerged at this level, I computed, but did not test, correlation coefficients between the highest recorded number of leaves of each plant, and the number of insect species found over it, during (a) the entire sampling period (April 1981 to February 1982), and (b) only the first 2 months of 1982 (when nearly all deciduous or frost-affected plants had already acquired a new set of leaves). By far, most of the coefficients were positive (27 out of 32, entire period; 21 out of 23, beginning of 1982).

Seasonal Faunal Buildup, Substitution, and Plant Phenology

For the purpose of analyzing faunal buildup and substitution, I defined the following four sampling periods: Period 1 (late autumn to early winter), 29 April to 8 July 1981, a very cold, dry period; Period 2 (late winter), 29 July to 16 September, a dry period with increasing daily temperatures; Period 3 (spring), 30

TABLE 24-5. Multiple Regression of Herbivore Species Richness on Plant Species Attributes[a]

Herbivore Species	Regression Model	R^2	F	df
Associated	$0.1175 + 0.0037$ LEAVES[b] $- 0.0278$ TOUGH[e] $- 0.0357$ HAIRS[b] $- 0.0015$ DRY[e]	0.46	5.52[d]	4,26
Potentially Associated	$0.1303 + 0.0037$ LEAVES[b] $+ 0.0288$ HAIRS[e] $- 0.0024$ DRY[b]	0.33	4.43[b]	3,27
A + PA	$0.2280 + 0.0076$ LEAVES[c] $- 0.0040$ DRY[c]	0.54	nt[c]	nt[c]
Leaf–stem sucking	$0.1740 + 0.0054$ LEAVES[d] $- 0.0261$ TOUGH[e] $- 0.0221$ ISOLN[b] $- 0.0016$ DRY[b]	0.61	10.33[d]	4,26
Leaf chewing	$0.0996 - 0.0214$ HAIRS[e] $+ 0.0276$ ISOLN[e] $- 0.0024$ DRY[d]	0.26	3.17[b]	3,27

[a]Statistical procedure: simultaneous step-down and step-up procedures; regression models accepted (i.e., best), are those which minimize bias and variance simultaneously (Wonnacott and Wonnacott 1981, 1986). In this procedure, some nonsignificant predictors may happen to be included in the final, accepted models. Variable names, Table 24-4.
[b]$0.05 > p > 0.01$.
[c]ntNot tested (see text).
[d]$p < 0.01$.
[e]nsNonsignificant.

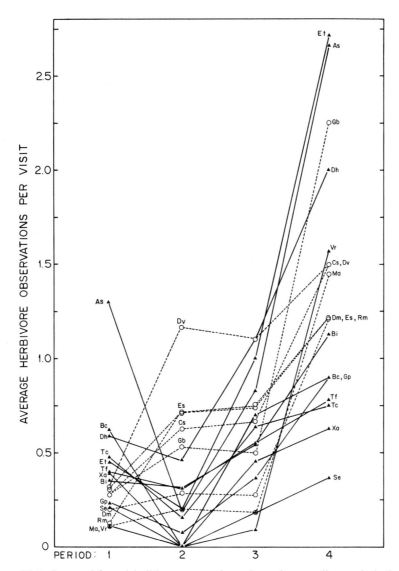

Figure 24-5. Seasonal faunal buildup: progression, along the sampling periods, in the average number of herbivore observations per plant visit. (1) Species whose largest numbers of herbivore observations occurred during the summer. Data are total number of herbivore observations on a given plant species, in a given period, divided by number of times plant species was visited. Periods as in Figure 24-2.

September to 18 November, a rainy period with high daily temperatures, and Period 4 (summer), 6 January to 22 February 1982, with high precipitation and daily temperatures.

In the majority of plant species, the number of insect observations per plant was greatest during the summer (Period 4, Figs. 24-5 and 24-6). During the winter and (early) spring, there were almost no insects on deciduous plants, the few insects observed being concentrated on those plants still with leaves, or already with leaf buds. Conversely, many insect observations were recorded on nondeciduous plants.

The number of insect observations on deciduous plants, or those affected by frost, greatly increased after these plants acquired new foliage. In some "late-leafing" species (e.g., *Miconia albicans* and *Vernonia rubriramea*), this increase was remarkable, even disproportionate, if we consider individual plant sizes. I

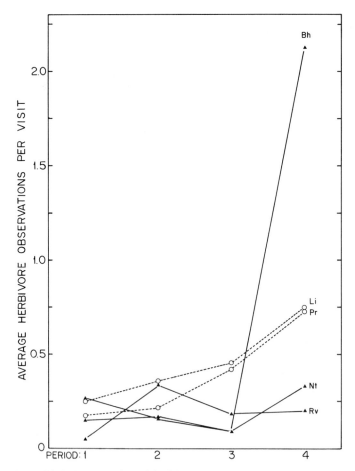

Figure 24-6. Seasonal faunal buildup. (II) See 24-5 for explanations.

hypothesize that these plants, with new, showy, tender foliage, might be very attractive for alighting (polyphagous?) herbivores, at a time when most of the other plants already have mature foliage.

In Lawton's (1983, p. 32) words, "Part of the seasonal progression of herbivores on a plant is dictated not by the absolute presence or absence of some critical resource but by marked seasonal changes in the chemistry, toughness, and general palatability of foliage, stems, etc...." This might be valid for generalist herbivores choosing among a variety of plants. It is worth noting that, of the 9 species whose faunal size decreased in 1982 (Fig. 24-7, plus *R. viburnioides* in Fig. 24-6), 7 have tough leaves; also, some of these plants grow continually throughout the year, others shed their leaves, but quickly acquired new foliage. Among the 22 species whose faunal size increased in 1982, only 7 have tough leaves, and many acquired new foliage relatively late.

Faunal substitution also varied markedly among plant species. This can be exemplified by considering two insect-rich plants, the deciduous *Casearia sylvestris*, and the nondeciduous *Myrcia albotomentosa*. In *C. sylvestris*, only two insect species are present for more than one period (both for two periods), while in

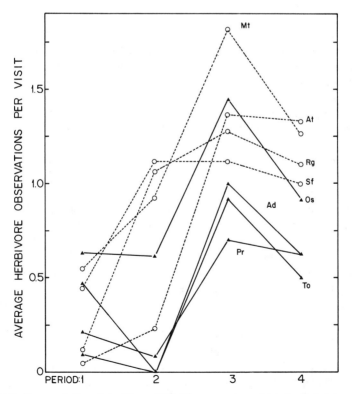

Figure 24-7. Seasonal faunal buildup. (III) Plant species which had their number of herbivore observations decreased in the summer. See 24-5 for explanations.

M. albotomentosa, five species appear at least in two periods (data not shown).

Faunal constancy was found to be correlated with plant phenology. Species with the least seasonal variation in the availability of green leaves (that is, nondeciduous plants), had more faithful insect faunas, corroborating in part a prediction made by Lawton (1983). Lawton also predicted that "plants with large seasonal changes in foliage characteristics should be attacked by more species of insects in toto than plants with more conservative seasonal strategies." This is not what happens in Campininha; leaf-shedding species stay leafless (or near) for a good portion of the year, and very few insects are found on them at this time. When they acquire new foliage, their faunal richnesses are (usually) compatible with the size of that foliage.

Naturally, we could ask whether nondeciduous plants might harbor, during the winter and early spring, insects otherwise found on deciduous plants. At the moment, the answer seems to be no; in fact, some nondeciduous species seem to have more exclusive faunas during this period than in more favorable periods, but this pattern is by no means general.

Size per se

Are the leaves × richness correlations I obtained evidence of the importance of size per se of the resource? Neuvonen and Niemelä (1981) found a relation between plant height and herbivore richness for Finnish trees and shrubs, and argue that the effect is of size per se, because they compared plants in the same category of growth form and only leaf-chewing herbivores (Lepidoptera).

The first argument might also apply to my results, especially if we contrast the plants I sampled, or those sampled by Neuvonen and Niemalä (1981), to those compared by Lawton and Schröder (1977): monocots and dicots, and from grasses to trees.

Like Neuvonen and Niemelä (1981), I also considered arthropods found on the foliage, but belonging to any taxon (though I often could not be affirmative about the insect × plant association).

Though I measured it, I did not feel confident about using plant height as an estimate of (plant) resource size. At Campininha, comparatively large individuals (large trees), can stand leafless for several months, while modest shrubs can produce leaves continually throughout the year. Also, canopy form can vary dramatically within many Cerrado plant species. As Cornell (1986) suggested, "height" may only be an index of the true parameter "size," with herbivores responding to better correlates of size such as form, structure, and size of the canopy.

As a further exploration of data, I forced in the regressions (without tests of hypotheses) the variable "height"—average height of plant species. While "number of leaves" still maintained a strong positive relation to herbivore richness, "height" was either noncorrelated, or could even be negatively correlated. Of course, one cannot extract much inference from skewed statistical procedures (in this case, colinearity), but I cannot refrain from thinking that some kind of index, the "leafiness" of a plant, or the amount of leaves per unit canopy volume, might be a good estimate of (resource) size.

Certainly, the definition of this index would be methodologically complex, and its use in regression subject to statistical pitfalls (see Green 1979). It is possible that the strong correlation between leaf number and richness in stem or leaf-sucking insects is a result of these insects responding to leafyness, if leafyness indicates more meristems, or more scattered phloem sites to suck from. In this case, once again it is difficult to separate resource size from complexity.

Lack of Correlation with Local Abundance

As striking as is the influence of plant size, is the apparent lack of influence of seemingly important variables such as local abundance and taxonomic isolation. Overall, the effects of local host abundance on the richness and composition of herbivore faunas do not seem to be clear. Fowler and Lawton (1982) concluded that "local abundance" was eliminated as a predictor of agromyzid richness, when the number of different habitats occupied by British umbellifers entered the regressions, and thus they argued against a direct effect of local abundance.

Gilbert and Smiley (1978) suggested that in monophagous herbivores variations in local host abundance should influence more the population densities than the number of species of specialized herbivores. On the other hand, Strong (1977a, b) found that locally more abundant Zingiberales in Costa Rica are utilized by greater numbers of specialist hispine beetles, while locally rarer Zingiberales are utilized exclusively by generalist herbivores.

Of course, this question is an exceedingly complex one, involving the use of locally adundant to locally rare hosts, belonging to many plant taxa, by herbivores ranging from strictly monophagous to highly polyphagous. The issue is not clear even at the level of individual herbivore populations: many populations are denser in larger resource patches (Root 1973), but many are not (McGarvin 1982).

I suggest that the lack of correlation between local host abundance and faunal richness is not unexpected, if the Campininha insect × plant × climate system is sufficiently patchy in space-time, and if the degreee of resource predictability is very low, a theme I will develop later on.

Lack of Correlation with Taxonomic Isolation

A first difficulty in interpreting the lack (or for that matter, the presence) of correlation between herbivore richness and plant taxonomic isolation resides precisely in the operational definition of isolation. Taxonomic isolation has many facets, one of them the degree of biochemical distinctiveness. Biochemical differences are very nonlinear, and exceedingly difficult or impossible to quantify (Strong et al. 1984).

There is no question that plant secondary compounds strongly affect insect × plant interactions. However, exactly how secondary compounds influence herbivore species richness or degree of herbivore specialization is still the subject of intense debates, as witnessed in the Forum section of the August 1988 *Ecology* issue. Unfortunately, I could not add to this debate, for Cerrado plants biochemically are almost entirely unknown, references being very scant (see Gottlieb et al. 1971).

The number of conspecific, congeneric, confamilial, or coordinal, and so on plants in a given site or region have alternatively been used by different authors as a measure of taxonomic isolation. Neuvonen and Niemelä (1981, p. 369) quote the use, by Lawton and Schröder (1977) of the genus level. In their own study, these authors chose plant order, stating that "Biochemical barriers to macro-lepidopteran species feeding on herbs seem to exist mainly between different plant families, but with woody plants such barriers are not so pronounced...." The same authors, when reanalyzing cicadellid data from Claridge and Wilson (1976, 1978), defined taxonomic isolation at the subclass level.

Taxonomical level notwithstanding, several authors have found correlations between taxonomic isolation and herbivore species richness, among them Strong (1977), who found more hispine beetles on Zingiberales belonging to better represented families than on Zingiberales from more modestly represented ones.

Taxonomically, more isolated plant species are hypothesized to be more slowly colonized by herbivores than less isolated ones (Cornell and Washburn 1979; Connor et al. 1980; McCoy and Rey 1983; Southwood and Kennedy 1983). The herbivores which colonize exotic (introduced) plants, for example, are known to be mainly polyphagous, ectophagous species, rather than endophagous specialists (Strong et al. 1984).

At Campininha, I did not find correlation between taxonomic isolation and total herbivore species richness. However, in the multiple regression, isolation was significantly negatively correlated with number of stem or leaf-sucking species (Table 24-5).

Why should Cerrado plants from families poorly represented in Campininha be richer is stem or leaf-sucking insects? One hypothesis that deserves further attention is that these plants are actually richer in oligophagous or polyphagous herbivores—if leaf or stem suckers tend to be less specialized than leaf chewers.

The rationale for this was given by Janzen (1973), who suggested that biochemical barriers against colonization may be greater for leaf-chewers than for phloem-feeders—phloem being comparatively poorer than leaves in second-ary compounds.

It is perhaps not surprising that, among the insect species which occurred in more than one plant species, about 70% were Homopterans or Hemipterans; that these families included by far the most abundant insects; and that, among the 16 most commonly observed insect families, 10 belonged to these two orders.

Competition?

Why do leafier Campininha plants have richer herbivore faunas? The answer might be that they are encountered by a greater numbers of herbivores. According to Island Biogeographic Theory, if plants were islands, which is arguable (Southwood and Kennedy 1983), larger plant species would tend to have, all else being equal, larger herbivore populations and richer faunas. The number of herbivore species would be in "equilibrium," and plant-islands would be "saturated" with herbivore species, the equilibrium numbers depending on island size (i.e., size, local abundance) and distance from mainland (taxonomic isolation, Janzen 1968).

Saturation implies competition, possible consequences being specialization on different resource parts, finer niche separation, and/or vertical and horizontal stratification of herbivores along the resource (e.g., vertically within the canopy), though stratification could result from other, noncompetitive causes (Lawton 1983; Cornell 1986).

I did not obtain much evidence for or against stratification, owing in part to the very low number of insect observations (an average of 0.5 observation per plant visit); most insect populations at Campininha seem to be very rarefied and patchy. However, interesting data came from the tree *Roupala montana*, represented in my study by 23 plants rather than the usual 11 or 12; small individuals clearly had a subsample of the herbivore fauna present in large individuals, but not a different fauna, a result comparable to that reported by Cornell (1986) for cynipines on oaks.

With so few insect observations per plant, I follow other authors (listed in Strong et al. 1984) in finding it hard to accept that interspecific competition caused, or is maintaining the plant size × insect richness correlations detected in Campininha, even through the automatic competition proposed by Janzen (1973), and criticized by Lawton and Strong (1981). The question of subtle interactions is a hotly debated one. In very recent papers on the importance of secondary compounds on the evolution of host specificity, Bernays and Graham (1988) try to exorcize the ghost of "little interaction, over a long time, causing large effects," but several other authors disagree with them (Ehrlich and Murphy 1988; Fox 1988; Rausher 1988; Schultz 1988; Thompson 1988).

COMMUNITY STRUCTURE

My results seem to follow closely those of Futuyma and Gould (1979) for a temperate forest, in that I did not find evidence for extensive component community structuring. Although on the average half of the herbivore fauna on a given plant species is not unique to it, that is, most of the plants do not seem to have highly specialized faunas, pairwise faunal similarity between plants is very low, and does not reflect well plant taxonomic affinity. Moreover, among the (few) insects shared by any two plant species, most are either polyphagous, or just happened to occur by chance on two plants contiguous in space-time (insect species appearing now and then, here and there).

Given that the insect populations seem to be very rarefied, and the system poorly structured, and given that I sampled mainly foliage-associated insects, a likely explanation for the plant size × herbivore richness correlation may involve a kind of passive-sampling (sensu Connor and McCoy 1979; Strong 1979) phenomenon.

PASSIVE-SAMPLING, ACTIVE CHOICE, AND EXPOSURE FREQUENCY

Suppose that, during the spring rains, a multitude of herbivores emerge (from larval stages, hiding places) seeking plants. At that time, a good portion of the Campininha vegetation is still dry or bare. Specialist foliage herbivores, emerging patchily in time and space, would have difficulties in locating adequate host plants, due to their high intraspecific phenological asynchrony, induced by climatic events and mediated by a number of site-specific factors. This assynchrony would not be common, however, at the core of Cerrado in Central Brazil, with its more tropical climate.

From the point of view of a specialist insect herbivore, then, Campininha may be a highly unpredictable system (see Kareiva 1986). Selection might foster the evolution of feeding generalization (Janzen 1973; Cates 1981; Lawton and Strong 1981), and hinder the evolution of specialization (at least of foliage feeders).

Suppose, next, that fairly generalized herbivorous insects emerge in a haphazard fashion, in favorable sites, amidst a highly diversified vegetation, looking for food plants. The best choices for alighting insects would be the locally most attractive plants. "Leafy" plants would be good candidates, perhaps because they reflect or transmit well those spectral hues used as cues by many insect herbivores (Cytrynowicz et al. 1982; Prokopy and Owens 1983). After the insect alights, whether or not the plant is eaten would depend on several plant attributes, among them palatability for that particular insect.

Thus, among adjacent plants, conspecific or not, those with a larger foliage size might be sampled by a larger proportion of the available (largely nonspecialized) herbivore fauna, than those with fewer leaves. Though this process resembles passive-sampling (larger canopies, more insect records), it would involve active alighting choice by herbivores. Because of the (very short-term) accumulation of observations on a plant species with large individuals, some insect species would be considered "associated," whether or not they actually feed on it. In the long run, however, they might really be able to use this plant as a host, among others to which the insect species would have been repeatedly exposed, as hypothesized by Southwood and Kennedy (1983).

In this way, plant size might correlate positively with species richness of herbivores found on the foliage, even in "unsaturated" herbivore communities, or without interspecific competition.

CAMPININHA AS A NONEQUILIBRATED SYSTEM?

With its lack of highly structured component communities, its diversified flora, its apparently rarefied herbivore populations, and its fairly unpredictable climate, Campininha may qualify as a good example of a nonequilibrated system, in the spectrum presented by Wiens (1984).

In closing, I want to draw some parallels between my study and the one by

Benson (1978), involving specialist butterflies (Heliconiinae) and a highly diversified plant taxon (Passifloraceae) at two tropical (Rincón, Costa Rica; Arima Valley, Trinidad) and one subtropical site (Rio de Janeiro). At the subtropical site, there are fewer heliconian species, which are much more generalized in their use of host-plant species, but exhibit marked habitat preferences (which, according to Benson, serve to reduce competition for the use of individual food items).

In Benson's (1978, p. 515) words "the unusual community structure at Rio de Janeiro is related to the seasonal growth patterns of the local passion vines, which results in strong cycles in plant biomass and fluctuations in the rank abundances of the different species. The best strategy for a short-lived, continuously breeding herbivore under these conditions would be to use the plant type that was currently more common...."

I am not arguing that interspecific competition between insect herbivores is universally unimportant, nor that all subtropical systems are not equilibrated (sensu Wiens 1984), as opposed to tropical ones. Benson (1978) and I arrive at different conclusions regarding competition, perhaps because of the kinds of herbivores and plants studied (specialists vs. nonspecialists). However, both studies might exemplify well (for highly specialized and for generalized insect herbivores) what happens with a highly diversified community of tropical origin, subject to a subtropical climate.

In central, tropical Cerrado areas, I would not be surprised to find stronger component community structuring, more specialization, greater intraspecific plant phenological synchrony, and stronger biotic regulation than I found in Campininha—and perhaps even interspecific herbivore competition.

ACKNOWLEDGMENTS

I thank Thomas Lewinsohn, Peter Price, Woodruff Benson, Keith Brown, and Geraldo Fernandes for continued encouragement and fruitful discussions. L. Knutson, V. Eastop, U. R. Martins, and V. Becker guided me in having specimens from the poorly known Campininha entomofauna identified. I thank Benedito Lopes and Helena C. de Morais for their friendship, and invaluable help in the field (while they were carrying on their own research). I also thank Universidade Federal de Santa Catarina, especially Silvio Coelho and Milton D. Muniz, for support and encouragement, and CNPq for a doctoral grant. My love, and gratitude to Eliana, Lúcia, Débora, Hadassah, and Heinrich, who shared some of the joys—and probably most of the pains—of my work with Cerrado herbivores.

REFERENCES

Bach, C. E. 1980. Effects of plant diversity and time of colonization on an herbivore–plant interaction. *Oecologia (Berl.)*, **44**: 319–326.

Benson, W. W. 1978. Resource partitioning in Passion vine butterflies. *Evolution*, **32**(3): 493–518.

Bernays, E. and M. Graham. 1988. On the evolution of host specificity in phytophagous arthropods. *Ecology*; **69**(4): 886–92.

Cates, R. G. 1981. Host plant predictability and the feeding patterns of monophagous, oligophagous, and polyphagous insect herbivores. *Oecologia* (Berl.), **48**: 319–26.

Claridge, M. F. and M. R. Wilson. 1976. Diversity and distribution patterns of some mesophyll-feeding leafhoppers of temperate woodland canopy. *Ecological Entomology*, **1**: 231–50.

Claridge, M. F. and M. R. Wilson. 1978. British insects and trees: A study in island biogeography or insect/plant coevolution? *American Naturalist*, **112**: 451–6.

Colwell, R. K. and D. W. Winkler. 1984. A null model for null models in biogeography. In *Ecological Communities*. D. R. Strong, Jr., D. Simberloff, L. G. Abele, and A. B. Thisle (Eds.). Princeton University Press, Princeton, NJ.

Connell, J. H. 1980. Diversity and coevolution of competitors, or the ghost of competition past. *Oikos*, **35**: 131–38.

Connor, E. F. and E. D. McCoy. 1979. The statistics and biology of the species-area relationship. *American Naturalist*, **113**: 791–833.

Connor, E. F., S. H. Faeth, D. Simberloff, and P. A. Opler. 1980. Taxonomic isolation and the accumulation of herbivorous insects: A comparison of introduced and native trees. *Ecological Entomology*, **5**: 205–11.

Cornell, H. V. 1985. Local and regional richness of cynipine gall wasps on California oaks. *Ecology*, **66**: 1247–60.

Cornell, H. V. 1986. Oak species attributes and host size influence cynipine wasp species richness. *Ecology*; **67**(6): 1582–92.

Cornell, H. V. and J. O. Washburn. 1979. Evolution of the richness-area correlation for cynipid gall wasps on oak trees: A comparison of two geographic areas. *Evolution*, **33**(1): 257–74.

Cytrynowicz. M. 1988. Determinantes da Riqueza Local de Espécies de Insetos Fitófagos Associados a Plantas de uma Área de Cerrado. Ph.D. thesis, UNICAMP, Campinas, Brasil.

Cytrynowicz, M., J. S. Morgante, and H. M. L. de Souza. 1982. Visual responses of South American fruit flies, *Anastrepha fraterculus* and Mediterranean fruit flies, *Ceratitis capitata*, to colored rectangles and spheres. *Environmental Entomology*, **11**: 1202–10.

Ehrlich, P. R. and D. D. Murphy. 1988. Plant chemistry and host range in insect herbivores. *Ecology*, **69**(4): 908–9.

Eiten, G. 1963. Habitat flora of Fazenda Campininha, São Paulo, Brazil. I. Introduction, species of the "cerrado", species of open wet ground. In *Simposio Sobre o Cerrado*. M. G. Ferri (Ed.). EDUSP, S. P.

Eiten, G. 1972. The Cerrado vegetation of Brazil. *Bot. Rev.* **38**(2): 201–341.

Ferri, M. G. 1969. *Plantas do Brazil: Espécies do Cerrado*. Edgard Blucher/EDUSP, S. P.

Feeny, P. 1976. Plant Apparency and Chemical Defense. *Recent Advances in Phytochemistry*, **10**: 1–40.

Fowler, S. V. and J. H. Lawton. 1982. The effect of host-plant distribution and local abundance on the species richness of agromyzid flies attacking British umbellifers. *Ecological Entomology*, **7**: 257–65.

Fox, L. R. 1988. Diffuse coevolution within complex communities. *Ecology*, **69**(4): 906–7.

Futuyma, D. J. and F. Gould. 1979. Associations of plants and insects in a deciduous forest. *Ecological Monographs*, **49**(1): 33–50.

Gibbs, P. E., H. F. Leitão Filho, and G. Shepherd. 1983. Floristic composition and community structure in an area of cerrado in SE Brazil. *Flora*; **173**: 433–49.

Gilbert, F. S. 1980. The equilibrium theory of island biogeography: Fact or fiction? *Journal of Biogeography*, **7**: 209–35.

Gilbert, L. E. and J. T. Smiley. 1978. Determinants of local diversity in phytophagous insects: Host specialists in tropical environments. In *Diversity of Insect Faunas*, L. A. Mound and N. Waloff (Ed.). Symposia of the Royal Entomological Society of London **9**: 89–104.

Goodland, R. and M. G. Ferri. 1979. *Ecologia do Cerrado*. EDUSP/Itatiaia, S. P.

Gottlieb, O. R., M. T. Magalhães, and W. B. Mors. 1971. Problemas e possibilidade da Fitoquímica no cerrado. In *Simposio sobre o Cerrado*. M. G. Ferri (Ed.). Edgard Blucher/EDUSP, S. P.

Green, R. H. 1979. *Sampling Design and Statistical Methods for Environmental Biologists*. Wiley-Interscience, New York.

Heringer, E. P. 1971. Propagação e sucessão de espécies arbóreas do cerrado em função do fogo, cupim, da capina e do aldrim. In *III Simpósio sobre Cerrado*. M. G. Ferri (Ed.). Edgard Blucher/EDUSP, S. P.

Heringer, E. P., G. M. Barroso, J. A. Rizzo, and C. T. Rizzini. 1977. A Flora do Cerado. In *IV Simposio sobre o Cerrado*. M. G. Ferri (Ed.). EDUSP/Itatiaia, S. P.

Janzen, D. H. 1968. Host plants as islands in evolutionary and contemporary time. *American Naturalist*, **102**: 592–95.

Janzen, D. H. 1973. Sweep samples of tropical foliage insects: Effects of seasons, vegetation types, time of day, and insularity. *Ecology*, **54**(3): 687–708.

Karban, R. and R. Ricklefs. 1983. Host characteristics, sampling intensity, and species richness of lepidopteran larvae on broad-leaved trees in Southern Ontario. *Ecology*, **64**(4): 636–41.

Kareiva, P. 1986. Patchiness, dispersal, and species interactions: Consequences for communities of herbivorous insects. In *Community Ecology*. J. Diamond and T. J. Case (Eds.). Harper & Row, New York.

Lawton, J. H. 1983. Plant architecture and the diversity of phytophagous insects. *Annual Review of Ecology and Systematics*, **28**: 23–39.

Lawton, J. H. and D. Schroeder. 1977. Effects of plant type, size of geographical range, and taxonomical isolation on number of insect species associated with British plants. *Nature*, **265**: 137–40

Lawton, J. H. 1984. Non-competitive populations, non-convergent communities, and vacant niches; The herbivores of bracken. In *Ecological Communities*. D. R. Strong, Jr., D. Simberloff, L. G. Abele, and A. B. Thistle (Eds.) Princeton University Press, Princeton, NJ.

Lawton, J. H. and D. R. Strong. 1981. Community patterns and competition in folivorous insects. *American Naturalist*, **118**(3): 317–38.

Lopes, B. C. 1984. Aspectos da ecologia de Membracideos (Insecta: Homoptera) em vegetação de Cerrado do Estado de São Paulo. Instituto de Biologia, UNICAMP. M.Sc.

MacArthur, R. H. and E. O. Wilson. 1967. *The Theory of Island Biogeography*. Princeton University Press, Princeton, NJ.

McCoy, E. D. 1984. Colonization by herbivores of *Heliconia* spp. plants (Zingiberales: Heliconiaceae). *Biotropica*, 16(1): 10–13.

McCoy, E. D. and J. R. Rey. 1983. The biogeography of herbivorous arthropods: Some species accrual on tropical crops. *Ecological Entomology*, 8: 305–13.

McGarvin, M. 1982. Species-area relationships of insects on host plants: Herbivores on rosebay willowherb. *Journal of Animal Ecology*, 51: 207–23.

Moran, V. C. 1980. Interactions between phytophagous insects and *Opuntia* hosts. *Ecological Entomology*, 5: 153–64.

Neuvonen, S. and P. Niemelä. 1981. Species richness of macrolepidoptera on Finnish deciduous trees and shrubs. *Oecologia (Berl.)*, 51: 364–70.

Opler, P. A. 1974. Oaks as evolutionary islands for leaf-mining insects. *American Scientist*, 62: 67–73.

Price, P. W. 1984. Communities of specialists: Vacant niches in ecological and evolutionary time. In *Ecological Communities*. D. R. Strong, Jr., D. Simberloff, L. G. Abele, and A. B. Thistle (Eds.). Princeton University Press, Princeton, NJ.

Prokopy, R. J. and E. D. Owen. 1983. Visual detection of plants by herbivorous insects. *Annual Review of Entomology*, 28: 337–64.

Ratter, J. A. 1971. Some notes on two types of cerradão occurring in north eastern Mato Grosso. In III *Simpósio sobre o Cerrado*. M. G. Ferri (Ed.). Edgard Blucher/EDUSP, S. P. Brasil.

Rausher, M. D. 1988. Is coevolution dead? *Ecology*, 69(4): 898–901.

Rodrigues, W. A. 1971. Plantas dos Campos do Rio Branco (Território de Rondonia). In III *Simpósio sobre o Cerrado*. M. G. Ferri (Ed.). Edgard Blucher/EDUSP, S. P.

Root, R. B. 1973. Organization of a plant–arthropod association in simple and diverse habitats: The fauna of collards (*Brassica oleracea*). *Ecological Monographs*, 43(1): 95–124.

Saxena, K. N. 1978. Role of certain environmental factors in determining the efficiency of host plant selection by an insect. *Entomologica Experimentalis et Applicata*, 24: 446–78.

Schultz, J. C. 1988. Many factors influence the evolution of herbivore diets, but plant chemistry is central. *Ecology*, 69(4): 896–97.

Silberbauer-Gottsberger, I., W. Morawetz, and G. Gottsberger. 1977. Frost damage of cerrado plants in Botucatu, Brazil, as related to the geographical distribution of the species. *Biotropica*, 9(4): 253–61.

Southwood, T. R. E. 1960. The abundance of the Hawaiian trees and the number of their associated insect species. *Proceeding of the Hawaii Entomological Society*, 17: 299–303.

Southwood, T. R. E. 1961. The number of the species of insect associated with various trees. *Journal of Animal Ecology*, 30: 1–8.

Southwood, T. R. E., V. C. Moran. and C. E. J. Kennedy, 1982. The assessment of arboreal insect fauna: Comparisons of knockdown sampling and faunal lists. *Ecological Entomology*, 7: 331–40.

Southwood, T. R. E. and C. E. J. Kennedy. 1983. Trees as islands. *Oikos*, 41: 359–71.

Strong, D. R. 1974a. Rapid asymptotic species accumulation in phytophagous insect communities: The pests of cacao. *Science*, 185: 1064–66.

Strong, D. R. 1974b. Nonasymptotic species richness models and the insects of British trees. *Proceedings of National Academy of Science, USA.,* **71**(7): 2766–69.

Strong, D. R. 1977a. Rolled-leaf hispine beetles (Chrysomelidae) and their Zingiberales host plants in Middle America. *Biotropica,* **9**(3): 156–69.

Strong, D. R. 1977b. Insect species richness: Hispine beetles of *Heliconia latispatha. Ecology,* **58**: 573–82.

Strong, D. R. 1979. Biogeographic dynamics of insect–host plant communities. *Annual Review of Ecology and Systematics,* **24**: 89–119.

Strong, D. R. 1984. Exorcizing the ghost of Competition past. In *Ecological Communities.* D. R. Strong, Jr., D. Simberloff, L. G. Abele, and A. B. Thistle (Eds.). Princeton University Press, Princeton, NJ.

Strong D. R., J. H. Lawton, and T. R. E. Southwood 1984. *Insects on Plants: Community Patterns and Mechanisms.* Blackwell, Oxford.

Thompson, J. N. 1988. Coevolution and alternative hypotheses on insect/plant interactions. *Ecology,* **69**(4): 893–95.

Wiens, J. A. 1984. On understanding a non-equilibrium world: Myth and reality in community patterns and process. In *Ecological Communities.* D. R. Strong, Jr., D. Simberloff, L. G. Abele, and A. B. Thistle (Eds.). Princeton University Press, Princeton, NJ.

Wonnacott, T. H. and R. J. Wonnacott. 1981. *Regression: A Second Course in Statistics.* Wiley, New York.

25 Contemporary Adaptations of Herbivores to Introduced Legume Crops

MARCOS KOGAN

Most regions of the world have suffered the effects of invasions by introduced crops. Plant invaders are often equated with weeds, even when they come to dominate the regional vegetation, as in the case of many crops (Baker 1984; Bazzaz 1984). The concept of invaders as used by ecologists presupposes the ability of a plant or an animal to compete with existing flora or fauna and replace or coexist with the native components of the community. Most agricultural crops are incapable of self-perpetuation and many are poor competitors, but these attributes are more than adequately compensated for by human intervention. Therefore, the introduction of an exotic crop into a new region constitutes a biological invasion that often has far-reaching ecological as well as social and economic consequences. If the new region was not previously in agriculture, the new crop often replaces climax vegetation, as was the case of vast forested areas in the state of Parana, Brazil after the expansion first of coffee and, since the 1960s, of soybean (Miyasaka and Medina 1981). The introduced-crop community undergoes dynamic changes, remaining for some time in a state of nonequilibrium. The recruitment of herbivores into these new crop communities has particular interest as a potential source of community instability (Kogan 1981). Among the factors that influence the rate of recruitment and the colonization success of the herbivores are (a) the reservoir of potential colonists—mainly the richness of the specialist fauna associated with native (wild) plants chemotaxonomically related to the exotic crop and the abundance of aggressive food generalists; (b) the synchronization of the crop phenology with the life history characteristics of potential colonizers; and (c) the characteristics of the extant antiherbivory defenses of the introduced crop and the degree to which these defenses are altered through the genotypic interaction with the new environment.

Many of the leguminous species ancestral to contemporary cultivars radiated

Plant-Animal Interactions: Evolutionary Ecology in Tropical and Temperate Regions, Edited by Peter W. Price, Thomas M. Lewinsohn, G. Wilson Fernandes, and Woodruff W. Benson. ISBN 0-471-50937-X © 1991 John Wiley & Sons, Inc.

from centers often distant from the major areas of current cultivation. Consequently, the cultivars are continuously exposed to faunas originally allopatric with the crop plant's ancestral species. The current status of these leguminous crop communities in regions of recent (100–200 years) introduction is that they lack highly adapted, oligophagous elements. On the other hand, crops generally deprived of broad spectrum defensive allomones (Kogan 1986) are vulnerable to colonization by species with broad host spectra. In this chapter, I discuss characteristics of fauna associated with grain legumes in temperate and subtropical Americas; soybean is used as a paradigm. By comparing these faunas with those of the Orient, I attempt to explain the absence in the Americas of entire guilds that are common and well adapted to soybean grown for long periods of time (> 1000 years) in regions closer to the centers of origin of the crop (China, Korea, and Japan) (Hymowitz 1970). Furthermore, comparisons among those faunas suggest that vacant feeding niches exist in many areas. Occasional records of new colonizers indicate that niche occupancy in these crops is evolving. Speculation on current patterns of niche occupancy is based on preliminary evidence on the action of inducible defenses in soybean.

THE LEGUMINOSAE

The family Leguminosae has provided humans with some of the most valuable food, oil, and forage crops. Collectively, these crops are planted annually on over 188.1 million ha, or about 12.7% of the total world cultivated land surface (FAO 1986). Two of these crops, the common bean (*Phaseolus vulgaris*) and cowpea (*Vigna unguiculata*), are staples in Central and South America and Africa. Together with soybean (*Glycine max*) in China, Japan, and Korea, and several pulses, lentils, and peas in India, these grain legumes are the main sources of protein for most of the people in the developing world. Forage crops, such as alfalfa (*Medicago sativa*) and clover (*Melilotus* spp.), are essential ingredients of animal feeds. Because of their nitrogen-fixing ability, many additional species of Leguminosae, together with grasses, are highly desirable components of the best pastures in every continent. By recent accounts, the Leguminosae have 505 genera with about 14,000 species (Allen and Allen 1981). The tropical flora is particularly rich in both arboreous and herbaceous genera and species of Leguminosae, but the family is well represented in the temperate zone as well. For example, a review of the species of Leguminosae of Illionis (Gambill 1953) recognized 110 species in 37 genera; interestingly, however, 29 species, or 26% of the flora, consisted of introduced species (invaders), mostly of European or Asian origin. Among those are economically important species in the genera *Trifolium, Melilotus, Vicia, Lathyrus,* and *Glycine.* The richness of species of economic importance in the Leguminosae is paralleled only in the family Gramineae. There are about 225 genera of Leguminosae with species of economic value (Table 25-1).

Many of the major legume crops have been translocated from their centers of

TABLE 25-1. Uses of the Leguminosae of Economic Value (Allen and Allen 1981)

Types of Crop	Number of Genera
Cover crops	40
Fiber crops	18
Forage crops	62
Food crops:	
Flour and meal	15
Flowers	6
Foliage	8
Fruit (pods)	26
Root tubers	16
Seeds	34

TABLE 25-2. Major Legume Crops and Their Purported Wild Relatives

Cultigen	Common Name	Wild Relative
Arachis hypogaea	Peanut	A. monticola
Cajanus cajan	Pigeon pea	Atylosia lineata
Cicer arietinum	Chickpea	C. reticulatum
Glycine max	Soybean	G. soja
Lens esculenta	Lentil	L. orientalis
Phaseolus lunatus	Lima bean	P. lunatus silvester
Phaseolus vulgaris	Common bean	P. aborigineus
Pisum sativum	Pea	P. humile
Vicia faba	Broad bean	V. galilaea
Vigna aureus	Green gram	—
Vigna mungo	Black gram	V. radiata sublobata
Vigna unguiculata	Cowpea	V. unguiculata dekindtiana

origin and have adapted well in their adoptive countries. The wild relatives of several of these crops are known (Table 25-2). Their study should provide valuable information on the evolution of insect–plant interactions, but detailed studies are scarce if existent at all. However, all of these native plants harbor an immensely diverse reservoir of arthropod species, and many are preadapted to invade new leguminous crops when their area of cultivation expands.

THE SOYBEAN CROP

The center of origin of soybean is placed in the northeastern regions of China, in the provinces of Heilongjian, Jilin, and Liaoning. References to cultivated soybean in China date back some 3000 years. Introduction to Korea and Japan probably dates to a period between 200 B.C.E. and 200 C.E. (Hymowitz 1970).

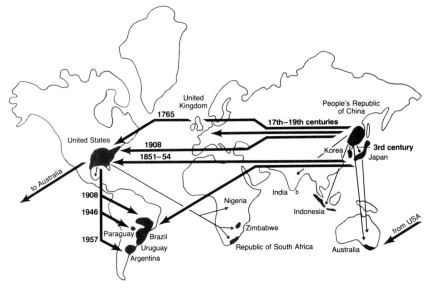

Figure 25-1. Soybean-producing regions of the world with approximate dates for the introduction of the plant into some major regions. (Reproduced from Kogan et al., 1988, by permission from the University of Illinois Agricultural Experiment Station.)

From the Far East soybean spread throughout the globe (Fig. 25-1) and it is now the most widely cultivated grain legume in the world, covering an area of about 52 million ha. The United States, Brazil, and Argentina account for 37.2 million ha, or 71% of the total cultivated area (American Soybean Association 1986).

Expansion of the crop occurred first in temperate regions at about the latitudes of Northeastern China. In the past 20 years, however, the crop expanded dramatically into the subtropics in cropping systems that involve such diverse crops as sugar cane, cotton, and coffee as well as the more traditional rice and corn. In addition, soybean has replaced previously important crops (e.g., cotton in the southern United States and coffee in the State of Parana, Brazil) and has been used as a pioneer crop in newly deforested areas of Brazil and freshly drained swamps in Louisiana. As might have been expected, the growth characteristics of the crop under such varied conditions are clearly diverse and have led to the establishment of regionally different herbivore communities associated with the crop.

MAJOR GUILDS OF PHYTOPHAGOUS ARTHROPODS ASSOCIATED WITH SOYBEAN

In both North and South America, species belonging to three major guilds account for about 75% of the economic injury to soybean. These guilds are (a) lepidopterous foliage and pod feeders, particularly in the families Noctuidae and

TABLE 25-3. Principal Species in Major Guilds of Phytophagous Insects Associated with Soybean in North and South America

Guilds	Host Specificity	North America[j]	South America[j]
	Foliage Feeders		
Lepidoptera			
Noctuidae			
Anticarsia gemmatalis	Oligophagous[a]	+ + +	+ + +
Pseudoplusia includens	Polyphagous[b]	+ + +	+ + +
Heliothis zea	Polyphagous[c]	+ + +	+
Heliothis virescens	Polyphagous[c]	+ +	−
Spodoptera exigua	Polyphagous	+	0
Spodoptera eridania	Polyphagous	+	+
Spodoptera latifascia	Polyphagous	0	+
Plathypena scabra	Oligophagous[d]	+	0
Coleoptera			
Chrysomelidae			
Cerotoma trifurcata	Oligophagous[e]	+ +	0
Cerotoma ruficornis	Oligophagous[e]	−	0
Cerotoma arcuata	Oligophagous[e]	0	+
Cerotoma fascialis	Oligophagous[e]	0	+
Colaspis brunnea	Oligophagous	−	0
Maecolaspis aeruginosus	Oligophagous	0	−
Diabrotica balteata	Polyphagous[f]	+	0
Diabrotica speciosa	Polyphagous[f]	0	+
	Pod Feeders		
Hemiptera			
Pentatomidae			
Nezara viridula	Polyphagous[g]	+ + +	+ + +
Acrosternum hilare	Polyphagous[h]	+ +	0
Acrosternum impicticorne	Polyphagous	0	+
Euschistus servus	Polyphagous	+	0
Euschistus heros	Polyphagous	0	+
Piezodorus guildinii	Polyphagous[i]	+	+ +

[a] Mostly Leguminosae (23 genera), but recorded also from five species of Begoniaceae, Gramineae, and Malvaceae (Herzog and Todd 1980).
[b] Recorded on species in 29 plant families (Herzog 1980).
[c] The two species have been reported feeding on 235 plant species in 36 families (Kogan et al. 1989).
[d] On 23 species of plants, mostly Leguminosae (Pedigo et al. 1973).
[e] Mostly on Leguminosae (Kogan et al. 1980).
[f] *D. balteata* larvae were recorded from Gramineae, Convolvulaceae, and Leguminosae; *D. speciosa* larvae were recorded from Gramineae, Leguminosae, and Cucurbitaceae. Adults of both species are highly polyphagous (Krysan 1986).
[g] On 145 species in 32 plant families (Kiritani et al. 1965); but American populations may prefer Leguminosae (Todd 1989) (see also Panizzi and Slansky 1985).
[h] No complete list available. Polyphagous habit deduced from host records in Schoene and Underhill (1933) and Miner (1966).
[i] On 41 species in 21 plant families (Panizzi and Slansky 1985).
[j] Key to symbols: + + + key pest; + + major pest; + occasional pest; − rarely a pest; 0 does not occur in the Continent.

Pyralidae; (b) coleopterous foliage or foliage and root feeders in the families Coccinellidae–Epilachninae and Chrysomelidae; and (c) hemipterous pod and seed suckers, mostly in the family Pentatomidae, but also Coreidae (Kogan and Turnipseed 1987). The principal species in these guilds are listed in Table 25-3.

The following are some of the most remarkable aspects of the guilds of soybean colonizers shown in Table 25-3.

1. The guild of lepidopterous defoliators is dominated by highly polyphagous species, most of which are typically r-strategists.
2. The most economically important species—*Heliothis zea*, *Anticarsia gemmatalis*, and *Pseudoplusia includens*—have a broad geographic range; they are widespread in North, Central, and South America.
3. The guild of coleopterous species is dominated by oligophagous species well adapted to feeding on native species of the leguminous subfamily Papilionoidae, with the exception of the polyphagous *Diabrotica* species.
4. The pod-feeding Pentatomidae are all highly polyphagous, and the dominant species—*Nezara viridula*—is cosmopolitan.

Kogan and Turnipseed (1987) and Kogan et al. (1988) provide a list of 106 of the most common arthropods associated with soybean in six regions of the world: North America; South and Central America; Africa; India and Indonesia; China, Japan, and Korea; and Australia and New Zealand. Eighty-three of the 106 species occurred predominantly in one region, 16 species in two regions, 6 species in three regions, 2 species in four regions, and only 1 species, *Nezara viridula*, occurred in all six regions (Fig. 25-2). This type of assessment, previously used by Strong et al. (1984), suggests that recruitment of colonizers of soybean fields has been mostly a local phenomenon. Ecological homologues native to the various regions tended to become established on the introduced crop and played similar ecological roles.

One remarkable aspect of the arthropod fauna associated with soybean in North and South America, however, is the general absence of guilds that are among the most host specific and damaging to soybean in the Orient. These

Figure 25-2. Number of species occurring in 1–6 soybean producing regions of the world. The six regions are: North America; Central and South America; Africa; Australia and New Zealand; China, Japan, and Korea; and India and Indonesia. (Based on a table in Kogan and Turnipseed 1987, and Kogan et al. 1988.)

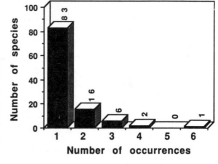

TABLE 25-4. Principal Species in Guilds of Phytophagous Insects Associated with Soybean in the Orient (China, Japan, Korea)[a]

Guilds	Host Specificity	Economic Importance[b]
Foliage Aphids		
Homoptera		
Aphididae		
Aphis glycines	Oligophagous	1
Aphis craccivora	Polyphagous	2
Aulacorthum solani	Polyphagous	2
Stem Borers		
Diptera		
Agromyzidae		
Melanagromyza sojae	Oligophagous	1
Melanagromyza dolichostigma	?	2
Ophiomyia shibatzuji	?	3
Pod and Seed Borers		
Lepidoptera		
Tortricidae		
Leguminivora glycinivorella	Oligophagous	1
Matsumuraeses phaseoli	Oligophagous	1
Gelechiidae		
Etiella zinckenella	Oligophagous	1

[a]Based on Kogan and Turnipseed 1987, Kogan et al. 1988.
[b]Economic importance based on Kobayashi (1977).

guilds include (a) soybean colonizing aphids, (b) agromyzid stem-boring flies, and (c) seed-feeding tortricids and gelechiids (Table 25-4). The lima bean pod borer *Etiella zinckenella* is also present in North and South America, but it causes only occasional injury to soybean. The South American tortricid, *Epinotia aporema*, seems to be the closest homologue of *Matsumuraeses fabivora*, in nature of injury and habits. No pod-feeding species in the New World compare with *L. glycinivorella* in adaptation to soybean phenology, host specificity, and nature of injury (Kogan and Turnipseed 1987). Agromyzid flies are represented in the New World by species of *Agromyza*, *Liriomyza*, and *Phytomyza* (Blickenstaff and Huggans 1974), but the habits and importance of the species on legumes are not well defined.

The guild of soybean-colonizing aphids offers the best support for the hypothesis advanced here. There are no known soybean-colonizing aphids in North and South America, although over 60 species of aphids, including *Aphis craccivova* and *Aulacorthum solani*, have been trapped in soybean fields (Halbert et al. 1981). Aphids are among the most serious pests of soybean in the Orient, either because of their direct injury to the plants or through the transmission of diseases. Damsteegt and Hewings (1986) compared Japanese, California, and New Zealand populations of *Aulacorthum solani* and concluded that the Japanese population colonized soybean in confined areas more readily than did the populations from either California or New Zealand. There were few morphological differences among the three populations.

THE INDUCIBLE DEFENSES OF SOYBEAN

Although grain legumes have a wide range of secondary metabolites, few constitutive compounds [with perhaps the exception of nonprotein amino acids (Harborne and Taylor 1984)] are very obvious candidates to account for the diversity of the interactions of herbivores with species within the Papilionoideae or to genotypes within the cultivars. On the other hand, species of Papilionoideae are known to respond to either biotically or abiotically induced stress with the *de novo* production of phytoalexins (Bailey and Mansfield 1982; Kogan and Paxton 1983). Since Hart et al. (1983) demonstrated that these same phytoalexins deterred Mexican bean bettle feeding, considerable research has been done on the induction of resistance in soybean to herbivorous insects. Soybean phytoalexins are pterocarpan isoflavonoids; the most active in induced defenses against plant pathogens are four isomers of glyceollin (Ingham 1982). The isoflavonoids coumestrol, phaseol, and afrormosin were extracted from soybean leaves and proved to be active feeding deterrents and growth inhibitors to the soybean looper *Pseudoplusia includens* (Caballero et al. 1986). Coumestrol was also identified as a soybean allomone for the Mexican bean beetle (Chiang et al. 1986). These compounds are products of the shikimic–polymalonic acid pathway via the biosynthesis of cinnamic acid or *p*-coumaric acid, which in turn are derived from phenylalanine or tyrosine. Key enzymes of the pathway are phenylalanine ammonia-lyase (PAL) or tyrosine ammonia-lyase (TAL). Activity of these enzymes was increased in plants subjected to resistance-inducing treatments (Chiang et al. 1986). Cinnamic and coumaric acids give rise to other simple phenolic acids and are used by plants to produce such structural and defensive chemicals as flavones, lignins, soluble polyphenols, phenylpropanoids, quinones, and tannins as well as the pterocarpan phytoalexins (Levin 1971). Phenolic acids were found in higher levels in the Mexican bean beetle resistant PI 171451 than in the susceptible cultivar "Forrest" (Hardin 1979). In Hardin's study, phenolic acid analysis was preceded by acid hydrolysis; thus, it is not known whether the acids were free or esterified. Fischer et al. (1990a) sprayed common bean leaves with acetone solutions of benzoic acid, cinnamic, and 21 simple phenolic acids (hydroxylated benzoic and cinnamic acids) and their methoxylated analogues. Discs from treated leaves and untreated controls were offered to Mexican bean beetle adults in dual choice preference tests. The most active compounds were hydroxylated benzoic and cinnamic acid derivatives with the hydroxyl group in position 2 or 6, that is, adjacent to the carboxylic acid group on the benzene ring. Gentisic acid (2,5 dihydroxy-benzoic acid) and salicylic acid (2-hydroxy benzoic acid) are reported to occur in soybean leaves. Using the same aerosol technique, Fischer et al. (1990b) tested the effect of pure glyceollins and found that sprayed leaves of common bean deterred the feeding of Mexican bean beetle but not the feeding of bean leaf beetle.

It seems that herbivory is capable of eliciting a phytoalexin response. Mexican bean beetle herbivory increased the level of antixenosis in soybean plants, which was also positively correlated with total phenolic content of the plants (Chiang et al. 1986). Previously, Reynolds and Smith (1985) showed that growth rates of soybean looper were reduced when larvae were fed foliage from previously injured soybean plants. Experiments conducted in our laboratory indicate that previous herbivory on the normally susceptible cv."Williams" soybean by either Mexican bean beetle or soybean looper larvae increases the levels of antixenosis in uninjured tissue of the same plants (Lin et al. 1990). Field experiments (N. Iannone, M. Kogan, D. Fischer, and C. Helm, in preparation) showed that following the removal of 20 or 35% of total leaf area by soybean looper herbivory, levels of antixenosis increased in newly developed leaves to a maximum of nearly 70% above the level of uninjured plants 17 days after injury onset. The level of antixenosis remained nearly 40% above the basal level for an additional 18 days. The effect was most accentuated in treatments with the highest defoliation level.

Although previously injured plants do not show a drastically altered phenolic compound chemistry, the concentrations of certain phenolics appear to be greatly increased (Kogan et al. 1991). The existence of these inducible defense mechanisms may explain, at least in part, why such obvious gaps among colonizers of soybeans in the New World occur.

A CHEMICAL ECOLOGICAL HYPOTHESIS

If previous herbivory increases levels of antixenosis (and probably also of antibiosis) in soybean, insects that tend to remain stationary (e.g., aphid apterae, mealybugs, whiteflies) and those that are confined within plant tissue (leaf miners, stem borers) must have the ability to withstand, tolerate, or detoxify the increasingly higher concentrations of phytoalexins induced by the feeding activity of potential colonists.

The chemical mechanisms of induced resistance to soybean pathogens have been studied with some detail (Bailey and Mansfield 1982). It is unknown whether herbivore injury disrupts plant cell walls in a fashion that triggers the protein synthesis associated with phytoalexin accumulations or whether the mechanical injury simply opens the way to pathogens that are the actual carriers of the inducing factors. Regardless of the exact means, continuous feeding, for example by a growing aphid colony, is likely to be a powerful inducer of phytoalexin responses. Because of their sedentary behavior, the aphids would receive the full impact of an increasingly higher rate of phytoalexin accumulation. If the aphids are not preadapted to accept these allomonal compounds in order to continue feeding and reproducing, fitness of the colony would be compromised. In actuality, New World aphids must be quite susceptible to even constitutive levels of as yet unknown soybean allomones because although they land on soybean and probe long enough to spread soybean mosaic virus (Irwin and Goodman 1981), they rarely remain long enough to establish colonies. Under certain conditions, however, other sedentary colonists (e.g., whiteflies) are occasionally capable of building large damaging colonies on soybean in the field. They often develop large populations in the greenhouse, but greenhouse-grown soybean plants often show signs of stress. Field populations seem to develop as a result of unusual environmental conditions, for example, prolonged hot, dry weather (Costa et al. 1973), that favor both the development of herbivores and, perhaps, the suppression or inhibition of the biochemical processes of induction.

ESCAPE FROM THE PHYTOALEXIN RESPONSE

The ability of aphids to colonize soybean in the Orient is probably the result of coevolution in the case of the oligophagous *Aphis glycines* (Fig. 25-3). Selection and adaptation over a long period of time may be the processes involved in the polyphagous *Aulacorthum solani* and *Aphis craccivora*. New World strains of these two species are frequent visitors of soybean fields, but no records of colonization exist.

Figure 25-3. Colony of *Aphis glycines* on soybean in Korea. Most colonies are tended by ants. (M. Kogan, photo).

Successful colonizers of soybean fields in North and South America display some of the following general behavioral patterns:

1. Eggs are usually laid singly and dispersed throughout the plant. Larvae of polyphagous lepidopterous species are fairly mobile. First instars merely scrape the leaf surface and consume very little leaf tissue, but later instars consume large areas of leaves and seldom remain on the same leaf for their entire larval development. They often move from one leaf to another, and this mobility may permit larvae to leave areas of high phytoalexin concentration for areas of low phytoalexin concentration. Larvae may also move to adjacent, sometimes uninjured plants. Under optimal conditions, life cycles are short (about 14 days) and outbreaks are explosive, probably overwhelming the capacity of the plant to "mobilize" its defenses.

2. Oligophagous chrysomelids (mainly species of *Cerotoma*) feed as larvae on small roots and inside N-fixing nodules. Although roots may concentrate phytoalexins (Kaplan et al. 1980), it is unlikely that nodules do so. Adult *Cerotoma* are very mobile and probably select feeding sites through probing bites. More importantly, however, our preliminary evidence suggests that *Cerotoma trifurcata* is unaffected by dosages of glyceollins that usually deter feeding by the Mexican bean beetle or the southern corn rootworm, *Diabrotica undecimpunctata howardi* (Fischer et al. 1990b). It is conceivable that *Cerotoma* spp. are preadapted to soybean inducible defenses through the evolution of *Cerotoma* associations with their South and North American leguminous hosts.

3. The polyphagous *Nezara viridula* offers the best example of a possible behavioral adaptation to an inducible plant defense mode. Eggs are laid in clusters of 30–50. The newly emerged nymphs remain aggregated and do not feed; second instars feed by sucking on leaves but rapidly molt (perhaps before a strong phytoalexin response occurs) and begin to disperse. Older nymphs disperse considerable distances from their original emergence sites (Panizzi et al. 1980) and switch to feeding on developing seeds. The mobility of the adults (strong fliers) probably permits them to select tissues least affected by induction.

In summary, the most successful herbivorous colonists of soybean in the New World are either indifferent to the phytoalexin response (*Cerotoma* spp.) or have life history and behavioral strategies that allow either escape or avoidance of plant tissues with high phytoalexin concentration. Insects that tend to form sedentary colonies or have sessile nymphal stages are unable to escape the phytoalexin response; hence, they fail to colonize soybean.

THE EXCEPTIONS

Several New World soybean colonizers may seem to contradict the hypothesis of the role of induced defenses in the limitation of host ranges of potential colonists. The arctiid moths (e.g., *Spilosoma virginica* in North America) oviposit clusters of eggs. Young instars remain aggregated and feed voraciously on a circular front; however, older instars are dispersive and often migrate over the ground surface. These are highly polyphagous species and whether they are affected at all by soybean phytoalexins is unknown.

Several species of hispine leaf miners have been recently recorded on soybean (Kogan and Kogan 1979, Buntin and Pedigo 1982). These leaf miners are oligophagous species well adapted to feeding on native Papilionoideae. Just as the *Cerotoma*, these hispines are probably insensitive to soybean phytoalexins.

The sporadic occurrence of whiteflies in North and South America and the incidence of mealybugs on soybean in Africa have been explained as a possible consequence of weather conditions that alter the phytoalexin response.

CONCLUDING REMARKS

The absence of entire herbivorous guilds from soybean arthropod communities may be explained by the induction of chemical defenses resulting from previous herbivory. Experimental testing of this hypothesis will be possible after we know more about the chemical nature of the induced phytoalexin response in the plant and the effect of phytoalexins on the various components of the herbivore fauna. In the long run, monitoring subtle changes in the faunal composition will allow detection of alterations either in the crop plant through breeding or through genotype x environment interactions, or in the characteristics of native herbivores through selection of adaptive traits that may lead to new potentially damaging associations.

ACKNOWLEDGMENTS

This paper is a contribution of the Office of Agricultural Entomology, College of Agriculture, University of Illinois at Urbana-Champaiign, and the Illinois Natural History Survey. This research was supported in part by grants from the American Soybean Association, the U.S. Department of Agriculture (CSRS-CGO Grant #86-CRCR-1-2174, and North Central Region IPM program, grant #88-34103-3376), and the Illinois Agricultural Experiment Station through Hatch project 12-324 and the Regional Project S-219.

Thanks are due to Peter Price and Thomas Lewinsohn for the organization of the symposium, Charles G. Helm, Dan Fischer, and Audrey Hodgins for technical and editorial reviews, and Jo Ann Auble for typing and word processing.

REFERENCES

Allen, O. N. and E. K. Allen. 1981. *The Leguminosae, a Source Book of Characteristics, Uses, and Nodulation.* University of Wisconsin Press, Madison, WI. 812 pp.

American Soybean Association. 1986. *Soya Bluebook '86.* St. Louis, MO. 278 pp.

Bailey, J. A. and J. W. Mansfield (Eds.). 1982. *Phytoalexins.* John Wiley and Sons, New York. 334 pp.

Baker, H. G. 1984. Patterns of plant invasion in North America, pp. 44–57. In H. A. Mooney and J. A. Drake (Eds.). *Ecology of Biological Invasions of North America and Hawaii.* Ecological Studies 58. Springer-Verlag, New York.

Bazzaz, F. A. 1984. Life history of colonizing plants: Some demographic, genetic, and physiological features, pp. 96–110. In H. A. Mooney and J. A. Drake (Eds.). *Ecology of Biological Invasions of North America and Hawaii.* Ecological Studies 58. Springer-Verlag, New York.

Blickenstaff, C. C. and J. L. Huggans. 1974. Soybean insects and related arthropods in Missouri. *Mo. Agr. Exp. Stn. Res. Bull. 803* (revised). 47 pp.

Buntin, G. D. and L. P. Pedigo. 1982. Foliage consumption and damage potential of *Odontota horni and Baliosus nervosus* (Coleoptera: Chrysomelidae) on soybean. *J. Econ. Entomol.* **75**: 1034–1037.

Caballero, P., C. M. Smith, F. R. Fronczek, and N. H. Fischer. 1986. Isoflavonoids from soybean with potential insecticidal activity. *J. Nat. Prod.* **49**: 1126–1129.

Chiang, H. S., D. Norris, A. Ciepiela, P. Shapiro, and A. Dosterwyk. 1986. Inducible versus constitutive PI 227687 soybean resistance to Mexican bean beetle, *Epilachna varivestis. J. Chem. Ecol.* **13**: 741–749.

Costa A. S., C. L. Costa, and H. F. G. Sauer. 1973. Surto de mosca branca em culturas do Paraná e São Paulo. *Ann. Soc. Entomol. Brasil* **2**: 20–30.

Damsteegt, V. D. and A. D. Hewings. 1986. Comparative transmission of soybean dwarf virus by three geographically diverse populations of *Aulacorthum (Acyrthosiphon) solani. Ann. Appl. Biol.* **109**: 453–463.

FAO. 1986. 1985 FAO production yearbook. *FAO Statistics Ser. 70(39).* 330 pp.

Fischer, D. C., M. Kogan, and J. Paxton. 1990a. Role of free phenolic acids in host plant resistance against the Mexican bean beetle, *Epilachna varivestis:* Behavioral evidence. *J. Entomol. Sci.* **25**: 230–238.

Fischer, D. C., M. Kogan, and J. Paxton. 1990b. Effect of glyceollin, a soybean phytoalexin, on feeding by three phytophagous beetles (Coleoptera: Coccinellidae and Chrysomelidae): Dose vs. Response. *Environ. Entomol.* 19 (in press).

Gambill, Jr., W. G. 1953. The Leguminosae of Illinois. *Ill. Biol. Monographs 23(4),* xii + 117 pp.

Halbert, S. E., M. E. Irwin, and R. M. Goodman. 1981. Alate aphid (Homoptera: Aphididae) species and their relative importance as field vectors of soybean mosaic virus. *Ann. Appl. Biol.* **97**: 1–9.

Harborne, J. B. and B. L. Turner. 1984. *Plant Chemosystematics.* Academic Press, London. 562 pp.

Hardin, J. M. T. 1979. *Phenolic acids of soybean resistant and nonresistant to leaf feeding larvae.* M.S. thesis. University of Arkansas, Fayetteville, AR. 50 pp.

Hart, S. V., M. Kogan, and J. D. Paxton. 1983. Effect of soybean phytoalexins on the herbivorous insects Mexican bean beetle and soybean looper. *J. Chem. Ecol.* **9**: 657–672.

Herzog, D. C. 1980. Sampling soybean looper on soybean, pp. 141–168. In M. Kogan and D. C. Herzog (Eds.). *Sampling Methods in Soybean Entomology.* Springer-Verlag, New York.

Herzog, D. C. and J. W. Todd. 1980. Sampling velvetbean caterpillar on soybean, pp. 107–140. In M. Kogan and D. C. Herzog (Eds.). *Sampling Methods in Soybean Entomology.* Springer-Verlag, New York.

Hymowitz, T. 1970. On the domestication of the soybean. *Econ. Bot.* **24**: 408–421.

Ingham, J. L. 1982. Phytoalexins from the Leguminosae, pp. 21–80. In J. A. Bailey and J. W. Mansfield. *Phytoalexins.* John Wiley, New York.

Irwin, M. E. and R. M. Goodman. 1981. Ecology and control of soybean mosaic virus, pp. 181–220. In K. Maramorosch and F. K. Harris (Eds.). *Plant Diseases and Vectors: Ecology and Epidemiology.* Academic Press, New York.

Kaplan, D. T., N. T. Keen, and I. J. Thomason. 1980. Studies on the mode of action of glyceollin in soybean: Incompatibility to the root knot nematode, *Meloidogyne incognita. Physiol. Plant Pathol.* **13**: 319–325.

Kiritani, K., N. Hokyo, K. Kimura, and F. Nakasuji. 1965. Imaginal dispersal of southern green stink bug, *Nezara viridula* L., in relation to feeding and oviposition. *Japan. J. Appl. Entomol. Zool.* **9**: 291–297.

Kobayashi, T. 1977. Insect pests of soybeans in Japan. Tech. Bull. ASPAC Food Fert. *Technol. Cent.* **36**: 1–24.

Kogan, J., M. Kogan, E. F. Brewer, and C. G. Helm. 1988. World Bibliography of Soybean Entomology, 2 vols. *Illinois Agricultural Experiment Station, Special Publication 73.* Volume I, xxix, 665 pp, Volume II, ix, 291 pp.

Kogan, M. 1981. Dynamics of insect adaptations to soybean: Impact of integrated pest management. *Environ. Entomol.* **10**: 363–371.

Kogan, M. 1986. Natural chemicals in plant resistance to insects. *Iowa St. J. Res.* **60**: 501–527.

Kogan, M. and D. C. Fischer. (1991). Inducible defenses in soybean against herbivorous insects. In M. J. Raupp and D. W. Tallamy (Eds.). *Phytochemical Induction by Herbivores.* John Wiley, New York.

Kogan, M. and D. D. Kogan. 1979 *Odontota horni*, a hispine leaf miner adapted to soybean feeding in Illinois. *Ann. Entomol. Soc. Amer.* **72**: 456–461.

Kogan, M. and J. Paxton. 1983. Natural inducers of plant resistance to insects, pp. 153–171. In P. A. Hedin (Ed.). *Plant Resistance to Insects.* Symposium on Plant Resistance to Insects at the 183rd meeting of the American Chemical Society, Las Vegas, NV, March 28-April 2, 1982. ACS Symposium Series Vol. 208. American Chemical Society, Washington, D.C. 375 pp.

Kogan, M. and S. G. Turnipseed. 1987. Ecology and management of soybean arthropods. *Ann. Rev. Entomol.* **32**: 507–538.

Kogan, M., C. G. Helm, J. Kogan, and E. F. Brewer. 1989. Distribution and economic importance of *Heliothis virescens* and *Heliothis zea* in North, Central and South America including a listing and assessment of the importance of their natural enemies and host plants, pp. 241–297. In E. King and R. Jackson, (Eds.). *Biological Control of Heliothis: Increasing the Effectiveness of Natural Enemies.* USDA-PL 480 India project. New Delhi, India.

Kogan, M., G. P. Waldbauer, G. Boiteau, and C. E. Eastman. 1980. Sampling bean leaf beetles on soybean, pp. 201–236. In M. Kogan and D. C. Herzog (Eds.). *Sampling Methods in Soybean Entomology.* Springer-Verlag, New York.

Krysan, J. L. 1986. Introduction: Biology, distribution, and identification of pest

Diabrotica, pp. 1–23. In J. L. Krysan and T. A. Miller (Eds.). *Methods for the Study of Pest Diabrotica.* Springer-Verlag, New York.

Levin, D. A. 1971. Plant phenolics: An ecological perspective. *Amer. Nat.* **105**: 157–181.

Lin, H., M. Kogan, and D. C. Fischer. 1990. Induced resistance in soybean to the Mexican bean beetle: Comparisons of induction factors. *Environ. Entomol.* (in press):

Miner, F. D. 1966. Biology and control of stink bugs on soybeans. *Ag. Exp. Stn., University of Arkansas, Fayetteville, Bull. 708,* 40 pp.

Miyasaka, S. and J. C. Medina (Eds.). 1981. *A soja no Brasil.* Instituto de Tecnologia de Alimentos, Campinas, São Paulo, Brazil. 1062 pp.

Panizzi, A. R., M. H. M. Galileo, H. A. O. Gastal, J. F. F. Toledo, and C. H. Wild. 1980. Dispersal of *Nezara viridula* and *Piezodorus guildinii* nymphs in soybeans. *Environ. Entomol.* 9: 293–297.

Panizzi, A. R. and F. Slansky, Jr. 1985. Review of phytophagous pentatomids (Hemiptera: Pentatomidae) associated with soybean in the Americas. *Fla. Entomol.* **68**: 184–214.

Pedigo, L. P., J. D. Stone, and G. L. Lentz. 1973. Biological synopsis of the green cloverworm in central Iowa. *J. Econ. Entomol.* **68**: 665–673.

Reynolds, G. W. and C. M. Smith. 1985. Effects of leaf position, leaf wounding, and plant age of two soybean genotypes on soybean looper (Lepidoptera: Noctuidae) growth. *Environ. Entomol.* **14**: 475–478.

Schoene, W. J. and G. W. Underhill. 1933. Economic status of the green stinkbug with reference to the succession of its wild hosts. *J. Agric. Res.* **46**: 863–866.

Strong, D. R., J. H. Lawton, and R. Southwood. 1984. *Insects on Plants: Community Patterns and Mechanisms.* Harvard University Press, Cambridge, MA. 313 pp.

Todd, J. W. 1989. Ecology and behavior of *Nezara viridula. Ann. Rev. Entomol.* **34**: 273–292.

26 Ecology of Tropical Herbivores in Polycultural Agroecosystems

MIGUEL A. ALTIERI

Tropical landscapes are usually characterized by a mosaic of agroecosystems representative of various stages of succession, with cropping systems that differ in age and diversity (Ewell 1986). Agroecosystems vary in structure and management and may be composed of seasonal crops in mixed or sole-cropping patterns and/or of perennial crops in plantation or agroforestry designs, devoted to subsistence based on low inputs or to commercial purposes under modern technology. All these systems are dynamic and subjected to different levels of management so that the crop arrangements in time and space are continually changing in the face of biological, cultural, socioeconomic, and environmental factors. Such landscape variations determine the degree of spatial and temporal heterogeneity characteristic of agricultural regions, which in turn may or may not benefit the pest protection of particular agroecosystems. Thus, one of the main challenges facing tropical agroecologists today is identifying the types of heterogeneity (either at the field or regional level) that will yield desirable agricultural results (i.e., pest regulation), given the unique environment and entomofauna of each area (Janzen 1973). This challenge can only be met by further analyzing the relationship between vegetation diversification and the population dynamics of tropical herbivore species, in light of the diversity and complexity of tropical agricultural systems. A hypothetical pattern in pest regulation according to agroecosystem temporal and spatial diversity is depicted in Figure 26-1. Whether this pattern holds across regions and systems is unclear; however, our current understanding has been significantly advanced in the last two decades by numerous studies that have explored the ways in which plant diversity in tropical agroecosystems influences the stability and diversity of the herbivore community (Altieri and Letourneau 1982, 1984).

Plant-Animal Interactions: Evolutionary Ecology in Tropical and Temperate Regions, Edited by Peter W. Price, Thomas M. Lewinsohn, G. Wilson Fernandes, and Woodruff W. Benson. ISBN 0-471-50937-X © 1991 John Wiley & Sons, Inc.

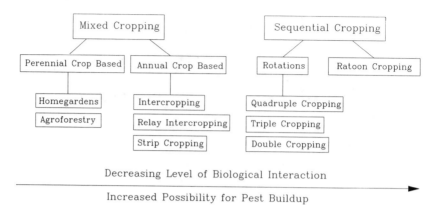

Figure 26-1. An hypothetical model describing the relationship between agroecosystem temporal and spatial complexity and pest-regulation trends in tropical areas.

HERBIVORE TRENDS IN POLYCULTURAL AGROECOSYSTEMS

Although herbivores vary widely in their response to crop distribution, abundance, and dispersion, the majority of agroecological studies show that structural (i.e., spatial and temporal crop arrangement) and management (i.e., crop diversity, input levels, etc.) attributes of agroecosystems influence herbivore dynamics (Fig. 26-2). Most experiments that mixed other plant species with the primary host of a specialized herbivore showed that in comparison with diverse crop communities, simple crop communities have greater population densities of specialist herbivores (Root 1973; Bach 1980a, b; and Risch 1981). In these systems, herbivores exhibit greater colonization rates, greater reproduction, less tenure time, less disruption of host finding and enhanced mortality by natural enemies.

So far, two hypotheses have been proposed to explain the commonly observed lower herbivore abundance in polycultures:

1. *The Resource Concentration Hypothesis.* This hypothesis states that crop monocultures represent a concentrated resource for specialized herbivores, which increases the attraction and accumulation of these species, the time they spend in the system and their reproductive success. Visual and chemical stimuli from host and nonhost plants in the polyculture affect the rate at which herbivores colonize polycultures and their behavior in these habitats. In a polyculture, nonhost species may mask the chemical attractants of the host, reduce the contrast between the host and its background or simply hide the plants from view (i.e., make them less apparent) (Bach 1984).

2. *The Enemies Hypothesis.* This hypothesis predicts an increased abundance of arthropod predators and parasitoids in polycultures due to the increased availability of alternate prey, nectar sources, and suitable microhabitats. The net

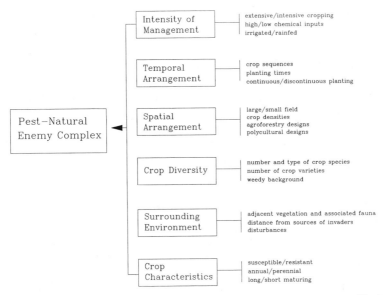

Figure 26-2. Structural and management features of tropical agroecosystems affecting the dynamics of pest–natural enemy complexes.

effect of natural enemies on pest abundance in polycultures will depend upon whether natural enemies are governed more in their behavior by prey density or by background plant density, although at times both factors simultaneously operate (Letourneau and Altieri 1983; Letourneau 1987).

There is general agreement that these hypotheses are not mutually exclusive since a particular herbivore population may be simultaneously affected by both the concentration of resources and by natural enemies. In fact, herbivore regulation in polycultures may involve other mechanisms not considered by the hypotheses such as microclimate, differences in levels of secondary compounds or in plant quality, and so on. The point is that it is important to identify key mechanisms, differences in mechanisms between cropping systems, and which plant assemblages enhance regulatory effects and which do not, and under what management and agroecological circumstances.

DILEMMAS OF THE HYPOTHESES

The most common studies used to evaluate the effects of vegetation diversity on herbivore populations have included:

1. Cultivating crop plants in monocultures and polycultures, including treatments that vary host-plant density in both systems, or vary the number of species of associated plants in the polyculture.

2. Varying the background in which crops are grown, usually involving cultivated versus weedy ground or soil covered by a living mulch.
3. Contrasting herbivore abundances on crop fields surrounded by plant communities of different ages and diversities.

Of these approaches, type 1 and 2 studies represent the overwhelming majority of studies. In their summary of 68 published experiments contrasting herbivore abundance between monocultures and weedy or intercropped systems, Risch et al. (1983) found that 79 herbivore species were more abundant in monocultures, 19 were more abundant in diversified systems, and 17 showed no difference.

One main criticism of the mentioned studies is that most have failed to separate the effects of several confounding variables, such as host-plant density and total plant density, both of which vary with diversity (Kareiva 1983). If the number of host plants is kept constant, adding nonhosts increases total plant density in the polyculture, as well as the level of interspecific competition, and the amount of leaf area upon which flying insects can alight. If total plant density is kept constant, then there will be more host plants per unit area in the monoculture than in the polyculture, and any differences observed may be pure host-plant density effects rather than responses to increased diversity. In both cases, differences in herbivore numbers could be due to differences in levels of plant competition (which may reduce size or quality of host plants) or differences in quantity of host plant available.

An additional complication arises from studies showing that a change in predator abundance does not necessarily result in a corresponding change in predator efficacy. This has led some authors to conclude that herbivore movement patterns are more important than the activities of natural enemies in explaining the reduction of herbivores in polycultures (Risch et al. 1983; Kareiva 1986). On the basis of currently available information it is not possible to rule out the ability of natural enemies to limit herbivore populations in polycultures. Part of the problem is that most comparative studies concentrated on insect species or life stages that exhibited a relatively low potential for population regulation by natural enemies (Sheehan 1986), and also most studies did not correlate predator population density and the herbivore mortality resulting from predator augmentation (Letourneau 1987).

Given the limitations mentioned above, a number of researchers have developed more sophisticated research methodologies aimed at more precisely elucidating the effects of agroecosystem diversity on herbivores and associated natural enemies. The recent studies of Bach (1980a,b; 1984), Risch (1980, 1981), Andow and Risch (1985) and Letourneau (1986, 1987) offer better approaches to separate the effects of density and diversity per se on herbivores and natural enemies by controlling each variable in the experimental plots. Despite the efforts described above, virtually no studies that explore ecological mechanisms have been conducted in actual farmers' fields that include polycultures. A notable exception is the survey of Trujillo (1987) of Tlaxcalan agroecosystems, where he observed that mixed untreated maize fields escaped some of the insect damage to give higher yields than those of unprotected sole crops. It follows then that the implications and projections of the two hypotheses to real farm situations, especially in relation to explaining herbivore dynamics in the prevalent polycultures of the tropics, are at this point highly speculative.

ECOLOGICAL THEORY AND INSECT MANIPULATION IN POLYCULTURES

The most vegetationally diverse tropical agroecosystems are those under extensive shifting cultivation and/or those under intensive subsistence farming. These systems are usually characterized by complex cropping systems (i.e., intercropping, agroforestry, rotations, etc.) with crop sequences and associations managed in a variety of ways in time and space.

Theoretically these systems contain built-in elements of natural pest control. The question is then: how can emergent ecological theories help in designing polycultures that offer even better or more effective herbivore protection features? Or in other words, is there anything that ecologists can do to help traditional farmers improve the polycultural systems they already have?

From a practical standpoint it is easier to design insect manipulation strategies in polycultures using the elements of the natural enemies hypothesis than those of the resource concentration hypothesis, mainly because we cannot yet identify the ecological situations or life history traits that make some pests sensitive (i.e., their movement is affected by crop-patterning) and others insensitive to cropping patterns (Kareiva 1986). The recognition of biological control practitioners that crop monocultures are difficult environments in which to induce efficient operation of beneficial insects, because those systems lack adequate resources for effective performance of natural enemies and because of the disturbing cultural practices often utilized in such systems, offer useful elements. It implies that because polycultures already contain certain specific resources provided by plant diversity, and are usually not disturbed with pesticides (especially when managed by resource-poor farmers who cannot afford high-input technology), they are more amenable for manipulation. Thus by replacing or adding diversity to existing systems, it may be possible to exert changes in habitat diversity that enhance natural enemy abundance and effectiveness by (Van den Bosch and Telford 1964; Altieri and Letourneau 1982; Powell 1986):

1. Providing alternative hosts–prey at times of pest host scarcity.
2. Providing food (pollen and nectar) for adult parasitoids and predators.
3. Providing refuges for overwintering, nesting, and so on.
4. Maintaining acceptable populations of the pest over extended periods to ensure continued survival of beneficial insects.

The specific resulting effect or the strategy to use will depend on the species of herbivores and associated natural enemies, as well as on properties of the vegetation, the physiological condition of the crop, or the nature of the direct effects of a particular plant species (Letourneau 1987). In addition, the success of enhancement measures can be influenced by the scale upon which they are implemented (i.e., field scale, farming unit or region), since field size, within-field and surrounding vegetation composition, and level of field isolation (i.e., distance from source of colonizers) will all affect immigration rates, emigration rates, and the effective tenure time of a particular natural enemy in a crop field.

Perhaps one of the best strategies to increase effectiveness of predators and parasitoids is the manipulation of nontarget food resources (i.e., alternate hosts–prey and pollen–nectar) (Rabb et al. 1976). Here it is not only important that the density of the nontarget resource be high to influence enemy populations, but that the spatial distribution and temporal dispersion of the resource be also adequate. Proper manipulation of the nontarget resource should result in the enemies colonizing the habitat earlier in the season than the pest, and frequently encountering an evenly distributed resource in the field, thus increasing the probability of the enemy to remain in the habitat and reproduce (Andow and Risch 1985). Certain polycultural arrangements increase and others reduce the spatial heterogeneity of specific food resources; thus particular species of natural enemies may be more or less abundant in a specific polyculture. These effects and responses can only be determined experimentally across a whole range of tropical agroecosystems. The task is indeed overwhelming since enhancement techniques must necessarily be site-specific.

NEW DIVERGENT EVIDENCE

So far, the two hypotheses provide the basis to predict herbivore response in polycultures if:

1. The herbivore is a specialist or exploits a narrow range of plants.
2. The polyculture is composed of a preferred host plant and one or more nonhost plants.
3. Host and nonhost plants overlap in time and space within the mixture.

Given the above limitations, not all results from studies of agricultural diversification leading to reduced pest populations can be adequately explained by the two hypotheses as they stand. One of such cases is a study by Gold (1987) on the effects of intercropping cassava with cowpeas on the population dynamics of the cassava whiteflies (*Aleurotrachelus socialis* and *Trialeurodes variabilis*) in Colombia. As expected, intercropping reduced egg populations of both species of whitefly on cassava (a 12-month crop), but the reductions were residual, persisting up to 6 months after the harvest of the cowpea, which only lasts 3–4 months (Fig. 26-3). The natural enemy hypothesis was rejected as a mechanism explaining reduced herbivore load, because predators were more abundant in monocultures than in polycultures, showing a numerical response. The resource

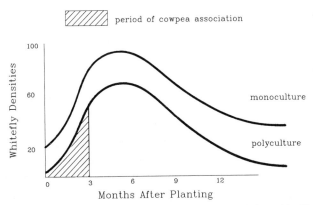

Figure 26-3. Expected population response (egg densities) of the whiteflies *Aleurotrachelus socialis* and *Trialeurodes variabilis* in cassava and cowpea mixtures in Colombia (after Gold 1987).

concentration hypothesis could explain lower whitefly densities during the period that cowpea was present, but cannot explain the reductions in whitefly populations observed long after the removal of the cowpea. Instead, it appeared that intercrop competition caused a reduction in cassava size or vigor that persisted through the remainder of the season. Thus, whitefly populations in polycultures were a function of host-plant selection and/or tenure time which, in turn, was related to host condition.

Another case is a study by Altieri and Schmidt (1986a), who found lower densities of the specialist herbivore *Phyllotreta cruciferae* on broccoli mixed with another crucifer host plant, the wild mustard *Brassica kaber*. Densities of *P. cruciferae* were greater on broccoli plants grown in monocultures than in polycultures, but not so on a per-plot basis. The differences in beetle abundance on a per-plant basis were basically due to the fact that the beetles concentrated more on wild mustard than on broccoli in the mixture. This preference has a chemical basis since wild mustard has higher concentrations of glucosinolates, a strong beetle attractant, than broccoli. Thus in this case the differences in beetle abundance caused by diversity resulted from differences in beetle-feeding preferences rather than from differences in colonization, reproduction, or predation. Obviously these trends do not conform to the assumptions of either of the two hypotheses.

These two case studies do not necessarily contradict the two hypotheses, but provide additional explanation to current theory and call for more caution and flexibility since the responses of herbivores to vegetational diversity are not uniform and cannot always be explained by diversity per se. In fact, differences in tenure time or movement patterns between mono- and polycultures are not always evident, although interplot differences in abundance persist. As suggested by the studies mentioned above, other factors such as visual cues, microclimatic changes, feeding preferences, or direct effects on host-plant vigor could influence habitat location and/or searching behavior of both herbivores and natural enemies.

THE EFFECTS OF SURROUNDING HABITAT DIVERSITY

Despite all the experimental work conducted on vegetational diversity at the field level, there is limited evidence to suggest whether at a more regional level (i.e., at the level of agroecosystem mosaics interspersed among natural vegetation) pest problems will diminish due to spatial and temporal heterogeneity of tropical agricultural landscapes. Although studies on temperate agroecosystems suggest that the vegetational settings associated with particular crop fields influence the kind, abundance, and time of arrival of herbivores and their natural enemies (Gut et al. 1982; Altieri and Schmidt 1986b), the role of primary and secondary forests and other natural vegetation remnants in harboring pests and other natural enemy complexes is virtually unexplained in the tropics.

In perennial orchards (i.e., pears and apple) at temperate latitudes, a diverse complex of predators usually develops early in the season in orchards surrounded by woodlands. In these systems, main pests (i.e., *Psylla pyricola, Cydia pomonella,* etc.) are quickly reduced and maintained at low levels throughout the season. In contrast, early season predators are absent in more extensive commercial orchards, and therefore pest pressure is more intense (Croft and Hoyt 1983). Assuming that these trends also occur in the tropics, one would expect that in certain tropical agroecosystems (i.e., shifting cultivation in the lowland tropics), forest and bush fallows have potential value in controlling pests. Clearing small plots in a matrix of secondary forest vegetation may permit easy migration of natural enemies from the surrounding jungle (Matteson et al. 1984). This positive potential role of natural vegetation on biological control is, however, expected to change in view of current deforestation rates and modernization trends toward commercial monocultures.

CONCLUSIONS

Although it is generally accepted that diversification of cropping systems often leads to reduced herbivore populations, the mechanisms explaining such reductions cannot always be adequately explained by the resource concentration and/or the natural enemies hypotheses. In addition, only limited work has been conducted in real tropical farming systems, so that the dynamics of herbivores and associated natural enemies, and the way they interact with other components of the agroecosystem, are virtually unknown, under the whole range of diversities and management intensities that polycultures exhibit under farmers' circumstances. If certain existing crop mixtures contain built-in elements of pest control, such elements should be identified and then retained in the course of modernization. In other cases, suboptimal interactions between plants, herbivores, and natural enemies could be improved (i.e., by adding or eliminating diversity) to enhance natural enemy effectiveness in regulating herbivore densities. To advance beyond merely cataloging the results of crop-patterning experiments, the mechanisms underlying pest responses to diversity must be identified (Kareiva 1986).

What is difficult is that each agricultural situation must be assessed separately, since herbivore–enemy interactions will vary significantly depending on insect species, location and size of the field, plant composition, the surrounding vegetation, and cultural management. One can only hope to elucidate the ecological principles governing herbivore dynamics in complex systems, but the polycultural designs necessary to achieve herbivore regulation will depend on the agroecological conditions and socioeconomic restrictions of each area. In this regard farmers' needs and preferences must be fully considered if adoption of new designs is expected. New designs will be most attractive if, in addition to pest regulation, polycultures offer benefits in terms of overyielding, increased soil fertility, decreased weed competition and diseases and evening out of labor demands.

Although studies of the relationships between landscape and agroecosystem diversity and insect herbivores are of great theoretical interest, the most important issue at hand is that this understanding of the ecology of insects in polycultures be applied to develop sustainable pest management systems that enable constant food production and conservation of the resource base, under low-input management, so that it benefits the large mass of resource-poor farmers who rely on polycultures for subsistence. Thus, agroecological research on polycultures has an important, yet unrealized, role in rural development, namely contributing to develop strategies that secure food self-sufficiency and conserve the natural resource base of tropical agricultural communities.

REFERENCES

Andow, D. and S. J. Risch. 1985. Predation in diversified agroecosystems: Relations between a coccinellid predator *Coleomegilla maculata* and its food. *Journal of Applied Ecology* **22**: 357–372.

Altieri, M. A. and D. L. Letourneau. 1982. Vegetation management and biological control in agroecosystems. *Crop Protection* **1**: 405–430.

Altieri, M. A. and D. L. Letourneau. 1984. Vegetation diversity and insect pest outbreaks. *CRC Critical Reviews in Plant Sciences* **2**: 131–169.

Altieri, M. A. and L. L. Schmidt. 1986a. Population trends and feeding preferences of flea beetles (*Phyllotreta cruciferae* Goeze) in collard-wild mustard mixtures. *Crop Protection* **5**: 170–175.

Altieri, M. A. and L. L. Schmidt. 1986b. The dynamics of colonizing arthropod communities at the interface of abandoned organic and commercial apple orchards and adjacent woodland habitats. *Agriculture, Ecosystems and Environment* **16**: 24–43.

Bach, C. E. 1980a. Effects of plant density and diversity on the population dynamics of a specialist herbivore, the striped cucumber beetle, *Acalymma vittata* (Fab.). *Ecology* **61**: 1515–1530.

Bach, C. E. 1980b. Effects of plant diversity and time of colonization on an herbivore–plant interaction. *Oecologia (Berlin)* **44**: 319–326.

Bach, C. E. 1984. Plant spatial pattern and herbivore population dynamics: Plant factors affecting the movement patterns of a tropical cucurbit specialist (*Acalymma innubum*). *Ecology* **65**: 175–190.

Croft, B. A. and S. C. Hoyt. 1983. *Integrated management of insect pests of pome and stone fruits*. Wiley, New York. 454 pp.

Ewell, J. J. 1986. Designing agricultural ecosystems for the humid tropics. *Ann. Rev. Ecol. Syst.* **17**: 245–271.

Gold, C. S. 1987. *Crop diversity and tropical herbivores: Effects of intercroping and mixed varieties on the cassava whiteflies, Aleurotrachelus socialis Bondar and Trialeurodes variabilis Quaintance, in Colombia*. Ph.D. Dissertation, University of California, Berkeley, CA.

Gut, L. J., C. E. Jochums, P. H. Westigard, and W. J. Liss. 1982. Variation in pear psylla (*Psylla pyricola* Foerster) densities in southern Oregon orchards and its implications. *Acta Hortic.* **124**: 101–111.

Janzen, D. H. 1973. Tropical agroecosystems. *Science* **182**: 1212–1219.

Kareiva, P. 1983. The influence of vegetation texture on herbivore populations: Resource concentration and herbivore movement. Pages 259–289 in R. F. Denno and M. S. McClure (Eds.). *Variable plants and herbivores in natural and managed systems*. Academic Press, New York.

Kareiva, P. 1986. Trivial movement and foraging by crop colonizers. In *Ecological Theory and Integrated Pest Management Practice*. M. Kogan (Ed.). John Wiley and Sons, New York, pp. 59–82.

Letourneau, D. K. and M. A. Altieri. 1983. Abundance patterns of a predator *Orius tristicolor* (Hemiptera: Anthocoridae) and its prey, *Frankliniella accidentalis* (Thysanoptera: Thripidae): Habitat attraction in polycultures versus monocultures. *Env. Ent.* **122**: 1464–1469.

Letourneau, D. K. 1986. Associational resistance in squash monoculture and polyculture in tropical Mexico. *Environmental Entomology* **15**: 285–292.

Letourneau, D. K. 1987. The enemies hypothesis: Tritrophic interactions and vegetational diversity in tropical agroecosystems. *Ecology* **68**: 1616–1622.

Matteson, P. C., M. A. Altieri, and W. C. Gagné. 1984. Modification of small farmer practices for better pest management. *Ann. Rev. Entomol.* **29**: 383–402.

Powell, W. 1986. Enhancing parasitoid activity in crops. In *Insect Parasitoids*. J. Waage and D. Greathead (Eds.). Academic Press, London. 389 pp.

Rabb, R. L., R. E. Stinner, and R. van den Bosch. 1976. Conservation and augmentation of natural enemies. In *Theory and Practice of Biological Control*. C. B. Huffaker and P. S. Messenger (Eds.). Academic Press, New York, pp. 233–254.

Risch, S. 1980. The population dynamics of several herbivorous beetles in a tropical agroecosystem: The effect of intercropping corn, beans and squash in Costa Rica. *J. Appl. Ecol.* **17**: 593–612.

Risch, S. 1981. Insect herbivore abundance in tropical monocultures and polycultures: An experimental test of two hypotheses. *Ecology* **62**: 1325–1340.

Risch, S., D. Andow, and M. Altieri. 1983. Agroecosystem diversity and pest control: Data, tentative conclusions and new research directions. *Environ. Entomol.* **12**(3): 625–629.

Root, R. B. 1973. Organization of a plant–arthropod association in simple and diverse habitats: The fauna of collards (*Brassica oleracea*). *Ecol. Monogr.* **43**: 95–124.

Sheehan, W. 1986. Response by specialist and generalist natural enemies to agroecosystem diversification: A selective review. *Env. Ent.* **15**: 456–461.

Trujillo, J. A. 1987. *The agroecology of maize production in Tlaxcala, Mexico: Cropping systems effects on arthropod communities.* Ph.D. Dissertation, University of California, Berkeley, CA.

Van den Bosch, R. and A. D. Telford. 1964. Environmental modification and biological control. In *Biological Control of Insect Pests and Weeds*. P. deBach (Ed.). Reinhold, New York, pp. 454–488.

27 Arthropods in a Tropical Corn Field: Effects of Weeds and Insecticides on Community Composition

MARIA ALICE GARCIA

Some of the theoretical issues under debate about processes and factors determining the arthropod fauna in a community are considered very important from a practical point of view, especially for agriculture. The idea that there exists a relationship between species diversity and stability in natural communities has been extrapolated to artificial systems and led to a questioning of the practice of community simplification usually adopted in agroecosystems (Murdoch 1975; Liss et al. 1986).

Alhough many tropical countries adopt polyculture in small areas, thus creating high-diversity systems as the common form of agriculture (see Altieri, this volume), in Brazil, especially in the eastern, western, and southern regions, monocuure is the general rule. Soybean, sugar cane, rice, and wheat crops are predominant in these regions.

According to Elton (1958) and MacArthur (1955) the stability of a community should be proportional to the number of trophic links between the component species. Revisions of the theoretical foundations of the relationship between diversity and stability such as those from Van Emden and Williams (1974) have, however, found no such obligate connection. The application of mathematical models (Maynard Smith 1974) reveals that with a certain threshold, the higher the diversity, the lower the stability of the populations. May (1973) demonstrated that a complex system with more trophic links can fluctuate more than a simpler system, and can therefore have lower stability. However, a higher species richness offers the possibility of multiple, interchangeable forms of organization (Holling 1973).

Goodman (1975) concludes his extensive revision of this subject cautiously

Plant-Animal Interactions: Evolutionary Ecology in Tropical and Temperate Regions, Edited by Peter W. Price, Thomas M. Lewinsohn, G. Wilson Fernandes, and Woodruff W. Benson. ISBN 0-471-50937-X © 1991 John Wiley & Sons, Inc.

suggesting that when there is an intense perturbation which practically eliminates interspecific interactions, a proportionally higher number of species can be lost in more diversified environments. But in the case of smaller perturbations it is difficult to establish any clear relationship between diversity and stability.

According to the diversity–stability principle, the problem of pest eruption and resurgence in agroecosystems should also be related to the adoption of monoculture as a generalized agricultural practice. In this system, with a low species diversity, population fluctuations would be greater, leading to demographic explosions and to local species extinction. Indeed, a large number of species in the higher trophic levels may contribute to maintain stability in lower levels (Southwood and Way 1970).

In practice, it is not only stability which is required for populations in an agroecosystem but the maintenance of the pest species below an economic damage level, and for this community diversity may be important (Dempster and Coaker 1974).

Another concept that has been extrapolated to agricultural systems is the theory of Island Biogeography of MacArthur and Wilson (1967). In this case a crop field could be considered as a potentially colonizable island with an immigration rate proportional to the distance to species-source areas, and species extinction rates inversely correlated to size. Size could also be related to variability of available habitats for colonizing species (Price 1983).

The application of the Island Biogeography theory in agroecosystems was questioned by Price (1976), especially because of their peculiarities of instability with time and space, and the difficulties of analyzing the effects of distance from the colonizer sources. Similarly, Rey and McCoy (1979) suggested caution when interpreting data from pest populations using this theory, since barriers that can differentially restrict the colonization by herbivores and predators are not predicted by it. The concept of the diversity–stability relationship has also been questioned, although these two general ideas are widely quoted in the literature on applied agroecosystem ecology. Some elements appear in two alternative, though not mutually exclusive, hypotheses, that Root (1973) discusses to explain the relationship between diversity in a plant community and the density of specialist and generalist herbivore populations in monoculture systems and in systems with invaders. The hypothesis that links resource concentration to higher numbers of specialist herbivores which may reach pest status represents in fact a situation of both low trophic diversity and of low numbers of species present in the environment. Such an event favors populational increase, which may reach eruption levels, among those herbivores that find abundant resources in an environment unsuitable for their natural enemies.

As to natural enemies, a more diverse environment with a higher variability of available habitats for colonizing species could be considered as a larger "island." This could allow the maintenance of generalist herbivores and natural enemy populations, which could act as regulators for the herbivore populations, keeping them at lower densities.

The mechanisms suggested by both of Root's hypotheses can be understood when the insects' responses to the various different attraction signals are analyzed. Different plant species in the habitat release chemicals and show visual signals that the vast majority of insects use as cues for resource location. The insect perception and the corresponding response to these signals is slightly different for each insect species (see Visser 1986). The establishment and population increase of each species will depend not only on the availability and distribution of resources, but also on the interactions that are established and on microclimatic factors.

The concept of an "associational resistance" that reduces the possibility of a population explosion of herbivores in environments with a higher taxonomic and microclimatic diversity is defended by Tahvanainen and Root (1972). Thus, the complexity maintained in agroecosystems may selectively favor or inhibit species of generalists, specialist herbivores, or colonizing natural enemies and also depend on the plant species present, their density and distribution, and on the arthropod species that may reach the area.

The colonization process and establishment of arthropods in monoculture agroecosystems or in unweeded crops in the tropics and subtropics may occur with patterns of special responses depending on each case. The understanding of these patterns is essential for the elaboration of proposals of pest management in these regions.

Searching for Response Patterns

Few studies on the arthropod community in tropical agroecosystems have used an approach based on ecological theories and hypotheses; see for instance Altieri and Letourneau (1982, 1984), Altieri et al. (1977), Letorneau (1986), McCoy and Rey (1983), Risch (1979, 1981), Risch et al. (1986) and Power et al. (1987).

However, the verification of the mechanisms underlying community organization is very useful for elaborating proposals concerning agroecosystems in general and especially tropical systems, where the problems of pests are added to those of soil fragility and losses due to erosion. The practice of no-tillage, which tends to reduce erosion, has been especially employed in southern Brazil but, in general, when misused it is associated with an intense use of herbicides and an increase in costs. It is recognized, however, that the maintenance of weed populations may also represent an efficient method of reducing losses due to erosion. This practice, according to the theories already presented, can contribute to the reduction of crop infestation by herbivorous pests. Comparing the responses of herbivores and natural enemies in monocultures with more complex environments is a way of assessing the efficiency of this practice for reducing pest incidence. The responses of monoculture or polyculture communities, or of a weedy crop, to natural perturbations like storms, or artificial ones like those caused by pesticide application, for example, may indicate their degree of stability, allowing forecasts necessary for the integration of management methods.

ARTHROPODS IN WEEDED AND UNWEEDED CORN FIELDS IN A TROPICAL REGION

During 1984 the colonization process by arthropods, and the effects of insecticide application in a cultivated field crop, comparing weeded with unweeded corn fields, were observed. The research was carried out in Campinas, São Paulo State, Brazil (22°47′ S, 47°02′ W), in an area of 7000 m², with eight 450-m plots (15 × 30 m). Four plots were hand-weeded and four plots were left unweeded. Corn was planted in the same density and row distance in all plots. Two weeded and two unweeded plots were sprayed with Endossulfan (175 g AI/ha), whereas the control areas were sprayed with water. Colonization by arthropods and insecticide effects were assayed by sweep net sampling. In each experimental plot, transects were carried out on four occasions: one month and one day before the insecticide application, and one day and one month after spraying. On each date, each plot was sampled with 3 replicates of 20 sweeps. Sampling started when average corn height was 1 m.

In the unweeded fields, the weed community was composed of 47 species of herbaceous and woody invaders, at an average density of 29 plants per square meter. The commonest species were Malvaceae (*Sida glaziovii, S. rhombifolia,* and *S. cordifolia*), Amaranthaceae (*Amaranthus hybridus* and *A. viridis*), and Asteraceae (*Ageratum conyzoides* and *Emilia sonchifolia*). At the start of sampling, invaders had reached a maximum height of 60 cm, and the soil was only visible as small patches among branches, stems, and leaves.

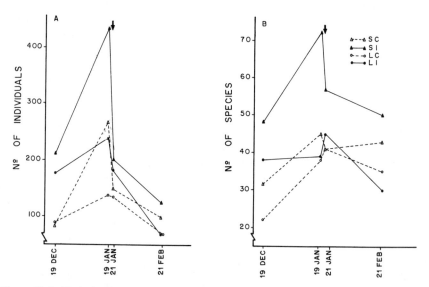

Figure 27-1. Variation in number of individuals (*A*) and species (*B*) of arthropods in corn fields at different times of the crop cycle. Corn field treatments included weeded (L), unweeded (S), treated (I), and untreated (C) with insecticides. The arrow indicates insecticide application.

Although annual or shorter-period cultures such as corn represent highly unstable systems, the current terminology used for similar studies on natural systems will be adopted here for the analysis of arthropod immigration and maintenance on this resource. It is considered here that in these systems a process of colonization and establishment of arthropod species occurs, even though many of them may be only feeding and not breeding.

In the study areas, the number of individuals was considerably higher in the unweeded than in the weed-free plots (Fig. 27-1A), but there are obvious oscillations in time, which are similar in all plots and treatments, implying that the "time" variable may be a significant component of the observed differences (Table 27-1). Concerning the variation in the number of species, there is, however, a tendency for a greater accumulation in the weedy fields that can be better explained by the differences between both fields (Table 27-2). Thus, an increase in individuals in the weeded plots seems to be entirely dependent on the establishment and reproduction of the initial colonizers, while in the unweeded plots the colonization process through new species input continues over a longer

TABLE 27-1. Kruskal–Wallis Test for Ranking Arthropod Numbers Collected in Weeded and Unweeded Corn Fields Submitted to Different Treatments and at Different Times (total Mean Square = 87.9516, df = 31)

Source of Variation	DF	SS	F	H = SS/MS	P > H
Time (T)	3	1368.4375	9.87	15.5590	0.005***
Treatment (TREAT)	1	78.7656	1.70	0.8956	0.500
FIELD	1	236.5313	5.12	2.6893	0.250
T × TREAT	1	19.1406	0.41	0.2176	0.750
T × FIELD	3	81.2813	0.59	0.9242	0.500
TREAT × FIELD	1	6.8906	0.15	0.0783	0.900
T × TREAT × FIELD	1	11.3906	0.25	0.1295	0.750

TABLE 27-2. Kruskal–Wallis Test for Ranking Arthropod Species Collected in Weeded and Unweeded Corn Fields Submitted to Different Treatments and at Different Times (total Mean Square = 87.4516, df = 31)

Source of Variation	DF	SS	F	H = SS/MS	P > H
Time (T)	3	606.1875	3.62	6.9317	0.100
Treatment (TREAT)	1	126.5625	2.27	1.4472	0.250
FIELD	1	621.2813	11.13	7.1043	0.010**
T × TREAT	1	3.0625	0.05	0.0350	0.900
T × FIELD	3	100.9813	0.60	1.1524	0.900
TREAT × FIELD	1	132.2500	2.37	1.5123	0.250
T × TREAT × FIELD	1	4.0006	0.07	0.0457	0.900

period, thus increasing progressively the differences in richness and composition between the plots. The higher similarity found between the arthropod communities in the first and second sampling dates in the weed-free fields (Morisita index = 0.62) when compared to the weedy fields (Morisita index = 0.23) lends support to this idea.

Colonization in a cultivated area occurs basically as a function of recruitment of individuals coming from neighboring areas, and of individuals dispersing through the air either actively or as "aerial plancton" (Price 1976) which can come from distant areas. The colonizing species are supposed to succeed in finding the culture. In this sense, the likelihood that a herbivore succeeds in arriving in a field where its host plant occurs is directly related to the perimeter of the field, or to its area in the case of air-dispersed insects.

This kind of generalization may be valid for colonizing arthropods, since natural enemies and herbivores initially must both be able to locate the habitat where their resources occur. Moreover, the stimulus–response mechanisms, related to insect orientation while searching for resources, may be highly specific, involving a combination of visual and chemical signals.

The greater complexity in polycultures or in weedy fields represents a more varied spectrum of signals to which a larger number of species and individuals may respond positively. On the other hand, the uniformity and intensity of signals coming from a monoculture would tend to elicit a positive response in a smaller group of potential colonizing species. One part of this assembly may indeed not respond positively to the same signal when there is a dilution, in part because of the presence of other plants in the same area, since their odors, colors, and shapes/forms represent "noise" in this communication system. So the differential attractiveness of fields with and without invaders to potential colonizing species may exert a central role in determining the richness and species composition of an arthropod community.

Colonization by Herbivorous Pests

An environment with a greater richness of plant species also represents a more variable habitat or a "larger island," which may imply a higher probability of an arthropod being able to settle successfully. Moreover, it is reasonable to expect that individuals more closely related to cultivated plants would tend to become established more easily in monocultures, where their resource is more abundant and more conspicuous, either visually or chemically (see Stanton 1983). There is some evidence that individual herbivores tend to remain longer in monocultures (Bach 1980).

It would also be reasonable to expect higher chances for survival and reproduction of these species in weed-free fields or in monocultures. However, during this study the presence of herbivores usually associated with corn fields was detected neither in the weeded nor in the unweeded plots. The absence of insects usually considered as pests may be a result of the peculiarities of the study area, either because it was located between a large cotton field and a small forest tract or because of the history of the area. Thus, these pest species would be absent in the species pool with the highest possibility of colonizing the fields. Besides, the neighboring vegetation may act as a barrier (Van Emden und Williams 1974) for potential colonizers. Liss et al. (1986) point out the necessity for understanding the array of the colonizing species pool, because of its great importance for the organization of the community that will be established in an area.

The possibility of interference by the greater higher plant diversity in the unweeded area on the location of the crop by the pest insects should not be overlooked, because of the proximity of the plots to each other. However, it is hard to envisage that, in the present study, natural enemies may account for the absence of these herbivores in either weed-free or in weedy plots, although there may be situations when the occurrence of natural enemies may constitute a limiting factor for the colonization of an area. This is the case for the chrysomelid beetle *Polyspila polyspila*, a rare species in the Campinas region but very common farther south, which feeds on several *Sida* species. In field experiments where I introduced a population of *P. polyspila* there was 100% mortality due to egg parasitism by *Erixestus* sp. (Hymenoptera, Pteromalidae). The parasite, which showed preference for *Polyspila* may be using other hosts in order to maintain its population in the area.

Spodoptera frugiperda, the most important insect pest of corn in the Campinas region, is a very common species, and although it is a host for some Ichneumonidae and Tachinidae, none is specific to it. Watt (1965) has analyzed the inefficiency of nonspecialist parasites as regulators of pest herbivore populations from his own and from Zwölfer's (1963) data. Watt concluded that, contrary to what could be expected from the trophic diversity–stability hypothesis, herbivore populations would not be more efficiently stabilized when there is a greater complexity at the higher trophic levels. Crawley (1986) stated that parasites and diseases are factors of secondary importance for the establishment of a species, while the presence of predators, which also add to complexity in the third trophic level, and chiefly generalist predators, are the main cause of failure. In this study, however, the absence of characteristic damage on the corn plants studied indicates that pest species probably did not reach the study area. Thus, failure to locate the habitat patch would be the main reason for their absence, although one cannot discard entirely the possibility of unsuccessful establishment.

Taxonomic Groups and Trophic Levels

Higher taxonomic categories are usually analyzed in data on sweep-net samples from arthropod communities (Risch 1979, 1981; Risch et al. 1986). The variation in the number of species and individuals of the different orders, however, does not reveal any tendency with respect to the structural complexity of the corn plots study (Fig. 27-2*A,B*). These variations occur through time and seem to be related to phenological aspects of the flora (Garcia 1988) and to habits and responses of the species which constitute each group. Clearer responses are seen in the

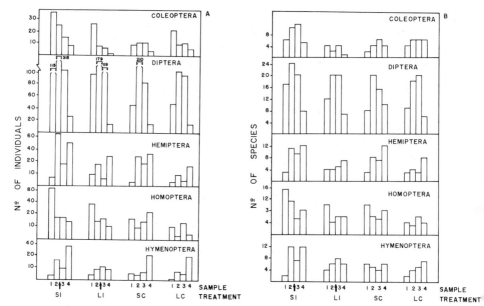

Figure 27-2. Distribution of number of individuals (*A*) and species (*B*) of the principal insect orders during the corn crop cycle in unweeded (S) and weeded (L) plots, and in areas where insecticide was applied after the second collection (I) and in control areas (C). The arrow indicates insecticide application. The four bars in each group correspond to sampling dates shown in Figure 27-1.

Homoptera, an exclusively phytophagous group, which was mainly represented by Cicadellidae. Greater homopteran densities and diversities were found at the beginning of the crop cycle, and from then on were gradually reduced. Such a tendency reflects the behavior of the Cicadellidae, which are strongly affected by shading (Saxena and Saxena 1974). The Hymenoptera, on the other hand, show an increase in density and species numbers in a more advanced phase of the agroecosystem development that coincides with the flowering peak of the weed community present at the site (Garcia 1988). This relationship is likely to be associated with the necessity for nectar and pollen, which accounts for a considerable part of the adult diet.

The Diptera were the best represented order, being the most abundant and richest both in the weedy and weed-free plot. Numbers of species and of individuals increase gradually until the end of the crop cycle, when they decrease markedly. In the remaining orders, the increase in individuals and species near the end of the cycle is directly related to fruit availability. At this time, pyrrochorid and rhopalid hemipterans proliferate, attracted to fruits of Malvaceae. The Coleoptera are among the first species to appear, with a predominance of defoliators; later on, diversity tends to increase in time largely because of the establishment of endophages such as stem and fruit borers.

The same orders are mentioned by Suttman and Barrett (1972) as the most common in oat fields; as in this study, Diptera make an overwhelming contribution. In the predator guild associated with *Brassica* sp., Root (1973) observed 50 species, 22 of which were dipterans. In the present study, from 165 morphospecies, 51 were to Diptera from 23 families, and they accounted for almost 48% of the total number of insects collected. Although feeding habits for many species have not yet been determined, the diversity and abundance of the mainly predatory Dolichopodidae and Chamaemyiidae are noteworthy. Unfortunately, Risch (1979), who studied arthropod communities associated with corn in Costa Rica, did not consider dipterans at all in his analysis, and his results can only be partially compared with mine. The relative contribution of hymenopterans, which constitute around 35% of the total number of species found in both weeded and unweeded plots after excluding Diptera, is very close to the values found by Risch (1979) for corn intercropped with sweet potatoes.

When one considers the assemblage of natural enemies observed, which includes especially Hymenoptera, Diptera, and, to a lesser extent, Hemiptera and Coleoptera, the number of individuals, but not the number of species, is significantly higher for the unweeded fields ($p = 0.0001$ for the second sample and $p = 0.0024$ for the third; χ^2-test, $df = 3$). But when parasites (mainly Hymenoptera) are separately analyzed, this difference disappears (Garcia 1988). The great mobility of adults from this group and the proximity between weeded and unweeded fields may help to account for this fact. The weedy areas may offer better microclimatic conditions and resources like nectar and pollen, but they may also represent a situation of structural complexity where hosts are more difficult to find. Sheehan (1986) argues that specialist natural enemies may respond negatively to agroecosystem diversification. In the present study, predators are the main reason for the differences between numbers of natural enemies in the weeded and unweeded plots. They are probably more favored by habitat complexity than parasites. The differences between the number of predator individuals in weedy and weed-free fields is still highly significant ($p = 0.0001$ for the first sample and $p = 0.0011$ for the third; χ^2-test, $df = 3$) while the difference in parasite numbers occurring between both treatments did not reach a significant level at any time. This information, however, does not answer a basic question: what is the efficiency of natural enemies in areas of monoculture and polyculture or weedy crops? To verify the effect of habitat complexity on the efficiency of natural enemies in reducing the herbivore population densities, detailed studies must be carried out in order to evaluate the differential rates of predation and parasitism under different habitat conditions, trying to eliminate other variables. Although there is a vast literature that seeks to compare these two situations, Kareiva (1983) considers these studies inconclusive because of their inability to eliminate other factors.

Responses to Disturbances

Resilient communities have been defined as those capable of rapid recovery of individuals or species which disappear temporarily because of disturbances. On the other hand, resistant communities are those with a high capacity to withstand disturbances because of the lower susceptibility of the component species to them (Goodman 1975).

The idea of stability in ecological systems can be associated with these concepts on a fixed-time scale. Annual crops are an example of a situation in which intense disturbances occur associated with the activities of planting, harvesting, and elimination of crop residues. In this case, the arthropod community which had been established is also eliminated, together with its substrate. A few species, usually associated with a particular crop, manage to keep themselves in the area generally in a quiescent state or in diapause. Most of the species which do not diapause bear adaptations which allow them to complete their cycle and migrate to other areas in search of host plants. In this case, the distance between cultivated areas can limit the maintenance of specialized herbivores in a certain region. Many species, however, can use ruderal plants on roadsides and forest margins as alternative hosts and as a bridge between resources with a discontinuous distribution in space or time.

There are also disturbances originated from pesticide application and unpredictable natural perturbations, such as storms and frost.

In what way might the arthropod fauna of a typical monoculture system and the fauna of a more complex system such as weedy crops react to these perturbations? The answer to this question may indicate the role of the structural complexity or of the floral diversity on the resilience or resistance of the arthropod community in agroecosystems.

In the weeded and unweeded plots where Endossulfan was applied as well as the controls, there was an overall reduction in the number of arthropod individuals. The number of species, however, was reduced only in the unweeded plots, while in the weeded fields there was a tendency for arthropod richness to increase after application (Fig. 27-3). Certainly, the general reduction in the number of individuals in all treatments may in part be due to a heavy rain during this period.

After applying another insecticide (Carbaryl) in oat fields, Suttman and Barrett (1979) also observed a decline in the number of individuals, both in treatment and in control plots. It seems likely that, in the case studied by Suttman and Barrett (1979) as well as in the present study, the mechanical perturbations caused by insecticide or water application may have led to an emigration of the insects from their sample sites.

The similarity between both weeded and unweeded fields continues to be high even after the insecticide spraying. However, there is a within-field variation, especially when analyzing the similarities before and after treatment and also when comparing with the controls. The weedy fields kept as a control were less affected by the disturbance, while the weeded fields showed the least similarity between treatment and control areas after insecticide application (Table 27-3). However, the similarity between the weeded field samples before and after application is very close to the one from weedy fields which received the same treatment. This suggests that the structural complexity may have some effect in

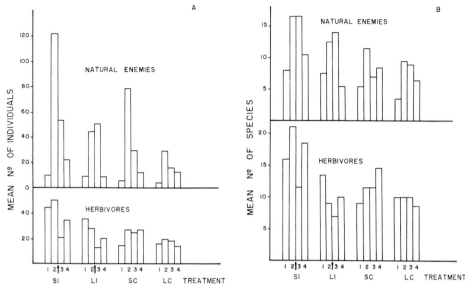

Figure 27-3. Variation in the number of individuals (*A*) and species (*B*) of herbivores and natural enemies during the corn crop cycle; in unweeded–insecticide treated (SI), weeded–insecticide treated (LI), unweeded–control (SC), and weeded–control (LC). The arrows indicate insecticide applications. The four bars in each group correspond to sampling dates shown in Figure 27-1.

compensating mechanical perturbations, although not against the effects of the insecticide. Hurd et al. (1971) have also observed that floral diversity does not buffer the herbivores and their natural enemies from the effects of insecticides.

The impact on this kind of system depends on the insecticide adopted (Root and Gowan 1978, Sukhoruchenko et al. 1981), and on the pest-management procedure. As an example, Fanchao and Changming (1986) observed that in rice fields where integrated pest management was being used, the richness of the parasite and predator species was, respectively, almost twice and three times that of areas where chemical control was exclusively used.

The responses of the herbivores may be more clearly related to the effects of the insecticide (Fig. 27-3*A* and *B*). However, in relation to the natural enemies, the response seems to be more related to a dispersal of individuals and species between the plots and to other sites outside the study area after the insecticide spraying.

Thus, the response observed in the herbivores, with a reduction in the number of individuals and species in fields treated with a selective insecticide, suggests a higher sensitivity of these insects to the insecticide used, as well as their more sedentary habits when compared to the habits of predators or parasites.

One month after the insecticide application, when the herbivorous fauna showed a tendency for recovery, the number of natural enemies decreased in all areas. Barrett (1968) reports a similar effect on the predators after application of

TABLE 27-3. Morisita's Similarity Index[a] Between the Arthropod Fauna of Unweeded (S) and Weeded (L) Corn Fields, Treated and Not Treated with the Insecticide Endossulfan

	Before	After	Treatment L	Treatment S	Control L	Control S	L Before	L After	S Before	S After
S × L	0.93	0.87								
Before × after application			0.75	0.70	0.64	0.88				
Treatment × control							0.86	0.68	0.99	0.90

[a]The Morisita index may range from 0 (no similarity) to 1.0 (identical), and refers to the probability of individuals drawn from each of the communities belonging to the same species in relation to the probability of a pair of individuals randomly collected from one of the communities belonging to the same species.

the selective insecticide Carbaryl in Georgia. In that case, and in the present study, this reduction may be either a result of the lower chances of establishment of migrant individuals in areas with low floral diversity, or a result of a delayed effect of the insecticide on the herbivores (hosts and prey), reducing the available resources or directly causing death among the usually more sedentary juvenile stages of their natural enemies.

CONCLUSIONS

The data obtained in this study indicate that a larger number of arthropod species and individuals are able to colonize and become established in corn field agroecosystems when weeds are present. The colonization process seems to be associated with the phenology of the plant species, which during their development represent different resource combinations for the herbivores and for adult natural enemies. In both weeded and unweeded plots, the herbivores normally associated with this crop did not occur. The absence of nearby corn fields and the maintenance of areas with higher diversity near to small areas of monoculture may represent an efficient form of reducing infestation by herbivorous pests.

The differential responses of parasites and predators, as a function of their necessity for food resources and specific microclimatic conditions may account for the greater accumulation of predators in the more diverse areas, while parasites showed an indifferent distribution. Detailed studies would be necessary to assess the efficiency of these groups in both conditions of habitat complexity. Such studies would be of great importance in order to verify the role of the floral diversity in the establishment of natural enemies and, consequently, their contribution to the regulatory mechanisms of herbivore populations.

Random dispersal seems to be a generalized response of the arthropods to disturbances, whether to natural events like rains or to insecticide or water application. This response results in a general reduction in arthropod density, but can provoke a temporary increase in species richness in monoculture areas located near more diverse systems if these are also disturbed.

According to the similarity analyses, unweeded fields tend to be less affected than weeded fields by mechanical perturbation, although both fields were equally affected by the Endossulfan application. Therefore, it seems that the structural complexity or the plant diversity may minimize the effects of mechanical, but not chemical, disturbances.

The herbivore populations showed the most obvious effects by the selective insecticide used. The response of the natural enemies may be better explained by the effect of random dispersal caused by perturbations in general. Habitat diversification may therefore represent a way of keeping a certain stability of the arthropod fauna in agroecosystems, besides being compatible with the practice of using selective defensive chemicals.

The maintenance of a higher floristic diversity in agroecosystems may be a productive cropping strategy in tropical and subtropical regions, since their composition and density do not represent a substantial competitive pressure on the crop species.

The higher plant diversity may reduce pest infestations, by interfering with the location of planted areas by colonists, and also by reducing the possibilities of expansion of the herbivorous populations, under conditions favorable for their natural enemies. Concomitantly, its compatibility with the use of selective defensive chemicals allows the use of curative methods when necessary. Moreover, this practice minimizes erosion processes, which are among the worst problems in tropical countries, and also results in the reduction of herbicide use.

Thus, for the tropics and subtropics, the maintenance of a high structural complexity and floral diversity may be part of a strategy for efficient crop production, not only for the small subsistence systems discussed by Altieri (Chapter 26), but also as a component of an alternative and effectively integrated pest-management system, especially in countries where monocultures represent an expensive and completely dependent method of production.

ACKNOWLEDGMENTS

I am very grateful to many friends who have helped me to clarify ideas during the elaboration of this chapter. My thanks go especially to Dr. Thomas M. Lewinsohn, Dr. John Winder, and Dr. Mohamed M. E. Habib for valuable suggestions; to Márcio Zikan who translated the manuscript; and to Lucia Paleari who prepared the drawings.

REFERENCES

Altieri, M. A. and D. K. Letourneau. 1982. Vegetation management and biological control in agroecosystems. *Crop Protection* **1**: 405–430.

Altieri, M. A., and D. K. Letourneau. 1984. Vegetation diversity and insect pest outbreaks. *CRC Critical Reviews in Plant Sciences* **2**: 131–169.

Altieri, M. A. and W. H. Whitcomb. 1980. Weed manipulation for insect pest management in corn. *Environmental Management* **4**: 1–7.

Altieri, M. A., A. van Schoohoven, and J. Doll. 1977. The ecological role of weeds in insect pest management systems. A review illustrated by bean (*Phaseolus vulgaris*) cropping systems. *PANS* **23**: 195–205.

Bach, C. E. 1980. Effects of plant diversity and time of colonization on an herbivore–plant interaction. *Oecologia* **44**: 319–326.

Barrett, G. W. 1968. The effects of an acute insecticide stress on a semi-enclosed grassland ecosystem. *Ecology* **49**: 1019–1035.

Bergelson, J. and P. Kareiva. 1987. Barriers to movement and the response of herbivores to alternative cropping patterns. *Oecologia* **71**: 457–460.

Crawley, M. J. 1986. The population biology of invaders. *Philosophical Transactions of Royal Society of London B* **314**: 711–731.

Dempster, F. P. and T. H. Coaker. 1974. Diversification of crop ecosystems as a means of controlling pests. In D. P. Jones, and M. E. Solomon (Eds.). *Biology in pest and disease control*, pp. 106–114. Blackwell Scientific Publications, Oxford.

Elton, C. A. 1958. *The ecology of invasions by animals and plants*. Methuen, London. 181 pp.

Fanghao, W. and C. Changming. 1986. Studies on the structure of the rice pest-natural enemy community and diversity under IPM area and chemical control area. *Acta Ecologica Sinica* **6**: 159–170.

Garcia, M. A. 1988. *Comunidades de plantas e artrópodes invasores em cultura de milho*. Ph.D. diss., Universidade Estadual de Campinas, São Paulo, Brazil. 272 pp.

Goodman, D. 1975. The theory of diversity–stability relationships in ecology. *The Quarterly Review of Biology* **50**: 237–266.

Holling, C. S. 1973. Resilience and stability of ecological systems. *Annual Review of Ecology and Systematics* **4**: 1–22.

Hurd, L. E., M. V. Mellinger, L. L. Wolf, and S. J. McNaughton. 1971. Stability and diversity at three trophic levels in terrestrial ecosystems. *Science* **173**: 1134–1136.

Kareiva, P. 1983. Influence of vegetation texture on herbivore populations: Resource concentration and herbivore movement. In R. F. Denno and M. S. McClure (Eds.). *Variable plants and herbivores in natural and managed systems*, pp. 259–289. Academic Press, New York, London.

Letourneau, D. K. 1986. Associational resistance in squash monocultures and polycultures in tropical Mexico. *Environmental Entomology* **15**: 285–292.

Liss, W. J., L. J. Gut, P. H. Westigard, and C. E. Warren. 1986. Perspectives on arthropod community structure, organization, and development in agricultural crops. *Annual Review of Entomology* **31**: 455–478.

MacArthur, R. H. 1955. Fluctuations of animal populations and a measure of community stability. *Ecology* **36**: 533–536.

MacArthur, R. H. and E. O. Wilson. 1967. *The theory of island biogeography*. Princeton University Press, Princeton, NJ. 203 pp.

May, M. L. and S. Ahmad. 1983. Host location in the Colorado potato beetle: Searching mechanisms in relation to oligophagy. In S. Ahmad (Ed.). *Herbivorous insects: Host-seeking behaviour and mechanisms*, pp. 173–200. Academic Press, New York, London.

May, R. M. 1973. Qualitative stability in model ecosystems. *Ecology* **54**: 638–641.

Maynard Smith, J. 1974. *Models in ecology*. Cambridge University Press, London.

McCoy, E. D. and J. R. Rey. 1983. The biogeography of herbivorous arthropods: Species accrual on tropical crops. *Ecological Entomology* **8**: 305–313.

Murdoch, W. W. 1975. Diversity, complexity, stability and pest control. *Journal of Applied Ecology* **12**: 795–807.

Power, A. G., P. M. Rosset, R. J. Ambrose, and A. J. Hruska. 1987. Population response of bean insect herbivores to inter- and intraspecific plant community diversity: Experiments in a tomato and bean agroecosystem in Costa Rica. *Turrialba* **37**: 219–226.

Price, P. W. 1976. Colonization of crops by arthropods: Non-equilibrium communities in soybean fields. *Environmental Entomology* **5**: 605–611.

Price, P. W. 1983. Hypotheses on organization and evolution in herbivorous insect communities. In R. F. Denno, and M. J. McClure (Eds.). *Variable plants and herbivores in natural and managed systems.* Pp. 559–596. Academic Press, New York, London.

Rey, J. R. and E. D. McCoy. 1979. Application of island biogeography theory to pests of cultivated crops. *Environmental Entomology* **8**: 577–582.

Risch, S. J. 1979. A comparison, by sweep sampling, of the insect fauna from corn and sweet potato monocultures and dicultures in Costa Rica. *Oecologia* **42**: 195–211.

Risch, S. J. 1981. Insect herbivore abundance in tropical monocultures and polycultures: An experimental test of two hypothesis. *Ecology* **62**: 1325–1340.

Risch, S. J., D. Pimentel, and H. Crover. 1986. Corn monoculture versus old fields: Effects of low levels of insecticides. *Ecology* **67**: 505–515

Root, R. B. 1973. Organization of plant arthropod associations in simple and diverse habitats: The fauna of collards (*Brassica oleracea*). *Ecological Monographs* **43**: 95–124.

Root, R. B. and J. A. Gowan. 1978. The influence of insecticides with differing specificity on the structure of the fauna associated with potatoes. *The American Midland Naturalist* **99**: 299–315.

Saxena, K. N. and R. C. Saxena. 1974. Patterns of relationship between leafhoppers and plants. II—Role of sensory stimuli in orientation and feeding. *Entomologia Experimentalis et Applicata* **17**: 493–503.

Sheehan, W. 1986. Response by specialist and generalist natural enemies to agroecosystem diversification: A selective review. *Environmental Entomology* **15**: 456–461.

Southwood, T. R. E. and M. J. Way. 1970. Ecological background to pest management. In R. L. Rabb, and F. E. Guthrie (Eds.). *Concepts of pest management*, pp. 6–29. North Carolina State University Press, Raleigh, NC.

Stanton, M. L. 1983. Spatial patterns in the plant communities and their effects upon insect search. In S. Ahmad (Ed.). *Herbivorous insects: Host seeking behaviour and mechanisms*, pp. 197–210. Academic Press, New York.

Sukhoruchenko, G. I., A. A. Smirnova, Y. E. V. Vikar, and A. I. Kapitan. 1981. The effect of pyrethroids on the arthropods of a cotton agrobiocenosis. *Entomological Review* **60**: 1–10.

Suttman, C. E. and G. W. Barrett. 1979. Effects of sevin on arthropods in an agricultural and an old field plant community. *Ecology* **60**: 628–641.

Tahvanainen, J. O. and R. B. Root. 1972. The influence of vegetational diversity on the population ecology of a specialized herbivore, *Phyllotreta cruciferae* (Coleoptera: Chrysomelidae). *Oecologia* **10**: 321–346.

Van Emden, H. F. 1965. The role of uncultivated land in the biology of crop pests and beneficial insects. *Scientific Horticulture* **17**: 121–136.

Van Emden, H. F. and G. F. WIlliams. 1974. Insect stability and diversity in agroecosystems. *Annual Review of Entomology* **19**: 455–475.

Visser, J. H. 1986. Host odor perception in phytophagous insects. *Annual Review of Entomology* **31**: 121–144.

Watt, K. E. F. 1965. Community stability and strategy of biological control. *The Canadian Entomologist* **97**: 887–895.

Zwölfer, H. 1963. Untersuchungen über die Structur von Parasitenkomplexen bei einigen Lepidopteren. *Zeitschrift für Angewandte Entomologie* **51**: 346–357.

INDEX